Springer Mineralogy

More information about this series at http://www.springer.com/series/13488

Michael Gaft • Renata Reisfeld • Gerard Panczer

Modern Luminescence Spectroscopy of Minerals and Materials

Second Edition

Springer

Michael Gaft
Laser Distance Spectrometry
Petach Tikva, Israel

Renata Reisfeld
The Enrique Berman Pr. of Solar Energy
Hebrew University
Jerusalem, Israel

Gerard Panczer
The Institute of Light and Matter (UMR5306)
Claude Bernard - Lyon 1 University
Lyon, France

ISSN 2366-1585 ISSN 2366-1593 (electronic)
Springer Mineralogy
ISBN 978-3-319-79674-1 ISBN 978-3-319-24765-6 (eBook)
DOI 10.1007/978-3-319-24765-6

Springer Cham Heidelberg New York Dordrecht London

Softcover re-print of the Hardcover 2nd edition 2015

Printed on acid-free paper

Springer International Publishing AG Switzerland is part of Springer Science+Business Media (www.springer.com)

Foreword

During the last decade, laser-induced luminescence spectroscopy has become an integral part of the routine analytical tools applied in virtually all sub-disciplines of the Earth sciences. The increased, and currently still even further increasing, use of laser-induced photoluminescence is favoured by a number of aspects, including the augmented availability of reliable spectrometer systems in many institutions worldwide. Laser-induced photoluminescence has the advantage of being a fully non-destructive technique (in contrast to, for instance, cathodoluminescence, whose implementation often causes samples to suffer surficial beam damage). Furthermore, luminescence excited with laser light allows one to not only analyse the surface of a given specimen with high spatial resolution on the order of the wavelength of light but also to perform in-depth analyses, on prepared as well as rough samples. By contrast, other luminescence techniques (such as cathodoluminescence and ionoluminescence), due to the limited penetration depth of the excitation, merely are surface analysis techniques that usually require the existence of a well-prepared sample surface.

A shortcoming, however, still exists, namely, the limited availability of comprehensive and dependable spectrum databases comprising photoluminescence emission spectra. Researchers often are obliged to do troublesome literature searches, in order to find reliable references backing up their own analytical findings and interpretations. Recognising the need of a compendium that summarises not only the emissions observed in certain geological materials but also the spectral peculiarities of a certain emission centre when incorporated in different host structures, Dr. Michael Gaft and co-authors have released in 2005 a book that contains a tremendous number of luminescence spectra of minerals. All three co-authors of the book are internationally recognised experts in the field of time-resolved luminescence spectroscopy. The reliability of the interpretation of spectra presented – a challenging and often controversial issue in visible-light spectroscopy – is backed up by the fact that many of the spectra presented have been obtained time resolved. That is, not only spectral information (i.e. energy of the light emitted, sharpness of bands, possibly the characteristic fine structure of an electronic transition and many more) but also information on the lifetime of the excited state is available. The data collection presented is backed up by a comprehensive list of relevant scientific articles, books and other sources.

Now, after 10 years past the appearance of the first edition of *Modern Luminescence Spectroscopy of Minerals and Materials*, Drs. Gaft, Reisfeld and Panczer release a new and updated edition of their book. Apart from the significantly extended number of spectra presented and minerals covered, the book has gained in usefulness as it now covers also the LIBS (laser-induced breakdown spectroscopy) and Raman scattering techniques, which many analysts actually conduct with the same analytical system as time-resolved photoluminescence. The book will prove useful for students as it provides the well-structured theoretical background of photoluminescence, with particular emphasis on rare earth elements, and general introductions to the basic experimental principles of other laser spectroscopy techniques. Furthermore, the three final chapters of the book are dedicated to the detailed presentation of diverse applications of laser-induced luminescence and related techniques in research and industry. Scientists working in diverse fields, from petrology over gemstone analysis and forensics to ore sorting, to name just a few examples, will recognise and appreciate this book as valuable basic provider of information that proves useful in the study of minerals and their emissions and subsequent interpretations and applications.

Institut für Mineralogie und Kristallographie
University of Vienna
Vienna, Austria

Lutz Nasdala

Epigraphs

Philosophy established as principle of things: water or abyss, dry substance or atoms or earth, spirit or air, and in the fourth place, light; Because these elements distinguished each other in the fact that they cannot exchange their nature but that all – here more, there less, here some of them only – meet and combine in a pleasant manner.

De Magia, Giordano Bruno (1590)

Great is the art to penetrate by intelligence in the subterranean deepness where nature forbids the access to our hands and to our eyes; to roam by thought in the Earth womb, to penetrate by reason into the sombre crevasses, to make the things and the activities hidden by the eternal darkness, coming to the day light.

Terrestrian strata, Mikhail Lomonossov (1763)

According to Alberto, the price of a lighter flint was equivalent to a ration of bread, that is, one day of life. (. . .) Total, one hundred and twenty stones, two months of life for me and two for Alberto and in two months the Russians would have arrived and liberated us; and finally the cerium would have liberated us, an element about which I knew nothing (. . .) and that it belongs to the equivocal and heretical rare-earth group family (. . .).

The periodic table, Primo Levi (1975)

You shall make a h'oshen of judgment (. . .). And you shall fill into it stone fillings, four rows of stones. One row: odem [ruby], pitdah [emerald], and bareketh [agate]; thus shall the one row be. The second row: nofeh' [aquamarine], sappir [sapphire], and yahalom [quartz]. The third row: leshem [lapis lazuli], shevo [amethyst], and ah'lamah [topaz]. And the fourth row: tarshish [chrysolite], shoham [onyx], and yashpheh [opal].

Exodus, Tetzaveh, 28

Comrade scientists, you, Newton dears
And precious Einstein, that we love to cry!
In the common earth will sleep our rotten rest;
Earth swallows everything: the apatite and the dung.

Comrade scientists, Vladimir Vissotski (1973)

Preface

In the broad sense the subject of our book belongs to the Physics of Minerals field, namely, to its spectroscopic part. The main features of our experimental technique are laser excitation of minerals and time-resolved detection of resulting emission and scattering. Laser sources allow study simultaneously as events of linear optics, such as luminescence and breakdown spectroscopies, and non-linear ones, such as Raman spectroscopy and Second Harmonic Generation. Time-resolved technique, namely, signal detection only during certain time gate after certain delay following the end of the laser pulse, results in drastically improved spectroscopic sensitivity and selectivity. It enables to collect new data, which contribute greatly to our fundamental knowledge of minerals and materials and may be practically used.

The Book is arranged as follows.

The **Chapter 1** contains the basic definitions of the main scientific terms, such as *spectroscopy, luminescence spectroscopy, luminescent mineral, luminescent center, luminescence lifetime, luminescence spectrum and excitation spectrum.* The state of the art in the *steady-state luminescence* of minerals field is presented. The main advantages of *laser-induced time-resolved technique* in comparison with steady-state one are shortly described.

The **Chapter 2** contains description of the main processes involved in luminescence, such as *absorption of the excitation energy*, and *radiative return* to the ground state. The resulting luminescence is considered based on *ligand field theory*, *configurational diagrams model*, *molecular orbitals theory* and *band scheme* approximation. The main attention is paid to *luminescence decay*, consisting of radiation decay, non-radiation decay and special types of decay, such as stepwise energy transfer, cooperative sensitization and cooperative luminescence.

The **Chapter 3** describes our experimental setup with the following main parts: *laser source* (Ar, excimer, Nd-YAG, nitrogen, dye, OPO), imaging monochromator, *gated detector* (Intensified Charge Coupled Device) and computer with corresponding software. The main features of the experimental devices are described, which enable to accomplish time-resolved detection.

The **Chapter 4** contains time-resolved luminescence spectra for more than 100 minerals and synthetic analogues. The following information is presented for each one, comprising literature and original results: short description of the *crystal structure, optically active colour centres and luminescence centres,* detected by steady-state and time-resolved techniques.

The **Chapter 5** is devoted to detailed explanation of luminescence centres interpretation process, which comprises investigation of spectral and kinetic parameters of excitation and emission at different temperatures from 4.2 to 300 K. In numerous cases the interpretation is confirmed by study of synthetic mineral analogues artificially activated by potential impurities, which may serve as emission centres. The following luminescence centres have been confidently detected and interpreted:

rare-earth elements (Ce^{3+}, Pr^{3+}, Nd^{3+}, Sm^{3+}, Sm^{2+}, Eu^{3+}, Eu^{2+}, Gd^{3+}, Tb^{3+}, Dy^{3+}, Er^{3+}, Ho^{3+}, Tm^{3+}, Yb^{3+}, Yb^{2+});
transition elements (Mn^{2+}, Mn^{3+}, Mn^{4+}, Mn^{5+}, Cr^{3+}, Cr^{4+}, Cr^{5+}, V^{2+}, Fe^{3+}, Ti^{3+}, Ni^{2+});
s^2 ions (Bi^{3+}, Bi^{2+}, Pb^{2+}, Pb^{+}, Sn^{2+}, Tl^{+}, Sb^{3+});
d^{10} ions (Ag^{+}, Cu^{+});
d^0 complex ions (WO_4, MoO_4, TiO_4, VO_4);
molecular centres (UO_2, S_2^-, O_2^-);
radiation-induced centres
diamond centres (N3, H3, H4, S2, S3);
reabsorption lines of O_2, H_2O, U^{4+}, Nd^{3+}.

The centres such as Mn^{6+}, Cr^{4+}, Cr^{5+}, V^{3+} and V^{4+} are described, which are not found in minerals yet but are known in synthetic analogues of minerals, such as apatite, barite, zircon and corundum.

The **Chapter 6** describes of laser-induced spectroscopies, which may be used in combination with laser-induced luminescence, namely, Breakdown, Raman.

The **Chapter 7** presents several examples of laser-based spectroscopies' application in remote sensing of minerals, rocks and foliage.

The **Chapter 8** describes several examples of laser-based spectroscopies' application in radiometric sorting and process control in mining industry.

The **Chapter 9** shortly describes laser-based spectroscopies' applications in gemology, micro-element mapping, waste storage, bio-minerals and LED. It contains Conclusions, Bibliography and Subject Index.

Laser Distance Spectrometry — Michael Gaft
Petach Tikva, Israel
Hebrew University — Renata Reisfeld
Jerusalem, Israel
Claude Bernard - Lyon 1 University — Gerard Panczer
Lyon, France

Acknowledgements

Particular thanks are due to Prof. **Georges Boulon** for his personal involvement in our lives as scientist and friend.

Our special thanks to Boris Gorobets and Harry Siegel who resently passed away for great contribution in minerals luminescence researh, advice and friendship.

We are very grateful for our colleagues and co-authors for long-time fruitful collaborations: **Philippe Blanc, Alain Baumer, Micheline Boudeulle, Bernard Champagnon, Dominique de Ligny, Elena Evdokimenko, Olivier Forni, Igor Gornushkin, Yoni Groisman, Victoria Gukova, Yannick Guyot, Patrick Jollivet, Christian Jörgensen, Yoram Kirsh, Svetlana Malinko, Arnold Marfunin, Nathalie Moncoffre, Vadim Moroshkin, Vincent Motto-Ros, Lev Nagli, Lutz Nasdalla, Nadege Ollier, Sylvain Peuget, Igor Prokofiev, Victor Rassulov, Ina Reiche, Alexander Rogojine, George R. Rossman, Irina Rudenkova, Anne-Magali Seydoux-Guillaume, Tatiana Shalashilina, Isamu Shinno, Shlomo Shoval, Arkadii Tarashchan, Riadh Ternane, Xiaochun Wang, Glenn Waychunas, Herb Yeates.**

Contents

Chapter 1
Introduction

Spectroscopy is the study of the interaction of electromagnetic radiation with the matter. There are three aspects of the spectroscopic measurement: irradiation of a sample with electromagnetic radiation; measurement of the absorption, spontaneous emission and scattering (Rayleigh elastic scattering, Raman inelastic scattering) from the sample; and analysis and interpretation of these measurements. The main subject of our book is spontaneous luminescence of natural minerals. We have used the term luminescence, which is the general term of the phenomenon. Luminescence may be subdivided to the fast fluorescence with spin-allowed transition and slow phosphorescence with spin-forbidden transition. Afterglow may be added, which is related to emission after trapping of an electron elsewhere.

Luminescence spectroscopy measures the energy levels of the luminescence centers. The energy level of a luminescence center is defined as its characteristic state, which is related to the physical nature of the center and to its energetic and dynamic processes that the center undergoes. The ground state is defined as the state with the lowest energy. States of higher energy are called excited states. A center possesses several distinct reservoirs of energy levels, including electronic, vibrational, rotational, transitional, and those associated with nuclear and electron spin. In minerals luminescence, the energy levels of interest are those that are associated with electronic and vibrational transitions.

A *luminescent mineral* is a solid, which converts certain types of energy into electromagnetic radiation over and above thermal radiation. The electromagnetic radiation emitted by a luminescent mineral is usually in the visible range, but can also occur in ultraviolet (UV) or infrared (IR). It is possible to excite luminescence of minerals by UV and visible radiation (photoluminescence), by beam of energetic electrons (cathodoluminescence), by X-rays (X-ray excited luminescence) and so on. The special case is so-called thermoluminescence, which is stimulation by heating of luminescence, preliminary excited in a different way.

The luminescent mineral consists of a host lattice and a *luminescent center*, often called an activator. The determination of the nature of the center responsible of luminescence is not generally a trivial task. Correlation of the observation of the

M. Gaft et al., *Modern Luminescence Spectroscopy of Minerals and Materials*,
Springer Mineralogy, DOI 10.1007/978-3-319-24765-6_1

specific luminescence with particular impurity concentration may give indication of the source of the emission but it is not proof of the origin, and can sometimes be misleading. Furthermore, it does not give any details about the precise nature of the center. Spectroscopic studies may provide a number of important parameters, which aid identification of the luminescence center.

Luminescence emission occurs as a result of a radiative electronic transition in which electron jumps from a higher energy state to a lower one, the difference in energy being released as a photon. Clearly the electron must first be excited into higher energy states by some means, for example, UV or visible light. After excitation the nuclei adjust their positions to the new excited situation, so that the inter-atomic distances equal the equilibrium distances belonging to the excited state. This process is called relaxation. During relaxation there is usually no emission. The system can return to the ground state spontaneously under emission of radiation from the lowest level of the excited state. The emission occurs at a lower energy than the absorption due to the relaxation process. The energy difference between the maximum of the lowest excitation band and that of the emission band is called *Stokes shift*.

The luminescence is usually characterized by its *quantum yield and lifetime*. Luminescence spectroscopy is the measurement and analysis of various features that are related to the luminescence quantum yield and lifetime. The quantum yield is the ratio of the number of photons emitted to the number absorbed. Luminescence centers with the largest quantum yields display the brightest emission. The lifetime is connected with the average time that the luminescence center spends in the excited state prior to its return to the ground state. It is defined as the time required for the luminescence intensity to drop to 1/e of its original value. The luminescence quantum yield and lifetime are modified by a number of factors that can increase or decrease the energy losses. The luminescence intensity of the luminescence centers is a function of their concentration, absorbing power at the excitation wavelength, and their quantum yield of the emission wavelength.

A *luminescence emission spectrum* represents the luminescence intensity measured over a range of emission wavelengths at a fixed excitation wavelength. On the other hand, a *luminescence excitation spectrum* is a plot of the luminescence intensity at a particular emission wavelength for a range of excitation wavelengths. A luminescence excitation-emission matrix is a two-dimensional contour plot that displays the luminescence intensities as a function of a range of excitation and emission wavelengths. Each contour represents points of equal luminescence intensity. Measurement of a luminescence lifetime requires excitation by a pulsed source, whose duration is negligible as compared to the lifetime of the emission process. The luminescence intensity is then recorded as a function of time. All spectral and kinetic measurements may be done at room temperature, or at liquid nitrogen or helium temperatures.

In many luminescence centers the intensity is a function of a specific orientation in relation to the crystallographic directions in the mineral. Even if a center consists of one atom or ion, such *luminescence anisotropy* may be produced by a compensating impurity or an intrinsic defect. In the case of cubic crystals this fact does not disrupt optical isotropy since anisotropic centers are oriented statistically uniformly over different crystallographic directions. However, in excitation of luminescence by polarized light the hidden anisotropy may be revealed and the orientation of centers can be determined.

Not every mineral shows luminescence. The reason is that the radiative emission process has a competitor, namely, the *nonradiative return* to the ground state. In that process the energy of the excited state is used to excite the vibrations of the host lattice, i.e. to heat the host lattice.

Many scientists have contributed to an essential part to our current knowledge in this field of mineral physics and its practical applications in earth sciences. In 1602, an Italian cobbler from Bologna, Vincencio Casiarolo found that by heating a heavy mineral so called « Bologna stone », in fact barite $BaSO_4$, it emitted light if it was previously exposed to the sun light or heated (Fig. 1.1 left). By such way (thermo-) luminescence was discovered. Despite the fact that the Arab alchemist, Alchid Bechil isolated Phosphorus since the twelfth century, the discovery of elementary phosphorus is attributed to Hennig Brand (1669) a German merchant and alchemist amateur who, while concentrating a great quantity of urine, succeeded to extract it as solid substance (Fig. 1.1 right). This substance was glowing in the dark and led to the *phosphorescence* term (from the Greek, φωσφορος *[fōsforos]*, which carry light). In 1786, the poet Goethe, a great collector of minerals, showed that, while sending colored lights (through filters) on an heated barite crystal, blue light only induced luminescence but not the red, yellow or green ones. Georges Stokes observed a luminescence phenomenon while studying fluorite and gave to this phenomenon the name of *fluorescence* (Stokes 1852). He developed the "Stokes Law of Fluorescence", which dictates that the wavelength of fluorescence emission must be greater than that of the exciting radiation. In 1859, Edmond Becquerel showed that *fluorescence* was in fact a *phosphorescence* phenomenon with very short duration, and that uranyl salts had phosphorescence 20 times more intense that calcium (greenockite) or zinc (sphalerite or blende) sulfides. He also emphasized that the red fluorescence of calcite was link to the presence of manganese, thus identifying for the first time the role of activators in fluorescence.

The detailed summary of some key elements in the history of luminescent studies of minerals is well presented in the following books and reviews: Marfunin 1979; Waychunas 1989; Gorobets and Walker 1994; Gorobets et al. 1995; Tarashchan and Waychunas 1995; Nasdala et al. 2004a. We want to mention specifically the excellent book of Tarashchan (1978), unfortunately not translated for English and remaining practically unknown to not Russian speaking scientists. We think that up today it remains one of the best in the field of mineral luminescence. Gorobets and Rogojine (2001) wrote Reference book "Luminescent spectra of minerals". It is the first systematic compilation of the luminescence spectra of minerals together with contemporary interpretation of the luminescence centers nature. This tremendous work summarizes the main achievements in this field up to 2000 years. Many luminescence properties and peculiarities of different minerals have been related to the crystallochemical and structural state of mineral species, varieties and individuals, searching at the same time for possibilities to associate the

Fig. 1.1 (*left*) Cabalistic representation image from the treaty «Il fosforo o vero la Pietra Bolognese » by Marc Antonio Cellio (1680) representing the light emission of heated barite; (*right*) Painting by Joseph Wright of Derby (eighteenth century) representing the discovery of the phosphorescence of the phosphorus extracted from urine by Hennig Brand in 1669

luminescence behavior with the different conditions of mineral formation and alteration. Different types of luminescent centers have been studied and identified using photo-, cathodo- and X-ray luminescence techniques frequently combined with optical absorption and EPR spectroscopy in minerals such as diamond, apatite, fluorite, scheelite, zircon, calcite etc., which are known as important genetic indicators in many rock- and ore-forming processes. The main minerals, which are luminescent under UV lamp excitation are summarized in Table 1.1.

Recently a lot of new information on minerals luminescent under UV lamp excitation appear (MacRae and Wilson 2008) and in several Internet sites: (http://www.fluomin.org/uk/contact.php, http://www.csiro.au/luminescence/about.html

At the same time, solid-state physicists intensively dealt with synthetic analogues of luminescent minerals as phosphors and quantum electronic materials, providing the theoretical and experimental background for further applications in high technology material sciences. These achievements are shortly summarized in Table 1.2.

These results were used also in geosciences, remote sensing, exploration, natural dosimetry, mineral processing etc. Starting from approximately 20 years ago, UV lasers have been used as luminescence of minerals excitation source. By such a way, not only luminescence spectra, but decay times also have been determined. The main motivation was to use laser-induced luminescence for remote sensing of mineral deposits (Kasdan et al. 1981; Seigel and Robbins 1985). Systematical

Table 1.1 Minerals luminescent under UV lamp excitation

Native	Diamond
Oxides	Brucite $Mg(OH)_2$, corundum Al_2O_3, ruby Al_2O_3, sapphire Al_2O_3
Halogenides	Fluorite CaF_2, halite NaCl, calomel HgCl
Silica	Agate SiO_2, opal SiO_2, quartz SiO_2
Phyllosilicates	Pyrophyllite $AlSi_2O_5OH$, talc $Mg_3Si_4O_{10}(OH)_2$, serpentine $(Mg, Fe)_3Si_2O_5(OH)_4$
Feldspar	Albite $NaAlSi_3O_8$, microcline $KAlSi_3O_8$
Pyroxenes	Diopside $CaMgSi_2O_6$, spodumene $LiAlSi_2O_6$
Amphiboles	Anthophyllite $(Mg, Fe)_7Si_8O_{22}(OH)_2$, tremolite $Ca_2Mg_5Si_8O_{22}(OH)_2$
Zeolites	Natrolite $Na_2Al_2Si_3O_{10} \times 2H_2O$, laumontite $CaAl_2Si_4O_{12} \times 4H_2O$, analcite $NaAlSi_2O_6 \times H_2O$
Feldspathoids	Sodalite $Na_4Al_3(SiO_4)_3Cl$
Other silicates	Zircon $ZrSiO_4$, datolite $CaBSiO_4(OH)$, danburite $CaB_2Si_2O_8$, eucryptite $LiAlSiO_4$, willemite Zn_2SiO_4, wollastonite $CaSiO_3$, sphene $CaTiSiO_5$, scapolite $Na_4(Al, Si)1_2O_{24}Cl$, manganaxinite $Ca_2(Mn, Fe, Mg)Al_2(BO_3OH)(SiO_3)_4$, benitoite $BaTiSi_3O_9$, thorite $(Th, U)SiO_4$
Carbonates	Aragonite $CaCO_3$, calcite $CaCO_3$, cerussite $PbCO_3$, magnesite $MgCO_3$, strontianite $SrCO_3$, witherite $BaCO_3$
Sulfides	Sphalerite (Zn, Fe)S, wurtzite (Zn, Fe)S
Sulphates	Barite $BaSO_4$, celestine $SrSO_4$, gypsum $CaSO_4 \times 2H_2O$, anhydrite $CaSO_4$, alunite $KAl_3(SO_4)_2(OH)_6$, zippeite $K_4(UO_2)_6(SO_4)_3OH_{10} \times 4H_2O$, ettringite $Ca_6Al_2(SO_4)_3(OH)_{12} \times 26H_2O$, hanksite $Na_{22}K(SO_4)_9(CO_3)_2Cl$
Phosphates	Apatite $Ca_5(PO_4)_3(OH,F,Cl)$, wavellite $Al_3(PO_4)_2(OH)_3 \times 5(H_2O)$, amblygonite $(Li, Na)AlPO_4(F, OH)$, pyromorphite $Pb_5(PO_4)_3Cl$, monazite $(Ce, La, Th, Nd, Y)PO_4$, autunite $Ca(UO_2)_2(PO_4)_2 \times 10H_2O$, xenotime$YPO_4$
Tungstates	Scheelite $CaWO_4$
Molybdates	Powellite $CaMoO_4$, wulfenite $PbMoO_4$
Borates	Colemanite $CaB_3O_4(OH)_3$-H_2O
Organics	Amber $C_{10}H_{16}O$

investigation of laser-induced luminescence of minerals has been started at the same time by one of the authors in the former Soviet Union (Gaft 1989). Luminescence centers established in minerals including decay times are summarized in Table 1.3.

So far presented results are connected with so-called *steady-state spectroscopy*. Conventionally studied steady state or CW (continuous-wave) luminescence is a process where the excitation sources pump the sample at constant intensity over time necessary to perform the measurement. The end result is emission spectrum, namely the distribution of energy emitted by an excited system in terms of the intensity of emitted optical photons as a function of wavelength or photon energy. Such spectroscopy in many cases is inadequate because the discriminatory power of the normal emission spectra is somewhat limited. Therefore, most of the previously provided emission spectra of minerals present overlapping features of several types of luminescence centers.

Table 1.2 Principal luminescent materials and their natural analogues (From Reisfeld and Jörgensen 1977; Moncorge et al. 1994; Boulon 1997)

Material	Synthetic	Natural analogues
Phosphors	(Ce^{3+}, Tb^{3+}):$LaPO_4$	Monazite
	Pr^{3+}:$CaWO_4$	Scheelite
	(Ag^{+}, Al^{3+}):ZnS	Sphalerite
	(Cu^{+}, Al^{3+}):ZnS; Mn^{2+}:ZnS	Sphalerite
	Ce^{3+}:$Y_5(SiO_4)_3F$	Britholite
	Tb^{3+}: $LaSiO_2F$	Wollastonite
	(Ce^{3+}, Tb^{3+}):$Gd_4(Si_2O_7)F_2$	Cuspidine
	Tb^{3+}:$CaYSi_3O_3F_4$	Melilite
Scintillators	Ce^{3+} : $CaWO_4$	Scheelite
	Ce^{3+} : $PbWO_4$	Stolzite
	Ce^{3+} : $YbPO_4$	Xenotime
Dosimeters	U^{6+} : CaF_2; REE^{3+} : CaF_2	Fluorite
	(Dy^{3+}, Tb^{3+}) : $CaSO_4$	Anhydrite
	(Cr^{3+}, Ti^{3+}) : Al_2O_3	Corundum
Laser materials	REE^{3+}: $CaWO_4$	Scheelite
	REE^{3+}: $CaMoO_4$	Molybdenite
	(Er^{3+}, Nd^{3+}): $Y_3Al_5O_{12}$	Garnet
	(Cr^{3+}, Cr^{4+}): $Y_3Al_5O_{12}$	Garnet
	(Yb^{3+}, Nd^{3+}): $Y_3Al_5O_{12}$	Garnet
	(Cr^{3+}, Ti^{3+}): $BeAl_2O_4$	Alexandrite
	REE^{3+}: $ZrSiO_4$	Zircon
	Cr^{4+}: Mg_2SiO_4	Forsterite (olivine)
	Cr^{4+}: $CaMgSiO_4$	Monticellite (olivine)
	Mn^{5+}: $Sr_5(PO_4)_3(Cl, F)$	F, Cl Apatite
	Yb^{3+}: $Sr_5(PO_4)_3(Cl, F)$	Apatite
	Mn^{2+}: $(Ca, Sr, Ba)_2(PO_4, VO_4)Cl$	Spodiosite (apatite)
	Cr^{3+}: $(Ca, Sr)(Y, Gd)_4Si_3O_{13}$	Apatite
	(Nd^{3+}, Ho^{3+}): $Ca_5(PO_4)_3F$	Fluorapatite

However, *luminescence lifetime*, which is a measure of the transition probability from the emitting level, may be effectively used. It is a characteristic and an unique property and it is highly improbable that two different luminescence emissions will have exactly the same decay time. The best way for a combination of the spectral and temporal nature of the emission can be determined by *laser-induced time-resolved spectra*. Time-resolved technique requires relatively complex and expensive instrumentation, but its scientific value overweights such deficiencies. It is important to note that there is simple relationship between steady-state and time-resolved measurements. The steady-state spectrum is an integral of the time-resolved phenomena over intensity decay of the sample, namely:

Table 1.3 Luminescence centers found in steady-state spectra of minerals

Luminescence center	Transition	Decay time
Rare earth ions (line emission): Nd^{3+}, Sm^{3+}, Sm^{2+}, Eu^{3+}, Gd^{3+}, Tb^{3+}, Dy^{3+}, Yb^{3+}	4f–4f	The parity does not change and the life time of the excited state is long (~10 μs–1 ms)
Rare earth ions (band emission): Ce^{3}	5d–4f	Transition is parity allowed and spin selection is not appropriate ⇒ fully allowed (10–100 ns)
Eu^{2+}		Emitting level contains octets and sextets, whereas the ground state level ^8S is an octet, so that the spin selection rule slows down the optical transition rate (~1 μs)
Yb^{2+}, Sm^{2+}		The spin selection rule is more important and decay times are of (100 μs–1 ms)
Transition metal ions	d–d	Decay time is long
Cr^{3+}	^2E–$^4A_2$	The parity selection rule as well as the spin selection rule apply (~1 ms)
	$^4T_2$–$^4A_2$	The parity selection rule (~10–100 μs)
Mn^{2+}, Fe^{3+}	$^4T_1$–$^6A_1$ $^4T_2$–$^6A_1$	The parity selection rule as well as the spin selection rule (~1 ms)
d^0 complex ions: $(VO_4)^{3-}$, $(WO_4)^{2-}$, $(MoO_4)^{2-}$, $(TiO_4)^{4-}$	Charge-Transfer	Emitting state is a spin triplet and decay time is long (10–100 μs)
$(UO_2)^{2+}$	Charge-Transfer	Considerably long decay time (~01–1 ms)

$$I_{(steady-state)} = \int I_0 e^{-t/\tau} dt = I_0 \tau \qquad (1.1)$$

where τ is the longest lifetime of the luminescence centers present in mineral. It turns out that much of the information available from luminescence spectra is lost during the time-averaging process. Time-resolved luminescence can separate overlapping features, which have different origins and therefore different luminescence lifetimes. The method involves recording the intensity in a specific time "window" at a given delay after the excitation pulse with carefully chosen delay time and gate width. The added value of the method is the energetic selectivity of a laser beam, which enables to combine time-resolved spectroscopy with individual monochromatic excitation.

Time-resolved luminescence spectroscopy may be extremely effective in minerals, many of which contain big amount of emission centers simultaneously. By steady state technique only the mostly intensive centers are detected, while the weaker ones remain unnoticed. Fluorescence in minerals is observed over time range of nanoseconds to milliseconds (Table 1.3) and this property was used in our research. Thus our main improvement is laser-induced time-resolved spectroscopy in the wide spectral range from 270 to 1500 nm, which enables to reveal new luminescence centers in minerals previously hidden by more intensive centers.

Time-resolved luminescence spectroscopy studies have been started approximately 20 years ago and the main results are presented by Gaft et al. (2005) and Panczer (2001). The task of our present paper is to present the main new data on luminescence spectroscopy of minerals, which have been received after this book publication both by our groups and by other researches.

References

Boulon G (1997) Broad band centers applied for laser materials: example of tetrahedrally coordinated centers. In: Di Bartolo B (ed) Spectroscopy and dynamics of collective excitations in solids. Plenum Press, New York, pp 561–573

Gaft M (1989) Luminescence of minerals under laser excitation. Ministry of Geology, Moscow (in Russian)

Gaft M, Reisfeld R, Panczer G (2005) Modern luminescence spectroscopy of minerals and materials. Springer, Berlin/New York

Gorobets B, Rogojine A (2001) Luminescent spectra of minerals. Handbook. RPC VIMS, Moscow, p 300

Gorobets B, Walker G (1994) Origins of luminescence in minerals: a summary of fundamental studies and applications. In: Marfunin A (ed) Advanced mineralogy 2, methods and instrumentation. Springer, Berlin/Heidelberg/New York, pp 138–146

Gorobets B, Portnov A, Rogojine A (1995) Luminescence spectroscopy of the earth. Radiat Meas 24:485–491

http://www.fluomin.org/uk/contact.php

http://www.csiro.au/luminescence/about.html

Kasdan A, Chimenti R, de Neufville J (1981) Selective detection of uranium by laser-induced fluorescence: a potential remote-sensing technique. 2: Experimental assessment of the remote sensing of uranyl geologic targets. Appl Opt 20:1279–1307

MacRae C, Wilson N (2008) Luminescence database I – minerals and materials. Microsc Microanal 14:184–204

Marfunin A (1979) Physics of minerals and inorganic materials. Springer, Berlin/Heidelberg/New York

Moncorgé R, Manaa H, Boulon G (1994) Cr^{4+} and Mn^{5+} active centers for new solid state laser materials. Opt Mater 4:139–144

Nasdala L, Götze J, Hanchar J et al (2004a) Luminescence techniques in Earth sciences. In: Biran A, Lubowitzky E (eds) EMU notes in mineralogy, spectroscopic methods in mineralogy, vol 6/2. Mineralogical Society of America, Chantilly, pp 43–91

Panczer G (2001) La photoluminescence résolue en temps, nouvel outil pour la minéralogie. Habilitation thesis, Universite Claude Bernard, Lyon (in French)

Reisfeld R, Jörgensen K (1977) Lasers and excited states of rare earths. Springer, Berlin/Heidelberg/New York

Seigel H, Robbins J (1985) Luminescence method-new method of air and ground exploration of ore deposits. ITC J 3:162–168

Stokes G (1852) Phil Trans Royal Soc London 142:463–562

Tarashchan A (1978) Luminescence of minerals. Naukova Dumka, Kiev (in Russian)

Tarashchan A, Waychunas G (1995) Luminescence of minerals. Interpretation of the luminescence spectra in terms of band theory, crystal field theory, sensitization and quenching. In: Marfunin A (ed) Advanced mineralogy 2, methods and instrumentation. Springer, Berlin, pp 124–135

Waychunas G (1989) Luminescence, X-Ray emission and new spectroscopies. In: Hawthorne FC (ed) Spectroscopic methods in mineralogy and geology. Rev Mineral. Mineralogical Society of America, Washington, DC, pp 638–698

Chapter 2
Theoretical Background

Theoretical data essential for understanding of luminescence phenomenon may be found in many books, but we believe that for specific field of minerals luminescence the fundamental books of Marfunin (1979a, b) are the best. Below we tried to present very shortly only the mostly essential data especially connected with kinetic considerations, which are the basis of time-resolved technique.

2.1 The s, p, d, f Atomic Orbitals

According to quantum mechanics laws, electrons in free atoms occupy so-called atomic orbitals. Each orbital is characterized by its energy and is determined by quantum numbers n, l, m_l, where n is the main quantum number, designated by numbers 1, 2, 3…, l is the orbital quantum number with 0, 1, 2 … (n − 1) values and m_l is the magnetic quantum number with $-l, -l+1, \ldots 0, \ldots l-1, l$ values.

For explanation of luminescence properties of the atomic orbitals, their symmetry and directional properties are most important. The shape of atomic orbitals in space is determined by the l quantum number, designated as s, p, d, f for $l = 0, 1, 2, 3$. The number of possible orientations of the orbitals is given by the number of possible values of the m_l magnetic quantum number and is equal for the s-orbitals to 1, for the p-orbitals to 3, for the d-orbitals to 5 and for the f-orbitals to 7, according to $(2l+1)$ rule. In a totally symmetric surrounding all orientations of the orbitals are equivalent and correspond to the same energy value. These orbitals are called degenerate, which means they are characterized by equal energy. Figure 2.1 represents real forms of the p, d and f orbitals.

Each element is characterized by specific scheme of energy levels. When individual elements are combined forming minerals the energy levels are changed. They have now different energies and form different system of energy levels typical

M. Gaft et al., *Modern Luminescence Spectroscopy of Minerals and Materials*, Springer Mineralogy, DOI 10.1007/978-3-319-24765-6_2

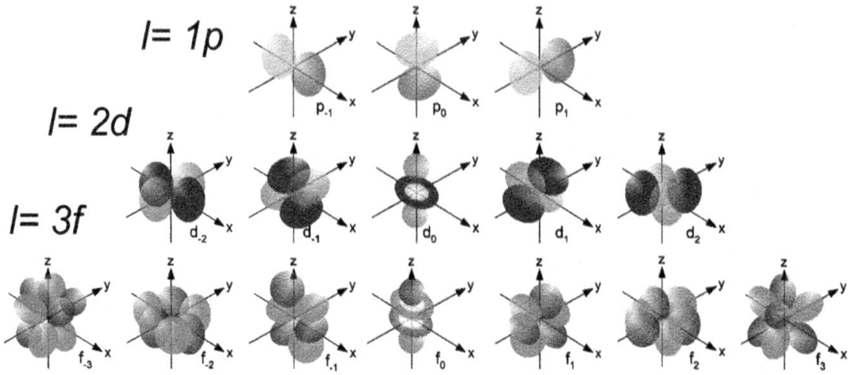

Fig. 2.1 The forms of the angular dependence functions for p, d, and f orbitals (Figgis 1966)

for mineral. The process of new energy levels system in minerals may be described using different theories of solid state physics, such as ligand field theory, molecular orbitals theory and zone theory, which are especially useful for absorption and luminescence properties explanation.

2.2 Point Group Symmetry

As we will demonstrate, luminescent properties, radiative transition characteristics as well as emission under site selective excitation depend of the local environment symmetry of the luminescent center. Therefore it is necessary to take into account and to describe the different local symmetry. There are two systems commonly used in describing symmetry elements of punctual groups:

1. The Schoenflies notation (1891) used extensively by spectroscopists, and
2. The Hermann-Mauguin (1935) or international notation preferred by crystallographers.

Schönflies notation is widely used to describe molecule or assemblages of atoms (polyhedron) such as local environment of an atom. Thus, it is widely used to describe the symmetry of structural sites. It is a more compact notation but less complete than Hermann-Mauguin one. It consists generally of one cap letter, followed by one subscript number and one final letter:

- Symmetry C_n for one rotational axis (cyclic); C_1: identity; C_n: the polyhedron has an n-fold axis of symmetry.
- Symmetry D_n for orthogonal axes (dihedral), n-fold axis of symmetry perpendicular to n twofold axes;
- Cubic symmetry: T for 4 axes (tetrahedral), O for 8 axes (octahedral) et I pour 20 axes (icosahedra, exists to describe molecules but not for crystals);

- Planar symmetry:
  - one subscript letter follows the rotation symmetry (by example C_{3h}); h indicates a planar symmetry (mirror) parallel to the rotation axis (horizontal), v a mirror perpendicular to the rotation axis (vertical) and d a diagonal mirror;
  - Sn, S indicates a mirror symmetry (Spiegel in German) and *n* invariance by a n-fold rotation around the perpendicular axis to the plan. An added superscript m number indicates sometimes that the symmetry operation is applied m times.

Table 2.1 presents the non-cubic crystallographic point groups with compared notation.

2.3 Absorption of the Excitation Energy

2.3.1 *Optical Absorption Spectroscopy*

It is a remarkable fact that the contemporary history of absorption and emission spectroscopy began simultaneously, from the simultaneous discoveries by Bunsen and Kirchhoff made in the middle of the nineteenth century. They observed atomic emission and absorption lines whose wavelengths exactly coincided. Stokes and Kirchhoff applied this discovery for the explanation of the Fraunhofer spectra. Nearly at the same time approximately 150 years ago, Stokes explained the conversion of absorbed ultraviolet light into the emitted blue light and introduced the term fluorescence. Apparently, the discovery of the Stokes shift marked the birth of luminescence as a science.

A rule due to Beer and Lambert is that the light absorbed per unit length in a sample depends at any wavelength only on the incident light intensity, i.e.,

$$\frac{dI(\lambda)}{dx} = -\alpha\ I(\lambda) \tag{2.1}$$

where α is a proportionality constant called the absorption coefficient, which also depends on wavelength.

Integrating Eq. 2.1, we find for a non-reflecting sample of thickness t:

$$I_T(\lambda) = I_o(\lambda)\exp(-\alpha t) \tag{2.2}$$

where I_T is the transmitted light intensity and I_o the incident light intensity. In these terms, the transmittance spectrum is

$$T(\lambda) = I_T(\lambda)/I_o(\lambda) \tag{2.3}$$

so that

Table 2.1 The non-cubic crystallographic point groups

	Hexagonal		Tetragonal		Trigonal		Orthorhombic		Monoclinic		Triclinic	
Cyclic	C_6	6	C_4	4	C_3	3			C_2	2	C_1	1
Cyclic + vertical planes	$\underline{C_{6v}}$	6 mm	$\underline{C_{4v}}$	4 mm	$\underline{C_{3v}}$	3 m	$\underline{C_{2v}}$	2 mm				
Cyclic + horizontal planes	$\mathbf{C_{6h}}$	6/m	$\mathbf{C_{4h}}$	4/m					$\mathbf{C_{2h}}$	2/m		
	C_{3h}	−6							$\underline{C_{1h}}$ $\underline{(C_s)}$	m (−2)		
Improper Rotation			S_4	−4	$\mathbf{S_6(C_{3i})}$	−3					$\mathbf{S_2}$ $\mathbf{(C_i)}$	−1
Dihedral	D_6	622	D_4	422	D_3	32	D_2 (V)	222				
Dihedral + horizontal planes	$\mathbf{D_{6h}}$	6/mmm	$\mathbf{D_{4h}}$	4/mmm			$\mathbf{D_{2h}(V_h)}$	2/mmm (mmm)				
	D_{3h}	−62 m										
Dihedral + planes between axes			D_{2d} (V_d)	−42 m	$\mathbf{D_{3d}}$	−32/m (−3 m)						

For cubic symmetry, T correspond to 23, T_h to m3, O to 432, T_d to 4–3 m, O_h to m3m and I to 532 (thus impossible for crystals, but possible for molecule) Column pairs give the Schönflies and equivalent Hermann-Mauguin international notation. *C* indicates cyclic, *D* for dihedral, *S* for spiegel (or “mirror”), with h, v and d meaning respectively horizontal, vertical and diagonal (referring to mirror plane locations). A minus before a number indicates a horizontal bar above that number. Brackets indicate equivalent alternative. The centrosymmetric point groups are shown in bold. The enantiomorphic point groups are shown in italic. The polar point groups are underlined

$$T(\lambda) = \exp(-\alpha t) \tag{2.4}$$

The absorption spectrum is given by an inversion of Eq. 2.4 i.e.,

$$\alpha(\lambda) = \frac{1}{t}\ell n\left(\frac{I_o}{I_T}\right) = \frac{1}{t}\ell n\left(T^{-1}\right) \tag{2.5}$$

The absorbance is defined as:

$$\log_{10}\left(\frac{I_c}{I_T}\right) \tag{2.6}$$

In optical spectroscopy a sample is illuminated by monochromatic light with a varying wavelength. The optically active centers rise from the ground to the excited states with resulting appearance of the absorption band in the optical spectrum. Useful information may be obtained from the absorption spectroscopy, which reveals the excited energy levels. Nevertheless, it has to be noted that absorption technique is much less sensitive compared to luminescence. The reason is that absorption is a subtractive process. One measures small differences in large numbers (intensity of incident light-intensity of transmitted light). This difference will be very small for very low absorbance from low concentrations. Best sensitivity is ~0.0005 OD (optical density), but most spectrophotometers are not nearly this good. On the other hand luminescence measures any detectable light against an essentially zero background. Modern detectors are capable of measuring single photons and thus luminescence can be extremely sensitive. There are a number of different processes, which may generate optical absorption in the visible and ultraviolet wavelength range. The following transitions are of primary importance for minerals (Platonov 1979; Rossman 1988; Matsyuk and Zintchouk 2001):

- d-d or f-f electronic transitions between ligand field split spectroscopic states of the free electronic levels of transition metals or rare earth ions. This process involves an ion with a partially filled d- or f-electron shell being incorporated into a crystalline environment whose symmetry produces a difference in energy between the different orbitals (see Sect. 2.4.1). Such effects are observed within compounds containing transition metal ions as Fe^{2+}, Fe^{3+}, Mn^{2+}, Ti^{3+}, Cr^{3+} and associated colors characterize a range of minerals. The intensities of d-d and f-f absorption bands can vary over four orders of magnitude: these are governed both by abundance of the absorbing cation, and by a combination of the symmetry of the cation environment and quantum mechanics selection rules (see Sect. 2.4.5).
- metal to metal charge-transfer between transition metal ions differing in valence state and accommodated in interconnected polyhedra of the mineral structure. This term used to describe absorption mechanism by transfer of electrons between transition metal ions in non-metals via an input of energy in the form of photon, especially those ions, which are able to adopt multiple valence states

in minerals, such as Fe^{2+} and Fe^{3+}, Ti^{3+} and Ti^{4+} and so on. Characteristically, these bands occur in the visible part of the spectrum, and are most intensive in those minerals, which have large quantities of transition metals, and relatively short metal-metal distances.

- ligand to metal charge transfer or an interband transitions, which involve transfer of electron density from ligand to the metal cation. Charge transfer transitions not only occur between transition metal ions, but also occur in excitation of electrons between cations and ions. Within oxides, such oxygen-to-metal charge transfer absorption bands typically occur at high energies in the ultraviolet, and are extremely intense: often three to four orders of magnitude more intense than d-d transitions. Effectively, such charge transfer between anions and cations is often associated with delocalization of an electron, or photoexcitation of an electron into the conduction band of a mineral. Within many minerals, such charge transfer bands appear as an absorption edge, rather than a discrete band: this edge is simply generated because all photons with energy above that of the edge will produce electron delocalization. Absorption generated by such cation-anion charge delocalization processes is common in ore minerals such as sulfides and arsenides.
- radiation induced centers from natural sources in rocks. Such electron-hole centers produced by ionizing radiation are mostly typical for smoky quartz, blue feldspar, green diamonds, and fluorite and calcite samples.

2.3.2 Luminescence Excitation Spectroscopy

A luminescent mineral will only emit radiation when the excitation energy is absorbed. It is possible to categorize luminescence mechanisms into two general classes. In the first the process of excitation is localized near an isolated center and occurs without ionization of any species in the mineral. This type of luminescence is called intra-centric. Excitation and emission occurs due to electronic transitions at the center. In certain luminescent minerals the situation is more complicated, because the activator does not absorb the exciting radiation, but it is absorbed by the other center, which subsequently transfers it to the activator. In this case the absorbing ion is called a sensitizer. The second mechanism occurs when the excitation is light of higher energy than the band gap, or consists of very high-energy radiation or particles. Ionization effects then occur, and the emission is produced by recombination of electrons or holes at ionized centers. Laser-induced luminescence mainly belongs to the first type direct excitation of centers of emission without their ionization. But during excitation with high-energy photons in semiconductors with a small inter-band spacing, an electron transfer of impurity ions of the ground substance to the conduction band takes place. In these crystals the energy levels of defects are divided according to their position in the forbidden band into donors (near the conduction band) and acceptors (in the vicinity of the valence band).

Absorption of the excitation energy resulting in luminescence is revealed by excitation spectra. Hear the emission results of the transition from the specific electronic level. The excitation spectrum is registered as the intensity of luminescence as a function of the excitation wavelength. The excitation spectrum is measured using two monochromators. One is used to separate the exciting frequency and the other is used for separating a specific region in the luminescence spectrum. The second monochromator is set to a specific frequency, whereas the first one scans the excitation frequency. The appearance of tunable lasers has made it possible to avoid using the first monochromator.

The excitation spectrum demonstrates that for an effective luminescence not only the presence of emitting level is important, but also of the upper levels with a sufficiently intensive absorption. The excitation spectra enable to choose the mostly effective wavelength for luminescence observation. The combination of excitation and optical spectroscopies enable to determine the full pattern of the center's excited levels, which may be crucial for luminescence center interpretation, energy migration investigation and so on. The main excitation bands and lines of luminescence in minerals are presented in the Table 2.2.

2.3.2.1 Selective Spectroscopy

There are a large number of cases when the spectra of luminescence center remain broad up to helium temperatures. In certain cases, this is explained by a strong electron-phonon interaction, but more often the inhomogeneous broadening, connected with several types of the same center presence, causes this. In such cases it is possible to simplify the spectrum by selective excitation of a specific center.

2.3.2.2 Two-Photon Absorption Spectroscopy

In most of luminescence experiments, at least in the minerals luminescence field, excitation is due to absorption of a single photon. However, it is also possible for a luminescence center to absorb two or more long-wavelength photons to reach the excited state. Two-photon excitation occurs by the simultaneous absorption of two lower-energy photons. Such excitation requires special conditions including high local intensities, which can only be obtained from laser sources.

Thus it was not observed until lasers were invented. In principal, one-photon and two-photon excitation follow different selection rules. For example, the inner shell one-photon transitions in transition metal, rare earth, and actinide ions are formally forbidden by parity selection rule. These ions have d- or f-shells and transitions within them are either even to even ($d \rightarrow d$) or odd to odd ($f \rightarrow f$). The electric dipole transition operator is equal to zero. The two-photon transition operator is a tensor whose components may be nonzero. Thus an important reason for doing two-photon spectroscopy is that it allows to observe the transitions directly as allowed transitions instead of indirectly as forbidden transitions as are al one-photon spectra of transition metal ions.

In single beam two-photon spectroscopy, an intense laser beam having a frequency $h\nu = 1/2(E_2–E_1)$ is passed through the crystal, and the attenuation of the

Table 2.2 Main luminescence excitation bands and lines in minerals

Center	Electronic transition	$\lambda_{excitation}$ (nm)	$\lambda_{luminescence}$ (nm)
Pr^{3+}	$^3H_4$-$^1S_0$ $^3H_4$-$^1I_1$ $^3H_4$-$^1I_0$	280 460 490	480,610, 650, IR
Nd^{3+}	$^4I_{9/2}$-$^4G_{7/2–5/2}$ $^4I_{9/2}$-$^2D_{7/2}$ $^4I_{9/2}$-$^2P_{1/2}$	580 320 380	880, 1060, 1340 380, 417 535, 600, 670
Sm^{3+}	Charge transfer $^6H_{5/2}$-$^6P_{5/2}$	230 407	540, 600, 675
Sm^{2+}	4f-5d	450, 580	650, 690, 700, 730, 780, 810
Eu^{3+}	Charge transfer $^7F_0$-$^5H_3$ $^7F_0$-$^5D_2$	220 325 532	590, 610, 650, 700
Eu^{2+}	$4f^7$-$4f^65d$	250, 310	380–450
Tb^{3+}	$^7F_6$-$^5D_3$	280	380, 417, 440, 550
Dy^{3+}	$^6H_{15/2}$-$^6P_{5/2}$	351	480, 575, 670, 760
Er^{3+}	$^4I_{15/2}$-$^4G_{11/2}$	370	550
Ho^{3+}	$^5I_7$-$^5G_2$	360	550, 640, 760
Tm^{3+}	$^3H_6$-$^3P_0$	360	360, 480, 780, 800
Ce^{3+}	4f-5d	280, 320–340	350–390
Yb^{2+}	$4f^{14}$-$4f^{13}d^1$ (k_1-k_4)	250, 280, 320	520–560
Yb^{3+}	$^2F_{7/2}$-$^2F_{5/2}$	920, 960	975, 1014
Gd^{3+}	$^8S_{7/2}$-$^6I_{7/2}$	270	300, 312
Mn^{2+}	$^6A_1(S)$-$^4E,A_1(G)$ $^6A_1(S)$-$^4T_1(G)$	380–400 560–575	480–700
Mn^{4+}	$^4A_2$-$^4T_1$ $^4A_2$-$^4T_2$	420 530	650–700
Mn^{5+}	$^3A_2$-$^3T_2$ $^3A_2$-^1E	400 590	1100–1500
Fe^{3+}	Charge transfer $^6A_1(S)$-$^4E(D)$ $^6A_1(S)$-$^4T_2(G)$	300 380 500	700–750
Cr^{3+}	$^4A_2$-$^4T_1$ $^4A_2$-$^4T_2$	410 550	690–1000
V^{2+}	$^4A_2$-$^4T_1$ $^4A_2$-$^4T_2$	410 560	750–850
Pb^{2+}	$^1S_0$-$^3P_1$	220–260	300–400
Bi^{3+}	$^1S_0$-$^3P_1$	280	400–650
Bi^{2+}	$^2P_{1/2}$-$^2S_{1/2}$	270	600–650

(continued)

Table 2.2 (continued)

Center	Electronic transition	$\lambda_{excitation}$ (nm)	$\lambda_{luminescence}$ (nm)
Ti^{3+}	$^2T_{2g}$-^2E_g	500	650–950
$(UO_2)^{2+}$	400	A_{1g}-E_{2u}	480–580
S_2^-	$3\Sigma_g$-$^2\Pi_{3/2}$	400	500–700
O_2^-	$3\Sigma_g$-$^2\Pi_{3/2}$	280	400–550
$(MeO_m)^{n-}$	$^1A_1$-$^1T_1$	250–340	480–520

beam is measured. To measure attenuation directly is extremely difficult and is better to measure the proportional quantity, while the fluorescence maybe the most useful. Two-photon spectroscopy was not used yet in luminescence of minerals field, but many relevant centers have been studied in artificial compounds, such as Cr^{3+}, Mn^{4+}, Cu^+, Ni^{2+}, Eu^{2+} (McClure 1990).

2.4 Radiative Return to the Ground State: Luminescence

The energy of an electronically excited state may be lost in a variety of ways. A radiative decay is a process in which a molecule discards its excitation energy as a photon. A more common fate is non-radiative decay, in which the excess energy is transferred into the vibration, rotation, and translation of the surrounding molecules. This thermal degradation converts the excitation energy into thermal motion of the environment (i.e., to heat). Two radiative processes are possible: spontaneous emission, just like radioactivity, which is a completely random process where the excited state decays:

$$M^* \rightarrow M + h\bar{\nu}c. \tag{2.7}$$

Rate of spontaneous emission in populating state

$$f = -N_f A_{fi}, \tag{2.8}$$

where A_{fi} is the Einstein coefficient for spontaneous emission (unit s^{-1}). It obeys 1st order kinetics:

$$\begin{aligned} \text{Rate} &= -d[M^*]/dt; \\ N_f[t] &= N_f[t=0]\exp(-kt), \end{aligned} \tag{2.9}$$

where k = rate constant for spontaneous emission = A_{fi} = $1/\tau$ (τ = natural lifetime of state f, in seconds). Stimulated (or induced) emission can occur in the presence of light of the correct transition energy, incident upon the sample, acting as a stimulant for it to emit a photon of the same energy:

$$M^* + h\bar{\nu}c \rightarrow M + 2\bar{\nu}c. \qquad (2.10)$$

Rate of induced emission in populating the state:

$$f = -N_f B_{fi} \rho(\bar{\nu}), \qquad (2.11)$$

where B_{fi} is the Einstein coefficient of induced emission.

2.4.1 Ligand Field Theory

2.4.1.1 The Concept of the Ligand Field

Most of the electronic spectroscopy in minerals can be interpreted by the well-known ligand field theory. The main luminescence centers in minerals are transition and rare-earth elements. The ground and excited levels in these cases are d and f orbitals, while d-d and d-f emission transitions are subjected to strong influence from nearest neighbors, so called ligands. The basic notion of a ligand field was first developed many years ago (Bethe 1929). It was supposed that the ions are undeformable spheres and the interactions, which take place between them, are due solely to the electrostatic potentials set by their charges. The charges are taken to be located at the center of the ions coincident with their nuclei. Summing potentials from the individual surrounding ions to give the total potential at any point near central ion, and then finding the effect of such a potential on the electrons of the central ion solve the problem. The fact that the original considerations were made on the potential developed near an ion, which was part of lattice, and that such a lattice could occur only in crystal, led to the term *crystal field* theory. However, later was found that few of the results of crystal field theory depend on the existence of a lattice, and it is possible to carry most of them over to the model, which is at the basis of coordination chemistry – that of the coordination cluster. By coordination cluster is meant a central metal ion associated with a number of attached ligands, the whole forming a distinguishable entity and possibly bearing a net electric charge. In the more general model of a coordination compound, the electrons of the central ion are subject to a potential, not necessarily of simple electronic origin, from the ligand atoms. The term *ligand field* theory has been employed to cover all aspects of the manner in which an ion or atom is influenced by its nearest neighbors. Ligand field theory, then, contains crystal field theory as a special case. A great many of the results of ligand field theory depend only on the approximate symmetry of the ligands distribution around the central metal ion rather than on the particular ligands or the details of their locations. Consequently, the results may be obtained in the first place from crystal field theory and then generalized to other models of the bonding. Once the mathematical formalism of the crystal field approach has been developed, calculations within this model are quite straightforward (Figgis 1966, Di Bartolo and Collins 2006).

The splitting of d orbital energies and its consequences are at the heart of crystal field theory. It is possible to examine qualitatively the effect of various distributions of ligand atoms around the central ion upon its d orbitals. The d orbitals in an isolated gaseous metal are degenerate. If spherically symmetric field of negative charges is placed around the metal, these orbitals remain degenerate, but all of them are raised in energy as a result of the repulsion between the negative charges on the ligands and in the d orbitals. On going from a spherical to an octahedral symmetry, all orbitals are raised in energy, relative to the free ion. However, not all d orbitals will interact to the same extent with the six point charges located on the +x, −x, +y, −y, +z and − z axes respectively. The orbitals which lie along these axes (i.e. $x^2 - y^2$, z^2) will be destabilized more that the orbitals which lie in-between the axes (i.e. xy, xz, yz). Δ_0 or alternatively 10Dq denote the extent to which these two sets of orbitals are split. The tetrahedral symmetry can be derived from a cubic symmetry where only four of the eight corners are occupied by point charges. In such a situation, it is the xy, yz, xz orbitals are destabilized as they point towards the incoming point charges, while the $x^2 - y^2$ and z^2 are stabilized. The crystal field splitting in a tetrahedral symmetry is intrinsically smaller than in the octahedral symmetry as there are only four ligands (instead of six ligands in the octahedral symmetry) interacting with the transition metal ion (Fig. 2.2).

The point charge model predicts that

$$\Delta_t = \frac{4}{9}\Delta_o. \tag{2.12}$$

After determination of the qualitative schemes for the term splitting in the crystal fields, it is possible to go to a quantitative estimation of separation between the crystal field levels. All the electronic interactions stages can be represented quantitatively by means of the few parameters:

$$H = H_0 + H_{ee}(B,\ C) + H_{CF}(D_q) + H_{LS}(\xi) \tag{2.13}$$

where B, C are inter-electronic repulsion (H_{ee}) parameters governing term separation due to Coulomb repulsion of electrons; D_q is the cubic crystalline filed parameter determining term splitting by cubic crystalline field (H_{CF}); ξ is spin-orbit coupling parameter, taking into account further splitting of ion levels in crystals due to an interaction between orbital and spin moments of an ion (H_{LS}).

Interelectron interactions depend on the size, namely the greater the ion size, the more distant the electrons from each other, the less repulsion between them. Hence B and C decrease with decreasing of oxidation state, from the first transition series to the second and third series and from the first to the last ions within each of the series. For an ion in crystal the overlapping of transition metal and ligand orbitals leads to decrease of B and C, namely the more covalent the bonding, the less B and C. For a given transition metal ion the bond covalency depends on the ligand ions, which are arranged in the order of B decreasing in so-called nephelauxetic series:

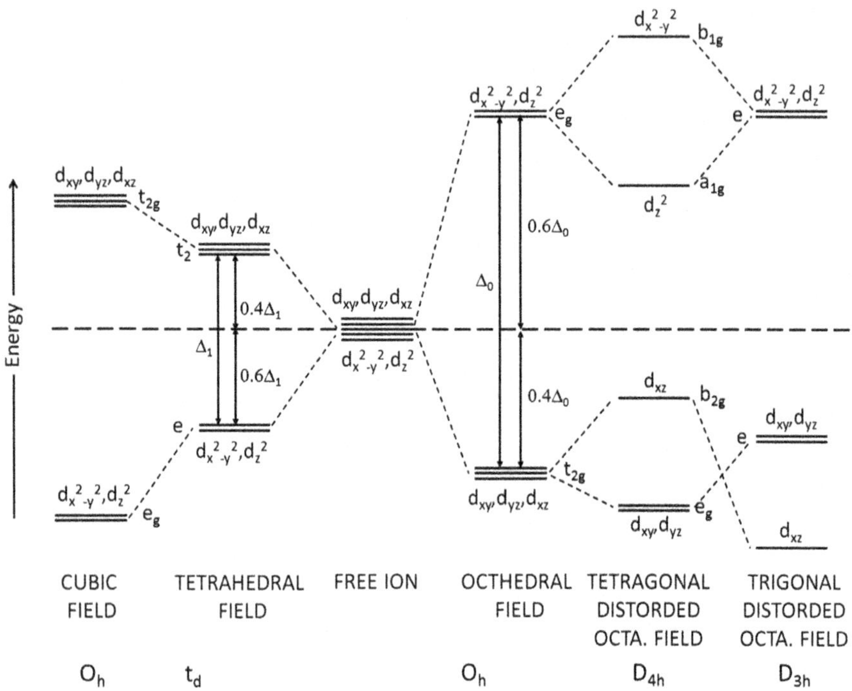

Fig. 2.2 Splitting of the five d-orbitals in various types of ligand fields (Bakhtin and Gorobets 1992)

$$F^- > O^{2-} > Cl^- > Br^- > S^{2-} > I^- > Se^{2-}.$$

The crystal field strength D_q is determined by the effective change of the ligands Q, by the average radius r of the d-orbital and by the ligand-metal distance R:

$$\Delta \sim \frac{(r^4)}{R^5} \tag{2.14}$$

This includes the dependence for D_q on the kind of transition metal ion, on the ligands arranges in the spectrochemical series in the order of increasing D_q ($I^- < Br^- < Cl^- < F^- < O^{2-} < H_2O$), on the metal-ligand distances and on the coordination (D_{qoct}: D_{qcub}: $D_{qtetr} = 1{:}8/9{:}4/9$).

It appears that the d^1 and d^6, the d^2 and d^7, the d^3 and d^8, and the d^4 and d^9 configurations have great similarity in the splitting of their levels with the crystal field.

After determination of the relative energies of the d orbitals in a transition-metal complex, the distribution of the electrons has to be considered. Degenerate orbitals are filled according to Hunds rules: one electron is added to each of the degenerate orbitals in a subshell before a second electron is added to any orbital in the subshell; electrons are added to a subshell with the same value of the spin quantum number until each orbital in the subshell has at least one electron. But such filling scheme is valid only for the high spin (or low field) configurations. If there is a large enough energy difference between the five d orbitals, as in the case if the ligand field is very large, the Hind's rule of maximum multiplicity no longer applies. As a result, the lower orbitals, for example the three t_{2g} orbitals in an octahedral system, will fill completely before any electrons enter the upper two e_g orbitals. This is low spin (or high field) configuration. Significant changes could occur in the octahedral $3d^4$ to $3d^7$ cases, and in the tetrahedral $3d^3$ to $3d^6$ cases.

2.4.1.2 Tanabe-Sugano Diagrams

Tanabe and Sugano taking the mutual interactions between the d-electrons as well as the crystal field into account have calculated the energy levels originating from such a configuration (Fig. 2.3). On the utmost left-hand side where crystal field $\Delta = 0$ we find the energy levels of the free ion. Many of these levels split into two or more levels for $\Delta = 0$. The lowest level, i.e. the ground state, coincides with the x-axis. For the free ion the levels are marked ^{2S+1}L, where S presents the total spin quantum number, and L the total orbital angular momentum. Values of L may be 0 (S), 1 (P), 2 (D), 3 (F), 4 (D), etc. The degeneracy of these levels is $2\,L + 1$and may be lifted by the crystal field. Crystal-field levels are marked ^{2S+1}X, where X may be A (no degeneracy), E (two fold degeneracy) and T (tree fold degeneracy). Subscripts indicate certain symmetry properties. The restrictions of these diagrams are connected with restrictions of the very model of crystalline field, the fact that the diagrams are calculated only for one B and C values, not speaking that the free ion values are taken, which differ considerably from corresponding values in crystals and the fact that splitting by cubic crystalline field only is shown without spin-orbital splitting.

In order to use Tanabe-Sugano diagram properly, the transition probabilities have to be taken into consideration. We have a ground state and we have many excited states, but not all transitions from ground to excited state and in the opposite direction are allowed. It is all a question of probability, while some have a high probability of occurrence, namely, allowed and intense transitions and some have a low probability, namely, forbidden and weak or very weak transitions.

The following selection rules are mostly important:

(i) There must be a change in parity (Laporte Rule) – transitions between states with the same parity are forbidden, thus all d-d transitions are forbidden
(ii) Transitions are allowed between states with the same spin and such transitions are seen only weakly in the 3d transition metal compounds, though more strongly in the 4d and 5d series.

The Laporte selection rule formally forbids all transitions within the d shell among all the energy levels. Nevertheless, the Laporte rule can be relaxed by two

Fig. 2.3 Tanabe-Sugano diagrams for d^4 transition elements

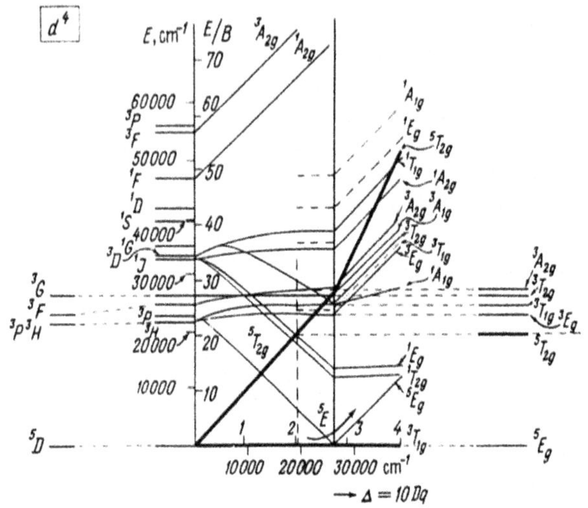

mechanisms. First, if the central ion is at a center of symmetry, there may be coupling of electronic and vibrational wave functions of opposite signs, the vibronic coupling, which lifts the symmetry and produces weakly allowed transitions. Second, in the absence of a center of symmetry, there can be a partial mixing of the 3d and 4p orbitals, which again results in weak transitions.

2.4.2 Configurational Coordinate Diagram

The energy level diagram of transition metal ions forms a starting point for the analyses of their luminescence properties. The configurational coordinate model is often used to explain optical properties, particularly the effect of lattice vibrations, of a localized center. In this model, a luminescent ion and the ions at its nearest neighbor sites are selected for simplicity. In most cases, one can regard these ions as an isolated molecule by neglecting the effects of other distant ions. In this way, the huge number of actual vibrational modes of the lattice can by approximated by a small number or a combination of specific normal coordinates. These normal coordinates are called the configurational coordinates. The configurational coordinate model explains optical properties of a localized center on the basis of potential curves, each of which represents the total energy of the molecule in its ground or excited state as a function of the configurational coordinate (Fig. 2.4). Here, the total energy means the sum of the electron energy and ion energy. To understand how the configurational coordinate model is built, one is first reminded of the adiabatic potential of a diatomic molecule, in which the variable on the abscissa is simply the inter-atomic distance. In contrast, the adiabatic potential of a platonic molecule requires a multidimensional space, but it is approximated by a single

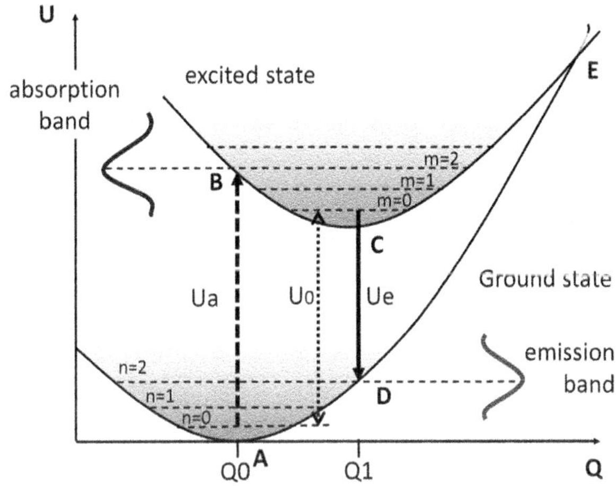

Fig. 2.4 General scheme for explanation of luminescence transition according to the Franck-Condon principle in a configurational coordinate diagram, showing the parameters energy (U) and configurational coordinate (Q). The excitation from the vibrational level n = 0 of the ground state to the excited state result in absorption band with the energy U_a. the relaxation of the system, i.e. from vibrational level m = 0 of the excited state to the ground state, causes an emission band with the energy U_e. The energy difference between the lowest possible vibrational levels of ground and excited state (n = 0 and m = 0) is U_0, with $U_a > U_0 > U_e$ (After Yacobi and Holt 1990 and Nasdala et al. 2004b)

configurational coordinate in the one-dimensional configurational coordinate model. In this model, the totally symmetric vibrational mode or the "breathing mode" is usually employed. Such a simple model can explain a number of facts qualitatively, such as:

- Stokes' law; i.e., the fact that the energy of absorption is higher than that of emission in most cases. The energy difference between the two is called the Stokes' shift;
- Widths of absorption or emission bands and their temperature dependence;
- Thermal quenching of luminescence.

It must be mentioned, however, that the one-dimensional model gives only a qualitative explanation of thermal quenching. A quantitatively valid explanation can be obtained only by a multidimensional model.

Following the path of the optical absorption transition, presume that Hooke's law expresses the bonding force between the luminescent ion and a nearest-neighbor ion. The deviation from the equilibrium position of the ions is taken as the configurational coordinate denoted as Q. The total energies of the ground state, U_g, and that of the excited state, U_e, are given by the following relations:

$$U_g = \frac{K_g Q^2}{2} \tag{2.15}$$

$$U_e = K_e(Q - Q_0)/2 + U_0 \tag{2.16}$$

where K_g and K_e are the force constants of the chemical bond, Q^o is the inter-atomic distance at the equilibrium of the ground state, and U_o is the total energy at $Q = Q_o$. The spatial distribution of an electron orbital is different between the ground and excited states, giving rise to a difference in the electron wavefunction overlap with neighboring ions. This difference further induces a change in the equilibrium position and the force constant of the ground and excited states, and is the origin of the Stokes' shift. In the excited state, the orbital is more spread out, so that the energy of such an electron orbital depends less on the configuration coordinate; in other words, the potential curve has less curvature. As Fig. 2.4 shows, the nucleus of an emitting ion stays approximately at the same position throughout the optical processes. This is called the Franck-Condon principle. This approximation is quite reasonable since an atomic nucleus is heavier than an electron by 10^3–10^5 times. The optical absorption proceeds from the equilibrium position of the ground state. The probability for an excited electron to lose energy by generating lattice vibration is 10^{12}–$10^{13}s^{-1}$, while the probability for light emission is at most $10^9\ s^{-1}$. Consequently, state B relaxes to the equilibrium position C before it emits luminescence. This is followed by the emission process C-D and the relaxation process D-A, completing the cycle. At finite temperature, the electron state oscillates around the equilibrium position along the configurational coordinate curve up to the thermal energy of kT. The amplitude of this oscillation causes the spectral width of the absorption transition. When two configurational coordinate curves intersect with each other, an electron in the excited state can cross the intersection E assisted by thermal energy and can reach the ground state nonradiatively. In other words, one can assume a nonradiative relaxation process with the activation energy ΔU, and with the transition probability per unit time N given by:

$$N = s\ \exp(-\Delta U/kT) \tag{2.17}$$

where s is a product of the transition probability between the ground and excited states and a frequency, with which the excited state reaches the intersection E. This quantity s can be treated as a constant, since it is only weakly dependent on temperature. It is called the frequency factor and is typically of the order of $10^{13}s^{-1}$. The luminescence efficiency η can by expressed as:

$$\eta = \frac{W}{W+N} = \left[1 + \frac{s}{W} exp \frac{-\Delta U}{kT}\right]^{-1} \tag{2.18}$$

If the equilibrium position of the excited state C is located outside the configurational coordinate curve of the ground state, the excited state intersects the ground state in relaxing from B to C, leading to a nonradiative process. As described above, the shape of an optical absorption or emission spectrum is decided by the Franck-Condon factor and also by the electronic population in the vibrational levels at thermal equilibrium. For the special case where both ground and excited states have the same angular frequency, the absorption probability can by calculated with harmonic oscillator wavefunctions in a relatively simple form:

$$W_{nm} = e^{-S}\left[\frac{m!}{n!}\right]S^{n-m}\left[L_m^{n-m}(S)\right]^2 \tag{2.19}$$

Here Laguerre's polynomial functions are used. The quantity S can by expressed as shown below, with K being the force constant of a harmonic oscillator and Q_o the coordinate of the equilibrium position of the excited state.

$$S = \frac{1}{2}\frac{k}{\hbar\omega}(Q - Q_0)^2 \tag{2.20}$$

Physically, S is the number of emitted phonons accompanying the optical transition. It is commonly used as a measure of electron-phonon interaction and is called the Huang-Rhys factor. At $m = 0$, the transition probability is given by the simple relation:

$$W_{n0} = S^n\frac{e^{-s}}{n!} \tag{2.21}$$

If $S < 1$ we are in the weak coupling mode, if $1 < S < 5$, in the intermediate coupling mode, if $S > 5$, in the strong coupling regime.

2.4.3 *Molecular Orbital Theory*

The further development of the ligand field concept takes place in Molecular Orbitals (MO) Theory. As an atomic orbital is a wave function describing the spatial probability density for a single electron bound to the nucleus of an atom, a molecular orbital is a wave function, which describes the spatial probability density for a single electron bond to the set of nuclei, which constitute the framework of a molecule.

The MO theory treats molecular bonds as a sharing of electrons between nuclei. Unlike the valence bond theory, which treats the electrons as localized balloons of electron density, the MO theory says that the electrons are delocalized. That means that they are spread out over the entire molecule. Now, when two atoms come together, their two atomic orbitals react to form two possible molecular orbitals. It emerges that the two types restrict all possible types of pair combination for s, p and d orbitals: σ and π molecular orbitals, each of them can be bonding and antibonding. Every pair of atomic orbitals forming the molecular orbital gives not one but always two molecular orbitals: bonding and antibonding. This is reflected in the formation of the two energy levels: a lower level corresponding to bonding and a higher one corresponding to antibonding. In the ground state, the bonding orbitals are usually completely occupied, while the antibonding orbitals are empty or partially occupied by d electrons. The σ- bonding orbitals made from identical atomic orbitals have a center of symmetry and are referred as even. The σ-antibonding orbitals without a center of symmetry are uneven. On the contrary, π-bonding orbitals are uneven, while π-antibonding orbitals are even. The parity in the MO theory determines some of the selection rules for optical transitions.

The molecule energy level diagrams show the energies of the molecular orbitals made from every possible combination of the atomic orbitals. To achieve this, one constructs the energy levels for each atoms and then forms from them the molecular orbitals under following conditions: (1) the atomic orbitals should have a comparable energy, (2) they must overlap appreciably, (3) they must have a same symmetry type, (4) each pair of atomic orbitals forming a bonding molecular orbital must give at the same time an antibonding molecular orbital, (5) each of the atomic orbitals contributes in greater or lesser degree to all the molecular orbitals of the same symmetry type. The electronic state of a molecule is represented by a set of all possible molecular orbitals and term symbols are designated: for linear molecules: $^{2S+1}\Lambda$ with $\Lambda = 0, 1, 2$, as Σ, Π, Δ; for nonlinear molecules and complexes: $^{2S+1}\Gamma$ where Γ is the symmetry type (A, B: one-dimensional, non-degenerate type; E: two-dimensional, doubly degenerate; T: three-dimensional, triply degenerate).

2.4.4 Discrete Variational Multi Electron Technique

The main deficiency of the ligand field theory is that its theoretical basis is mainly based on semi-empirical methods. For this reason it is not possible to discuss quantitatively the chemical difference between different ions, since the magnitude of the adjustable parameters may be determined only from experimental data. Recently the general method for the calculation of electronic structures including many-electron interactions from first principles (DV-ME–Discrete Variational Multi Electron) has been developed. By this technique, the multiplet energies are calculated in two steps: in the first step, one electron molecular orbitals (MO) and MO energies are calculated by the DV-X_α molecular orbital method; in the second step, the many-electron Hamiltonian, in which the interactions between two electrons are exactly described, is diagonalized within a subspace spanned by the Slater determinants made up of the obtained one electron data. This method enables to calculate multiplet structures without any restrictions on the symmetry of the crystal. It was revealed that the DV-ME method not only well reproduced the peak positions in optical spectra from first principles, but also was useful to discuss the effect of covalency and trigonal distortion of impurity-state wave functions on the multiplet structure. It was successfully applied to analyze the multiplet structures of $3d^2$ and $3d^3$ transition metals, f-d transitions of heavy metal ions and rare-earth ions in solids (Ishii et al. 1999, 2002).

2.4.5 *Luminescence Decay*

The mechanisms of luminescence decay from an optical center are of critical importance. In particular we have to know if there are any processes internal to the center or external to it, which reduce the luminescence efficiency. It is possible to define two decay times, τ_r, the true radiative decay time which a transition would have in absence of all non-radiative processes, and τ, the actual observed decay time, which may be temperature dependent, as will usually occur when there are internal non-radiative channels, and which may also be specimen dependent, as when there is energy transfer to other impurities in mineral. The quantum yield may be close to unity if the radiationless decay rate is much smaller than the radiative decay.

2.4.5.1 Radiative Decay

For allowed emission transitions the true lifetime is short, namely 10^{-7}–10^{-8} s, for strongly forbidden transitions in solids it is much longer, a few 10^{-3} s. For the two-level system (excited state and ground state) the population of the excited state decreases according to

$$dN/dt = -N_eP_{eg}. \tag{2.22}$$

The value of N_e gives the number of luminescent ions in the excited state after an excitation pulse, t the time, and P_{eg} the probability for spontaneous emission from the excited to the ground state. Integration yields

$$N_e(t) = N_e(0)e^{-t/\tau}, \tag{2.23}$$

where τ is the radiative decay time.

In the absence of non-radiative decay processes the experimentally observed decay time equals the radiative decay time. When non-radiative processes are present, the experimental value is reduced by a factor equal to the quantum efficiency of the luminescence. There are many factors, which affect the decay time. One is due competing non-radiative processes, which shorten the measured decay time. We will consider the latter first. The experimentally observed decay time of the luminescence is given by

$$\tau = 1/(K_r + k_i) \tag{2.24}$$

where k_r is the probability of radiative decay and k_I the probability of non-radiative decay processes from the same state. If k_I is much large than k_r not only will decay time be shortened appreciably, but the luminescence intensity will be very weak. Non-radiative deactivation within a center occurs by interaction with the vibrating lattice, which depends on temperature. Thus k_I is temperature dependent.

2.4.5.2 Non-radiative Decay

Radiative return from the excited state to the ground state is not the only possibility of completing the cycle. The alternative is nonradiative return, i.e. a return without emission of radiation. All the energy absorbed by the mineral, which is not emitted as radiation is dissipated to the lattice (radiationless process). Actually there are many centers, which do not emit at all. Absorption and emission transitions are possible with Stokes shift. The relaxed excited state may, however, reach the crossing point of the two parabolas if the temperature is high enough and to return to the ground state in a nonradiative manner. The excitation energy is then completely given up as heat to the lattice. This model accounts for the thermal quenching of luminescence.

Another possibility to return to the ground state is by transfer of the excitation energy from the excited center (S*) to another center (A). The energy transfer may be followed by emission from A and species S is then said to sensitize species A. However, A* may also decay nonradiatively and in this case species A is sad to be a quencher of the S emission. The mostly important quenchers in minerals are Fe^{2+}, Co^{2+} and Ni^{2+}, which have intense charge-transfer bands.

Let consider two centers, S and A, separated in mineral by distance R, which is so short that the centers S and A have a non-vanishing interaction with each other. If S is in the excited state and a in the ground state, the relaxed excited state of S may transfer its energy to A. Energy transfer can occur only if the energy differences between the ground and excited states of S and A are equal (resonance condition) and if a suitable interaction between both systems exists. The interaction may be either an exchange interaction (if we have wave functions overlap) or an electric or magnetic multipolar interaction. In such case, if the excitation spectrum of the A emission is measured, the excitation bands of S will be found as well, since excitation of S yields emission from A via energy transfer. If S is excited selectively, the luminescence of A also presents. Finally, the decay time of the S emission should be shortened by the presence of nonradiative energy transfer, since it shortens the life time of the excited state S*.

If the spectral overlap consists of a considerable amount of overlap of an emission band and an allowed absorption band, there can be a considerable amount of radiative energy transfer: S* decays radiatively and the emission band vanishes at the wavelengths where A absorbs strongly.

If we consider now transfer between two identical ions the same considerations can be used. If transfer between S ions occurs with a high rate, in a lattice of s ions there is no reason why the transfer should be restricted to one step. This can bring the excitation energy far from the site where the absorption took place. If in this way, the excitation energy reaches a site where it is lost nonradiatively (quenching site), the luminescence will be quenched. This phenomenon is called concentration quenching.

An additional delay between excitation and emission is often introduced by metastable electron states known as electron traps, which are filled during excitation. Once am electron has become trapped in such a state it requires certain energy to release it but this can be provided thermally. Emission, which follows the emptying of such traps at a fixed temperature, is known as phosphorescence.

2.4.5.3 Special Cases of Decay

The processes involved are stepwise energy transfer, cooperative sensitization of luminescence and cooperative luminescence. As an example of stepwise energy transfer a system containing Er^{3+} and Yb^{3+} may be considered. The latter ion absorbs at 970 nm (10,300 cm^{-1}) and in phonon assisted an Er ion is excited to its $^4I_{11/2}$ state. Then a second photon is absorbed by another Yb ion and produces $^4I_{11/2}$-$^4F_{7/2}$ in the excited Er by resonance energy transfer. This successive double energy transfer accounts for the fact that the excitation spectrum of such a system agrees with the reflection spectrum of Yb^{3+} and that the intensity of the resulting green emission $^4S_{3/2}$-$^4I_{15/2}$ (after radiationless $^4F_{7/2}$-$^4S_{3/2}$) of erbium is proportional to the square of the flux of infra-red photons. Two excited ($^2F_{5/2}$) Yb ions can cooperatively excite a third ion of another rare-earth element, such as terbium, if the sum of the energies of the donor system corresponds to the excited level of the acceptor system. Cooperative luminescence is the opposite process of cooperative absorption and can be described as two excited ions that simultaneously make transitions downwards, emitting one photon having the sum of the two energy differences. A similar case where an ion emits one photon and simultaneously exciting another ion is also known.

A related phenomenon is the conversion of single visible photons with the result the quantum efficiency can be higher than 100 %. If, for example, 0.1 % of Pr^{3+} is incorporated in YF_3 and excited with the mercury spectral line at 185 nm, the electron from 4f5d states decay non-radiatively to the $^1S_0$ ($4f^2$) state. This system is able to generate two visible photons by $^1S_0$-$^1I_6$ followed by non-radiative decay to the closely adjacent $^3P_0$ and by transition another photon is emitted by transitions to one of the six J-levels of ^3H or ^3F. A condition for this cascade process is that the nephelauxetic effect for inter-shell transitions is sufficiently weakly pronounced for the lowest 4f5d state to be above $^1S_0$ (Reisfeld and Jörgensen 1977).

2.4.6 *Luminescence in the Band Scheme*

Atoms in minerals are very close and influence strongly on each other. As a result in certain minerals the energy levels of individual atoms are combined and form the energy zones filled by electrons (Fig. 2.5). The inner atom orbitals of the individual atoms form inner energy zones totally filled by electrons. The outer atom orbitals filled by electrons form the outer energy zones filled by electrons. The last filled zone is called valence band (VB) because valence electrons take part in its occupation. The higher energy zones, following the valence zone, are empty because

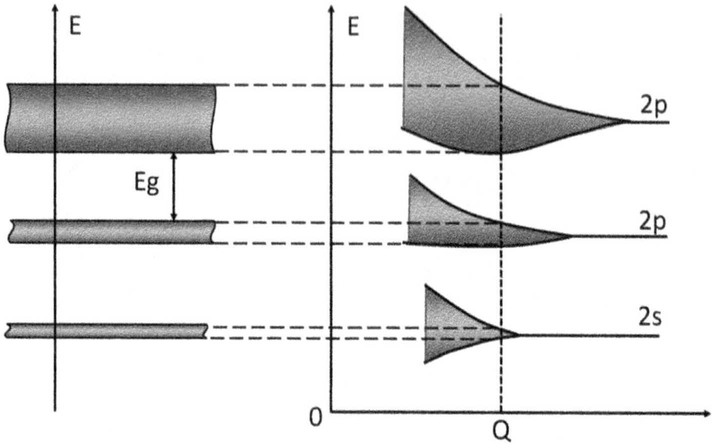

Fig. 2.5 Energy zones formation scheme (Bakhtin and Gorobets 1992)

they are formed by empty electron orbitals. The lowest of these zones is called conductivity band (CB). The energy interval between the valence zone and conductivity zone is called forbidden zone and designated as E_g.

With small dimensions of the forbidden band the electron transfer of the impurity or of the main substance to the conduction band may take place. Most important luminescent minerals of such kind are ZnS and silver bromides. With the interband spacing of 3–4 eV a UV irradiation with a wavelength of less than 300 nm has enough energy to detach electrons and transfer them from filled valence band into an empty conduction band. Thus we have a free electron in the conduction band and a free hole in the valence band. The certain defects may catch a free electron, forming an electron center, or a free hole forming a hole center. Upon disappearance of such centers, as result of recombination with the opposite charge carrier, recombination energy serves as excitation source for luminescence center and emission appears (Fig. 2.6). The following processes are possible:

- "Intrinsic" luminescence where an electron is excited from the valence band to the conduction band, so-called interband transition. Recombination of this electron with a hole in the valence band generates a photon, the energy of which corresponds to the energy difference of the band gap (Fig. 2.6a);
- The excited electron may also recombine with an activator, with the following luminescence, or a trap, with the following electron capturing, within the forbidden gap. Traps and activator energy levels are caused by defects in the crystal lattice (Fig. 2.6b);
- Luminescence generation through direct recombination of electrons from a trap with an activator (tunneling) or with a hole in the valence band (Fig. 2.6c);
- Stimulated release from an electron from the trap to the collection band, followed by emissive recombination with an activator. This process is called

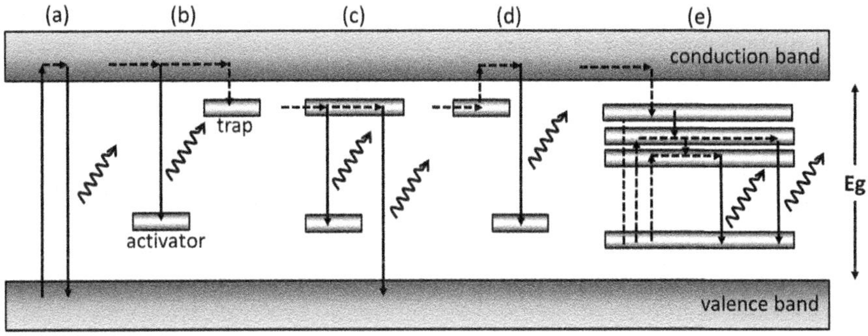

Fig. 2.6 (**a–e**) Simplified sketch of electron band structure of an semiconductor mineral, showing the processes of excitation (energy absorption), non-radiative energy transfer and generation of luminescence (After Nasdala et al. 2004b)

thermoluminescence (electron release stimulated by heating) and optically stimulated luminescence (electron release stimulated by light) (Fig. 2.6d);

– "Extrinsic" luminescence, where after being excited, electrons of defect ions recombine with the ground state with luminescence emission (Fig. 2.6e).

References

Bakhtin A, Gorobets B (1992) Optical spectroscopy of minerals and ores and its use in geology. Kazan University Publishing, Kazan (in Russian)

Bethe G (1929) Splitting of terms in crystals. Ann Phys 3:133–206

Di Bartolo B, Collins J (2006) Luminescence spectroscopy. In: Vij D (ed) Handbook of applied solid state spectroscopy. Springer, New York, pp 509–575

Figgis B (1966) Introduction to ligand fields. Wiley, New York

Ishii T, Ogasava K, Tanaka A, Adachi H (1999) Theoretical calculation for multiplet structure of chromium ion pair in ruby. Mater Trans JIM 40:416–419

Ishii T, Ogasava K, Adachi H (2002) First-principles analysis for the multiplet structures of tetrahedrally and octahedrally oxo-coordinated 3d2 and 3d3 transition metals. J Chem Phys 116:471–479

Marfunin A (1979a) Physics of minerals and inorganic materials. Springer, Berlin/Heidelberg/New York

Marfunin A (1979b) Spectroscopy, luminescence and radiation centers in minerals. Springer, Berlin/Heidelberg/New York

Matsyuk S, Zintchouk N (2001) Optical spectroscopy of the upper mantle minerals. Nedra, Moscow (in Russian)

McClure D (1990) Techniques and applications of two-photon spectroscopy of inorganic solids. Standard Publication 1066 (American Society for testing and materials), pp 21–35

Nasdala L, Götze J, Hanchar J et al (2004b) Luminescence techniques in Earth sciences. In: Biran A, Lubowitzky E (eds) EMU notes in mineralogy, spectroscopic methods in mineralogy, vol 6/2. Mineralogical Society of America, Chantilly, pp 43–91

Platonov A (1979) Color of minerals. Naukova Dumka, Kiev (in Russian)

Reisfeld R, Jörgensen K (1977) Lasers and excited states of rare earths. Springer, Berlin/Heidelberg/New York

Rossman GR (1988) Optical spectroscopy. In: Hawthorne FC (ed) Spectroscopic methods in mineralogy and geology, vol 18, Reviews in mineralogy. Mineralogical Society of America, Washington, DC, pp 207–254

Yacobi C, Holt D (1990) Cathodoluminescence microscopy of inorganic solids. Plenum, New York

Chapter 3
Experimental Techniques

3.1 Lasers

When lasers first emerged from the laboratory and became commercially available in the late 1960s, they have found application in diverse areas. Spectrometric analysis was among the technology's earliest applications. Unlike established continuum sources, the monochromatic, coherent beams from lasers delivered high energy density. The result was a quantum leap in spatial and temporal resolution, sensitivity and speed of acquisition. For more than 40 years, spectrometric techniques implemented sources with narrower wavebands, shorter pulses and tunable emission.

The lasers are used in the luminescence of minerals for the following reasons:

- The concentration of luminescence centers is usually extremely low and we need high-energy sources.
- Several luminescence centers often present simultaneously in one mineral and we need selective monochromatic excitation in broad spectral range.
- Decay time of many luminescence centers in minerals is extremely short and for time-resolved spectroscopy we need the pulsed source with very short pulse width.
- For practical application in remote sensing we need high directional, very intense and powerful source.

The only source, which combines all those parameters, is monochromatic pulsed laser.

M. Gaft et al., *Modern Luminescence Spectroscopy of Minerals and Materials*,
Springer Mineralogy, DOI 10.1007/978-3-319-24765-6_3

3.1.1 The Basic Theory of Laser Action

Einstein's laws of absorption and emission describe the operation of lasers. Luminescence of minerals, considered in this book, is spontaneous emission where the luminescence is independent of incident radiation. In stimulated emission the relaxation is accomplished by interaction with a photon of the same energy as the relaxation energy. Thus the quantum state of the excited species and the incident photon are intimately coupled. As a result the incident and the emitted photons will have the same phase and propagation direction. The emitted light of stimulated emission is therefore coherent as opposed to the random distribution of propagation direction and phase found in spontaneous emission.

The energy levels systems used in laser action contain at least three levels. Population inversion is achieved by stimulated absorption through a spin allowed transition from the ground state into a state with high energy, which is then nonradiatively relaxed to a lower state, which has symmetry and spin characteristics that hampers fast spontaneous decay to the ground state. It is from this lower metastable state that the laser action occurs. Lasing is accomplished by placing the laser material between two mirrors. It is initiated either by a spontaneous transition from metastable level or by an external photon of the right energy. The beam thus created is passed back and forth between the two mirrors, so that it on each pass through the laser media initiates more stimulated emission and increases in power.

3.1.2 Short Description of the Lasers Used for Luminescence of Minerals Excitation

(a) Argon ion CW laser is one of the earliest types of lasers. We used it in steady-state spectroscopy. It consists of capillary tube where an electrical discharge generates high current density plasma of argon. The argon ions created in the discharge are the laser medium. The argon laser is capable of emitting several different wavelengths, one at a time or even simultaneously, such as 514.5, 496.5, 488.0, 476.5 and 457.9 nm.

(b) Excimer laser contain an excited complex consisting of two or more different atoms or molecules, which exists only in an excited state, while the ground state is repulsive and after radiative de-excitation it falls apart again. An electric pulsed discharge is used within the laser tube to supply the energy to the laser process. The gas filling consists of three major constituents: rare gas (Ar, Kr, He), a halogen compound (F, Cl, Br) and a buffer gas (He, Ne). Within the discharge the excited complexes are formed, e.g. XeCl. Sometimes it is also possible to use nitrogen as laser gas. The major advantages of the excimer laser are high pulse energies, high repetition rates, and direct generation of the UV light. Selecting the appropriate gas mixture may choose several different wavelengths (ArF-193 nm, KrCl-222 nm, KrF-248 nm, XeCl-308 nm, N_2-337 nm).

The main disadvantages are poor beam quality, relatively long pulse durations and toxic gases.

(c) Nd-YAG laser (neodymium doped yttrium aluminum garnet) can work in CW mode and in pulsed version. In a first step electric energy is stored in capacitors. This energy is discharged into flash lamps within approximately 100 μs. The emitted optical energy is subsequently stored in the laser crystal and is opened by means of an electro-optical cell (Q-switch) as a single pulse of 0.1–1 J within approximately 6–10 ns. The primary IR radiation is at 1064 nm, which may be frequency doubled to 532 nm. By additional optically non-linear frequency conversion wavelengths of 355 and 266 nm are attainable.

(d) The dye laser is the best laser for spectroscopy because it may be tunable over comparatively large range of wavelengths. The laser media are molecules of an organic dye dissolved in a suitable solvent. A so-called pump laser, an excimer laser in our case, provides the energy input.

(e) OPO lasers (Optical Parametric Oscillator) are optical devices pumped by a laser to produce two wavelengths, called the signal and idler, different from those of the pump laser. They are more complicated than laser oscillators due to the nonlinear nature of the gain medium. Within an optically non-linear crystal an incoming photon is broken up into two photons, while the cutting line between the two photon energies can be chosen arbitrarily. OPOs can be used to shift the wavelength of fixed-frequency lasers to a desired wavelength.

3.2 Steady-State Luminescence Spectroscopy

3.2.1 Experimental Apparatus

Firstly we shall consider the various forms of conventionally determined spectra, as distinct from time-resolved spectra, which can be obtained using continuous excitation. The measurement of emission spectra enables parameters such as the spectral position, spectral width, and spectral intensity to be determined. Moreover, the variation of these parameters with temperature is very important in deciding the origin of spectral bands. Sometimes the luminescence from single crystals is polarized and this can also give information as to the nature of the center.

The traditional steady-state experimental setup has a UV lamp as excitation source. It is equipped with two monochromators to select both the excitation and emission wavelengths. To record an excitation spectrum, the emission monochromator is set at the desired wavelength, generally the emission maximum. The excitation monochromator is then scanned through the excitation bands of the luminescence center. To improve the specificity of the technique is to employ a tunable laser as the active component of the system. In an ideal laser-based spectroscopic system the exciting wavelength would be tunable through the large fraction of the ultraviolet and visible regions of the electromagnetic spectrum. Emission spectra are recorded by selection of an appropriate excitation wavelength and scanning wavelength with the emission monochromator. The luminescence is quantified with the corresponding electronic devices. Band-pass excitation filters have to be used to remove second order diffraction from the excitation source. Polarizes may be added to both the excitation and emission paths. Sample holder usually contains cryostat enabling spectral measurements at low temperatures.

The observed excitation spectrum is distorted because the light intensity of the excitation source is a function of the wavelength and the transmission efficiency of the excitation monochromator is a function of wavelength. The emission spectra are distorted by the wavelength-dependent efficiency of the emission monochromator and the photomultiplier tubes (PMP). Thus both spectra have to be corrected for the system response. It is well known that for gratings the maximum efficiency is controlled by the blaze applied during manufacture of the grating. In the same way, PMP are strongly wavelength dependent. Consequently, luminescence of a given apparatus represents biased information about the original luminescence of the sample. The recorded signal differs greatly from that of the original emission. Determination of the system response is essential to correct spectra for discrepancies due to the method of acquisition so that they may be compared among laboratories. The calibration of wavelength and intensity may be achieved with reference lamps whose emission spectra are known.

In our research the steady-state photoluminescence spectra in the UV-visible range were investigated under UV lamps and lasers excitations and analyzed by grating monochromator equipped with continuous-flow helium cryostat at temperatures from 20 to 300 K. The steady-state luminescence in the range of 200–900 nm was measured by photomultiplier tube and a range of 900–1600 nm was detected by a Ge-detector. The steady-state cathodoluminescence (CL) spectra were recorded with a CL spectrometer on a Jeol JSM-840 scanning electron microscope. The recording system consists of a parabolic mirror, a silica window allowing the passage of UV emission, a Jobin-Yvon H-10 spectrometer and a Hamamatsu R636 photomultiplier, which allows detection from 200 to 900 nm. Steady-state spectroscopy has been combined with lifetime measurements. We used time-domain technique, where the sample is excited with a pulse of laser light, while width of pulse is much shorter than the measured decay time. The time-dependent intensity is measured following the excitation pulse, and decay time is calculated from the slope of a plot of log I(t) versus t, or from time at which the intensity decreases to 1/e of a value at $t = 0$. Decay information is extremely important, but if you found several decay components at the same λ, it is not possible to know if they connected with individual luminescence center or with several ones.

3.2.2 *Spectral Deconvolution*

Natural minerals may contain simultaneously up to 20–25 luminescence centers, which are characterized by strongly different emission intensities. Usually one or two centers dominate, while others are not detectable by steady-state spectroscopy.

In certain cases deconvolution of the luminescence spectra may be useful, especially in the case of broad emission bands. It was demonstrated that for deconvolution of luminescence bands into individual components, spectra have to be plotted as a function of energy. This conversion needs the transposition of the Y-axis by a factor λ^2/hc (Townsend and Rawlands 2000). The intensity is then expressed in arbitrary units. Deconvolution is made with a least squares fitting algorithm that minimizes the difference between the experimental spectrum and the sum of the Gaussian curves. Based on the presumed band numbers and wavelengths, iterative calculations give the band positions that correspond to the best fit between the spectrum and the sum of calculated bands. The usual procedure is to start with one or two bands and to increase the band numbers until the deconvolution does not significantly improve. Band positions can be constrained or assumed for calculation. Such deconvolution can reveal different activators and establish their relative importance, particularly where others overlap some luminescence centers with broader or more intensive emission bands. Deconvolution must be carried out with care and at any case not all bands present in steady-state spectrum may be assigned to a specific activator by deconvolution into Gaussian curves. Our experience with deconvolution evidences that it may be effective only when the exact number of the bands and lines and their spectral positions are already known. In such case the relative intensity of each one may be evaluated. Deconvolution of the broad bands of unknown origin for several components and their physical interpretation may lead to mistakes, as it was in our attempts with zircon and barite emissions.

3.3 Time-Resolved Luminescence Spectroscopy

The different decay times of luminescence centers in minerals, changing from ns to ms (Table 1.3), make time resolved spectroscopy possible and very effective. Let suppose that we have three different luminescence centers with different decay times emitting on the same spectral range. In traditional spectra, the time-integrated intensity is measured on each wavelength from all three centers, namely the resulting signal will be the sum of the three areas under decay curves. This procedure will be repeated on each wavelength with certain steps and the resulting spectrum will be complicated mixture of the three overlapped spectral features.

Figure 3.1 schematically represents time-resolved experimental setup used in our experiments. The excitation sources were pulsed lasers, such as excimer XeCl (308 nm), nitrogen (337 nm), three harmonics of Nd-YAG (266, 355 and 532 nm), and tunable dye and OPO, which deliver pulses of $\approx$10 ns duration. The spectra observed at the geometry of 90° are analyzed by intensified CCD matrix. Image intensifiers comprise three main components: a photocathode, microchannel plate (MCP) and phosphor screen. The standard operation of these devices starts when the incident photons become converted into electrons at the photocathode. The electrons then accelerated towards the MCP where they are multiplied to an amount

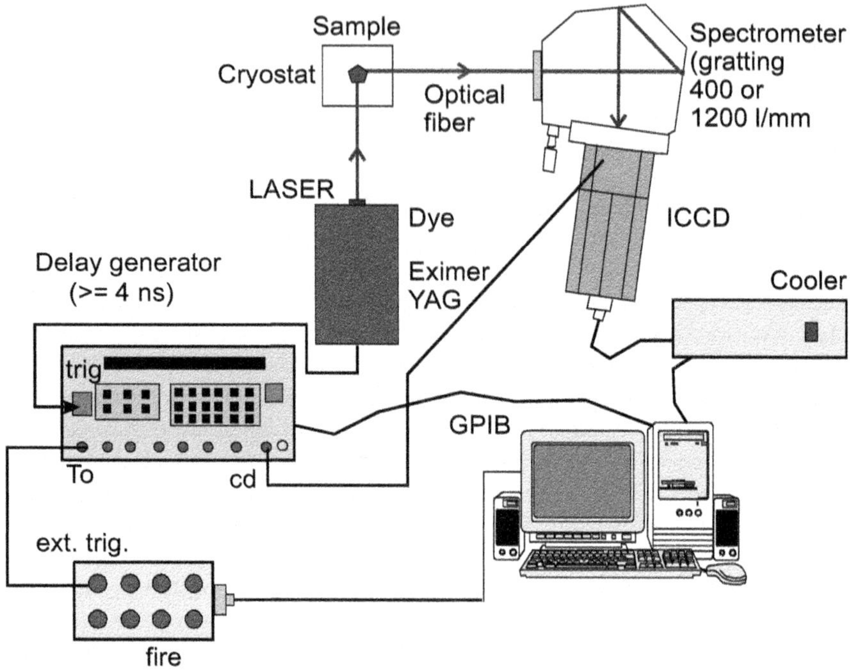

Fig. 3.1 Schematic experimental setup for time-resolved luminescence spectroscopy

dependent on the gain voltage across the MCP. After MCP multiplication, the electrons are accelerated further toward the phosphor screen where they are converted back into photons ready for the CCD to detect. CCD is two-dimensional array, which comes in a semiconductor "chip" package. Each pixel is overlaid with a small voltage-carrying element known as an electrode. During illumination of the chip, charge accumulates in the pixel. To collect the data, a sequence of voltages is applied across the electrodes to move the charge row by row down the vertical dimension of the chip and into a shift register at the bottom of the array. This charge is then moved similarly across the shift register to the output node where it converted to a digital form processing. The most significant benefit of this readout mode is that the associated readout noise is very low. Typically, CCDs have sensitivity similar to a photomultiplier tube; however, unlike photomultiplier tube a CCD is not damaged by over-exposed to bright lights. The second purpose of the image intensifier is electronic ultra-short shutter function, which allows a so-called gated mode of operation crucial for time-resolved spectroscopy.

In recent years the use of CMOS (complementary metal oxide semiconductor) technology has received attention for the implementation of image sensors, e.g. for general-purpose imagers and vision systems. CMOS imaging systems have several advantages, which include lower cost and higher frame rate, which is important for signal/noise ratio when luminescence intensity is very low.

Laser pulses synchronize an iStar detector (ANDOR V). It enables to accomplish spectral measurements in certain time windows, which are determined by delay time, namely the time between the end of the laser pulse and the beginning of the measurement, and gate width, namely the time between the beginning and the end of the measurement. Simply speaking, detector is closed during delay time and open during the gate width. After that the system is waiting for other laser pulse and accomplishes the cycle one again. The number of pulses is determined according to signal/noise ratio and depends on luminescence intensity. Delay time and gate width may be changed from 1 ns to 19 ms. Spectral resolution is determined by the number of measuring pixels and the monochromator gratings. We use 1024 channels and gratings from 300 to 1200 lines/mm. Gratings were of two kinds: ruled where grooves are physically formed into a reflective surface with a diamond on a ruling machine and holographic, produced from laser constructed interference patterns and a photolithographic process.

For delivering the luminescence to monochromator we used quartz fiber optics, which enables the spectral measurement in the range from 200 to 900 nm. Thanks to the need for fiber optics in the communication technology, low absorption silica fibers have been developed. Similar fibers can be used as measurement fibers to transport light from the sample to the optical bench of the spectrometer. The easy coupling of fibers allows a modular build-up of a system that consists of light source, sampling accessories and fiber optic spectrometer. Advantages of fiber optic spectroscopy are the modularity and flexibility of the system.

The additional improvement used in the system compared to traditional luminescence methods used in luminescence of minerals is so-called multichannel advantage. Traditionally, spectroscopy has involved using a scanning monochromator and a single element detector, for example photomultiplier tube, placed at exit slit. In such a scanning system, a complete spectrum is built up point by point, by moving the grating to select each wavelength. By contrast, in multichannel spectroscopy the grating is fixed and the exit slit and single detector are replaced with an array of detectors, each viewing a different wavelength. In our system 1024 26 μm detectors are used, so a complete spectrum is recorded in the same time it takes to record one wavelength point with a scanning system. Array based spectrometers also have potential for better stability and reproducibility since they have no moving parts.

Now let suppose that three centers with similar spectral characteristics have different decay times of 10 ns, 1 μs and 1 ms (Fig. 3.2). If we use delay time of 10 μs and gate width of 19 ms, two centers with decay times of 10 ns and 1 μs will be already quenched, while the third one with the longest decay time remains. The resulting spectrum will represent only emission of the long-lived center. If we use delay time of 1 ns and gate width of 100 ns, only the first center with the shortest decay time will be registered, while the other two are still in excited state during the measuring time and do not participate in emission. If we use delay time of 100 ns and gate width of 10 μs the first center will be already quenched before the beginning of the measuring time, while the third one will be mainly still in excited state during the measuring time and the resulting spectrum represents only the

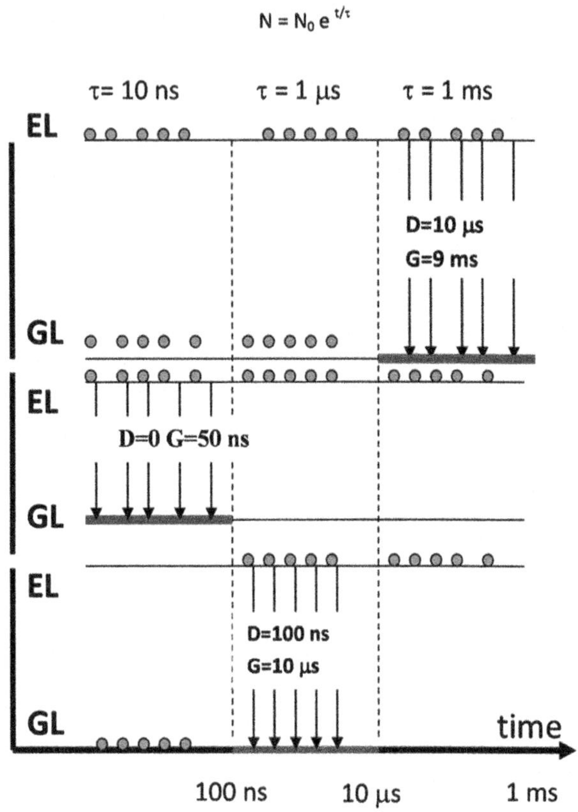

Fig. 3.2 Scheme of simultaneous emission of three luminescence centers in the same spectral range with different decays, which may be resolved by time-resolved spectroscopy

second center with intermediate decay time. If we use delay time of 1 ns and gate width of 50 ns, two long-lived centers will be still in excited state during the measuring time and the resulting spectrum represents only the first center with the shortest decay time.

One example demonstrates the advantage of time-resolved technique compared to steady-state one. Time-integrated cathodoluminescence spectrum of apatite enables to detect only two dominant luminescence centers, Mn^{2+} and Ce^{3+}. Laser-induced luminescence spectrum with minimal delay time of several nanoseconds and maximum gate width of several milliseconds, which is practically steady-state spectroscopy, also demonstrate the same centers, while their relative intensities are different (Fig. 3.3a, b). After short delay of 100 ns the second Ce^{3+} centers is detected, which has slightly longer decay time compared to the first one (Fig. 3.3c). Using delay time of 500 ns in order to quench the short-lived Ce^{3+} centers and narrow gate of 500 ns in order to quench long-lived centers the band of Eu^{2+} dominates the spectrum (Fig. 3.3d). After delay of 10 μs when luminescence of Eu^{2+} also disappears, the lines of REE^{3+} with very long decay times, such as Tb^{3+}, Dy^{3+} and Eu^{3+} are confidently detected (Fig. 3.3e), while changing excitation from 266 to 337 nm the second type of Eu^{3+} luminescence center appears (Fig. 3.3f).

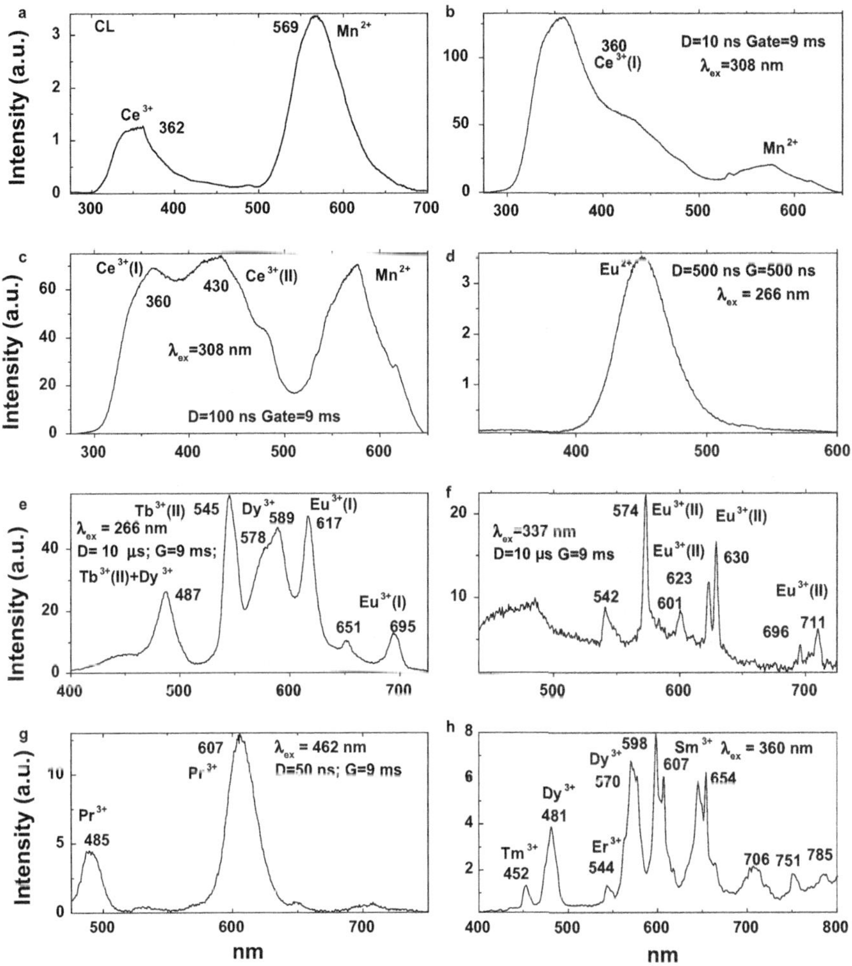

Fig. 3.3 (**a–h**) Cathodoluminescence (**a**) and laser-induced time-resolved luminescence (**b–h**) spectra of the same apatite sample

Excitations with 480 and 360 nm enable to see luminescence lines of Pr^{3+}, Tm^{3+}, Er^{3+} and Sm^{3+} (Fig. 3.3g, h).

The optical setup for time-resolved micro-luminescence measurements is based around an Axiotech 100 HD Zeiss microscope, modified to allow laser injection and fluorescence collection. The sample is observed either under transmission or reflection of polarized white light, or under UV illumination (HBO lamp). A set-up consisting on a dichroic mirror, for the selection of the excitation wavelength, and an objective (Epiplan Neofluar obj. >350 nm; Ealing/Coherent reflection obj. <350 nm) is used to focus the laser beam on the sample (spatial resolution over 5 μm with a ×50 objective). The limitation of the spatial resolution is mainly caused by the divergence of the laser beam. The same objective collects emitted fluorescence and feeds it back in a pure silica glass fiber (transparent in the UV, visible and near IR range) (Fig. 3.4).

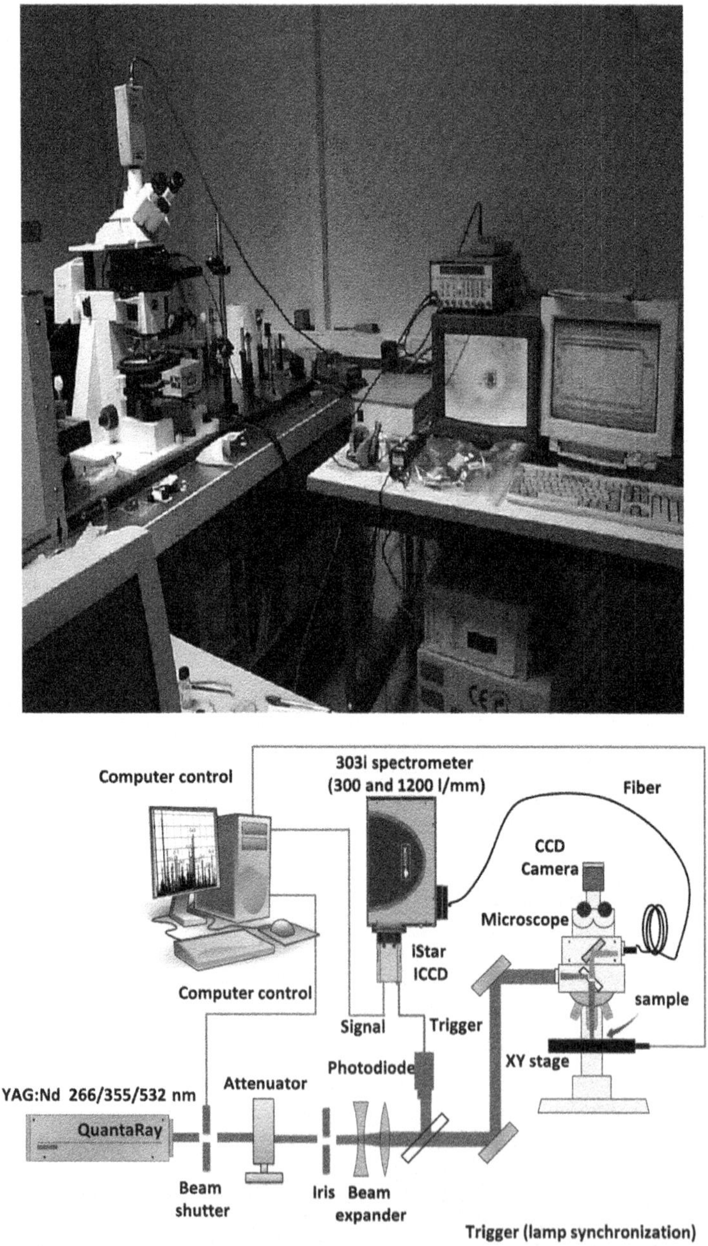

Fig. 3.4 Laser-induced time-resolved micro-luminescence system

3.4 Other Laser Based Spectroscopic Techniques

The experimental setup presented on Fig. 3.1 may be easily transformed for investigation of other spectroscopic properties under laser excitation, such as laser-induced breakdown spectroscopy (LIBS) and Raman spectroscopy. For the LIBS the focusing lens has to be added between the laser and the mineral to create enough power to form plasma. For Raman investigation holographic Notch filter has to be added between the mineral and the monochromator. They are fabricated by recording interference patterns formed between two mutually coherent laser beams unlike conventional interference filters, which are made by vacuum evaporation techniques. Since all layers are recorded simultaneously within a thick stack, the optical density of the notch filter is high and its spectral bandwidth can be extremely narrow. Also, since the layering profile is sinusoidal instead of square wave, holographic notch filters are free from extraneous reflection bands and provide significantly higher laser thresholds.

Reference

Townsend P, Rowlands A (2000) Information encoded in cathodoluminescence emission spectra. In: Pagel M, Barbin V, Blanc P, Ohnenstetter D (eds) Cathodoluminescence in geosciences. Springer, Berlin/Heidenberg/New York, pp 41–58

Chapter 4
Luminescent Minerals

Gorobets and Rogojine (2001) gave the general systematization of minerals according to their luminescent properties (Table 4.1). In the case considered, mineral groups comprise phases with similar luminescence properties determined by luminescence and quenching centers. At the first level, all minerals are divided on dielectric, semiconductors, and metals together with intermetallic compounds. From the physical point of view, it means classification by the energy band gap E_g. Minerals with $0 \leq Eg \leq 1$ are incapable of luminescence, because they are opaque to optic photons. Other minerals are principally luminescent if they are not saturated with quenching centers.

At the second level, the type of chemical compounds subdivides luminescent minerals: homoatomic compounds, sulfides, halides and oxygen-bearing compounds. Such level is mainly important for recombination induced luminescence and practically is not considered in our book.

At the third level, the most detailed partition of luminescence minerals is carried out on the basis of metals in the mineral formulae. In rare cases we have minerals with host luminescence, such as uranyl minerals, Mn minerals, scheelite, powellite, cassiterite and chlorargyrite. Much more often luminescent elements are present as impurities substituting intrinsic cations if their radii and charges are close enough. Thus, for example, Mn^{2+} substitutes for Ca^{2+} and Mg^{2+} in many calcium and magnesium minerals, REE^{3+} and REE^{2+} substitutes for Ca^{2+}, Cr^{3+}substitutes for Al^{3+} in oxygen octahedra, Fe^{3+} substitutes for Si^{4+}in tetrahedra and so on. Luminescence centers presently known in solid-state spectroscopy are summarized in Table 4.2 and their potential substations in positions of intrinsic cations in minerals in Table 4.3.

Thus, for a mineral to be luminescent the following three conditions must be satisfied at once: (1) a suitable type of crystal lattice favorable to form emission centers; (2) sufficient content of luminescence centers and (3) a small amount of quenchers. We arrange the luminescent minerals in our book according to main major element, the substitution of which by luminescence centers determines the emission properties of mineral.

M. Gaft et al., *Modern Luminescence Spectroscopy of Minerals and Materials*,
Springer Mineralogy, DOI 10.1007/978-3-319-24765-6_4

Table 4.1 Minerals assemblage according to luminescence properties (Gorobets and Rogojine 2001)

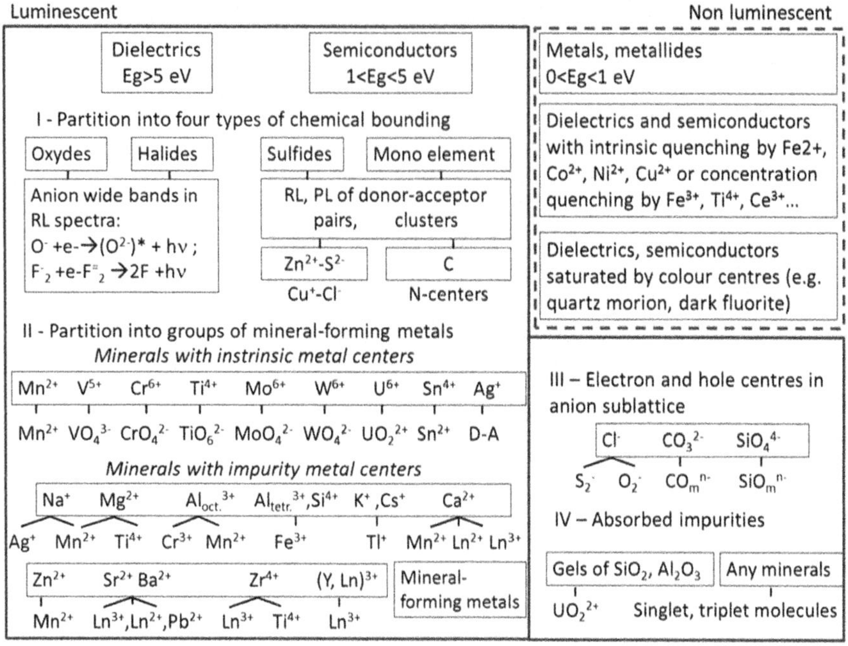

Table 4.2 Luminescent activator ions and valence states ordered to electronic configurations (column numbers correspond to the periodic table columns)

4	5	6	7	8	9	10	11	12	13	14	15	16	17	18
Transitions elements $3d^n$							d^{10}		Heavy ions $4\text{-}6s^2$ $4\text{-}6s\,p$					
Ti^{3+}	V^{2+}	Cr^{2+} Cr^{3+} Cr^{4+} Cr^{5+}	Mn^{2+} Mn^{4+} Mn^{5+}	Fe^{3+}	Co^{2+}	Ni^{2+}	Cu^{+}		Ga^{+}	Ge^{2+}	As^{3+}			
							Ag^{+}		In^{+}	Sn^{2}	Sb^{3+}			
							Au^{+}		Tl^{+}	Pb^{2+}	Bi^{3+}			
	Lanthanides $4f^n\text{-}4f^{n-1}5d$													
La	Ce^{3+}	Pr^{3+}	Nd^{3+}	Pm^{3+}	Sm^{3+} Sm^{2+}	Eu^{3+} Eu^{2+}	Gd^{3+}	Tb^{3+}	Dy^{3+} Dy^{2+}	Ho^{3+} Ho^{2+}	Er^{3+} Er^{2+}	Tm^{3+} Tm^{2+}	Yb^{3+} Yb^{2+}	Lu
	Actinides $5f^n$													
Ac	Th^{4+}	Pa^{4+}	U^{6+}	Np^{3+}	Pu^{3+}	Am^{3+}	Cm^{3+}	Bk^{3+}	Cf^{3+}	Es^{3+}	Fm^{3+}	Md^{3+}	No^{3+}	Lr

Table 4.3 List of some elements which can be substituted and potential replacement ions in function of charge, ionic radii and coordinence (in brackets) (*IR*: infrared; *R*: red; *P*: pink; *Y*: yellow; *G*: green; *B*: blue; *UV*: ultraviolet)

Element	Ionic radii	Transition elements	Ionic radii pm	Absorption (color)	Emission	Rare earths	Ionic radii	Emission
Al^{3+}	0.67 (6)	Cr^{3+}	0.75 (6)	R-G	R	Eu^{3+}	1.09 (6)	R
		Ti^{3+}	0.81 (6)	(B)	R			
		V^{3+}	0.78 (6)	G				
		Fe^{3+}	0.69 (6)	Y	R			
		Mn^{2+}	0.81 (6)	P	G-R			
		Mn^{3+}	0.72 (6)	R-G				
		Mn^{4+}	0.67 (6)					
		V^{2+}	0.93 (6)					
		V^{4+}	0.72 (6)					
Si^{4+}	0.40 (4)	Cr^{4+}	0.55 (4)		IR			
		Fe^{3+}	0.63 (4)	Y	R			
Ca^{2+}	1.14 (6)	Mn^{2+}	0.81 (6)	P	G-R	Eu^{2+}	1.31 (6)	B-UV
		Ni^{2+}	0.83 (6)	G-Y	G-R-IR	Eu^{3+}	1.09 (6)	R
		Cu^{2+}	0.87 (6)	B-G	B	Yb^{2+}	1.06 (6)	
						Sm^{2+}	1.41 (8)	
	1.26 (8)	Mn^{2+}	1.10 (8)	P	G-R	Eu^{2+}	1.39 (8)	B-UV
		Fe^{2+}	0.74 (8)	G		Eu^{3+}	1.21 (8)	R
Ba^{2+}	1.49 (6)	Ag^{+}	1.29 (6)			Eu^{2+}	1.31 (6)	B-UV
		Cu^{2+}	0.87 (6)			REE^{3+}	1.00–1.17 (6)	
		Bi^{3+}	1.17 (6)					
		Bi^{2+}	0.59 (6)					
P^{5+}	0.31 (4)	Mn^{5+}	0.47 (4)	B	IR			
Zr^{4+}	0.98 (6)	Cr^{5+}	0.71 (8)	G	IR	REE^{3+}	1.00–1.17 (6)	
		Cr^{4+}	0.69 (6)					

(continued)

Table 4.3 (continued)

Element	Ionic radii	Transition elements	Ionic radii pm	Absorption (color)	Emission	Rare earths	Ionic radii	Emission
		Mn^{2+}	0.81 (6)	P	G-R			
		Fe^{3+}	0.69 (6)	R	R			
Zn^{2+}	0.88 (6)	Mn^{2+}	0.81 (6)	P	G-R			
		Pb^{2+}	1.33 (6)					
Na^{+}	1.16 (6)	Mn^{2+}	0.81 (6)		G-R			
Ti^{4+}	0.74 (6)	Cr^{3+}	0.75 (6)					

Besides luminescence properties, we also added the short data on color of corresponding minerals, because, as it was already mentioned, such information is often useful for luminescence nature understanding. The data on ionic radii of different elements and the main structural data are taken from Internet sites. The Figures contain the following symbols: (a.u.) – arbitrary units, D-delay, G-gate.

4.1 Calcium (Ca^{2+}) Bearing Minerals

The main substituting luminescence centers are Mn^{2+}, REE^{2+} (Sm, Eu, Yb), and REE^{3+} (Ce, Pr, Nd, Sm, Eu, Gd, Tb, Er, Ho, Tm, Yb). Ionic radii of Ca^{2+} are 1.14 Å in 6-coordinated form and 1.26 Å in 8-coordinated form, while ionic radius of Mn^{2+} is 0.97 Å, ionic radii of Sm^{2+}, Eu^{2+} and Yb^{2+} are 1.36 Å, 1.31 Å and 1.16 Å, correspondingly, and ionic radii of REE^{3+} are changing from 1.00 Å to 1.15 Å from Ce to Yb, correspondingly.

4.1.1 *Apatite* $Ca_5(PO_4)_3(F,Cl,O,OH)$

Ninety-five percent of the phosphorus on Earth belongs to the minerals of the apatite group. Apatites are inorganic constituents of bones and teeth of vertebrate and man, as well as a basis of many pathologic solid formations. Minerals of the apatite group are the main raw materials in the production of phosphorus fertilizers, fodder and technical phosphates, elementary phosphorus, and phosphor-organic compounds. The mineral is sometimes substantially enriched in rare-earth elements (REE) making their extraction possible.

There are apatites with violet, blue, green, yellow orange and brown colors. The mostly abundant are blue and green and despite numerous investigations the origin of these colors are ambiguous. According to last investigation, violet-blue color is connected with broad absorption band of SO_3^--radical, which is result of incorporation of S and U(Th) in the apatite structure. The reason of blue color is Mn^{5+} incorporation substituting for P^{5+}. Green color is mainly connected with complex defect SO_3-TR(Ce), but the influence of Pr, Nd and also Cr and Fe is also possible (Gilinskaya and Mashkovtsev 1995).

It has been known for a long time that natural apatite is fluorescent. Its luminescent properties have found many applications, such as detection and identification of mineral in rocks, ores, and outcroppings by luminescent Lidar. The synthetic halo-apatite was one of the first phosphors applied for fluorescent lamps. The fluorescence permits also to establish genetic relationships between mineral samples and their source (mantle or earth's crust) and the type of rock. An enrichment of apatite ores by luminescent sorting is possible (Gorobets 1968; Portnov and Gorobets 1969; Gorobets et al. 1997; Gorobets and Rogojine 2001; Waychunas

2002). Synthetic apatite activated by Yb^{3+} and Mn^{5+} are used as potential laser materials (Payne et al. 1994; Moncorge et al. 1994).

Apatite is considered for a barrier that will prevent the leakage of radioactive nuclei from the radioactive waste storage. Because of the similarity in the chemical and spectral features REE have been chosen as a model of the fission products of the actinides. For this reason it is of importance to recognize whether the elements are incorporated in the bulk of the barrier, or adsorbed on the surface where they can be subjected to leaching out (Martin et al. 1999, 2000).

The composition of the ideal fluor-apatite corresponds to the formula $Ca_{10}(PO_4)_6F_2$. The structure of the apatite is hexagonal, with the symmetry group P63/m. There are two different sites for the Ca^{2+} in this structure: 40 % of Ca^{2+} ions are associated with the Ca(I) sites and 60 % are associated with Ca(II) sites. The point group symmetry of the Ca(I) site is C_3, with each Ca having six oxygen nearest neighbors that form a distorted triangular prism about the Ca^{2+} ion. The Ca (II) site has C_s symmetry with the Ca^{2+} ions sitting at the corners of equilateral triangles with an F^- ion in the center (Fig. 4.1).

Trivalent rare-earth (TR^{3+}) ions that have taken the Ca(II) position are characterized by abnormally high values of Stark splitting. This permits an easy separation of the spectra of these ions from those of REE^{3+} ions in Ca(I) positions (Voron'ko et al. 1991). The luminescence of REE in synthetic artificially activated apatites may be subdivided in two groups according to activation conditions. A number of RE have been studied after activation in oxidative conditions with the main type always in the low-symmetry Ca(II) site. In nature, REE can easily substitute for Ca, becoming luminescence centers in a crystallographic environment. Luminescence of natural apatites has been the subject of numerous investigations. With steady-state luminescence spectroscopy it was discovered that Eu^{2+}, Ce^{3+}, Mn^{2+}, Dy^{3+}, Nd^{3+}, Sm^{3+} and Sm^{2+} mainly determine apatite luminescence. It was established that REE^{3+} have been detected only in the high symmetry Ca(I) position (Gorobets 1968; Portnov and Gorobets 1969; Tarashchan 1978; Gorobets and Rogojine 2001; Waychunas 2002). Laser-induced luminescence of apatite was investigated by nitrogen laser excitation and the main luminescence center being Ce^{3+} with decay time of ~20 ns accompanying by Mn^{2+} with decay time of ~ 1.5 ms (Gaft 1989).

The natural fluor-apatites in our study consisted of 75 samples from a variety of geologic environments. Concentrations of potential luminescence impurities in several samples are presented in Table 4.4.

For the correct interpretation of the luminescent bands, artificial apatite standards have been investigated, as nominally pure, as activated by different potential luminogen impurities. Natural carbonate-fluor-apatites not containing REE were heated with 1–5 wt% of oxides of Eu, Pr, Sm and Dy at 900 °C in air and in vacuum. By changing the activation conditions the differentiation between isomorphous substitutions in different Ca-sites has been achieved. Under vacuum the compensation of the excessive positive charge by substitution of F^- by O^{2-} is impossible and the luminescence centers in Ca(II) sites may be less preferential. After heating

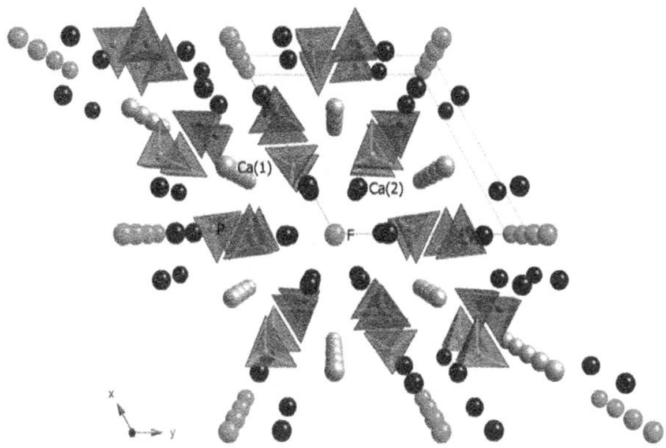

Fig. 4.1 Crystal structure of fluorapatite showing the two different calcium sites Ca(1) and Ca(2)

Table 4.4 Concentrations of rare-earth elements and manganese (ppm) in several apatites with different colors

	Pr	Sm	Eu	Nd	Dy	Tb	Er	Tm	Ho	Gd	Ce	Mn
Red	156	134	37	583	72	23	46	4	13	208	1367	
Green	63	50	12	237	41	7	24	4	9	40	519	
Blue	424	199	28	1470	59	14	25	3	11	133	4052	
Florida	5	4	1	19	7	1	6	1	2	6	45	130
Ivory	0.03	0.07	0.1	0.18	0.3	0.07	1.2	0.4	0.2	1.0	0.3	550

at this temperature carbonate-fluor-apatite loses its carbonate content and becomes very similar to natural fluor-apatite.

Time-resolved luminescence spectra enable to detect the following emission centers in natural apatite: Eu – Eu^{2+} in Ca(II) site, Eu^{3+} in Ca(I) and Ca(II) sites; Sm^{3+} in Ca(I) and Ca(II) sites; Ce^{3+} in Ca(I) and Ca(II) sites; Pr^{3+} – in Ca(I) site; Nd^{3+}, Dy^{3+}, Tb^{3+}, Er^{3+} and Tm^{3+} in Ca(I) site; Yb^{3+} possibly in Ca(II) site, Mn^{5+} in P^{5+} site, Mn^{2+} in Ca(I), Ca(II) sites and adsorption on carbonate-fluor-apatite surface; reabsorption lines of molecular oxygen and water have been found and ascribed (Gaft et al. 1996; Gaft et al. 1997a, b, c) (Figs. 4.2, 4.3, 4.4, 4.5, 4.6 and 4.7).

Under cw laser excitation at 514 nm several narrow lines have been found in IR part of the spectrum (Fig. 4.8). Evidently they may be connected with different types of Nd^{3+} centers, such as Nd^{3+} in Ca II site, Nd^{3+} in Ca I site, and Nd^{3+} in Ca II site combined with oxygen substituting for F (Tarashchan 1978). Participation of Pr^{3+} is also possible with the main line peaking at 871 nm (Morozov et al. 1970). Similar luminescence lines have been found also under cw 780 nm excitation (http://rruff.info).

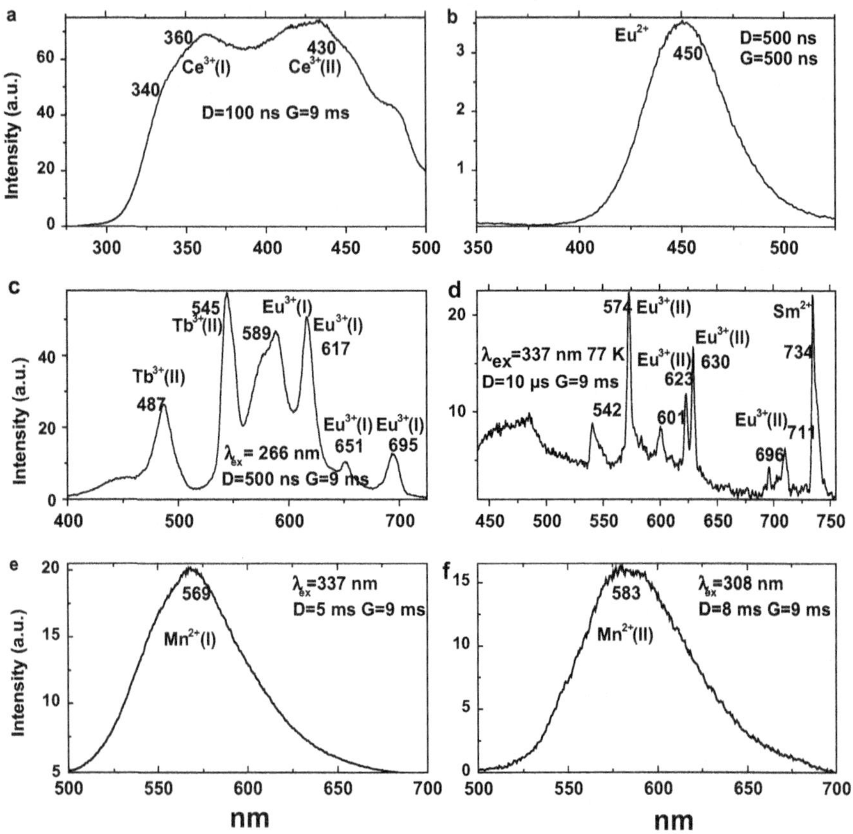

Fig. 4.2 (**a–f**) Laser-induced time-resolved luminescence spectra of apatite demonstrating two Ce^{3+}, Tb^{3+}, Eu^{3+} and Mn^{2+} centers in Ca(I) and Ca(II) structural positions. Besides that, Eu^{2+}, Sm^{2+} and Dy^{3+} are detected

4.1.2 Scheelite $CaWO_4$ and Powellite $CaMoO_4$

Scheelite is a calcium tungstate, $CaWO_4$, with a tetragonal structure ($I4_1/a$) where irregular tetrahedra WO_4 share edges with CaO_8 polyhedra. It is ore mineral of W and a common accessory mineral in various kinds of rocks and ore deposits. REE geochemistry of scheelite may give information about the source of the ore matter, and about the physico-chemical conditions of ore transport/or precipitation. In nature, REE can easily substitute Ca, and become luminescent centers in such a crystallographic environment.

The method of luminescent spectroscopy is widely used to study the rare-earth elements distribution in natural scheelite (Brugger et al. 2000). Comparisons with synthetic pure and doped phosphors showed that WO_4^{2-} are responsible for blue emission in scheelite (Tiede and Schleede, quoted in Pringsheim and Vogel 1946).

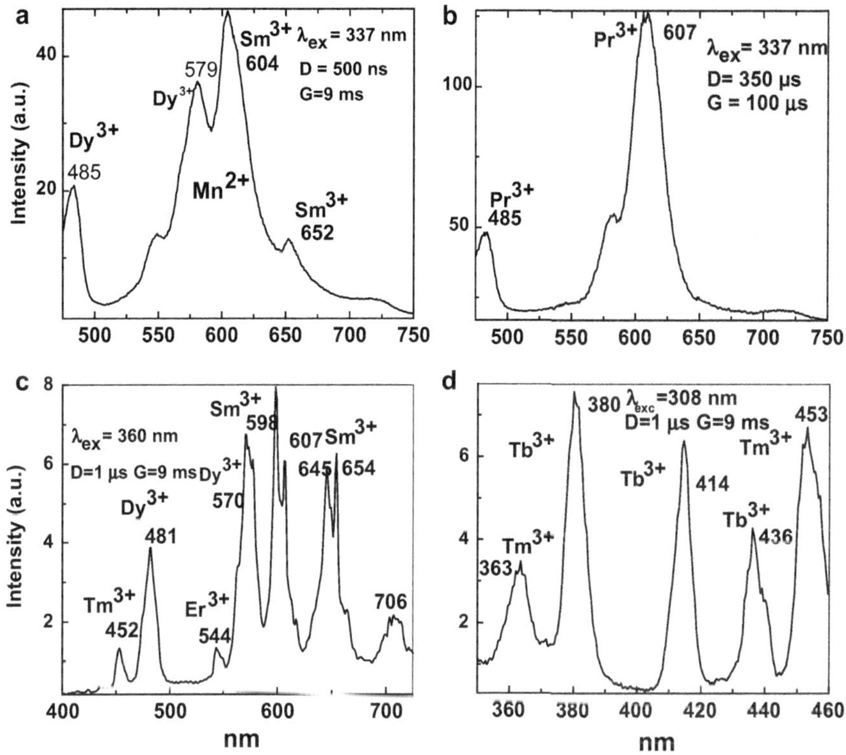

Fig. 4.3 (**a–d**) Laser-induced time-resolved luminescence spectra of apatite demonstrating Pr^{3+}, Tm^{3+} and two Sm^{3+} centers. Besides Mn^{2+} and Dy^{3+} are detected

Systematic spectroscopic investigation enabled to detect and interpret trivalent rare-earth elements emission (Gorobets and Kudrina 1976; Tarashchan 1978). Some regularity appears in the luminescence spectra of scheelite. These regularities are related to the genetic type of mineralisation; in particular, it is possible to recognize a scheelite from a skarn/calcsilicate rock, from a molybdenite-vein, from a metamorphic deposit, from a hydrothermal deposit or from Au-deposit (Uspensky et al. 1989; Gorobets and Walker 1994). Nevertheless in many cases the possibilities of luminescence spectroscopy are limited because scheelite is characterized by broad luminescent bands of intrinsic (WO_4)-groups and impurity (MoO_4)-groups (Tarashchan 1978; Blasse 1980; Gaft 1989). Such strong bands prevent in many cases the detection of lines of rare-earth elements, especially Tm, Er and Ho, which have weak luminescence in the corresponding spectral range. The thermal X-ray excited luminescence enables to solve this problem in certain cases using the fact that while the intensity of broad band luminescence quickly decreases with increasing temperature, the intensities of the REE lines remains nearly constant (Uspensky and Aleshin 1993). Nevertheless, the lines of certain REE may be hidden by

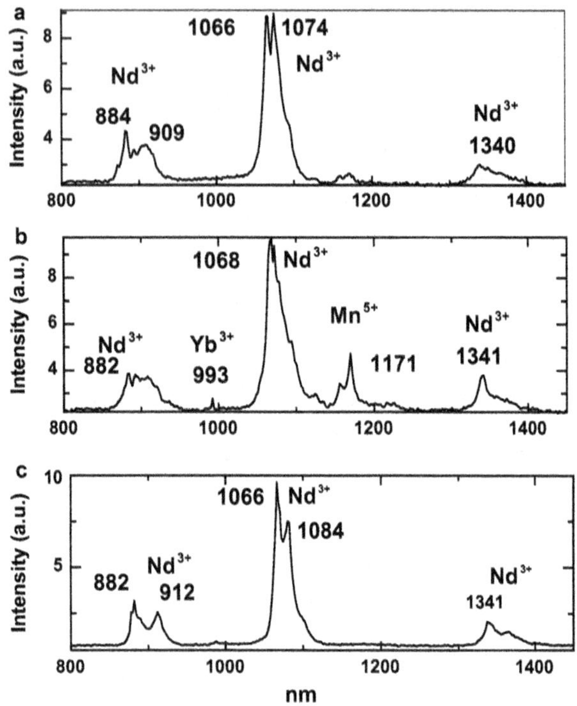

Fig. 4.4 (a–c) Laser-induced time-resolved IR luminescence spectra of apatite demonstrating Nd^{3+}, Yb^{3+} and Mn^{5+} centers

stronger luminescence of others REE. For example, luminescence of Pr^{3+} is difficult to detect because its radiative transitions are hidden by the lines of Sm^{3+}, Dy^{3+} and Nd^{3+}, luminescence of Tm^{3+} is concealed by Tb^{3+} and so on.

The natural scheelite in our study consisted of 90 samples from a variety of geologic environments. Concentrations of potential luminescence impurities in several samples are presented in Table 4.5.

For comparison, steady-state cathodoluminescence spectra (Fig. 4.9) are presented from two scheelite samples with different rare-earth elements concentrations (Table 4.5). It is clearly seen that only broadband emissions are detected, while the narrow lines of several rare-earth elements, mostly Sm^{3+} are extremely weak.

For the correct interpretation of the luminescent bands, artificial standards have been investigated, as nominally pure, as activated by different potential luminogen impurities. Stoichiometric mixtures of $Na_2WO_4 \times 2H_2O$ and $CaCl_2$, together with the REE chloride of interest, were kept for 2 h at 900 °C in open quartz test-tubes filled with molten NaCl, and then where slowly cooled during 20 h. After washing with distilled water, the colorless crystals of scheelite were obtained, 0.01–0.5 mm in size. By using laser-induced time-resolved spectroscopy we were able to detect and ascribe various trivalent rare-earth impurities, such as Sm^{3+}, Dy^{3+}, Eu^{3+}, Pr^{3+}, Er^{3+}, Tm^{3+}, Ho^{3+}, Nd^{3+} and Yb^{3+} (Gaft et al. 1999a, b) (Figs. 4.10 and 4.11).

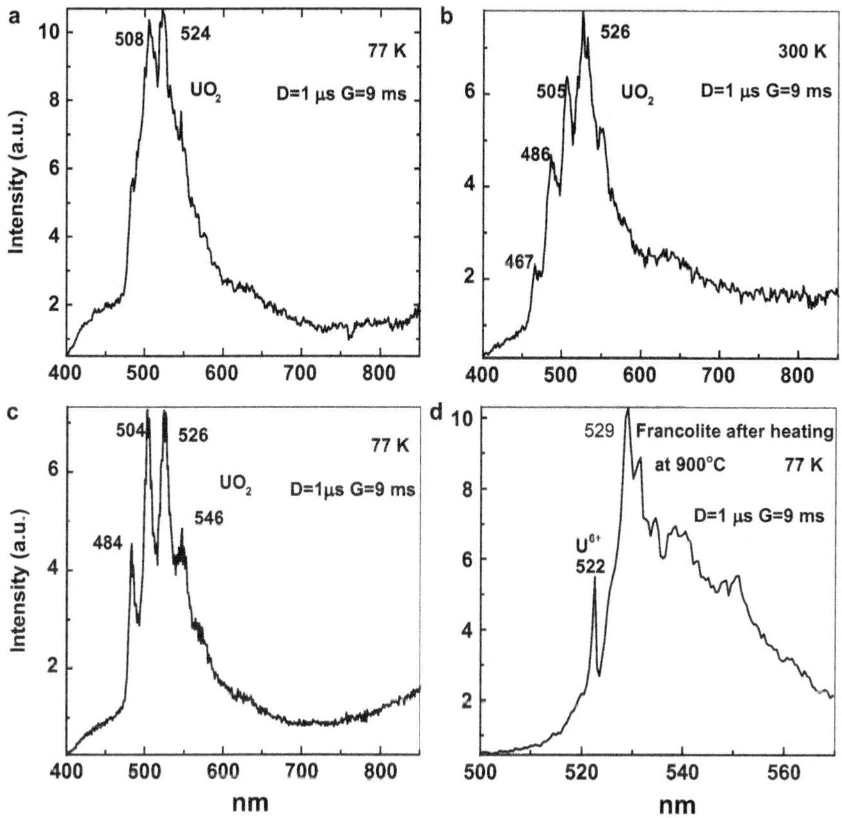

Fig. 4.5 (**a–d**) Laser-induced time-resolved luminescence spectra of apatite with uranyl (**a–c**) and U^{6+} luminescence (**d**) after oxidizing heating (possibly in fluorite inclusions)

Under cw laser excitation at 532 and 780 nm several narrow lines have been found in red and IR part of the spectrum (Fig. 4.12). Evidently they may be connected with Nd^{3+} and possibly Sm^{3+} centers.

Powellite $CaMoO_4$ is a tetragonal crystalline calcium molybdate, which is characterized by well-studied intrinsic $(MoO_4)^{2-}$ luminescence (Tarashchan 1978; Gorobets and Rogojine 2001) with relatively long decay time of 200 μs. Besides powellite was found associated with pitchblende (UO_2) and molybdenite (MoS_2) in South Kazakhstan (Bota-Burum Mo-U deposits). These samples contain traces of U, Pb and rare earth elements (mainly Nd) substituting Ca and are supposed to have undergone self-irradiations by the radionuclides incorporated inside the structure during the Devonian mineralization approximately 360 My ago. The effects of self-irradiation were analyzed by Raman spectroscopy, electronic microprobe, LA-ICP-MS, cathodo- and photo-luminescence spectroscopies. Under 532 nm cw excitation the luminescence of Pr^{3+}, Er^{3+}, Nd^{3+} as well as $[UO_2]^{2+}$ were revealed (Fig. 4.13). The combination of these techniques enables

Fig. 4.6 (**a–c**) Laser-induced time-resolved luminescence spectra of sedimentary apatite (francolite) (**a**) and francolite activated by Pr (**b**, **c**) after heating demonstrating uranyl and Pr^{3+} centers

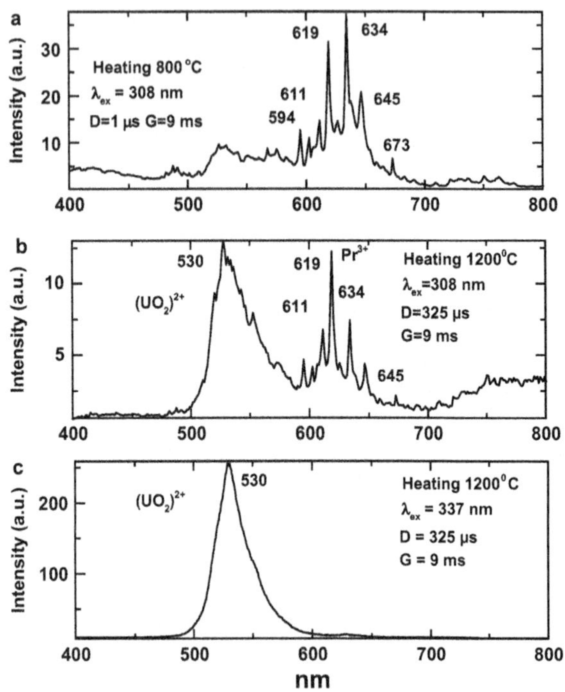

Fig. 4.7 Reabsorption lines of molecular oxygen and water in laser-induced time-resolved luminescence spectra of sedimentary apatite (francolite)

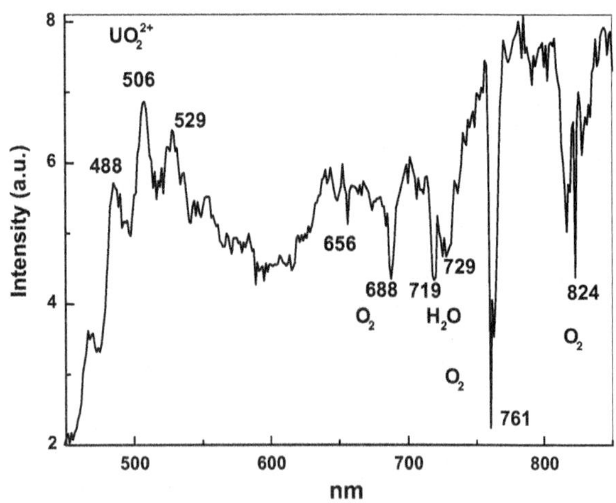

the determination of the sample composition, the local and medium range order and the valence number of emitting centers such as uranium and rare earth elements. The polycrystallinity and zonation of the sample was revealed by BSE SEM, cathode- and photo-luminescence mapping. Natural powellite structure can then incorporate a wide range of elements on the calcium site (up to 8 % in case)

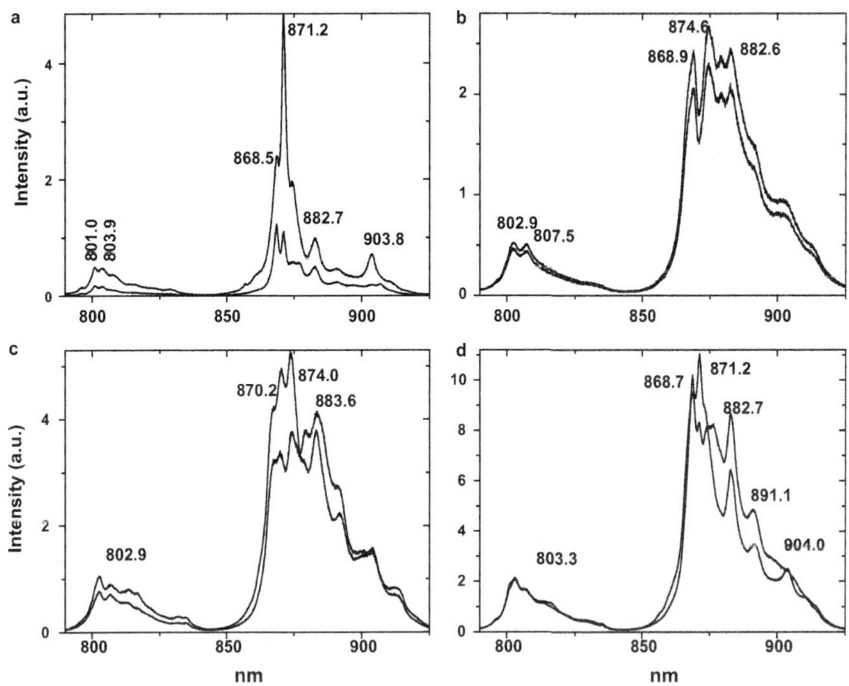

Fig. 4.8 (**a–d**) Laser induced cw spectra of apatite under 514 nm excitation with different Nd^{3+} centers

Table 4.5 Concentrations of rare-earth elements in several scheelites (ppm)

	Pr	Sm	Eu	Nd	Dy	Tb	Er	Tm	Ho	Gd	Ce	Yb
ZRK	129	134	11	589	41	9	20	2	8	72	832	–
FRS	11	10	35	42	13	2	8	1	2	11	100	–
BR7	20	120	75	160	120	25	25	2	15	175		8

including the whole rare earth elements range, Pb and Sr. Up to 7000 ppm of uranium was also incorporated by the sample and photoluminescence experiments confirm the presence of $[UO_2]^{2+}$ uranyl complexes substituting to molybdenum (Panczer et al. 2012).

4.1.3 Fluorite CaF_2

Fluorite is calcium fluoride, CaF_2, with cubic face-centered lattice, while each fluorine ion is at the center of one of the smaller cubes obtained by dividing the

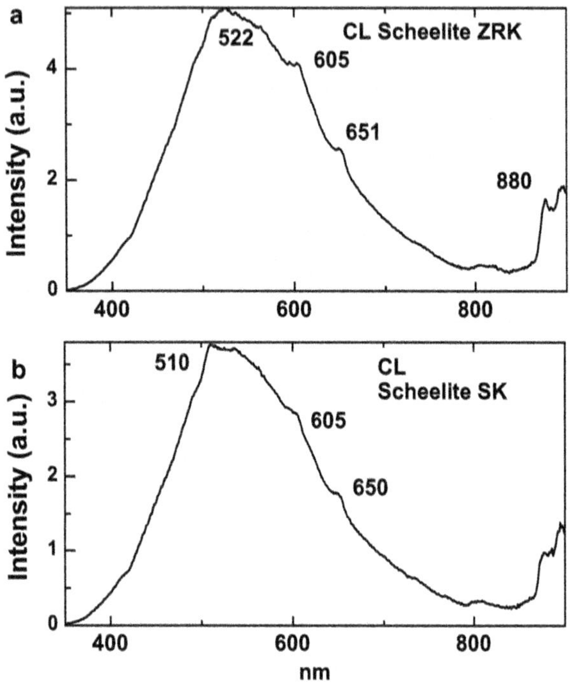

Fig. 4.9 (**a**–**b**) Steady-state cathodoluminescence spectra of scheelite samples

unit cube into eight parts. Each Ca is coordinated by eight F ions and each F is surrounded by four Ca ions arranged at the corners of a regular tetrahedron.

All varieties of color are mainly connected with two main absorption bands in the violet and yellow parts of the spectrum. The secondary bands are also present – in blue and green diapasons. The main absorption bands are connected with F and M-centers. The first one is anion vacancy, which traps electron and the second is two neighboring anion vacancy with two trapped electrons. The short-wave band in fluorite is generated by mutual absorption of F and M-centers, while the long-wave band is connected with M-center absorption only. In the green varieties the REE (Sm^{2+}, Yb^{2+} and Dy^{2+}) is also appreciable. Besides that, the centers O_{2-}, O_{3-} and $(Y,TR)O_2$ sometimes influence with resulting yellow and pink colors (Platonov 1979; Krasilschikova et al. 1986).

Fluorite was the first material where Stokes proved that the blue emission excited by UV light had a longer wavelength and was not diffuse light. For this reason he called the blue light fluorescence (Stokes 1852; Pringsheim and Vogel 1946). Fluorite is a reservoir for many of the rare earths and their luminescence has been carefully studied (Haberland et al. 1934; Haberland and Köhler 1939; Tarashchan 1978; Krasilschikova et al. 1986; Kemper et al. 2002). Under UV excitation fluorite usually emit violet fluorescence, which is connected with impurity of Eu^{2+}. Such emission is extremely strong and other centers are not detected. In very rare cases when Eu concentration is unusually low, other centers dominate,

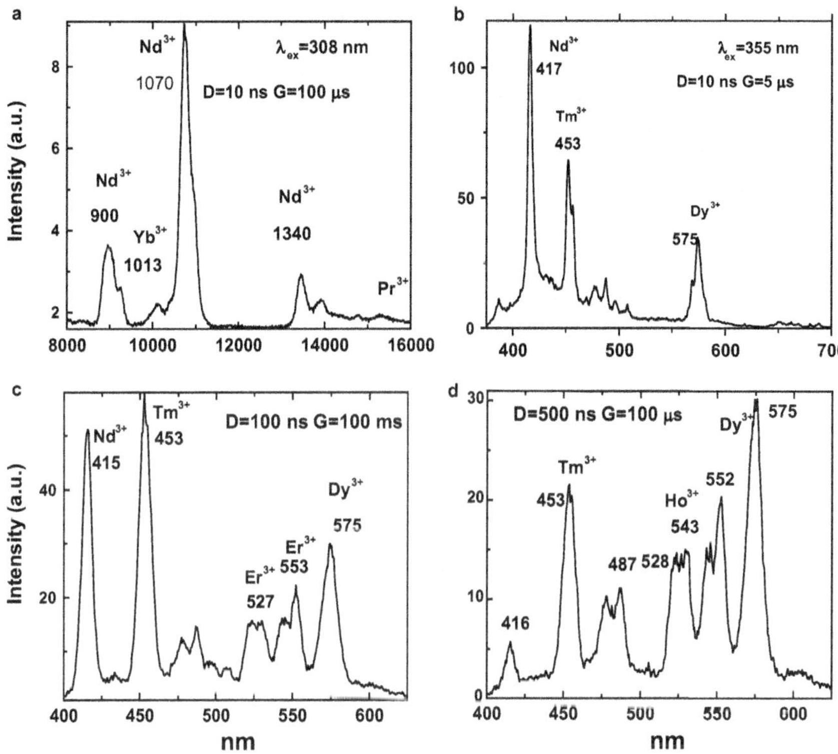

Fig. 4.10 (**a–d**) Laser-induced time-resolved luminescence spectra of scheelite demonstrating Nd^{3+}, Yb^{3+}, Tm^{3+}, Ho^{3+}, Er^{3+} and Dy^{3+} centers

for example, Dy^{3+} and emission color becomes yellow (Sidike et al. 2003). Nevertheless, decay time of Eu^{2+} is 600–800 ns and a delay of 1–5 μs makes it possible to quench it and to see the characteristic lines of other rare-earth elements previously undetected.

Recently the luminescence properties of Pr^{3+}, Nd^{3+}, Tm^{3+} and Yb^{3+} ions in fluorite have been obtained by steady-state measurements. In addition, the luminescence spectra of Ce^{3+}, Sm^{2+}, Sm^{3+}, Dy^{3+}, Er^{3+} and Yb^{3+} were measured. It was pointed out that $\lambda_{ex} = 415$ nm is most suitable for measuring the Ho^{3+} emission beside the Er^{3+}. The emission of trivalent holmium and erbium ions was measured independently using time-resolved measurements and tentative assignment of luminescence lines to C_{3v} and C_{4v} symmetry sites was proposed. Besides for natural fluorite crystal, the transition between Stark energy levels of lanthanide ions were presented (Czaja et al. 2012).

The fluorite in our study consisted of 40 samples from different environments. Concentrations of luminescence impurities in several samples are given in Table 4.6. By using laser-induced time-resolved spectroscopy we were able to detect and ascribe the following emission centers: Eu^{2+}, Ce^{3+}, Gd^{3+}, Sm^{3+}, Dy^{3+},

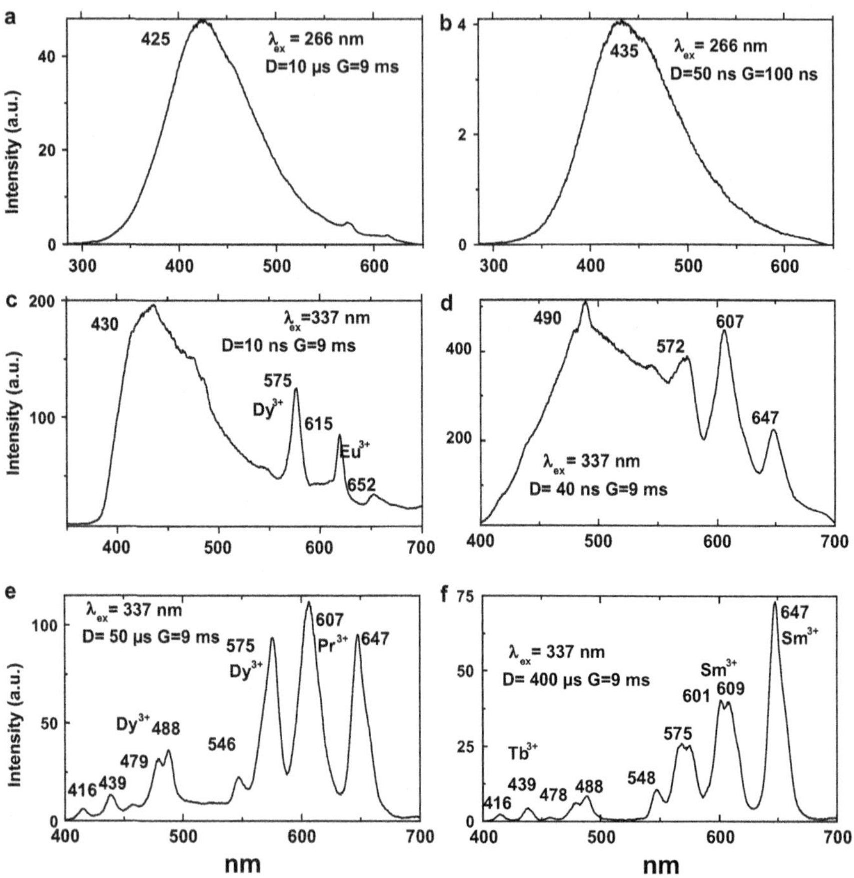

Fig. 4.11 (**a–f**) Laser-induced time-resolved luminescence spectra of scheelite demonstrating intrinsic, Eu^{3+}, Tb^{3+}, Pr^{3+} and Sm^{3+} centers

Eu^{3+}, Pr^{3+}, Er^{3+}, Tm^{3+}, Ho^{3+}, Nd^{3+}, Mn^{2+} and M-center (Gaft et al. 1998) (Figs. 4.14, 4.15 and 4.16).

Under cw laser excitation at 780 and 785 nm several narrow lines have been found in IR part of the spectrum (Fig. 4.17). Evidently they may be connected with different types of Nd^{3+} centers, such as Nd^{3+}-I and Nd^{3+}-II (Krasilschikova et al. 1986). Emission of Er^{3+} at 862 nm may be also supposed.

4.1.4 Calcite, Island Spar and Aragonite $CaCO_3$

Calcite is a calcium carbonate, $CaCO_3$, with a rhombohedral cell consisting of the large planar CO_3 groups, which contain a Ca ion at the center of an equilateral

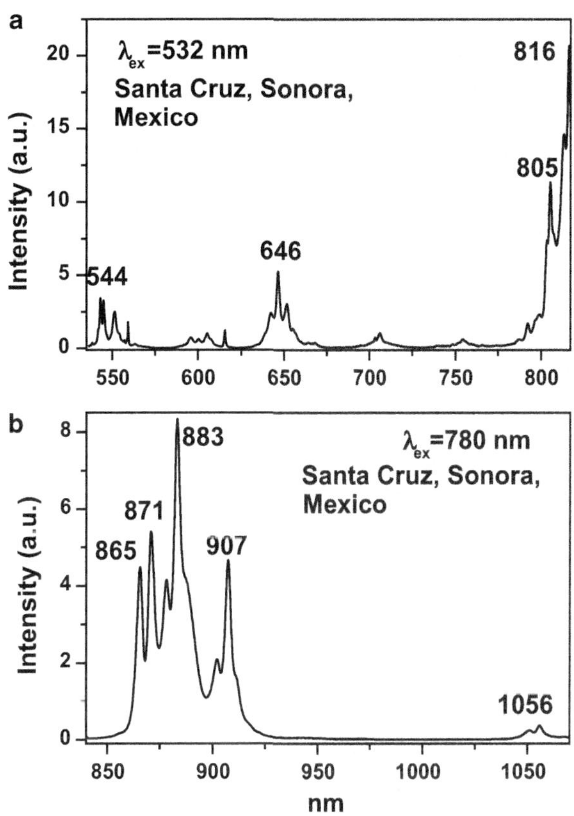

Fig. 4.12 (**a**–**b**) Laser induced cw luminescence spectra of scheelite under 532 and 785 nm excitations with Nd^{3+} centers (http://rruff.info)

triangle of oxygens (symmetry 3 2/m). The structure of calcite is analogous to that of halite NaCl if we consider the unit cube of halite to be shortened along one trigonal axis.

The color of calcite can be due to its iron content, which is present as Fe^{2+}, to Mn^{2+}, to radiation damage centers, which often involve rare earth or other elements, or occasionally to other elements such as Co^{2+}. The color of rhodochrosite is due to its Mn^{2+} content. Most minerals with Mn^{2+} in six-coordination are pale pink. Because Mn^{2+} does not absorb light strongly, a mineral must have a high Mn^{2+} concentration to be strongly colored by Mn^{2+} (Platonov 1979).

In nature, RE and Mn^{2+} can substitute for Ca, becoming luminescence centers in the crystallographic environment. Most calcite luminescence is attributed to Mn^{2+}, while Mn^{2+} hides rare earth emission, and so far their luminescence in calcite is rather poorly characterized (Blasse and Aguilar 1984; Pedone et al. 1990). The lines of Sm^{3+} and Dy^{3+} have been confidently established using a hot cathode cathodoluminescence method (Haberman et al. 1996). Besides that, luminescence of Pb^{2+} (Tarashchan 1978) and radiation-induced CO_3^{3-} are known (Kasyanenko and Matveeva 1987). The natural calcite in our study consisted of 30 samples from a

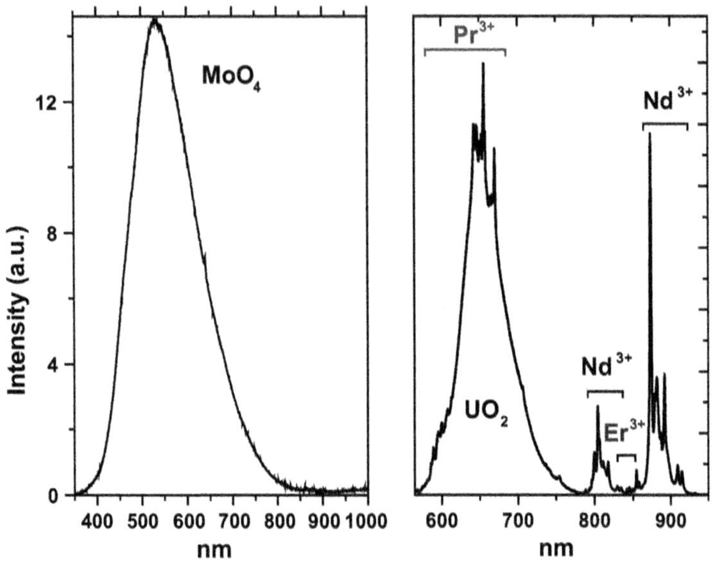

Fig. 4.13 Laser induced cw luminescence spectra of powellite from South Kazakhstan (Bota-Burum Mo-U deposits) with MoO_4, uranyl and trivalent REE centers

Table 4.6 Concentrations of rare-earth elements in several fluorites (ppm)

	Pr	Sm	Eu	Nd	Dy	Tb	Er	Tm	Ho	Gd	Ce	Yb
Russia	0.3	0.3	0.3	1.0	0.4	0.1	0.2	–	0.1	0.3	2.6	–
DNT	5.0	12.0	6.5	26.0	27.0	4.0	14.0	1.7	5.5	22.0	33	
FLS	57	200	28	370	350	55	220	35	70	280	280	260
UCB	0.1	0.3	–	0.5	0.3	–	0.2	–	0.1	0.3	0.3	0.1

variety of geologic environments. Concentrations of potential luminescence impurities in several samples are presented in Table 4.7. By using laser-induced time-resolved spectroscopy we were able to detect the following emission centers: Pb^{2+}, Mn^{2+}, Ce^{3+}, Sm^{3+}, Dy^{3+}, Eu^{3+}, Tb^{3+}, Tm^{3+}, Nd^{3+} and radiation-induced center (Figs. 4.18 and 4.19).

One specific variety of calcite is known as Terlingua-type (TT) because of its original discovery in large quantities at Terlingua, Texas. It is characterized by unique intense blue emission under short-wave UV lamp excitation with an extremely long decay time, accompanied by pink-orange luminescence under long-wave UV excitation with very short decay time (Fig. 4.20). TT calcite samples contain negligible quantities of impurities, which may be potentially connected to such luminescence, and it was supposed that violet luminescence center is radiation-induced (Gaft et al. 2008a).

Unusual luminescence was also found in Iceland spar, transparent variety of calcite. It is dominated by two relatively broad luminescence bands peaking at

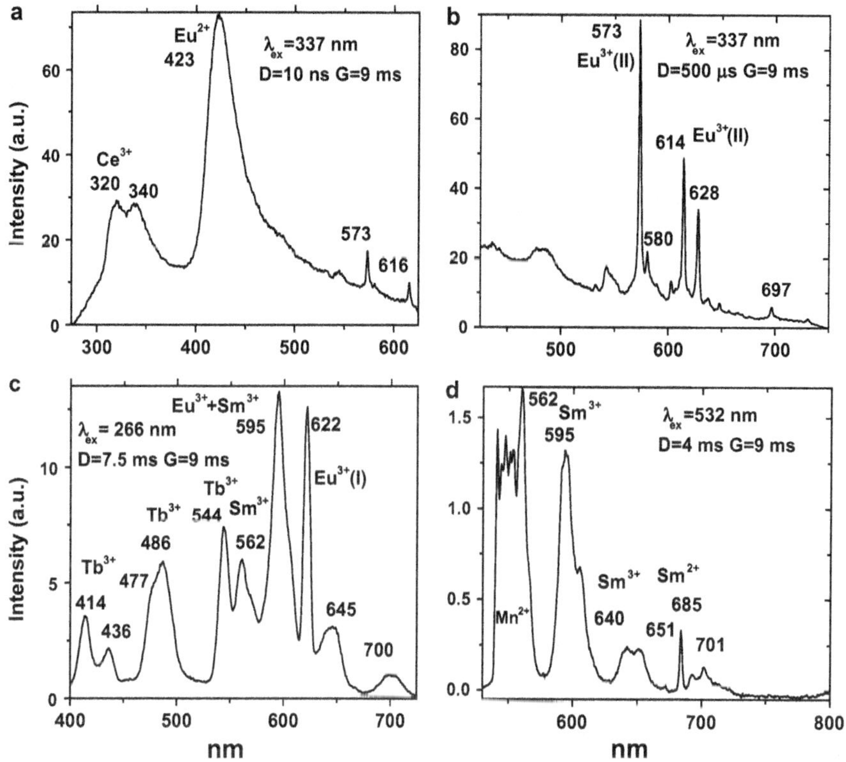

Fig. 4.14 (**a–d**) Laser-induced time-resolved luminescence spectra of fluorite demonstrating Ce^{3+}, Eu^{2+}, two types of Eu^{3+}, Sm^{3+}, Sm^{2+}, Tb^{3+} and Mn^{2+} centers

420 and 550 nm (Fig. 4.21). Because Iceland spar usually contains fewer impurities than calcite, it supports the possible connection of luminescence with radiation induced centers.

Aragonite is a polymorph of calcite, which means that it has the same chemistry as calcite but it has a different structure, and more importantly, different symmetry and crystal shapes. Aragonite's more compact structure is composed of triangular carbonate ion groups (CO_3), with a carbon at the center of the triangle and the three oxygens at each corner. Aragonite has an orthorhombic symmetry (2/m2/m2/m) instead of calcite's "higher" trigonal (bar 3 2/m) symmetry.

The natural aragonite in our study consisted of 12 samples from a variety of geologic environments. By using laser-induced time-resolved spectroscopy we were able to detect evidently Mn^{2+}, Ce^{3+}, Eu^{3+} and broad luminescence band peaking at 580 nm with very short decay time of approximately 20 ns, which by analogy with Terlingua-type calcite may be preliminary ascribed to radiation-induced luminescence center (Fig. 4.22).

Under cw laser excitation at 532 and 780 nm several bands and narrow lines have been found in visible and IR parts of the spectrum (Fig. 4.23). Evidently they

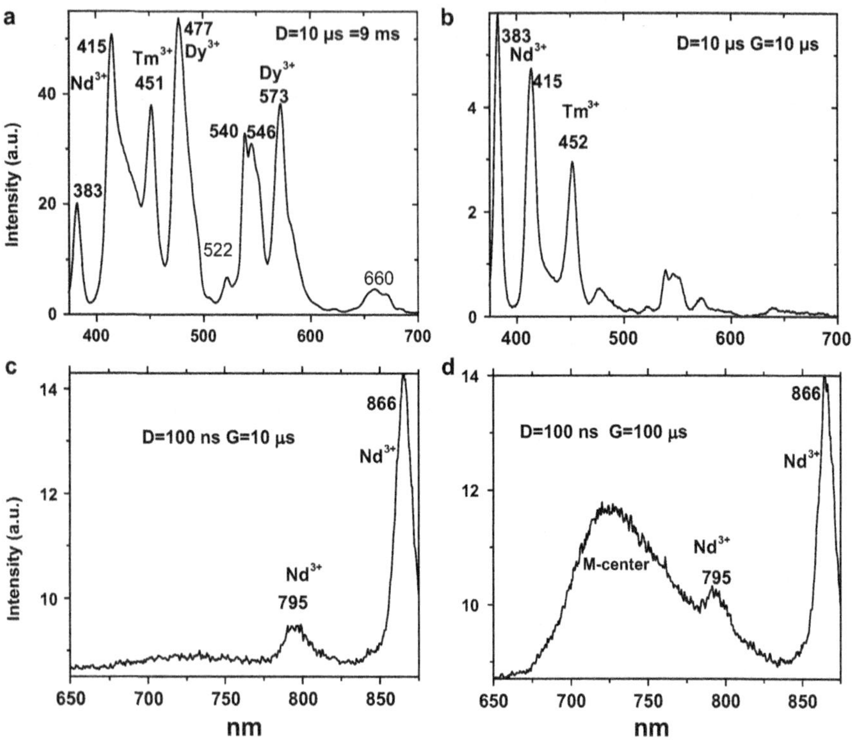

Fig. 4.15 (**a–d**) Laser-induced time-resolved luminescence spectra of fluorite demonstrating Tm^{3+}, Dy^{3+}, Nd^{3+} and M centers

may be connected with another type of Mn^{2+}, trivalent REE, such as Sm^{3+}, Eu^{3+} and different types of Nd^{3+} centers. The origin of the broad band peaking at approximately 700 nm needs further study.

4.1.5 Dolomite (Ca,Mg)(CO$_3$)$_2$

Dolomite is an anhydrous magnesium and calcium bearing carbonate (Ca,Mg)$(CO_3)_2$. It crystallizes in the trigonal-rhombohedral system and forms white, tan, gray, or pink crystals. Sometimes it is luminescent and the following emission centers have been found by steady-state spectroscopy: two different Mn^{2+} centers substituting for Ca^{2+} and Mg^{2+} (Fig. 4.24) and rare-earth elements such as Ce^{3+}, Gd^{3+}, Tb^{3+}, Dy^{3+} and Sm^{3+} substituting for Ca^{2+} (Gorobets and Rogojine 2001). By using laser-induced time-resolved spectroscopy we were able to detect the following emission centers in dolomite from Kovdor Peninsula: Mn^{2+} substituting for Ca^{2+} and Mg^{2+}, Ce^{3+}, Sm^{3+}, Dy^{3+}, Eu^{3+} and Tb^{3+} (Fig. 4.25).

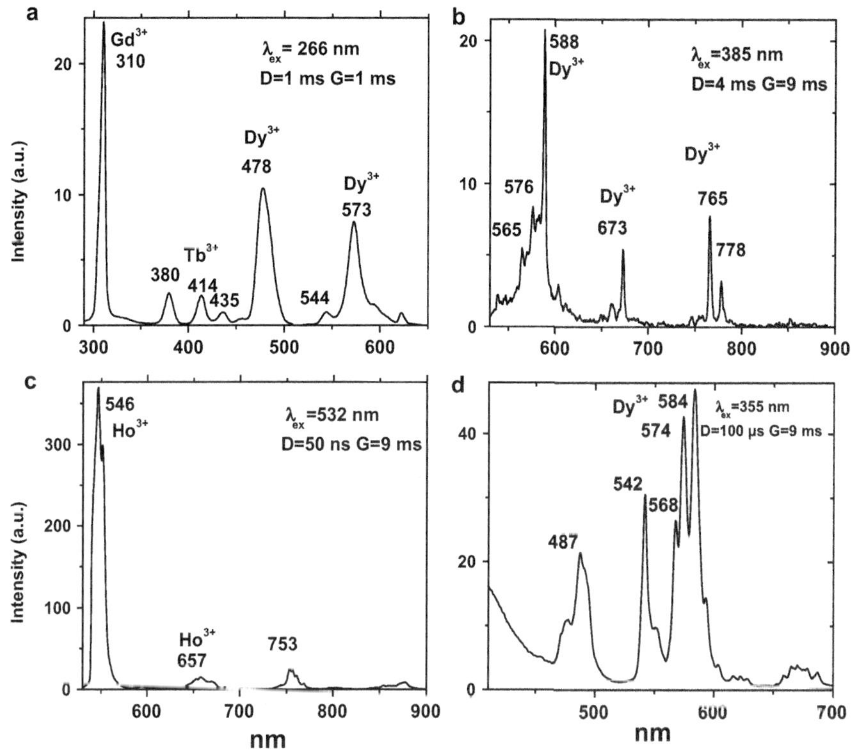

Fig. 4.16 (**a–d**) Laser-induced time-resolved luminescence spectra of fluorite demonstrating Gd^{3+}, Dy^{3+}, Sm^{3+} and Ho^{3+} centers

Excitation by CW laser with 532 and 780 nm revealed several luminescence lines and bands, supposedly connected to Mn^{2+} substituting for Mg^{2+} and different kinds of Nd^{3+} (Fig. 4.26).

4.1.6 *Danburite $CaB_2(SiO_4)_2$*

Danburite is calcium borosilicate with orthorhombic structure (2m/2m/2m) and space group Pnam. The natural danburite in our study consisted of five samples from a variety of geologic environments. Concentrations of potential luminescence impurities in one sample are presented in Table 4.8.

Luminescence centers Eu^{2+}, Yb^{2+}, Ce^{3+}, Dy^{3+} and Sm^{3+} characterize steady-state spectra of danburite (Gaft et al. 1979). By using laser-induced time-resolved spectroscopy we were able to detect and the following emission centers: Ce^{3+}, Eu^{2+}, Eu^{3+}, Sm^{3+}, Dy^{3} (Fig. 4.27).

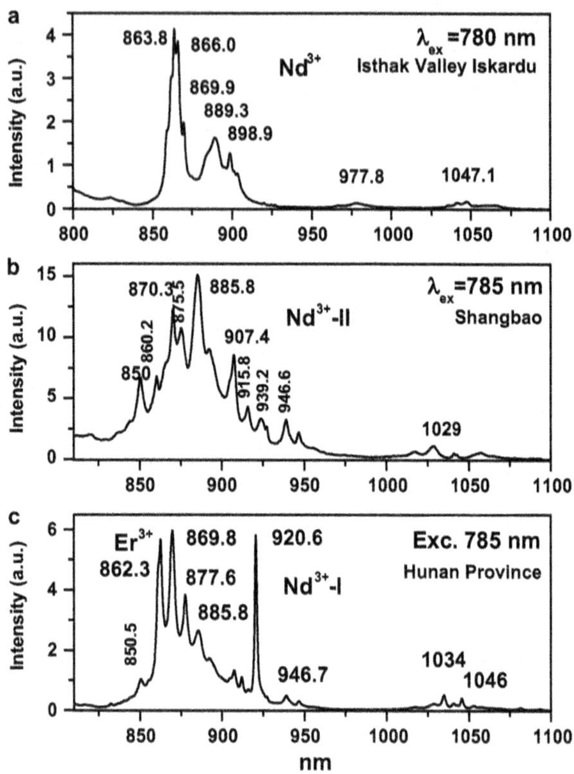

Fig. 4.17 (a–c) Laser induced cw fluorite spectra under 780 nm excitation with different Nd^{3+} centers (http://rruff.info)

Table 4.7 Concentrations of rare-earth elements and manganese in calcites (ppm)

	Pr	Sm	Eu	Nd	Dy	Tb	Er	Tm	Ho	Gd	Ce	Mn	Pb
ST1E	5	4	22	18	6	1	2	0.5	1	6	41	6300	–
Franklin	3.3	3.9	0.7	15	7.8	1	–	0.7	2	6	22	6700	450
Red	28	17	2.5	92	15	3	9	1.2	3	21	250	113	9.4

Excitation by CW laser with 532 and 780 nm revealed several luminescence lines, supposedly connected to Eu^{3+} and several types of Nd^{3+} (Fig. 4.28).

4.1.7 *Datolite* $CaB(SiO_4)(OH)$

Datolite is calcium borosilicate with a monoclinic structure (2/m) and space group $P2_1/c$. It consists of superimposed complex sheets of linked oxygen and O, OH tetrahedra around silicon and boron atoms respectively: the SiO_4 and $B(O,OH)_4$ tetrahedra alternate, forming rings of four and eight tetrahedra. The natural datolite

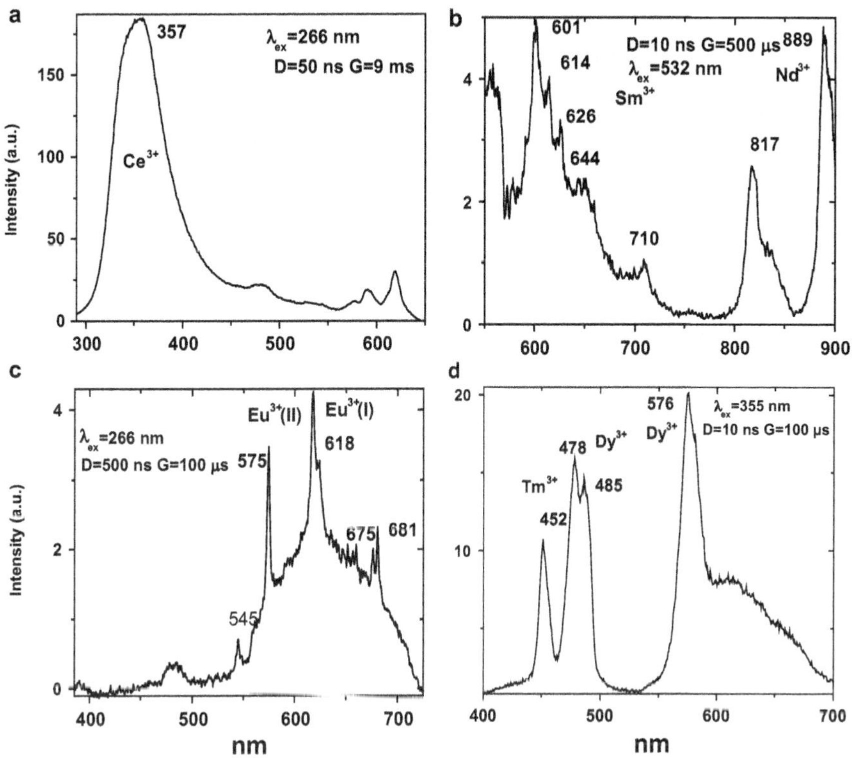

Fig. 4.18 (**a–d**) Laser-induced time-resolved luminescence spectra of calcite demonstrating Ce^{3+}, Eu^{3+}, Tm^{3+}, Nd^{3+} and Dy^{3+} centers

in our study consisted of seven samples, from a variety of geologic environments. Concentrations of potential luminescence impurities in one sample are presented in Table 4.8.

Excitation by CW laser with 532 and 780 nm revealed several luminescence lines, supposedly connected to Eu^{3+} and several types of Nd^{3+} (Fig. 4.28).

Luminescence centers Eu^{2+}, Yb^{2+} and Mn^{2+}, characterize steady-state spectra of datolite (Gaft et al. 1979). By using laser-induced time-resolved spectroscopy we were able to detect and the following emission centers: Mn^{2+}, Ce^{3+}, Eu^{2+}, Eu^{3+}, Sm^{3+}, Dy^{3} (Fig. 4.29).

4.1.8 *Anhydrite* $CaSO_4$

Anhydrite is a calcium sulfate, $CaSO_4$, with an orthorhombic structure (mmm) where sulfur atoms, which are at the centers of tetrahedra of oxygens, and calcium atoms lay on the lines of intersection of mirror planes (100) and (010). In nature,

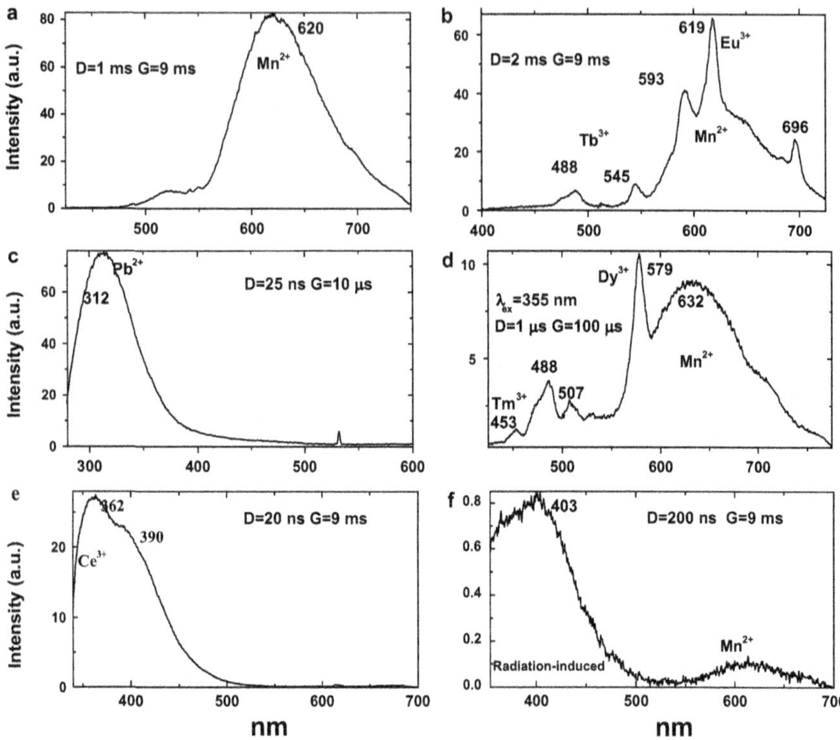

Fig. 4.19 (**a–f**) Laser-induced time-resolved luminescence spectra of calcite demonstrating Mn^{2+}, Pb^{2+} and radiation-induced centers

REE and Mn can easily substitute for Ca, becoming luminescence centers in such a crystallographic environment. X-ray excited and CL spectroscopies have established the luminescence of Gd^{3+}, Ce^{3+}, Eu^{2+}, Er^{3+}, Sm^{3+}, Pr^{3+}, Tb^{3+} and Dy^{3+}, while PL spectroscopy has established Sm^{2+} (Tarashchan 1978; Baumer et al. 1997).

The natural anhydrite in our study consisted of five samples from a variety of geologic environments. Concentrations of potential luminescence impurities in one sample are presented in Table 4.9.

By using laser-induced time-resolved spectroscopy we were able to detect and the following emission centers: Mn^{2+}, Sm^{2+}, Sm^{3+}, Dy^{3+}, Eu^{2+} and Eu^{3+} (Figs. 4.30 and 4.31).

Excitation by CW laser with 785 nm revealed several luminescence lines, supposedly connected to Nd^{3+} (Fig. 4.32).

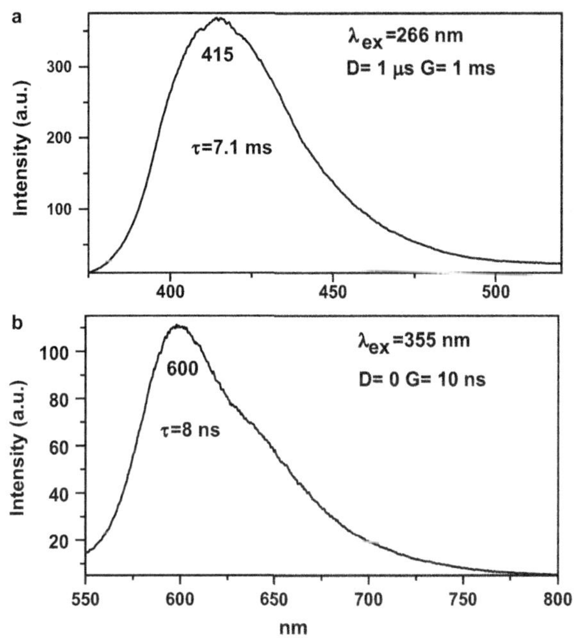

Fig. 4.20 (**a–b**) Time-resolved luminescence spectra of blue (**a**) and orange (**b**) radiation induced centers in calcite

4.1.9 Apophyllite $Ca_4KF(Si_8O_{20})\cdot 8H_2O$

Apophyllite is an uncommon mineral, which is of interest largely because of its unusual atomic structure: this bears some relationship to that of the micas since a basic part of it is a sheet of composition Si_8O_{20}. Instead of forming an approximately hexagonal network, however, the (Si-O) tetrahedra are arranged in four-fold and eight-fold rings, and alternate rings of four tetrahedra point in opposite directions. Between the sheets of tetrahedra lie K, F, Ca ions and water molecules. The structure is characterized by eight-fold tetragonal-prismatic potassium site and seven-fold capped trigonal-prismatic calcium site (Fig. 4.33).

Two different Mn^{2+} luminescence centers have been found in steady-state spectra of apophyllite: in Ca position with orange luminescence peaking at 620 nm and in K position with green emission peaking at 500 nm (Tarashchan 1978). The apophyllite in our study consisted of three samples from different environments. Laser-induced time-resolved technique enables to detect the following emission centers: Ce^{3+}, Mn^{2+} with orange emission and possibly $(UO_2)^{2+}$ (Fig. 4.34).

Excitation by CW laser with 532 and 780 nm revealed several luminescence bands in visible and IR spectral ranges (Fig. 4.35) which need additional study.

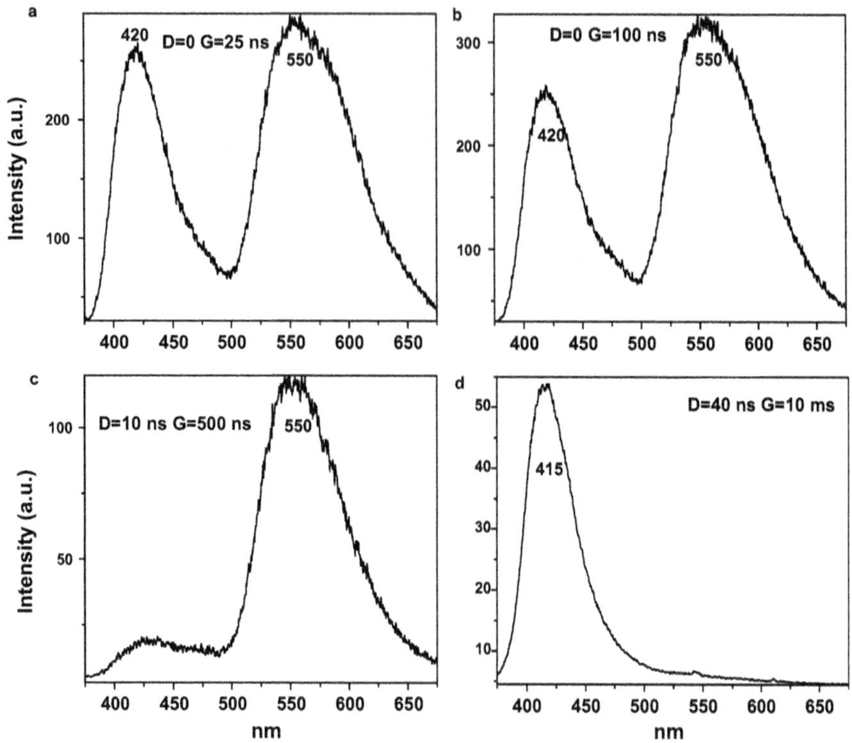

Fig. 4.21 (**a–d**) Time-resolved luminescence spectra of blue and green bands with very short decay time (**a–c**) and violet band with very long decay time (**d**) of supposedly radiation induced origin in Iceland spar

4.1.10 Hardystonite $Ca_2ZnSi_2O_7$

Hardystonite is a calcium zinc silicate of the melilite group. The crystal structure is tetragonal ($P42_1m$) consisting of [$Zn_2Si_2O_7$] sheets within which the nature of the bonding is dominantly covalent, with the adjacent sheets being held together by Ca^{2+} ions. The melilite group is composed of minerals of the general formula $X_2YZ_2O_7$ where X is Ca or Na, Y is Al or Mg and Z is Si. X is a large 8-coordinated site. Luminescence properties of hardystonite have been not studied.

The natural hardystonite in our study consisted of three samples. Concentrations of potential luminescence impurities in one sample are presented in Table 4.10. Laser-induced time-resolved technique enables to detect the following emission centers: Gd^{3+}, Ce^{3+}, Mn^{2+}, Pb^{2+}, Tb^{3+}, Tm^{3+} and Dy^{3+} (Fig. 4.36).

Excitation by CW laser with 532 and 780 nm revealed several luminescence lines in IR spectral ranges (Fig. 4.37) evideny connected with Nd^{3+} centers.

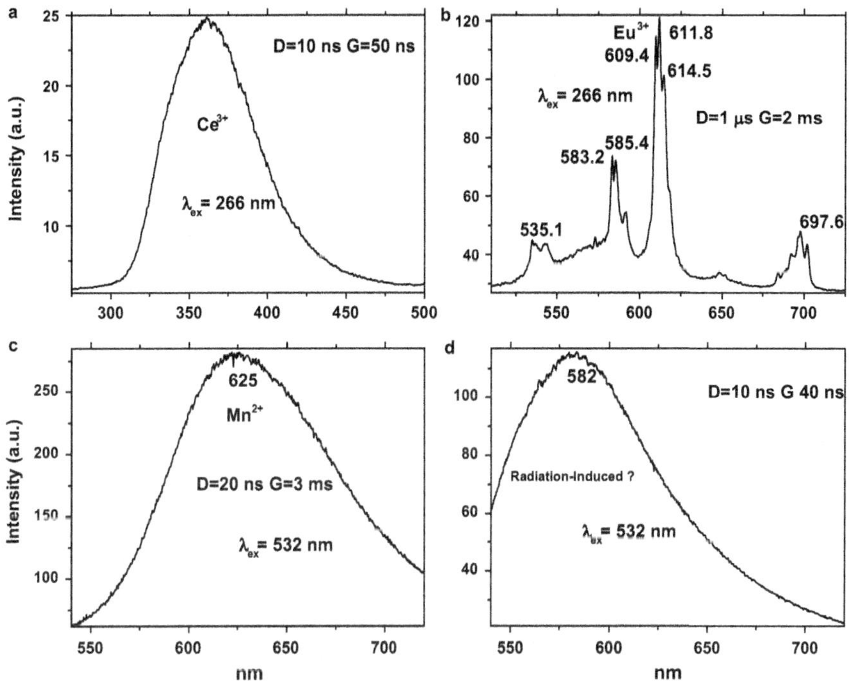

Fig. 4.22 (**a–d**) Time resolved luminescence spectra of aragonite under 266 and 532 nm excitations with Ce^{3+} (**a**), Eu^{3+} (**b**) Mn^{2+} (**c**) and possibly radiation induced (**d**) emission spectra

4.1.11 *Esperite* $Ca_3PbZn_4(SiO_4)_4$

Esperite is a calcium lead zinc-silicate mineral. The crystal structure is monoclinic-prismatic (P21/m) with B21/m group. Steady-state laser-induced luminescence of esperite was ascribed to Mn^{2+}in Zn and Ca positions, accompanied by Dy^{3+} and Sm^{3+} lines. Besides that reabsorption lines of Nd^{3+} and U^{4+} have been found (Gorobets and Rogojine 2001). Laser-induced time-resolved technique enables to detect Ce^{3+} and green band of Mn^{2+} in evidently Zn position (Fig. 4.38).

4.1.12 *Charoite* $K_2NaCa_5(Si_{12}O_{30})Fx3H_2O$

The structure of charoite is monoclinic-prismatic (2/m) with space group P*/4. Luminescence centers Ce^{3+}, Eu^{2+} and Mn^{2+} characterize steady-state spectra of charoite (Gaft 1989; Gorobets and Rogojine 2001).

The natural charoite in our study consisted of one sample. Laser-induced time-resolved technique enables to detect the Ce^3 and Eu^{2+} luminescence centers (Fig. 4.39).

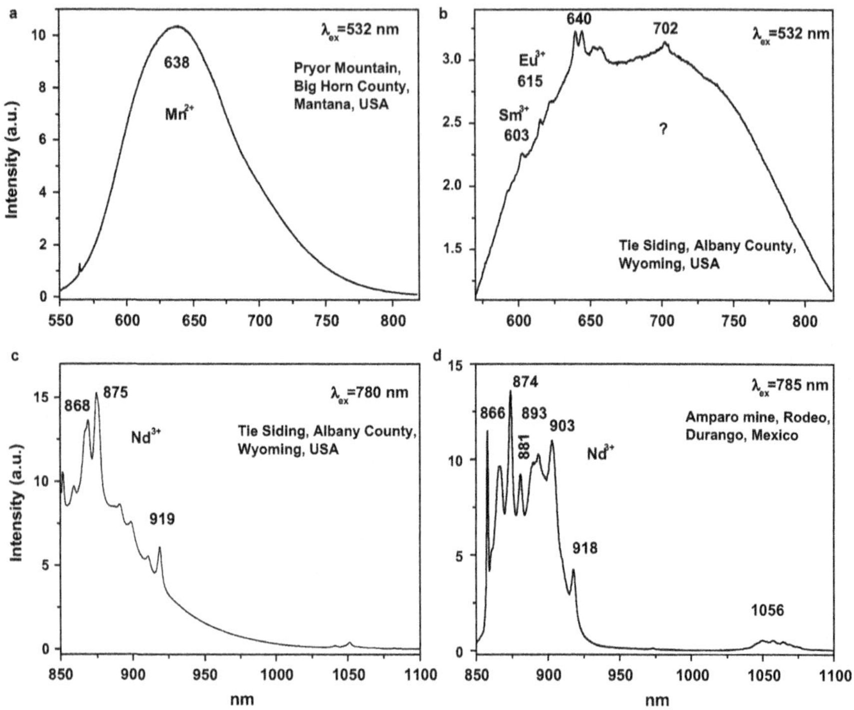

Fig. 4.23 (**a–d**) Laser induced cw spectra of calcite samples under 532 and 780 nm excitation with possibly Mn^{2+} (**a**) and trivalent REE elements (**b–d**) centers (http://rruff.info)

4.1.13 *Prehnite $Ca_2Al_2Si_3O_{10}(OH)_2$*

Prehnite is sheet silicate with orthorhombic structure (mm^2). Luminescence properties of prehnite have not been studied. The natural prehnite in our study consisted of two samples. Laser-induced time-resolved technique enables to detect emission center Pb^{2+} (Fig. 4.40).

4.1.14 *Pectolite $NaCa_2Si_3O_8(OH)$*

Pectolite is hydrous pyroxenoid with triclinic-pinacoidal structure with space group P-1. The structure of pectolite is based on single Si-O chains in a sequence of alternate single and double tetrahedral groups similar to that found in wollastonite. The Ca atoms are coordinated by oxygen octahedra sharing edges to form a lath-like strip, while the Na atoms have trigonal pyramidal coordination and these pyramids also share edges with those of the Ca octahedra. Luminescence center Mn^{2+} characterizes steady-state spectra of pectolite (Gorobets and Rogojine 2001).

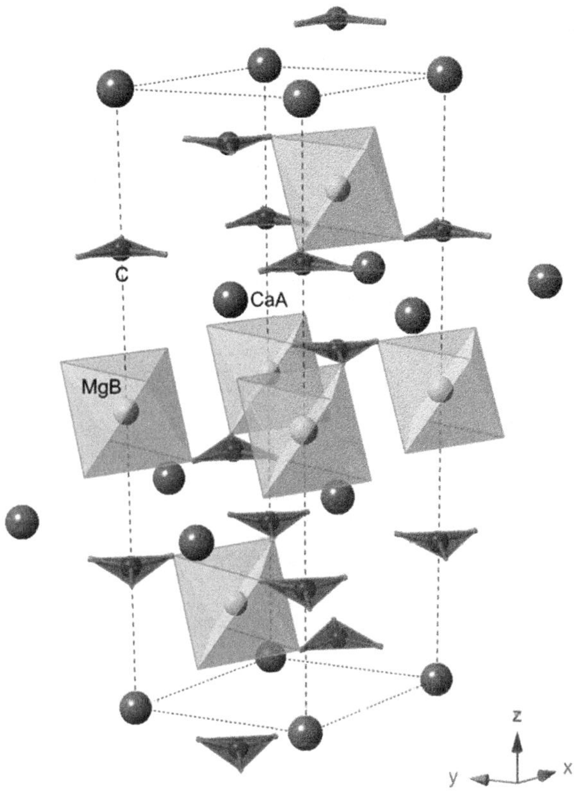

Fig. 4.24 Crystal structure of dolomite showing the CO_3 carbonate groups, Ca and Mg sites (CN6)

The natural pectolite in our study consisted of one sample. Laser-induced time-resolved technique enables to detect Mn^{2+} emission center (Fig. 4.40).

Excitation by CW laser with 532 and 780 nm revealed several luminescence lines in pectolite visible and IR spectral ranges (Fig. 4.41) evidently connected with Mn^{2+}, Fe^{3+} and Nd^{3+} centers. Excitation by CW laser with 532 nm revealed several luminescence lines in prehnite visible spectral range (Fig. 4.42) possibly connected with Mn^{2+}, Fe^{3+} and Cr^{3+} centers.

4.1.15 *Pyrochlore $(Ca,Na)_2Nb_2O_6(OH,F)$*

The structure of pyrochlore is considered to be an anion deficient derivative of the fluorite structure type. Ca atoms are in eight-fold coordination, while Nb atoms are in six-fold coordination. Steady-state luminescence spectra of pyrochlore revealed emission of REE, such as trivalent Dy and Nd (Gorobets and Rogojine 2001). The natural pyrochlore in our study consisted of four sample. Laser-induced time-resolved technique enables to detect Sm^{3+}, Dy^{3+}, Tb^{3+}, Eu^{3+} and Nd^{3+} emission

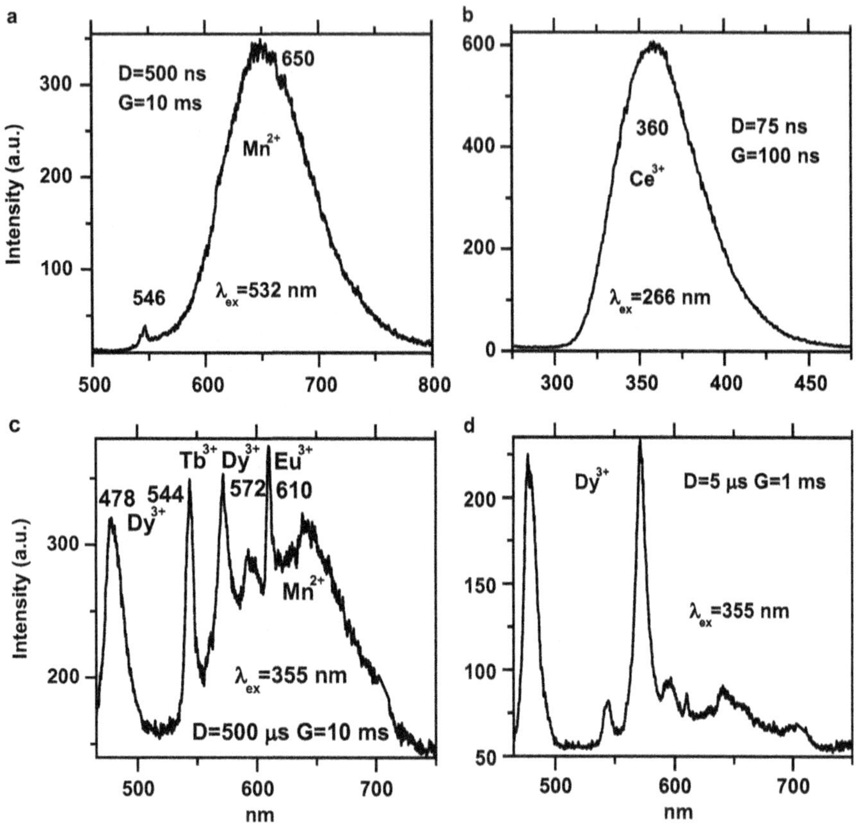

Fig. 4.25 (**a–d**) Laser-induced time-resolved luminescence spectra of dolomite demonstrating Mn^{2+}, Ce^{3+}, Eu^{3+}, Tb^{3+}, Sm^{3+} and Dy^{3+} centers

centers (Fig. 4.43). Excitation by CW laser with 532 and 780–785 nm revealed several luminescence bands and lines in pyrochlore in visible and IR spectral ranges (Fig. 4.44) connected with Eu^{3+} and Nd^{3+} and unidentified centers.

4.1.16 *Leucophane NaCaBe(Si$_2$O$_6$)F*

Leucophane is a relatively rare beryllium silicate. Of interest are the trace amounts of rare earth elements in its chemistry, especially cerium which substitutes for some calcium. Its true symmetry is triclinic, pedion class which is the lowest symmetry possible in a three dimensional system. The only symmetry element is translational shift as it lacks any mirrors, rotations, or even a center. Ce^{3+}, Eu^{2+}, Sm^{3+}, Dy^{3+}, Tb^{3+}, Nd^{3+} and Mn^{2+} centers characterize steady-state luminescence spectra of leucophane (Gorobets and Rogojine 2001).

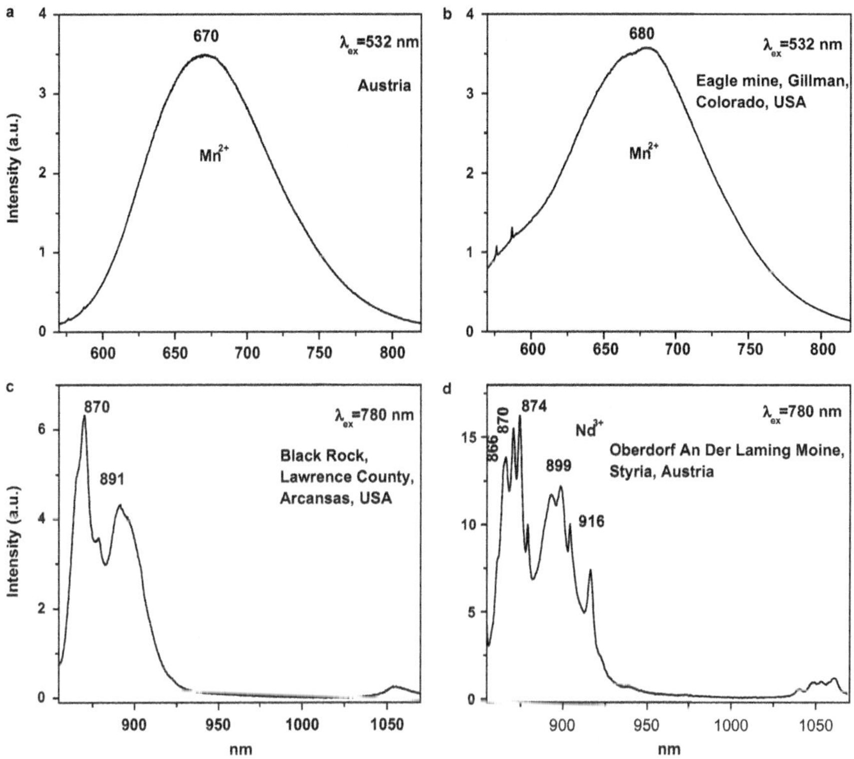

Fig. 4.26 (**a–d**) Laser-induced CW luminescence spectra of dolomite samples demonstrating Mn^{2+}, and Nd^{3+} centers (http://rruff.info)

Table 4.8 Concentrations of rare-earth elements in danburite and datolite (ppm)

	Pr	Sm	Eu	Nd	Dy	Tb	Er	Tm	Ho	Gd	Ce
Danburite	0.3	0.2	1.0	1.0	0.1	–	–	–	–	0.1	3
Datolite	3.0	2.0	0.5	11.0	1.0	0.2	0.7	0.1	0.2	1.7	22.0

The application of multiple forms of excitation (Friis et al. 2011) revealed that the UV-Blue emission in leucophanite and meliphanite $Ca_4(Na, Ca)_4Be_4AlSi_7O_{24}(F,O)_4$ consists of more than one emission centre and is therefore more complex than previously thought. The most likely centers are defects related to the structure, e.g. in connection with the tetrahedral sites, and a Ce^{3+} centre. The difference in Na/Ca ratio between the two minerals make it possible for REE to substitute into two sites in meliphanite contrary, to just one in leucophanite. The lower total REE content in meliphanite compared to leucophanite is consistent with a model whereby the formation of clusters is reduced and consequently generates sharper line emissions. The small difference in the average bond-length of the Ca coordination between the two minerals is consistent with the shift observed in the position of the Mn^{2+} emission. The excitation of specific REE in both leucophanite

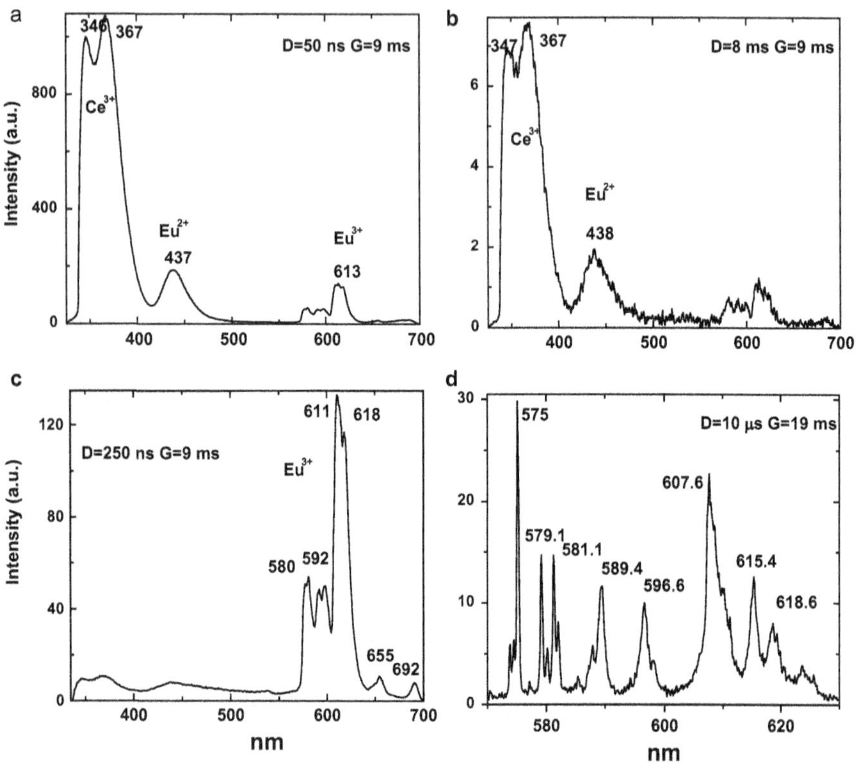

Fig. 4.27 (**a–d**) Laser-induced time-resolved luminescence spectra of danburite demonstrating Ce^{3+}, Eu^{2+} and Eu^{3+}

and meliphanite show that very little energy transfer occurs between different REE. Conversely, excitation through the lattice excites all emission centers. Both leucophanite and meliphanite have excellent luminescence properties, especially at low dopant levels. Furthermore, the structural difference between the two related species indicates that meliphanite might be a promising phosphor.

Time-resolved luminescence spectra in our study contain additionally Eu^{3+} and Tm^{3+} centers (Fig. 4.45). Excitation by CW laser with 532 and 785 nm revealed several luminescence bands and lines supposedly connected to Mn^{2+} and Nd^{3+} (Fig. 4.46).

4.1.17 *Milarite* $K_2Ca_4Al_2Be_4Si_{24}O_{60}\times H_2O$

Milarite group minerals crystallize in space group p6/mcc with six-fold Si tetrahedra surrounding each K^+ ion (CN 12) while Ca^{2+} (CN7) are centered by 3 Be tetrahedra. Milarite crystals are often colored a muted green or yellow. Ce^{3+}, Eu^{2+}

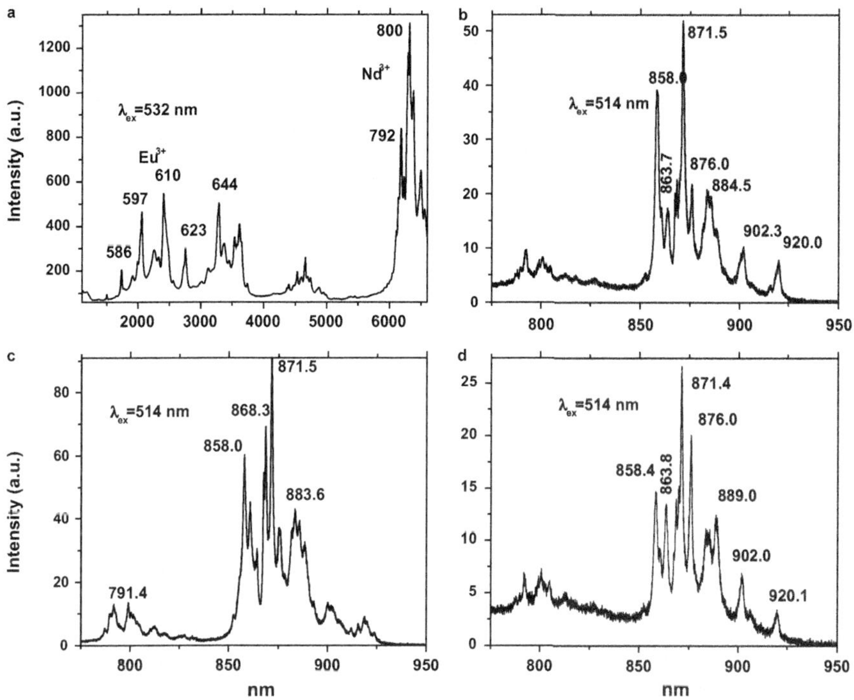

Fig. 4.28 (**a–b**) Laser-induced CW luminescence spectra of danburite demonstrating Eu^{3+} and Nd^{3+} centers

and Mn^2 centers characterize steady-state luminescence spectra of milarite (Gorobets and Rogojine 2001). Time-resolved luminescence spectra are characterized by short lived Ce^{3+}, Eu^{2+}, and possibly Pb^{2+} (Fig. 4.47).

Excitation by CW laser with 532 and 785 nm revealed several luminescence bands and lines supposedly connected to Mn^{2+}, Fe^{3+} and Nd^{3+} (Fig. 4.48).

4.1.18 *Walstromite $Ca_2BaSi_3O_9$*

Walstromite is related to $CaSiO_3$ group of minerals, with the substitution of the 1/3 of Ca atoms (CN 6-7) by Ba (CN 8) atoms. Its luminescence was not studied by steady-state spectroscopy. Time-resolved luminescence spectra (Fig. 4.49) are characterized by Tm^{3+}, Eu^{2+} and Mn^{2+} centers (Gaft et al. 2013).

Excitation by CW laser with 532 and 785 nm revealed several luminescence bands and lines supposedly connected to Mn^{2+} and Nd^{3+} (Fig. 4.50).

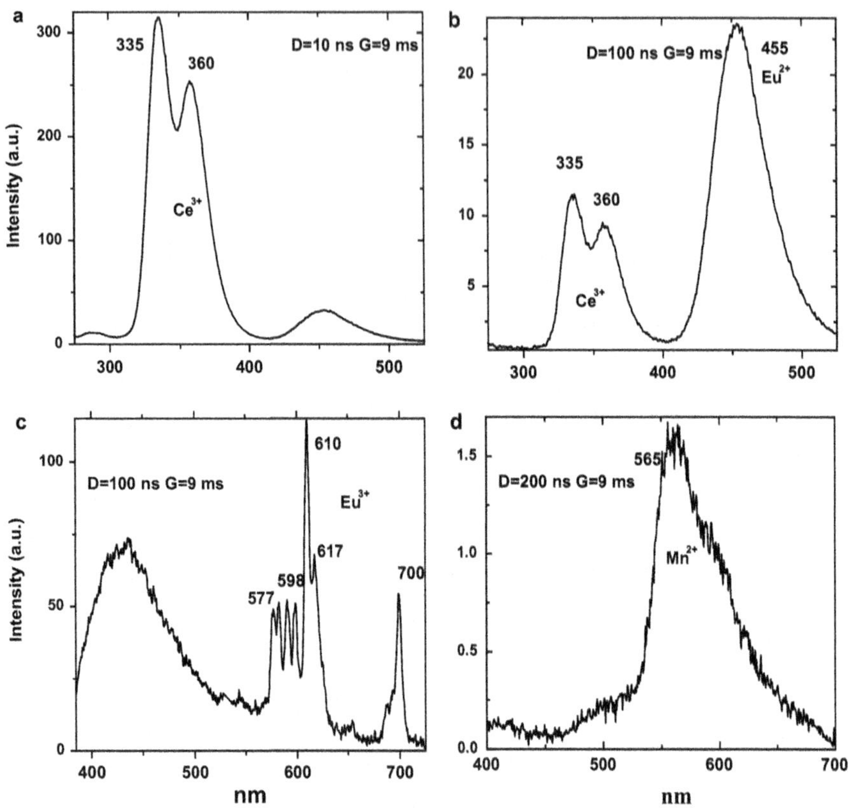

Fig. 4.29 (**a–d**) Laser-induced time-resolved luminescence spectra of datolite demonstrating Mn^{2+}, Ce^{3+}, $Eu^{2+}$and Eu^{3+} centers

Table 4.9 Concentrations of rare-earth elements in anhydrite (ppm)

Pr	Sm	Eu	Nd	Dy	Tb	Er	Tm	Ho	Gd	Ce
0.3	0.6	0.3	1.7	0.7	0.2	0.4	0.1	0.2	0.8	2.3

4.1.19 Bavenite $Ca_4Be_2Al_2Si_9O_{26}(OH)_2$

Natural fibrous crystals of bavenite collected in intra-granitic pegmatite bodies of Bustarviejo (Madrid, Spain) have been examined (Garcia-Guinea et al. 2005). The strong luminescence emissions has been recorded. The bavenite luminescence takes place in the 5d electron shell that interacts strongly with the crystal field; the spectra bands assignment are Gd^{3+} (319 nm), Sm^{3+} (562 and 594 nm), Dy^{3+} (572 nm) and Tb^{3+} (495 nm). A Mn^{2+} band at about 578 nm in Ca^{2+} sites is present as a broad band that overlaps with the Dy^{3+}, Sm^{3+} and Tb^{3+} bands. The ICP-MS analyses show

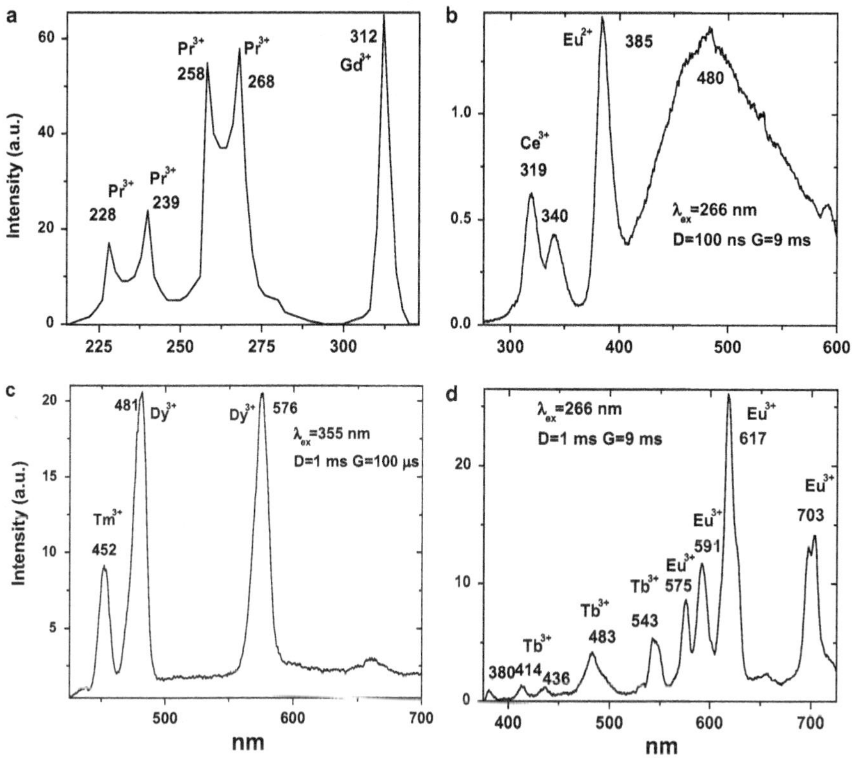

Fig. 4.30 (**a–d**) Laser-induced time-resolved luminescence spectra of anhydrite demonstrating Gd^{3+}, Ce^{3+}, Eu^{2+}, Eu^{3+}, Pr^{3+}, Tm^{3+}, Tb^{3+} and Dy^{3+} centers

concentrations of Yb – 29.7, Dy = 22.7, Sm = 9.45, Nd = 8.95 and Gd = 8.15 ppm in the bavenite lattice.

Excitation by CW laser with 532 nm revealed two luminescence bands evidently belonging to Mn^{2+} and lines at 656 and 659 nm (Fig. 4.51). Additional study is needed to ascribe those lines including their decay times.

4.1.20 Diopside $CaMgSi_2O_6$

Diopside is a mineral belonging to the pyroxene family; in particular it is an inosilicate with a single calcium and magnesium chain. Natural diopside displays a large variety of luminescence centers: at room temperature main features are a 450 nm band, that can be ascribed to the TiO_6 complex, an intense bands at 585 and 690 nm that can be ascribed to Mn^{2+} ions in M2 (Ca^{2+}) and M1 (Mg^{2+}) sites, lines near 684 nm belonging to Cr^{3+} and band peaking at 690 nm ascribed to Fe^{3+} (Gorobets and Rogojine 2001).

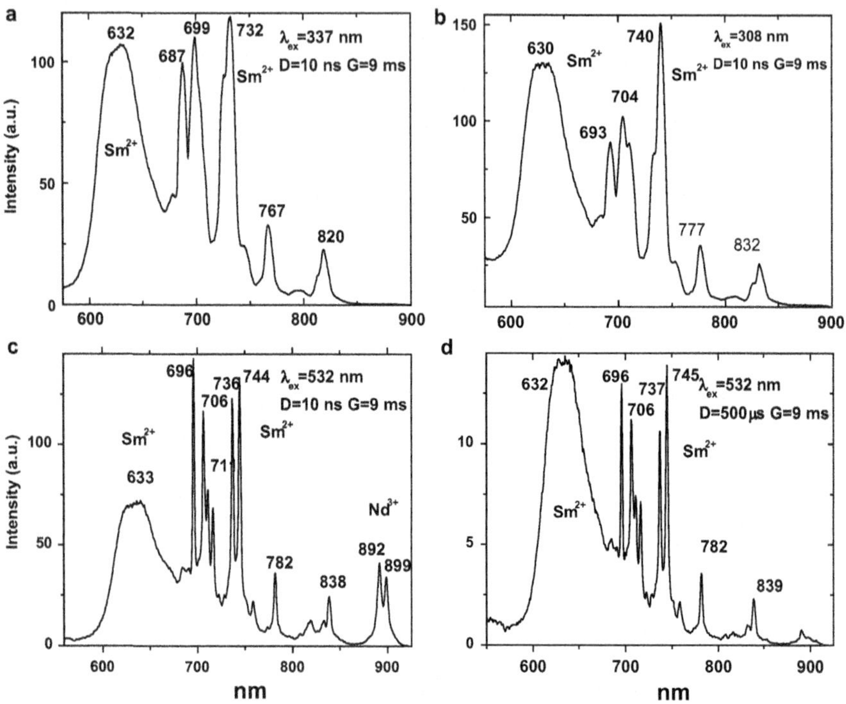

Fig. 4.31 (a–d) Laser-induced time-resolved luminescence spectra of anhydrite demonstrating Sm^{2+} centers

Fig. 4.32 Laser-induced CW luminescence spectra of anhydrite demonstrating Nd^{3+} centers (http://rruff.info)

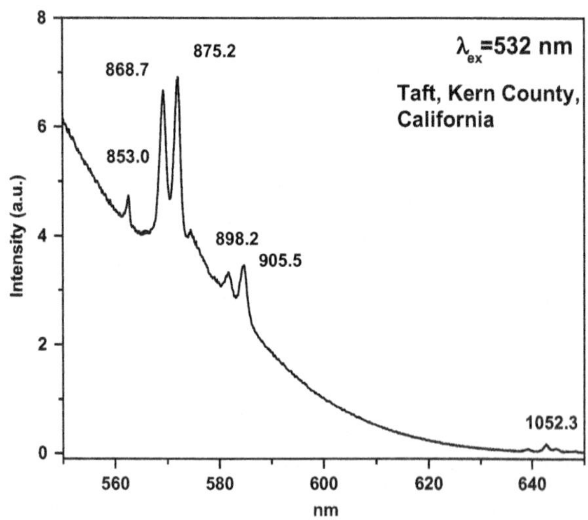

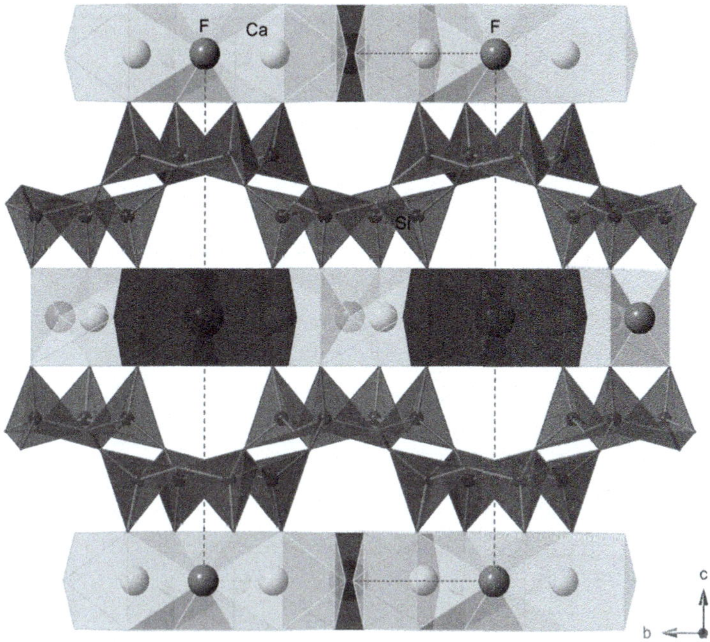

Fig. 4.33 Crystal structure of apophyllite showing the four-fold Si tetrahedra rings as well as Ca (CN7) and K (CN8) sites

By time resolved spectroscopy we found the broad band peaking at approximately 785 nm with relatively short decay time of 5 μs (Fig. 4.52a) evidently connected to Cr^{3+} center in weak crystal field. Excitation by CW laser with 532 and 780 nm revealed many luminescence bands and lines (Fig. 4.52b–f), which may be ascribed to Mn^{2+}, Fe^{3+}, Nd^{3+} and the broad structured band peaking at approximately 700 nm or to Cr^{3+} in average crystal field or to V^{2+}.

4.1.21 *Eudialite* $Na_{15}Ca_6(Fe,Mn)_3Zr_3SiO(O,OH,H_2O)_3$

Excitation by CW laser with 532 and 780 nm revealed narrow luminescence lines possibly belonging to trivalent REE (Nd, Eu) and Cr^{3+} and two luminescence bands peaking at 650 and 700 nm (Fig. 4.53). Additional study is needed to ascribe those emission centers including their decay times and excitation spectra.

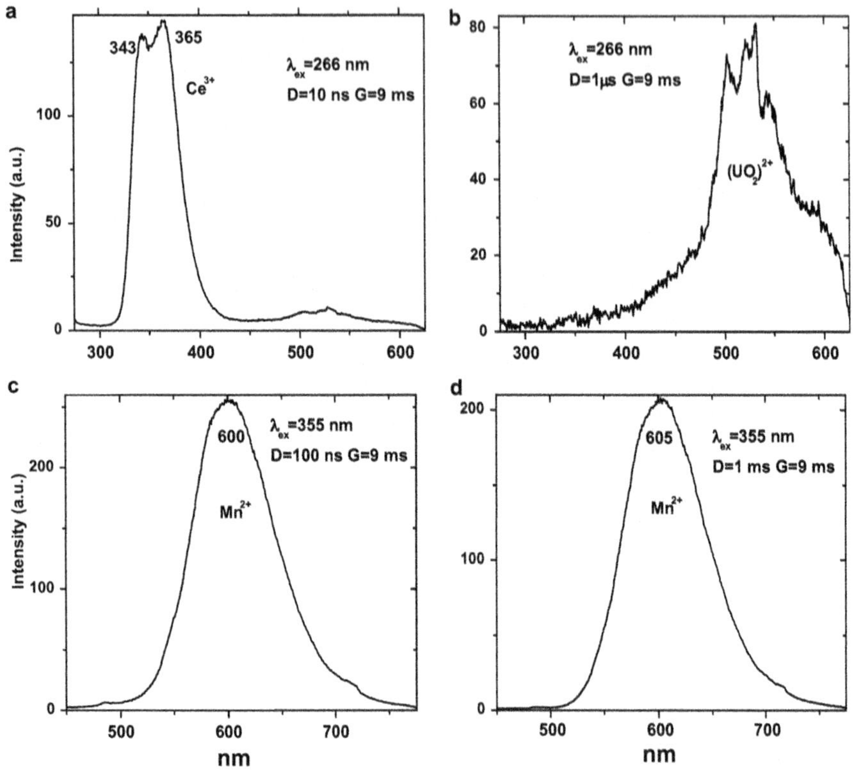

Fig. 4.34 (**a–d**) Laser-induced time-resolved luminescence spectra of apophyllite demonstrating Ce^{3+}, Mn^{2+} and $(UO_2)^{2+}$ centers

4.1.22 Shortite $Na_2Ca_2(CO_3)_3$

Shortite was studied by steady-state luminescence spectroscopy and luminescence of trivalent REE, such as Gd, Dy, Sm and Tb was found (Gorobets and Rogojine 2001). Excitation by CW laser with 532 and 780 nm revealed narrow luminescence lines possibly belonging to Nd^{3+} and two luminescence bands peaking at 650 and 705 nm (Fig. 4.54). Additional study is needed to ascribe those emission centers including their decay times and excitation spectra.

4.1.23 Burbankite $(Na,Ca)_3(Sr,Ba,Ce)_3(CO_3)_5$

Burbankite was studied by steady-state luminescence spectroscopy and luminescence of trivalent REE, such as Dy, Sm and Nd, was found (Gorobets and Rogojine 2001).

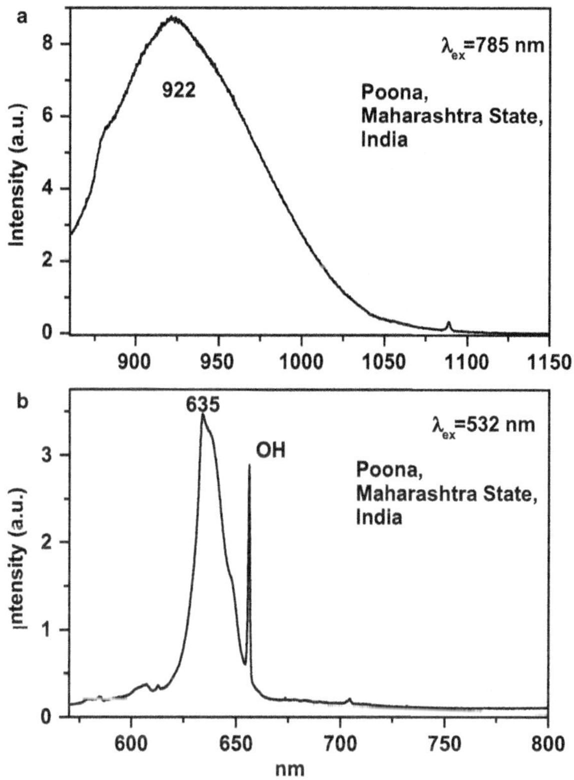

Fig. 4.35 (**a–b**) Laser-induced CW luminescence spectra of apophyllite with unknown centers (http://rruff.info)

Table 4.10 Concentrations of rare-earth elements in hardystonite sample (ppm)

	Pr	Sm	Eu	Nd	Dy	Tb	Er	Tm	Ho	Gd	Ce	Yb
NJ	9.7	9.2	1.3	43.5	7.5	1.4	2.5	0.2	0.1	11	55	0.8

Laser-induced photoluminescence was studied for burbankite group minerals-hexagonal burbankite, calcioburbankite, khanneshite and monoclinic remondite-(Ce), remondite-(La), petersenite-(Ce), as well as for orthorhombic carbocemaite and strontianite from carbonatites and hydrothennalites of alkaline complexes from various regions of the world. In burbankites of both genetic types, color centers probably do not correspolld to luminescence centers. All REE-rich minerals of burbankite group ($REE_2O_3 \sim 10.5$ wt%) are characterized by violet luminescence and their photo- and X-ray luminescel1ce spectra always contaill four bands of Sm^{3+}. Low-REE hydrothermal burbankites ($REE_2O_3 < 3$ wt%) are distinguished by different luminescence colors (from whitish to violet) and the low intensive spectra without Sm^{3+} bands. The essential differences in luminescence properties of indicated burbankites are determined by concentration ratio of the luminogene donor (Ce^{3+}) and the luminogene acceptor (Sm^{3+}). Photoluminescence spectra of

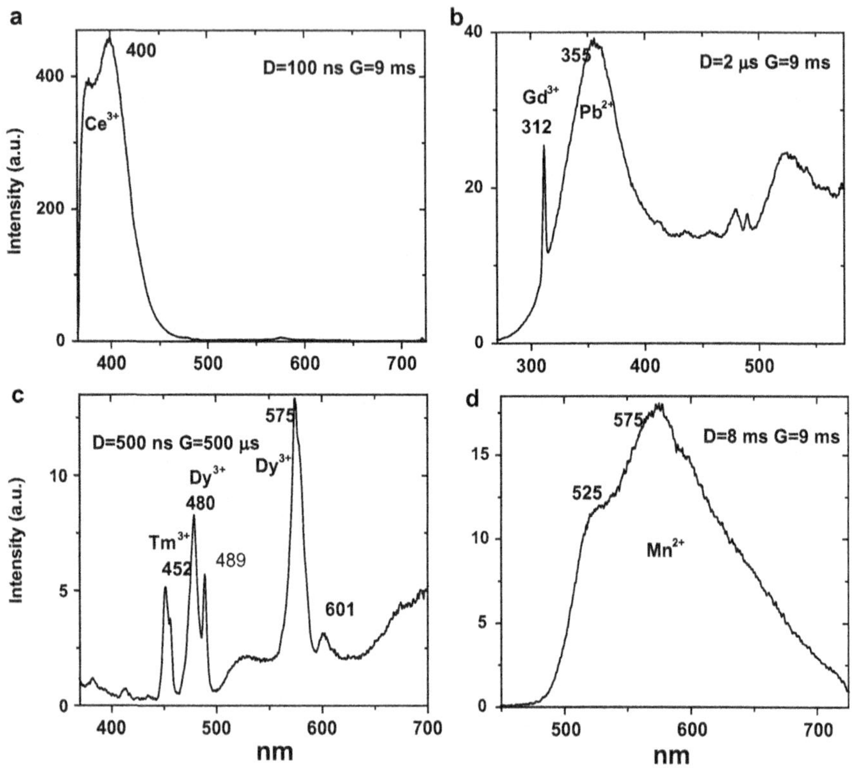

Fig. 4.36 (**a**–**d**) Laser-induced time-resolved luminescence spectra of hardystonite demonstrating Ce^{3+}, Pb^{2+}, Tm^{3+}, Dy^{3+} and Mn^{2+} centers

carbocernaite and strontianite contain also four bands of Sm^{3+} but, in the whole, their intensity is below.

Obtained data testify for insignificant influence of matrix and crystal symmetry upon ability of Sm^{3+} to manifest itself as luminescence center in the studied carbonates (Sokolov and Rassulov 2004).

Excitation by CW laser with 532 and 780 nm revealed narrow luminescence lines possibly belonging to Sm^{3+}, Eu^{3+} and Nd^{3+} (Fig. 4.55).

4.1.24 *Prosopite $CaAl_2(F,OH)_8$*

Prosopite was studied by steady-state luminescence spectroscopy and luminescence of trivalent REE, such as Gd, Ce, Dy, and Sm, was found (Gorobets and Rogojine 2001). Excitation by CW laser with 532 nm revealed narrow luminescence lines possibly belonging to Eu^{3+}, Sm^{3+} and Cr^{3+} and luminescence band peaking at 720 nm (Fig. 4.56).

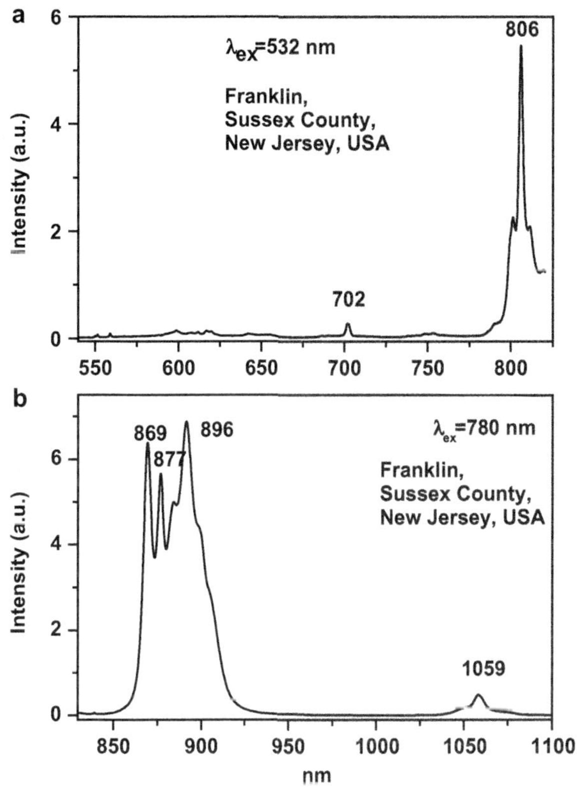

Fig. 4.37 (**a–b**) Laser-induced CW luminescence spectra of hardystonite with Nd^{3+} centers (http://rruff.info)

4.1.25 *Svabite* $Ca_5(AsO_4)_3F$

Svabite was studied by steady-state luminescence spectroscopy and luminescence of Mn^{2+} was found (Gorobets and Rogojine 2001). Excitation by CW laser with 532 and 780 nm revealed band of Mn^{2+} and narrow luminescence lines of evidently Nd^{3+} (Fig. 4.57).

4.1.26 *Anorthite* $CaAl_2Si_2O_8$

Anorthite has a triclinic crystal system where are six crystallographically independent cation sites, namely, four Ca^{2+} sites, one Al^{3+} site, and one Si^{4+} site. Al and Si atoms both occupy tetrahedral sites with 4 coordinated oxygen atoms, while one type of Ca^{2+} ion occupies an octahedral site with six oxygen atoms and other Ca^{2+} ions occupy three kinds of polyhedral sites with 7 coordinated oxygen atoms with different Ca-O bond distance.

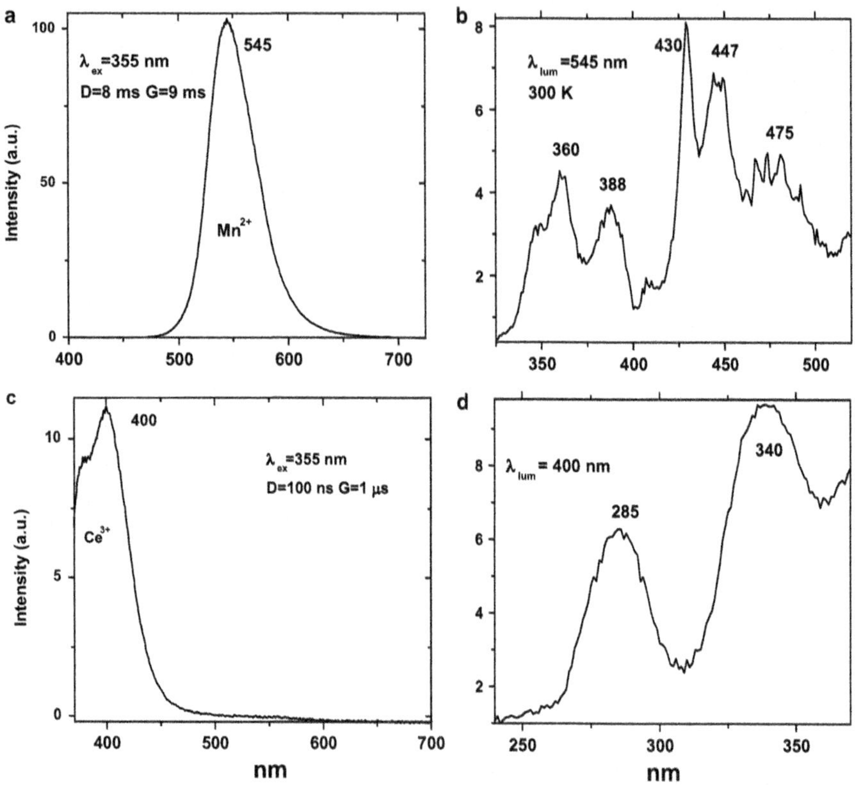

Fig. 4.38 (**a–d**) Laser-induced time-resolved luminescence spectra of esperite demonstrating Ce^{3+} and Mn^{2+} centers. (**b**, **d**) Excitation spectra of Mn^{2+} and Ce^{3+}

Excitation by CW laser with 532 and 780 nm revealed several broad bands and narrow luminescence lines (Fig. 4.58). Narrow lines peaking at 588 and 611 nm (Fig. 4.58b) evidently belong to Eu^{3+} luminescence as in synthetic anorthite artificially activated by Eu (Yu et al. 2012). The mostly probable substitution is for one of the Ca^{2+} sites. Other narrow lines in visible and NIR spectral ranges may be connected with Dy^{3+}, Sm^{3+} and Nd^{3+} (Fig. 4.58b, d and f). Two broad emission bands evidently present peaking at 700 (Fig. 4.58a) and 752 (Fig. 4.58e) and often they present together resulting in emission with intermediate maximum (Fig. 4.58c). Those bands interpretation needs decay time study, but preliminary it may be supposed that they are connected to Fe^{3+} in tetrahedral coordination in Si and Al positions.

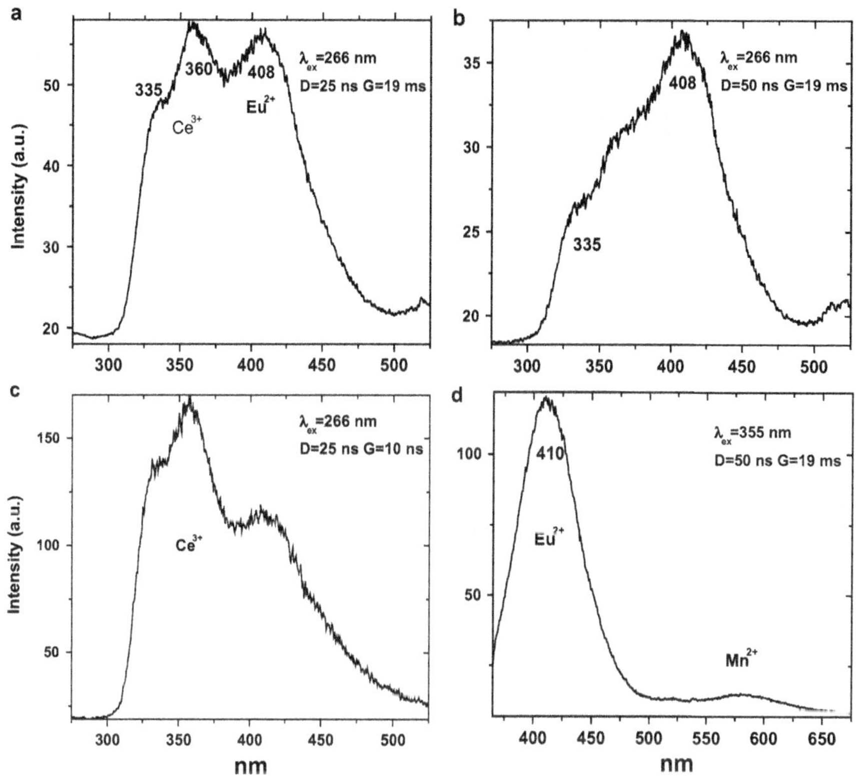

Fig. 4.39 (**a–d**) Laser-induced time-resolved luminescence spectra of charoite demonstrating Ce^{3+} and Eu^{2+} centers

4.2 Lead (Pb^{2+}) Bearing Minerals

Minerals of leas are principally capable to intrinsic luminescence. The main substituting luminescence centers are REE^{2+} (Eu), REE^{3+} (Ce, Sm, Eu, Dy). Ionic radii of Pb^{2+} are 1.33 Å in 6-coordinated form and 1.43 Å in 8-coordinated form, while ionic radius of Eu^{2+} is 1.31 Å and ionic radii of REE^{3+} are changing from 1.00 to 1.13 Å.

4.2.1 *Pyromorphite* $Pb_5(PO_4)_3Cl$

Pyromorphite belongs to the apatite group. This group is divided into two series, the apatite series and pyromorphite series. These minerals are isostructural, however the unit cell volume of pyromorphite is about one-fifth larger than that of apatite.

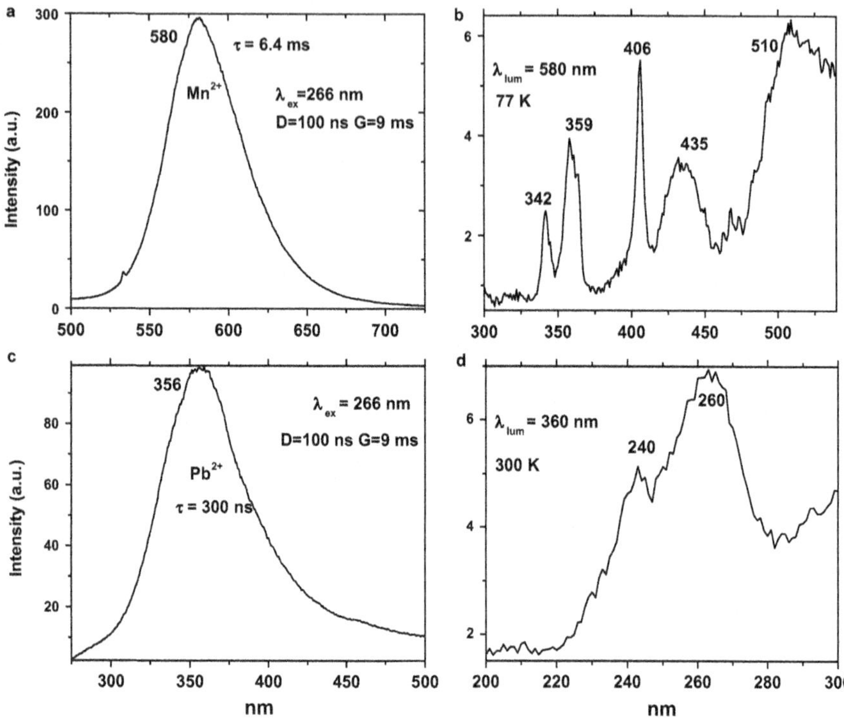

Fig. 4.40 (**a–d**) Laser-induced time-resolved luminescence spectra of Mn^{2+} in pectolite and Pb^{2+} in prehnite. (**b**, **d**) Excitation spectra of Mn^{2+} and Pb^{2+}

Luminescence centers O_2^- (Tarashchan 1978) and evidently $(VO_4)^{3-}$ (Gaft 1984) characterize steady-state spectra of pyromorphite. The natural pyromorphite in our study consisted of two samples. Laser-induced time-resolved technique enables to detect $Ce^{3+,}$ Eu^{3+}, Sm^{3+} and Tb^{3+} emission centers (Fig. 4.59).

Excitation by CW laser with 532 and 780 nm revealed narrow luminescence lines possibly belonging to trivalent REE (Nd, Eu, Sm) and two luminescence bands peaking at 625 and 675 nm (Fig. 4.60). Additional study is needed to ascribe those emission centers including their decay times and excitation spectra.

4.2.2 *Vanadinite* $Pb_5(VO_4)_3Cl$

Vanadinite was studied by steady-state luminescence spectroscopy and luminescence of $(VO_4)^{3-}$ metal-oxygen complexes was found (Gorobets and Rogojine 2001). Excitation by CW laser with 532 and 780 nm revealed narrow luminescence lines possibly belonging to Nd^{3+} and luminescence band similar to ascribed to $(VO_4)^{3-}$ (Fig. 4.61).

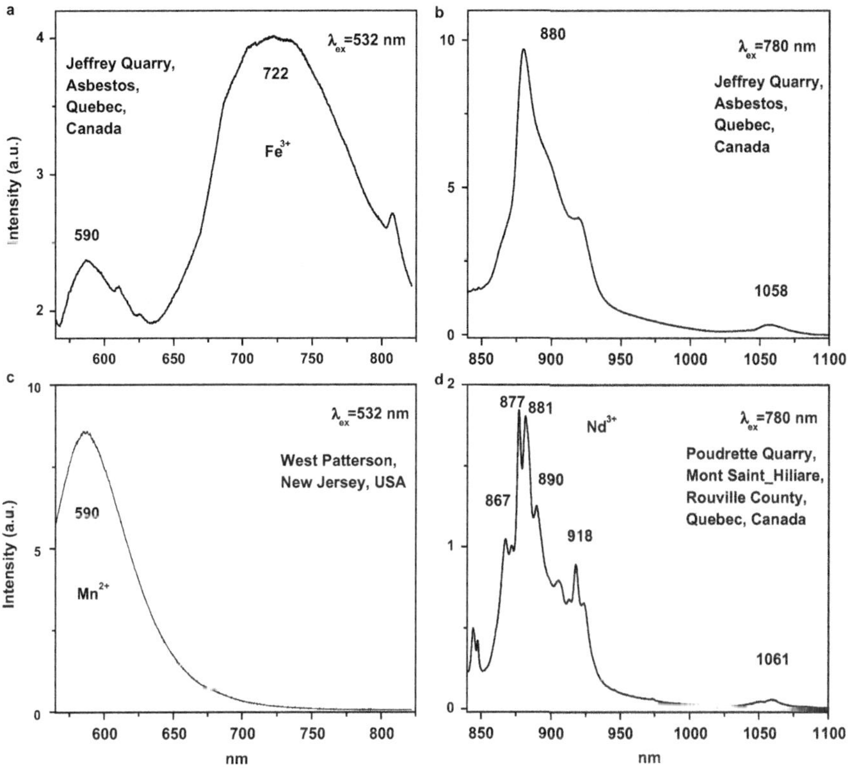

Fig. 4.41 (**a–d**) Laser-induced CW luminescence spectra of pectolite with Mn^{2+}, Fe^{3+} and Nd^{3+} centers (http://rruff.info)

4.2.3 *Wulfenite $PbMoO_4$*

Wulfenite was studied by steady-state luminescence spectroscopy and luminescence of $(MoO_4)^{3-}$ metal-oxygen complexes was found (Gorobets and Rogojine 2001). Excitation by CW laser with 532 and 780 nm revealed narrow luminescence lines possibly belonging to Nd^{3+} and several broad luminescence bands of unknown origin (Fig. 4.62).

4.2.4 *Margarosanite $Pb(Ca,Mn)_2Si_3O_9$*

The unit cell of margarosanite ideally contains two ($PbCa_2Si_3O_9$). The margarosanite structure has the following principle features: planes of tetrahedral sites alternate with planes of calcium sites between sheets of close packed oxygen atoms parallel to (101); Pb and Ca are ordered; Ca(l) polyhedra form an infinite

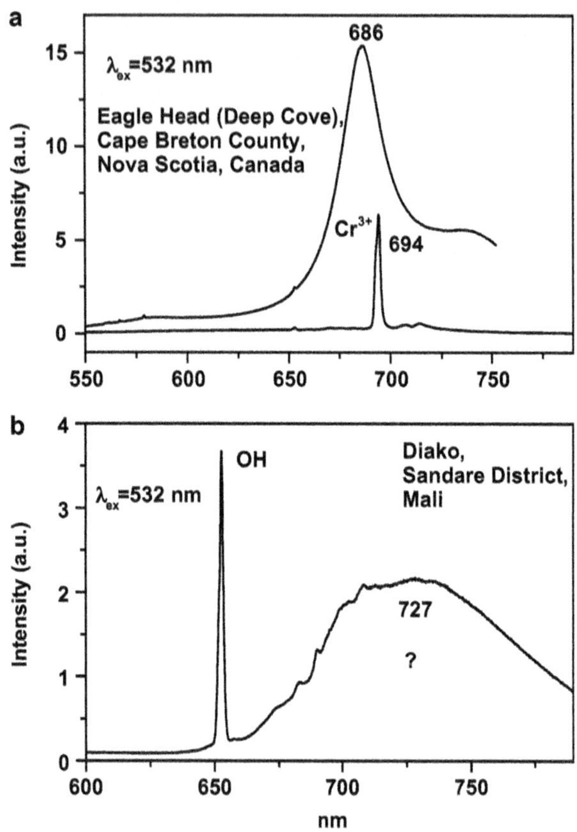

Fig. 4.42 (**a–b**) Laser-induced CW luminescence spectra of pectolite with possibly Mn^{2+}, Fe^{3+} and Cr^{3+} centers (http://rruff.info)

edge-sharing chain parallel to [101] and Pb and Ca(2) sites alternate along the edge of this chain; tetrahedral sites form three membered rings which are linked to the Ca (1) polyhedral chain; the two Ca atoms have six-fold coordination and the Pb atom has seven-fold coordination (Fig. 4.63)

It is a rare lead silicate which was not studied by steady-state luminescence spectroscopy. Laser-induced time-resolved technique (Figs. 4.64 and 4.65) enables to detect two types of Pb^{2+}, two types of Mn^{2+} and Ce^{3+} emission centers (Gaft et al. 2013a).

Excitation by CW laser with 532 and 780 nm revealed narrow luminescence lines possibly belonging to Nd^{3+} and broad luminescence band of Mn^{2+} (Fig. 4.66).

4.2.5 Cerussite $PbCO_3$

Excitation by CW laser with 532 and 780 nm revealed several luminescence bands and lines (Fig. 4.67). IR lines under 780 nm excitation may be evidently ascribed to

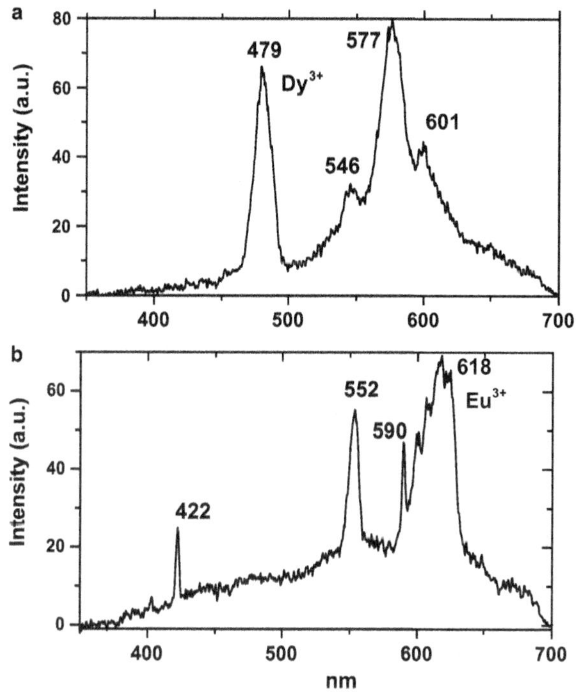

Fig. 4.43 (**a–b**) Laser-induced time-resolved luminescence spectra of pyrochlore demonstrating Dy^{3+} and Eu^{3+} centers

Nd^{3+} and other trivalent REE luminescence present under 532 nm excitation, while the bands origin needs further investigation.

4.3 Tin (Sn^{4+}) Bearing Minerals

Minerals of tin are capable for intrinsic luminescence, possibly connected with defect centers containing Sn^{2+}. The ionic radius of Sn^{4+} is of 0.83 Å and the possible substituting luminescence center is Ti^{4+} with ionic radius of 0.75 Å.

4.3.1 Cassiterite SnO_2

The structure of cassiterite resembles that of rutile TiO_2, each tin ion being surrounded by six oxygen ions approximately at the corners of a regular octahedron, and each oxygen having three tin ions around it forming a nearly equilateral triangle.

The considerable zonation was found in CL of cassiterite with respect to W and Ti. It was concluded that W is responsible for blue luminescence, while Ti for the

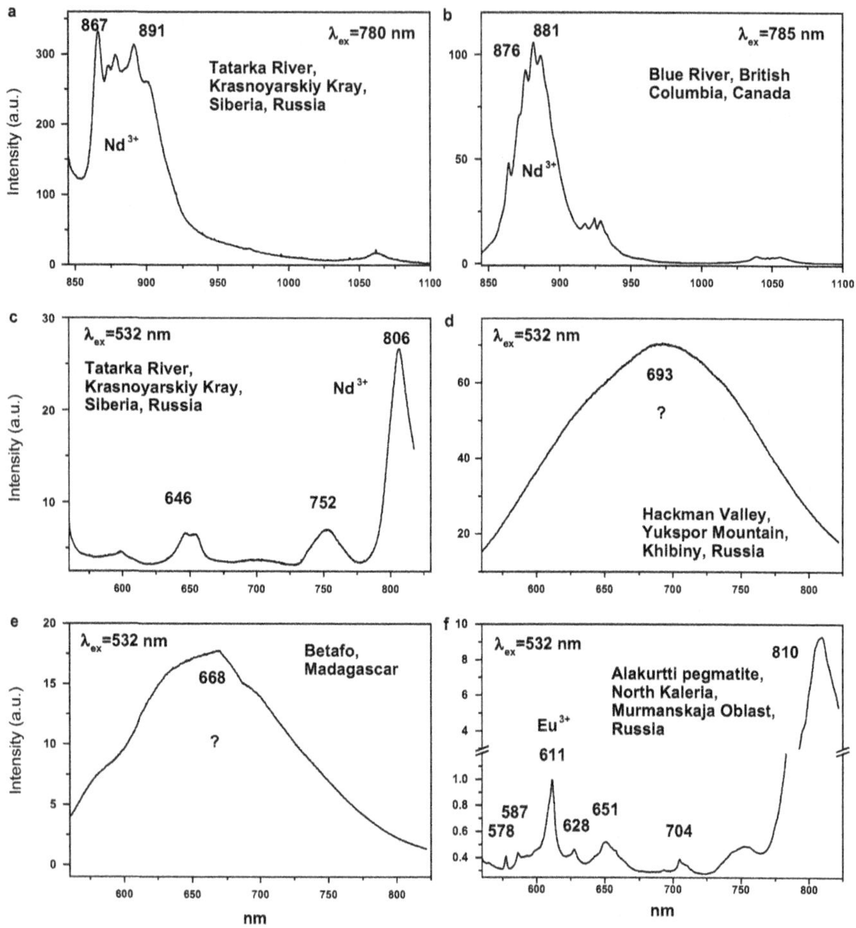

Fig. 4.44 (**a–f**) Laser-induced CW luminescence spectra of pyrochlore with Eu^{3+}, Nd^{3+} and unknown centers (http://rruff.info)

yellow one (Waychunas 1989). Under UV lamp excitation cassiterite samples are luminescent only at low temperatures. The luminescence center was found to be related to intrinsic centers, namely to Sn^{2+} and Sn^{3+}. Under powerful UV laser excitation cassiterite has luminescence even at 300 K with spectrum similar to those at 77 K with lamp excitation. Decay time of luminescence has two components, namely 100 ns and 280 ns (Gaft and Vorontsova 1982; Gaft et al. 1988).

The natural cassiterite in our study consisted of six samples. Laser-induced time-resolved technique detected emission centers similar to those in sw one (Fig. 4.68).

4.3.2 Malayaite CaSn(SiO$_5$)

Other tin minerals with luminescence are malayaite $CaSn(SiO_5)$ and sorensenite $Na_4SnBe_2Si_6O_{18}x2H_2O$, where emission centers (Fig. 4.69) are evidently Ti

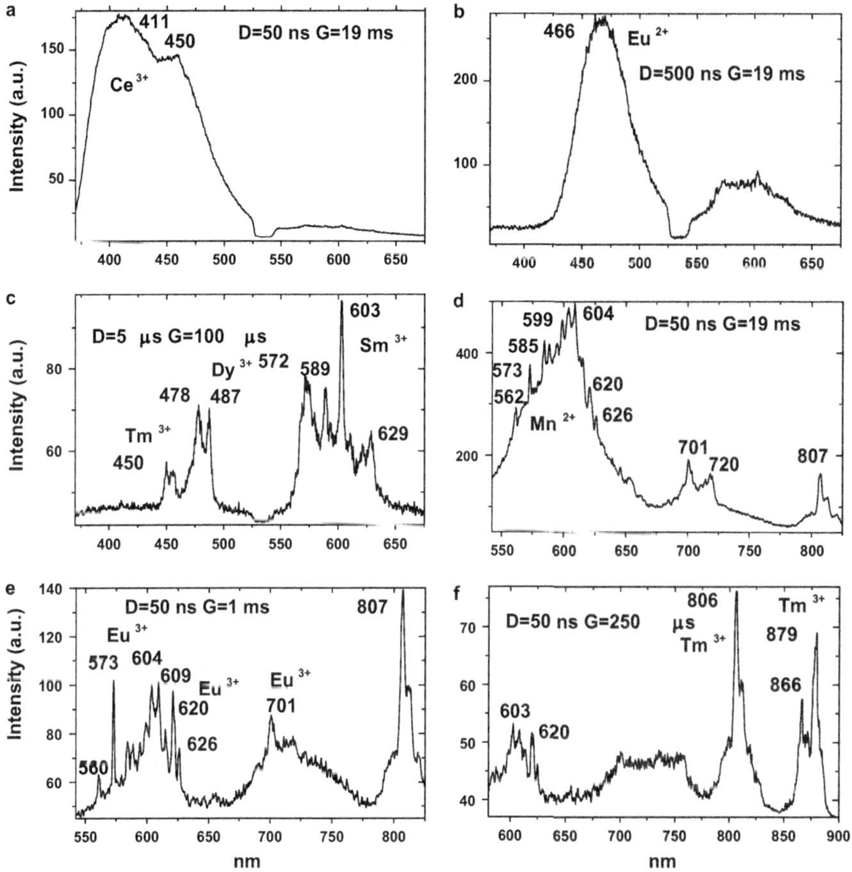

Fig. 4.45 (**a–f**) Laser-induced time-resolved luminescence spectra of leucophane under 355 nm (**a–c**) and 532 nm (**d–f**) excitations demonstrating Ce^{3+}, Eu^{2+}, Eu^{3+}, Tm^{3+}, Dy^{3+}, Sm^{3+} and Mn^{2+}

impurities (Gaft et al. 1982). Excitation by CW laser with 532 and 780 nm revealed several luminescence bands and lines (Fig. 4.70). IR lines under 780 nm excitation may be evidently ascribed to Nd^{3+} and other trivalent REE luminescence, such as Sm^{3+}, also present. The bands origin needs further investigation.

4.4 Barium (Ba^{2+}) Bearing Minerals

Ionic radii of Ba^{2+} are of 1.49 Å in 6-coordinated form and 1.56 Å in 8-coordinated form. The main substituting luminescence centers are Bi^{3+} with ionic radius of 1.49 and 1.56 Å in 6-coordinated and 8-coordinated forms, correspondingly, Ag^{+} with ionic radii of 1.29 and 1.42 Å in 6-coordinated and 8-coordinated forms.

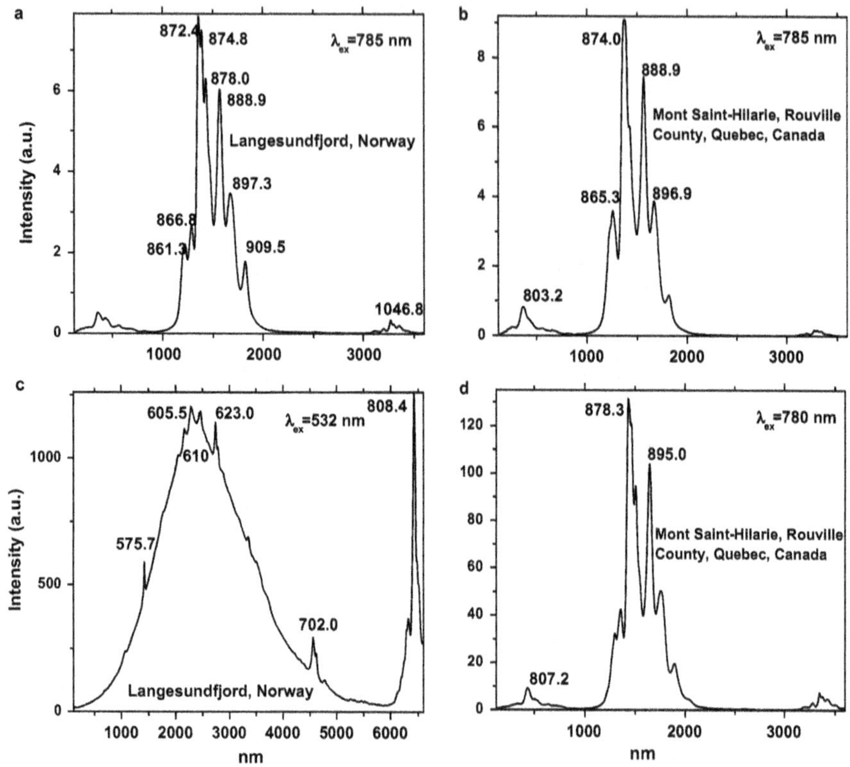

Fig. 4.46 (**a–d**) Laser-induced CW luminescence spectra of leucophane demonstrating Mn^{2+}, and Nd^{3+} centers (http://rruff.info)

4.4.1 Barite $BaSO_4$

Barite is a barium sulfate, $BaSO_4$, with orthorhombic structure (2/m2/m2/m) where the sulfur is situated in tetrahedral coordination with oxygen, and barium in 12-fold coordination with oxygen. Mineral barite is one of the first luminescent materials from which the famous "Bologna stone" was obtained. Nevertheless, up today the understanding of natural barite luminescence is very scarce. It has been known for a long time that some specimens of barite are fluorescent under UV exposure and emit white, yellow, green or orange light. In steady-state luminescence spectra of barite different luminescence bands from the UV to the red part of the spectrum have been detected. However, only UO_2^{2+} and Eu^{2+} luminescence centers have been confidently identified (Tarashchan 1978; Gaft et al. 1985).

Barite would be colorless is pure, but radiation damage centers commonly because color. For example, O^- (in some blue crystals), SO_3^- (in some honey-yellow and blue crystals), and SO_2^- (in yellow crystals) centers have been identified in barite.

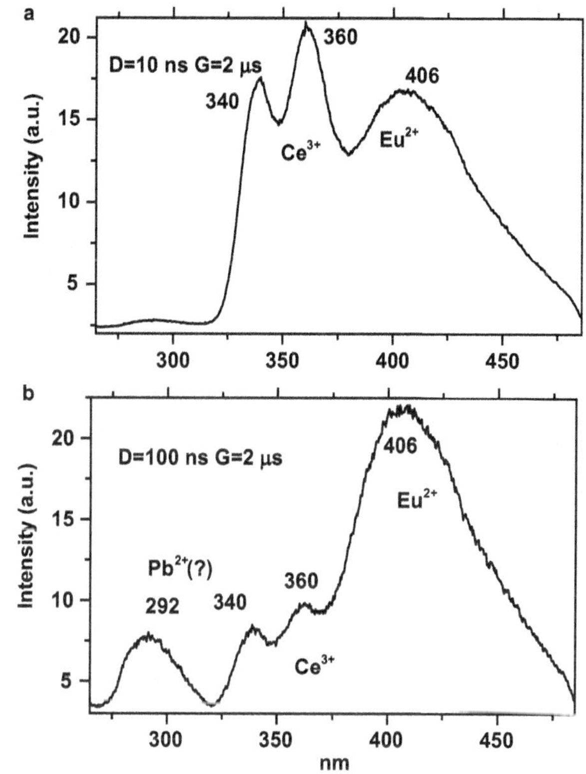

Fig. 4.47 (**a–b**) Laser-induced time-resolved luminescence spectra of milarite demonstrating Ce^{3+}, Eu^{2+}, and possibly Pb^{2+}

The natural barite in our study consisted of 25 samples of different origin. Concentrations of potential luminescence impurities in several samples are presented in Table 4.11. For the correct interpretation of the luminescent bands, artificial barite standards have been investigated, as nominally pure, and activated. Laser-induced time-resolved technique enables to detect Ag^+, Bi^{2+}, Bi^{3+}, Eu^{2+}, Eu^{3+}, Ce^{3+}, Nd^{3+}, $(UO_2)^{2+}$ and several still not identified emission centers (Gaft et al. 2001, 2008b) (Figs. 4.71, 4.72, 4.73 and 4.74).

4.4.2 *Witherite $BaCO_3$*

Witherite was studied by steady-state luminescence spectroscopy and trivalent REE, such as Gd, Dy and Eu, have been found (Gorobets and Rogojine 2001). Excitation by CW laser with 532 and 780 nm revealed several luminescence lines and one broad band (Fig. 4.75). IR lines under 780 nm excitation may be evidently ascribed to Nd^{3+}. The band is somewhat similar to orange emission of barite ascribed to Ag^+ luminescence center. Excitation spectra and decay time have to be studied.

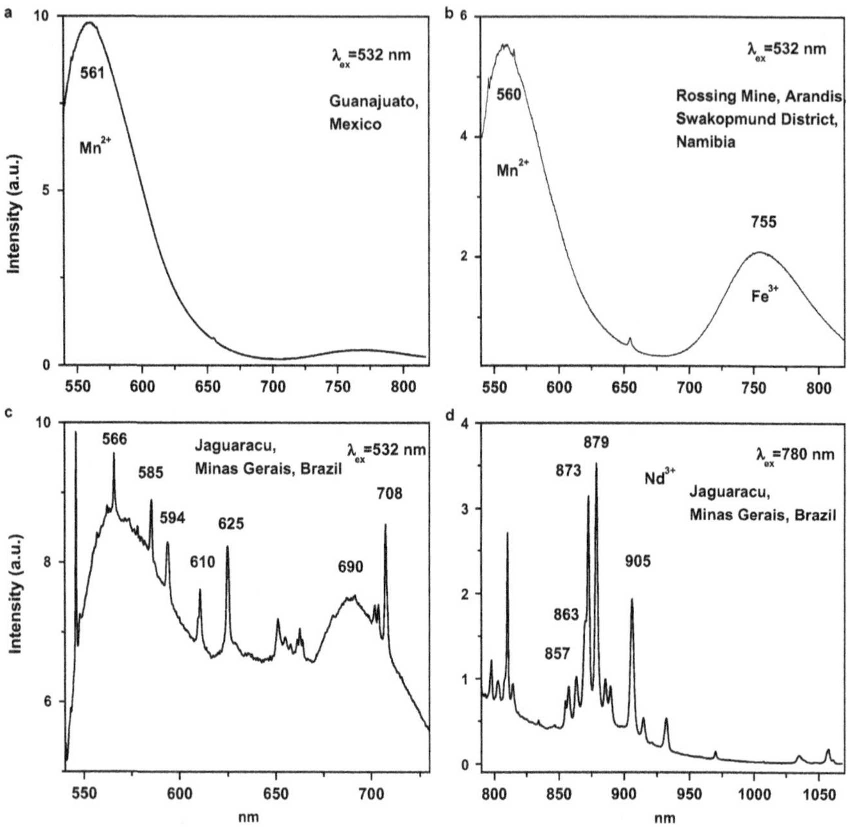

Fig. 4.48 (**a–d**) Laser-induced CW luminescence spectra of milarite demonstrating Mn^{2+}, Fe^{3+} and Nd^{3+} centers (http://rruff.info)

4.5 Strontium (Sr^{2+}) Bearing Minerals

The ionic radius of Sr^{2+} is 1.32 Å in 6-coordinated form. The possible substituting elements are Eu^{2+} with ionic radius of 1.31 Å and Ce^{3+} with ionic radius of 1.15 Å.

4.5.1 Celestine $SrSO_4$

The structure of celestine is similar to that of barites, with Sr taking the place of Ba. Steady-state luminescence of celestine is characterized by broad bands, which are associated with adsorption of water-organic complexes (Tarashchan 1978). The natural celestine in our study consisted of five samples. Laser-induced time-

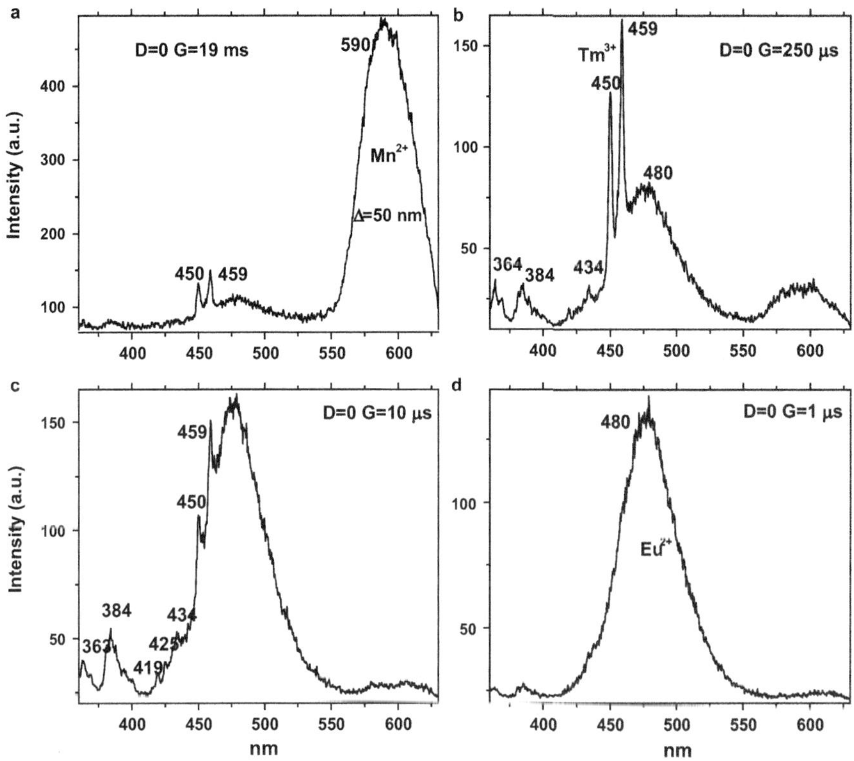

Fig. 4.49 (**a–d**) Laser-induced time-resolved luminescence spectra of walstromite demonstrating Tm^{3+}, Eu^{2+}, and Mn^{2+}

resolved technique enables to detect strong emission bands bur their interpretation is not finished yet (Fig. 4.76).

4.5.2 *Strontianite* $SrCO_3$

Strontianite has been studied by steady-state luminescence spectroscopy and trivalent REE, such as Dy, Sm and Ce, were found together with Mn^{2+} and uranyl (Gorobets and Rogojine 2001). Excitation by CW laser with 532 and 780 nm revealed several luminescence lines and one broad band (Fig. 4.77). IR lines under 780 nm excitation may be evidently ascribed to Nd^{3+}. The bands may be ascribed to Mn^{2+}.

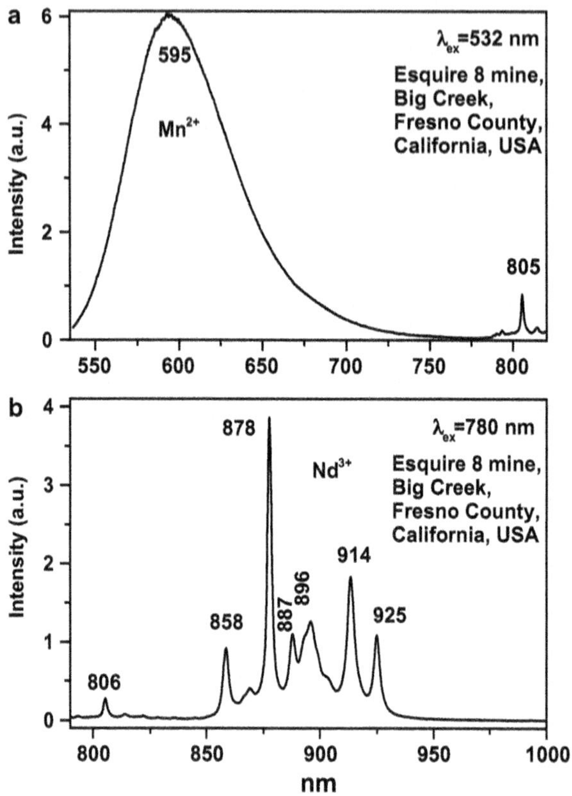

Fig. 4.50 (**a–b**) Laser-induced CW luminescence spectra of walstromite demonstrating Mn^{2+}, and Nd^{3+} centers (http://rruff.info)

4.6 Titanium (Ti^{4+}) Bearing Minerals

Minerals of Ti are capable for intrinsic luminescence, which is connected with $(TiO_6)^{8-}$ groups and Ti^{3+} center. The ionic radius of Ti^{4+} is 0.75 Å in 6-coordinated form. The possible substituting element is Cr^{3+} with ionic radius of 0.76 Å.

4.6.1 Titanite $CaTiSiO_5$

Titanite (formerly called "sphene") is an orthosilicate mineral with the chemical formula $CaTiOSiO_4$, where approximately 20 % of the oxygens can be partially replaced by OH (hydroxyl) and F. Titanite has a monoclinic symmetry with a space group $P2_1/a$. It has optically positive character with $\alpha = 1.84 - 1.95$, $\beta = 1.87 - 2.034$, $\gamma = 1.943 - 2.11$, very high refractive indices (1.843-1.950) and extreme birefringence (0.100–0.192). The structure of titanite consists of chains of octahedral sites occupied by Ti^{4+}, cross-linked by isolated SiO_4 tetrahedra. Large

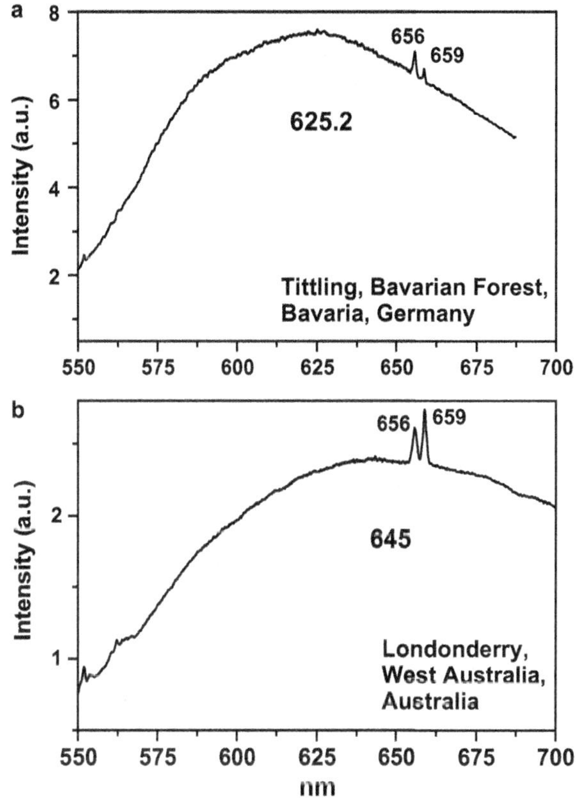

Fig. 4.51 (**a–b**) Laser-induced CW luminescence spectra of bavenite demonstrating Mn^{2+}, centers (http://rruff.info)

cavities in this structure provide the seven-fold coordinated site occupied by Ca^{2+} (Fig. 4.78).

Titanite is very interesting luminescent material. Potential luminescent centers include intrinsic TiO_6 and Ti^{3+} and different impurities such as trivalent and divalent rare-earth elements (REE), Pb^{2+} and Mn^{2+} substituting for Ca^{2+}, Cr^{3+} and Mn^{4+} substituting for Ti^{4+}, and Cr^{4+}, Cr^{5+} and Fe^{3+} substituting for Si^{4+}. Besides that, titanite can incorporate minor amounts of radioactive impurity components (particularly U and Th) that affect the crystal structure through α- and β-decay events. Thus radiation-induced luminescence centers are also possible. Nevertheless, steady-state luminescent spectroscopy of titanite under UV and X-ray excitations did not reveal characteristic bands and lines (Gorobets and Rogojine 2001) and only ionoluminescence spectrum of titanite exhibits various narrow lines of Sm^{3+}, Eu^{3+} and Nd^{3+} (Yang 1995).

The natural titanite in our study consisted of nine samples. Concentrations of potential luminescence impurities in one sample are presented in Table 4.12. Trace element concentrations in natural titanite samples have been analyzed by Lenz et al. 2015 and it was found that chromium and REE concentrations vary

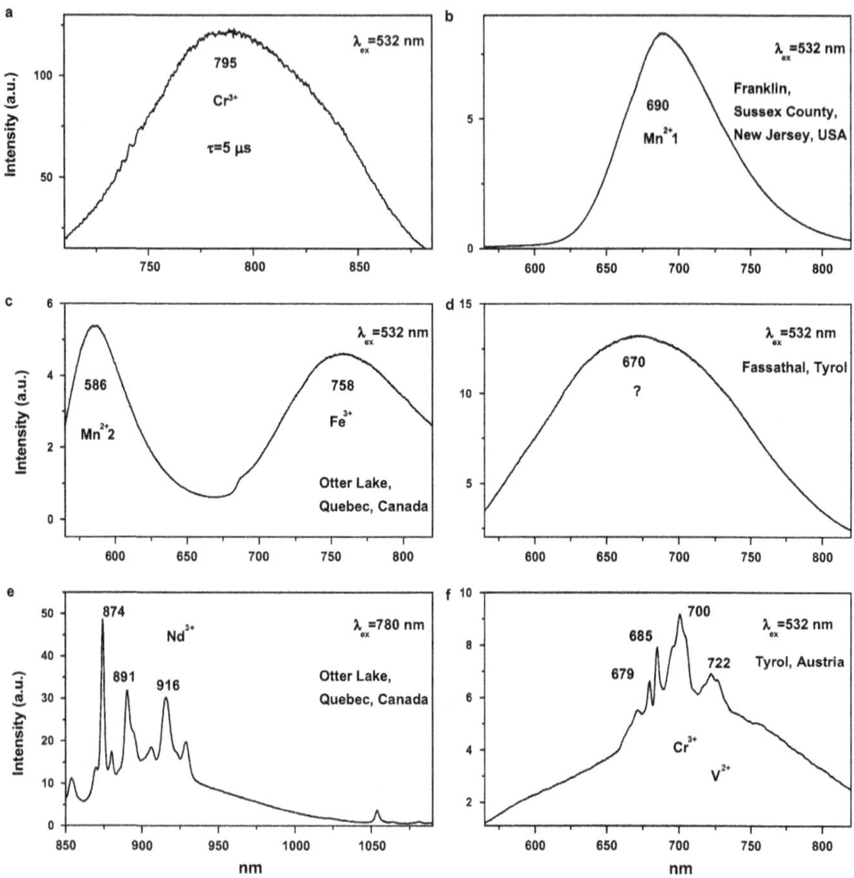

Fig. 4.52 (**a–f**) Laser-induced time-resolved spectrum of diopside demonstrating Cr^{3+} (**a**) and CW luminescence spectra of diopside demonstrating evidently Cr^{3+}, Fe^{3+}, Mn^{2+}, Nd^{3+} and possibly V^{2+} centers (http://rruff.info)

appreciably, depending on sample origin. Certain samples are characterized by relatively low Cr (<100 ppm) and elevated REE concentrations (ranging from ca. 1100 ppm to 16,600 ppm), while other yielded elevated Cr (ca. 160–1930 ppm) and low REE concentrations (less than 200 ppm) or be rich in both, Cr and REE. Chromium and other metals, such as Nb, Ta, V, Mn, Mg, Sn, Al, and Fe, are generally considered to be incorporated at the six-fold coordinated Ti-site whereas REEs substitute Ca on its large, seven-coordinated site. Laser-induced time-resolved technique enables to detect Sm^{3+}, Nd^{3+}, Tm^{3+}, Pr^{3+}, Eu^{3+}, Er^{3+} and Cr^{3+} emission centers (Gaft et al. 2003b) (Figs. 4.79 and 4.80).

Excitation by CW laser with 514 nm revealed several IR luminescence (Fig. 4.81). They may be evidently ascribed to Nd^{3+} emission. Many luminescence lines in the IR part of the spectrum have been found under excitation with 785 nm

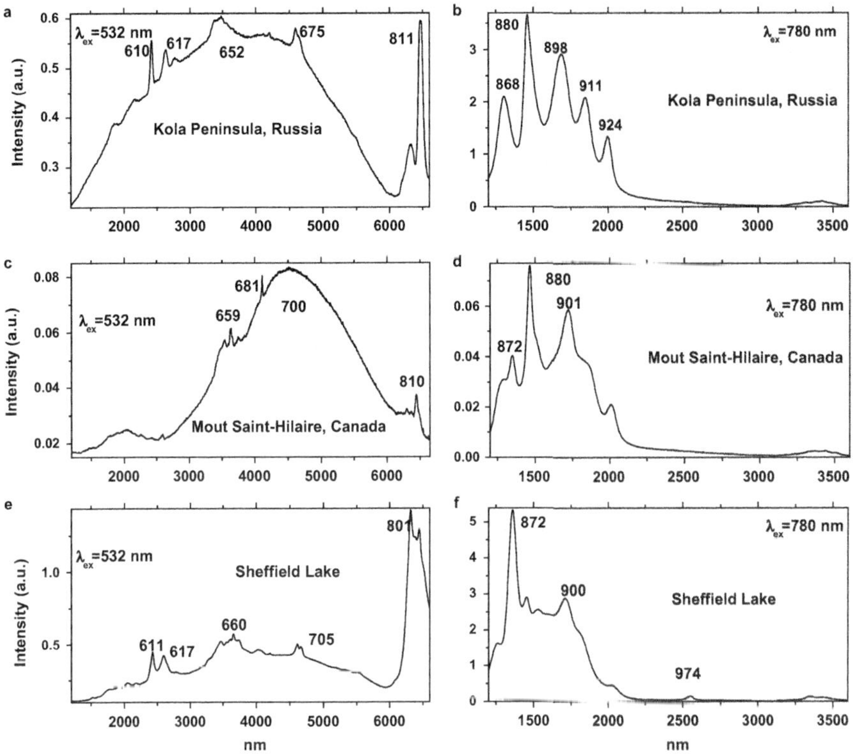

Fig. 4.53 (**a–f**) Laser-induced CW luminescence spectra of eudialite (http://rruff.info)

and ascribed to REE^{3+}, such as to Nd^{3+} (870 and 890 nm), Pr^{3+} (950 and 870 nm), Ho^{3+} (960 nm) and Er^{3+} (850 nm) (Jasinevicius 2009).

Detailed research of titanite REE luminescence under different cw laser excitations was done by Lenz et al. (2015) including the study of luminescence and its excitation of artificial titanite samples activated by different REE, such as Sm, Nd, Pr and Eu. Relative emission intensities of individual REEs depend strongly on the excitation wavelength. The Raman spectra of titanite obtained using a 473 nm laser excitation shows emissions of Pr^{3+}, Sm^{3+} and Nd^{3+}, whereas green excitation (532 nm) excites preferentially the PL of Sm^{3+}, and Nd^{3+}, red excitation (633 nm) predominantly Cr^{3+} and Nd^{3+}, and NIR excitation (785 nm) Nd^{3+} only. Those results have been confirmed by excitation spectroscopy of artificially activated titanite samples. Under 785 nm excitation, Nd^{3+} emissions are exceptionally strong (whereas Raman scattering is weaker under NIR excitation when compared to visible excitation). Therefore, Raman spectra of titanite samples obtained with IR excitation typically are obscured vastly by Nd^{3+} emissions.

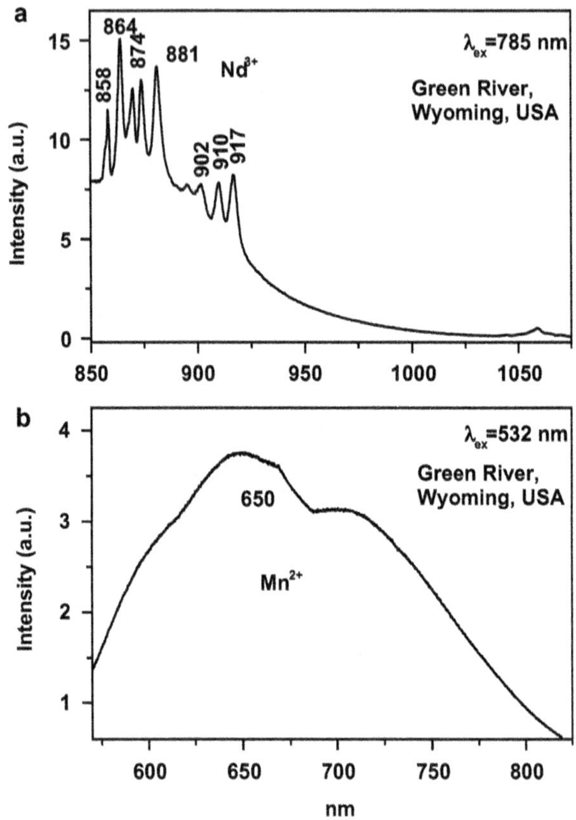

Fig. 4.54 (**a–b**) Laser-induced CW luminescence spectra of shortite (http://rruff.info)

4.6.2 Benitoite $BaTiSi_3O_9$

Benitoite belongs to cyclosilicates subclass. It contains rings of linked SiO_4 tetrahedra. The tetrahedra are linked in the same manner as in the single-chain inosilicates, each SiO_4 group sharing two oxygens with adjoining tetrahedra on either side, giving the same overall formula $(SiO_3{}^{-2})_n$. However, instead of forming straight chains the tetrahedra are joined at angles, which result in the formation of the rings. The minimum number of tetrahedra to form a ring is three and a few three-member-ring minerals are known, the best known being benitoite. It is the only known mineral to crystallize in the bar 6 m^2 class called the ditrigonal-dipiramidal symmetry class.

Benitoite is rare, strongly dichroic, blue mineral used as a gemstone. In spite of much effort to its study, the origin of color in benitoite has not been definitively established. Because traces of Fe are found, ideas proposed include the Fe^{2+}-Ti^{4+} or the Fe^{2+}-Fe^{3+} inter-valance charge transfer. While most benitoite is colorless when viewed down the c-axis, there are a very small number of exceedingly rare stones, which are pink in this direction (Rossman 1988).

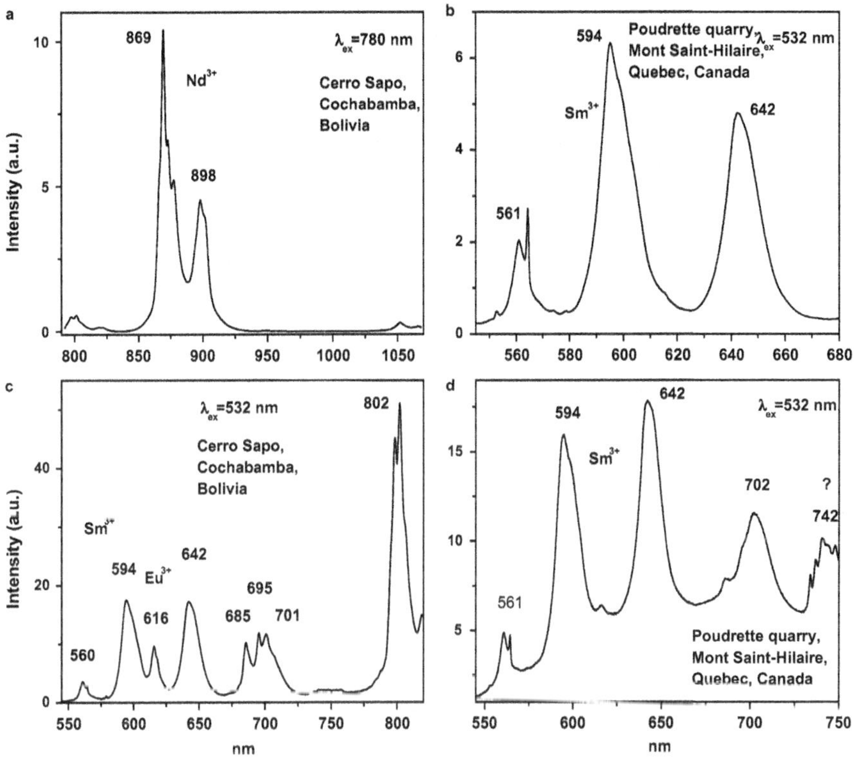

Fig. 4.55 (**a–d**) Laser-induced CW luminescence spectra of burbankite (http://rruff.info)

Fig. 4.56 Laser-induced CW luminescence spectra of prosopite (http://rruff.info)

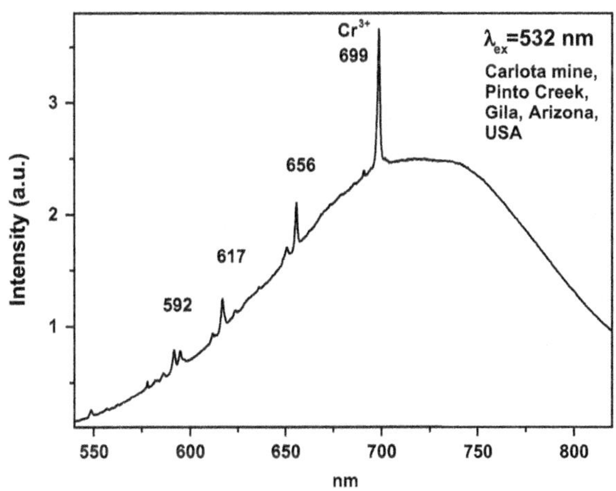

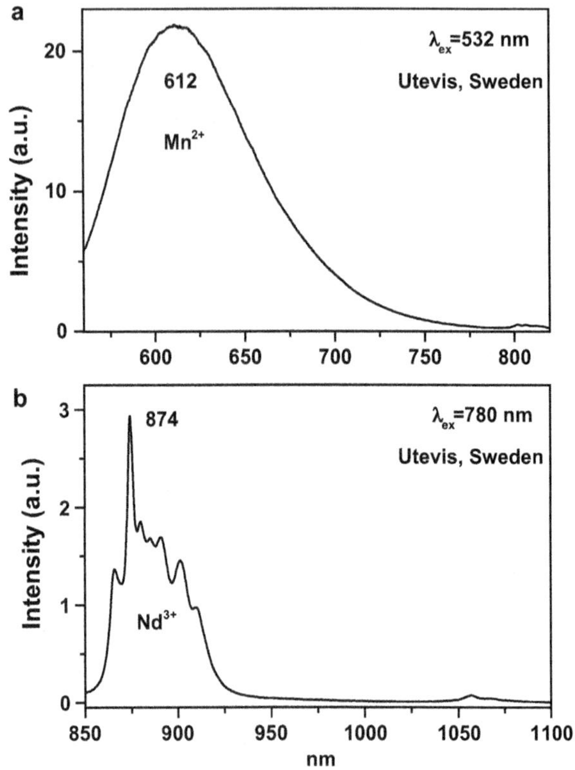

Fig. 4.57 (**a**–**b**) Laser-induced CW luminescence spectra of svabite (http://rruff.info) with Mn^{2+} and Nd^{3+} emission

Benitoite is characterized by very intensive blue luminescence (White 1990). Laser-induced time-resolved technique enables to detect three broad bands and one narrow line, connected with TiO_6, Ti^{3+} and Cr^{3+} or Mn^{4+} luminescence centers (Gaft et al. 2004) (Fig. 4.82).

4.6.3 *Baghdadite $Ca_3(Zr,Ti)Si_2O_9$*

Baghdadite from Fuka, Okayama Prefecture, Japan shows a bright yellow fluorescence under UV (Hg 253.7 nm) excitation (Sidike et al. 2006) which was ascribed to the existence of titanium. Our study of decay time revealed that it is equal of approximately 10 μs, which corresponds to TiO_6 emission properties (Fig. 4.83).

4.6.4 *Other Titanium Bearing Minerals*

Groups of Ti bearing minerals have been studied by Gaft et al. (1981a) and the broad bands in the blue-green part of the spectrum have been connected with

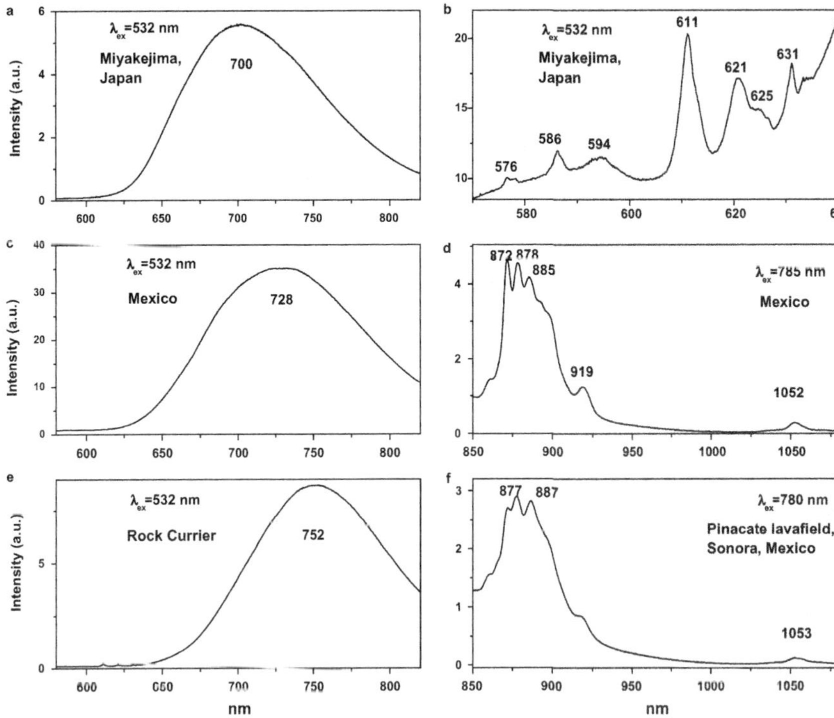

Fig. 4.58 (a–f) Laser-induced CW luminescence spectra of anorthite (http://rruff.info) with evidently, Fe^{3+}, Cr^{3+} and Nd^{3+} emission

individual TiO_6 luminescence centers and their clusters, while the yellow band ascribed to Ti^{3+} center (Fig. 4.84).

4.7 Zinc (Zn^{2+}) Bearing Minerals

Ionic radii of zinc are of 0.74 Å in 4-coordinated form and 0.88 Å in 6-coordinated form. The main substituting luminescence centers are Mn^{2+} and Pb^{2+}.

4.7.1 *Hydrozincite $Zn_5(CO_3)_2(OH)_6$*

Hydrozincite is anhydrous carbonates. The crystalline system is monoclinic-prismatic with the space group C2/m. The structure is composed of Zn in both octahedral and tetrahedral coordination, in the ratio 3:2. The octahedral Zn atoms

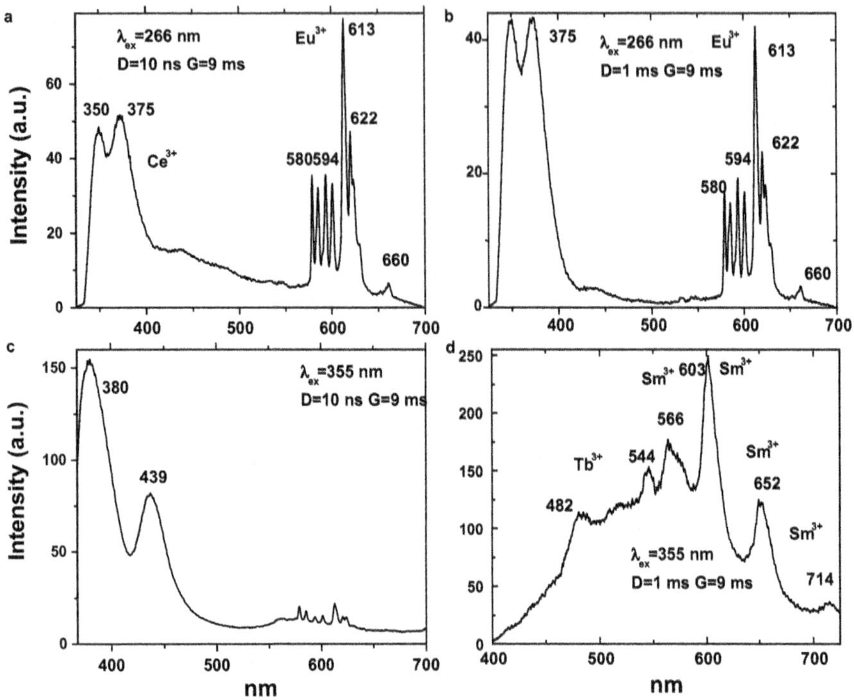

Fig. 4.59 (**a–d**) Laser-induced time-resolved luminescence spectra of pyromorphite demonstrating Ce^{3+}, Eu^{3+}, Tb^{3+}, Sm^{3+} and possibly Pb^{2+} centers

form part of C6 type sheet with holes. The octahedral Zn atoms occur above and below these holes.

The natural hydrozincite in our study consisted of three samples. Concentrations of potential luminescent impurities are presented in Table 4.13. Laser-induced time-resolved technique enables to detect Pb^{2+} center (Gaft et al. 2002a) (Fig. 4.85a, b).

4.7.2 *Willemite Zn_2SiO_4*

Willemite is trigonal silicate with well-known green luminescence. It is connected with Mn^{2+} luminescence center (Tarashchan 1978). The natural willemite in our study consisted of three samples. Laser-induced time-resolved technique enables to detect Mn^{2+} emission center (Fig. 4.85c, d).

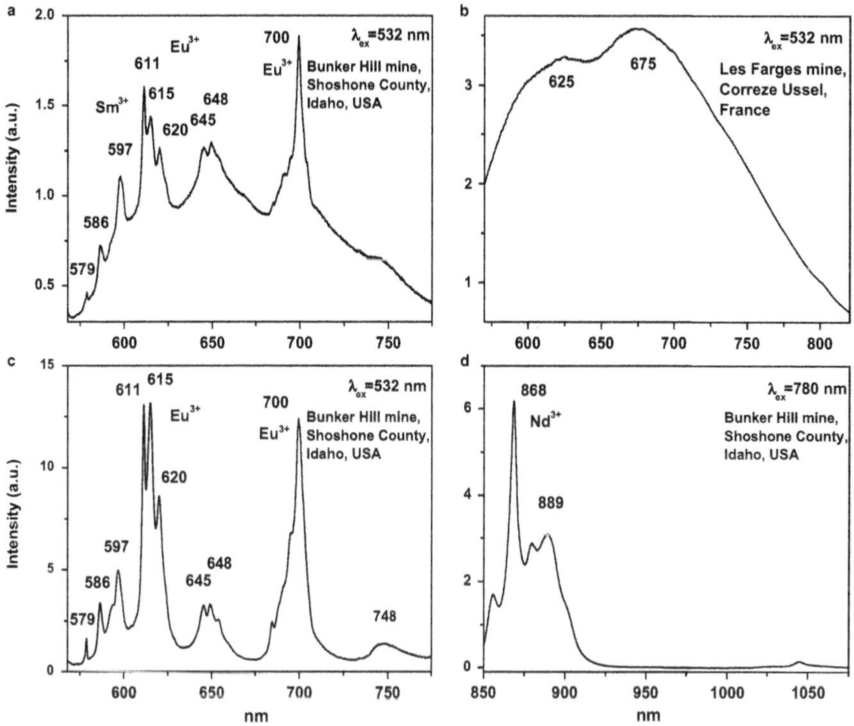

Fig. 4.60 (**a–d**) Laser-induced CW luminescence spectra of pyromorphite (http://rruff.info)

4.7.3 Gahnite $ZnAl_2O_4$

Gahhite was studied by steady-state luminescence spectroscopy and Mn^{2+} luminescence was found peaking at approximately 515 nm (Goirobets and Rogojine 2001). Excitation by CW laser with 532 nm revealed several two additional bands peaking at 647 and 732 nm and narrow lines peaking at 676, 686 and 698 nm (Fig. 4.86). Narrow lines evidently may be ascribed to Cr^{3+} while the broad bands need further study.

4.7.4 Smithonite $ZnCO_3$

Smithsonite has been studied by steady-state luminescence spectroscopy and evidently Mn^{2+} was found (Gorobets and Rogojine 2001). Excitation by CW laser with 532 nm revealed one band peaking at 680 nm (Fig. 4.87) which is different from green emission previously ascribed to Mn^{2+}.

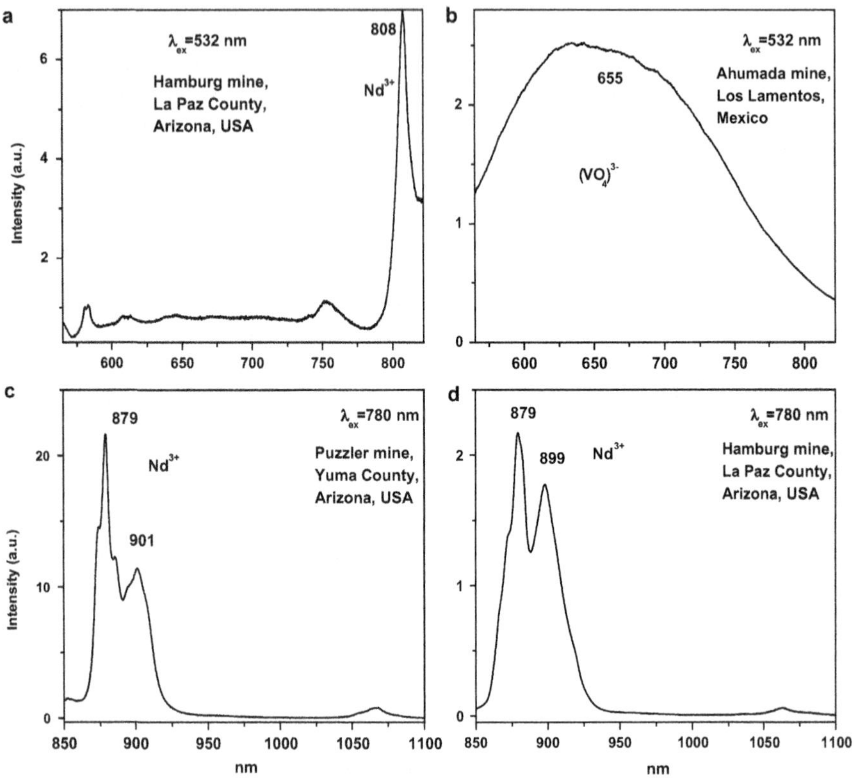

Fig. 4.61 (**a–d**) Laser-induced CW luminescence spectra of vanadinite (http://rruff.info)

4.7.5 *Genthelvite $Be_3Zn_4(SiO_4)_3S$*

Genthelvite has been studied by steady-state luminescence spectroscopy and Mn^{2+} was found (Gorobets and Rogojine 2001). Excitation by CW laser with 532 nm revealed one band peaking at 740 nm (Fig. 4.88) which is possibly connected to Fe^{3+}.

4.8 Zirconium (Zr^{4+}) Bearing Minerals

Ionic radii of zirconium are of 0.73 Å in 4-coordinated form and 0.86 Å in 6-coordinated form. The possible substituting luminescence centers are Ti^{4+} with ionic radius of 0.75 Å in 6-coordinated form, REE^{3+}, Cr^{5+}, Cr^{4+}, Mn^{2+} and Fe^{3+}. Impurities of U and Th are also possible, which may radiatively decay with formation of radiation induced luminescence centers.

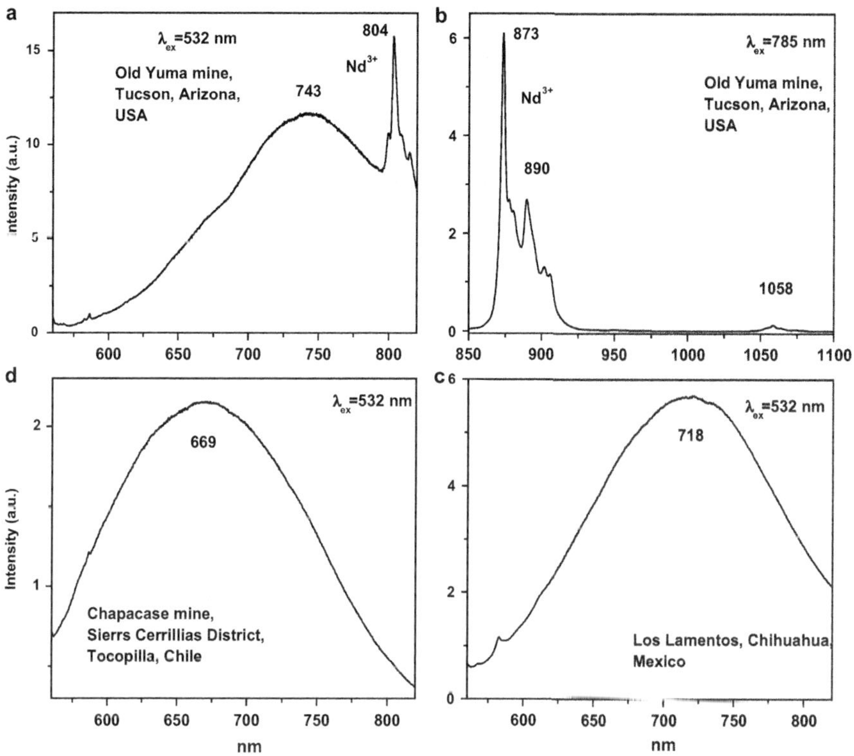

Fig. 4.62 (**a–d**) Laser-induced CW luminescence spectra of wulfenite (http://rruff.info)

4.8.1 *Zircon* $ZrSiO_4$

Zircon, is a zirconium silicate, $ZrSiO_4$, with a tetragonal structure ($I4_1/amd$) where in the unit cell there are four SiO_4^{4-} and four ZrO_8^{12-} groups (Fig. 4.89). In nature, zircon is an accessory mineral almost always found in igneous, sedimentary and metamorphic rock. Zircon's crystal chemistry strongly favors the incorporation of REE in Zr^{4+} site. The REE impurities become luminescent in a crystallographic environment of the lattice. This property, coupled with the ability to form wave-guides in this material by the technique of ion implantation, makes zircon of interest as a potential host material for laser waveguide cavities. Thus spectroscopic analysis of zircon is needed to identify possible laser transitions.

Zircon may show a large variety of colors, depending on the content of transition metals and radiation-induced color centers. Reported colors range for crystalline to moderately radiation-damaged zircons from colorless to pale brown, olive, yellow, orange, green, and blue. The color in zircon comes from both uranium ions and radiation damage. U^{4+} enters the zircon structure because, like Zr^{4+}, it has a large ionic radius. U^{4+} substituting in zircon causes the blue color of heat-treated zircon.

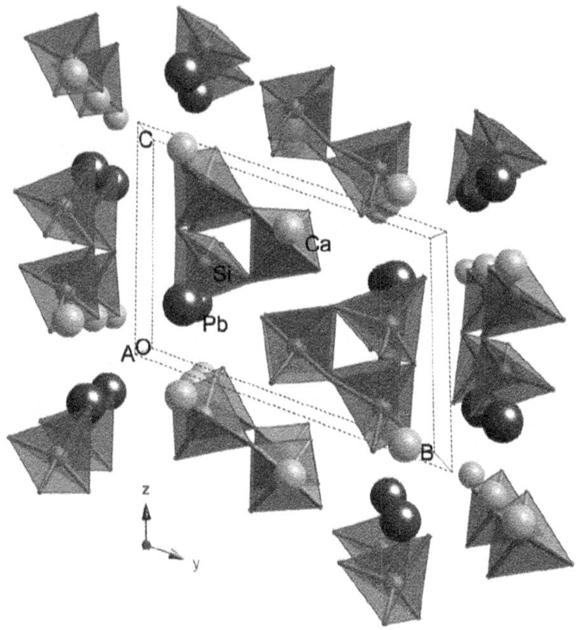

Fig. 4.63 Margarosanite structure showing Ca (CN 6 and 8) and Pb (CN 7) sites.

Over geologic time, the uranium undergoes radioactive decay and the resulting radiation damage centers cause a range of red-brown and amber colors. Heat treatment removes the radiation damage centers and restores the blue color with which zircon presumably originally crystallized. Green colors are associated with a mixture of the blue color from uranium and red-brown color from radiation damage centers. Colorless zircons have little uranium (Platonov et al. 1984). Chromium activated zircon exhibited bluish to greenish color, which was originally interpreted on the basis of local D_{2d} symmetry for Cr^{4+} substituting for Si^{4+} (Beletti et al. 1995). Alternative explanation was double substitution Cr^{3+}-Cr^{5+} within two Zr^{4+}-Si^{4+} neighbors of the zircon lattice (Gaft et al. 2000b). Synthetic zircon doped with vanadium exhibits a blue color, which was attributed to V^{4+} substituting for Zr^{4+}. The red color of zircon was attributed to Nb^{4+} substituting for Zr^{4+}.

X-ray Fluorescence Spectrometry and Inductively Coupled Plasma analysis reveal the presence in the zircons of all existing REE. The steady-state luminescence in natural zircons is dominated by broad emission arising from radiation-induced centers and narrow emission lines of Dy^{3+} (Trofimov 1962; Tarashchan 1978). These emissions obscure the spectra of other REE. The thermal treatment enables to solve this problem in certain cases using the fact that the intensity of broad band luminescence quickly decreases after heating at 700 °C–800 °C, while the intensities of the REE lines remain nearly constant (Shinno 1986, 1987). Even after heating the samples not all the REE can be identified by steady-state spectroscopy since the weaker luminescence lines of certain REE are obscured by stronger luminescence of others. For example, luminescence of Pr^{3+} is difficult to

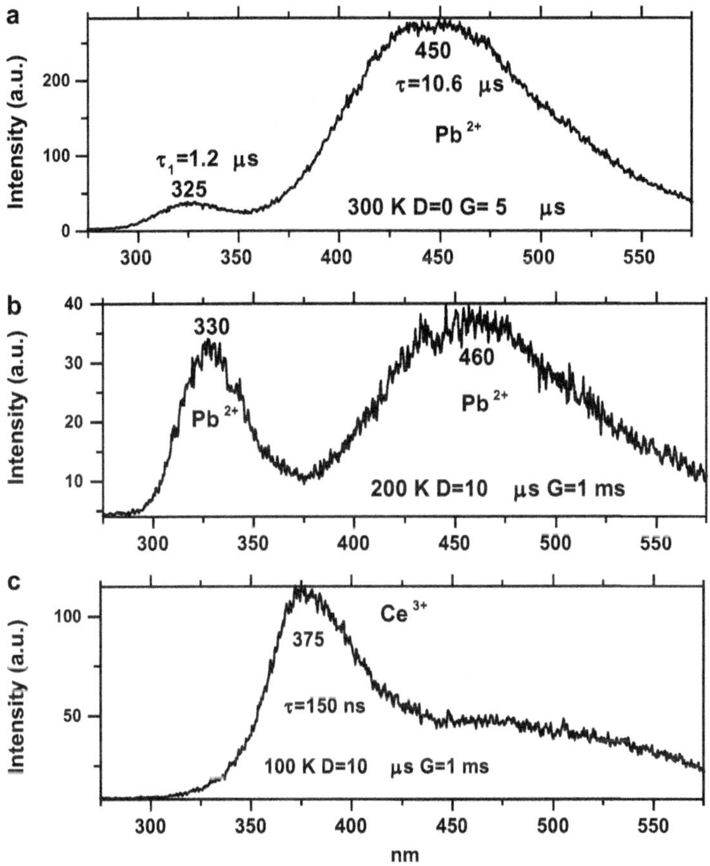

Fig. 4.64 (**a–c**) Laser-induced time-resolved luminescence spectra of margarosanite demonstrating two types of Pb^{2+} and Ce^{3+} centers

detect because the lines of Sm^{3+}, Dy^{3+} and Nd^{3+} hide its radiative transitions. In turn, Tb^{3+} conceals luminescence of Tm^{3+} and so on.

Besides REE, broad spectral bands characterize the luminescence of zircon. They are structureless down to 4.6 K, which makes difficult the correct interpretation of the nature of the luminescent centers. Different suppositions are made in previous studies and even the question about yellow luminescence connection with intrinsic or impurity defect remains open. For example, yellow band ("C-band") was ascribed to SiO_m^{n-} -defects (Votyakov et al. 1993; Krasnobayev et al. 1988) while the same emission ("band VII") was explained by impurity luminescence, namely by Yb^{2+} created by radioactive reduction of Yb^{3+} (Kempe et al. 2000).

It was proposed that yellow zircon luminescence is connected not with one, but with many centers, which have similar luminescence and excitation spectra, but different decay times and thermal stability (Gaft et al. 1986, 2002b; Shinno 1987;

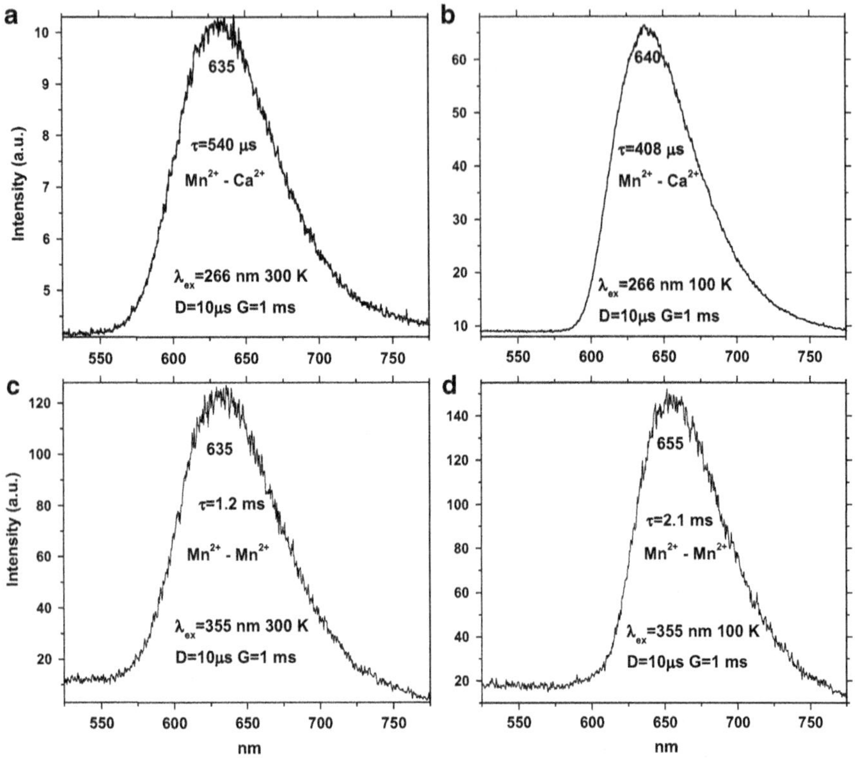

Fig. 4.65 **(a–d)** Laser-induced time-resolved luminescence spectra of margarosanite demonstrating two types of Mn^{2+} centers

Nasdala et al. 2003). For these reason spectroscopic methods alone are not enough for the separation of the broad spectra into individual bands.

It is interesting that yellow zircon luminescence is very specific and different from other Zr-bearing minerals such as catapleite, keldyshite, vlasovite, khibinskite and others, which are usually characterized by blue luminescence evidently connected with titanium impurity, namely TiO_6 complexes (Gaft et al. 1981a).

The photoluminescence properties of synthetic $ZrSiO_4$, doped with REE^{3+} (REE = Pr, Sm, Eu, Gd, Dy, Ho, Er) was investigated using combined excitation and emission spectroscopy (Friis 2009; Friis et al. 2010). All samples showed luminescence characteristics of intra-ion energy transitions, similar to other lanthanide doped materials. However, differences caused by the different mode of excitation were observed as well as evidence for the lanthanides occurring in more than one structural site. Energy transfer between different lanthanides was observed in some samples with more than one lanthanide present, indicating that zircon is a possible host for luminescent materials with quantum efficiency >100 %. Furthermore, emissions that were unassigned in previous studies of these samples could be assigned to specific lanthanides based on excitation spectra.

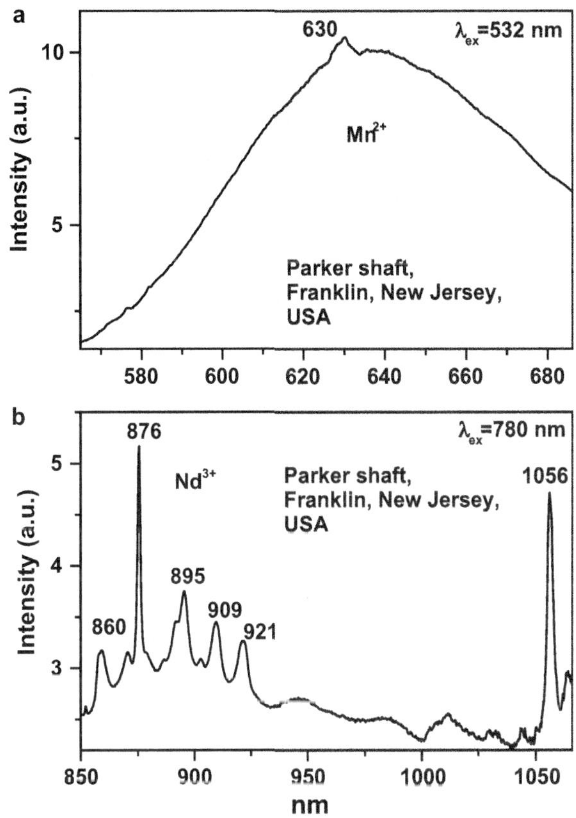

Fig. 4.66 (**a–b**) Laser-induced CW luminescence spectra of margarosanite (http://rruff.info)

Approximately 50 natural zircons have been investigated together with synthesized analogs, as nominally pure and activated by potential luminogens. Concentrations of potential impurities in several zircon samples are presented in Tables 4.14 and 4.15. Laser-induced time-resolved technique enables to detect the following emission centers: radiation induced; trivalent rare-earth elements such as Gd^{3+}, Ce^{3+}, Tb^{3+}, Tm^{3+}, Er^{3+}, Ho^{3+}, Dy^{3+}, Eu^{3+}, Sm^{3+}, and Nd^{3+}; $(UO_2)^{2+}$; Fe^{3+} and Cr^{3+} (Gaft et al. 2000a, b, c) (Figs. 4.90 and 4.91).

Excitation by CW laser with 532 and 780 nm revealed multiple narrow lines (Fig. 4.92). Recently cw luminescence spectra of zircon were studied under different laser excitations including artificially activated samples (Lenz et al. 2015). It was found that Raman spectra of zircon may be obscured strongly by emissions of Tm^{3+} and Nd^{3+} if spectra were recorded using 785 nm laser-excitation. Emissions of Er^{3+} typically interfere with Raman spectra obtained with green excitation. In the 532 and 514 excitations the $^4S_{3/2}$-$^4I1_{5/2}$ transition of Er^{3+} is observed as group of bands at different relative Raman shifts.

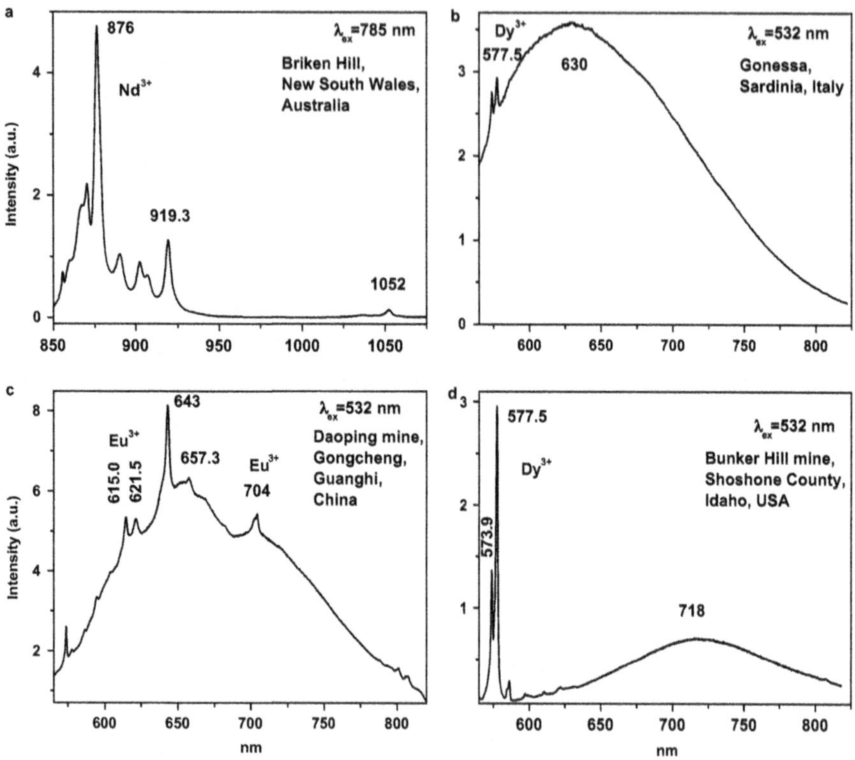

Fig. 4.67 (**a–d**) Laser-induced CW luminescence spectra of cerussite samples demonstrating evidently Nd^{3+} and other trivalent REE centers (http://rruff.info)

4.8.2 *Baddeleyite* ZrO_2

Baddeleyite has monoclinic structure with space group P21/c. Zr^{4+} ion has seven-fold coordination, while the idealized ZrO_7 polyhedron is close to tetrahedral orientation, where one angle in the structure is different significantly from the tetrahedral value. Natural baddeleyite is a raw material for zirconium. In industry ZrO_2, named usually zirconia, is important in areas such as surface chemistry, where its activity as a red ox material and its acid-based functions are important. As a ceramic material, zirconia can resist very high temperatures and its stabilized form, yttrium-stabilized zirconium, shows remarkable mechanical properties.

Steady-state spectra of baddeleyite are characterized by bright green-blue luminescence which was ascribed to Ti impurity (Gaft et al. 1981a). Besides that, emission lines of trivalent Dy and Sm have been found (Eremenko and Khrenov 1982). Laser induced time resolved luminescence of baddeleyite from Kovdor Peninsula enables to detect the same luminescence centers with decay time of 0.7–2.0 μs often accompanied by trivalent REE with longer decay (Fig. 4.93). Excitation by CW laser with 532 and 780 nm revealed several luminescence bands (Fig. 4.94) which need further study.

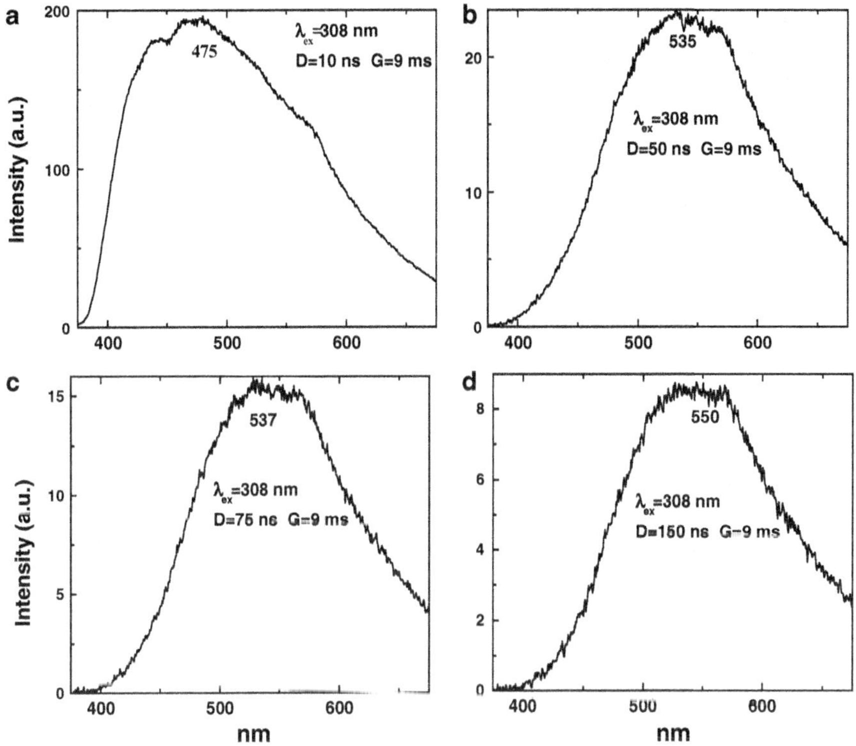

Fig. 4.68 (**a–d**) Laser-induced time-resolved luminescence spectra of cassiterite with intrinsic, possibly Sn^{2+} centers

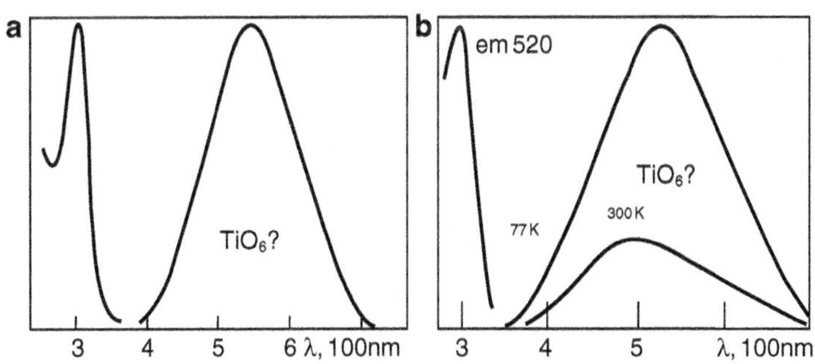

Fig. 4.69 Luminescence (*right*) and excitation (*left*) spectra of malayaite (**a**) and sorensenite (**b**) (Gaft et al. 1982)

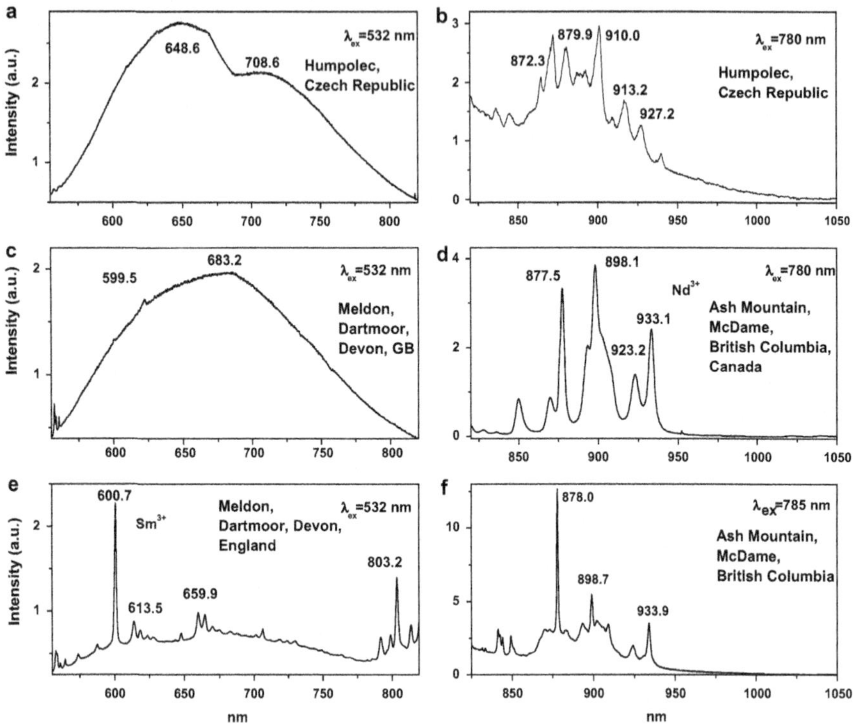

Fig. 4.70 (**a–f**) Laser-induced CW luminescence spectra of several malayaite samples demonstrating evidently Nd^{3+} and other trivalent REE centers (http://rruff.info)

Table 4.11 Concentrations of rare-earth elements in barite samples (ppm)

Origin	Ag	Bi	Cu	Ti	Fe	Mn	Al	Pb	Sn	U	Sr
Spain	–	–	3.5	0.5	75	6.0	20	1.0	–	0.2	1180
France 1	1.5	–	3.3	2.7	950	3.5	5.0	62	–	0.1	760
France 2	21	–	5.0	–	45	2.1	120	1.5	–	–	685
France 3	4	–	3.6	1.3	230	6.5	60	0.8	–	0.2	590
Romany	0.4	–	1.2	1.0	70	2.2	–	1.0	1.5	–	10,300
England	0.6	–	5.7	1.7	120	13.0	200	9500	–	2.0	390

4.9 Silicon (Si^{4+}) Bearing Minerals

Ionic radius of silicon in tetrahedral coordination is of 0.4 Å. The main substituting luminescence center is Fe^{3+} with ionic radius of 0.63 Å in tetrahedral coordination.

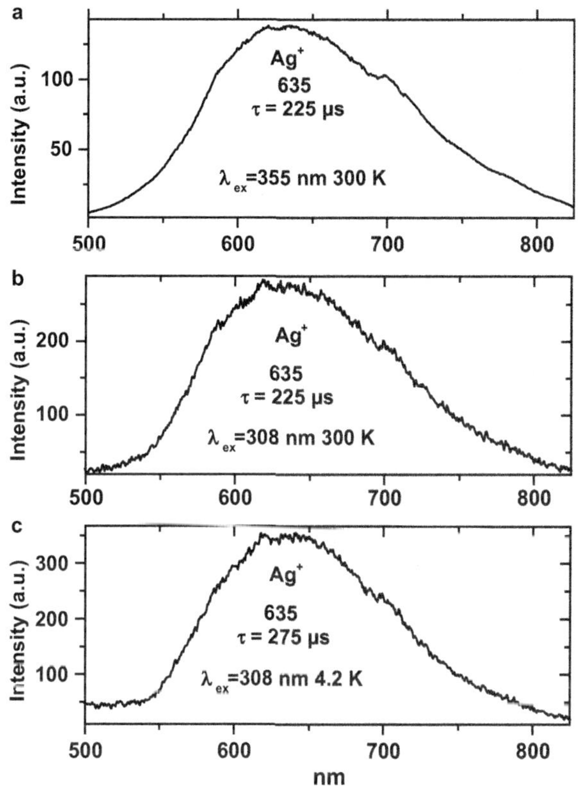

Fig. 4.71 (**a–c**) Laser-induced time-resolved luminescence spectra of barite with Ag^{+} centers

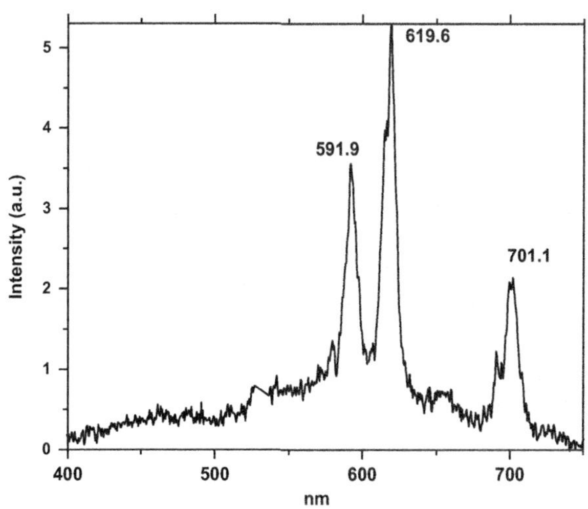

Fig. 4.72 Laser-induced time-resolved luminescence spectra of barite demonstrating Eu^{3+}

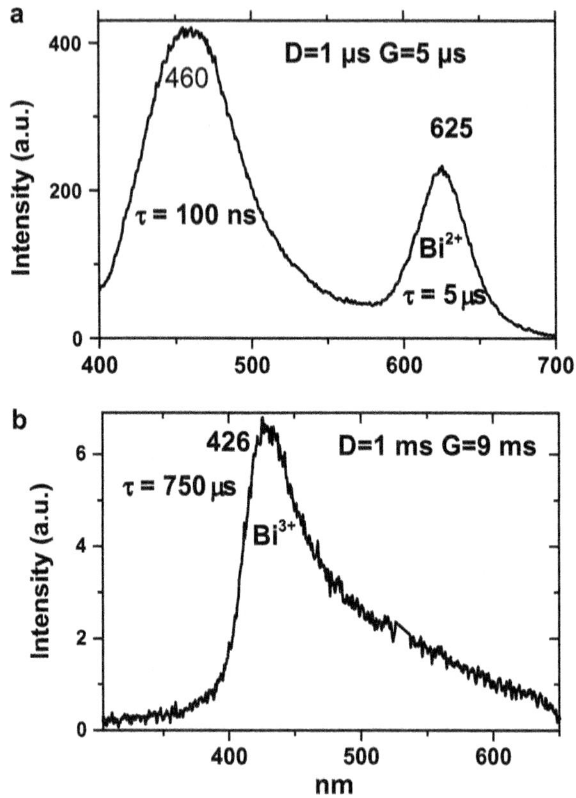

Fig. 4.73 (**a–b**) Laser-induced time-resolved luminescence spectra of barite with Bi^{2+} and Bi^{3+}

4.9.1 Wollastonite $CaSiO_3$

Wollastonite is calcium silicate with triclinic crystal system (P21). It has infinite-chain structure, with three tetrahedra per unit sell arranged parallel to y, this repeat unit consisting of a pair of tetrahedra joined apex to apex as in the [SiO_7] group, alternating with a single tetrahedron with one edge parallel to the chain direction. Steady-state luminescence of wollastonite has been previously studied and luminescence of Mn^{2+}, Fe^{3+} and supposedly Cr^{3+} has been proposed (Min'ko et al. 1978).

The natural wollastonite in our study consisted of three samples. Laser-induced time-resolved technique enables to detect Mn^{2+}, Fe^{3+} and possibly Cr^{3+} emission centers (Fig. 4.95).

Excitation by CW laser with 532 and 780 nm revealed narrow luminescence lines possibly belonging to trivalent REE (Nd) and two luminescence bands peaking at 590 and 620 nm (Fig. 4.96). Additional study is needed to ascribe those emission centers including their decay times and excitation spectra.

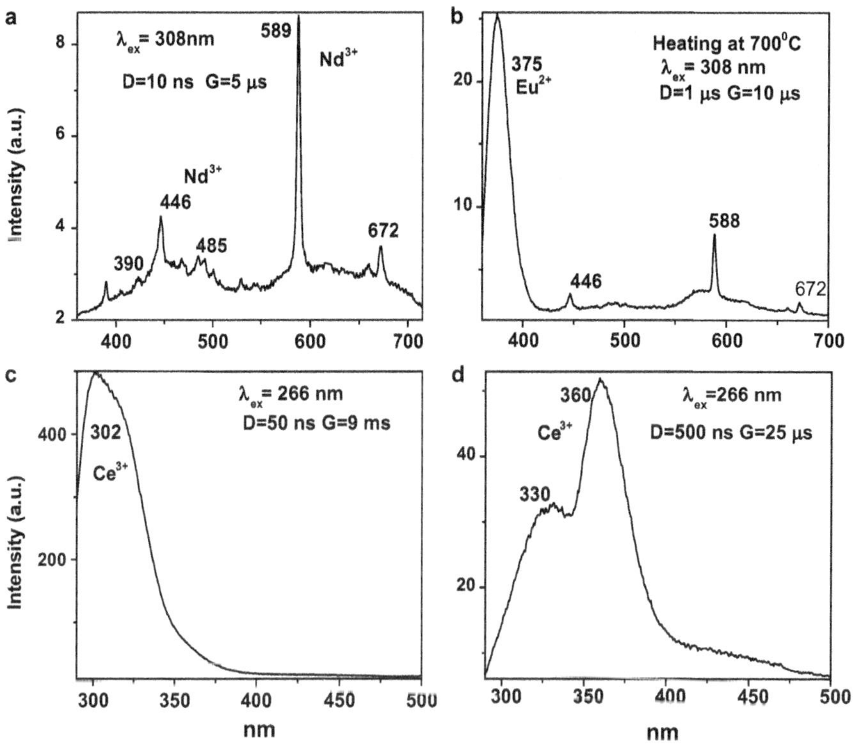

Fig. 4.74 (a–d) Laser-induced time-resolved luminescence spectra of barite demonstrating Eu^{2+}, Ce^{3+} and possibly Nd^{3+} centers

4.9.2 *Feldspars*

Feldspars are mainly MT_4A_8 alumosilicates whose structures are composed of corner-sharing AlO_4 and SiO_4 tetrahedra (T sites) linked in an infinite three-dimensional array. Charge compensating cations (K^+, Na^+, Ca^{2+}, Ba^{2+}) occupy large, irregular cavities in the tetrahedral framework (M sites). Most natural feldspars occur in the K-Na-Ca ternary diagram (orthoclase-albite-anorthite). In general, feldspars have a disordered structure with respect to Al and Si at high temperatures and ordered structures at low temperatures. Monoclinic sanidine is the most disordered K-feldspar polymorph. If the rocks are annealed over geologic times at lower temperatures, Al migrates and monoclinic orthoclase and ordered triclinic microcline are formed. In K-feldspars large cations enter the M-sites, whereas the number of substituents in plagioclases is limited because of the smaller cations in the M site. Small cations are incorporated into the tetrahedral T sites. Some of these substituting elements in natural feldspars can act as luminescence centers.

The pale yellow color in feldspar is due to Fe^{3+} in tetrahedral Si/Al site. This color is often masked by the pervasive turbidity of common feldspars. Smoky color, the

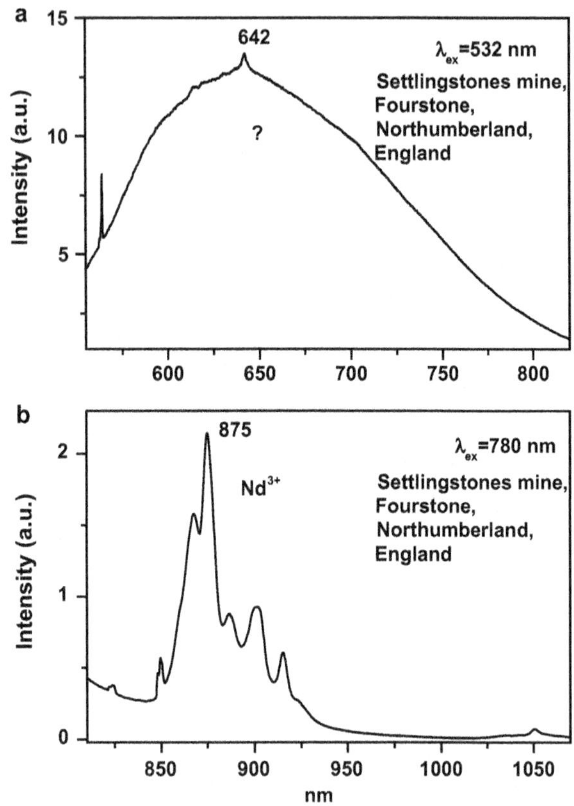

Fig. 4.75 (**a–b**) Laser-induced CW luminescence spectra of witherite demonstrating evidently Nd^{3+} and possibly Ag^{+} centers (http://rruff.info)

result of radiation damage from the decay of K-40, is also common but often masked. Blue color in the amazonite variety of potassium feldspar (and pale-blue albite) is from the interaction of trace amounts of Pb^{2+} in the feldspar with ionizing radiation. Lead-containing feldspars with a higher degree of Al/Si disorder than occurs in microcline become green from radiation damage. There are also rare varieties of plagioclase feldspar colored green and red from Cu ions (Platonov 1979).

Steady-state luminescence of feldspars is well studied. The following impurity centers have been found: Tl^{+}, Pb^{+}, Pb^{2+}, REE^{3+} (Ce, Dy, Sm, Tb, Nd), Eu^{2+}, Mn^{2+}, Fe^{3+}, Cr^{3+} (Tarashchan 1978; Gorobets and Rogojine 2001; Götse 2000; Waychunas 1989; Kuznetsov and Tarashchan 1988; Bakhtin and Moroshkin 1986; Krbetschek et al. 2002; White et al. 1986).

The natural feldspars in our study consisted of twelve samples. Concentrations of rare-earth elements in one of them are presented in Table 4.16. Laser-induced time-resolved technique enables to detect Pb^{2+}; Gd^{3+}, Ce^{3+}, Eu^{2+}, Eu^{3+}, Tb^{3+}, Er^{3+}, Dy^{3+}, Sm^{3+}, Nd^{3+}; Mn^{2+}, Fe^{3+} and possibly Cr^{3+} emission (Figs. 4.97, 4.98 and 4.99).

Luminescence of monovalent Pb was firstly proposed for explanation of IR luminescence band peaking at 860 nm in emission spectrum of feldspars

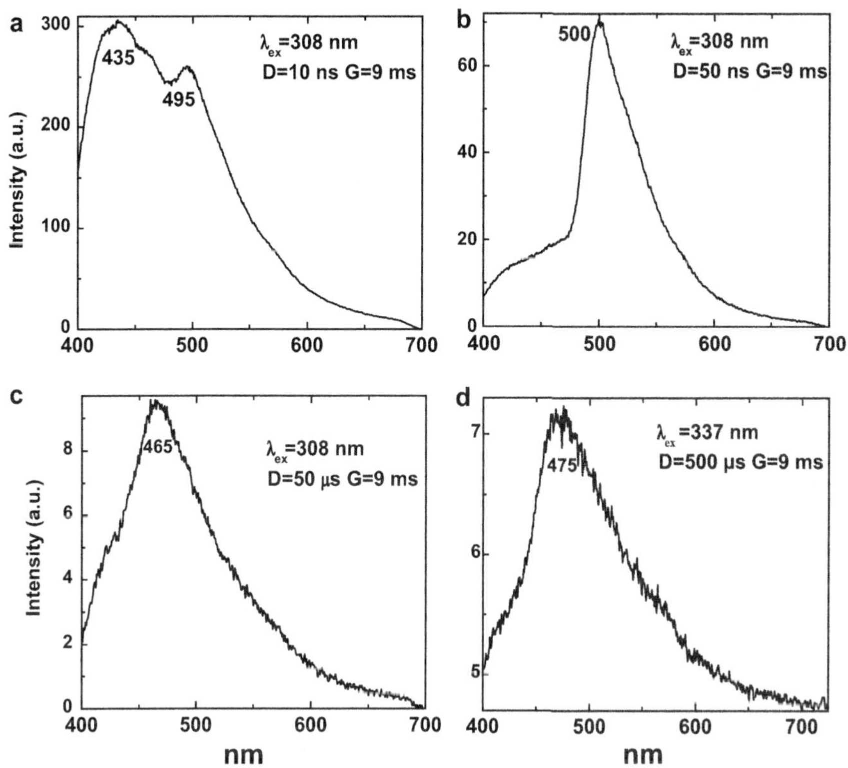

Fig. 4.76 (**a–d**) Laser-induced time-resolved luminescence spectra of celestine demonstrating possibly Pb^{2+} centers

(Kusnetsov and Tarashchan 1988). Later new results have been reported on the characteristics of photo- and radioluminescence emissions at 865 nm and 910 nm emitted by potassium-rich feldspars (Erfurt 2003). Under laser excitation using 355 nm, similar band peaking at 850 nm appears at 300 K with relatively short decay time of 1 μs (Gaft and Panczer 2013) (Fig. 4.100) Such IR band was detected only in potassium feldspar with elevated Pb concentration.

Excitation by CW laser with 780 nm revealed several luminescence lines in albite (Fig. 4.101) which may be evidently ascribed to Fe^{3+}, Nd^{3+} and Yb^{3+}, while microcline revealed mainly emission bands under excitation by 532 nm peaking at 715, 720, 865 and 870 nm (Fig. 4.102.) The origin of those bands needs additional study including decay times and excitation spectra.

4.9.3 *Quartz* SiO_2

Silica makes up 12.6 mass-% of the Earth crust as crystalline and amorphous forms. It was found that both modifications show similar main luminescence bands,

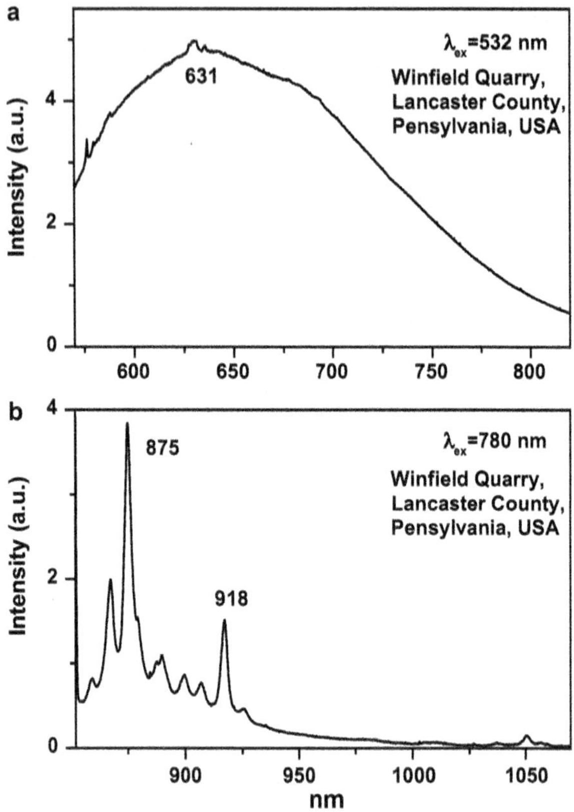

Fig. 4.77 (**a–b**) Laser-induced CW luminescence spectra of strontianite demonstrating Nd^{3+} and possibly Mg^{2+} centers (http://rruff.info)

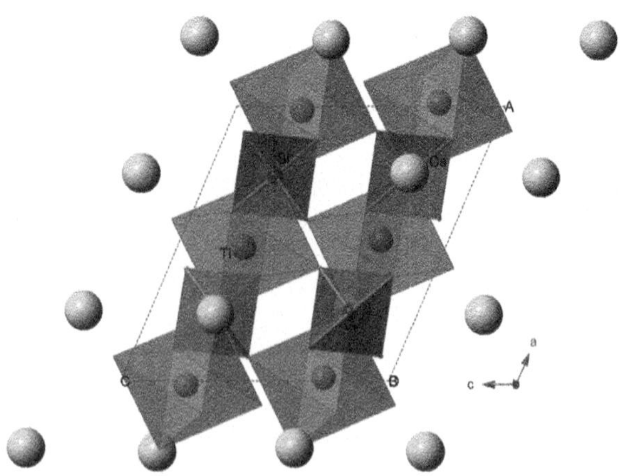

Fig. 4.78 Crystal structure of titatite showing the isolated SiO_4 tetrahedra cross-linked by the Ti^{4+} octahedral sites and the Ca^{2+} seven-fold coordinated sites

Table 4.12 Concentrations of rare-earth elements in titanite sample (ppm)

	Pr	Sm	Eu	Nd	Dy	Tb	Er	Tm	Ho	Gd	Ce
Green	90	210	275	555	290	50	165	22	60	280	390

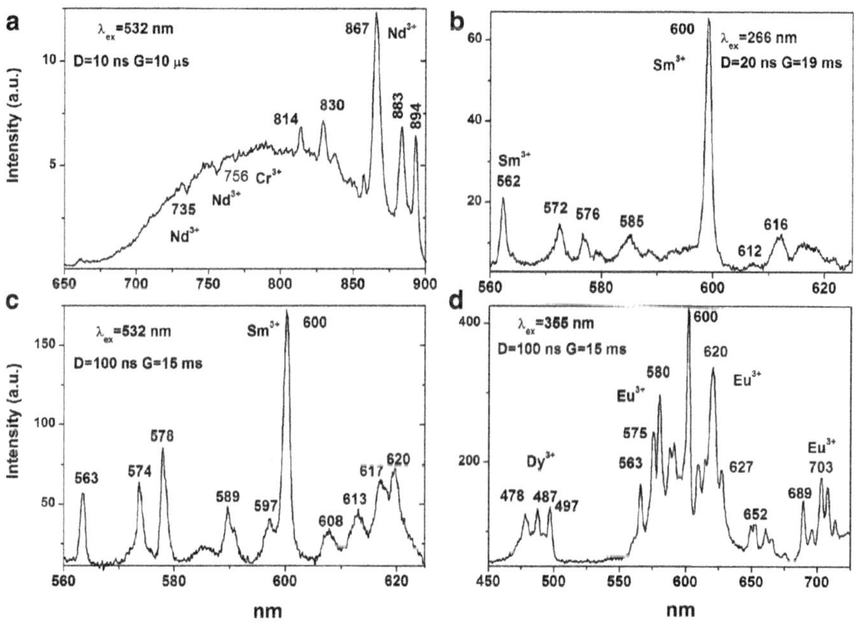

Fig. 4.79 (**a–d**) Laser-induced time-resolved luminescence spectra of titanite demonstrating Cr^{3+}, Eu^{3+}, Pr^{3+} and Sm^{3+} centers

namely blue one centered at 450 nm ascribed to Al^{3+} which substitutes for Si^{4+}, red centered at 650 nm linked with non-bridge O*, and dark-red at 700–730 nm linked with Fe^{3+} (Getze et al. 2015). Time resolved luminescence of quartz is characterized by green emission of uranyl with relatively long decay time of 30–110 μs and typical vibrational structure, often accompanied by orange structureless band peaking at 656 nm with shorter decay of 10 μs (Fig. 4.103), evidently connected to non-bridge O*. Several other structureless bands exist in UV-violet, blue-green and yellow-orange (Fig. 4.104) spectral ranges. The origin of those bands needs further study, but probably they have organic nature.

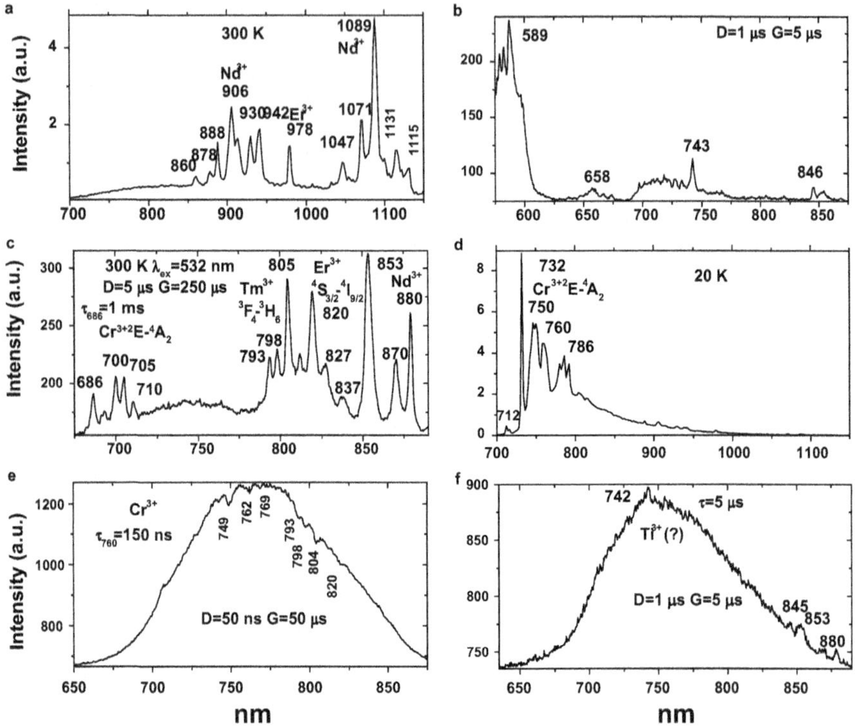

Fig. 4.80 (**a–f**) Laser-induced time-resolved luminescence spectra of titanite demonstrating Cr^{3+}, Nd^{3+}, Tm^{3+} and possibly Ti^{3+} centers

Fig. 4.81 Laser-induced CW 514 nm luminescence spectra of titanite demonstrating Nd^{3+} centers (V vertical; H horizontal polarization)

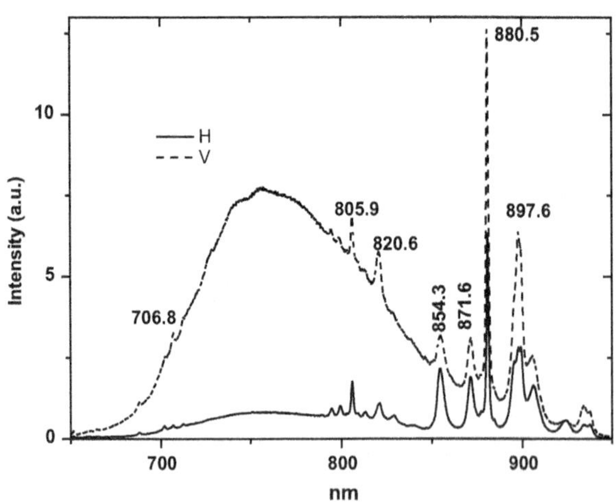

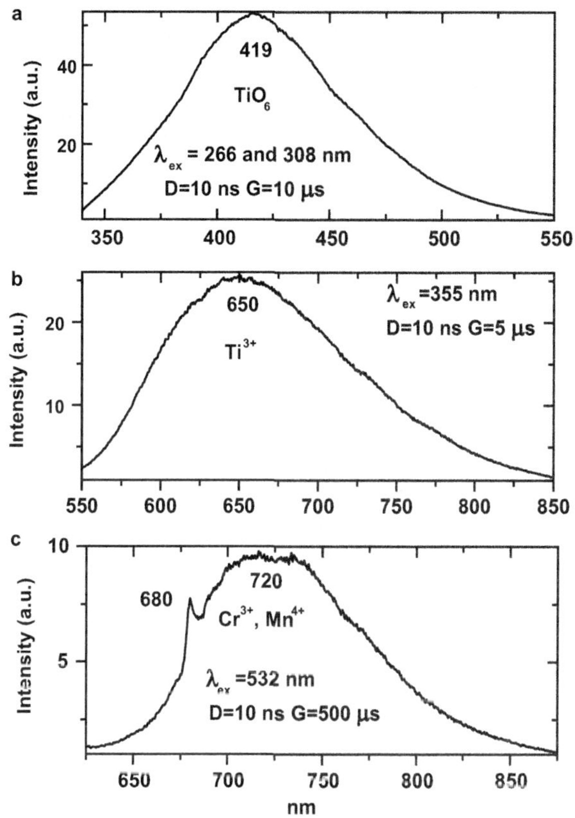

Fig. 4.82 (**a–c**) Laser-induced time-resolved luminescence spectra of benitoite demonstrating TiO_6, Ti^{3+} and Cr^{3+} or Mn^{4+} centers

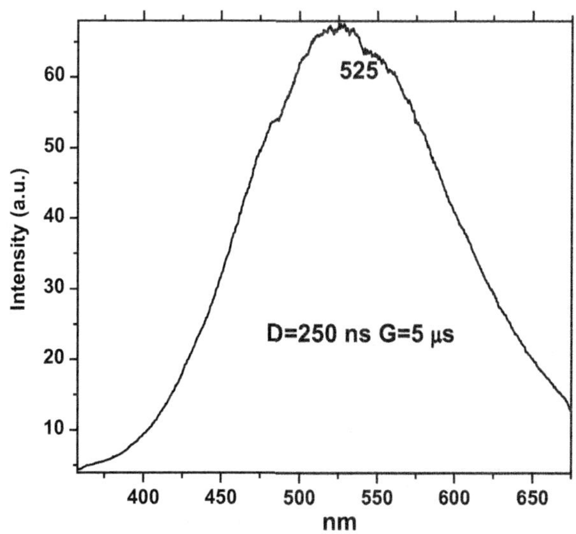

Fig. 4.83 Laser-induced time-resolved luminescence spectra of baghdadite demonstrating TiO_6 centers

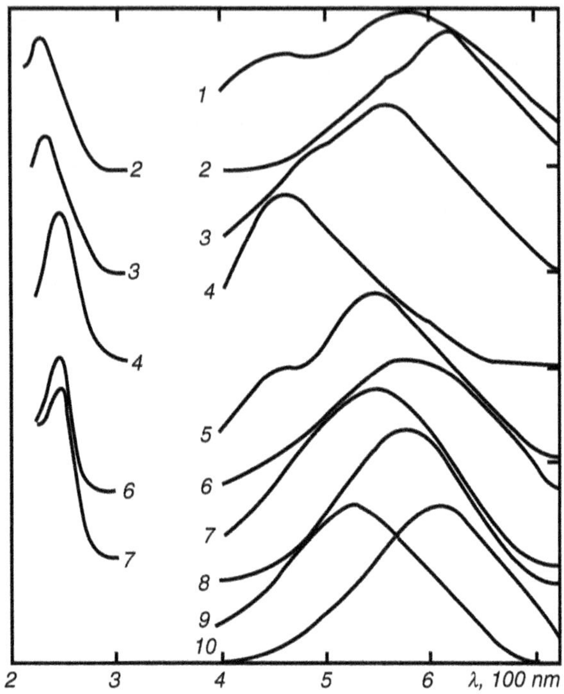

Fig. 4.84 Luminescence (*right*) and excitation (*left*) spectra of titanite (*1*), vuonnemite (*2*), epistolite (*3*), natisite (*4*), penkvilsite (*5*), Ti-silicate (*6*), vinogradovite (*7*), leucosphenite (*8*), Ti-carbonatesilicate (*9*), fersmanite (*10*) (Gaft et al. 1981a)

Table 4.13 Concentrations of rare-earth elements and other potential luminescent impurities in hydrozincite samples (ppm)

	Pr	Sm	Eu	Nd	Dy	Tb	Tm	Ho	Gd	Ce	Pb	Mn
1	3.6	2	0.5	13.5	2.3	0.4	0.2	0.5	3	1	2000	–
2	0.4	0.3	0.5	1.5	0.2	0.04	0.02	0.04	0.35	3.5	35	200
3	0.2	0.2	0.006	1	0.45	0.06	0.06	0.12	0.3	0.5	600	10

4.9.4 *Phenakite Be_2SiO_4*

Phenakite was studied by steady-state luminescence spectroscopy and supposedly Fe^{3+} emission was found (Gorobets and Rogojine 2001). Excitation by CW laser with 532 and 780 nm revealed several luminescence lines which may be evidently ascribed to Nd^{3+} and band peaking at 670 nm of unknown origin (Fig. 4.105).

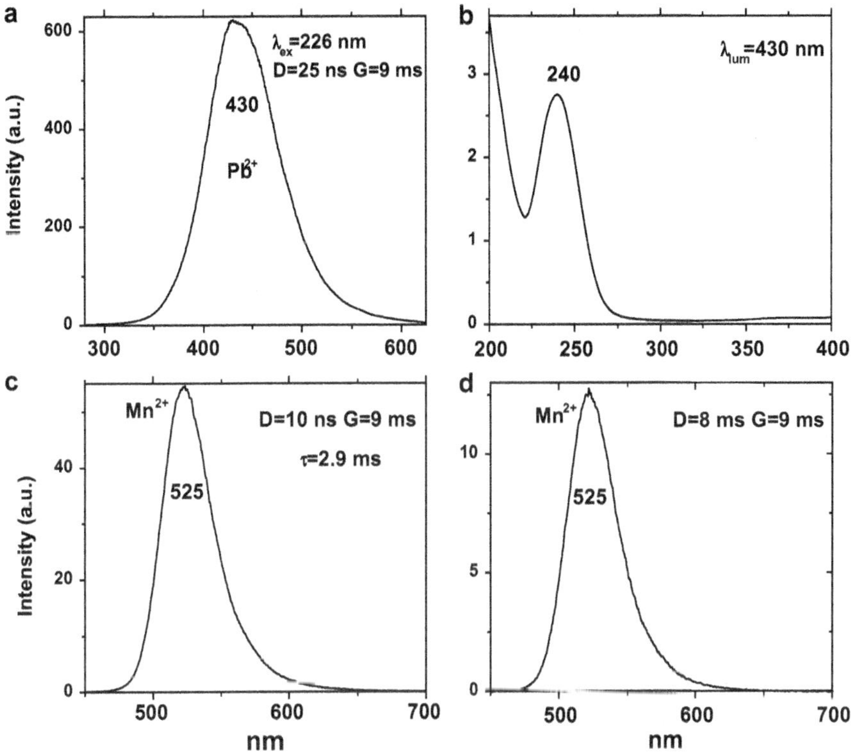

Fig. 4.85 (**a–d**) Laser-induced time-resolved luminescence and excitation spectra of hydrozincite (**a**, **b**) and luminescence of willemite (**c**, **d**) demonstrating Pb^{2+} and Mn^{2+} centers, correspondingly.

4.10 Aluminum (Al^{3+}) Bearing Minerals

Ionic radius of aluminum in octahedral coordination is of 0.67 Å. The main substituting luminescence centers are Cr^{3+} with ionic radius of 0.75 Å in octahedral coordination, Mn^{2+} and Mn^{4+} with ionic radii of 0.81 and 0.67 Å in octahedral coordination and V^{2+}, V^{3+} and V^{4+} with ionic radii of 0.93, 0.78 and 0.72 Å in octahedral coordination correspondingly.

4.10.1 Kyanite, Sillimanite and Andalusite $Al_2O[SiO_4]$

Kyanite is one of the three polymorphs of Al_2SiO_5, the others being sillimanite and andalusite. It has four inequivalent Al^{3+} sites, which may be substituted by Cr^{3+}, all of them with low symmetry: two have symmetry close to tetragonal and two are characterized by substantial rhombic distortion. The sizes of all sites are of similar

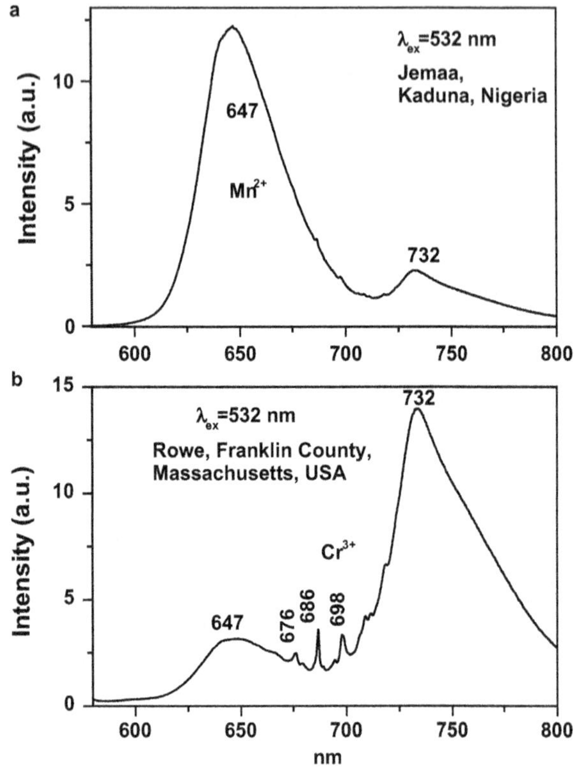

Fig. 4.86 (**a–b**) Laser-induced CW luminescence spectra of gahnite (http://rruff.info) demonstrating evidently Cr^{3+} lines and unknown bands

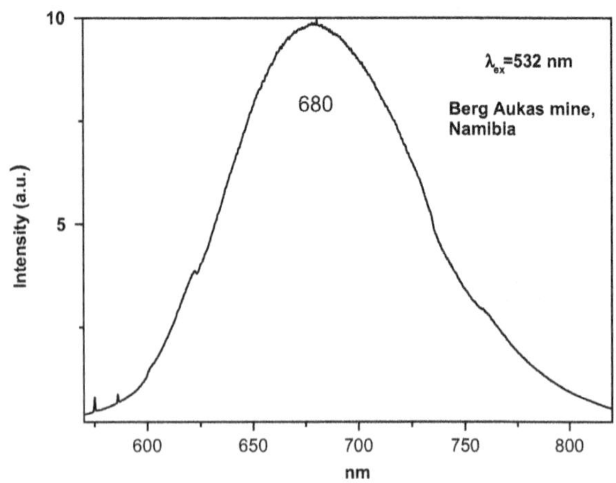

Fig. 4.87 Laser-induced CW luminescence spectra of smithsonite (http://rruff.info)

average size (average Al-O distances are 1.902, 1.912, 1.919 and 1.896 Å, correspondingly). Andalusite and sillimanite both belong to the orthorhombic system but differ in the space groups that are Pnnm and Pbnm, respectively. The two Al^{3+} sites

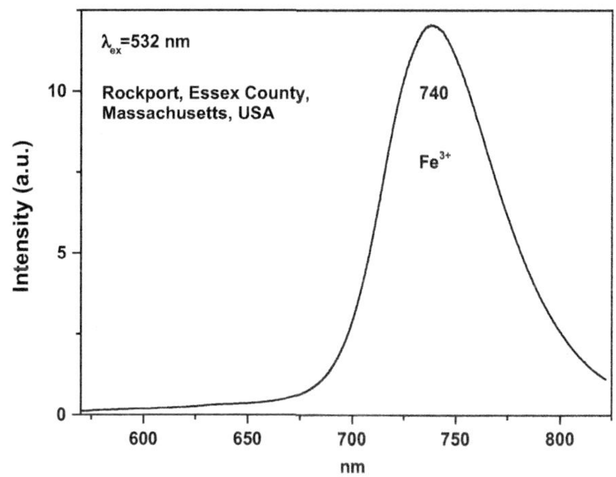

Fig. 4.88 Laser-induced CW luminescence spectra of genthelvite (http://rruff.info)

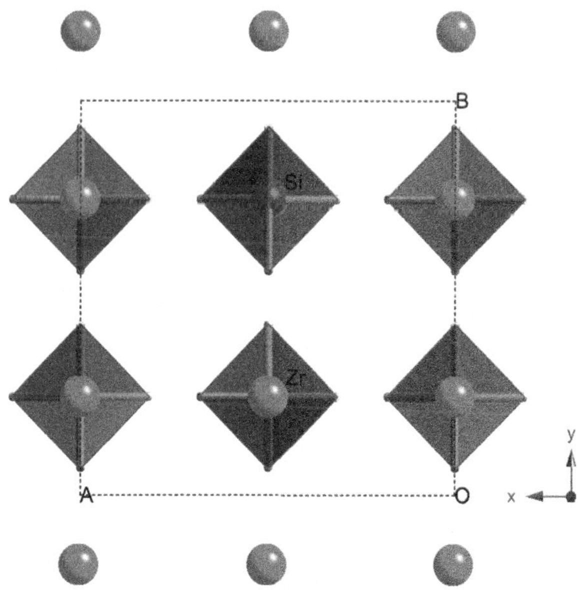

Fig. 4.89 Crystal structure of zircon with isolated Si tetrahedra and the Zr octahedra sites

Table 4.14 Concentrations of rare-earth elements (ppm) in several zircon samples

	Pr	Sm	Eu	Nd	Dy	Tb	Er	Tm	Ho	Gd	Ce	Yb
Kola	10	48	25	63	352	40	300	56	96	143	68	350
Nrw	2.5	7	5	12	120	9	220	50	45	30	120	15
GQ	3	4	1.3	12	56	6	100	32	22	12	28	300
Ural	2.5	7.5	4	13	86	8	125	35	30	25	38	300

Table 4.15 Concentrations of potential luminescent impurities (ppm) in zircon

	Mn	Fe	Ti	Mo	V	Nb	Ni	Pb	Ta	U
Australia	10	400	1200	0.5	90	20	10	15	4	550
Norway	150	475	120	0.5	3	100	15	8	15	120
Canada	100	2200	75	5	250	20	75	?	20	700
Kola	40	400	40	2	150	150	30	?	2.5	40

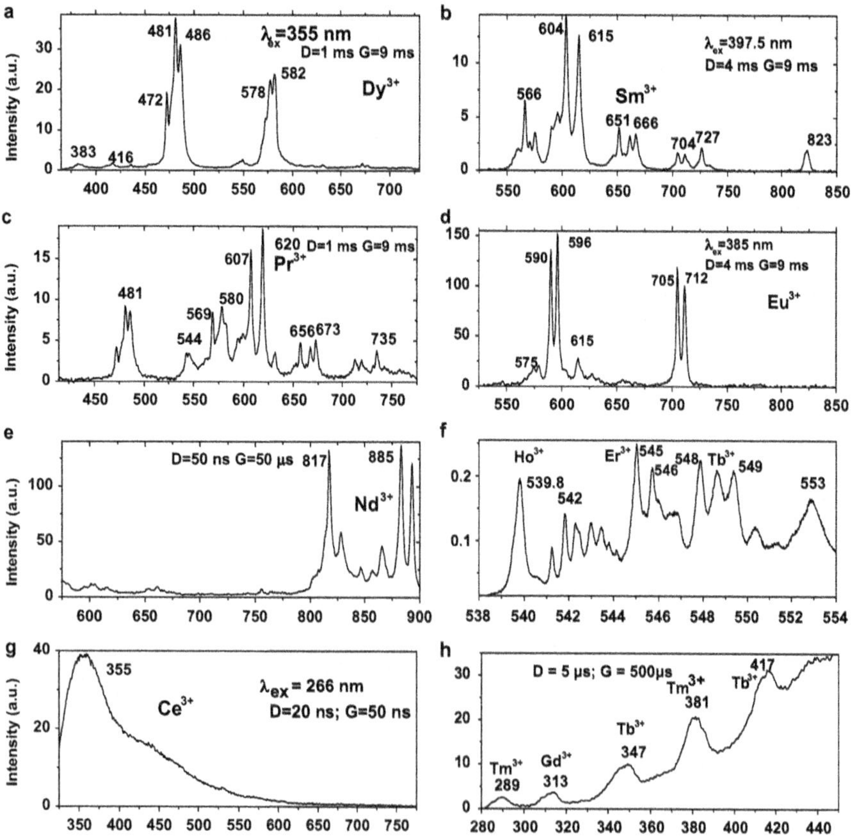

Fig. 4.90 (**a–h**) Laser-induced time-resolved luminescence spectra of zircon demonstrating Dy^{3+}, Sm^{3+}, Eu^{3+}, Pr^{3+}, Tm^{3+}, Er^{3+}, Ce^{3+}, Gd^{3+}, Tb^{3+} and Nd^{3+} centers

have C_2 and C_s point symmetry with distorted octahedral and trigonal bipyramid coordination in andalusite whereas C_i and Cs point symmetry with distorted octahedral and tetrahedral coordination in sillimanite (Fig. 4.106).

Kyanite can be regarded as a model matrix demonstrating the whole range of D_q covering the very sensitive region of doublet-quartet crossing. Substitional Cr^{3+} is

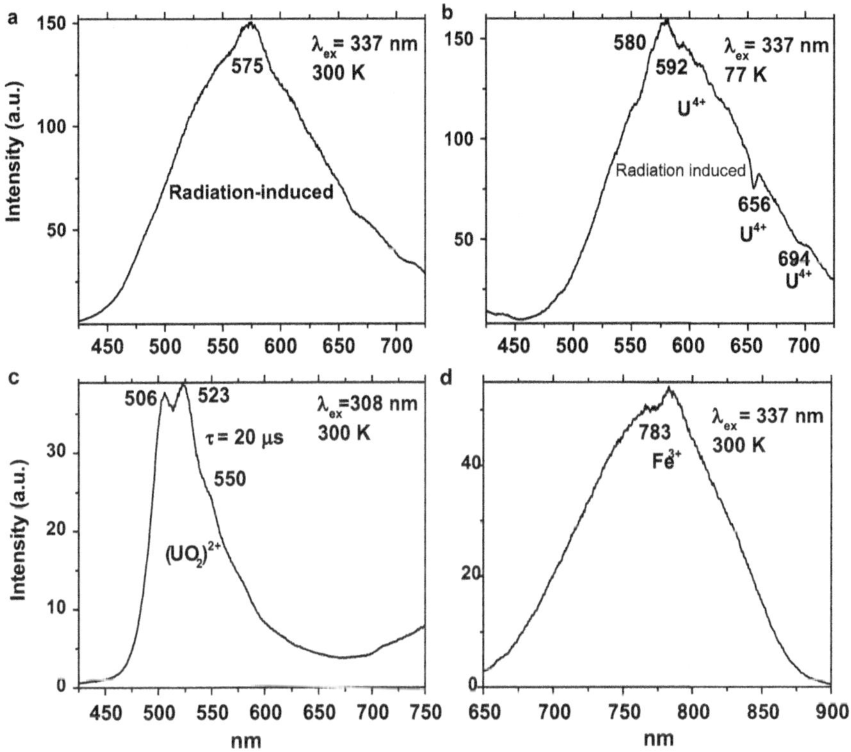

Fig. 4.91 (**a–d**) Laser-induced time-resolved luminescence spectra of zircon demonstrating intrinsic radiation induced, uranyl and Fe^{3+} center

an ordinary constituent in kyanite from many different localities. The study of Cr^{3+} luminescence in kyanite has long history. Four different luminescence lines $\lambda_1 = 704.9$, $\lambda_2 = 703.4$, $\lambda_3 = 689.3$ and $\lambda_4 = 687.6$ nm have been found, which were ascribed to two Cr^{3+} doublets, the first with λ_1 and λ_2 and the second with λ_3 and λ_4. The following study including absorption and emission spectroscopy at low temperature changed such interpretation, namely the first doublet with λ_1 and λ_3 and the second with λ_2 and λ_4 (Deutschbein 1932). The second doublet appearance was attributed to Stark's splitting of the first doublet lines in the kyanite crystal field. Later it was found that the lines λ_1 and λ_4 at 300 K have long decay time of 700 and 900 μs, correspondingly, while the lines λ_2 and λ_3 have much shorter decay times of 80 μs. The lines λ_3 and λ_4 disappeared at low temperatures, while the lines λ_1 and λ_2 remain and are characterized by much longer but still different decay times of 3.1 ms and 590 μs, correspondingly. It was concluded that the first doublet contains λ_1 and λ_4 lines, while the second one the λ_1 and λ_3 lines and their existence is explained by Cr-Al substitutions in different Al sites of kyanite structure (Tolstoy and Shinfue 1960). Tarashchan (1978) described the same four emission lines. He ascribed them to Cr^{3+} doublets in different structural positions, Al_I and Al_{II}, with

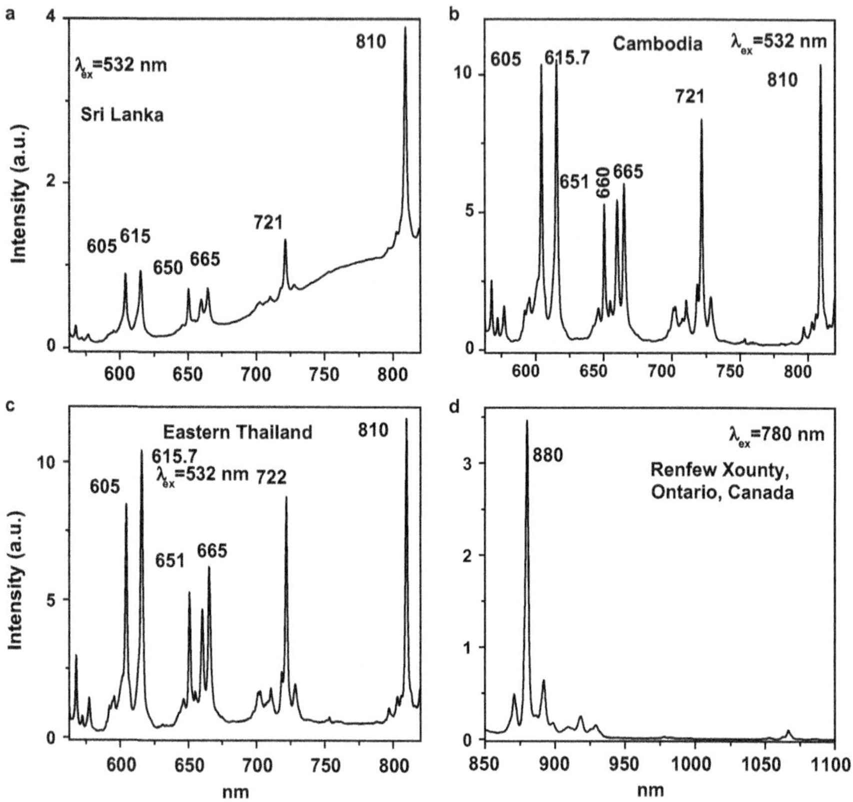

Fig. 4.92 (**a–d**) Laser-induced CW luminescence spectra of different zircons (http://rruff.info)

different Cr-O length and similar crystal fields of $D_q \sim 1700\ cm^{-1}$, but returned to the first Deutschbein's interpretation, namely λ_4 and λ_3 for the first one and λ_2 and λ_1 for the second. The splitting for those doublets was calculated as 36 and 30 cm^{-1}, correspondingly. The distribution of Cr^{3+} to those sites is not equal and may change from sample to sample. Wojtowicz (1991), using absorption, excitation and emission spectra (steady-state and time resolved) as well as decay measurements, identified three different emissions originating in three different sites. At 26 K the spectrum consists of the two clearly resolved sharp lines at 703.7 (Q) and 705.2 (R) nm, a highly structured complex band around 720 nm (S) and a weakly structured broad band at 800 nm (T). At 300 K a new sharp line at 688.5 nm (P) can be observed, while those around 705 nm merge together. It was found that line Q has a relatively short decay time changing from 620 μs at 26 K to 67 μs at 300 K, while line R has much longer decay times, changing from 3250 to1920 μs in the same temperatures interval. Line P also has a long decay time, changing from 2260 μs at 200 K, where it was initially detected to 980 μs at 300 K. Band T has short decay time of 12–14 μs at all temperatures. The following assignments have

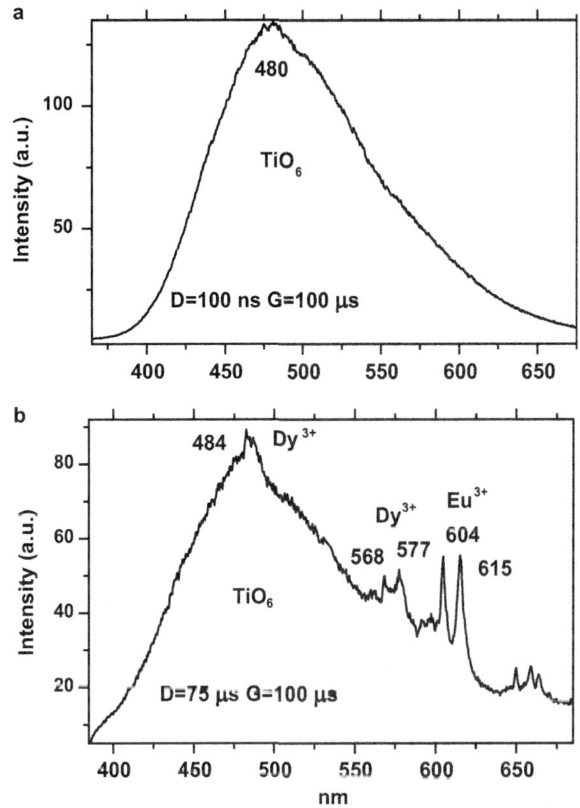

Fig. 4.93 (**a–b**) Laser-induced time resolved luminescence of baddeleyite with TiO_6, Dy^{3+} and Eu^{3+} centers

been proposed: site A: R-lines P + R without the quartet band; site B: only one component of R-lines Q and the quartet broad band T′, which shows up only at higher temperatures; Site C: quartet broad band T. Theoretically calculated D_q for sites A, B and C were 1715, 1638 and 1552 cm^{-1}, correspondingly. It was concluded that the range of D_q spanned by the sites is much larger than expected on the basis of the average bond lengths and the electrostatic theory. The wide variation in the doublet-quartet coupling among different sites of kyanite manifests itself in doublet radiative lifetimes which span the range 100 to 3000 μs. The decay time of both R-lines were found to be similar within the experimental error for the whole range of temperatures as expected in the case of two thermalized levels. Platonov et al. (1998) studied a selected set of five different kyanite samples with Cr_2O_3 concentration from practically zero to 2.58 wt%. The room temperature spectra revealed two narrow bands at 687.9 nm (R_2, $\lambda_1 + \lambda_2$) and at 706.0 nm (R_1, $\lambda_3 + \lambda_4$) as well as a broad band centered near 780 nm. On lowering the temperature to 77 K, both R-lines split into two components. Excitation spectra revealed maxima at 590 and 410 for λ_4 and 610, 490 and 450 nm for λ_3. Based on optical and luminescence spectroscopy it was concluded that the Cr^{3+} is in two greatly

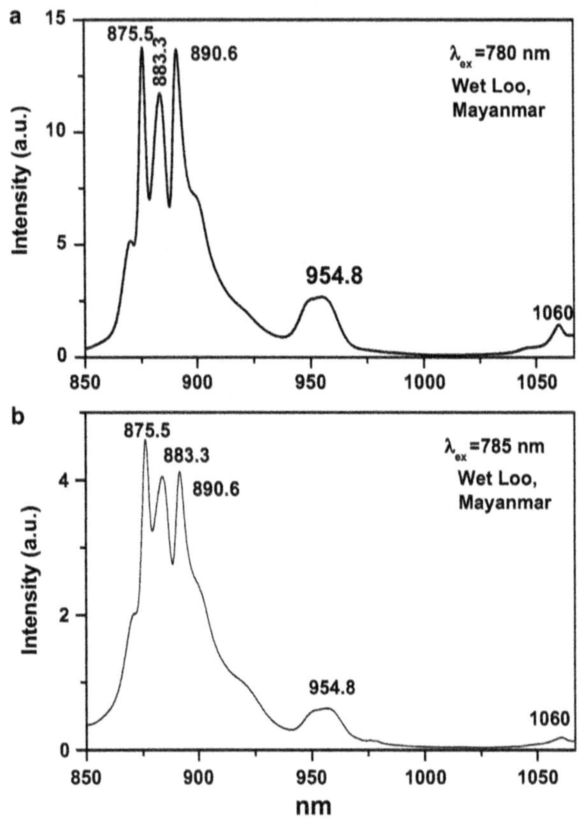

Fig. 4.94 (**a–b**) Laser-induced time resolved luminescence of baddeleyite with under CW 532 and 780 nm excitation (http://rruff.info)

differing crystal fields "a" and "b" ($D_q = 1730$ and $1600\ cm^{-1}$, correspondingly). It was stated that these great differences in the crystal field parameters of environments cannot be explained by a distribution between two or more of the four crystallographically different octahedral sites in the kyanite structure, because the mean M-O distances and the degree of distortion of these octahedral are too close to each other. Thus the type "a" was interpreted as caused by Cr^{3+} substituting for Al^{3+} in some or all of the (M1) to M(4) octahedral of the kyanite structure, which are crystallographically different but close in their mean Al-O distances and slight degrees of distortion. The "b" type resembles those of Cr^{3+} in oxides, and may be interpreted by the assumption that a fraction of the total chromium contents might be allocated in a precursor of a corundum type solution.

The natural kyanite in our study consisted of 30 samples. Concentrations of several potential impurities luminogens are presented in Table 4.17. Laser-induced time-resolved technique enables to detect all lines and bands previously found evidently connected to different kinds of Cr^{3+} luminescence (Fig. 4.107). Mn bearing orange kyanite was studied by laser induced time resolved spectroscopy (Fig. 4.108). Broad band peaking at 760 nm excited by 355 nm has $\tau = 550\ \mu s$. It

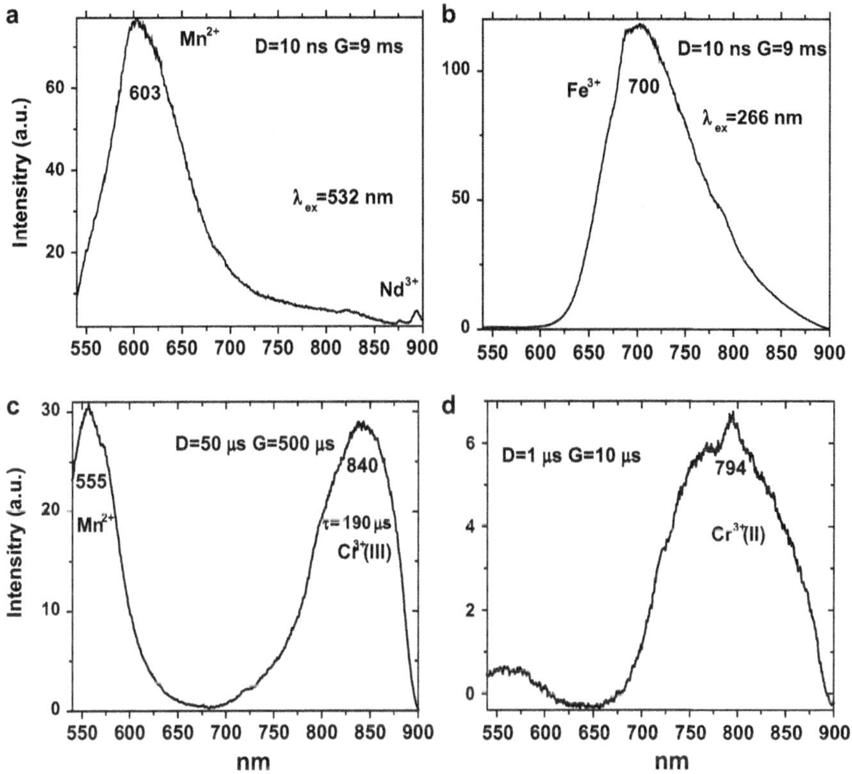

Fig. 4.95 (a–d) Laser-induced time-resolved luminescence spectra of wollastonite demonstrating Mn^{2+}, Fe^{3+} and possibly Cr^{3+} centers

may be ascribed to Mn^{3+} luminescence. Narrow lines at 697, 694, 678 and 677 nm have been connected in Mn^{4+} in different structural positions (Gaft et al. 2011, 2013c).

Sillimanite luminescence has been much less studied than kyanite. The broad-band luminescence with relatively long decay time of 160 μs peaking at 790 nm was ascribed to Cr impurities (Wojtowicz and Lempicki 1988). The natural sillimanite in our study consisted of 12 samples. Concentrations of several potential impurities luminogens are presented in Table 4.18. Laser-induced time-resolved technique enables to detect Fe^{3+}, Cr^{3+} and possibly V^{2+} and Mn^{4+} luminescence (Gaft et al. 2012) (Fig. 4.109).

Under UV excitation kyanite and sillimanite sometimes exhibit broad luminescence bands in blue-green-orange spectral range with decay time in 1–10 μs range. They need further research but in very preliminary form may be ascribed to metal-oxygen complexes (Fig. 4.110).

Andalusite luminescence spectra have not been studied. By laser induced time-resolved spectroscopy using excitation by 266 nm we found band emission peaking at approximately 480 nm with long decay time of 6.1 ms (Fig. 4.111a). Under excitation by 532 nm, very weak red emission takes place peaking at approximately

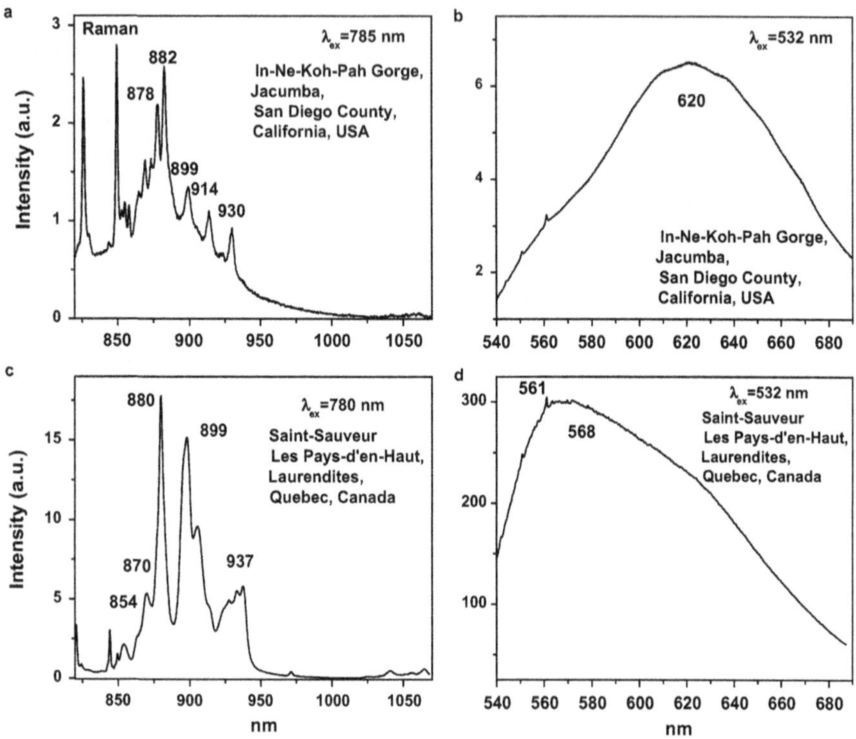

Fig. 4.96 (**a–d**) Laser-induced CW luminescence spectra of two wollastonite samples (http://rruff.info)

Table 4.16 Concentrations of rare-earth elements (ppm) in two different feldspar samples

Pr	Sm	Eu	Nd	Dy	Tb	Er	Tm	Ho	Gd	Ce
5.0	9.0	0.3	30.0	11.0	1.9	6.4	0.8	3.4	14.0	32.0
1.1	3.3	1.4	7.0	4.0	0.8	1.8	0.3	0.6	4.5	6.2

700 nm while after heating at 500 K, the first band remains unchanged and the second one becomes much stronger and has Cr^{3+} like structure (Fig. 4.111b). The origin of this bands needs further study. Excitation by CW laser with 532 and 780 nm revealed additional luminescence bands at 628 (Fig. 4.112a) and 924 nm (Fig. 4.112b).

4.10.2 *Topaz* $Al_2SiO_4F_2$

A chain like structure of connected irregular octahedrons controls topaz structure. These octahedrons have Al in the middle surrounded by four O atoms. Above and below the Al are the (OH) or F ions. The chains of octahedrons are held together by

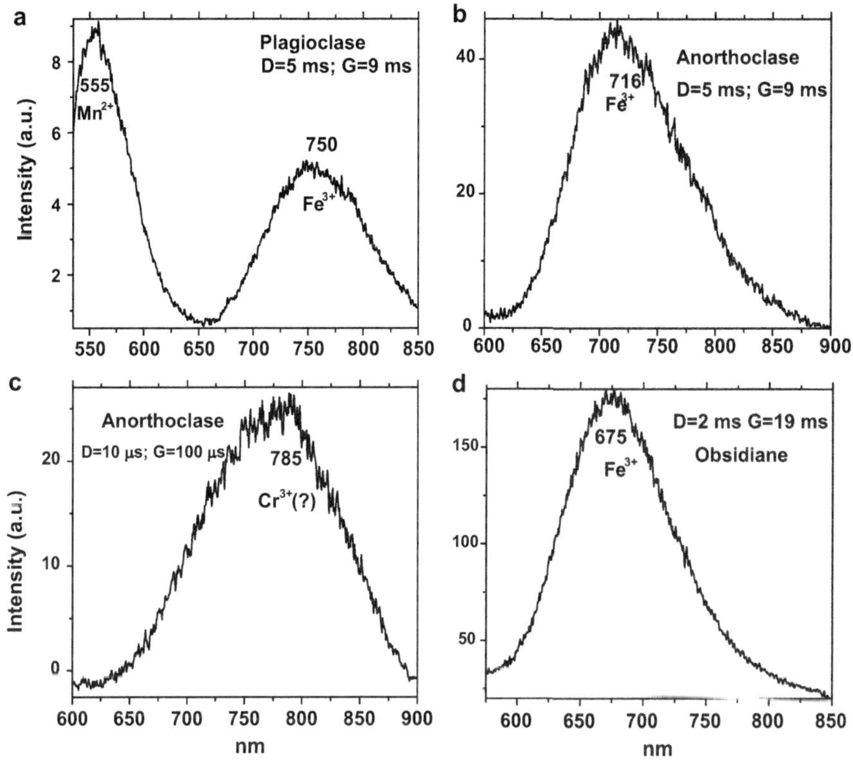

Fig. 4.97 (**a–d**) Laser-induced time-resolved luminescence spectra of various feldspars and obsidian demonstrating Mn^{2+}, Fe^{3+} and Cr^{3+} centers

individual Si tetrahedrons. The crystal system is orthorhombic (2/m2/m2/m) (Fig. 4.113).

The color is clear, yellow, orange, red, blue and green, while the main color centers are radiation induced. Violet and violet-red colors in Cr-containing topaz are generated by two absorption bands in the visible part of the spectrum, which are connected with Cr^{3+} substituting for Al^{3+}. Yellow topaz besides Cr^{3+} contains radiation-induced O^- hole-center (O^{2-} with trapped hole) with corresponding strong ultra-violet absorption. Orange-red topaz besides Cr^{3+} contains F-center (F-vacancy with trapped electron) with strong absorption in the violet part of the spectrum. The origin of the color in topaz without Cr is radiation induced. Yellow color is connected with absorption in the blue-violet part of the spectrum generated by F-center. Red-brown color is connected with absorption in the blue part of the spectrum created by the combination of F-center and O^--center. Blue color is connected with broad absorption band in the red part of the spectrum generated by so-called R-centers (two F-vacancies with two trapped electrons) (Platonov et al. 1984; Schott et al. 2003).

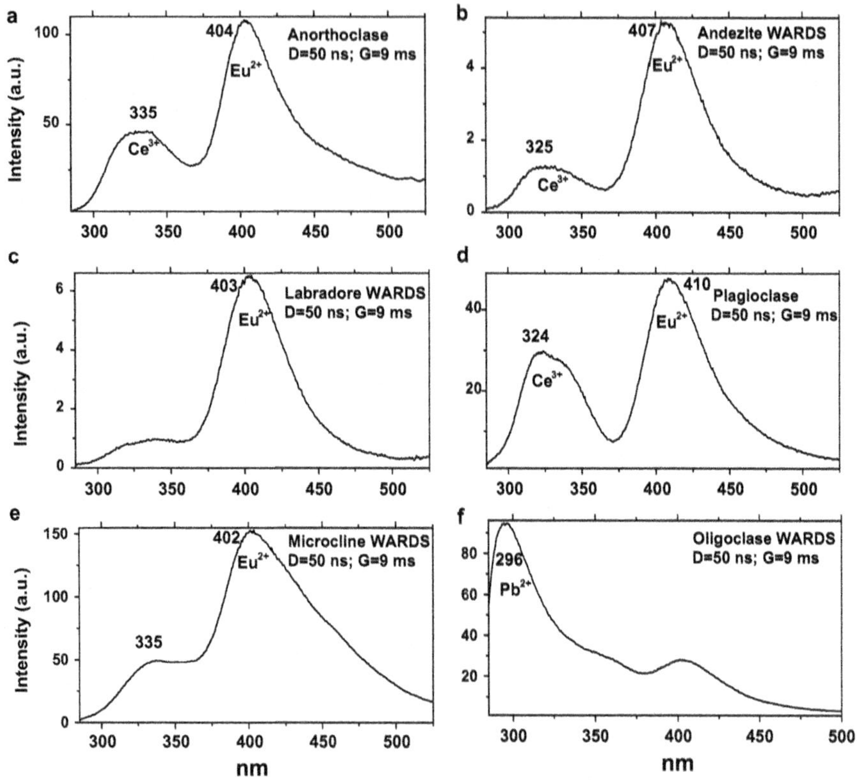

Fig. 4.98 (**a–f**) Laser-induced time-resolved luminescence spectra of various feldspars demonstrating Ce^{3+}, Eu^{2+} and Pb^{2+} centers

Steady-state luminescence of topaz from Ouro Preto, Brazil has been studied in (Tarashchan 1978; Tarashchan et al. 2005; Taran et al. 2003). It was established that all photoluminescence characteristics of variously colored topazes are due to three structurally non-equivalent Cr^{3+} centers isomorphically substituting Al^{3+} in the topaz structure and forming $[CrO_4F_2]^{7-}$, $[CrO_4OH,F]^{7-}$, and $[CrO_4(OH)_2]^{7-}$ complexes. Kinetics of thermal annealing indicates different thermal stability of these centers. Less stable $[CrO_4OH, F]^{7-}$- and $[CrO_4(OH)_2]^{7-}$ complexes diminish in temperature range 950°C–1100 °C, accompanied by appearance of corundum phase. The most stable $[CrO_4F_2]^{7-}$ centers completely decay at 1250 °C and luminescence spectrum of product obtained becomes identical to that of Cr^{3+} in mullite indicating that topaz completely transforms to mullite.

The natural topaz in our study consisted of seven samples. Concentrations of several potential impurities luminogens are presented in Table 4.19. Laser-induced time-resolved technique enables to detect three different Cr^{3+} and possibly Mn^{4+} and (TiO_4) emission centers (Gaft et al. 2003a) (Figs. 4.114 and 4.115).

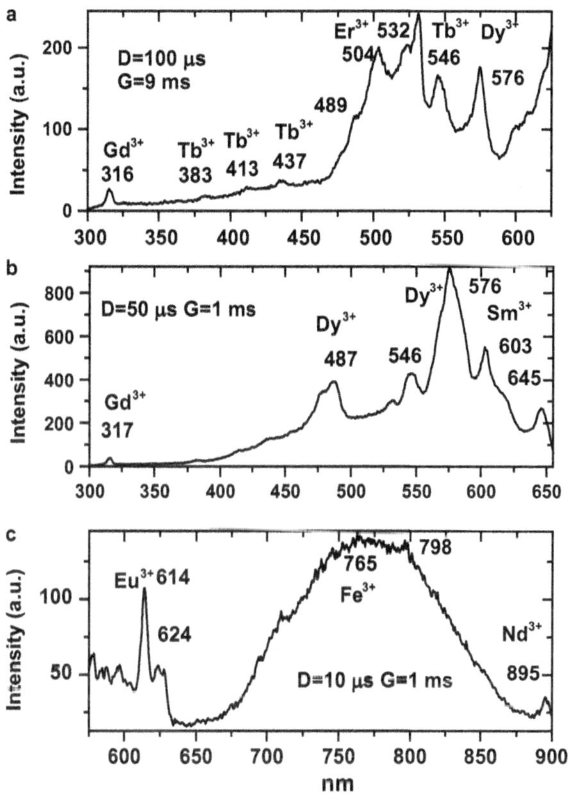

Fig. 4.99 (**a–c**) Laser-induced time-resolved luminescence spectra of feldspars demonstrating Gd^{3+}, Tb^{3+}, Er^{3+}, Dy^{3+}, Sm^{3+}, Eu^{3+}, Nd^{3+} and Fe^{3+}

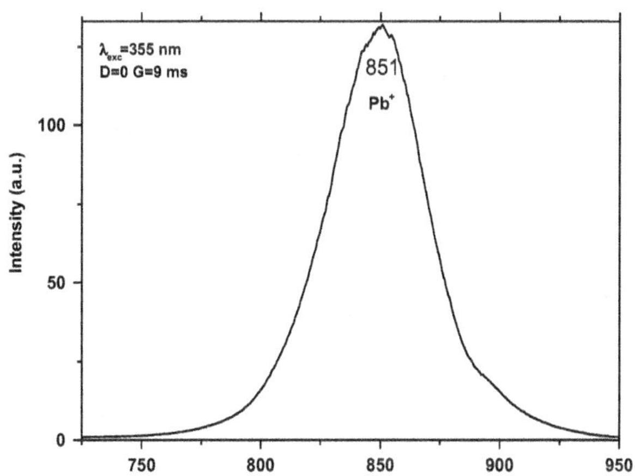

Fig. 4.100 Laser-induced time-resolved luminescence spectra of feldspar demonstrating Pb^{+}

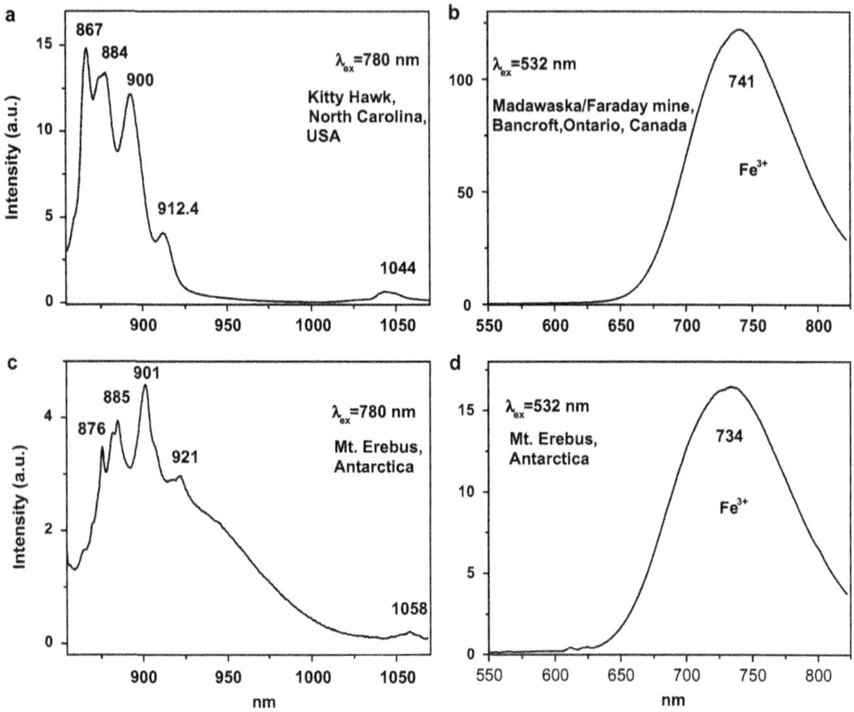

Fig. 4.101 (**a–b**) Laser-induced cw luminescence spectra of albite demonstrating Nd^{3+} and Fe^{3+} (http://rruff.info)

4.10.3 Corundum Al_2O_3

Corundum has trigonal structure. The oxygen ions are arranged in approximately hexagonal closed packing. Between the oxygen layers there are sites for cations octahedrally coordinated by six oxygen ions, but in corundum only two-thirds of the available positions are filled. Groups of three oxygen ions form a common face of two neighboring octahedra and thus the groups are linked to a pair of Al ions (Fig. 4.116).

The main colored varieties are ruby (red) and sapphire (blue). The corresponding color centers may present in different quantitative and qualitative proportions with resulting varieties of colors (up to 2000 different tints). The main color center in ruby is impurity of Cr (up to 4 %). Ion Cr^{3+} generates two strong absorption bands in the visible part of the spectrum, which explain the red color. Different tints of red, which influence strongly on the commercial value, are connected with impurities of Fe^{3+} and V^{3+}. Under UV excitation, ruby may exhibit red luminescence of Cr^{3+}, which makes the red color of ruby stronger under sun illumination. The main color centers in sapphire are impurities of Fe and Ti. Strong absorption bands in the green-blue sapphires are connected with Fe^{2+} and Fe^{3+}. Usually the Fe absorption is

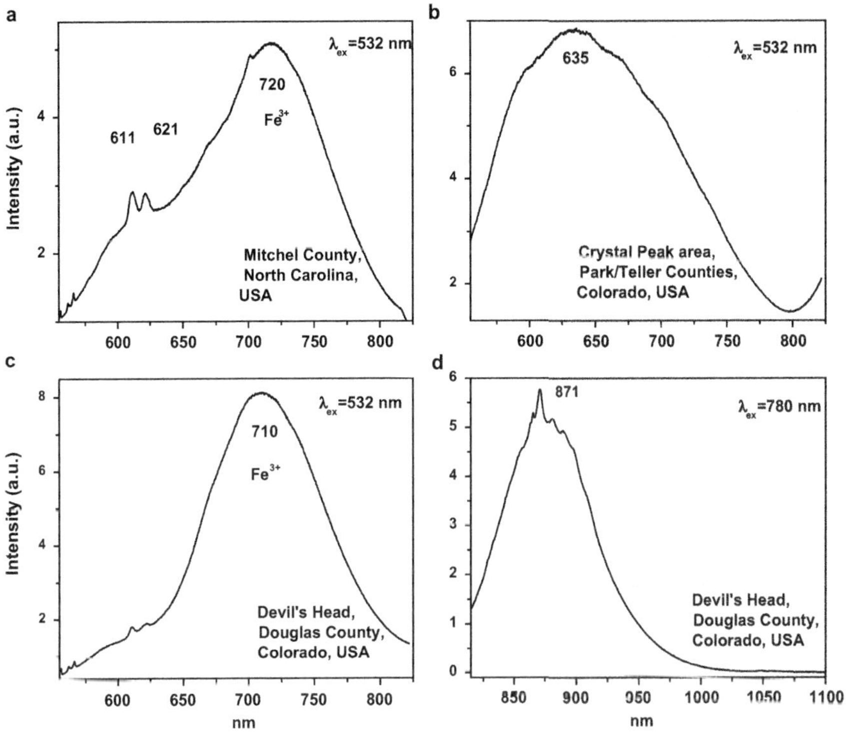

Fig. 4.102 (a–d) Laser-induced cw luminescence spectra of microcline samples demonstrating Nd^{3+} (http://rruff.info)

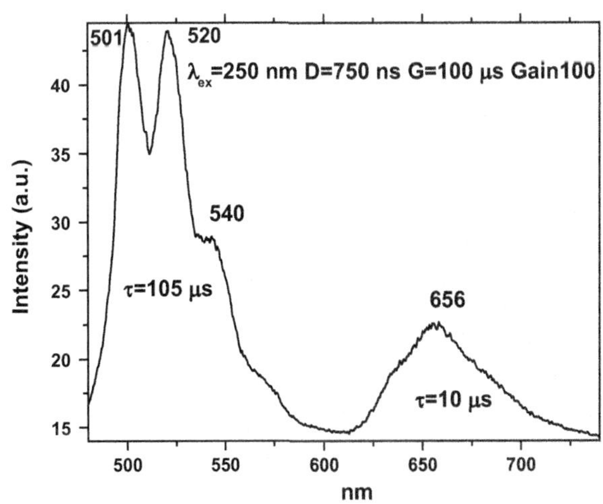

Fig. 4.103 Laser-induced time-resolved luminescence spectra of quartz demonstrating uranyl green band

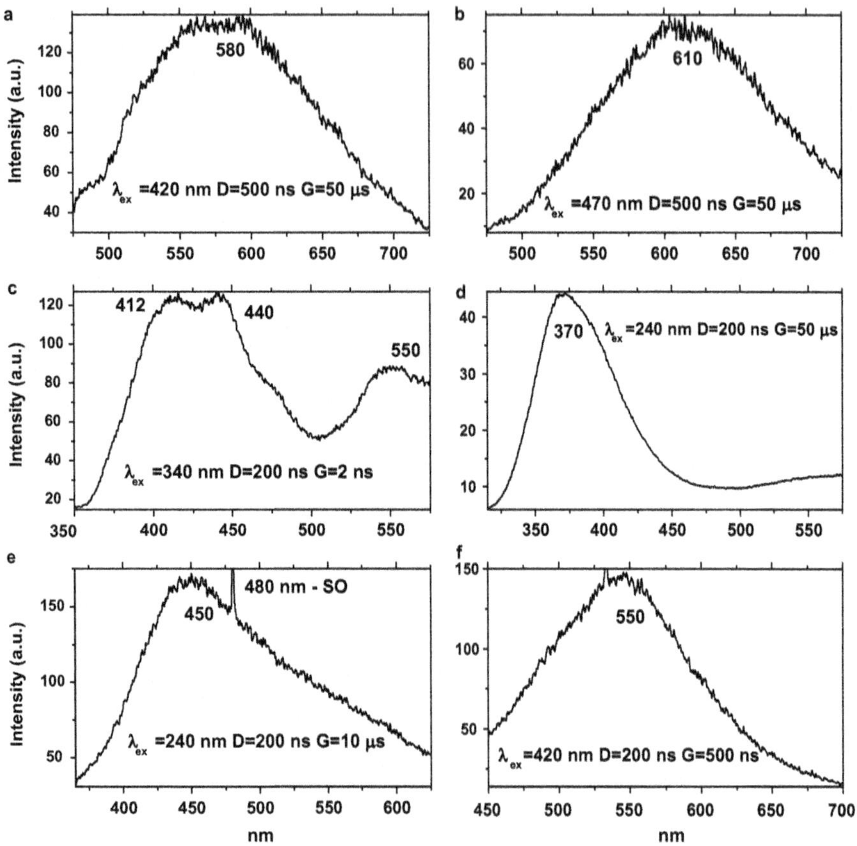

Fig. 4.104 (**a–f**) Laser-induced time-resolved luminescence spectra of quartz

relatively weak, but in sapphire the cations Fe^{2+} and Fe^{3+}, are structurally paired and exchange interaction between ions provides mechanisms by which the intensity of absorption increases dramatically. Absorption bands of blue sapphire are connected with Ti^{4+}-Fe^{2+} and Fe^{2+}-Fe^{3+} structural pairs (Platonov et al. 1984).

Steady-state luminescent properties of natural corundum Al_2O_3 were carefully investigated. Strong well-known $^2E \rightarrow {}^4A_2$ lines of Cr^{3+} with long decay time characterize their photoluminescence spectra. Besides that, much weaker narrow lines present, which are connected with Cr-pairs and more complicated complexes (so called N-lines) (Tarashchan 1978). Broad blue band of CL peaking at approximately 480 nm of undetermined origin (Ponahlo 2000) and broad red band peaking at 690 nm preliminary ascribed to Ti^{3+} (Gorobets and Rogojine 2001) have been also detected. The natural corundum in our study consisted of four samples. Laser-induced time-resolved technique enables to detect Cr^{3+} and possibly Fe^{3+} emission centers (Fig. 4.117).

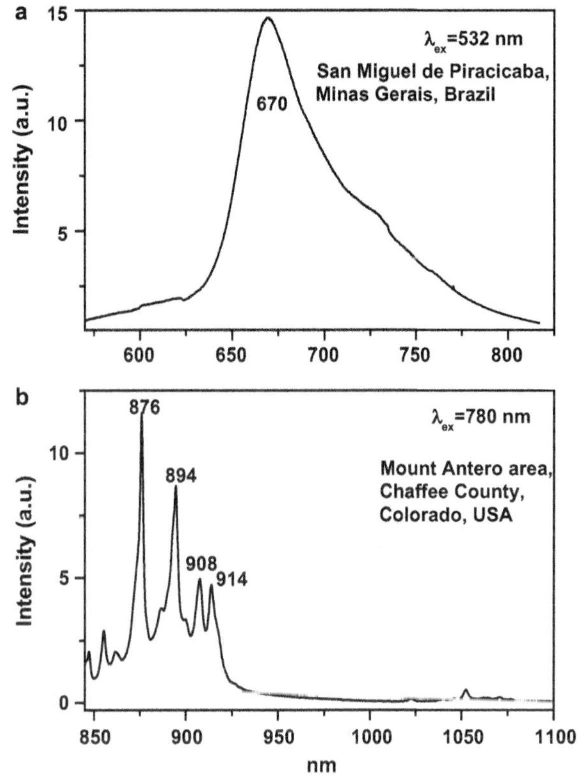

Fig. 4.105 (**a**–**b**) Laser-induced time-resolved luminescence spectra of phenakite (http://rruff.info)

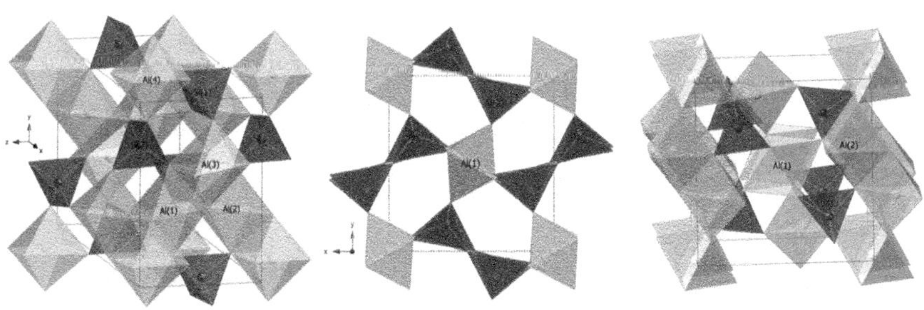

Fig. 4.106 Crystal structures of kyanite, sillimanite and andalusite showing the various Al and Si sites

An interesting feature exists in "classical" Cr^{3+} luminescence in corundum. The weaker peaks from the long wavelength side from those lines are also known and ascribed to so call N-lines, namely Cr^{3+} pairs. The lines from the short wavelength side are less famous (Fig. 4.118). The peak at 658 nm is corrected to combination of R_3' and R_2' lines from $^4T_1$ level, at 668 nm to R_1' from the same level, at 675 and

Table 4.17 Trace impurities (ppm) – potential luminescence centers in different kyanites

Sample	Ti	V	Cr	Mn	Fe	Ni
Big white	352.8	1348	279	7.6	7218	28.7
5025	6096	1.6			115	
White broad	450	1395	68	5.5	7136	19.8
Jem	276	923	344	6.2	7328	22.7
Transparent	534	457	181	168	6894	22.7
Blue	302.5	77	565	2.1	7471	8.7
Australia slab		118	67	1.7		
Sinai slab		101	66	1.5		

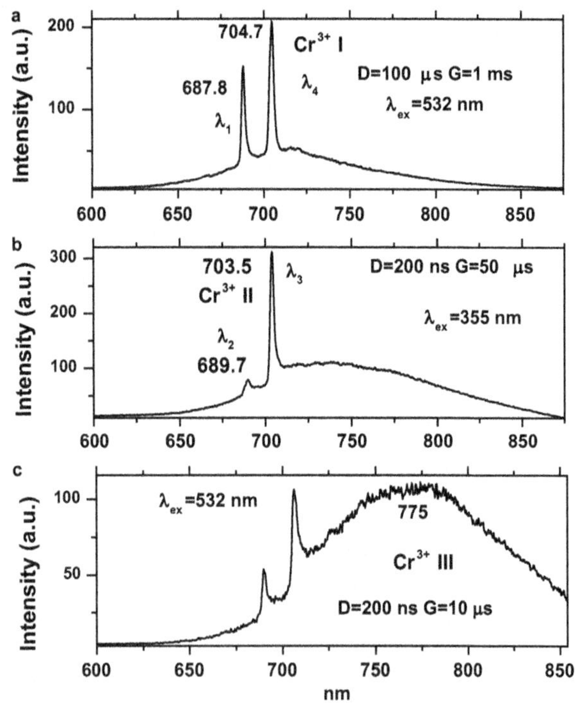

Fig. 4.107 (**a–c**) Laser-induced time-resolved luminescence spectra of kyanite demonstrating different Cr^{3+} centers

680 nm are anti-Stokes phonon repetitions as result of superposition of the energies of electron transitions connected to R-lines with the energies of atomic vibrations (Nelson and Sturge 1965).

4.10.4 *Spinel* $MgAl_2O_4$

The spinel structure is a cubic close packing of anions bound together by interstitial cations. The lattice is face-centered cubic, space group Fd3m, with eight $MgAl_2O_4$

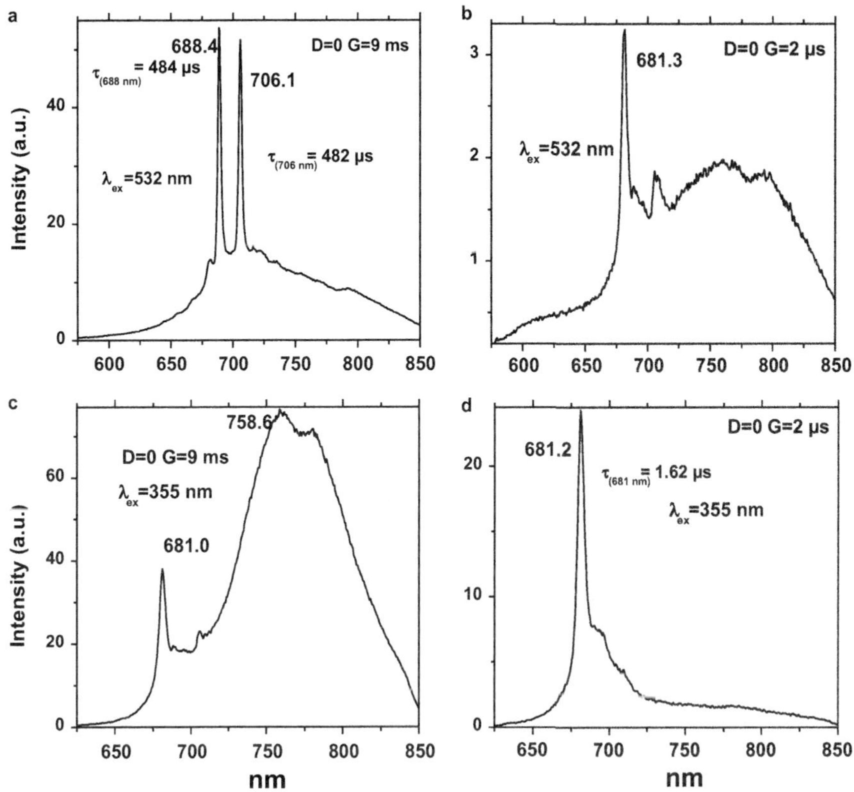

Fig. 4.108 (**a**–**d**) Laser-induced time-resolved luminescence spectra of orange Mn bearing kyanite demonstrating Mn^{3+} and different Mn^{4+} centers

Table 4.18 Trace impurities (ppm) – potential luminescence centers in different sillimanites

Sample	Ti	V	Cr	Mn	Fe
1924	107	656	936	1	1723
2020	115	623	114	2	1532
284	82	151	23	1	9340
518	86	108	894	1	3286
282	18	77	104	3	8718
273	36	107	24	-	8814
509	104	296	29	1	1933

formula units per face-centered cell. One of the spinel peculiarities is cation inversion responsible for its disordered structure. There are two cation positions in the spinel lattice, one tetrahedrally coordinated and one with octahedral coordination. Among them, two kinds of cations A^{2+} and B^{3+} are distributed. The so-called normal spinel is described by the formula $A^{\{4\}}B2^{\{6\}}O_4$, where A is in tetrahedral and B in octahedral coordination. In the diverse spinel, the divalent

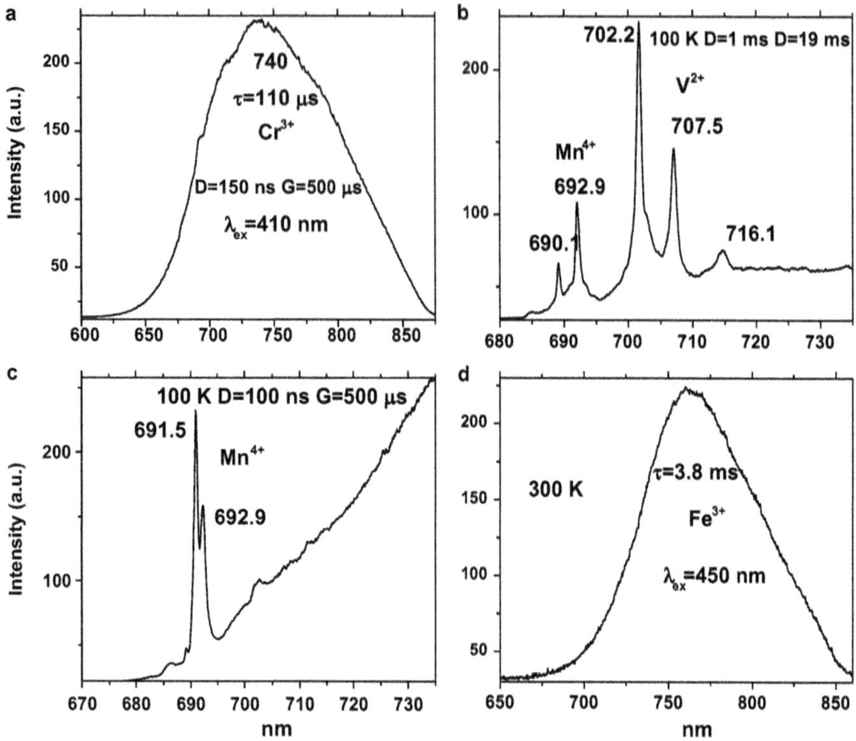

Fig. 4.109 (**a–d**) Laser-induced time-resolved luminescence spectra of sillimanite demonstrating Fe^{3+}, Cr^{3+} and possibly V^{2+} and Mn^{4+} luminescence

cations A are placed in the octahedral positions $B^{\{4\}}(AB)^{\{6\}}O_4$. Thus in a normal spinel, the Al^{3+} ions are in octahedral coordination with local symmetry D_{3d}, while the Mg^{2+} ions are in regular tetrahedral coordination, point symmetry T_d.

The colored varieties are red, pink-red, orange-red and violet-red. The main color centers are Cr^{3+}, Fe^{2+}, Co^{2+} and Fe^{3+}, which give all tints of color from violet to red. The mutual feature of pink and red spinel is two broad absorption bands in the visible part of the spectrum, which are connected with Cr^{3+} substituting for Al^{3+}. The set of the narrow absorption bands, which give the violet color are connected with Fe^{2+} impurity substituting for Mg^{2+}. Absorption spectra of the blue spinels are similar to those of the violet ones, but the absorption in the red and violet parts of the spectrum are higher. This is connected with additional influence of Fe^{2+} and Co^{2+} substituting for Al^{3+} and Mg^{2+}, correspondingly. Mostly clear the coloring action of these impurities is seen in green and blue spinels (Platonov et al. 1984).

The emission of natural spinel exhibits a great deal of fine structure, which was ascribed to Cr^{3+} luminescence (Tarashchan 1978; Mohler and White 1995). The natural spinel in our study consisted of three samples. Laser-induced time-resolved

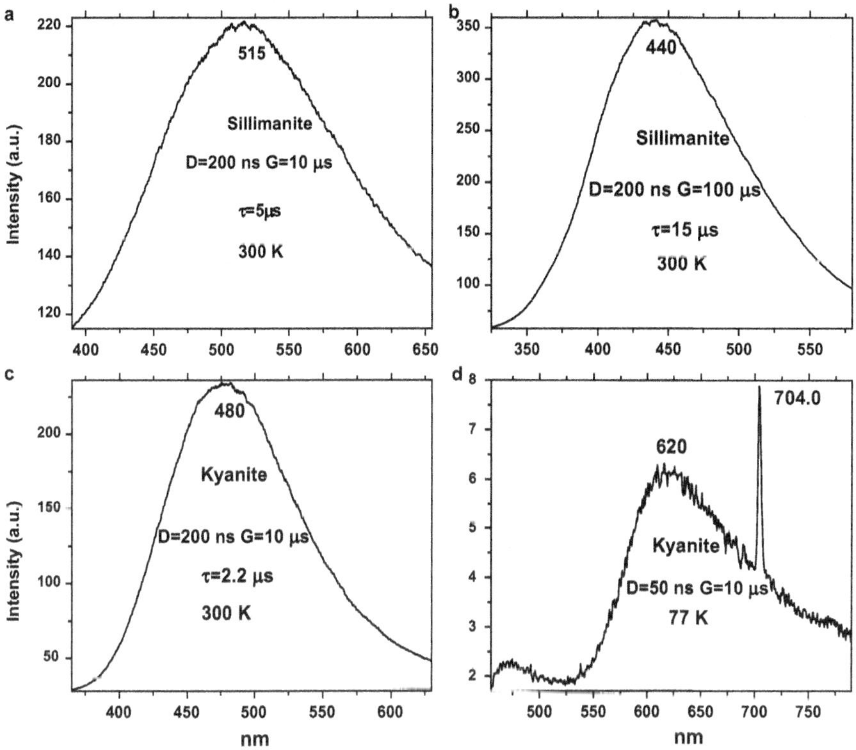

Fig. 4.110 (**a–d**) Laser-induced time-resolved luminescence spectra of sillimanite and kyanite demonstrating UV excited luminescence possibly connected to metal-oxygen complexes

technique enables to detect Cr^{3+} and possibly Mn^{2+} emission centers (Platonov et al. 1984).

The emission of natural spinel exhibits a great deal of fine structure, which was ascribed to Cr^{3+} luminescence (Tarashchan 1978; Mohler and White 1995). The natural spinel in our study consisted of three samples. Laser-induced time-resolved technique enables to detect Cr^{3+} and possibly Mn^{2+} emission centers (Fig. 4.119).

Excitation by CW laser with 532 and 780 nm revealed narrow luminescence lines belonging to Cr^{3+} and three luminescence bands peaking at 660, 720 and 895 nm (Fig. 4.120). Aditional study is needed to ascribe those emission centers including their decay times and excitation spectra.

4.10.5 *Beryl and Emerald $Be_3Al_2Si_6O_{18}$*

The dominant features in the structure are the hexagonal rings of six Si-O tetrahedra, these rings forming hollow columns parallel to the z-axis. Within the rings two

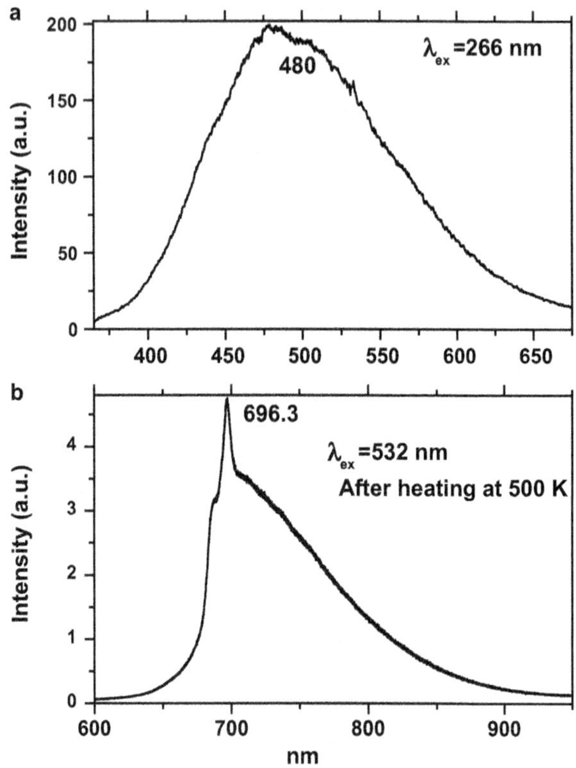

Fig. 4.111 (**a–b**) Laser-induced time-resolved luminescence spectra of andalusite under 266 and 532 nm excitation and after thermal treatment (500 K)

of the oxygen atoms in each SiO_4 group are shared by SiO_4 groups on either side, thus giving the meta-silicate ratio. Between the rings lie the Al and Be atoms, each Al coordinated with an octahedral group of six oxygen atoms, and each Be surrounded by four oxygen atoms on a distorted tetrahedron (Fig. 4.121).

The main colored varieties are emerald (Green), heliodor (Yellow) and aquamarine (Blue). The main color center in emerald is Cr^{3+} substituting for Al^{3+}, which generates two strong absorption bands in the visible part of the spectrum. An additional impurities which influence also on the color are Fe^{2+} in two different structural positions: instead of Be and instead of Al. They generate absorption bands in the red and near IR parts of the spectrum and give the yellowish tints to the color. Green beryl is known which does not contain Cr^{3+}. The corresponding absorption bands are connected with V^{3+} substituting for Al^{3+} and the name emerald is controversial in this case. Two different color centers are known in heliodor, both connected with Fe^{3+} impurities, but in different structural positions: instead of Be^{2+} (yellow-green color) and instead of Al^{3+} (yellow-orange color). In both cases the corresponding absorption bands are connected with charge transfer electron transitions in the O^{2-}-Fe^{3+} centers. Three different color centers are known in aquamarine, all connected with Fe impurities. The first type of absorption bands is connected with electronic transitions inside Fe^{2+} substituting for Al^{3+}. The

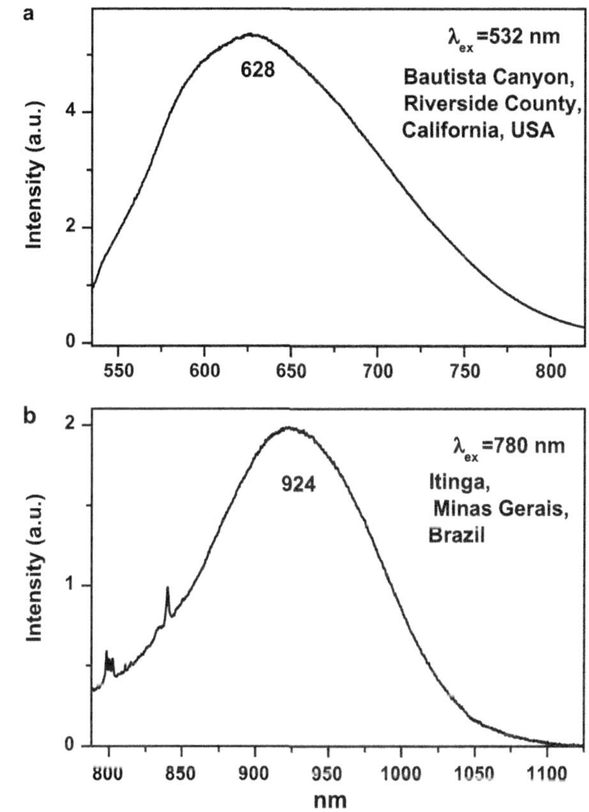

Fig. 4.112 (**a–b**) Laser-induced time-resolved luminescence spectra of andalusites under CW 532 (**a**) and 780 (**b**) nm laser excitation (http://rruff.info)

second color center is connected with double substitution: two Al^{3+} are substituted by Fe^{2+} on the place of Al^{3+} and Fe^{3+} in interstitial position (between two positions of Al^{3+}). The corresponding absorption bands are connected with charge transfer electronic transitions between Fe^{2+} and Fe^{3+}. The last color center has the same nature as the previous one, but Fe^{3+} substitutes for Al^{3+} position (Platonov et al. 1984).

Steady-state emission of beryl was previously studied. The broad band at 720 nm is connected with Fe^{3+}, while the relatively narrow bands at 480 and 570 nm are ascribed to Mn^{2+} in tetrahedral and octahedral coordination, correspondingly. Cr^{3+} emission was connected with narrow R-lines at 680 and 682 nm (Tarashchan 1978; Kuznetsov and Tarashchan 1988).

The Cr^{3+} luminescence properties in natural beryl minerals have been studied as a function of the Cr content as well as impurities such as Fe and V. It appears that the Cr^{3+} crystal field is linked to the Cr amount and decreases when Cr increases. A competition between Cr and V was noticed for very low Cr concentration (Ollier et al. 2015).

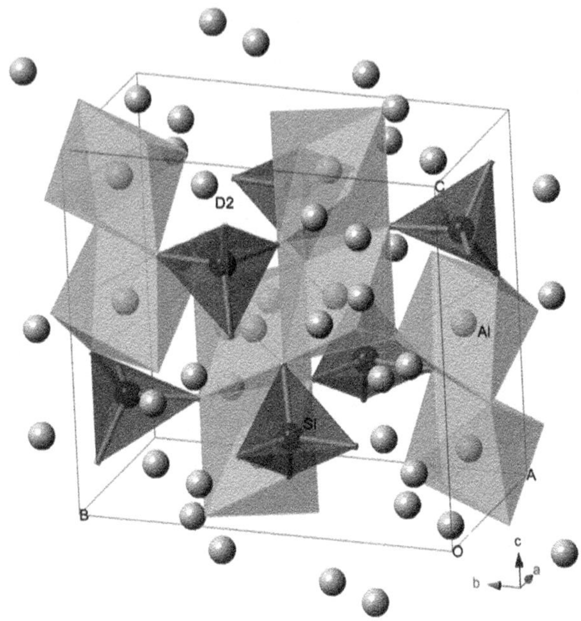

Fig. 4.113 Crystal structure of topaz showing tetrahedral Si and octahedral Al sites

Table 4.19 Concentrations of Mn, Fe, Cr, V and Ti (ppm) in topaz samples

	Mn	V	Cr	Ti
Yellow 1	<10	25	25	160
Yellow 2	<10	20	55	130
Yellow 3	<10	25	180	110
Yellow 4	<10	30	60	90
Red	125	30	500	130

The natural beryl and emerald in our study consisted of seven samples. Laser-induced time-resolved technique enables to detect Cr^{3+}, Fe^{3+} and possibly Mn^{4+} and V^{2+} emission center (Figs. 4.122 and 4.123). Excitation by CW laser with 532 and 780 nm revealed additional luminescence line at 657.4 nm, bands peaking at 685 and 755 nm (Fig. 4.124) and IR bands peaking at 880, 927, 945 and 952 nm (Fig. 4.125).

4.10.6 *Chrysoberyl $BeAl_2O_4$*

Alexandrite is the common name for Cr-doped chrysoberyl, a crystal whose unit cell contains four formula units $BeAl_2O_4$, forming an orthorhombic structure of space group Pnma. Synthetic $BeAl_2O_4$:Cr^{3+} is widely used as a material for solid state tunable lasers and is well studied. The Al^{3+} ions are octahedrally coordinated by the oxygen ions and occur in two not equivalent crystal field sites in the lattice.

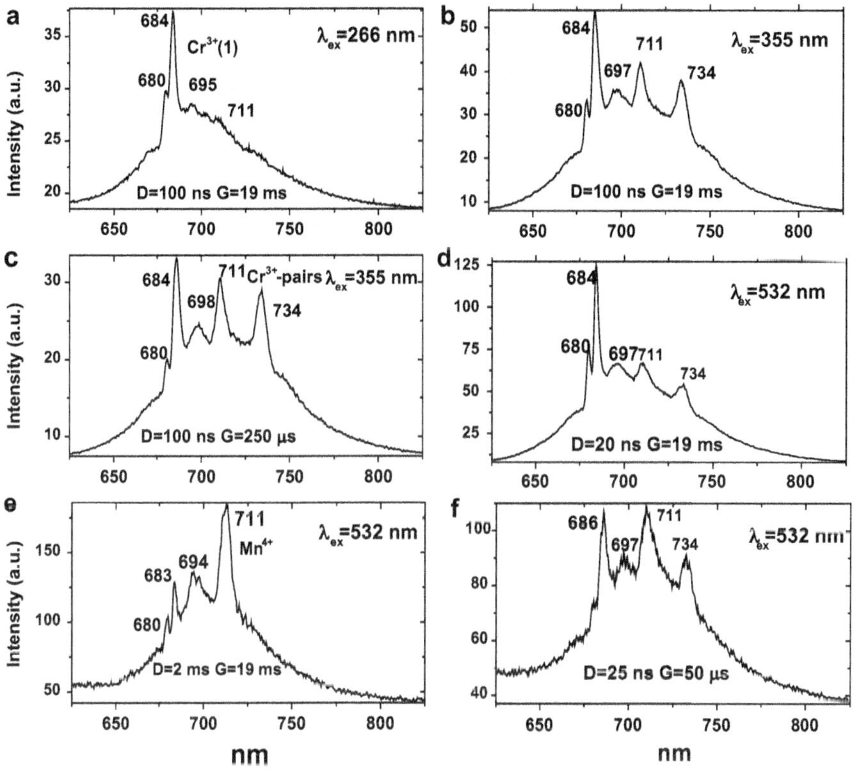

Fig. 4.114 (**a–f**) Laser-induced time-resolved luminescence spectra of topaz demonstrating different Cr^{3+} and possibly Mn^{4+} centers

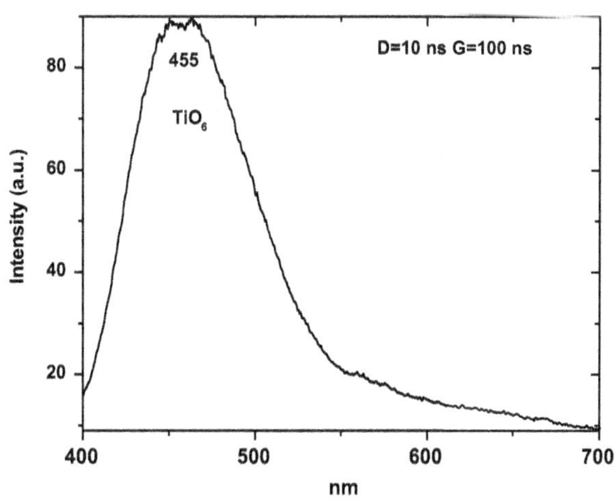

Fig. 4.115 Laser-induced time resolved luminescence spectra of topaz demonstrating possibly TiO_4 luminescence

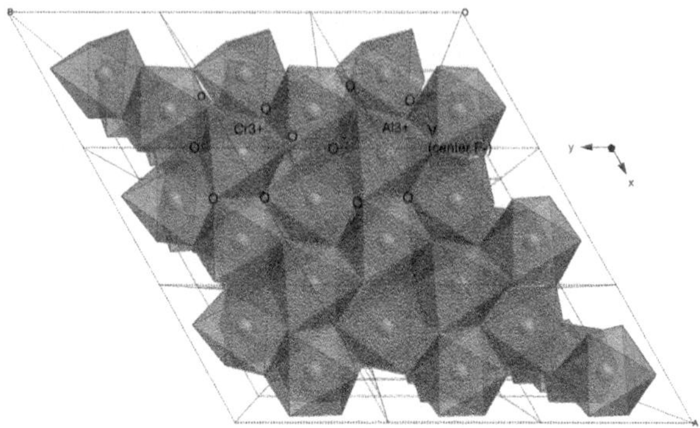

Fig. 4.116 Crystal structure of corundum showing substituting Cr^{3+} and F^{-} vacancy center

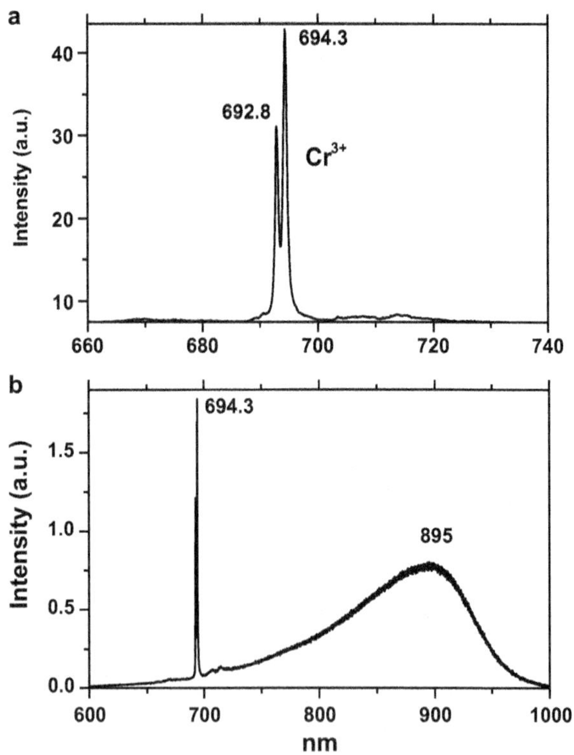

Fig. 4.117 (**a–b**) Laser-induced time-resolved luminescence spectra of ruby (**a**) and steady-state spectra of sapphire (**b**) demonstrating Cr^{3+} and possibly Fe^{3+} centers

The Al^{3+} sites lying in the mirror-symmetry planes of the lattice have the site symmetry of the C_s point group, while the other Al^{3+} sites possess inversion symmetry and belong to the C_i point group. The Cr^{3+} ions enter the crystal substitutionally for the Al^{3+} ions, 78 % replacing Al^{3+} ions in the mirror sites and the rest going into the inversion sites (Fig. 4.126).

Fig. 4.118 Laser-induced time-resolved luminescence spectra of ruby demonstrating short wavelength lines of Cr^{3+} centers

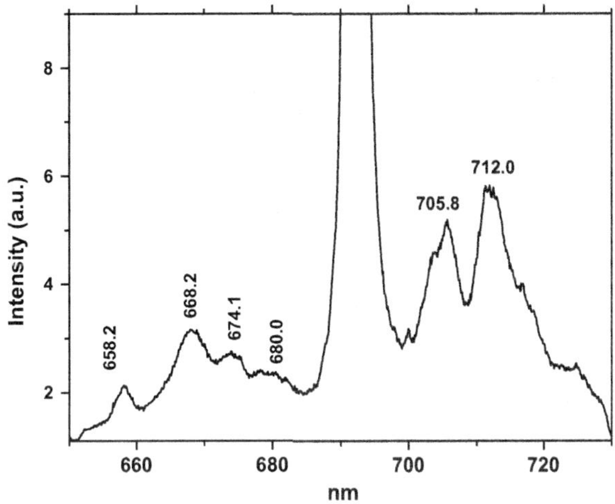

Fig. 4.119 (**a–b**) Laser-induced time-resolved (**a**) and steady-state (**b**) luminescence spectra of spinel demonstrating Cr^{3+} and possibly Mn^{2+} centers

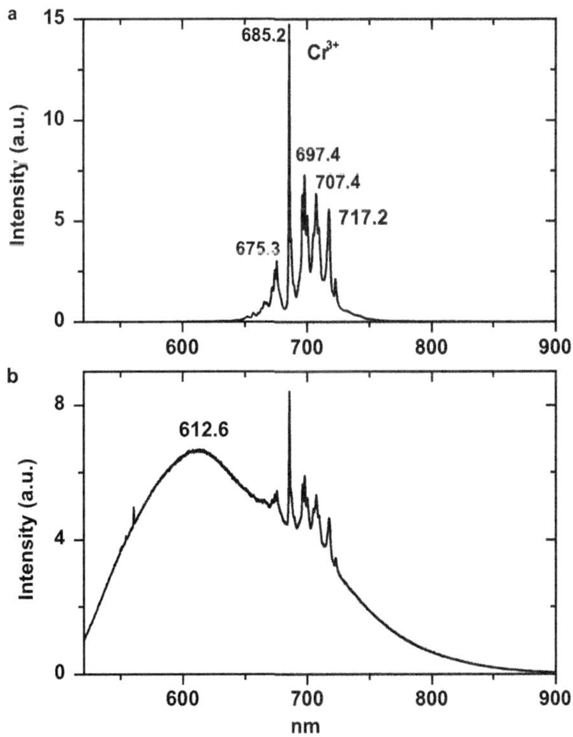

It was found that the major features of the spectra at 300 K are the two sharp lines R_{1m} and R_{2m} peaking at 680 nm and the broad, structured band peaking at lower energies. The latter is due the superposition of low-energy vibronic emission transitions from the components of the ^2E_g level and emission from the relaxed excited state of the $^4T_{2g}$ level which is Stokes-shifted to lower energy compared to

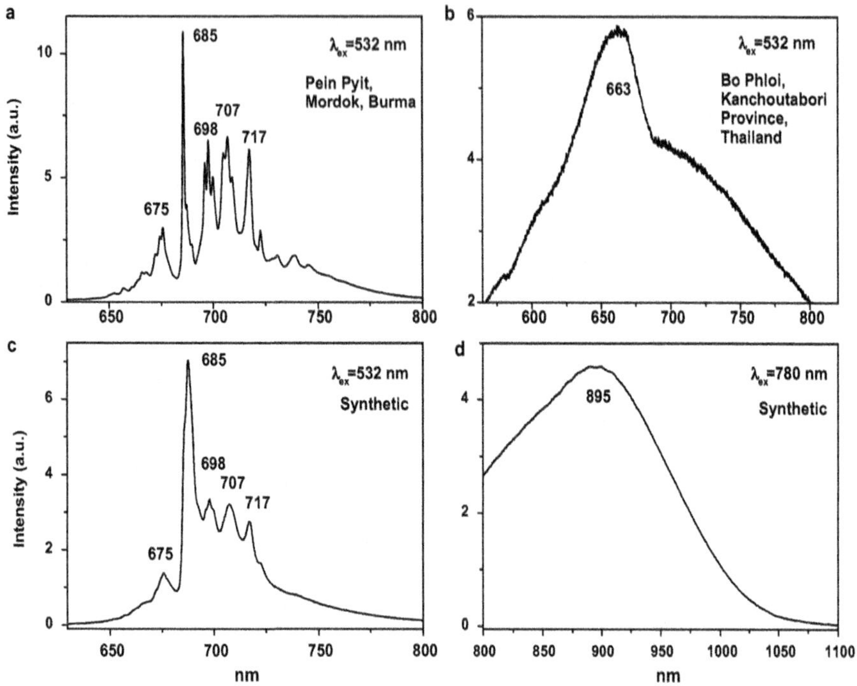

Fig. 4.120 (a–d) Laser-induced CW luminescence spectra of natural and synthetic spinels (http://rruff.info)

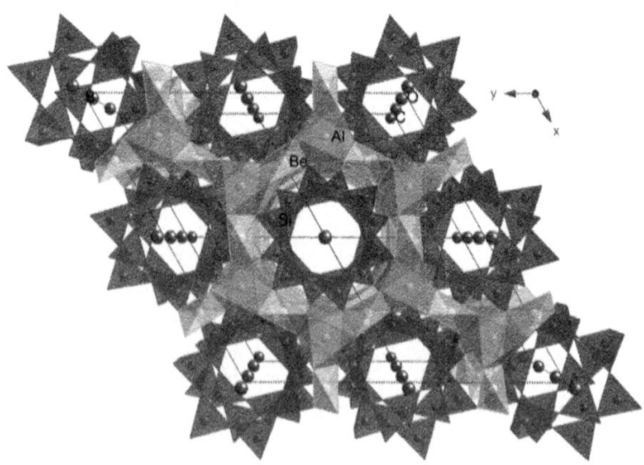

Fig. 4.121 Crystal structure of beryl demonstrating the octahedral Al sites and tetrahedral Be ones

the absorption transition involving this level. The structure in this sideband is a mixture of specific one-phonon vibronic peaks, and zero-phonon lines associated with transitions from the $^4T_{2g}$ and $^2T_{1g}$ levels. At certain excitation wavelengths, such as 488.0 nm, the R_{1i} and R_{2i} peaking at 690.0 and 695.0 nm lines associated with Cr_i ions in inversion sites appear (Powell et al. 1985). The 689.5 and 695.2 nm

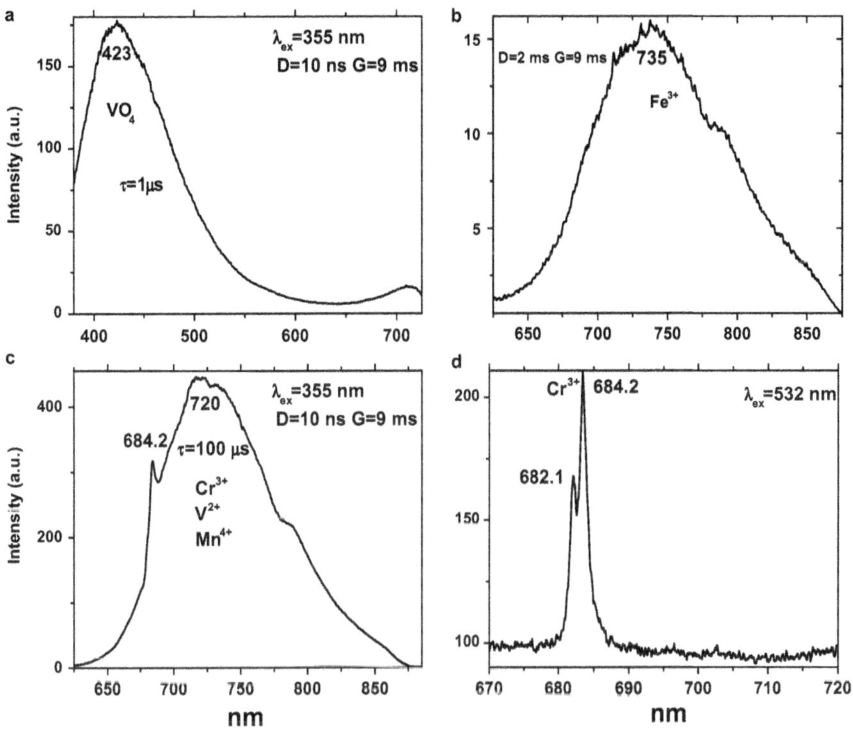

Fig. 4.122 (a–d) Laser-induced time-resolved luminescence spectra of beryl demonstrating different Cr^{3+}, Fe^{3+} and possibly Mn^{4+}, V^{2+} and VO_4 centers

decay time is much larger than that for the 677.9 and 679.5 nm emission (Suchoki et al. 1987). The excitation spectrum of C_m coincides with the absorption spectrum of alexandrite peaking in the visible range at 415–430 and 570–600 nm ranges orientation dependent, while for C_i one band peaking at 490 nm takes place. The crystal field strength Dq was determined and the result obtained indicates an increase in Dq when passing from 1740 cm^{-1} for C_m to 2053 cm^{-1} for C_i. The parameters of the interelectronic interaction for Cr(Ci) were also determined and they are as follows: $B = 680\ cm^{-1}$, $C = 3108\ cm^{-1}$ (Eliseev et al. 1988).

Two types of Cr^{3+} luminescence centers have been found in natural alexandrite, characterized at 300 K by R-lines at approximately 679.0 with 677.3 nm and 694.4 with 691.7 nm, accompanied by very many N-lines of Cr-Cr pairs. Those centers have been identified as connected with substitutions of Al^{3+} in different structural sites (Tarashchan 1978). In another study, two doublets at 680.4 with 678.5 nm and 695.8 with 689.9 nm were attributed to the same centers (Scalvi et al. 2003; Trindade et al. 2011). It was found that natural alexandrites with very rare exceptions are characterized by very low CL intensities (Ponahlo 2000). Pulse CL study revealed that the spectrum consists of relatively broad red band peaking at

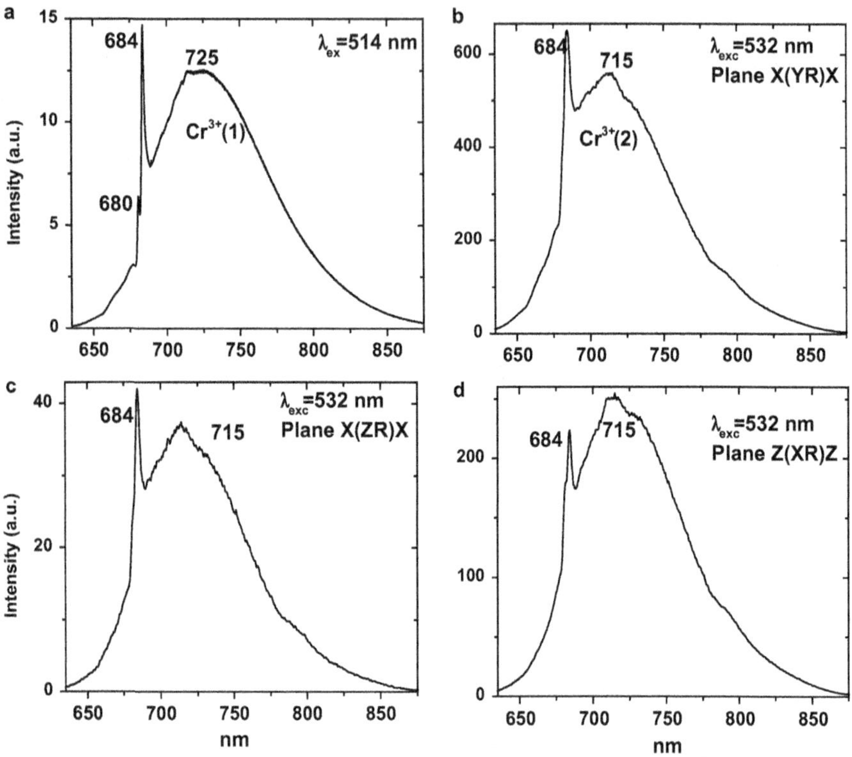

Fig. 4.123 (**a–d**) Laser-induced time-resolved luminescence spectra of oriented emerald demonstrating different Cr^{3+} centers

685–695 nm, accompanied by narrow lines with the strongest one at 679 nm and the weaker ones at 650, 655, 664, 700, 707 and 716 nm. All lines and bands have been ascribed to several Cr^{3+} centers (Solomonov et al. 2002; Shoval et al. 2003).

The Cr^{3+} luminescence properties in natural chrysoberyl minerals have been studied as a function of the Cr content as well as impurities such as Fe and V. A competition was found between Cr and V for very low Cr concentration with the vanishing of Cr^{3+} emission from Cr^{3+} ions located in inversion site. The Fe^{3+} ions substitute in mirror site efficiently with a strong impact on the Cr^{3+} lifetime of mirror site (Ollier et al. 2015).

The natural chrysoberyl in our study consisted of ten samples. Concentrations of several potential impurities luminogens are presented in Table 4.20. Laser-induced time-resolved technique enables to detect two different Cr^{3+} centers (Fig. 4.127).

Figure 4.128 presents additional luminescence spectra detected in chrysoberyl. Their origin needs further investigation but most probably they are connected to metal-oxygen complexes.

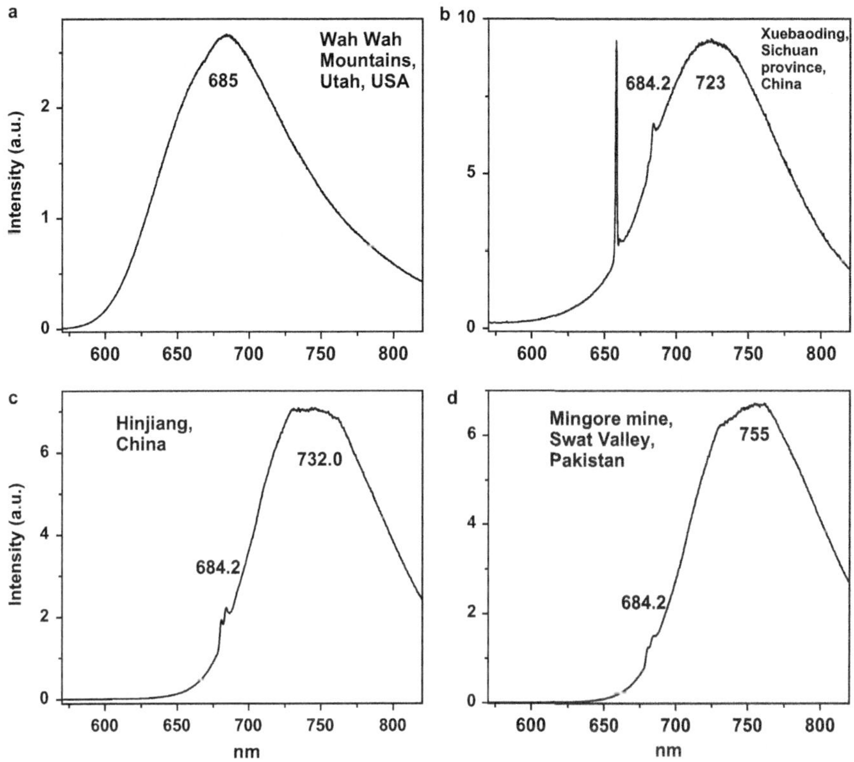

Fig. 4.124 (**a–d**) Laser-induced cw luminescence spectra of beryls under 532 nm laser excitation (http://rruff.info)

4.10.7 Grossular $Ca_3Al_2(SiO_4)_3$

Grossular has a chemical composition $Ca_3Al_2(SiO_4)_3$ and is the Ca-Al species of the six common anhydrous minerals of the garnet group. The structure of grossular is composed of SiO_4 tetrahedra and AlO_6 octahedra joint to each other by their corners. Ca cations are situated in this cavities surrounded by eight oxygen atoms in the form of a strongly distorted cube (Fig. 4.129).

The main color center is Cr^{3+} substituting for Al^{3+}. It generates two absorption bands in the visible part of the spectrum with resulting red color. The larger intensity of the ultra-violet absorption, resulting in red tint increasing, is connected with the presence of Fe^{3+} impurity. Besides that, orange-red color of pyrope may be connected with Fe^{2+} and Ti^{4+} impurities. Rhodolite has complex pyrope-almandine composition and its red color is connected with Fe^{2+} center. The main color centers in spessartite are Mn^{2+} and Fe^{2+}. Mn^{2+} generates absorption bands in the violet part of the spectrum with corresponding yellow and orange tints. Green color of grossular is mainly connected with Fe^{3+} impurity substituting for Al^{3+}. Orange

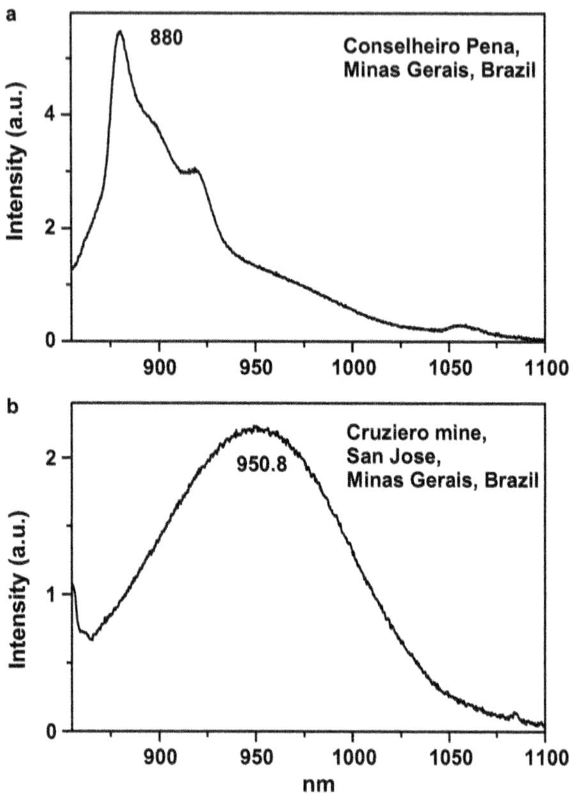

Fig. 4.125 (a–d) Laser-induced cw luminescence spectra of beryls under 780 nm laser excitation (http://rruff.info)

color is connected with additional absorption in the violet part of the spectrum. The reason is the influence of Fe^{3+} impurities substituting for Si^{4+}. Another reason for the green color may be V^{3+} impurity with two absorption bands in the visible part of the spectrum (Platonov et al. 1984).

Natural grossular luminescence was studied by continuous wave (CW) luminescence technique and has been connected with Cr^{3+} centers, which substitutes for Al^{3+} and occupy a site with distorted octahedral symmetry (which in fact is trigonal C_{3i}). Trivalent chromium in grossular occupies high crystal field position, the vibronic $^4T_{2g}$ level is found at energies higher than ^2E_g state. The emission can occur from both levels at room and liquid helium temperatures. Based on optical absorption spectra, spectral field strength D_q and Racah parameters B and C have been calculated: $10D_q = 17{,}460\ cm^{-1}$, $B = 670\ cm^{-1}$ and $C/B = 4.78$. The luminescence spectra of Cr^{3+} in grossular at room temperature contain a strong broad band peaking at 720 nm and three sharp lines centered at 697, 700 and 701 nm. The decay times of both R-lines were found to be equal within the experimental error for the whole range of temperatures as expected in the case of two thermalized levels. This indicates that the non radiative transitions between them are much faster than any other radiative transitions in the system (Czaja and Mazurak 1994). Later

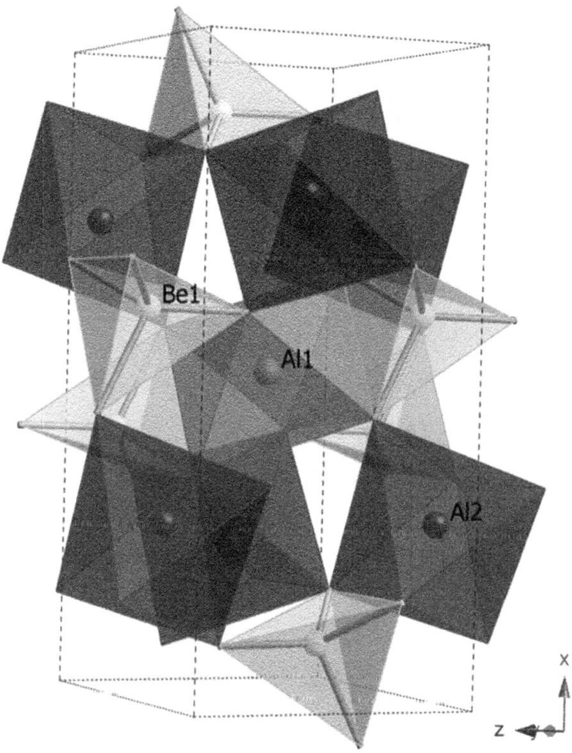

Fig. 4.126 Crystal structure of chrysoberyl showing the two Al (CN6) and Be (CN4) sites

Table 4.20 Concentrations of Mn, Fe, Cr, V, Fe, Ni and Ti (ppm) in chrysoberyl

Info	Ti	V	Cr	Mn	Fe	Ni
Siberia	2300	200	7150	255	31,650	870
13.11.	3400	145	600	560	21,000	770
11.11	5500	220	340	225	39,700	1130
5.04.	2600	225	520	255	16,000	1150
Brasilia	5500	185	1170	165	21,000	1530
US	9100	55	450	180	23,000	1130
4.3	7470	740	6600	210	58,000	2400
Synt + V	12,000	22,000	4400	535	11,000	4500

luminescence of green grossular, tsavorite variety, was studied (Mazurak and Czaja 1996), where vanadium content is usually substantially higher than Cr and whose color was ascribed to V^{3+} absorption (http://minerals.gps.caltech.edu). The same luminescence of Cr^{3+} in strong crystal field position was found, accompanied by broad IR band peaking at 1205 nm, which was ascribed to V^{3+}.

Luminescence of artificial garnets activated by different emission centers, such as Mn^{3+} (Kuck et al. 1998) and Mn^{4+} (Brenier et al. 1992), have been well studied. Besides, optical absorption spectra of grossular connected with Mn and V impurities have been analyzed and interpreted (http://minerals.gps.caltech.edu; Geiger

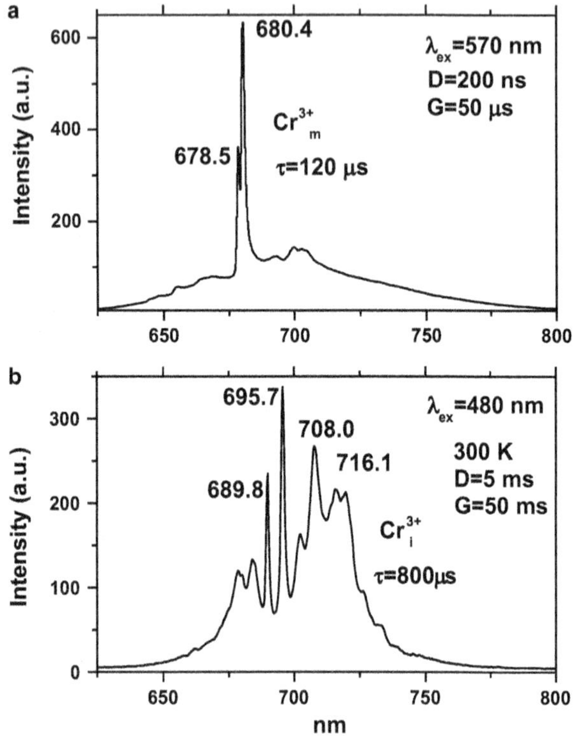

Fig. 4.127 (**a–b**) Laser-induced time-resolved luminescence spectra of chrysoberyl demonstrating different Cr^{3+} centers

et al. 1999). It represents a substantial basis for natural grossular luminescence research enabling comparison between emission parameters detected in minerals with artificial garnet activated by specific impurity.

The natural grossular in our study consisted of ten samples with different colors. Concentrations of several potential impurities luminogens are presented in Table 4.21. Laser-induced time-resolved technique enables to detect trivalent REE, Mn^{2+}, Mn^{3+}, Mn^{4+}, and possibly Cr^{3+}, V^{2+} and Ni^{2+} emission centers (Gaft et al. 2013b) (Figs. 4.130 and 4.131).

Excitation by CW laser with 785 nm revealed IR luminescence lines which evidently may be ascribed to trivalent REE, such as Pr, Ho, Nd, Er and Yb (Fig. 4.132). Excitation by CW laser of uvarovite (Cr-grossular) with 532 and 780–785 nm revealed several luminescence lines which evidently may be ascribed to trivalent REE, such as Pr, Ho, Nd, Er and Yb and red bands peaking at 720–740 and 900 nm (Fig. 4.133).

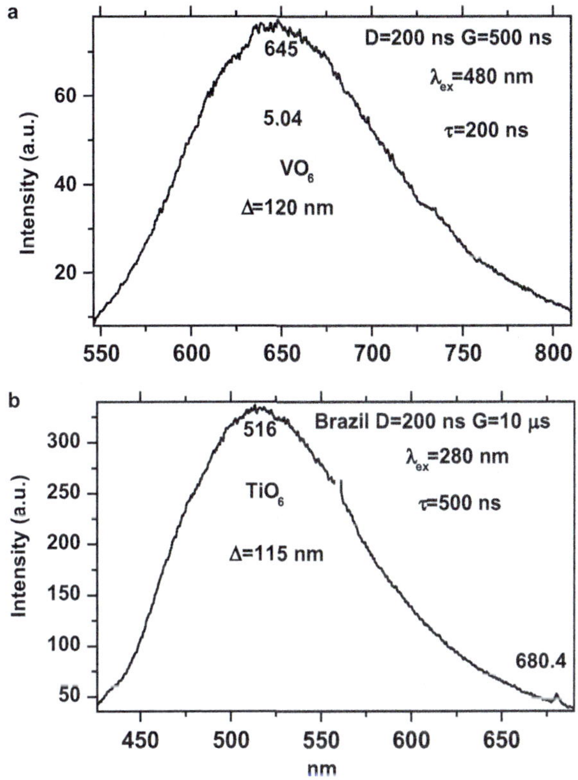

Fig. 4.128 (**a–b**) Laser-induced time-resolved luminescence spectra of chrysoberyl demonstrating possibly metal-oxygen complexes

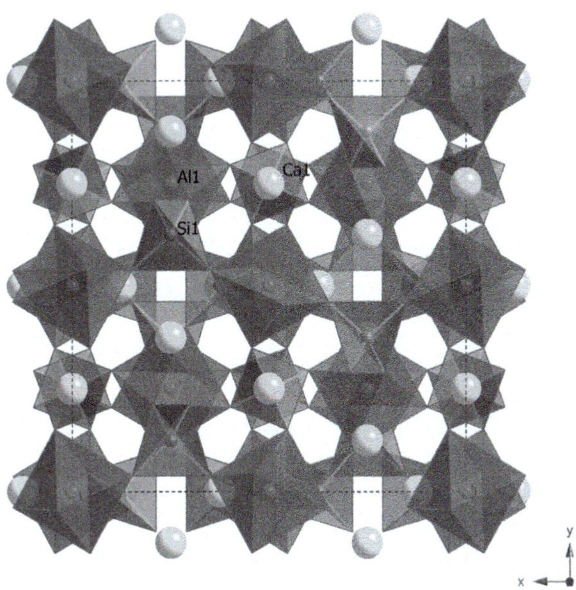

Fig. 4.129 Crystal structure of grossular showing the Al octahedral sites and Ca (CN8) ones

Table 4.21 Concentrations of Mn, Fe, Cr, V, Fe, Ni and Ti (ppm) in grossular

Info	Ti	V	Cr	Mn	Fe	Ni
Mexico	140	2.2	4.7	400	585	17
Unknown	170	26	23	135	140	26
2.77	175	33	23	190	125	20

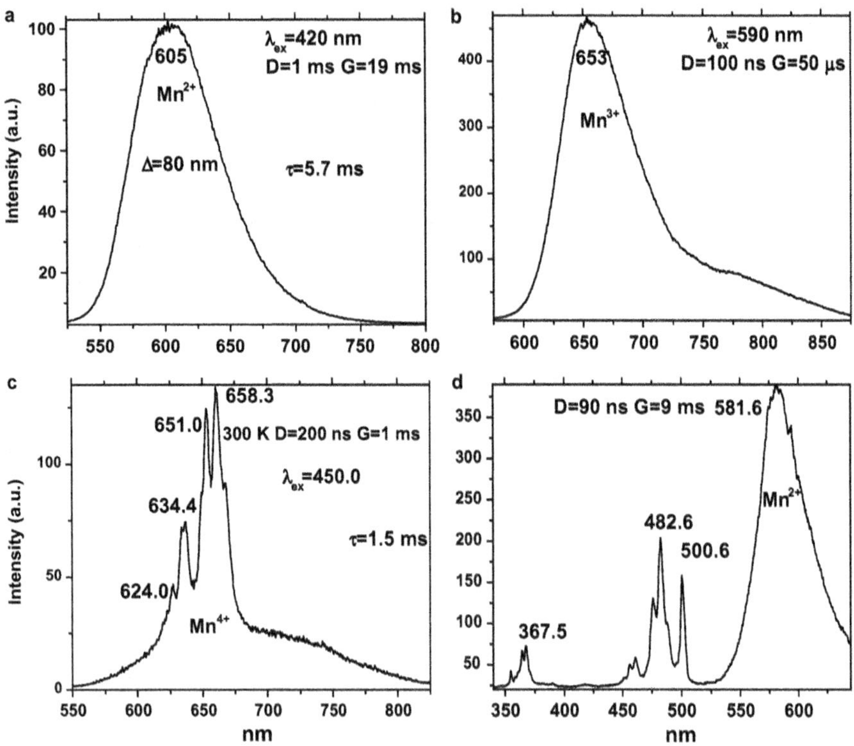

Fig. 4.130 **(a–d)** Laser-induced time-resolved luminescence spectra of grossular demonstrating REE^{3+}, Mn^{2+}, and possibly Mn^{4+} and Mn^{3+} centers

4.10.8 *Tourmaline (Na,Ca)(Mg,Al, Li)$_3$Al$_6$ × (BO$_3$)$_3$Si$_6$O$_{18}$OH*

Tourmaline has a rhombohedral unit cell. The six silicon atoms are each surrounded tetrahedrally by four oxygen atoms, each tetrahedron sharing two of its oxygens with the neighboring tetrahedra to form a six-membered ring of composition Si_6O_{18}: the Al ions are coordinated to six oxygen atoms in a very distorted octahedron.

Under lamps excitation tourmaline is practically non-luminescent, while under X-ray excitation it exhibits impurity luminescence from Fe^{3+} centered at

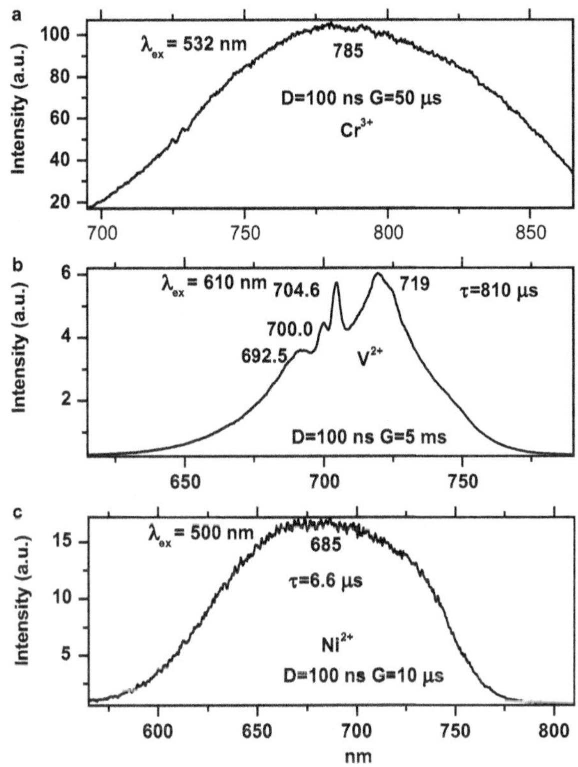

Fig. 4.131 (**a–c**) Laser-induced time-resolved luminescence spectra of grossular demonstrating possibly Cr^{3+} (**a**), V^{2+} (**b**) and Ni^{2+} (**c**) centers

700–750 nm and Mn^{2+} centered at 560–570 nm (Kusnetsov and Tarashchan 1988). The natural tourmaline in our study consisted of four samples. Laser-induced time-resolved technique enables to detect Cr^{3+} emission centers (Fig. 4.134).

Excitation of elbaite by CW laser with 532 nm revealed additional luminescence line at 652 and 658 nm and bands peaking at 617, 635, 684, 704 and 738 nm (Fig. 4.135).

Excitation of tourmaline by CW laser with 532 nm revealed luminescence lines at 682 nm and bands peaking at 735 and 765 nm (Fig. 4.136).

4.10.9 *Epidote (Zoisite) $Ca_2AlAl_2O(OH)(Si_2O_7)(SiO_4)$*

Epidote is the group name for a family of minerals of general composition $Ca_2(Fe^{3+}$, Al, $Mn^{3+})Al_2O[SiO_4][Si_2O]$-(OH). Epidote (octahedral Fe^{3+}) and clinozoisite (Al) represent the most common compositions among the epidote group. Zoisite is equivalent to clinozoisite, but it has a different crystalline system. Rare epidote-clinozoisites abundant in Cr, V and Pb also exist. The epidote structure consists of

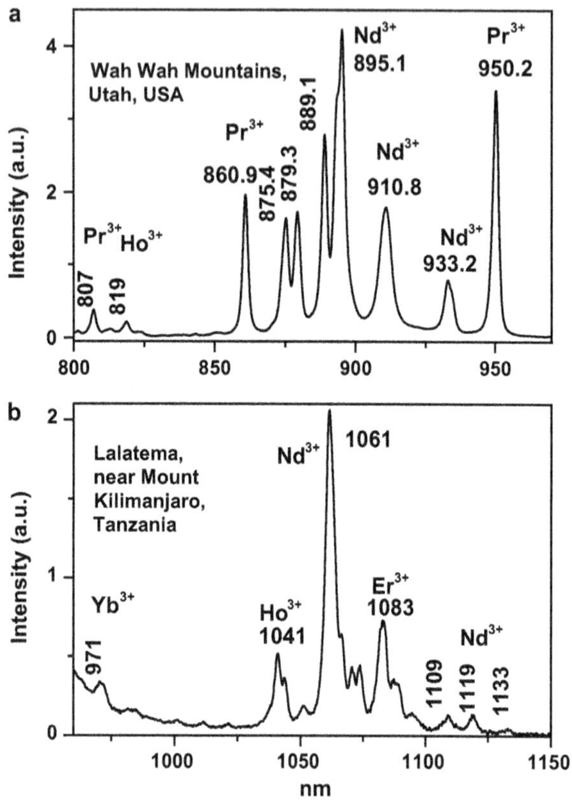

Fig. 4.132 (**a–b**) Laser-induced cw luminescence spectra of grossular under 785 nm laser excitation (http://rruff.info)

chains of octahedra (M sites) elongated parallel to the *y* crystallographic axis and cross-linked by the silicate groups SiO_4 and Si_2O_7. There are three distinct kinds of octahedra (M1, M2, and M3) arrayed along two types of chains. One chain type consists entirely of edge-sharing octahedra (M2), whereas the other contains M1 octahedra alternating with M3 octahedra along its length. The resulting framework structure contains large cavities (A sites) where the Ca^{2+} cations are housed in 9–10-fold coordination. Like the chains of octahedra, epidote crystals themselves are also elongated parallel to the *y*-axis. In typical epidote-clinozoisites, Fe^{3+} and Mn^{3+} substitute for Al principally in M3 sites; M2 sites house only Al. M1 sites are usually dominated by Al but may contain Mn^{3+}. The explanation for the preference of these cations for the M3 site is twofold. First, both cations are larger than Al^{3+}, and the M3 octahedron is the largest in the structure. Second, the M3 sites are relatively distorted, and this characteristic may permit Mn^{3+} to gain additional stabilization energy here, over the other M sites (Fig. 4.137).

The optical absorption and luminescence properties of zoisite have been studied (Koziarsca et al. 1994). It was found that the strong absorption bands around 750 and 585 nm are connected with transitions between the ground $^3T_1$ state and first excited state $^3T_2$ of V^{3+} ions at Al(II) and Al(I) sites, respectively. The band

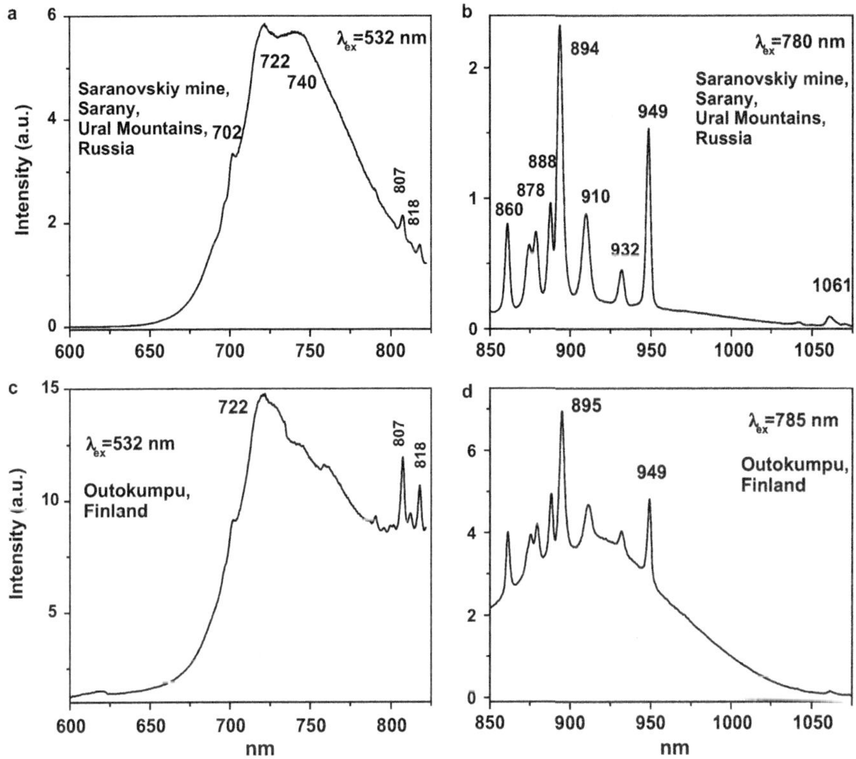

Fig. 4.133 **(a–d)** Laser-induced cw luminescence spectra of uvarovite under 532 and 780–785 nm laser excitations (http://rruff.info)

peaking at 545 nm was associated with transition from the $^3T_1$ ground state to the $^1A_1$ states for both of these crystallographic sites. The possible presence of V^{2+} ions was also proposed. The strong line at 709 nm in the low-temperature luminescence spectra has been assigned to transitions between the ground and the lower component of the first excited level ^2E (R_1-line) in the $3d^3$ shell. The line at 692 nm, observed in the luminescence at higher temperatures was associated with luminescence from the upper component (R_2 line) of the ^2E level. At room temperature the luminescence from ^2E level overlaps with a broad band of vibronic origin. The other line at 704 nm was also observed and connected to vibronic origin. Decay time of those lines at room temperature was 260 μs. The possible ions with $3d^3$ shell were supposed as Cr^{3+} and V^{2+}. The weak low-temperature IR emission band was connected with V^{3+} luminescence. The natural zoisite in our study consisted of one sample. Laser-induced time-resolved technique enables to detect Eu^{2+}, Dy^{3+}, Tb^{3+}, Sm^{3+} and possibly V^{2+} emission centers (Fig. 4.138). Excitation of zoisite and epidote by CW laser with 532 and 785 nm revealed luminescence IR luminescence lines evidently belonging to Nd^{3+}, red lines belonging to Cr^{3+} and broad red bands of unknown origin (Fig. 4.139).

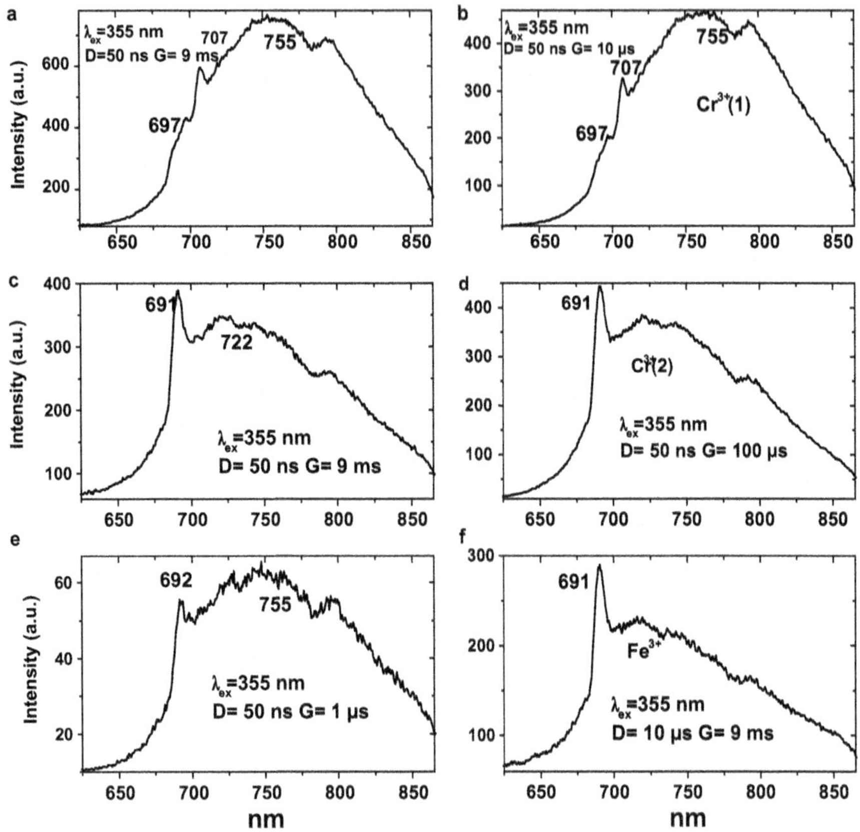

Fig. 4.134 (a–f) Laser-induced time-resolved luminescence spectra tourmaline demonstrating Cr^{3+}, and Fe^{3+} centers

4.10.10 *Eosphorite $AlPO_4Mn(OH)_2H_2O$*

Mineral eosphorite has been studied by Vergara et al. (1990). The crystal structure of eosphorite contains two kind of octahedral. One of these is composed of $MnO_4(OH)_2$ sharing opposite O-O edges and the other of $AlO_2(OH)_2(H_2O)_2$ sharing opposite H_2O corners. Absorption spectra are characterized by a number of bands, which have been associated with transitions of Mn^{2+}. All photoluminescence was connected with Cr^{3+} ions (Fig. 4.140).

Excitation of eosphorite by CW laser with 532 nm revealed several emission bands and lines peaking at approximately 646 nm (Fig. 4.141). Luminescence of Cr^{3+} and Mn^{4+} may be supposed.

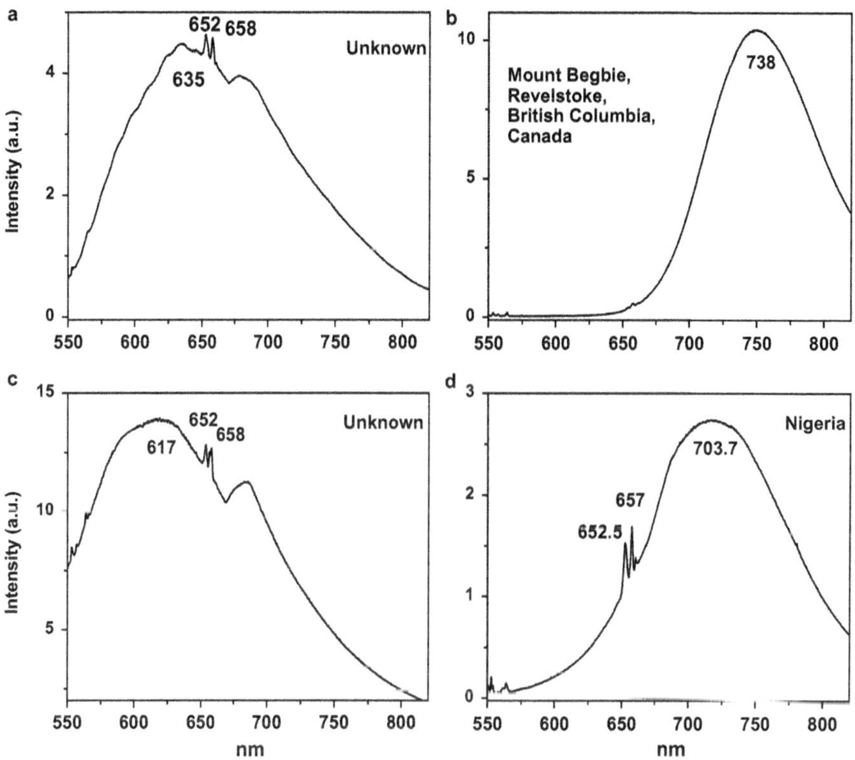

Fig. 4.135 (**a–d**) Laser-induced cw luminescence spectra of elbaite under 532 nm laser excitation (http://rruff.info)

4.10.11 Spodumen $LiAlSi_2O_6$

Spodumen is a monoclinic pyroxene, space group C_{2h}^6(C2/c), with two not equivalent metal cation sites M1 and M2. The aluminum occupies the smaller M1 site, which is approximately octahedral (actual symmetry C_2) with an average metal-oxygen distance of 1.92 A. The M2 site, occupied by Li, is also six-fold coordinated with an average metal-oxygen distance of 2.23 Å. Both Al and Li sites may be substitutionally replaced by ions of the transitional metals in various proportions. Both Mn^{2+} and Cr^{3+} centers have been identified in luminescence spectra by steady-state spectroscopy (Tarashchan 1978; Walker et al. 1997). At room and lower temperatures only one emission band of Mn^{2+} occurs and the excitation spectra taken for the different wavelengths of the luminescence bands are always the same. So it is very probable that Mn^{2+} ions in the spodumen matrix present only in one site. The calculated values of 10 Dq and B are consistent with the occupation of larger M2(Li) weak-field site. Mn^{2+} is mainly in Li-sites rather than Al-sites.

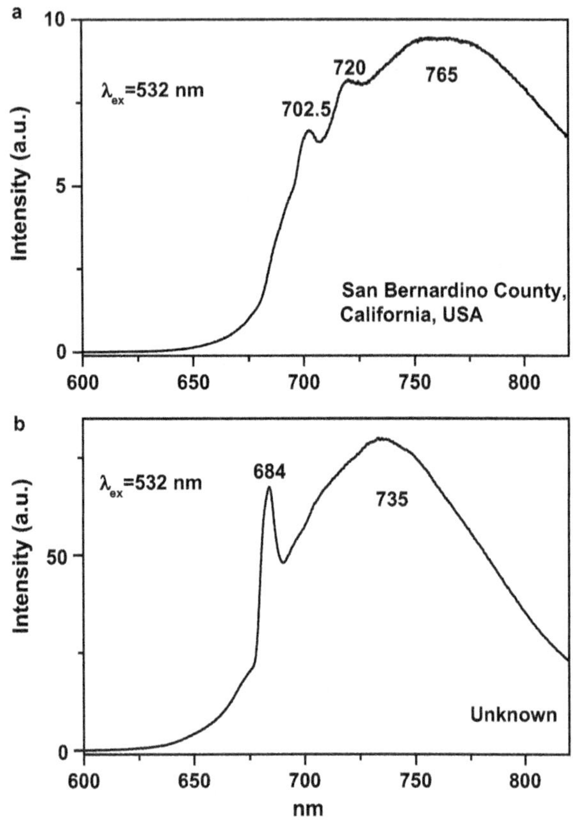

Fig. 4.136 (**a–b**) Laser-induced cw luminescence spectra of tourmaline under 532 nm excitation (http://rruff.info)

Color of spodumen can vary from pale pink for the kunzite variety to a deep green for the hiddenite variety, but colorless crystals also exist. Green color of hiddenite is connected with Cr^{3+} substituting for Al^{3+}, while V^{3+} also takes part in optical absorption spectrum formation. Green-yellow color of spodumen is connected with Fe^{3+} in M1 position and Fe^{2+} in M2 position. Pink color of kunzite is connected with Mn^{3+} impurities in different structural positions M1 and M2 (Platonov et al. 1984).

The spodumen in our study consisted of three samples. Laser-induced time-resolved technique enables to detect Mn^{2+}, Cr^{3+} and supposedly TiO_6 molecular emission. Besides, long lived deep red luminescence was found supposedly generated by Fe^{3+} (Fig. 4.142).

4.10.12 *Boehmite and Diaspor AlOOH*

The structure of boehmite contains double sheets of octahedra with Al ions at their centers, and sheets themselves are composed of chains of octahedra. In diaspor the

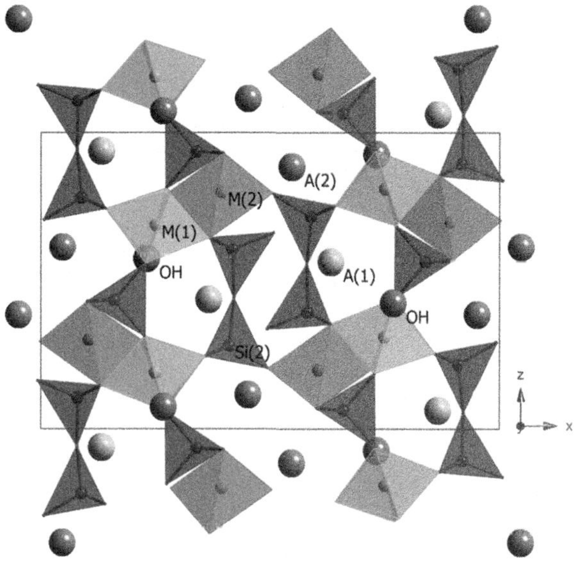

Fig. 4.137 Crystal structure of zoisite (tanzanite) showing the two Al (M1 and M2) sites and the two Ca ones (A1 and A2)

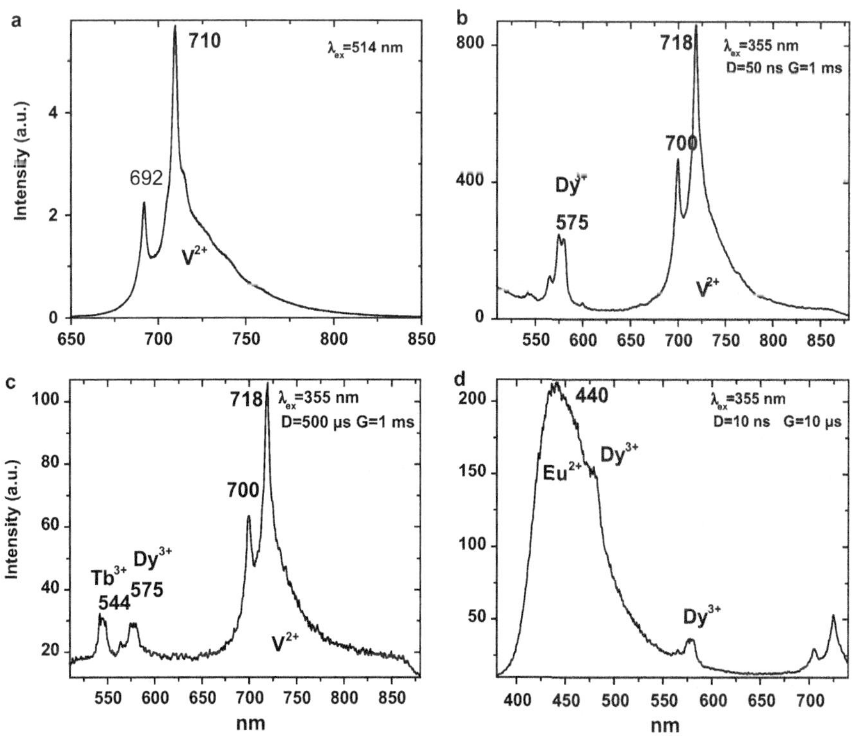

Fig. 4.138 (**a–d**) Laser-induced steady-state (**a**) and time-resolved (**b–d**) luminescence spectra of zoisite demonstrating Cr^{3+}, Eu^{2+}, Tb^{3+}, Dy^{3+} and possibly V^{2+} centers

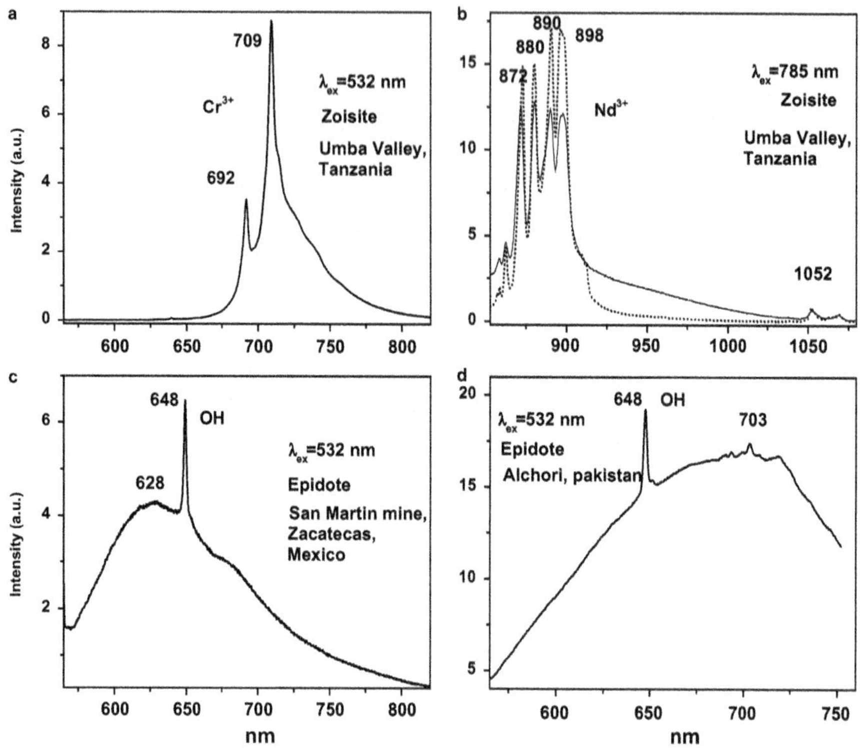

Fig. 4.139 (**a–d**) Laser-induced cw luminescence spectra of zoisite and epidote under 532 and 785 nm excitations (http://rruff.info)

oxygens are in hexagonal close packed layer; those within the double octahedral layers in boehmite are in cubic packing relationship. Luminescence center Cr^{3+} characterizes steady-state spectra (Solomonov et al. 1994; Shoval et al. 2003). The natural boehmite and diaspor in our study consisted of 12 samples. Laser-induced time-resolved technique enables to detect Cr^{3+} emission centers (Figs. 4.143 and 4.144).

4.10.13 Chlorite $(Mg,Al,Fe)_{12}[(Si,Al)_8O_{20}](OH)_{16}$

The chlorites are a group of minerals with layered structure, which in many respects resemble the micas. The monoclinic structure of chlorite is one of regularly alternating talc-like $Y_6Z_8O_{20}(OH)_4$ and brucite-like $Y_6(OH)_{12}$ sheets.

For chromium containing lilac chlorite two types of luminescence were observed at 15 K: phosphorescence at 14,518 and 1547 cm^{-1} with decay time of 60 μs and fluorescence band at about 13,850–13,500 cm^{-1} with decay time of

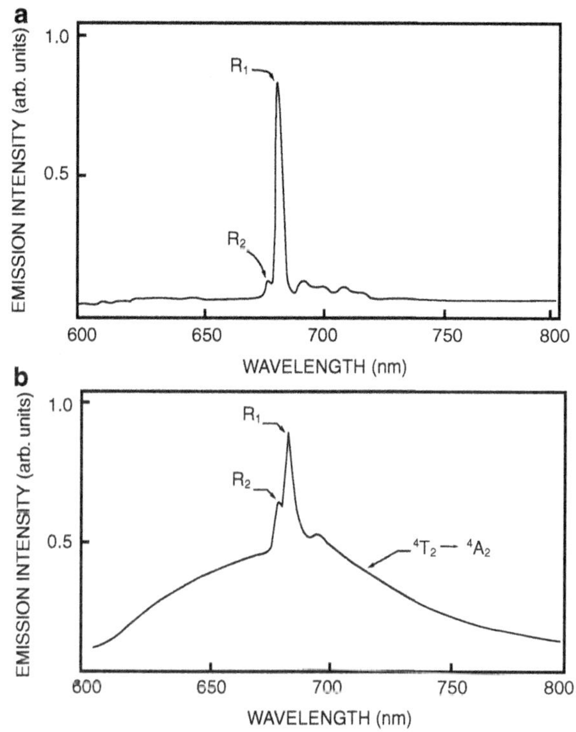

Fig. 4.140 (**a–b**) Laser-induced luminescence of eosphorite at (**a**) 12 K and (**b**) 300 K (Vergara et al. 1990)

several microseconds. For green chlorite weak fluorescence at 13,900 cm^{-1} and phosphorescence at 14,320 and 14,665 cm^{-1} were observed at room temperature. At low temperature phosphorescence lines at 14,580 and 14,676 cm^{-1} were observed on the short wave range of vibronic band (Fig. 4.145). Such emission has been connected with Cr^{3+} present in different sites in the brucite-like layers (Czaja 1999). It is very interesting that iron ions (1.54 wt%) did not quench Cr^{3+} emission. The same was detected in iron containing chlorites under X-ray excitation (Kusnetsov et al. 1991).

4.10.14 Clintonite $Ca(Mg,Al)_3(Al_3Si)O_{10}(OH)_2$

Excitation by CW laser with 532 nm revealed luminescence lines at 681.3, 685.b, 719.3 and 723.6 nm, which may be ascribed to Cr^{3+} and band peaking at 760 nm preliminary connected to Fe^{3+} (Fig. 4.146).

4.10.15 Cordierite $Mg_2Al_3[AlSi_5O_{18}]$

Steady state luminescence of cordierite was connected to Mn^{2+} substituting for Mg^{2+}, Fe^{3+} substituting for Al^{3+} or Si^{4+} and Cr^{3+} substituting for Al^{3+} (Gorobets and

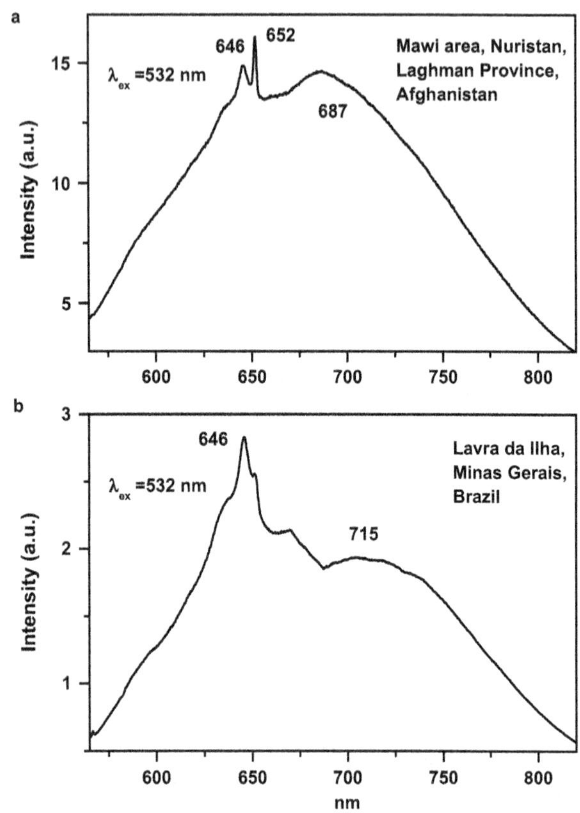

Fig. 4.141 (**a–b**) Laser-induced cw luminescence spectra of eosphorite under 532 nm excitation (http://rruff.info)

Rogojine 2001). Recently it was found that the photoluminescence spectrum of natural cordierite shows a structured band in the red spectral region with a main component centered at 14,420 cm^{-1} and a weaker one at about 12,400 cm^{-1}, by exciting at two different energies within the pleochroic band (at 15,978 and 20,492 cm^{-1}). The emission intensity has a clear dependence on the direction of the exciting beam polarization. Specifically, the emission band is more intense when the polarization of the excitation within the pleochroic absorption band is directed so as to give a larger absorption. Time resolved PL measurements give a decay time of about 5 ms, varying from 5.0 to 5.2 ms rotating the polarization within the plane orthogonal to the c-axis from the condition of maximum to minimum PL intensity, respectively. This band was ascribed to manganese ions in cordierite, specifically to Mn^{3+} in octahedral sites (Spinolo et al. 2012).

Luminescence band peaking at 650 nm range has been found under 532 nm laser excitation (http://rruff.info) (Fig. 4.147), which tentatively may be connected to Mn^{2+}.

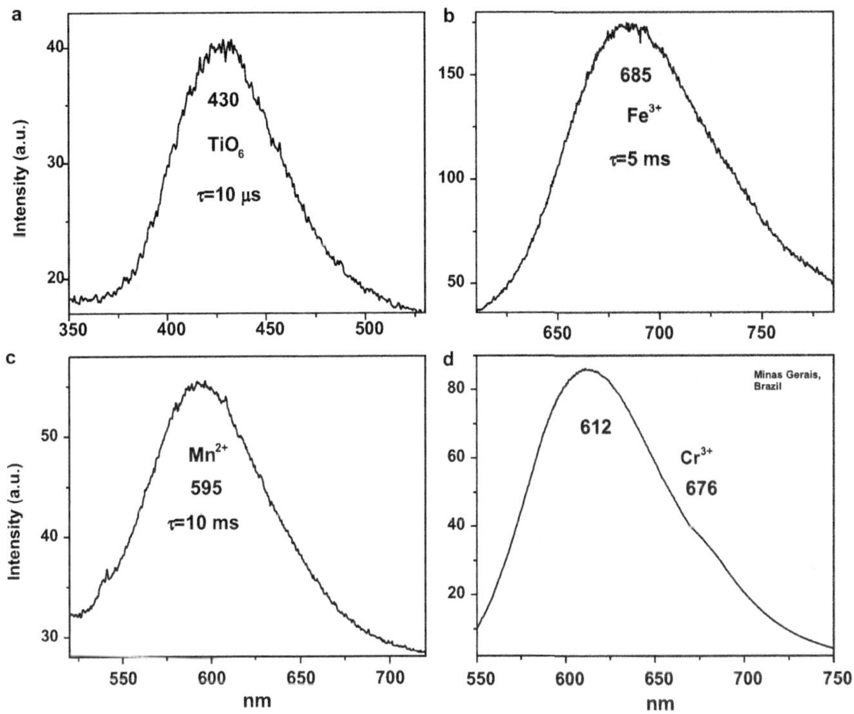

Fig. 4.142 (**a–d**) Laser-induced time resolved spectra of spodumen demonstrating TiO_6 (**a**), Fe^{3+} (**b**) and Mn^{2+} (**c**) centers. CW spectrum with weak Cr^{3+} emission (**d**) under 532 nm excitation (http://rruff.info)

4.10.16 Nepheline (Na,K)(AlSi)$_2$O$_4$

Time resolved luminescence spectra under excitation by 266 nm are characterized by the band peaking at 695 nm with decay time of 100 μs (Fig. 4.148). Principally it may be connected to Fe^{3+} or Cr^{3+} centers.

4.10.17 Jadeite NaAlSi$_2$O$_6$

Excitation by CW laser with 514 nm revealed luminescence bands peaking at 752 and 710 nm (Fig. 4.149). Under 532 and 780 nm cw excitations another two bands were found peaking at 768 and 950 nm, correspondingly (http://rruff.info).

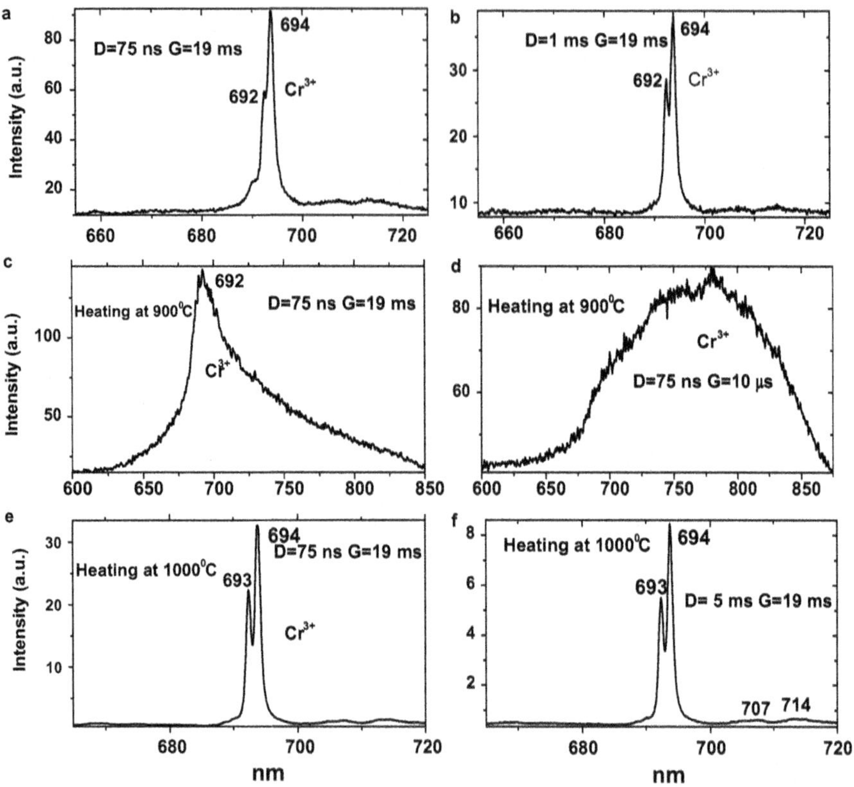

Fig. 4.143 (**a–f**) Laser-induced time-resolved luminescence spectra of diaspore demonstrating Cr^{3+}centers and its change with heating

4.10.18 Phlogopite $KMg_3AlSi_3O_{10}(F,OH)_2$

Steady state luminescence of phlogopite was connected to Mn^{2+} substituting for Mg^{2+} and Mn^{2+} at unknown site (Gorobets and Rogojine 2001). Excitation by CW laser with 532 nm revealed luminescence band peaking at 765 nm evidently belonging to Fe^{3+} center and lines at 661 and 663 nm which may be ascribed to Cr^{3+} (Fig. 4.150).

4.10.19 Clinochlore $(Mg,Fe)_5AlSi_3AlO_{10}(OH)_8$

Steady state luminescence of clinochlore was connected to Mn^{2+} and Fe^{3+} (Gorobets and Rogojine 2001). Excitation by CW laser with 532 nm revealed narrow band luminescence band which may be ascribed to Cr^{3+} (Fig. 4.151).

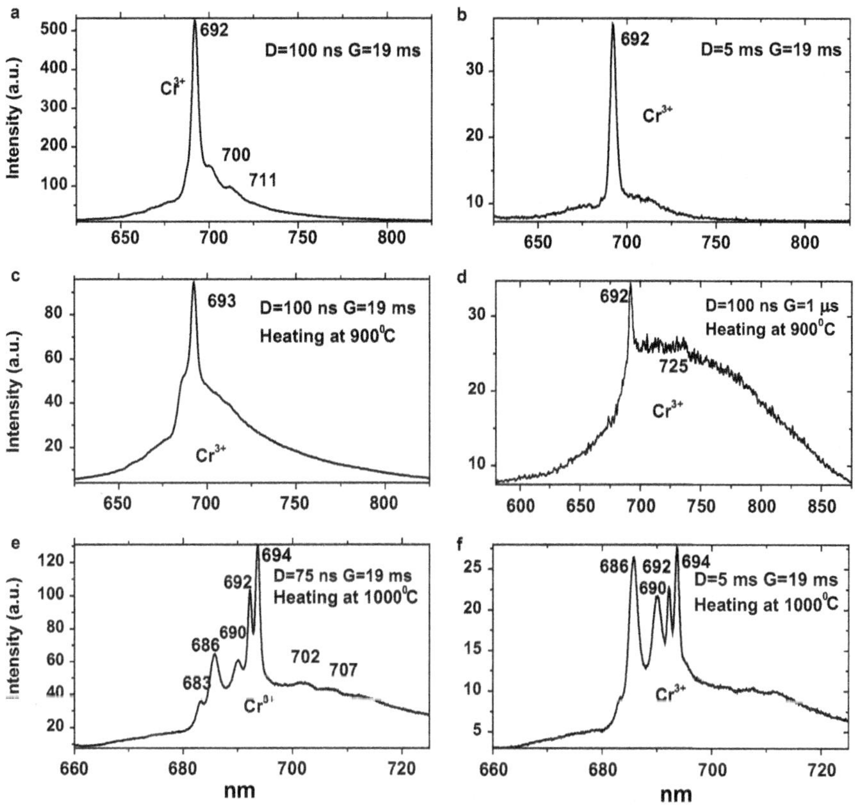

Fig. 4.144 (**a–f**) Laser-induced time-resolved luminescence spectra of boehmite demonstrating Cr^{3+}centers and its change with heating

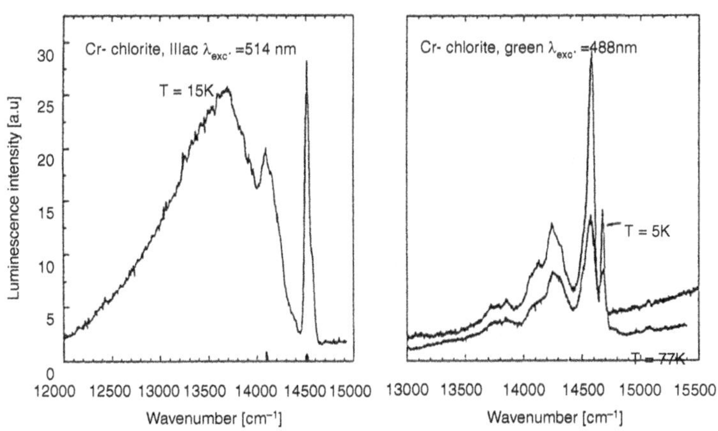

Fig. 4.145 Laser-induced luminescence of chlorite with Cr^{3+} emission (Czaja 1999)

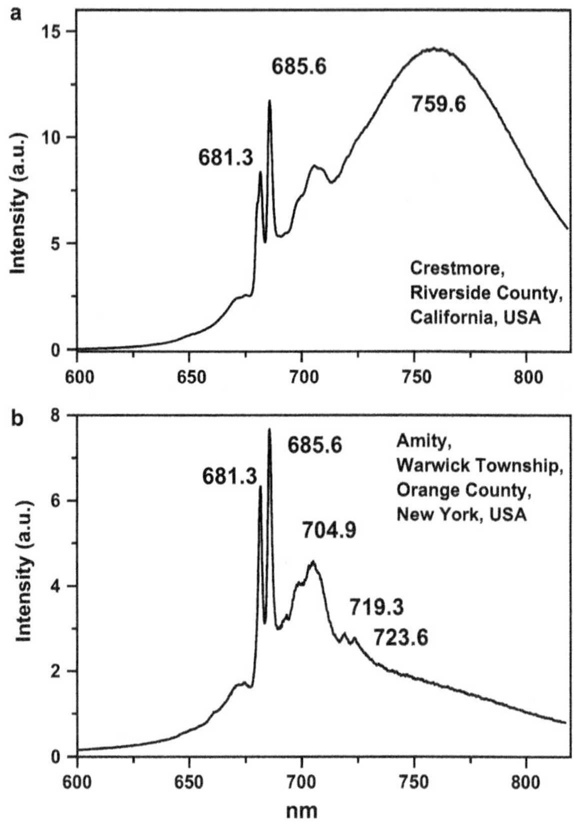

Fig. 4.146 (**a–b**) Laser-induced cw luminescence spectra of clintonite under 532 nm laser excitation (http://rruff.info) with evidently Cr^{3+} and Fe^{3+} centers

4.10.20 Clinohumite $(Mg,Fe)_9(SiO_4)_4(F,OH)_2$

Steady state luminescence of clinohumite was connected to Mn^{2+} and possible TiO_6 centers (Gorobets and Rogojine 2001). Excitation by CW laser with 532 nm revealed two emission bands which may be ascribed to Mn^{2+} and Fe^{3+} (Fig. 4.152).

4.10.21 Euclase $BeAlSiO_4(OH)$

Excitation by CW laser with 532 nm revealed narrow line and broad band which may be ascribed to Cr^{3+} and Fe^{3+} (Fig. 4.153).

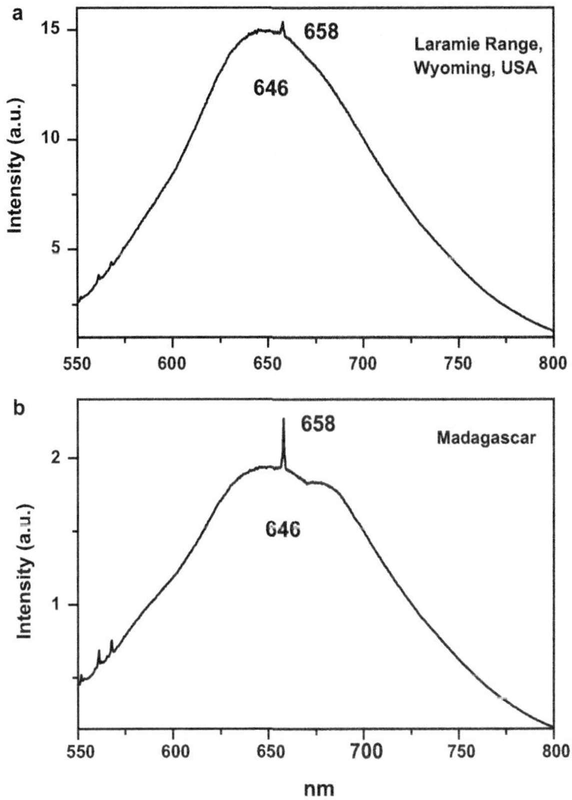

Fig. 4.147 (**a–b**) Laser-induced cw luminescence spectra of cordierite under 532 nm laser excitation (http://rruff.info) with evidently Cr^{3+} and Mn^{2+} centers

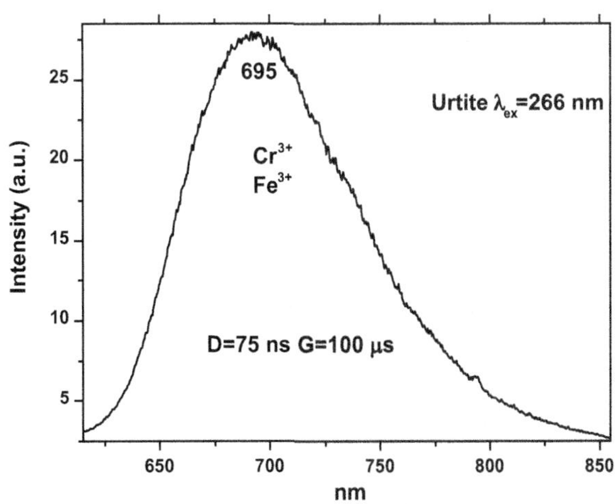

Fig. 4.148 Laser-induced luminescence of nepheline with Cr^{3+} or Fe^{3+} emission

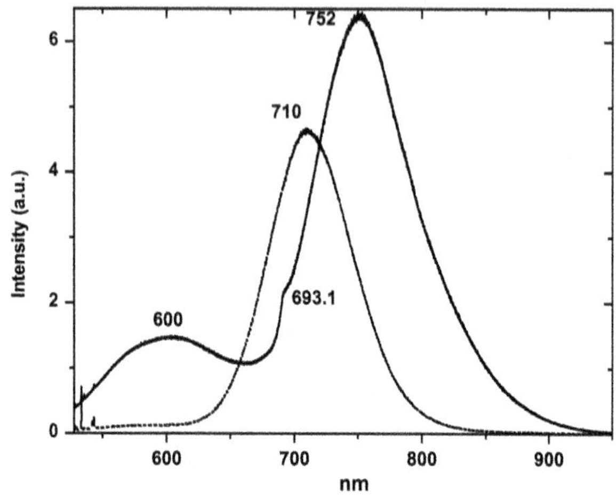

Fig. 4.149 (**a–b**) Laser-induced cw luminescence spectra of jadeite under 514 nm laser excitation

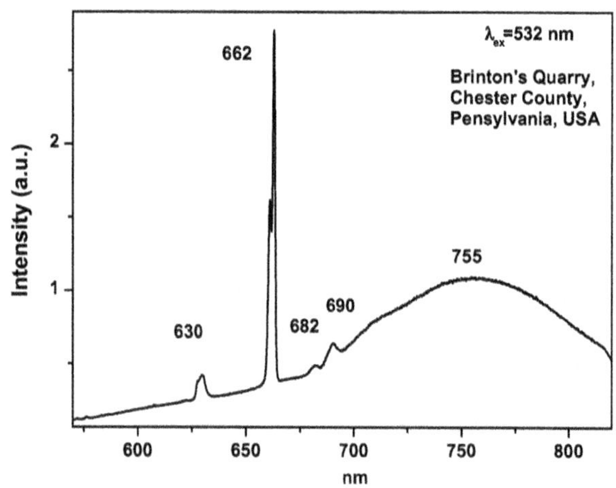

Fig. 4.150 (**a–b**) Laser-induced cw luminescence spectra of phlogopite under 532 nm laser excitation (http://rruff.info) with evidently Cr^{3+} and Fe^{3+} centers

4.10.22 Dickite $Al_2Si_2O_5(OH)_4$

Excitation by CW laser with 532 nm revealed narrow lines and broad band which may be ascribed to Cr^{3+} and Fe^{3+} (Fig. 4.154).

4.11 Sodium (Na^+) Bearing Minerals

The main substituting luminescence center are Mn^{2+}, Fe^{3+} and REE (ionic radius of $Na^+ = 1.16$ Å). It is interesting to note that Na bearing minerals are very often characterized by luminescence connected with S_2^- and O_{2-} impurities.

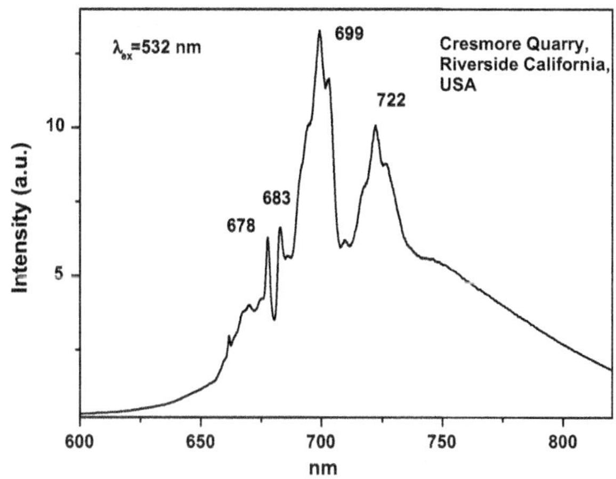

Fig. 4.151 Laser-induced cw luminescence spectra of clinochlore under 532 nm excitation (http://rruff.info) with evidently Cr^{3+} centers

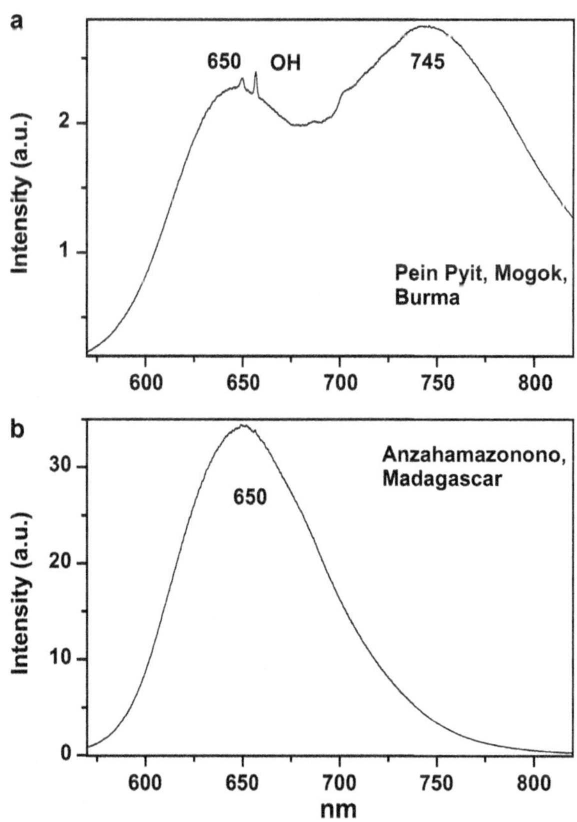

Fig. 4.152 (**a–b**) Laser-induced cw luminescence spectra of clinohumite under 532 nm excitation (http://rruff.info) with evidently Mn^{2+} and Fe^{3+} centers

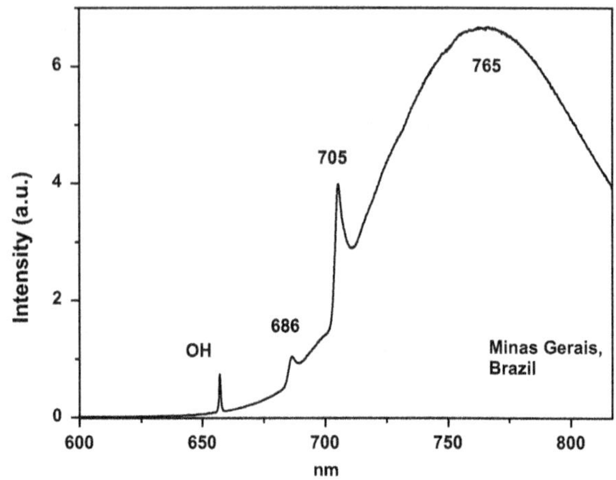

Fig. 4.153 Laser-induced cw luminescence spectra of euclase under 532 nm excitation (http://rruff.info) with evidently Cr^{3+} and Fe^{3+} centers

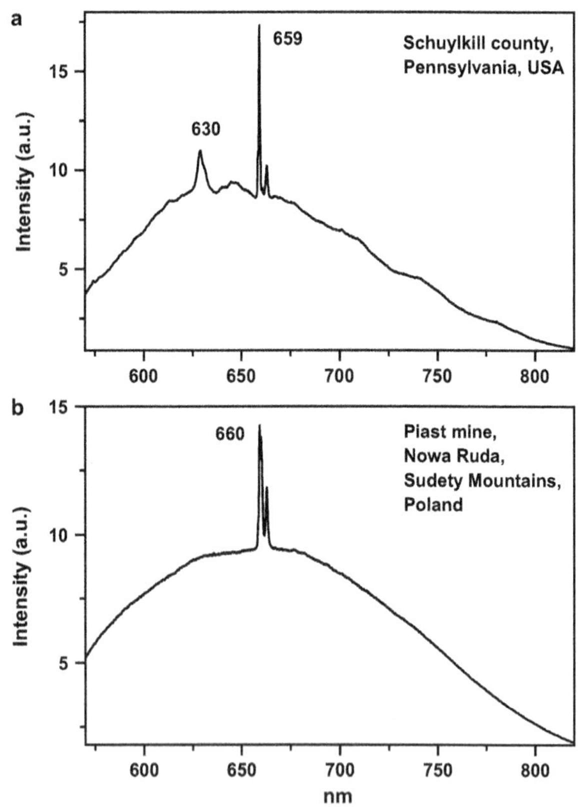

Fig. 4.154 (**a–b**) Laser-induced cw luminescence spectra of dickite under 532 nm excitation (http://rruff.info) with evidently Cr^{3+} and Fe^{3+} centers

4.11.1 *Sodalite $Na_8[Al_6Si_6O_{24}]Cl_2$, Hackmanite $Na_8[Al_6Si_6O_{24}](Cl_2,S)$ and Tugtupite $Na_4AlBeSi_4O_{12}C$*

Sodalite and tugtupite are members of the silica deficient feldspathoid mineral group, while 'hackmanite' is sulfur bearing photochromic variety of sodalite. These minerals may change color when exposed to sunlight or UV light, an unusual optical property known as tenebrescence, or reversible photochromism. It has been known for a long time that natural sodalite and tugtupite are fluorescent under UV excitation.

Despite the very diverse luminescence properties of tugtupite and sodalite, their spectroscopic study is relatively limited compared to other famous luminescent minerals, such as apatite, fluorite, scheelite, calcite and zircon. Luminescence data on tugtupite are extremely scarce. Actually only one spectrum of tugtupite from Ilı'maussaq at 77 K was published (Povarennykh et al. 1971), which later was reproduced by Tarashchan (1978), Gorobetz and Rogojine (2001). It is generally accepted that the yellow-orange emission of tugtupite with weak vibrational structure at 300 K, which becomes more prominent at low temperatures, is connected to a S_2^- luminescence center (Povarennykh et al. 1971). The research history of the yellow-orange luminescence of sodalite was well presented by Sidike et al (2007). It may be summarized that published data on the yellow-orange luminescence spectra of sodalite are more abundant compared to tugtupite. It was also ascribed to S_2^-, which was confirmed by the study of synthetic analogs activated by sulfur. Nevertheless, Sidike et al. (2007) noted that the peak wavelengths in the yellow orange band reported by many investigators for sodalite and ascribed to S_2^- do not agree. It was never reported by single author on several types of S_2^- luminescence spectra from the same deposit, which may be explained by different chemical composition and impurities, but different spectra were ascribed to the S_2^- luminescence center in sodalite samples from the same deposit. The values for luminescence maxima were reported to range from 630 nm to 708 nm. A similar situation takes place in scapolite, where the peak wavelengths of S_2^- emission bands reported by different authors do not agree (Sidike et al. 2008). It was proposed that the main reason for these discrepancies is probably the unsuitable correction of the measured spectra. Nevertheless, it is difficult to assume that such strong discrepancies may be explained only by the absence of proper calibration. A possible explanation for the differences between published spectra may be the presence of several different S_2^- centers, or the presence of the luminescence centers with vibrational structure of other types.

The crystal structure of sodalite is an ordered framework of linked AlO_4 and SiO_4 tetrahedra in which Si and Al alternate on the tetrahedral sites. The overall linkage of $(Al,Si)O_4$ tetrahedra results in cubo-octahedral cavities, which contain a centrally placed anion coordinated tetrahedrally to four cations. The flexibility of the sodalite structure allows a wide range of cations and anions with potential luminescence ability to be substituted into it. Thus, many impurities with potential luminescence ability besides S_2^- -may be present in the sodalite structure, such as

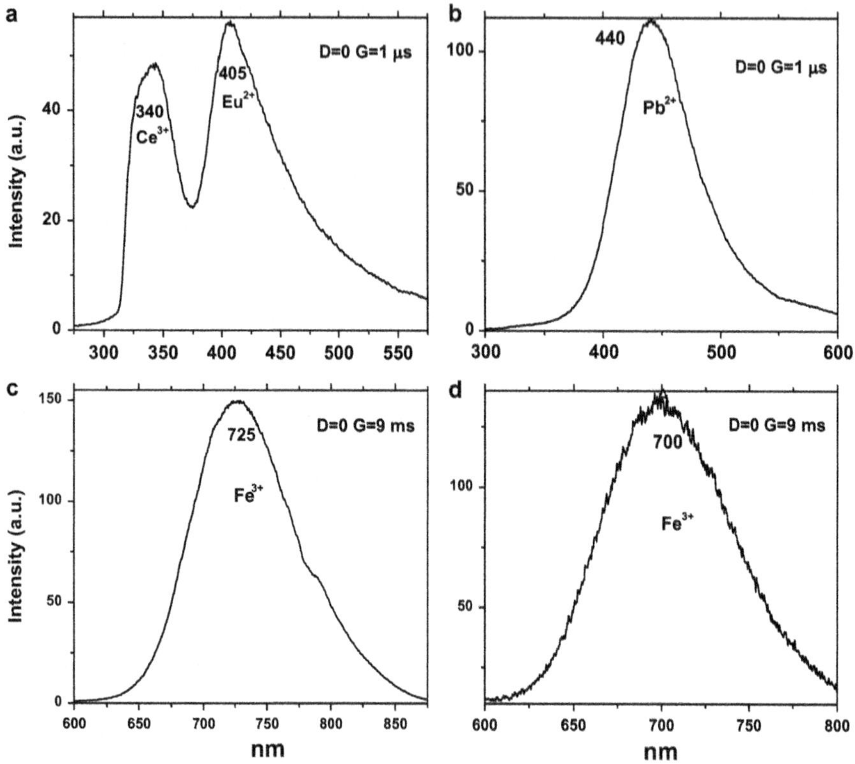

Fig. 4.155 (**a–d**) Laser-induced time-resolved luminescence of Eu^{2+} and Ce^{3+}, two different Fe^{3+} and possibly Pb^{2+} in sodalite

Fe^{3+} in Al and Si positions, Mn^{2+} in Na or Be positions, Pb^{2+} or Tl^{+} in the Na position and so on. Indeed we found that tugtupite and sodalite are characterized by red, green, blue and violet luminescence colors, besides the yellow-orange ascribed to a S_2^- luminescence center.

The natural sodalite in our study consisted of 15 samples. Laser-induced time-resolved technique enables to detect Eu^{2+} and Ce^{3+}, two different Fe^{3+} and possibly Pb^{2+} (Gaft et al. 2009) (Fig. 4.155) and S_2^- emission centers (Fig. 4.156).

Excitation by CW laser revealed luminescence bands peaking at 700 and 710 nm with excitation by 532 nm and 920 and 945 nm excited by 780 nm (Fig. 4.157).

The natural hackmanite in our study consisted of one sample. Laser-induced time-resolved technique enables to detect Fe^{3+} or Cr^{3+}, possibly Pb^{2+} (Fig. 4.158) and S_2^- emission centers (Fig. 4.159).

The natural tugtupite in our study consisted of ten samples. Laser-induced time-resolved technique enables to detect Eu^{2+} and Ce^{3+}, two different Mn^{2+} and S_2^- emission centers (Fig. 4.160).

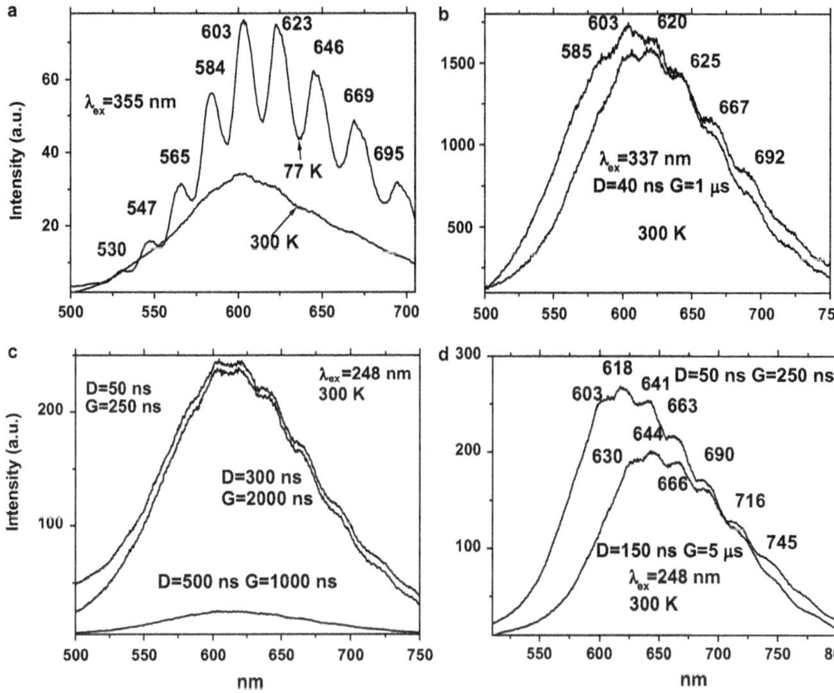

Fig. 4.156 (a–d) Laser-induced time-resolved luminescence of S_2^- in sodalite

4.11.2 Lazurite $Na_6Ca_2(Al_6Si_6O_{24})(SO_4,S_3,S_2,Cl,OH)_2$

Lazurite CW luminescence was firstly studied many years ago and emission of S_2^- centers was with vibrational structure even at room temperature was identified by Tarashchan (1978). Our study of S_2^- revealed that this center has very fast decay of approximately 200–500 ns enabling to quench it by relatively small delay time and to detect other previously hidden luminescence centers. This approach revealed, besides S_2^- emission two broad bands not definitely identified yet, but the last one is evidently belongs to Mn^{2+} based on its spectral position and form and long decay time (Fig. 4.161) Excitation by CW laser at 532 and 780 nm revealed luminescence evidently Nd^{3+}, Fe^{3+}, $S_2^-$and Fe^{3+} centers (Fig. 4.162).

4.11.3 Scapolite $Na_4Al_3Si_9O_{24}Cl$

The emission and excitation spectra of yellow luminescence due to S_2^- in scapolite were observed at 300, 80 and 10 K (Sidike et al. 2008). Emission and excitation bands at 10 K showed vibronic structures with a series of maxima spaced 15–30 and

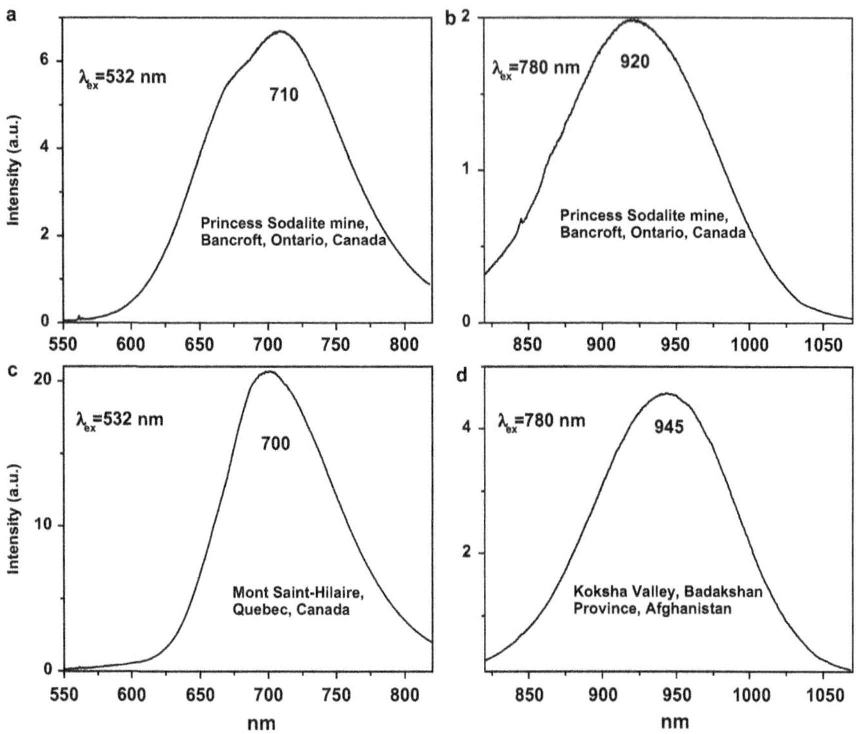

Fig. 4.157 (**a–d**) Laser-induced cw luminescence spectra of sodalite under 532 nm laser excitation (http://rruff.info)

5–9 nm, respectively. The relative efficiency of yellow luminescence from one scapolite sample was increased up to 117 times by heat treatment at 1000 °C for 2 h in air. The enhancement of yellow luminescence by heat treatment was ascribed to the alteration of SO_3^{2-} and SO_4^{2-} to S_2^- in scapolite.

Excitation by CW laser at 532 nm revealed several luminescence bands which may be supposedly connected to Mn^{2+} and Fe^{3+} centers (Fig. 4.163).

4.11.4 *Analcime* $NaAlSi_2O_6 \cdot (H_2O)$

Laser induced time-resolved luminescence spectra of analcime revealed slightly structured orange band peaking at 603 nm with very short decay time evidently belonging to S_2^-, red band peaking at 725 nm with very long decay time evidently belonging to Fe^{3+} and blue band peaking at 450 nm with short decay time with unknown origin (Fig. 4.164).

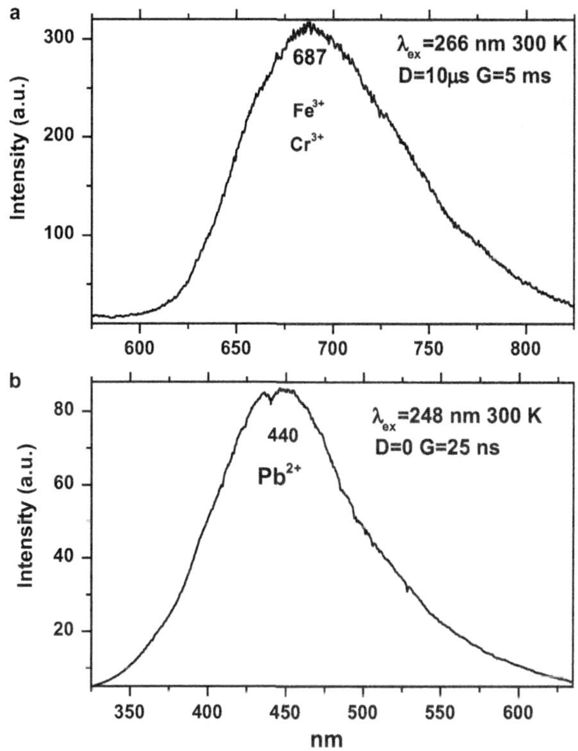

Fig. 4.158 (**a–b**) Laser-induced time-resolved luminescence of Fe^{3+} or Cr^{3+} and probably Pb^{2+} in hackmanite

4.11.5 *Delhayelite* $(Na,K)_{10}Ca_5Al_6Si_{32}O_{80}(Cl_2,F_2,SO_4)_3 \cdot 18(H_2O)$

Delhayelite was studied by steady-state luminescence and S_2^-, trivalent REE, such as Ce, Dy, Sm, and Mn^{2+} were found. Excitation by CW laser at 532 and 785 nm revealed trivalent REE, mainly Eu and Nd, and evidently S_2^- band (Fig. 4.165).

4.12 Silver (Ag^+) Bearing Minerals

Silver halides are capable to intrinsic luminescence.

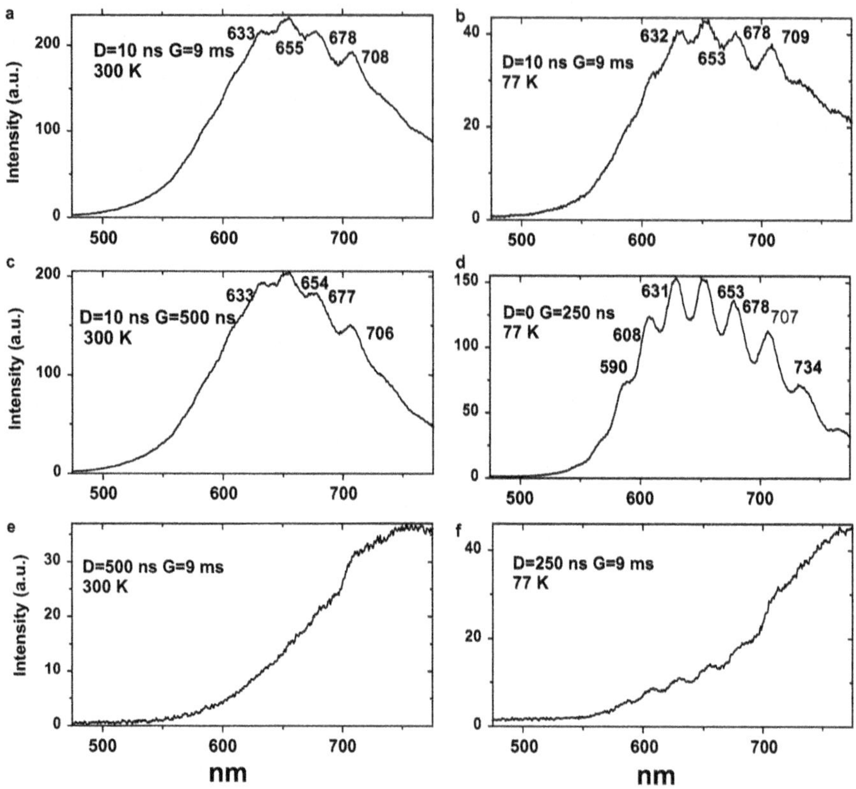

Fig. 4.159 (**a–f**) Laser-induced time-resolved luminescence of S_2^- in hackmanite

4.12.1 Chlorargyrite AgCl and Embolite Ag(Cl,Br)

It was found by steady-state luminescence spectroscopy that silver halides, such as chlorargyrite and embolite, formed in oxidized zones of silver ores demonstrate luminescence at low temperatures. Such emission was explained in energy band scheme by donor-acceptor recombination. By analogy with synthetic AgCl and AgBr samples the found bands have been connected in energy bands scheme with recombination of electrons with auto-localized holes and holes associated with metal-vacancy complexes. Under powerful laser excitation such luminescence was detected even at room temperature. The spectra are evidently connected with the same luminescence centers (Gaft et al. 1989). Laser-induced time-resolved luminescence spectrum of chlorargyrite demonstrate relatively narrow green band with short decay time of ~2–3 μs, which is especially strong in spectra with narrow gate (Fig. 4.166).

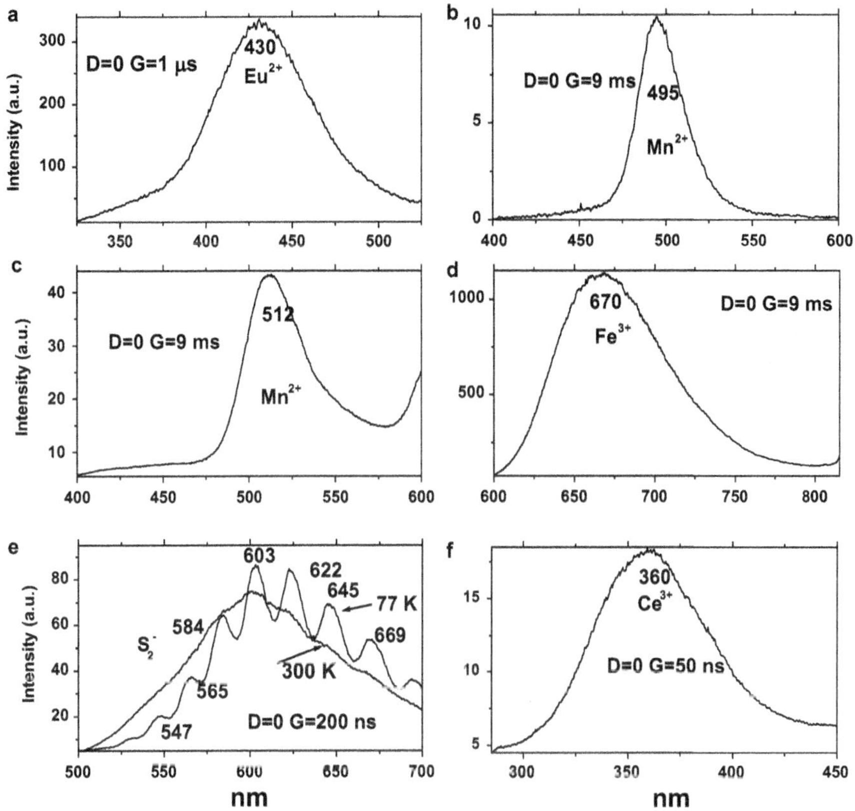

Fig. 4.160 (a–f) Laser-induced time-resolved luminescence of Eu^{2+}, Ce^{3+}, two different Mn^{2+} and S_2^- tugtupite

4.13 Manganese (Mn^{2+}) Bearing Minerals

Minerals of manganese are capable to intrinsic luminescence.

4.13.1 *Rhodonite $CaMn_4Si_5O_{15}$, Rhodochrosite $MnCO_3$, Axinite $Ca_2MnAl_2BSi_4O_{15}OH$*

It was established by steady-state luminescence spectroscopy that minerals of Mn, such as rhodonite, rhodochrosite, helvine, triplite, Mn-apatite, Mn-milarite and others, show dark red luminescence, mainly at 77 K, which is uncommon to impurity Mn^{2+}. The excitation center proved to be regular Mn^{2+}, while the emission

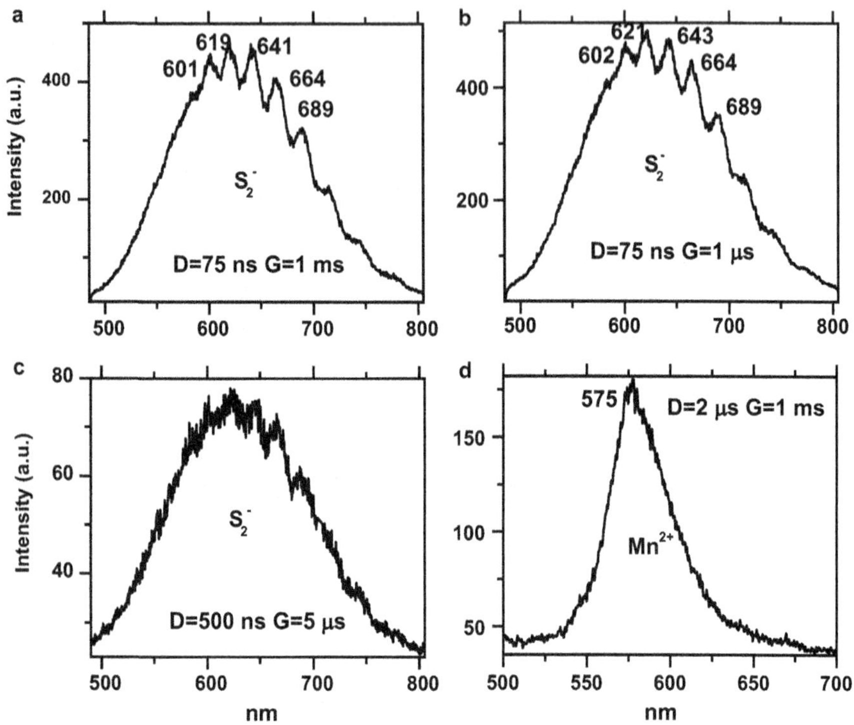

Fig. 4.161 (**a–d**) Laser-induced time-resolved luminescence of Mn^{2+} and S_2^- in lazurite

center is Mn^{2+}, situated near some lattice defect (Gorobets et al. 1978, 1981b, Gaft and Gorobets 1979; Gaft et al. 1981b).

The Mn minerals in our study consisted of seven sample. Laser-induced time-resolved technique enables to detect impurity and intrinsic luminescence of Mn^{2+} and Nd^{3+} (Fig. 4.167).

Excitation of rhodonite by CW laser revealed luminescence bands peaking at 660–680 nm with excitation by 532 nm and 890 and 970 nm excited by 780 nm (Fig. 4.168). IR lines of evidently Nd^{3+} are also detected.

4.13.2 *Helvine* $Mn_4(BeSiO_4)_5S$

Excitation of helvine by CW laser revealed luminescence bands peaking at 727 nm with excitation by 532 nm (Fig. 4.169) evidently connected with Mn^{2+}.

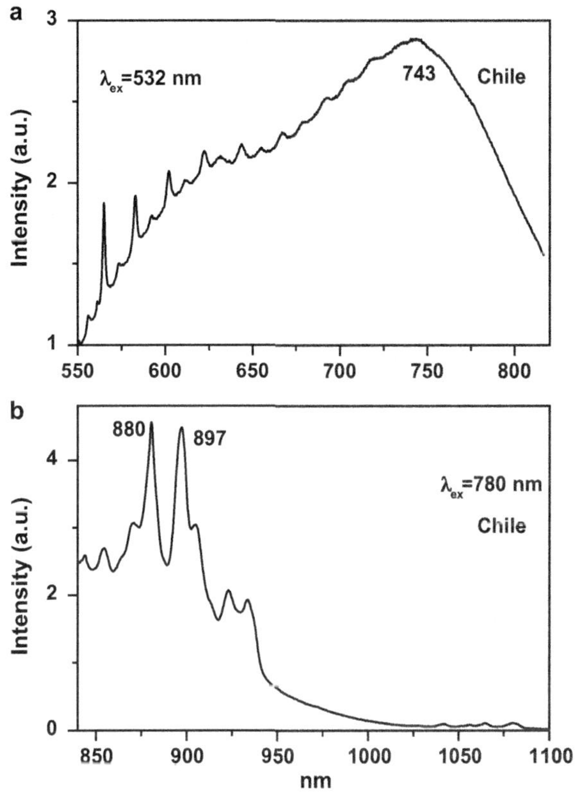

Fig. 4.162 (**a–b**) Laser-induced cw luminescence spectra of lazurite under 532 nm excitation (http://rruff.info)

4.14 Uranium (U^{6+}) Bearing Minerals

Minerals of uranium are capable to intrinsic luminescence, which is connected with uranyl emission. It is well known that a certain minerals which contain uranium as the uranyl ion, UO_2^{2+} exhibit fluorescence with distinctive spectral and temporal characteristics. The photoluminescence of nearly 40 uranyl minerals was investigated at 300 and 77 K, mainly under UV lamps, but also under UV lasers excitation. The highly fluorescent uranyl ion may also be present in other geologic materials, such as siliceous matrixes, calcites and barites (Gororbets and Sidorenko 1974; Tarashchan 1978; Kasdan et al. 1981; Gorobets and Rogojine 2001). This luminescence is effectively excited only by the short UV and its decay time is in the several hundred μs range. Three different models of uranyl ion interrelation with host mineral were proposed: surface films of uranyl-containing minerals; presence in the host matrices that are relatively transparent at the excitation and luminescence wavelength of UO_2^{2+}. Uranium mineralization was the first object for luminescence LIDAR.

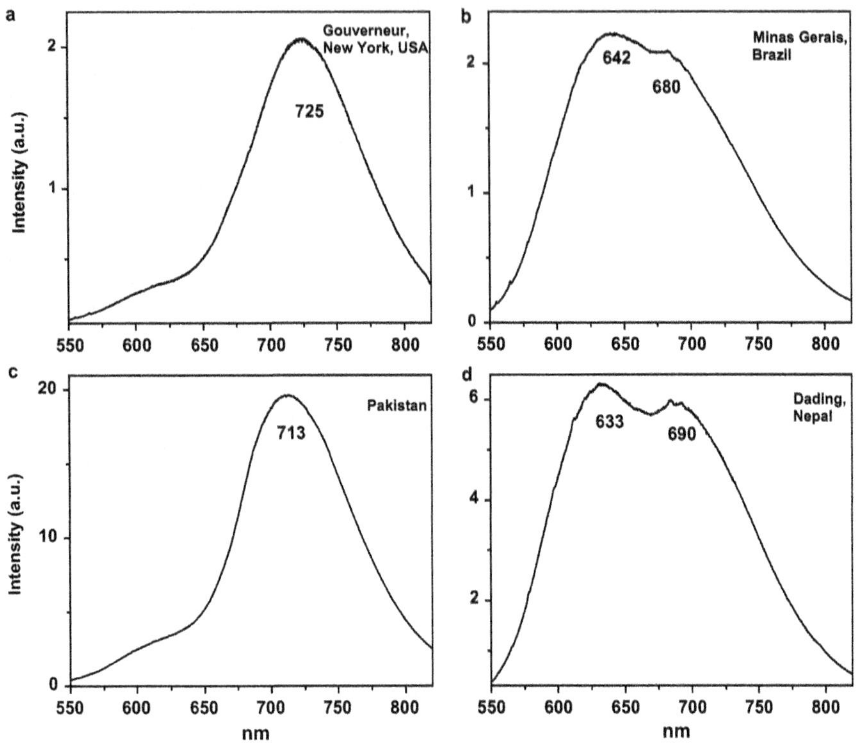

Fig. 4.163 (**a–d**) Laser-induced cw luminescence spectra of scapolite under 532 nm excitation (http://rruff.info)

Representative time-resolved luminescence spectrum of zippeite, autunite, zeunerite, agate, orangite, thorite, monazite and xenotime are given in Fig. 4.170.

4.15 Magnesium (Mg^{2+}) Bearing Minerals

Mn^{2+} is the mostly probable luminescence center suitable for substitution for Mg^{2+} in natural process of crystallization because of their close ionic radii and charges. The next possible candidates are Cr^{3+} replacing Mg^{2+} and Cr^{4+} replacing Si^{4+}.

4.15.1 Forsterite Mg_2SiO_4

Forsterite is a mineral of the olivine group, which possess orthorhombic symmetry and a structure consisting of independent SiO_4 tetrahedra linked by divalent atoms

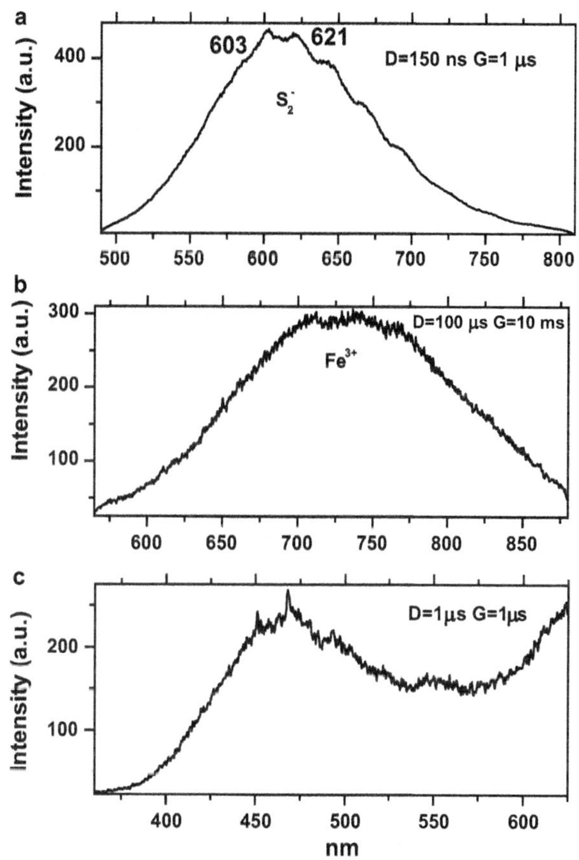

Fig. 4.164 (a–c) Laser-induced time-resolved luminescence of S_2^- and Fe^{3+} in analcime

in octahedral coordination. There are two non-equivalent distorted octahedral sites occupied by Mg^{2+} ions, one with inversion symmetry (M1) and one with mirror symmetry (M2). Forsterite has space group Pnma. The oxygen atoms are arranged in approximately hexagonal packing and lie in sheets parallel with the (100) plane. The SiO_4 tetrahedra point alternatively either way in both the x and y direction; half the Mg atoms are located at centers of symmetry and half on reflection planes.

Visible luminescence of many centers, such as Cr^{3+}, Mn^{2+} and Ni^{2+}, has been found and interpreted in the visible range in synthetic forsterite excited with electrons and laser. The polarized absorption spectra and time-resolved luminescence spectroscopy of Cr^-activated forsterite led to the discovery of at least two different kinds of Cr^{3+} emitting centers (Moncorge et al. 1991) that correspond to Cr^{3+} ions in two different sites of six-fold coordination. There are Cr^{3+} ions, probably substituting for Mg^{2+} ions in both the M1 and M2 sites, which form medium-field systems giving rise to ^4E-$^4T_2$ sharp line emissions at low temperatures with millisecond lifetimes around 700 nm and $^4T_2$, ^4E-$^4A_2$ thermalized short-lived broad band at high temperatures between about 690 and 750 nm. The second

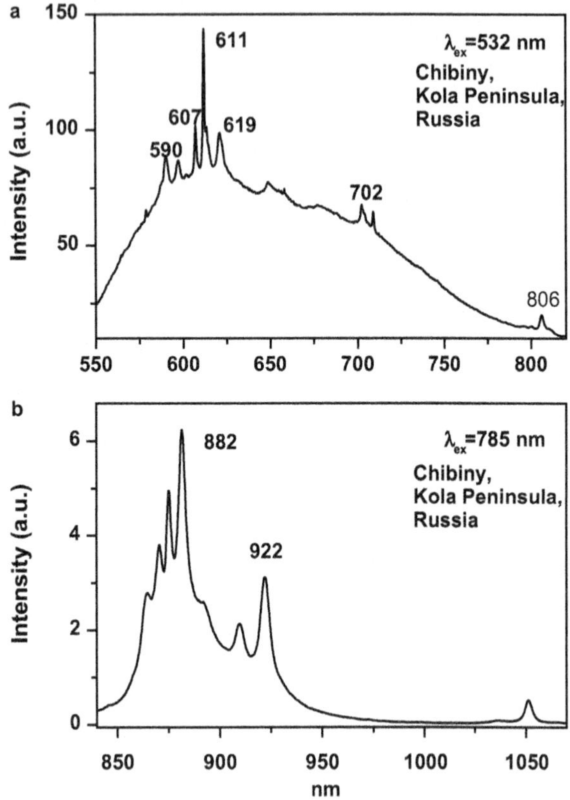

Fig. 4.165 (**a–b**) Laser-induced cw luminescence spectra of delhayelite under 532 and 785 nm excitation (http://rruff.info)

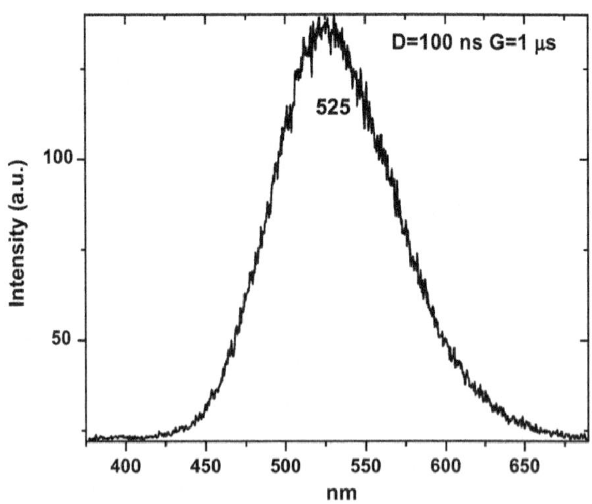

Fig. 4.166 Laser-induced time-resolved luminescence spectrum of chlorargyrite under 355 nm excitation

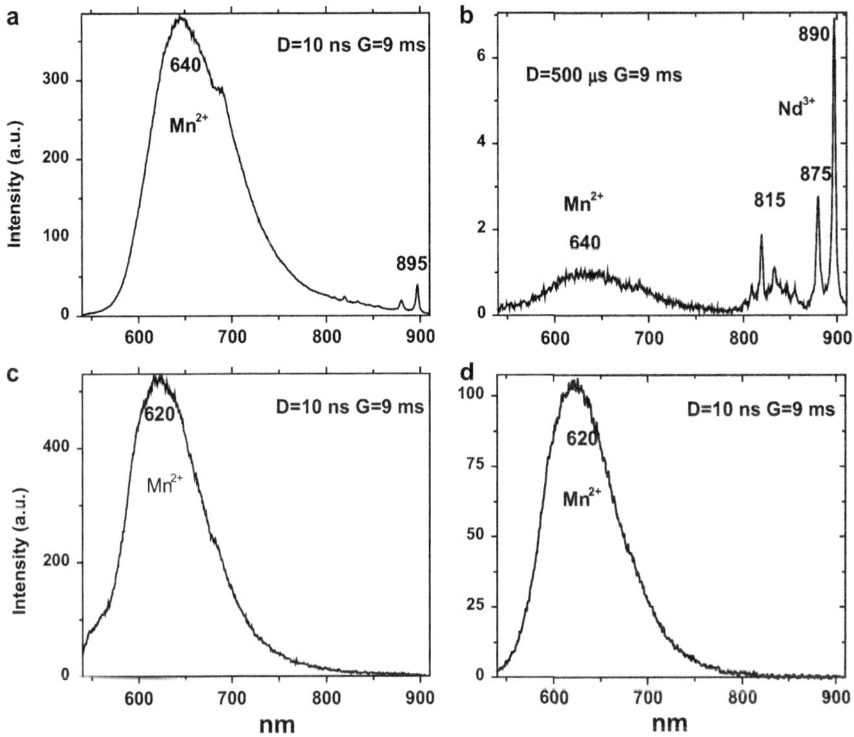

Fig. 4.167 (**a**–**d**) Laser-induced time-resolved luminescence spectra of rhodonite and rhodochrosite demonstrating Mn^{2+} and Nd^{3+} centers (**a**–**c**) and axinite (**d**) with Mn^{2+} emission

lower-energy ligand field Cr^{3+} center is characterized by a broadband emission peaking around 805 nm with a lifetime ranging between 12.5 and 18 μs. Low temperature spectroscopy of visible Cr-activated forsterite luminescence with high spectral resolution detected the main narrow R_1 Cr^{3+} lines at 692.7, 699.5 and 716.0 nm, while at room temperature the broad band peaking at 715 nm is observed (Glynn et al. 1991). Such emissions correspond to three paramagnetic centers Cr^{3+} has been found in chromium-doped forsterite: Cr^{3+}(M1), Cr^{3+}(M2) and Cr^{3+} (M1)-$V_{Mg}{}^{2+}$ formed by a Cr^{3+} ion substituting for Mg^{2+} at the M1 structural position with a nearest-neighbor Mg^{2+} vacancy at the M2 position (Ryabov 2012).

It was found that Mn^{2+} is characterized by a luminescence band peaking at 643 nm with a typical excitation spectrum and a long decay time of 29 ms. Low temperature study revealed a pure electronic transition at 615.4 nm which is the energy of the lowest split component of the $^4T_1(G)$ state above the ground state. Phonon replicas of this transition are evident showing that a particular phonon mode of 180 cm^{-1} is dominantly involved (Green and Walker 1985).

At 300 K there is the visible emission in the green at low temperature that appears to consist of two overlapping bands with maxima at about 510 and 560 nm

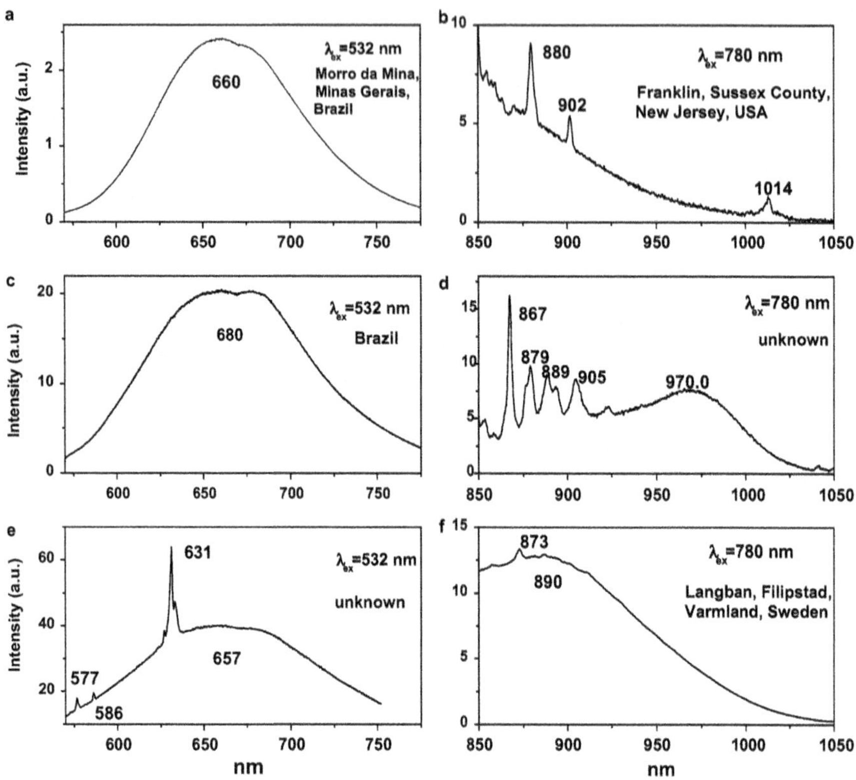

Fig. 4.168 (a–f) Laser-induced cw luminescence spectra of rhodonite under 532 and 780 nm excitation (http://rruff.info)

Fig. 4.169 Laser-induced cw luminescence spectra of helvine under 532 nm excitation (http://rruff.info)

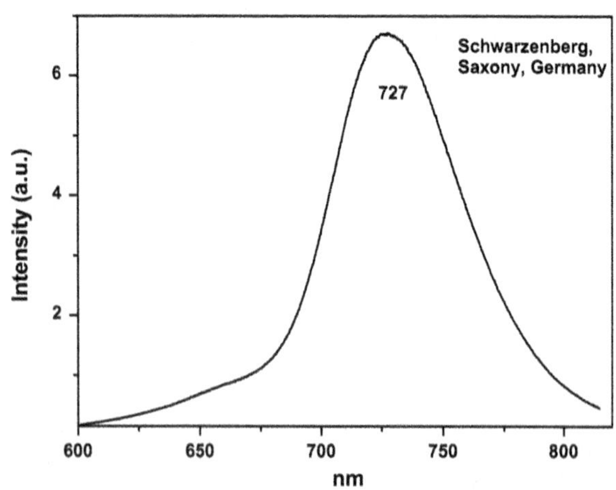

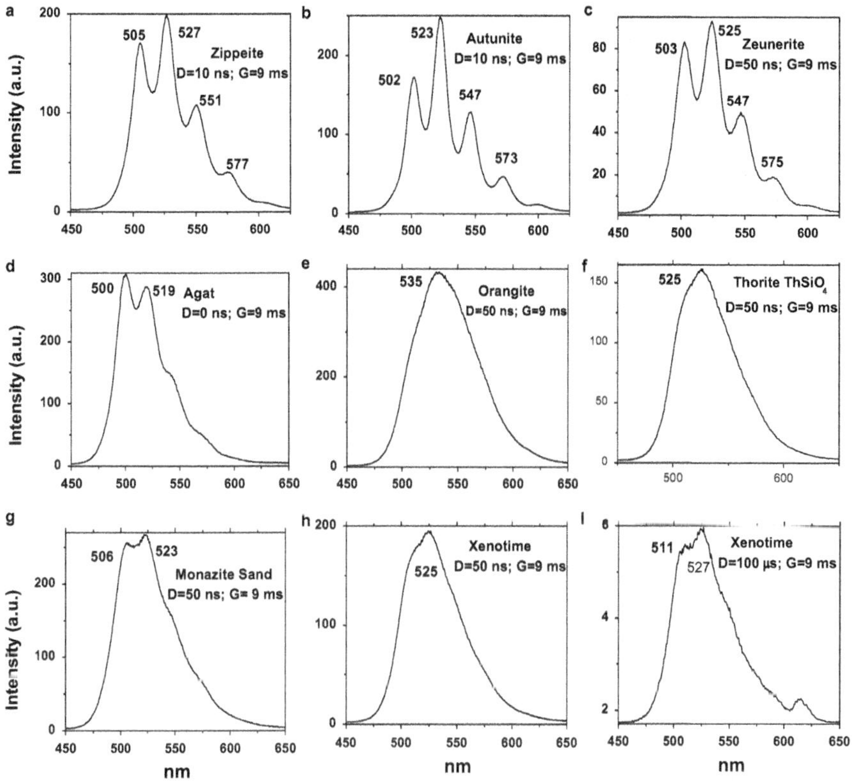

Fig. 4.170 (**a–i**) Laser-induced time-resolved luminescence spectra of uranyl minerals (zippeite, autunite, zeunerite, agate, orangite, thorite, monazite, xenotime)

(Walker et al. 1994). The fluorescence spectra of Ti-activated forsterite show a broad band covering the wavelength range of 560–1440 nm and peaking at approximately 850 nm. The intensity and the half-width of fluorescence band do not change much with the temperature (Shenjun et al. 1994). Cathodoluminescence spectra of meteoritic and Cr- and Mn-activated forsterite at 300 and 77 K revealed narrow lines peaking near 700 nm at 77 K assigned to Cr^{3+} in octahedral coordination and the broad band at 800 nm assigned to Cr associated with structural defects, while the band at 640 nm was assigned to Mn^{2+} (Benstock et al. 1997). Natural iron-free forsterite demonstrates luminescence of Mn^{2+} peaking at 640–660 nm and Fe^{3+} peaking at 700–710 nm identified by excitation spectra (Bakhtin et al. 1995).

Numerous spectral data appeared on natural forsterite as result of its continuous wavelength (CW) Raman spectra study under 532 and 780 nm laser excitations where Raman signals are accompanied by multiple luminescence lines and bands. Those luminescence spectral data exhibit much more emission centers than the

Table 4.22 Concentrations of Mn, Fe, Cr, V, Fe, Ni and Ti (ppm) in forsterite samples

Info	Ti	V	Cr	Mn	Fe	Ni
Balsam	25	2	5	24	1200	95
Suppat	15	0.6	19	67	3500	160
Trudy	21	0.9	18	52	3100	120
China	20	0.9	22	96	5500	210
Fen	17	0.7	12	38	1800	120

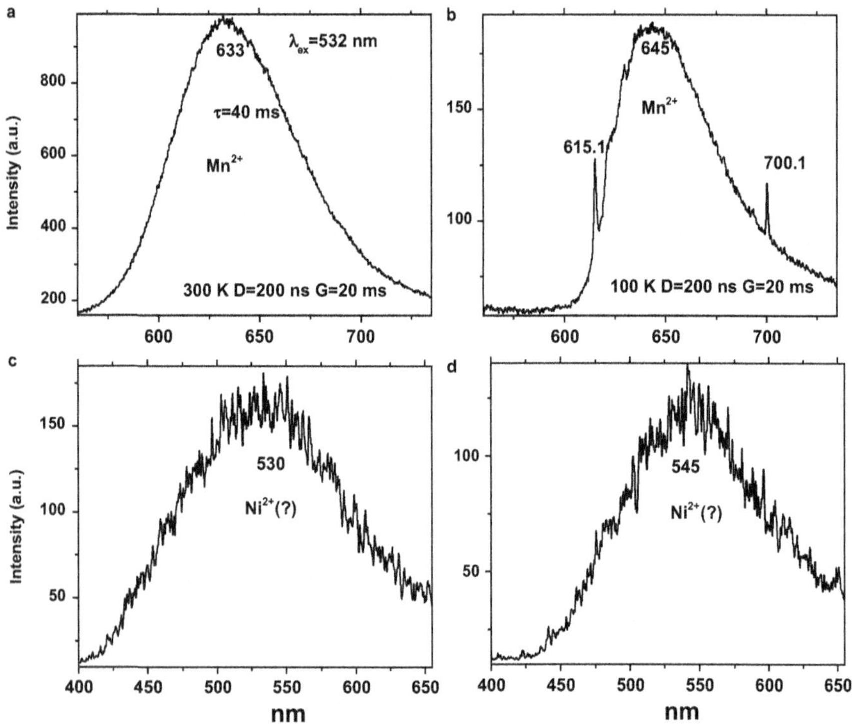

Fig. 4.171 (**a–d**) Laser-induced time-resolved luminescence spectra of Mn^{2+} with zero phonon line at 100 K (**a–b**) and possibly Ni^{2+} (**c–d**) in forsterite

Mn^{2+} and Fe^{3+} presently known in natural forsterite. Luminescence spectra detected only at room temperature without excitation spectra and decay times are very difficult to interpret.

The forsterite minerals in our study consisted of nine sample. Concentrations of several potential impurities luminogens are presented in Table 4.22. Laser-induced time-resolved technique enables to detect Mn^{2+} with zero phonon line at low temperatures (Fig. 4.171a, b), possibly Ni^{2+} (Fig. 4.171c, d), Cr^{3+} in strong CF (Fig. 4.172a–c), Cr^{3+} in weak CF (Fig. 4.172d, e) and possibly V^{2+} (4.172f).

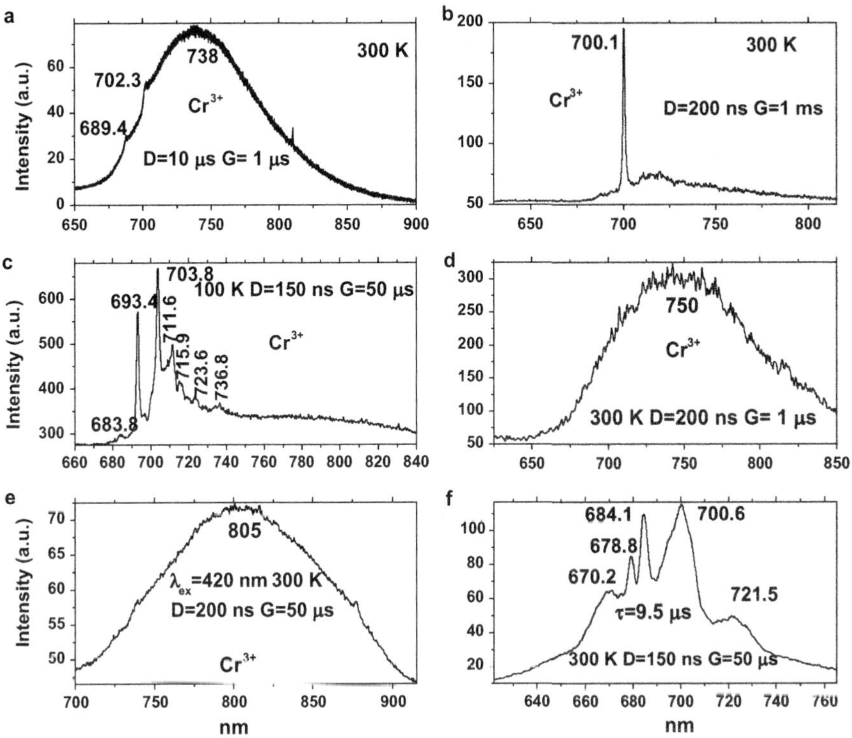

Fig. 4.172 (**a–f**) Laser-induced time-resolved luminescence spectra of Cr^{3+} in strong CF (**a–b**), weak crystal field (**c–e**) and possibly V^{2+} (**f**) in forsterite

Excitation of helvine by CW laser at 532 nm revealed multiple luminescence bands (Fig. 4.173) evidently connected with Mn^{2+}, Cr^{3+} and presently not identified.

4.15.2 *Enstatite $MgSiO_3$*

Enstatite is the magnesium end member of the pyroxene silicate mineral series enstatite ($MgSiO_3$) – ferrosilite ($FeSiO_3$). Enstatite is orthorombic and has space group Pbca. The basic enstatite structural unit can be visualized as infinitely long SiO_3 chains (||c) based on SiO_4 tetrahedra sharing two vertices and having all of their apical oxygens pointing in a common d direction(|| a, or -a) perpendicular to the length of the chain. Neighboring chains possess apices pointing in opposite directions. The result is that an a–c projection consists of "plates" containing silicon and oxygen. The larger cations – Mg for pure enstatite but Fe, Ca, Al, Mn for other pyroxenes-lie between the "plates", in octahedral coordination to the oxygen within the plates.

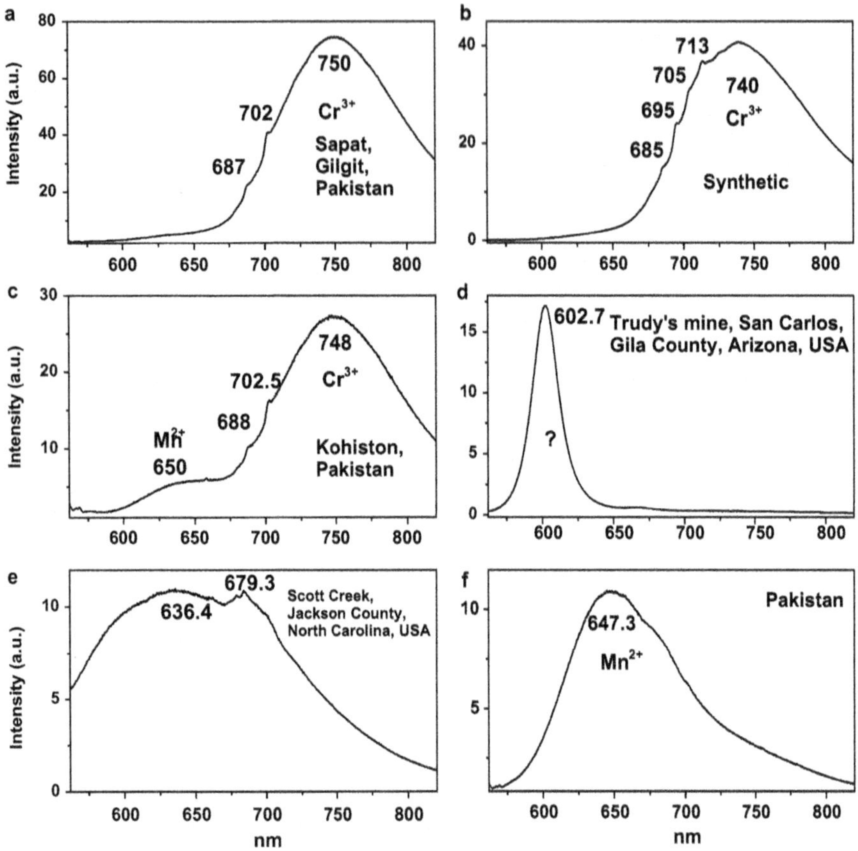

Fig. 4.173 **(a–f)** Laser-induced cw luminescence spectra of several forsterite samples under 532 nm excitation (http://rruff.info)

Emission and excitation spectra and luminescence decay curves of enstatite single crystals nominally doped with Ni and Cr have been measured at temperatures down to 10 K. For all the crystals under investigation, the emission band peaking at about 1520 nm was assigned to the ${}^3T_2$-${}^3A_2$ transition of octahedral Ni^{2+} center, while the luminescence around 800 nm was assigned to the ${}^4T_2$-${}^4A_2$ transition of octahedral Cr^{3+} centers (Moncorge et al. 1999).

Natural iron free enstatite exhibits mainly two orange-red luminescence bands peaking at 660 and 590 nm ascribed to Mn^{2+} centers substituting for Mg in different structural positions (Gorobets and Rogojine 2001). Luminescence bands peaking at 620–628 nm range have been found under 532 nm laser excitation (http://rruff.info) (Fig. 4.174).

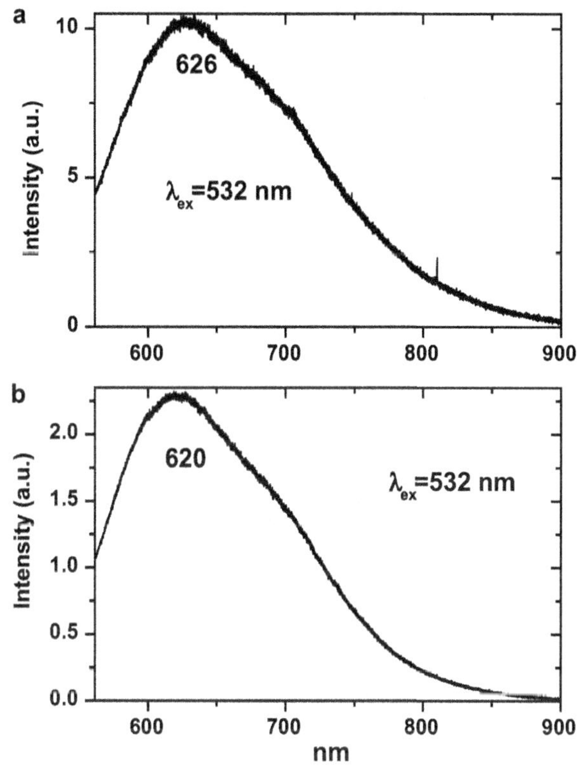

Fig. 4.174 (**a–b**) Laser-induced sw luminescence spectra of enstatite

4.15.3 *Magnesite $MgCO_3$, and Phlogopite $KMg_3AlSi_3O_{10}(F,OH)_2$*

Those minerals have been studied by steady-state luminescence spectroscopy and Mn^{2+} emission was found in magnesite peaking at 660 nm and at 580 and 630 nm in phlogopite (Gorobets and Rogojine 2001). Under cw laser excitation the bands peaking at 673 nm were found in magnesite, while in phlogopite evidently Cr^{3+} and Fe^{3+} present (Fig. 4.175).

4.15.4 *Clinochlore $Mg_5Al(Si_3Al)O_{10}(OH)_8$*

The luminescence spectra of Cr^{3+} centers in two chlorite crystals (Fig. 4.176) have been presented (Czaja et al. 2014). Chromium ions occupy the strong crystal-field site M4 in the brucite sheet and the intermediate crystal-field site in the inner

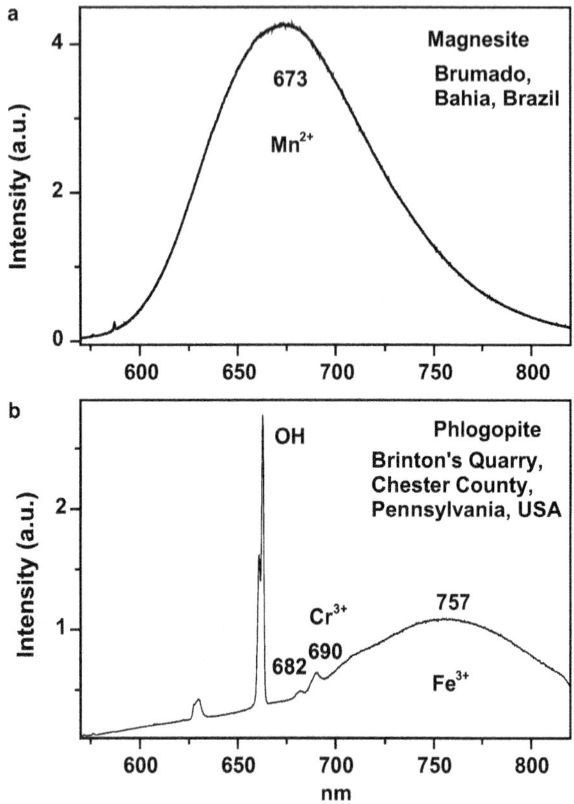

Fig. 4.175 (**a–b**) Laser-induced SW luminescence spectra of magnesite and phlogopite (http://rruff.info)

octahedral sheet for purple and green chlorite, respectively. The influence of an effective positive charge on the Cr^{3+} ion and an effective negative charge of ligands on the differences in the values of the Dq and B parameters were discussed. It was concluded that the presence of Fe^{2+} ions and other point defects, as well as concentration quenching, causes the very short luminescence lifetimes of chromium ions.

4.15.5 Sinhalite $MgAlBO_4$

Under cw laser excitation the structured emission spectrum appear (Fig. 4.177) which preliminary may be ascribed to V^{2+} luminescence.

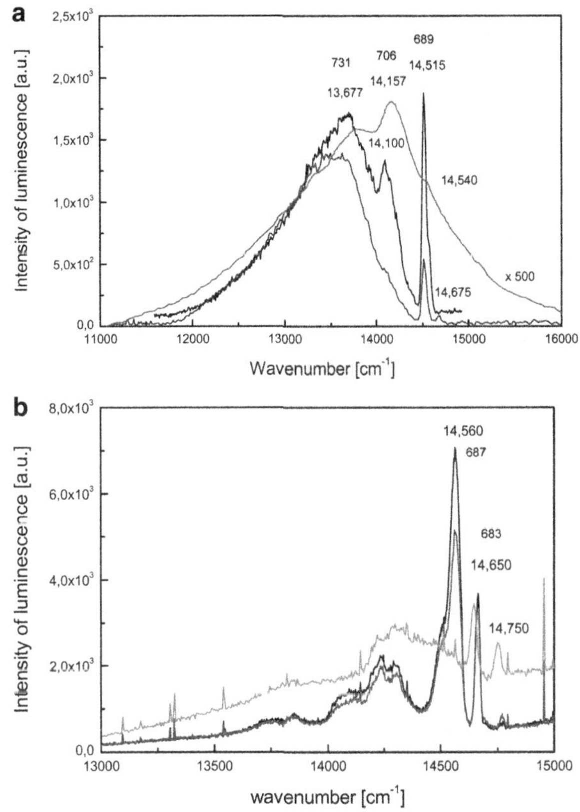

Fig. 4.176 (**a–b**) The luminescence spectra of the Cr^{3+} ion in purple chlorite measured at different temperatures; the *pink line* – T = 300 K, the *purple line* – T = 77 K, the *black line* – T = 15 K (**a**). The luminescence spectra of the Cr^{3+} ion in green chlorite measured at different temperatures; the *green line* – T = 300 K, the *olive line* – T = 77 K, the *black line* – T = 20 K (**b**) (Czaja et al. 2014)

4.16 Rare-Earth Bearing Minerals

Mineral matrix is formed chiefly by La^{3+} or Ce^{3+}. The last one is widely met as luminescence impurity center, but in rare-earth bearing minerals it is subjected to concentration quenching because of strong exchange interaction of outer 5d orbitals of neighboring Ce^{3+} ions. Hereby its luminescence is not observed in REE minerals. From the other side, other trivalent REE can achieve some concentrations without quenching of luminescence, because outer electron shells, such as $5s^2$ and $5p^6$, shield their inner 4f-4f transitions.

4.16.1 Monazite (Ce,La)PO_4

The monazite structure consists of distorted PO_4 tetrahedra with each metal atom roughly equidistant from nine oxygen atoms. Minor amounts of other rare-earth elements may occur. Steady-state luminescence under X-ray excitation of monazite

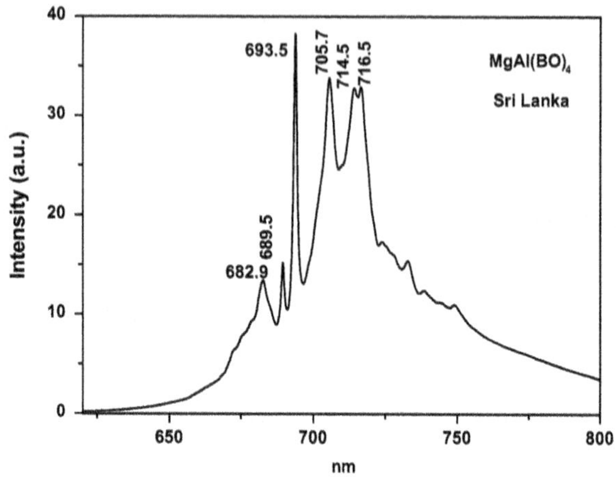

Fig. 4.177 Laser-induced sw luminescence spectra of sinhalite (http://rruff.info)

revealed emission of Gd^{3+}, Tb^{3+}, Dy^{3+} and Sm^{3+} (Gorobets and Rogojine 2001). Laser-induced time-resolved luminescence enables to detect Sm^{3+}, Eu^{3+} and Nd^{3+} emission centers (Fig. 4.178b–d).

Luminescence under steady-state laser excitation with 780–785 nm revealed several narrow lines in IR range of the spectrum (Fig. 4.179). Natural monazite luminescence under cw laser excitations have been studied together with synthetic samples activated by different REE (Lenz et al. 2015). It was proved that the luminescence lines are mostly connected to trivalent REE such as Nd, Er, Pr, Sm. The broadening of the Stark sublevels correlates well with Th content and the degree of amorphization of the various monazite (Fig. 4.180).

4.16.2 *Thorite $ThSiO_4$*

Thorite and orangite (orange thorite) have a tetragonal structure and are isostructural with zircon. Steady-state spectra under X-ray and laser (337 nm) excitations are connected with REE^{3+}, namely Sm^{3+}, Tb^{3+}, Dy^{3+} and Eu^{3+}. Reabsorption lines of Nd^{3+} have been also detected (Gorobets and Rogojine 2001). Laser-induced time-resolved luminescence enables to detect Eu^{3+} and uranyl emission centers (Fig. 4.178a).

4.16.3 *Xenotime YPO_4*

Luminescence under steady-state laser excitation with 532 and 780 nm revealed several narrow line in IR range of the spectrum. Natural xenotime luminescence

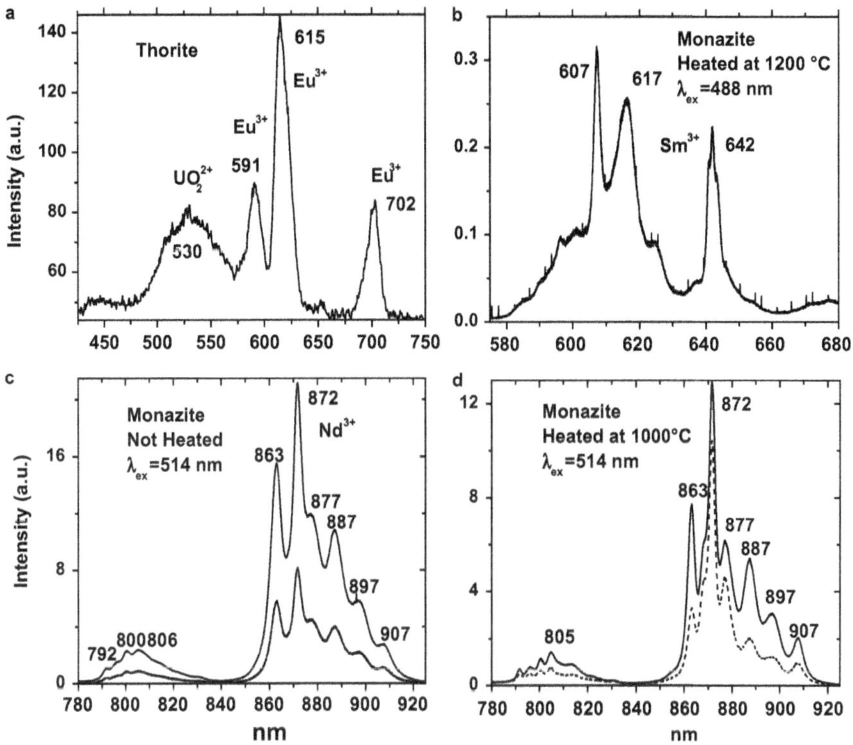

Fig. 4.178 (**a**–**d**) Laser-induced steady-state luminescence spectra of thorite (**a**) and monazite (**b**–**d**) natural and heated demonstrating uranyl, Eu^{3+}, Sm^{3+} and Nd^{3+} centers. Straight line-vertical polarization, dashed line – horizontal polarization

under cw laser excitations have been studied together with synthetic samples activated by different REE (Lenz et al. 2015). It was proved that the luminescence lines are mostly connected to trivalent REE such as Tm, Ho, Nd, Er, Pr, Sm, Eu (Fig. 4.181).

4.16.4 Bastnasite $CeCO_3F$

Photoluminescence studies of bastnasite confirmed the presence of at least four trivalent rare earths, such as Sm, Eu, Dy and Tb in trace quantities. The emission spectrum and decay time data were evaluated. It was observed that, the rare earth ions are situated at distorted sites in the system surrounded by defect centers (Mohapatra et al. 2014).

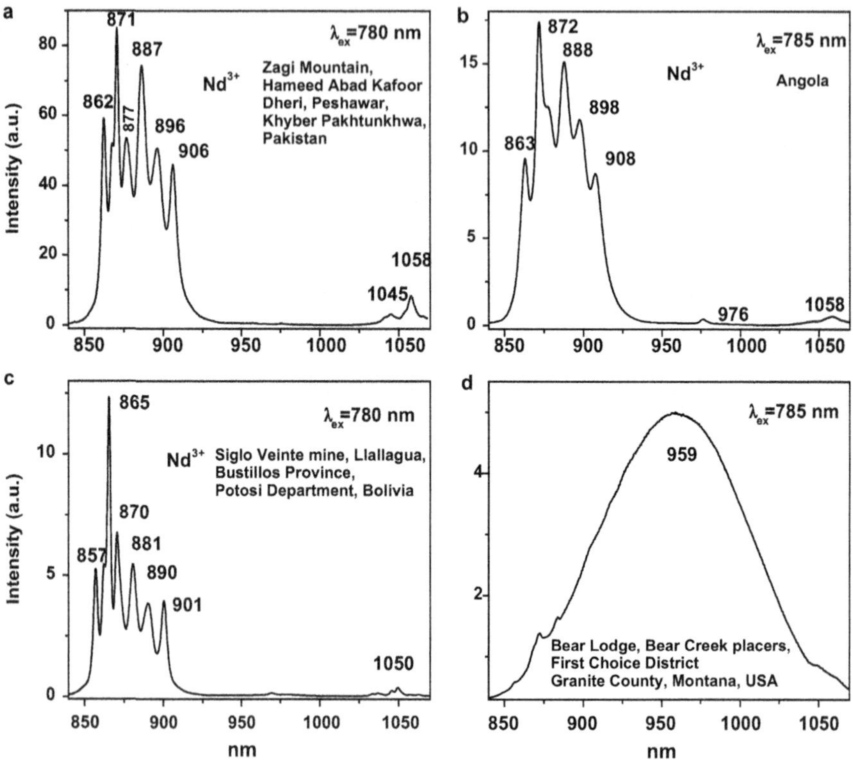

Fig. 4.179 (**a–d**) Laser-induced SW luminescence spectra of monazite (http://rruff.info)

Fig. 4.180 Nd^{3+} luminescence spectra in different natural monazite samples with different Th content and the degree of amorphization under cw excitation by 514 nm

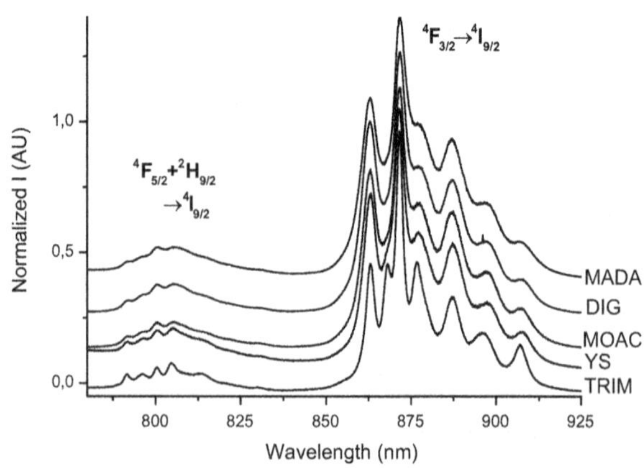

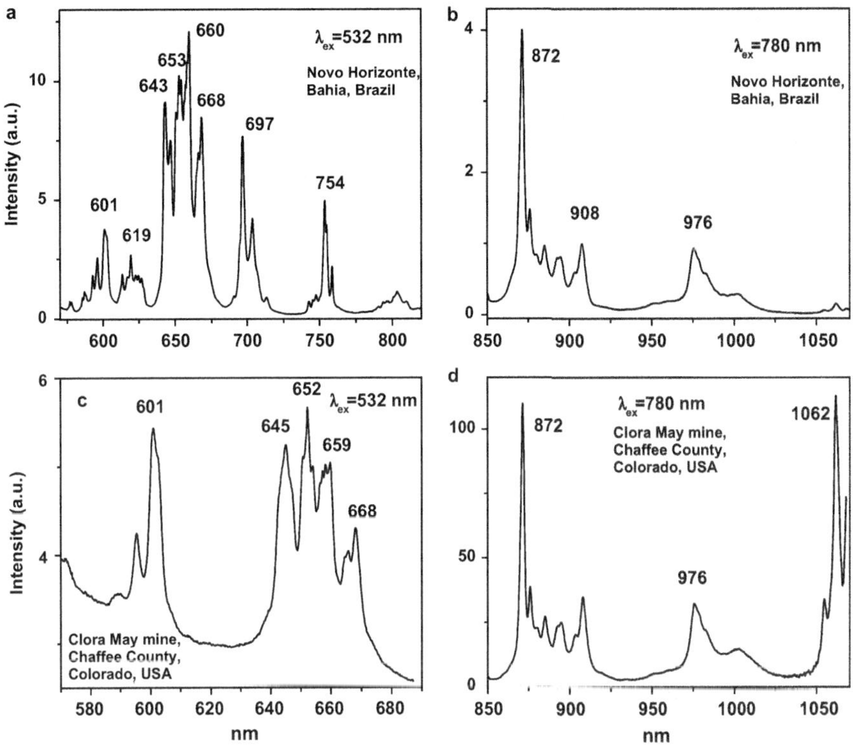

Fig. 4.181 (**a–d**) Laser-induced sw luminescence spectra of xenotime (http://rruff.info)

Luminescence under steady-state laser excitation with 532 and 780 nm revealed several narrow line in IR range of the spectrum evidently connected to trivalent REE such as Nd, Sm, Pr. Besides narrow "negative" lines and bands evidently may be ascribed to Nd^{3+} reabsorption (Fig. 4.182).

4.16.5 *Cerite $(Ca,La,Ca)_9(Mg,Fe)(SiO_4)_6(SiO_3OH)(OH)_3$*

Luminescence under steady-state laser excitation with 532 and 785 nm revealed several narrow line in IR range of the spectrum which evidently may be connected to Nd^{3+} (Fig. 4.183).

4.16.6 *Parisite $Ca(Ce,La)(CO_3)_3F_2$*

Luminescence under steady-state laser excitation with 532 and 785 nm revealed several narrow line in IR range of the spectrum which evidently may be connected to Nd^{3+} (Fig. 4.184).

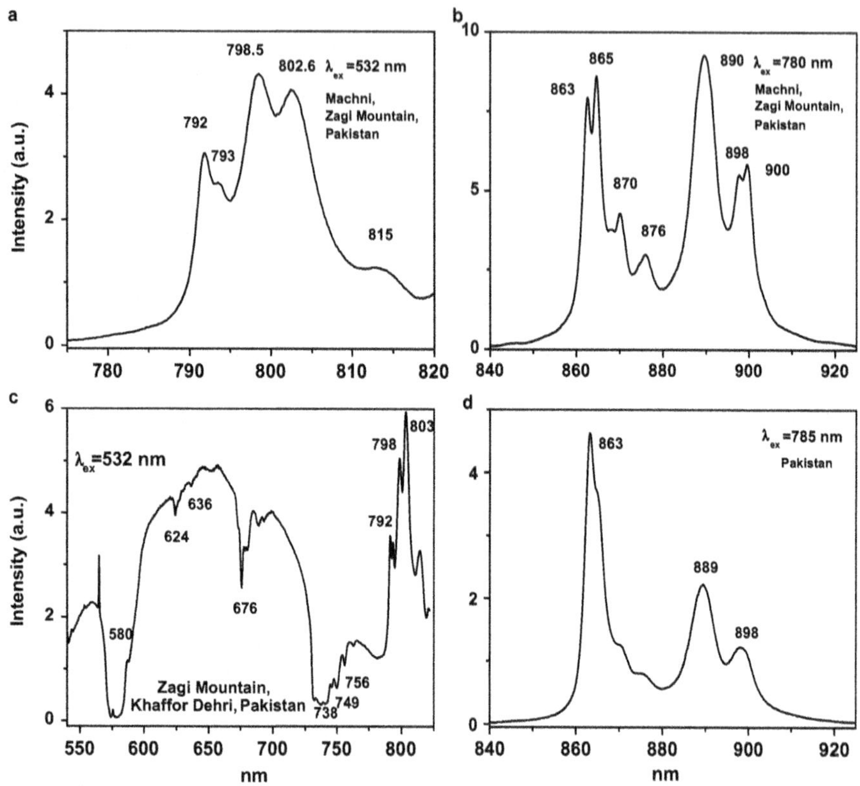

Fig. 4.182 (**a–d**) Laser-induced SW luminescence spectra of bastnäsite (http://rruff.info)

4.16.7 *Shomiokite $Na_3Y(CO_3)_3$•3(H_2O)*

Luminescence under steady-state laser excitation with 532 and 785 nm revealed several narrow line in IR range of the spectrum which evidently may be connected to Eu^{3+}, Pr^{3+} and Nd^{3+} (Fig. 4.185).

4.16.8 *Fergusonite (Y, RE)NbO_4*

Photoluminescence spectra, which are dominated by emissions of centers related to rare-earth elements, such as Er^{3+} and Ho^{3+}, have been studied upon gradual reconstitution of the crystallinity. General sharpening of bands at higher annealing temperatures was found due to increasing crystal field effects and indicates recovery of the fine structure (Rushel et al. 2007).

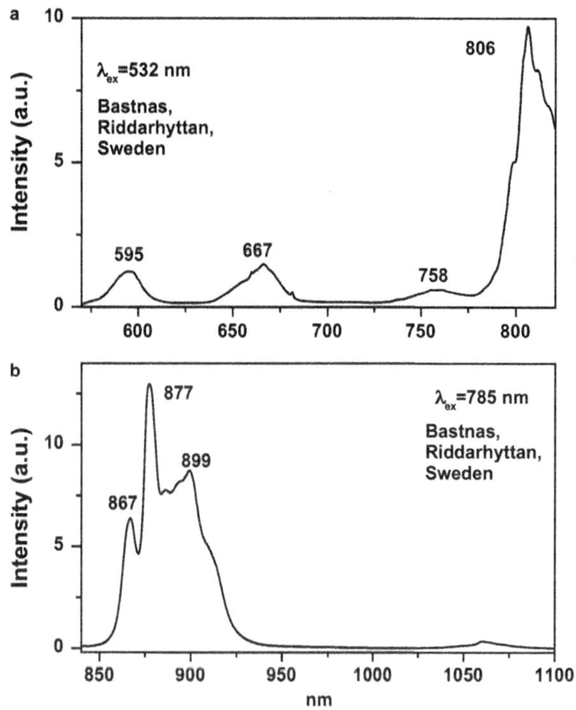

Fig. 4.183 (**a–b**) Laser-induced sw luminescence spectra of cerite (http://rruff.info)

4.17 Cesium (Cs^+) Bearing Minerals

4.17.1 Pezzottaite $Cs(Be_2Li)Al_2(Si_6O_{18})$

Pezzottaite is not isostructural with beryl $Be_3Al_2Si_6O_{18}$. However, the arrangement of the atoms in each structure is very similar, so the structures can be considered as related. In beryl, the three Be atoms indicated in the formula occupy three symmetrically related positions. In pezzottaite, these three positions are occupied in an ordered fashion by 2 Be and one Li rather that 3 Be ions as in beryl. As a result, pezzottaite differs from other members of the beryl group by having rhombohedral symmetry rather that hexagonal and a larger unit cell. Lithium substituting Be^{2+} in the tetrahedral linking the six-membered Si_6O_{18} rings are charge balance by Na^+ and Cs^+ present in the channels passing through the centres of the vertically stacked rings. Sodium is accommodated in the centres of the individual rings of silicate tetrahedra, but the larger Cs^+ is located halfway between these centres along the c-axis. The increase in charge caused by the incorporation of alkali ions into the channels compensate the Be replacement by Li in tetrahedral site. Vis-NIR absorption spectrum of purplish pink pezzottaite is dominated by a band centered at 494 nm, with a distinct shoulder at 476 nm. A second absorption band at 563 nm appears in the same orientation, along with a broad band centered at 820 nm. The spectral positions of the bands at 476 nm and 563 nm, together with its sensitivity to

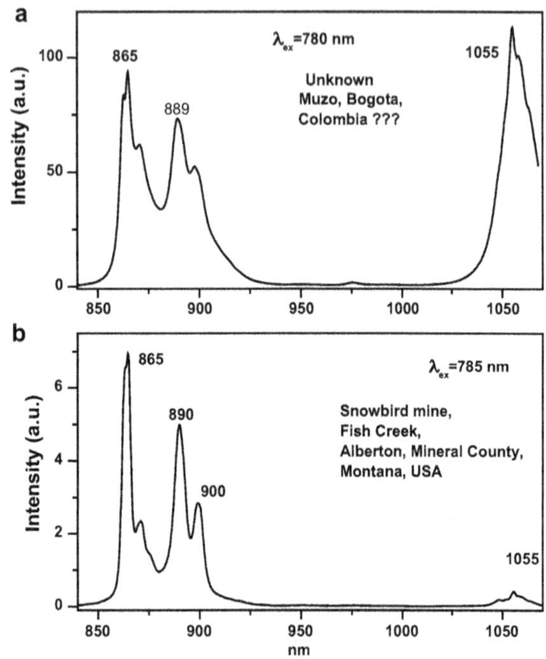

Fig. 4.184 (**a**–**b**) Laser-induced SW luminescence spectra of parisite (http://rruff.info)

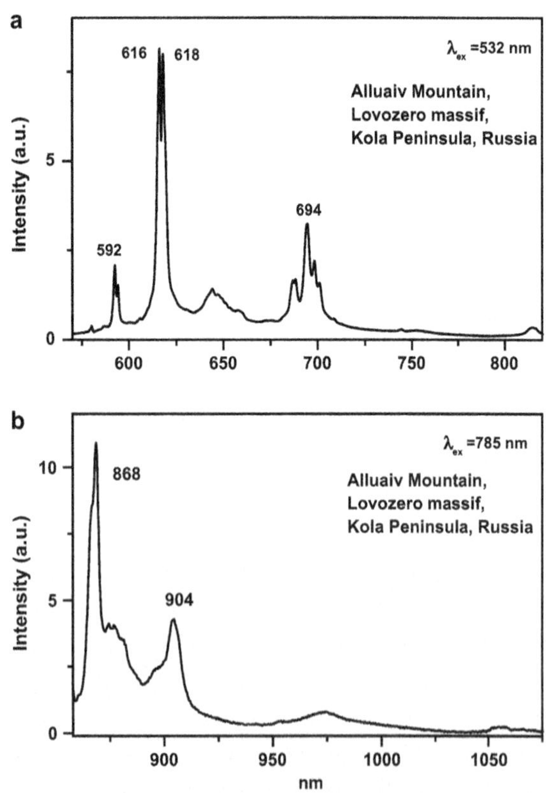

Fig. 4.185 (**a**–**b**) Laser-induced sw luminescence spectra of shomiokite (http://rruff.info)

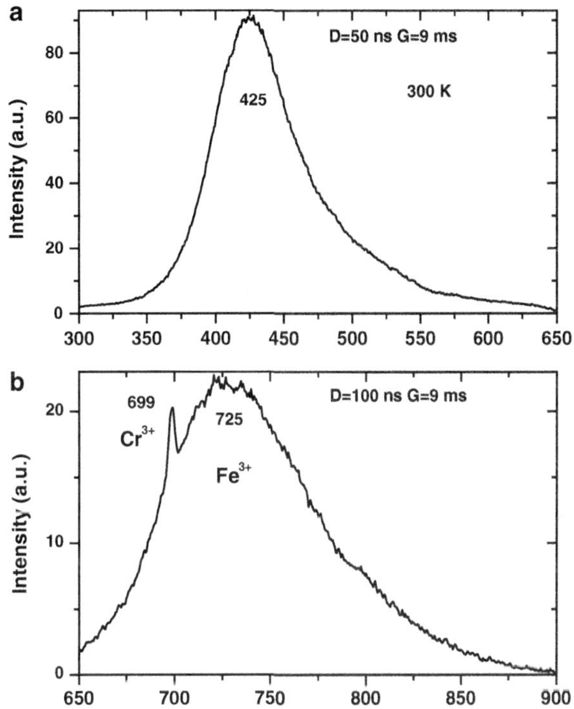

Fig. 4.186 (**a–b**) Laser-induced time-resolved luminescence spectra of Tl^{+}, Fe^{3+} and Cr^{3+} in pezzottaite

heat treatment and irradiation, evidence that they should be connected with spin-allowed transitions $^{5}E_{g}$-$^{5}T_{2g}$ in radiation-induced center involving Mn^{3+}. The band at 820 nm, which is strongly σ-oriented, is very similar to those in beryl and ascribed to Fe^{2+} in tetrahedral coordination (Chtihambo et al. 1995; Solntsev and Bukin 1997; Laurs et al. 2003).

No fluorescence under long and short-wave ultraviolet excitation was reported. The pezzottaite mineral in our study consisted of two samples. Laser-induced time-resolved technique enables to detect Tl^{+}, Fe^{3+} and Cr^{3+} (Panczer et al. 2010) (Fig. 4.186).

4.18 Antimony (Sb^{3+}) Bearing Minerals

4.18.1 Swedenborgite $NaBe_4SbO_7$

No fluorescence under long and short-wave ultraviolet excitation was reported. The swedenborgite mineral in our study consisted of two samples. Laser-induced time-resolved technique enables to detect supposedly Sb^{3+} luminescence (Gaft et al. 2013a) (Fig. 4.187).

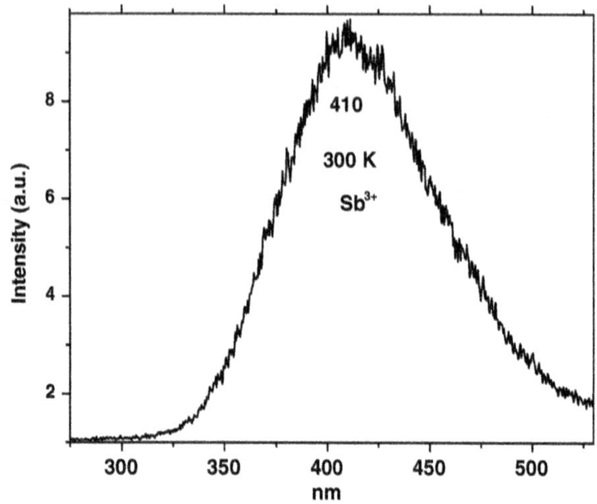

Fig. 4.187 Laser-induced time-resolved luminescence spectra of supposedly Sb^{3+} in swedenborgite

4.18.2 Diamond C

The unique situation exists in the case for diamonds, when the detailed spectroscopic descriptions of the centers are received, but the models are proposed only for the several. From more than 100 detected centers only for seven the models are determined, mainly based on EPR interpretations. The model includes identification of the impurity, vacancy, interstitial atom, their aggregations and their crystallochemical position together with quantum-chemical and spectroscopic description.

Luminescence of diamonds is related to various defects in its structure. Almost always, luminescence centers in diamonds are related to N atoms. It is logical, because the atomic radii of C and N are nearly equal (approximately 0.77 Å). Luminescence spectroscopy has proven to be a most widely used method in studies of diamonds even in comparison with optical absorption, ESR, IR and Raman spectroscopies. Hundreds of spectra have been obtained, fluorescence characteristics enter into diamond quality gemological certificates, wide range of electronic and laser applications are based on diamond optical properties in excited states, nitrogen center aggregation is controlled by residence time of diamond in the mantle, distinction between natural and synthetic, irradiated, enhanced diamonds and identification of any particular diamond can be made by detailed luminescence studies, CL topography reflects growth parameters (Walker 1979; Bokii et al. 1986; Plotnikova 1990; Lupashko et al. 1996; Zaitsev 2005; Panczer et al. 2000; Nasdala et al. 2013).

Determination of tint is the essential and the most intricate aspect in diamonds evaluation and may be useful for luminescence interpretation. The diamond coloration is a reflection of its complex structural peculiarity. To describe a slight shade of color, one has to use a lot of physics, crystallography and analytical tools. The following types of diamond coloration are generally distinguished.

Colorless to yellowish: a common color of diamonds of gem quality given to crystal by nitrogen impurity and certain types of nitrogen centers. Brown to yellow Australian (Argyle) called Cognac and Champagne due to dislocation related centers. In rare blue diamonds: a color is connected with boron impurity. In green diamonds the color is radiation induced (resulting from either natural or artificial irradiation) and is connected with vacancies (vacancies formation is possible not only after irradiation, but during growth and as result of certain transformation in the solid state). Rare Australian rose or purple: the color is connected with dislocation related centers of a different nature comparing with the centers of brown and yellow colors. Yellow, where individual nitrogen atoms substitute for carbon. Black and ink violet from epigenetic finely dispersed carbon. Certain raw diamonds acquire or change colors when subjected to radiation and further changing color when heated. This dynamics is traced to earlier crystal's history. As much as the diamonds colors are extraordinary their absorption spectra are monotonous. Their absorption curve bends slightly from IR to UV.

Twenty five specimens have been chosen from the collection of approximately 200 characteristic crystals taken from some thousands of Yakutian diamonds. Representative time-resolved luminescence spectra are given in Figs. 4.188, 4.189 and 4.190.

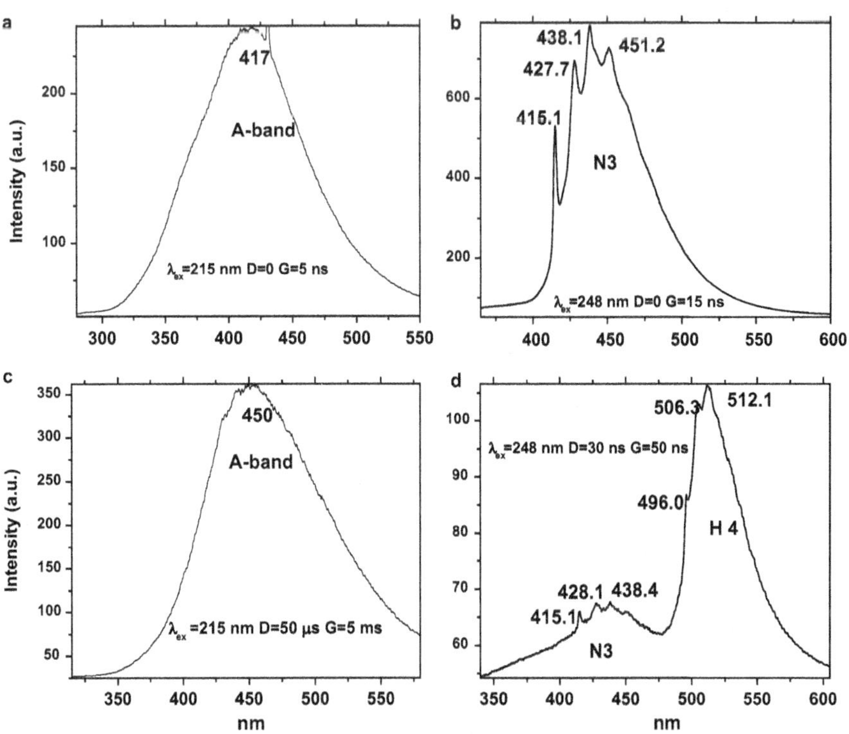

Fig. 4.188 (**a–d**) Laser-induced time-resolved luminescence spectra of diamonds with N3, A and H4 centers

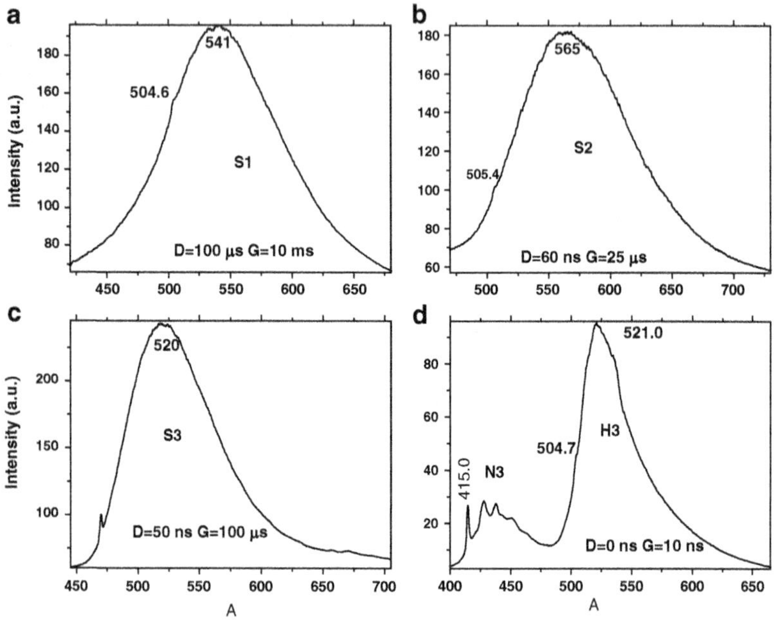

Fig. 4.189 (**a–d**) Laser-induced time-resolved luminescence spectra of diamonds demonstrating S1, S2, S3 and H3 centers

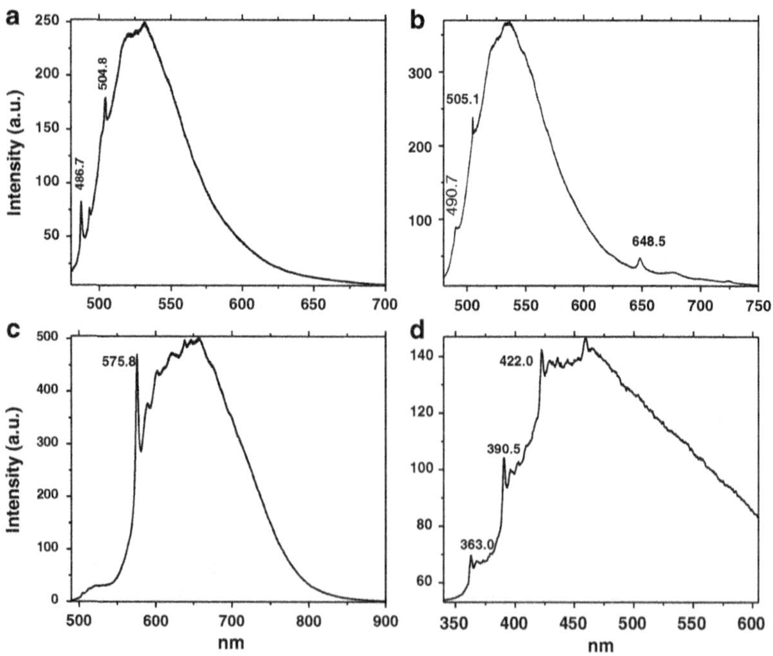

Fig. 4.190 (**a–d**) Laser-induced time-resolved luminescence spectra of diamonds demonstrating 486.7, 490.7, 575 and 391 centers

References

Bakhtin A, Moroshkin V (1986) Luminescence of Cr^{3+} in plagioclase as indication of under core material source. Geochemistry 10:1514–1515 (in Russian)

Bakhtin A, Denisov I, Lopatin O (1995) Photoluminescence of hole centers in oilivine crystals. Opt Spectrosc 79:773–777

Baumer A, Blanc P, Cesbron F, Ohnenstetter D (1997) Cathodoluminescence of synthetic (doped with rare-earth elements) and natural anhydrites. Chem Geol 138:73–80

Beletti A, Borromei R, Oleari L (1995) Absorption spectra of zircon crystals doped with Cr(IV): $ZrSiO_4$:Cr^{4+}. Inorg Chim Acta 235:349–362

Benstock E, Buseck P, Steele I (1997) Cathodoluminescence of meteoritic and synthetic forsterite at 296 and 77 K using TEM. Am Mineral 82:310–315

Blasse G (1980) Luminescence and energy transfer. Struct Bond 42:1–41

Blasse G, Aguilar M (1984) Luminescence of natural calcite ($CaCO_3$). J Lumin 29:239–241

Bokii G, Bezrykov G, Kliyev Yu et al (1986) Natural and synthetic diamonds. Nauka, Moscow (in Russian)

Brenier A, Suchocki A, Pedrini C, Boulon G, Made C (1992) Spectroscopy of Mn^{4+} -doped calcium-substituted gadolinium gallium garnet. Phys Rev B 46:3219–3227

Brugger J, Bettiol A, Costa S et al (2000) Mapping rare earth element distribution in scheelite using luminescence. Miner Mag 64:891–903

Chithambo M, Raymond S, Calderon T, Townsend P (1995) Low temperature luminescence of transition metal-doped beryls. J Afr Earth Sci 20:53–60

Czaja M (1999) Luminescence properties of Cr^{3+} ions in some phyllosilicates (chlorites). Excited States of Transition Elements, Duszniki Zdroj, Abstracts, Polish Academy of Science, PO9

Czaja M, Mazurak Z (1994) Crystal-field analysis of Cr^{3+} in grossular. Opt Mater 3:95–98

Czaja M, Bodył-Gajowska S, Lisiecki R, Meijerink A, Mazurak Z (2012) The luminescence properties of rare-earth ions in natural fluorite. Phys Chem Miner 39:639–648

Czaja M, Kadziołka-Gaweł M, Lisiecki R, Bodył-Gajowska S, Mazurak Z (2014) Luminescence and other spectroscopic properties of purple and green Cr-clinochlore. Phys Chem Miner 41:115–126

Deutschbein O (1932) Die linienhafte emission und absorption der chromphosphore. Ann Phys 14. I. 14,712-14,728; II:729–754

Eliseev A, Yurkin A, Samoilova E (1988) Cr^{3+} energy levels in alexandrite. Phys Stat Solid (a) 105:K169–K173

Eremenko G, Khrenov A (1982) Luminescence of baddeleyite. Miner Mag 4:93–95 (in Russian)

Erfurt G (2003) Infrared luminescence of Pb^{+} centres in potassium-rich feldspars. Phys Stat Sol (a) 200:429–438

Friis H (2009) Luminescence spectroscopy of natural and synthetic REE-bearing minerals. PhD Thesis, University of St. Andrews

Friis H, Finch A, Williams C, Hanchar J (2010) Photoluminescence of zircon ($ZrSiO_4$) doped with REE^{3+} (REE = Pr, Sm, Eu, Gd, Dy, Ho, Er). Phys Chem Miner 37:333–342

Friis H, Finch A, Williams C (2011) Multiple luminescent spectroscopic methods applied to the two related minerals, leucophanite and meliphanite. Phys Chem Miner 38:45–57

Gaft M (1984) Luminescence of vanadium containing centers in natural phosphates. Miner J 4:83–86 (in Russian)

Gaft M (1989) Luminescence of minerals under laser excitation. Ministry of Geology, Moscow (in Russian)

Gaft M, Panczer G (2013) Laser induced time resolved luminescence spectroscopy of minerals – a powerful tool for studying the nature of emission centers. Miner Petrol 107:363–372

Gaft M, Vorontsova L (1982) Luminescence of cassiterite and the possibilities of its practical use. Miner J 4(5):75–78

Gaft M, Gorobets B, Malinko S (1979) Studies of luminescence of boron minerals. Docl Acad Nauk SSSR 244(5):171–174

Gaft M, Gorobets B, Homyakov A (1981a) On luminescence nature of titanium and zirconium minerals. Docl Acad Nauk SSSR 263(3):1234–1237
Gaft M, Gorobets B, Naumova (1981b) Relation between luminescent properties and crystallochemistry of manganese minerals. Miner J N3:80–90 (in Russian)
Gaft M, Gorobets B, Marshukova N, Pavlovskii A (1982) Tin minerals luminescence and its use in the prospecting of tin ore deposits. Docl Acad Nauk SSSR 266(1):217–220
Gaft M, Bershov L, Krasnaya A, Yaskolko V (1985) Luminescence centers in anhydrite, barite, celestite and their synthesized analogs. Phys Chem Miner 11:255–260
Gaft M, Rassulov V, Zukova V, Rakov L (1986) Nature of zircon photoluminescence. Miner J 8 (3):74–78 (in Russian)
Gaft M, Gorobets B, Marshukova N, Pavlovskii A (1988) Diagnostics of cassiterite in the place of natural occurrence by laser induced luminescence. Docl Acad Nauk SSSR 299(1):176–179
Gaft M, Scorobogatova N, Rassulov V, Moroshkin V (1989) The use of natural silver halogens luminescence for mineral prospecting. Miner J 11:58–64 (in Russian)
Gaft M, Reisfeld R, Panczer G et al (1996) Luminescence of Eu^{3+}, Pr^{3+} and Sm^{3+} in carbonate-fluor-apatite. Acta Phys Polonica A 90:267–274
Gaft M, Reisfeld R, Panczer G et al (1997a) Accommodation of REE and Mn by apatite. Opt Mater 8:149–156
Gaft M, Reisfeld R, Panczer G et al (1997b) Luminescence of Eu^{3+} in high symmetry Ca position in apatite structure. J Lumin 72–74:572–574
Gaft M, Reisfeld R, Panczer G, Champagnon B (1997c) Reabsorption lines of molecular oxygen and water in natural apatite. Opt Mater 8(1-2):143–149
Gaft M, Reisfeld R, Panczer G et al (1998) Laser-induced time-resolved luminescence of minerals. Spectrochim Acta A 54:2163–2175
Gaft M, Reisfeld R, Panczer G et al (1999a) Luminescence of Pr^{3+} in minerals. Opt Mater 13:71–79
Gaft M, Panczer G, Uspensky E, Reisfeld R (1999b) Laser-induced time-resolved luminescence of rare-earth elements in scheelite. Mineral Mag 63:199–210
Gaft M, Panczer G, Reisfeld R, Shinno I (2000a) Laser-induced luminescence of rare-earth elements in natural zircon. J Alloys Comp 300–301:267–274
Gaft M, Boulon G, Panczer G et al (2000b) Unexpected luminescence of Cr-doped zircon crystals $ZrSiO_4$. J Lumin 87–88:1118–1121
Gaft M, Panczer G, Reisfeld R, Shinno I (2000c) Three centers of Eu^{3+} luminescence in zircon $ZrSiO_4$. J Lumin 87–89:1032–1035
Gaft M, Reisfeld R, Panczer G et al (2001) The nature of orange luminescence of mineral barite. Opt Mater 16(1-2):279–290
Gaft M, Seigel H, Panczer G, Reisfeld R (2002a) Laser-induced time-resolved luminescence of lead Pb^{2+} in Franklin, NJ, minerals. Eur J Miner 14:1041–1048
Gaft M, Panczer G, Reisfeld R et al (2002b) Laser-induced time-resolved spectroscopy of broad luminescence bands in zircon $ZrSiO_4$. Miner Petrol 76:235–246
Gaft M, Nagli L, Reisfeld R, Panczer G, Brestel M (2003a) Time-resolved luminescence of Cr^{3+} in topaz Al SiO (OH, F). J Lumin 102–103:349–356
Gaft M, Nagli L, Reisfeld R, Panczer G (2003b) Laser-induced time-resolved luminescence of titanite. Opt Mater 24:231–241
Gaft M, Nagli L, Waychunas G (2004) The nature of blue luminescence of natural benitoite $BaTiSi_3O_9$. Phys Chem Miner 31:365–373
Gaft M, Nagli L, Panczer G et al (2008a) The nature of unusual luminescence of natural calcite $CaCO_3$. Am Miner 93:158–167
Gaft M, Reisfeld R, Panczer G, Dimova M (2008b) UV-visible luminescence of Nd^{3+} in minerals. J Alloy Comp 451:56–61
Gaft M, Nagli L, Panczer G, Yeates H (2009) Laser-induced time resolved luminescence of tugtupite, sodalite and hackmanite. Phys Chem Miner 36:127–141
Gaft M, Nagli L, Panczer G et al (2011) Laser-induced time-resolved luminescence of orange kyanite Al_2SiO_5. Opt Mater 23:1476–1480

Gaft M, Strek W, Nagli L et al (2012) Laser-induced time-resolved luminescence of natural sillimanite Al_2SiO_5 and artificial Al_2SiO_5 activated by Cr. J Lumin 132:2855–2862

Gaft M, Nagli L, Panczer G, Yeates H (2013a) Laser-induced time resolved luminescence of natural margarosanite Pb(Ca, Mn)2Si_3O_9, swedenborgite $NaBe_4SbO_7$ and walstromite $BaCa_2Si_3O_9$. Eur J Mineral 25:71–77

Gaft M, Yeates H, Nagli L, Panczer G (2013b) Laser-induced time resolved luminescence of natural grossular $Ca_3Al_2(SiO_4)_3$. J Lumin 137:43–53

Gaft M, Nagli L, Panczer G, Rossman G (2013c) Long-lived laser induced time-resolved luminescence of Cr^{3+}in kyanite Al_2SiO_5. J Spectr Dynam 3:22–29

Garcia-Guinea J, Correcher V, Quejido A et al (2005) The role of rare-earth elements and Mn^{2+} point defectes on the luminescence of bavenite. Talanta 65:54–61

Geiger C, Stahl A, Rossman GR (1999) Raspberry red grossular from Mexico. Eur J Miner 11:1109–1113

Gilinskaya L, Mashkovtsev R (1995) Blue and green centers in natural apatites by ERS and optical spectroscopy data. J Struct Chem 36:89–101 (in Russian)

Glynn T, Imbusch C, Walker G (1991) Luminescence from Cr^{3+} centers in forsterite Mg_2SiO_4. J Lumin 48&49:541–544

Gorobets B (1968) On the luminescence of fluorapatite doped with rare earth elements. Opt Spectrosc 25:292–294 (in Russian)

Gorobets B, Kudrina M (1976) Typomorphic features of scheelite as revealed by their rare-earth elements luminescence spectra. Const Svoy Miner, Kiev, Naukova Dumka 10:82–88 (in Russian)

Gorobets B, Rogojine A (2001) Luminescent spectra of minerals. Handbook. RPC VIMS, Moscow

Gorobets B, Sidorenko G (1974) Luminescence of secondary uranium minerals at low temperature. At Energy 36:6–13

Gorobets B, Walker G (1994) Origins of luminescence in minerals: a summary of fundamental studies and applications. In: Marfunin A (ed) Advanced mineralogy 2, methods and instrumentation. Springer, Berlin/Heidelberg/New York, pp 138–146

Gorobets B, Gaft M, Laverova L (1978) Photoluminescence of manganese minerals. J Pricladn Spectrosk 28:1100–1102

Gorobets B, Litvintsev E, Rogojine A (1997) Luminescence sorting of nonmetallic raw materials. In: ICAM'96, 5th international on congress applied mineralogy, Warsaw, Programme and Abstract, 2–5 June 1996, pp 229–233

Götze J (2000) Cathodoluminescence microscopy and spectroscopy in applied mineralogy. Technische Universität Bergakademie Freiberg, Freiberg

Götze J, Gaft M, Möckel R (2015) Uranium and uranyl luminescence in agate/chalcedony. Miner Mag 79:985–995

Green G, Walker G (1985) Luminescence excitation spectra of Mn^{2+} in synthetic forsterite. Phys Chem Miner 12:271–278

Haberland H, Köhler A (1939) Uber die blaue Fluoreszen: von naturlichen Silikaten in UV lichte und iib er syntetischen verzeiche an silikatschmelzen mit eingebautem zweiwertigem. Naturwiss 27:275–281 (In German)

Haberland H, Karlik B, Przibram K (1934) Zur Fluoreszen: des Fluoriten II. Sitzber Akad Wiss Wien, Abt Iia 143:151–161 (in German)

Haberman D, Neuser R, Richter D (1996) Low limit of Mn^{2+} activated cathodoluminescence of calcite: state of the art. Sed Geol 101:1–7

Jasinevicius R (2009) Characterization of vibrational and electronic features in the Raman spectra of gemstones. MSc thesis, Department of Geosciences, University of Arizona, Tucson

Kasdan A, Chimenti R, de Neufville J (1981) Selective detection of uranium by laser-induced fluorescence: a potential remote-sensing technique. 2: Experimental assessment of the remote sensing of uranyl geologic targets. Appl Opt 20:1279–1307

Kasyanenko E, Matveeva O (1987) Ультрафиолетовое поглощение и люминесценция исландского шпата. J Prikladn Spectrosc 46:943–949

Kempe U, Torsten G, Nasdala L, Wolf D (2000) Relevance of cathodoluminescence for the interpretation of U-Pb zircon ages, with an example of an application to a study of zircons

from the Saxonian granulate Complex, Germany. In: Pagel M, Barbin V, Ohnenstetter D (eds) Cathodoluminescence in geosciences. Springer, Berlin/Heidenberg/New York, pp 425–457
Kemper U, Plötze M, Brachmann A, Böttcher R (2002) Stabilisation of divalent rare earth elements in natural fluorite. Miner Petrol 76:213–234
Koziarsca B, Godlewski M, Suchoki A et al (1994) Optical properties of zoisite. Phys Rev B 50:12297–12300
Krasilschikova O, Tarashchan A, Platonov A (1986) Color and luminescence of natural fluorite. Naukova Dumka, Kiev (in Russian)
Krasnobaev A, Votyakov S, Krohalev V (1988) Problems of the applied spectroscopy of minerals. Nauka, Ekaterinburg (in Russian)
Krbetschek M, Götze J, Irmer G et al (2002) The red luminescence emission of feldspar and its wavelength dependence on K, Na, Ca-composition. Miner Petrol Miner Petrol 76:167–177
Kuck S, Hartung S, Hurling S et al (1998) Emission of octahedrally coordinated Mn^{3+} in garnets. Spectrochim Acta A 54:1741–1749
Kusnetsov G, Tarashchan A (1988) Luminescence of minerals from granitic pegmatites. Naukova Dumka, Kiev (in Russian)
Kusnetsov G, Silaev V, Lupashko T, Melnikov V (1991) Rentgenoluminescence of chlorite and its typomorpohical meaning. Zap Vses Miner Ob-va 4:70–77 (in Russian)
Laurs B, Simmons W, Rossman GR et al (2003) Pezzottaite from Ambatovita: a new gem mineral. Gems & Gemology, Madagascar, pp 284–301
Lenz C, Nasdala L, Talla D et al (2015) Laser-induced REE^{3+} photoluminescence of selected accessory minerals – an "advantageous artefact" in Raman spectroscopy. Chem Geol
Lupashko T, Tarashchan A, Kvasnytsya V et al (1996) Luminescence, EPR and morphology of diamond crystals from placer deposits. In: Abstracts of third European meeting "Spectroscopic Methods in Mineralogy", Kiev, 10–13 p 27
Martin P, Carlot G, Chevarier A et al (1999) Mechanisms involved in thermal diffusion of rare earth elements in apatite. J Nucl Mater 275:268–278
Martin P, Chevarier A, Panczer G (2000) Diffusion under irradiation of rare earth elements in apatite. J Nucl Mater 278:202–206
Mazurak Z, Czaja M (1996) Optical properties of tsavorite $Ca_3Al_2(SiO_4)_3$:Cr^{3+}, V^{3+} from Kenya. J Lumin 65:335–340
Min'ko O, Bahtin A, Shepelev Yu (1978) Luminescence spectra of wollastonite as typomorphic feature of wollastonite rocks origin. Abstracts of USSR conference, Tallinn, pp 81–83 (in Russian)
Mohapatra M, Natarajan V, Rajeswari B et al (2014) On the use of bastnasite ore as a phosphor material. J Lumin 145:105–109
Mohler R, White W (1995) Influence of structural order on the luminescence of oxide spinels: Cr^{3+} activated spinels. J Electrochem Soc 142:3923–3927
Moncorge R, Cormier G, Simkon D, Capobianco J (1991) Fluorescence analysis of chromium doped forsterite (Mg_2SiO_4). IEEE J Quantum Electron 27:114–120
Moncorgé R, Manaa H, Boulon G (1994) Cr^{4+} and Mn^{5+} active centers for new solid state laser materials. Opt Mater 4:139–144
Moncorgé R, Bettinelli M, Gyot Y et al (1999) Luminescence of Ni^{2+} and Cr^{3+} centers in $MgSiO_3$ enstatite crystals. J Phys Condens Matter 11:6831–6841
Morozov A, Morozova L, Trefilov A, Feofilov P (1970) Spectral and luminescent characteristics of fluorapatite single crystals activated by rare earth ions. Opt Spectrosc 29:590–596
Nasdala L, Zhang M, Kempe U et al (2003) Spectroscopic methods applied to zircon. In: Hanchar J, Hoskin P (eds) Zircon. Review in mineralogy and geochemical 53. Mineralogy Society American, Washington, DC, pp 427–467
Nasdala L, Grambole D, Wildner M et al (2013) Radio-colouration of diamond: a spectroscopic study. Contrib Mineral Petrol 165:843–861
Nelson G, Sturge M (1965) Relation between absorption and emission in the region of the R-lines of ruby. Phys Rev A 137:1117–1130

Ollier N, Fuchs Y, Cavani O et al (2015) Influence of impurities on Cr^{3+} luminescence properties in Brazilian emeralds and alexandrite minerals. Eur Mineral J. doi:10.1127/ejm/2015/0027-2484
Panczer G, Gaft M, Marfunin A (2000) Systems of interacting luminescence centers in natural diamonds: laser-induced time-resolved luminescence and cathodoluminescence spectroscopy. In: Pagel M, Barbin V, Blanc P, Ohnenstetter D (eds) Cathodoluminescence in geosciences. Springer, Berlin, pp 359–373
Panczer G, Gaft M, de Ligny D, Champagnon B (2010) Luminescence characteristics of the Cs bearing "beryl", pezzottaite Cs(Be2, Li)Al2 (SiO3)6H2O. Eur Mineral J 22:605–612
Panczer G, De Ligny D, Mendoza C et al (2012) Raman and fluorescence in earth sciences. In: Dubessy J, Caumon M, Rull F (eds) Raman spectroscopy applied to earth sciences and cultural heritage. Chapter 2, vol 12. EMU Notes in Mineralogy, London, pp 61–80
Payne S, DeLoach L, Smith L et al (1994) Ytterbium-doped apatite-structure crystals: a new class of laser materials. J Appl Phys 76:497–503
Pedone V, Cercone K, Burrus R (1990) Activators of photoluminescence in calcite: evidence from high resolution, laser excited luminescence spectrometry. Chem Geol 88:183–190
Platonov A (1979) Color of minerals. Naukova Dumka, Kiev (in Russian)
Platonov A, Taran M, Balizkii B (1984) The nature of color of gems. Nedra, Moscow (in Russian) 196p
Platonov A, Tarashchan A, Langer K et al (1998) Electronic absorption and luminescence spectroscopic studies of kyanite single crystals: differentiation between excitation of FeTi charge transfer and Cr^{3+} dd transitions. Phys Chem Miner 25:203–212
Plotnikova S (1990) Classification and selection of diamonds for electronics. In: Kvaskov V (ed) Diamonds in electronic technology. Energoatomizdat, Moscow
Ponahlo J (2000) Cathodoluminescence as a tool in gemstone identification. In: Pagel M, Barbin V, Ohenstetter D (eds) Cathodoluminescence in geosciences. Springer, Berlin, pp 479–500
Portnov A, Gorobets B (1969) Luminescence of apatite from different types of rocks. Dokl Akad Nauk SSSR Seriya Geol 184:199–202
Povarennykh A, Platonov A, Tarashchan A, Belichenko V (1971) The colour and luminescence of tugtupite (beryllosodalite) from Ilîmaussaq, South Greenland. Medd Gronl 181:1–12
Powell R, Xi L, Gang Xu, Quarles G (1985) Spectroscopic properties of alexandrite crystals. Phys Rev B 32:2788–2797
Pringsheim P, Vogel M (1946) Luminescence of liquids and solids. Interscience, New York
Rossman GR (1988) Optical spectroscopy. In: Hawthorne FC (ed) Spectroscopic methods in mineralogy and geology, vol 18, Reviews in mineralogy. Mineralogical Society of America, Washington, DC, pp 207–254
Rushel K, Nasdala L, Gaft M et al (2007) Photoluminescence recovery upon annealing of fergusonite. Goldschmidt conference abstracts p A861
Ryabov I (2012) EPR study of chromium-doped forsterite crystals: Cr^{3+} (M1) with associated trivalent ions Al^{3+} and Sc^{3+}. Phys Chem Miner 39:725–732
Scalvi R, Li M, Scalvi L (2003) Annealing effects on optical properties of natural alexandrite. J Phys Condens Matter 15:7437–7443
Schott S, Rager H, Schurmann K, Taran M (2003) Spectroscopic study of natural gem quality "Imperial" – topazes from Ouro Preto, Brasil. Eur J Miner 15:701–706
Shenjun L, Lin L, Zulun W et al (1994) Growth and characteristics of Mg_2SiO_4:Ti crystal. J Cryst Grow 139:327–331
Shinno I (1986) Three types of photo-luminescence in natural zircon. J Japan Assoc Miner Pet Econ Geol 81:433–445
Shinno I (1987) Color and photo-luminescence of rare earth element-doped zircon. Miner J 21:119–130
Shoval S, Gaft M, Panczer G (2003) Luminescence of Cr^{3+} in natural and calcined diaspore. J Therm Anal Calorim 71:699–706

Sidike A, Kusachi I, Yamashita N (2003) Natural fluorite emitting yellow under UV light. Phys Chem Miner 30:478–485
Sidike A, Kusachi I, Yamashita N (2006) Yellow fluorescence from baghdadite and synthetic $Ca_3(Zr, Ti)Si_2O_9$. Phys Chem Miner 32:665–669
Sidike A, Sawuti A, Wang X-M et al (2007) Fine structure in photoluminescence spectrum of S_2^- center in sodalite. Phys Chem Miner 34:477–484
Sidike A, Kusachi I, Kobayashi I et al (2008) Photoluminescence spectra of S_2^- center in natural and heat treated scapolites. Phys Chem Miner 35:137–149
Sokolov S, Rassulov V (2004) A luminescence characteristics of the burbankite group minerals. Zapiski VMO 6:73–88
Solntsev V, Bukin G (1997) The nature of beryl color from rare metal pegmatite. Russian Geol Geophys 38(10):1661–1668
Solomonov V, Michailov S, Osipov V et al (1994) Application of impulse catodoluminescence to study of minerals with Cr^{3+}. Zapiski Vses Miner Obs 6:39–50 (in Russian)
Solomonov V, Mikhailov S, Lipchak A et al (2002) Impurity luminescence of alexandrite crystals. J Appl Spectrosc 69:423–429
Spinolo G, Gabrielli S, Palanza V et al (2012) Mn sites in cordierite – electron paramagnetic resonance, luminescence, and optical absorption analysis. Eur J Mineral 24:447–458
Stokes G (1852) On the Change of Refrangibility of Light. Phil Trans R Soc London 142:463–562
Suchocki A, Gilliland G, Powell R, Bowen J (1987) Spectroscopic properties of alexandrite crystals II. J Lumin 37:29–37
Taran M, Tarashchan A, Rager H et al (2003) Optical spectroscopy study of variously colored gem-quality topazes from Ouro Preto, Minas Gerais, Brazil. Phys Chem Miner 30:546–555
Tarashchan A (1978) Luminescence of minerals. Naukova Dumka, Kiev (in Russian)
Tarashchan A, Taran M, Rager H, Iwanuch W (2005) Luminescence spectroscopic study of Cr^{3+} in Brazilian topazes from Ouro Preto. Phys Chem Miner 32(10):679–690
Tolstoy N, Shinfue L (1960) Kinetick of emission of chromium luminophores. Opt Spectrosc 9:415–419
Trindade N, Tabata A, Scalvi R, Scalvi L (2011) Temperature dependent luminescence spectra of synthetic and natural alexandrite ($BeAl_2O_4$:Cr^{3+}). Mater Sci Appl 2:284–287
Trofimov A (1962) The nature of linear luminescence spectrum in zircon. Geochemistry 11:972–975 (in Russian)
Uspensky E, Aleshin P (1993) Patterns of scheelite mineralization in the Muruntau gold deposit, Uzbekistan. Int Geol Rev 35:1037–1051
Uspensky E, Novgorodova M, Mineeva R et al (1989) On europium anomaly in scheelite from gold bearing deposits. Docl Acad Nauk SSSR 304:55–59
Vergara I, Sole J, Hoyos M, Calderon T (1990) Fluorescence properties of Cr^{3+} in natural eosphorite crystals. Solid State Commun 76:289–292
Voronko K, Maksimova G, Sobol A (1991) Anisotropic luminescence centers of TR^{3+} ions in fluorapatite crystals. Opt Spectrosc 70:203–206 (in Russian)
Votyakov S, Krasnobaev A, Krohalev V (1993) Problems of applied spectroscopy of minerals. Nauka, Ekaterinburg, 233 p (in Russian)
Walker J (1979) Optical absorption and luminescence in diamond. Rep Progr Phys 42:1605–1659
Walker G, Kamaluddin B, Glynn T, Cherlock R (1994) Luminescence of Ni^{2+} centers in forsterite (Mg_2SiO_4). J Lumin 60&61:123–126
Walker G, Jaer A, Sherlock R et al (1997) Luminescence spectroscopy of Cr^{3+} and Mn^{2+} in spodumene ($LiAlSi_2O_6$). J Lumin 72–74:278–280
Waychunas G (1989) Luminescence, X-Ray emission and new spectroscopies. In: Hawthorne FC (ed) Spectroscopic methods in mineralogy and geology. Rev Mineral. Mineralogical Society of America, Washington, DC, pp 638–698
Waychunas G (2002) Luminescence of natural and synthetic apatites. In: Kohn M, Rakovan J, Hughes J (eds) Reviews in mineralogy and geochemistry. Phosphates: geochemical, geobiological, and material importance 48. Mineralogical Society of America, Washington, DC, pp 701–742

White W (1990) Photoluminescence, candoluminescence, and radical recombination luminescence of minerals. In: Coyne L, McKeever W, Blake D (eds) Spectroscopic characterization of minerals and their surfaces. American Chemical Society, Washington, DC, pp 118–134
White W, Masako M, Linnehan et al (1986) Absorption and luminescence of Fe^{3+} in single-crystal orthoclase. Am Miner 71:1415–1419
Wojtowicz A (1991) Luminescence of Cr^{3+} in kyanite. J Lumin 50:221–230
Wojtowicz A, Lempicki (1988) Luminescence of Cr^{3+} in sillimanite. Phys Rev B 39:8695–8701
Yang C (1995) Ionoluminescence techniques for geological applications. Department of Nuclear Physics, Lund Institute of Technology, Lund
Yu X, Xu X, Yang P et al (2012) Photoluminescence properties and the self-reduction process of $CaAl_2Si_2O_8$: Eu phosphor. Res Bull 47:117–120
Zaitsev A (2005) Optical properties of diamond. Springer, Berlin, 502 p

Chapter 5
Interpretation of Luminescence Centers

The interpretation of luminescence in minerals begins with the characterization of the luminescence centers, including the identity of the ions involved, their locations in the crystal structure, their energetic interactions, and their modes of energy transfer with each other, with other ions in the structure, or with vibrational states (Jørgensen and Judd 1964; Reisfeld 2015 and references within).

5.1 Rare-Earth Elements (REE)

The geochemical rare-earth (REE) data provide major evidence to test, support, and constrain various theories and processes proposed for the origin of the major rock groups in the Earth. Our understanding of the various rock-forming processes and geochemical evolution of the Earth has been considerably aided by data provided by the geochemical studies of the rare-earth elements. Neutron-activation analysis, isotope dilution mass spectrometry, inductive coupled plasma spectrometry, optical emission spectrometry, and X-ray fluorescence spectrometry are used to determine the RE contents in geological materials. The naturally occurring members of the lanthanide series make up a significant class of activator ions in minerals. The most common oxidation state for lanthanide ions is trivalent, but several members are known to exist in the divalent state. The optical properties of rare earths in minerals may be analyzed in a manner similar to that of impurities in crystals.

5.1.1 Introduction

The rare earths (REE) also named lanthanides (Ln) are a fascinating group of elements the optical properties of which arise from the inner f-electrons which are starting with one in Cerium and terminate with thirteen in Ytterbium. The

M. Gaft et al., *Modern Luminescence Spectroscopy of Minerals and Materials*,
Springer Mineralogy, DOI 10.1007/978-3-319-24765-6_5

transition probabilities within the for f-orbital are forbidden by Laporte rule and become partially allowed either by mixing of the for f with 5 d-orbital or with a charge transfer states of the neighboring ligands. The basic theory of the electronic spectra can be found in references. Since the beginning of the twentieth century, the radiative transitions of lanthanide ions have received academic and industrial attention. The characteristic absorption and emission spectra of lanthanide compounds in the visible, near-ultraviolet, and infrared are attributed to transitions between *4f* levels due to the fact that they present sharp line with oscillators strengths typically of the order of 10^{-6}. These transitions are electric dipole forbidden but became allowed as forced electric dipole transitions.

5.1.2 Electronic Levels of 4f Elements

The energy levels of the triply charged ions may be obtained with the highest accuracy from the analysis of the free ion spectra. Another way of obtaining information on the free ion states is to calculate the energy levels theoretically. Such calculations are practically a necessity for guiding the interpretation of the empirical results. Moreover, the calculations furnish the only way to obtain eigenvectors. The object of the theory is the calculation of the energy levels and the wave functions of atoms even of such complexity as the rare earth ions. As in all quantum mechanical calculation of complex systems, the theory starts with a system which is sufficiently simplified by the omission of certain interactions for its properties to be obtained without undue complications. This is the so-called zero-order approximation, characterized by the Hamiltonian H^0 with the energies $E_n{}^0$ and wavefunctions U_n. The zero order approximation considers the outer electrons in a central field produced by the nucleus and the 54 electrons in the completed xenon-like shell but disregards the electrostatic repulsion between the outer electrons and any influence of their spins. All outer electrons are, therefore, completely independent of each other and the energy cannot depend on the relative orientations of their orbits. In this approximation, all states of a configuration have the same energy. The state of an individual single electron is characterized by four quantum numbers n, l, m_l, m_s. A configuration of N electrons is the assembly of states in which the principal quantum number n and the orbital angular momentum l for each electron are given, but m_l and m_s can take all possible values. This can also be expressed by saying that the size and shape of each electron orbit is given but its orientation in space and the spin orientation are arbitrary. For the condensed phase spectra of the rare earths, the configuration where all outer electrons are in 4f orbits, the $4f^N$ configuration is of greatest importance.

The degeneracy of a configuration in the zero order approximation is given by the product of $2(2l_i + 1)$ for all outer electrons when no equivalent electrons are involved. For the $4f^N$ configuration which consist entirely of equivalent electrons, the degeneracies are given by the binomial coefficient $\binom{14}{N}$ which are 1, 14, 91, 364,

1001, 2002, 3003, 3432 or N = 0–7. For the second half of the group, the degeneracies are the same as in the first half, in the opposite order in accordance with the principle that N 4f electrons are equivalent to 14 – N holes in the completed shell and a configuration with a certain number of holes has the same number and kind of states as a configuration with the same number of electrons. If the interaction between the electrons and between the orbital and spin angular moment are introduced by appropriate terms H^1, H^2, etc. in the Hamiltonian, some of the degeneracy is removed. When the ion is then placed in a condensed matrix field, all of it may be removed so that only non-degenerate states remain. The modification of the state through the influence of the interaction can be calculated from conventional perturbation theory. There are more than 3000 different wavefunctions to deal with in one configuration near the middle of the rare earths group. This shows that the mere writing down of the wavefunctions and interaction matrices alone would be a nearly impossible task. For this reason, systematic methods have been developed, chiefly by Racah, which reduce the labor of computation considerably. The chief interactions in the free ion are the electrostatic repulsion between the outer electrons expressed by:

$$H^1 \sum_{ij} \frac{e^2}{r_{ij}} \tag{5.1}$$

Summed over all electron pairs, and the spin-orbit interaction by

$$H^2 = \sum_{i=1}^{n} \zeta_i (l_1 * s_i) \tag{5.2}$$

Where ζ_1 is the spin-orbit interaction constant for the ith electron. For a Coulomb field with an effective nuclear charge Z'',

$$\zeta_i = \frac{\alpha^4 R Z^n}{r_i^s} \tag{5.3}$$

with α fine structure constant and R the Rydberg constant. The effective charge Z'' is, in general, different from the effective charge Z' occurring in the electrostatic interactions.

In the interaction between the states within one configuration is taken into consideration and, thus interaction between states of different configurations are left out, this is the approximation usually employed. It is reasonably good for the $4f^N$ configuration of trivalent rare earth ions because this configuration is isolated from all others with which it could interact. Because of symmetry considerations (parity), the nearest configuration that can interact with $4f^N$ is $4f^{N-1}$ or $4f^{N-1}5d$ or $4f^{N-1}6p$ or $4f^{N-1}6s$ which are of the order of 10^5 cm^{-1} above it. Broken shell configurations such as $5p^5 4f^{N+1}$ can be of importance even though the energy difference is much larger, since the overlap integral may be quite large. According to the radial integral calculations, the contribution from $5p^5 4f^{N+1}$ is even more

Table 5.1 Maximum number of interacting levels under electrostatic H^1 and spin orbit H^2 interaction

N=	1	2	3	4	5	6	7
H^1	1	1	2	4	7	9	10
$H^1 + H^2$	1	3	7	19	30	46	50
Total number of levels	2	13	41	107	198	295	327
Multiplets	1	7	17	47	73	119	119

important. When only the influence of the electrostatic interaction is considered, we have the following situation. From the original states in which each electron is characterized by the four quantum numbers: n, l, m_e, m_s, (where only m_e and m_s are of importance for further consideration, as n and l are fixed for the whole configuration), we can by standard methods obtain linear combinations expressing the wave functions of states characterized by the quantum numbers L, S, $L_Z = M_L$, $S_Z = M_S$. L and S correspond as usual to the total orbital and spin angular moment, respectively, and M_Z and M_S to the projection of these vectors on a fixed axis, usually the Z-axis. The energy does not depend on M_Z and M_S. The energy actually does not depend directly on the numerical value of the total spin S either. In this approximation, S serves merely to identify a symmetry property of the state. Each level characterized by L and S is (2 L + 1)(2S + 1) – fold degenerate. Electrostatic interaction (mixing of states) can take place only between two levels with the same L and S. Such interacting levels do not exist for one- and two-electron configurations, but interacting pairs become increasingly common as N increases. The maximum number of interacting levels for a particular value of L and S is given in the second row of Table 5.1.

In order to obtain energies and eigenvectors, it is necessary to solve secular determinants of order equal to the number of interacting states that that is at most of the order of the number given in Table 5.1.

5.1.3 *Matrix Elements of Operators*

Calculation of the energy levels of an atom or ion normally processed by first figuring the matrix elements of the electrostatic perturbation potential V, defined as:

$$V = \sum_{i=1}^{N} \left[-\frac{Ze^2}{r_i} - U(r_i) \right] + \sum_{i<j}^{N} \frac{e^2}{r_{ij}} \tag{5.4}$$

The first term is purely radial and contributes to energy shifts that are the same for all levels belonging to a given configuration without affecting the energy-level of the configuration. The repulsive Coulomb interaction of the electrons will be different for different states of the same configuration. The summations in Eq. (5.4) are over the coordinates of all the electrons. However, the term energies of a configuration which contains closed shells are the same, apart from a constant energy shift of all the terms, as those of the configuration in which the closed shell have been omitted. Thus in considering the energy-level structure of a configuration, we shall restrict the summation in Eq. (5.4) to those electrons in incomplete shells. Hence in calculation the energy-level structure of a configuration, produced by the repulsive Coulomb interaction, we must calculate matrix elements of the type

$$\tau SLJM\left|\sum\nolimits_{i<j}\frac{e^2}{r_{ij}}\right|\tau' S' L' J' M' \tag{5.5}$$

Since the electrostatic Hamiltonian commutes with the angular momentum operators corresponding to L^2, S^2, J^2 and M the matrix elements will be diagonal in L and S (although not it τ) and independent of J and M. Calculation of the matrix elements of Eq. (5.5) commences by first expanding the interaction between each pair of electrons in Legendre polynomials of the cosine of the angle w_{ij} between the vector from the nucleus to two electrons.

$r<$ indicates the distance from the nucleus (the point of origin) to the nearer electron and $r>$ indicates distance from the nucleus to the further away electron.

The energies and wavefunctions can be expressed in terms of certain integrals

$$F^K\int_0^\infty\int_0^\infty\frac{r_<^k}{r_>^{k+1}}R_i^2(r_i)R_j^2(r_j)r_i^2r_j^2dr_idr_j \tag{5.6}$$

Where $r<$ is the smaller and $r>$ is the greater of r_i and r_j. The Slater integrals are usually expressed in terms of a subscripted quantity F_k where for f-electrons the relationships are $F_2=F^2/225$; $F^4=F^4/1089$; $F_6=F^6/7361$. If the wavefunctions of the central field approximation are known, the integrals F_k can be calculated. For the $4f^N$ configuration only the integrals F_0, F_2, F_4, F_6 need be considered. Here, a further simplification can be made which for the $4f^N$ configuration probably introduced only small errors in comparison to those present because of neglect of the inter-configuration interaction. In the central field is a Coulomb one, it can be characterized by a single constant, the effective nuclear charge Z'. In that case, the F_k must all be functions of Z' or of F_2 one can write $F_4=0.145\ F_2$; $F_6=0.0164\ F_2$ and (for the triply ionized rare earth) $F_2=12.4Z'=12.4\ (Z\text{-}34)$ approximately. The parameter F_0 is a common additive constant to all levels and for this reason, is of no importance for the structure of the configuration.

Recent investigations have shown that the wavefunctions, in fact, deviate considerably from being hydrogenic, but in spite of this, the ratios remain relatively stable. As more experimental details of the crystal and first ion levels become available, these problems will become increasingly important. The simplified theory has nevertheless proved very useful in clarifying the level assignment. Although the theoretical calculations are of limited validity they give a reasonably good account of the observed energies.

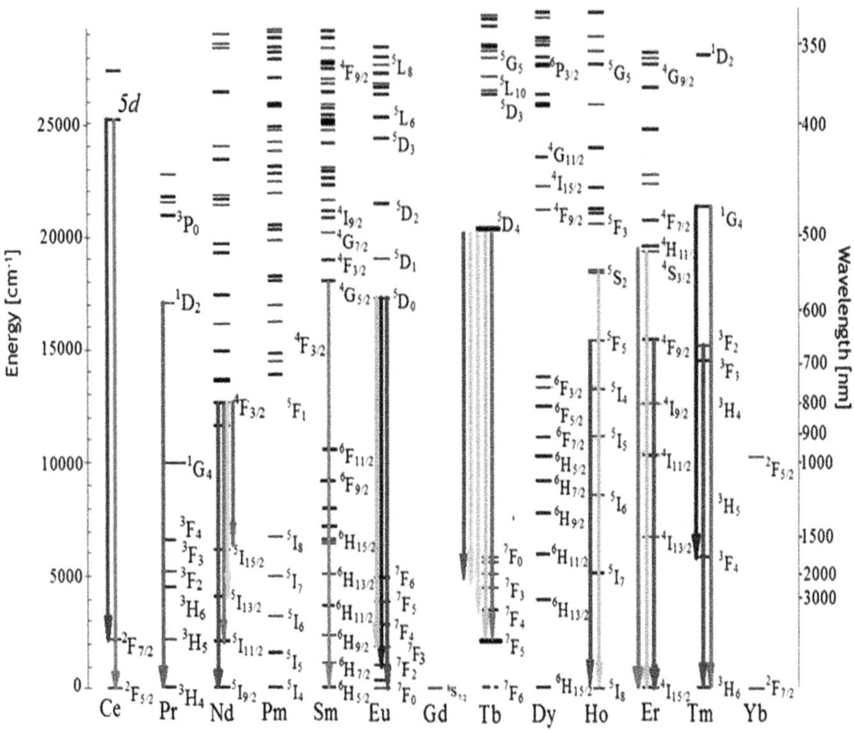

Fig. 5.1 Energy level of the trivalent lanthanides

The calculated levels are, however, good enough for the identification of the empirical levels, at least the lower ones, and ones identified, the latter can, of course, be determined with any desired accuracy. The energies and the assignments of the grand and exited levels can be found in (Reisfeld and Jorgensen 1977) and in Figure 5.1 where the electronic levels are presented. Figure 5.2 presents the excitation spectra (Shinoya and Yen 1999).

5.1.4 Results of Theoretical Calculation

Most detailed theoretical calculations of energy levels and wavefunctions have been carried out on the $4f^N$ configuration of the trivalent ions and are of direct interest for the crystal spectra. In all cases the calculations for $4f^N$ and $^{4fN+1}$ are the same but with a positive ζ for $N < 7$. Some examples for the simple cases are $4f^1$ and $4f^{13}$: This is the case of (Ce^{3+}, Yb^{3+}). There is no problem here, as there is only one electron. For $4f^2$ and $4f^{12}$: Pr^{3+}, Tm^{3+} here the relatively small number of levels and the fact that nothing more complicated than the solution of a cubic equation makes this a problem suitable even for a desk calculator. The electrostatic

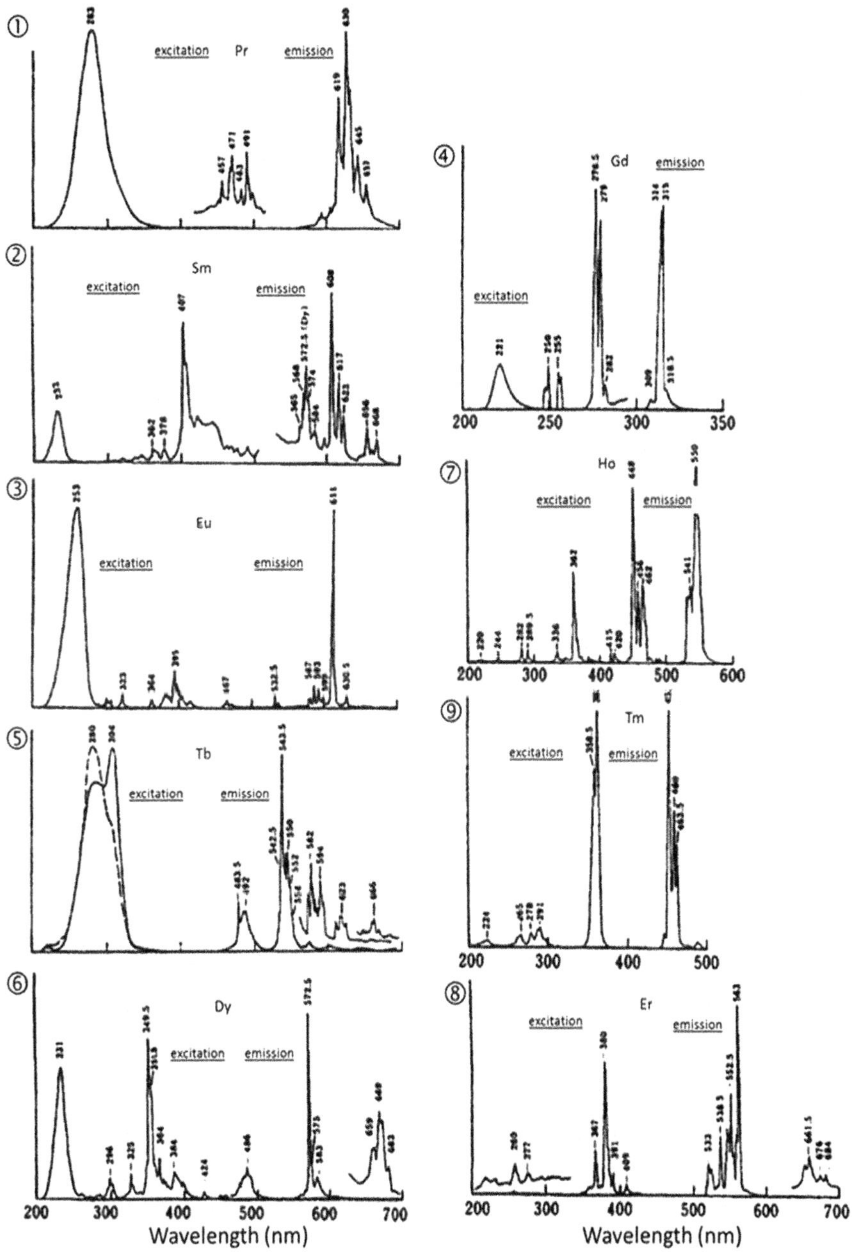

Fig. 5.2 Excitation spectra of trivalent REE (Shinoya and Yen 1999)

interaction involves no non-diagonal matrix elements, and the positions of the various multiplets can be directly given:

$$
\begin{aligned}
{}^3H &: E_1 = 0\\
{}^3F &: E_2 = 15F_2 + 18F_4 - 273F_6 = 13.133F_2\\
{}^1G &: E_3 = -5F_2 + 148F_4 + 91F_6 = 17.952F_2\\
{}^1D &: E_4 = 44F_2 - 48F_4 + 728F_6 = 48.979F_2\\
{}^3P &: E_5 = 70F_2 + 84F_4 - 127F_6 = 61.286F_2\\
{}^1I &: E_6 = 50F_2 + 60F_4 + 14F_6 = 58.923F_2\\
{}^1S &: E_7 = 85F_2 + 249F_4 + 1729F_6 = 149.461F_2
\end{aligned}
$$

The last column gives the results in a Coulomb field.

5.1.5 Transition Probabilities of 4 f-elements and the Judd-Ofelt Theory

The optical transitions so typical of the spectra of rare earth (RE) within condensed phase usually correspond to intra- f^N transitions of predominantly electric-dipole character. For a free ion, electric-dipole transitions between states of the same configuration are strictly parity forbidden, and thus any explanation of the observed spectra of glasses must concern itself with non-centro-symmetric interactions that lead to a mixing of states of opposite parity. This mixing may result from several distinct mechanisms. One of the most obvious mechanisms is simply the coupling of states of opposite parity by way of the *odd* terms in the crystal field expansion.

The systematic theoretical approach explaining the contribution of the next configuration to the *f* configuration via odd terms in the condensed phase expansion and the energy difference between the two configurations together with the radial integrals can be found in the literature.

Several selection rules follow in application of the triangular conditions to the 3-j and 6-j symbols contained in the theoretical equations. These are:

$$\Delta L = \pm l;\ \Delta S = 0;\ \Delta L \le 21;\ \Delta J \le 21$$

The $\Delta L = \pm 1$ selection rule means that for f^N configurations the perturbing n figurations may differ only by the single substitution of n'd or n'g-electrons. Spin-orbit interaction will usually lead to a breakdown of the selection rules on S and L. The selection rule $\Delta J \le 6$ will be valid as long as J is a "good" quantum number. This selection rule is seen in the usual absence of the transitions ${}^5D_0 \rightarrow {}^7F_3$ and ${}^7F_5$ in Eu^{3+} spectra. In general, the oscillator strength f or a transition may be expressed as

$$
\begin{aligned}
f_e &= \frac{8\pi^2 mc}{3he^2}\bar{\upsilon}\left(A\middle|P_\rho^{(1)}\middle|B\right)^2\frac{(n^2+2)^2}{9}\\
&= \bar{\upsilon}\textstyle\sum_{\lambda,q}\zeta(\lambda,q,\rho)\left(l^N\alpha SLJJ_Z\middle|U_{q+\rho}^{(\lambda)}\middle|l^N\alpha' S'L'J'J'_Z\right)^2
\end{aligned}
\tag{5.7}
$$

$$\zeta(\lambda, q, \rho) = \frac{8\eta^2 mc}{3he^2} \frac{(n^2+2)^2}{9} Y^2(\lambda, q, \rho) \tag{5.8}$$

Eq. (5.7) in a simplified form is

$$\begin{aligned} f_m &= \frac{8\pi^2 mc}{3h^2} \overline{\upsilon} \left| \frac{-e}{2mc} \left(\alpha SLJJ_Z | L + g_S S | \alpha SLJJ'_Z \right) \right|^2 n \\ &= 4.028 \times 10^{-11} \overline{\upsilon} \left(\alpha \alpha SLJJ_Z | L + g_S S | \alpha SLJJ'_Z \right)^2 n \end{aligned} \tag{5.9}$$

Where $\overline{\upsilon}$ is in units of cm^{-1} and n is the refractive index. The matrix of $L + g_s S$ will be given by

$$(\alpha LJJ_Z | L + g_S S | \alpha LJJ_Z) = (-1)^{J-J_Z} \begin{pmatrix} J & 1 & J' \\ -J_Z & \rho & J_Z' \end{pmatrix} \times (\alpha SLJ \| L + g_S S \| \alpha SLJ') \tag{5.10}$$

where $\rho = 0$ gives the z-components of S and L, which will correspond to the absorption or emission of σ-polarized light, and $\rho = \pm 1$ gives the $x \pm iy$-corresponding to π-polarized light. The above equations hold for a perturbing configuration of the type $nl^{N-1}n'l'$. Perturbing configurations of the type $n'l'^{4l+1}nl^{N+1}$ will differ in the angular factors by not more than a phase factor. The summation index is limited to even values (that is, 2, 4, and 6 for f-electrons).

The following selection rules may be readily established:

$\Delta l = 0$	$\Delta S = 0$	$\Delta L = 0$	$\Delta J = 0, \pm 1$	(not $0 \leftrightarrow 0$)
$\Delta J_Z = 0$			$\Delta \mu = 0$	(σ-polarization)
$\Delta J_Z = \pm 1$			$\Delta \mu = \pm 1$	(π-polarization)

In the trivalent REE the transitions of interest take place between states within the [Xe] $4f^1 5d^0 6s^0$ configuration (with $n = 1$ for Ce^{3+} to $n = 13$ for Yb^{3+}). As a consequence, they are parity forbidden and consist mainly of weak magnetic dipole (MD) and induced electric dipole (ED) transitions. The intensities of the MD transitions are practically not influenced by the chemical surroundings of the ion, whereas those of the ED transitions are quite sensitive to it. Judd-Ofelt theory has been very successful in understanding and predicting the spectral intensities of the latter, especially for ions in inorganic glasses and crystals. Also, the absorption spectra of lanthanide complexes in solution and molecular crystals have been subjected to analyses on the basis of Judd-Ofelt theory, but it has rarely been used to predict the emissive properties of lanthanide complexes. A wide variety of lanthanide complexes has been synthesized and studied over the past decades, some of which are highly luminescent. One of the attractions of luminescent lanthanide complexes is that the organic ligand can be endowed with specific physical and chemical properties through chemical design and synthesis.

5.2 Radiative Transitions in Rare Earth Elements

Radiative relaxation from an excited state ψJ of a lanthanide ion usually occurs in various lower lying state ψ′J′, giving rise to several lines in the emission spectrum. For example, the red luminescence of Eu^{3+} is a result of transitions from its $^5D_0$ state to all of the lower lying ^7F_J levels. The spontaneous emission probability, A, of the transition ψJ → ψ′J′ is related to its dipole strength according to

$$A(\psi J, \psi' J') = \frac{64\pi^4 \upsilon^{\sim S}}{3h(2J+1)} \left[\frac{n(n^2-2)^2}{9} D_{ED} + n^3 D_{MD} \right] \qquad (5.11)$$

Here is υ is the average transition energy in cm^{-1}, h is Planck's constant (6.63×10^{-27} erg s), and 2 J + 1 is the degeneracy of the initial state (1 for $^5D_0$). D_{ED} and D_{MD} are the electric and magnetic dipole strengths (in esu^2 cm^2) respectively. The factors containing the medium's refractive index, *n*, result from local field corrections that convert the external electromagnetic field into an effective field at the location of the active center in the dielectric medium. The transitions from $^5D_0$ to $^7F_{0,3,5}$ ($J' = 0, 3, 5$) are forbidden both in magnetic and induced electric dipole schemes (D_{ED} and D_{MD} are zero). The transition in $^7F_1$ ($J' = 1$) is the only magnetic dipole contribution. Magnetic dipole transitions in lanthanide ions are practically independent of the ion's surroundings and can be well calculated by theory ($D_{MD} = 9.6 \times 10^{-12}$ esu^2 $cm^2 = 9.6$ 10^{-6} $debye^2$). The remaining transitions (J = 2, 4, 6) are purely of induced dipole nature. According to the Judd-Ofelt theory, the strength of all induced dipole transitions (absorption and emission) of a lanthanide ion in a certain matrix can be calculated on the basis of only three parameters Ω_λ, using the expression

$$D_{ED} = e^2 \sum\nolimits_{\lambda=2,4,6} \Omega_\lambda \left| \left\langle J \| U^{(\lambda)} \| \right\rangle J' \right|^2 \qquad (5.12)$$

The elementary charge *e is* 4.803×10^{-10} esu. The Ω_λ, parameters belong the particular combination of the RE ion and its coordinating environment, end are usually determined experimentally. $|\langle J \| U^{(\lambda)} \| \rangle J'|^2$ are the square reduce dmatrix elements whose values are independent of the chemical environment of the ion. A set of values of all matrix elements for all lanthanide ions is now available in the literature, but smaller sets can be found. The matrix elements that may be used in Eq. (5.12) for Eu^{3+} are tabulated in Table 5.2.

Table 5.2 Squared reduced matrix elements for calculation of the dipole strengths of the allowed induced ED transitions in the emission spectrum of Eu^{3+}

$^5D_0 \rightarrow$	$\|\langle J \| U^{(2)} \| J' \rangle\|^2$	$\|\langle J \| U^{(4)} \| J' \rangle\|^2$	$\|\langle J \| U^{(6)} \| J' \rangle\|^2$
$^7F_2$	0.0032	0	0
$^7F_4$	0	0.0023	0
$^7F_6$	0	0	0.0002

Transitions arising from the ^5D_o level of the $4f^6$ electronic configuration of Eu^{3+} are intensities between the magnetic dipolar $^5D_0 \rightarrow {}^7F_1$ ($\approx$ 590 nm) and the electric dipolar $^5D_0 \rightarrow {}^7F_2$ ($\approx$ 610 nm) transitions. The higher this ratio, the closer the local symmetry around Eu^{3+} is to an inversion center. In the standard theory, the spontaneous emission of a integrated coefficient of the transition between two manifolds J and J' is given by

$$A_{JJ'} = \frac{4e^2\omega^S}{3\hbar c^S}\left[\frac{n(n^2-2)^2}{9}S_{ed} + n^3 S_{md}\right] \tag{5.13}$$

Where ω is a angular frequency of the transition, e is the electronic charge, c is the velocity of light, $\hbar$ is Planck's constant over 2π and n is the refractive index of the medium. The electric and magnetic dipole strengths, respectively, S_{ed} and S_{md} (in units of e^2), are given by

$$S_{ed} = \frac{1}{2J+1}\sum\nolimits_{\lambda=2,4,6}\Omega_\lambda\left\langle \alpha' J\|U^{(\lambda)}\|\alpha J\right\rangle^2 \tag{5.14}$$

Where the quantities Ω_λ are the Judd-Ofelt intensity parameters as described before and m is the electron mass.

$$S_{md} = \frac{\hbar^2}{4cm_e c^2}\left\langle \alpha' J\|L+2S\|\alpha J\right\rangle^2 \frac{1}{2J+1} \tag{5.15}$$

The reduced matrix elements appearing in Eq. (5.14) are evaluated in the intermediate coupling scheme, and the angular momentum operators L and S are in the units of $\hbar$. The corresponding expression for the oscillator strength may be obtained from the relation

$$P_{JJ'} = \frac{2J=1}{2J'+1}\frac{mc^2}{2\omega^2 e^2 n^2}A_{jj'} \tag{5.16}$$

The intensity parameters Ω_λ depend on both chemical environment and the lanthanide ion. Theoretically, they are given by

$$\Omega_\lambda = (2\lambda+1)\sum\nolimits_{t.p}\frac{|B_{\lambda tp}|^2}{2t+1} \tag{5.17}$$

Where $B_{\lambda p}$ depends on the radial wave function, and the odd crystal parameters and energy difference between the $4f$ and next excited configuration of the opposite parity.

Here we are discussing only the static model and omitting the vibronic interaction.

$$B_{\lambda t p} = \frac{2}{\Delta E}\langle rt+1\rangle\theta(t,\lambda)\gamma_{p}^{t} - \left[\frac{(\lambda+1)(2\lambda+3)}{(2\lambda+1)}\right]^{\frac{1}{2}}\langle r^{\lambda}\rangle(-\sigma_{\lambda})\left\langle 3\|C^{(\lambda)}\|3\right\rangle\Gamma_{p}^{t}\sigma_{t,\lambda+1} \tag{5.18}$$

Where ΔE is the energy difference between the barycenters of the excited $4f^{N-1}5d$ and ground $4f^{N}$ configurations, $\langle r^{\lambda}\rangle$ is a radial expectation value, $\theta(t,\lambda)$ is a numerical factor, σ_{λ} is a screening factor, $C^{(\lambda)}$ is a Racah tensor operator of rank λ, and $\sigma_{t,\lambda+1}$ is the Kronecker delta function. The first term in the right-hand-side of Eq. 5.18 corresponds to the forced electric dipole mechanism as expressed by the average energy denominator method, and the second term corresponds to the dynamic coupling mechanism within the point dipole isotropic ligand polarizability approximation. The measured intensity of an absorption band is related to the probability (P) for absorption of radiant energy (oscillator strength) by the expression

$$P = \frac{2303\ cm^{2}}{N\pi 2^{2}}\int \varepsilon_{i}(\upsilon)d\upsilon = 4.318 * 10^{9}\int \varepsilon_{i}(\upsilon)d\upsilon \tag{5.18a}$$

where ε is the molar absorptivity at the energy υ (cm^{-1}) and the other symbols have their usual meanings.

5.3 Relaxation Process in Trivalent Rare Earths

The luminescence intensity of emission from a given level depends on the radiative and non-radiative probabilities. The quantum yields is expressed as

$$QY = \frac{\Sigma Ar}{\Sigma Ar + Wnr} \tag{5.19}$$

where Ar are the radiative transition probabilities and Wnr the non-radiative ones. The several processes can affect the quantum efficiency due to additional non-radiative relaxations. These can arise from multiphonon decay, and concentration quenching, in addition to the energy transfer to neighboring ions. All these processes will lower the quantum efficiency. A multiphonon decay mechanism is responsible for the non-radiative decay in most cases. The rates as functions of the energy gaps are straight lines as in Figs. 5.3 and 5.4. The ab initio calculation of the transition rate between two electronic states with the emission of p phonons involves a very complicated sum over phonon modes and intermediate states. Due to this complexity these sums are extremely difficult to compute: However, it is just this complexity which permits a very simple phenomenological theory to be used. There are an extremely large number of ways in which p phonons can be

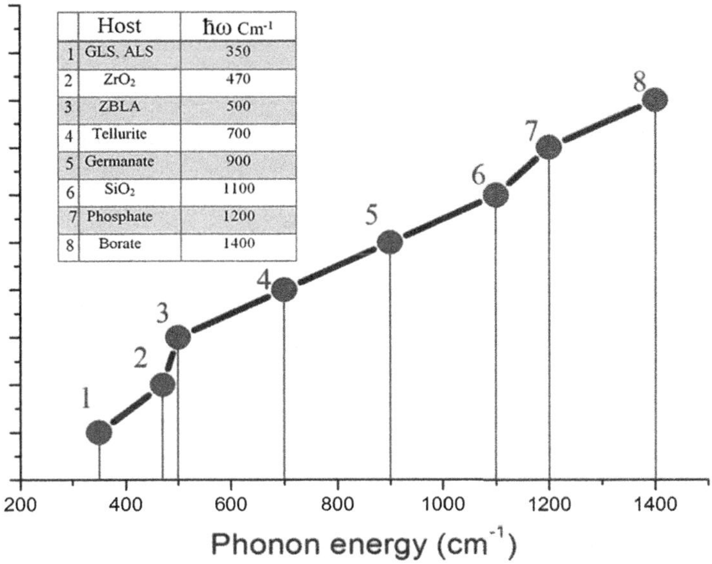

Fig. 5.3 Relative nonradiative relaxation of the highest phonon energy as a function of the host materials: * GLS, ALS: $3Ga_2S_3{*}La_2S_3$ (GLS), $3Al_2S_3{*}La_2S_3$ (ALS) ** ZBLA: 57 % ZrF_4 – 34 % BaF_2 – 4 % AlF_3 - (5-X) % LaF_3 – X % PrF_3 with $0 < X < 5$

Fig. 5.4 Energy gap to next lower level

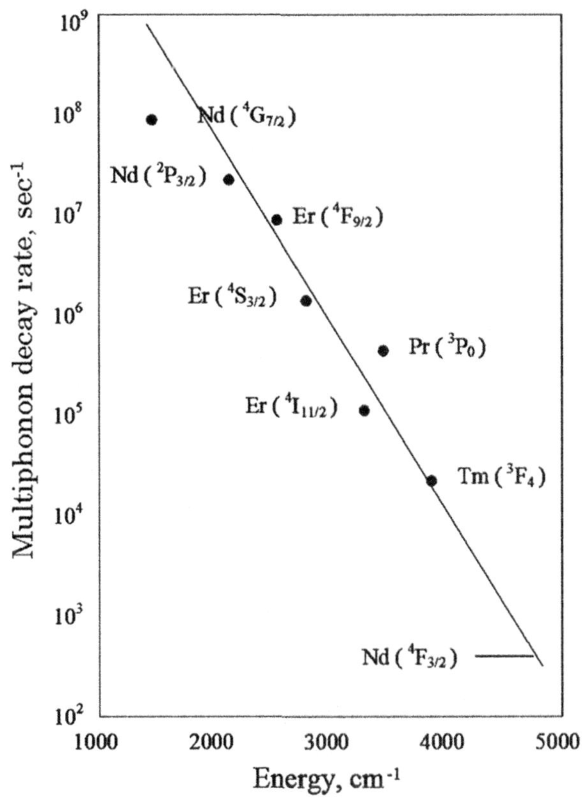

emitted and the sums over phonon modes and intermediate states are essentially a statistical average of matrix elements. In the phenomenological approach it is assumed that the ration of the pth and $(p-1)^{th}$ processes will be given by a coupling constant characteristic of the matrix in which the rear earth is situated and not depending on the rare earth electronic states. For a given lattice at low temperatures the spontaneous relaxation rate is given by:

$$W(0) = Be^{\alpha \Delta E} \tag{5.20}$$

Where B and α are characteristics of the host (a is negative). Experimental data have shown that the approach is very good for a large variety of hosts. In this way all multiphonon rates can be inferred from a few measured rates.

The dominant emission process is the one which requires the least number of phonons to be emitted. The minimum number of phonon required for a transition between states separated by an energy gap ΔE is

$$p = \frac{\Delta E}{\hbar \omega_{\text{max}}} \tag{5.21}$$

Where $\hbar \omega_{\text{max}}$ is the maximum energy of optical phonons. With increasing temperature the simulated emission of phonons by termal phonons increases the relaxation rate W according to

$$W(T) = W(0)(1 + \bar{n}(\hbar \omega_{\text{max}}))^p \tag{5.22}$$

where n is the average occupation number of phonons at energy $\hbar \omega_{\text{max}}$.

The non-radiative relaxations in the rare earth ions are related to their excited state populations and are governed by the energy difference between the emitting level and the next lower level, separated by the number of phonons of the host. In the weak-coupling case, for the rare earth ions, the temperature dependence of the non-radiative rate is given by:

$$W(T) = \beta \exp[-(\Delta \text{E} - 2h\nu_{\text{max}})\alpha] \left(\left[\exp\left(\frac{h\nu}{kT} \right) - 1 \right]^{-1} + 1 \right)^p \tag{5.23}$$

where $p = {}^{\Delta \text{E}}/_{h\nu}$, ΔE is the energy difference between the levels involved, α and β are constants and ν_{max} is the highest available vibrational frequency of the surroundings of the rare earth ion. For example, Eu^{3+} may not only emit from $^5D_0$, but also from $^5D_1$ and $^5D_2$. The Tb^{3+} ion may not only emit from $^5D_4$ (green), but also from $^5D_3$ (blue). However, this depends critically upon the host lattice.

5.4 Energy Transfer Between Two Rare Earth Ions

The above sections summarize the ways in which the electronic f levels of RE ions are populated and depopulated by radiative and non-radiative transitions. As already mentioned the population density of the luminescent levels is generally weak because of the low transition probabilities. However there are other possibilities to increase the luminescence mainly by creating strongly absorbing species that will transfer the energy to the RE ion. The three main ways of such excitation are:

1. Exciting closed shell transition metal complexes such as tungstates, molybdates, vanadates, titanates, etc. This way of excitation and its transfer to the RE has been known since the late 1950s. The method has been applied to phosphors for luminescent lamps and television screens. There is a vast literature and many reports by Phillips, General Electric, Westinghouse, etc.
2. Energy transfer from strongly absorbing RE due to f-d transitions or charge transfer.
3. Energy transfer between the ligands and the RE ion in complexes.

Rare earth ions are especially suitable for energy transfer studies due to their well-defined and narrow electronic levels, to which absorption occurs and from which fluorescence is observed. Symbolically, energy transfer can be written as

$$2(D) \rightarrow 1(D) \rightarrow 1(A) \rightarrow 2(A) \text{ or } D^* + A \rightarrow D + A^* \tag{5.24}$$

The donor system returns from the excited state 2(D) to the ground state 1(D) and the energy releases is used to bring the activator system from the ground state 1 (A) to its exited state 2(A). In the case of the rare earths, the transfer is a non-radiative one, i.e. no photon will appear in the system during transfer. Förster predicted that for the organic system, the rate of energy is proportional to the overlap of the donor emission and the acceptor absorption spectra and to R^{-6}, where R is the distance between the donor and the acceptor. The probability of energy transfer by dipole-dipole (dd) interaction in a simple case where the Born-Oppenheimer approximation holds is given as:

$$P_{da(dd)} = \left(\frac{3h^4c^4Q_a}{4\pi R^6 n^4 \tau_d}\right)\left(\frac{\varepsilon}{k^{1/2}\varepsilon_c}\right)^4 \times \int \left[\frac{f_d(E)f_a(E)}{E^4}\right] dE \tag{5.25}$$

Where R is the separation of nuclei of donors and acceptors, ε_c is the electric field within the crystal, ε is the electric field in vacuum and $f_d(E)$ is the observed shape of the emission band normalized to unity;

$$\int f_d(E)dE = 1.$$

E is the energy; the subscripts d and a refer to donors and acceptors, respectively; $f_a(E) = \sigma(E)$ is the normalized function of the acceptor absorption

$$Q_a = \int \sigma(E)dE,$$

The measured area under the absorption band, and

$$\int F_a(E)dE = 1.$$

τ_d is the decay constant of the pure donor; and σ is the absorption cross section. The equation may also be written as follows:

$$P_{da(dd)} = \frac{3hc^2 Q_a Q_d g_d}{4\pi^3 n R^6 g_{d'}} \left(\frac{\varepsilon}{k^{1/2}\varepsilon_c}\right)^4 \frac{f_d(E)F_a(E)}{E^2} dE \tag{5.26}$$

where g are the degeneracies of the initial an excited state of the donor and Q are the areas of the acceptor and donor absorption curves.

In addition to dipole-dipole energy transfer we may observe energy transfer due to exchange interaction. The probability p(exc) of these can be written according as

$$p(exc) = \frac{2\pi}{h} Z^2 f_d(E)F_a(E)dE \tag{5.27}$$

In this formula, Z is not accessible to direct optical measurement and Z^2 varies with distance R as exp(-R/L), where L is an effective average Bohr radius of donor and acceptor ions in excited and unexcited states. While the former two interactions are electrostatic in origin, the exchange interaction arises from the antisymmetry requirements of the electronic wavefunction for a system consisting of a donor and an acceptor.

5.5 The Treatment of Optical Properties of Divalent REE

Two specific features determine the similarity and difference of energy levels in di- and trivalent ions of rare-earth elements. First, isoelectronic configurations of TR^{2+} and TR^{3+} of the next elements in the periodic system determines a qualitatively similar pattern of terms and multiplet levels, namely for TR^{2+} the order of the $4f^k$ – configuration levels and of excited $4f^{k-1}5d$-configuration levels is the same as for trivalent ions isoelectronic to them. Second, difference in the charge of the nuclei in isoelectronic TR^{2+} and TR^{3+} ions, on the one hand, in an essential but relatively small and approximately equal fall of energies of all levels derived from $4f^k$ – configuration and, on the other hand, it brings about a sharp drop in the energy of levels derived from mixed $4f^{k-1}5d$-configurations. It is just this fall of the $4f^{k-1}5d$ levels that causes a sharp difference in the absorption and luminescence spectra of the TR^{2+} and TR^{3+} ions (Marfunin 1979a).

Transitions onto the levels of mixed $4f^{k-1}5d$-configuration result in the appearance of broad intensive absorption bands. These specific features are due, first, to the fact that the said transitions occur between states with dissimilar electron configurations and, therefore, are parity-allowed. In these transitions the oscillator strength is by 3–4 orders higher than for the forbidden *f-f* transitions and characterized by much shorter emission lifetimes. Second, in these transitions d-electrons are involved, unshielded against interactions with the lattice, which makes for greater splitting by the crystal field, greater half-width of the lines and greater shifting of these bands in different crystals (Blasse and Grabmaier 1994).

A relative ratio between the $4f^k$ and $4f^{k-1}5d$ configuration levels energies specifies a sharply distinctive position of broad bands in the spectra of trivalent and divalent rare-earth ions. In the TR^{3+} spectra, with the exception of Ce^{3+}, broad bands fall into a relatively far UV region and they yield only line spectra in the visible and adjacent regions. In the TR^{2+} spectra broad bands fall into the visible and near-UV regions. Thus in the case of TR^{2+} the f-d and f-f transitions lie close to each other and overlap. Three individual cases are distinguished in the TR^{2+} luminescence spectra, namely broad bands due to d-f transitions, line IR spectra and combination of bands and lines.

Emission of d-f type in the divalent lanthanides is usually quenched by multiphonon relaxation from the 5d level to levels of the $4f^n$ configuration. These are a few exceptions. When the red shift and Stokes shift place the 5d level of Sm^{2+} near to or below the $^5D_0$ state (Fig. 5.5) – df emission may occur. The same applies for Eu^{2+} when the level is shifted to near to or below $^6P_{7/2}$. When for Tm^{2+} the 5d level is not shifted too forwards the $^2F_{5/2}$ level, df emission can also be observed. Finally, for Yb^{2+} there is no multiphonon relaxation path and df emission cannot be quenched in that way. Figure 5.6 illustrates different emission mechanisms with a configuration coordinate diagram. Initially the system is in the $4f^n$ configuration state, indicated by point A on parabola a. After excitation to the 5d level, point B on parabola b is reached. Subsequent lattice relaxation brings the system to point C and several routes can be followed from here. The usual possibility is normal df emission (arrow CD) with a small Stokes shift. Nevertheless, for example, the system may relax to the impurity-trapped exciton state indicated by point E on parabola d. From here anomalous emission (arrow EF) takes place with high Stokes-shifted emission band. Another possibility is relaxation to point G on parabola e with the following anomalous emission (arrow GF). Parabola c represents an excited state of the $4f^n$ configuration. Via the crossing point with parabola d, the anomalous emission is quenched and the excited $4f^n$ state becomes populated. The existence of the $^7F_6$ and $^2F_{5/2}$ levels in Sm^{2+} and Tm^{2+} is very likely the reason that anomalous emission has never been observed for these lanthanides. It has only been detected for Eu^{2+} and Yb^{2+}, precisely those lanthanides where excited $4f^n$ levels, that may quench anomalous emission, are absent (Dorenbos 2003).

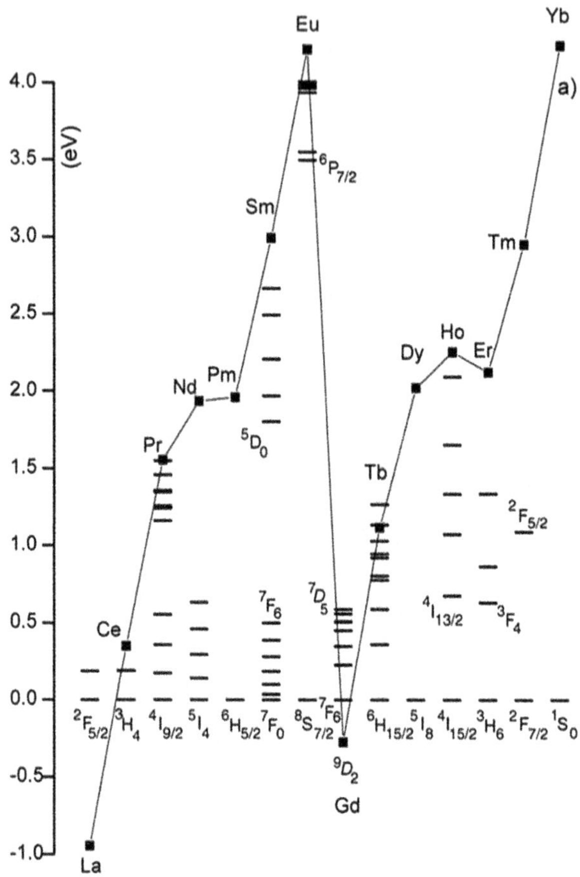

Fig. 5.5 The energy levels scheme of the free divalent lanthanides or REE. The solid curve (*a*) connects the locations of the first $4f^{n-1}5d$ level (Dorenbos 2003)

5.6 Rare-Earth Elements Luminescence in Minerals

The promise of luminescent methodology is based on many types of information that can be derived from mineralogical samples. These include REE from Ce to Yb, identities down to the ppb range, the valence states of the REE, the nature of the sites at which REE reside and the ways of compensating the charge, and features related to the presence of other ions (donors, activators). All this information can be used to determine the chemical, thermal, and deformational history of the material. Steady state and laser-induced time resolved luminescence spectroscopy is an excellent tool for REE study in minerals (Czaja et al. 2013; Gaft et al. 2001c, 2005).

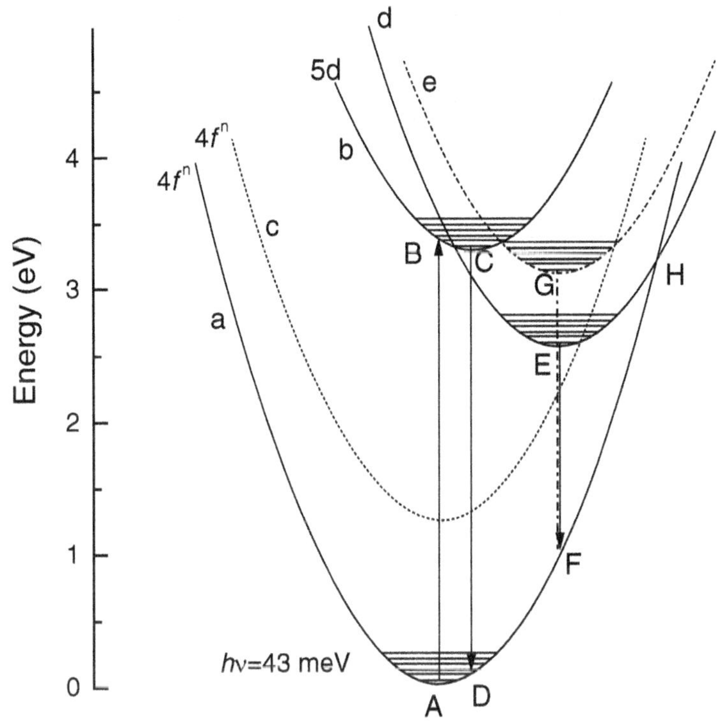

Fig. 5.6 The configuration diagram illustrating normal df emissions and anomalous emission (energy values realistic for SrF_2 and BaF_2 (Dorenbos 2003))

5.6.1 Ce^{3+}

In minerals cerium can occur in a trivalent state i.e. by losing its two 6s electrons and one of its 4f electrons. When cerium enters a solid, the expansion of the electron shells decreases the electrostatic interaction between the electrons resulting in a reduction of the energy of the excited states from their free ion values. This nephelauxetic shift increases with the degree of covalence of the cerium-anion bond. The spin-orbit interaction splits the 2f ground into two J states separated by ~2200 cm^{-1}. The (2 J + l)-fold degeneracy of these states is reduced by the ligand field. Because the 4f electron is shielded from the ligand field by the closed 5s and 5p electron shells, the overall splitting of the 2J states is small, typically only a few hundred cm^{-1}. When the 4f electron is excited to the outer 5d state, however, it is subjected to the effect of the ligands. Depending upon the site symmetry, the degeneracy of the 5d state is partially or completely removed. The overall splitting of the 5d manifold is typically in the order of 5000–10,000 cm^{-1}.

Electric-dipole transitions between the 4f ground state and the 5d excited state of Ce^{3+} are parity and spin allowed and have a large oscillator strength and very short decay time. Contrary to a long lived emission of many $4f^q$ j-levels of several lanthanides in glasses and crystals, corresponding to very low (10^{-6}) oscillator strength, Ce^{3+} is the only trivalent lanthanide which in the UV spectrum region shows high oscillator strength (0.01–0.1). Due to configurational admixing, normally forbidden transitions to the 6s state can also be observed in wide band gap materials. Optical excitation in the Ce-containing materials is normally performed in the range between 300 and 350 nm. The excitation arises from optical inter-shell Ridberg transitions between 4s and 5d levels.

5.6.1.1 Apatite

Two types of Ce^{3+} centers have been found in apatite by steady-state spectroscopy, connected with substitution in two different Ca sites (Tarashchan 1978). Both centers are well detected by time-resolved spectroscopy (Fig. 4.2a). The narrow UV band with two shoulders at 340 and 360 nm in apatite is connected with ^2D–$^2F_{3/2,5/2}$ electron transitions in Ce^{3+} in higher symmetry Ca(I) position. Very short decay time of this band of 25 ns is typical for Ce^{3+}. An additional band peaking at 430 nm is also characterized by short decay time, only slightly longer compared to the former one. It is considerably shifted to the long wavelength side from excitation bands, and two bands corresponding to transitions from the excited state to two levels of the ground multiplet are absent. This luminescence is connected with Ce^{3+} in the Ca(II) site. Since the 4f electron is shielded from the ligand field by the closed 5s and 5p electron shells, the overall splitting of the ^2F_J states is small. This is due to spin-orbit coupling, which may be lower in the Ca(II) position, leading to the absence of the double-band shape. When the 4f electron is excited to the outer 5d state, however, it experiences the full effect of the ligands. Depending upon the site symmetry, the degeneracy of the 5d state is partially or completely removed. The overall splitting of the 5d manifold is much larger in the Ca(II) site, which explains the larger Stokes shift. The longer decay time is consistent with the fact that for Ce^{3+} the decay time is longer if the emission is at longer wavelength (Blasse and Grabmaier 1994).

5.6.1.2 Barite and Anhydrite

A narrow band with a main shoulder at 302 nm with a very short decay time of ≈ 25 ns, and another with shoulders at 330 and 360 nm with a longer decay of ≈ 75 ns (Fig. 4.74c, d) in time-resolved spectra have spectral-kinetic parameters suitable for Ce^{3+}. It is known that narrow bands near 300 nm are especially strong in $(Ba,Sr)SO_4$ (baritocelestine), while in barite they are situated at a longer wavelength (Gaft et al. 1985). The ionic radius of Ce^{3+} is 128 pm and a possible accommodation is isomorphic substitution for Ba^{2+} (156 pm) or Sr^{2+} (140 pm). Thus, two types of Ce^{3+} centers may be connected with Sr impurity, the presence of which in barite samples is confirmed by ICP analysis (450–720 ppm). Luminescence of Ce^{3+} in anhydrite is seen under excitation at 266 nm (Fig. 4.30b) and it is disappeared after delay of 75–100 ns because of its short decay time (Gaft et al. 1985; Baumer et al. 1997).

5.6.1.3 Calcite, Aragonite and Dolomite

Two types of Ce^{3+} centers in calcite were detected by steady-state spectroscopy (Kasyanenko and Matveeva 1987). The first one has two bands at 340 and 370 nm and is connected with electron-hole pare Ce^{3+}-$CO_3{}^{3-}$. The second one has a maximum at 380 nm and was ascribed to complex center with Ce^{3+} and OH^- or H_2O as charge compensators. Such center becomes stronger after ionizing irradiation and disappears after thermal treatment. The typical example of Ce^{3+} luminescence in time-resolved luminescence of calcite, aragonite and dolomite consists of narrow band at approximately 350–360 nm with very short decay time of 30 ns, which is very characteristic for Ce^{3+} (Figs. 4.18a, 4.22a and 4.25b). It was found that Ce^{3+} excitation bands occur also in the Mn^{2+} excitation spectrum, demonstrating that energy transfer from Ce^{3+} to Mn^{2+} occurs (Blasse and Aguilar 1984).

5.6.1.4 Feldspars

Narrow bands at 320–335 nm with very short decay time of 20–30 ns may be confidently ascribed to Ce^{3+} luminescence (Fig. 4.98). In steady–state spectra different bands in this spectral range without decay time analyses, especially under X-ray and electron beam excitations may be mistakenly considered as Ce^{3+} emissions (Götze 2000).

5.6.1.5 Danburite and Datolite

Steady state luminescence of Ce^{3+} in danburite and datolite was found under X-ray and UV laser excitations (Gaft et al. 1979). The narrow band with two maxims at 346 and 367 nm in danburite (Fig. 4.27a), with short decay time component of 30 ns is evidently connected with Ce^{3+}. After a long delay, the luminescence of Ce^{3+}still dominant (Fig. 4.27b), showing the presence of a long decay-time component. The change in the lifetime is unambiguously indicative of a resonance radiationless mechanism of the energy transfer. The possible mechanism is Gd^{3+}-Ce^{3+} energy migration, because the emission of Gd^{3+} $^6P_{7/2}$-$^8S_{7/2}$ in the UV region coincides with excitation of Ce^{3+}.

The narrow band with two maxims at 335 and 360 nm in time-resolved emission spectra of datolite (Fig. 4.29a), with short decay time of 30 ns is connected with Ce^{3+}.

5.6.1.6 Pyromorphite

At $\lambda_{ex} = 266$ nm excitation, with 10 ns delay and 9 ms gate (practically steady state), there is a UV band with two maxims at 350 and 375 nm in the emission spectrum (Fig. 4.59a). This band looks exactly like a Ce^{3+} center. However, its luminescence is not quenched after a delay of 100 ns, which would be typical for Ce^{3+}, or after a delay of 1 ms, when even luminescence of Eu^{3+} with forbidden f-f transitions starts to reduce its intensity (Fig. 4.59b). In addition, the intensity of the maximum at 375 nm decreases with increasing delay time more quickly than that at 350 nm, thus indicating a somewhat shorter decay time. This behavior is not consistent with the existence of two transitions from one excited state into split 2f ground states of Ce^{3+}, and rather confirms the presence of two different luminescence centers, or two excited states. At $\lambda_{ex} = 355$ nm excitation, the band with two maxims at 380 and 439 nm in the spectrum with 10 ns delay also resembles a Ce^{3+} (Fig. 4.59c). It has a short decay component of approximately 67 ns and a much longer one of several ms. The presence of long component in Ce^{3+} decay may be connected with energy migration from the long-lived centers, for example Pb^{2+} or Gd^{3+}.

5.6.1.7 Esperite

The narrow band with two maxims at 378 and 400 nm (Fig. 4.38c), with decay time of 300 ns is evidently connected with Ce^{3+}, while decay time is unusually long for this center. Excitation spectrum (Fig. 4.38d) exhibits short-waved band ad 280 nm, which may be connected with Pb^{2+} excitation with the following energy migration from Pb to Ce. It may be the reason of relatively long decay component in Ce^{3+} luminescence.

5.6.1.8 Sodalite and Tugtupite

A UV bands peaking at 340 and 360 nm, with a short decay time of 50–70 ns, were detected in tugtupite and sodalite (Figs. 4.160f and 4.155a). Spectral and kinetic properties of this band are typical for Ce^{3+} luminescence. Cerium presence in sodalite up to 800 ppm quantities was detected, but it is not easy to find its proper place in the sodalite and tugtupite lattices. Its substitution for Na^+ is difficult because of the big charge difference, while substitution for Al^{3+} or Si^{4+} is difficult because of big radii difference (ionic radii of Ce^{3+} 115 pm in octahedral site and 128 pm in 8-coordinated site; Al^{3+}: 53 pm in tetrahedral site and 67.5 pm in octahedral site; Si^{4+}: 40 pm in tetrahedral site and 54 pm in octahedral site).

5.6.1.9 Other Minerals

Luminescence of Ce^3 is clearly seen under excitation at 266 nm in apophyllite (Fig. 4.34a), charoite (Fig. 4.39a, c), fluorite (Fig. 4.14a), leucophane (Fig. 4.45a), milarite (Fig. 4.47a), hardystonite (Fig. 4.36a). The photoluminescence of Ce^{3+} was found in chabazite (Kobayashi et al. 2011).

5.6.2 *Pr^{3+}*

Luminescence of Pr^{3+} consists of many multiplets as follows: 480–500 ($^3P_0$-$^3H_4$), 650–670 ($^3P_0$-$^3F_2$), 750–770 ($^3P_0$-$^3F_4$), 610–630 ($^1D_2$-$^3H_6$), 400–410 ($^1S_0$-$^1I_6$), and UV (5d-4f) transitions. The emission color of Pr^{3+} depends strongly on the host mineral lattice. The radiative decay time of the $^3P_0$-^3H_j or ^3F_j emissions is approximately 10–5 μs, which is the shortest lifetime observed in 4f-4f transitions. The short decay time of Pr^{3+} is ascribed to the spin-allowed character of the transition. It also may be explained by nonradiative relaxation due to the presence of high frequency vibrations in the lattice. And other possible reason is that the 4f orbitals are probably more spread out in the lighter rare earths (with lower nuclear charge), facilitating the mixing with opposite-parity states (Blasse and Grabmaier 1994; Reisfeld and Jörgensen 1977).

The steady-state luminescence of Pr^{3+} in minerals was found only in scheelite, where the line near 480 nm has been ascribed to this center (Gorobets and Kudrina 1976) and possibly in fluorite (Krasilschikova et al. 1986). The luminescence of Pr^{3+} in minerals is difficult to detect because its radiative transitions are hidden by the stronger lines of Sm^{3+} in the orange range of 600–650 nm, Dy^{3+} in the blue range of 470–490 nm and Nd^{3+} in the near IR (870–900 nm). In order to extract the hidden Pr^{3+} lines time-resolved luminescence was applied. The fact was used that Pr^{3+} usually has relatively short decay time compared to its competitors Dy^{3+}, Sm^{3+} and Nd^{3+}, especially from $^3P_0$ level. In order to correct identification of Pr^{3+} lines in minerals several of them were synthesized and artificially activated by Pr (Fig. 5.7). Besides, comparison has been made with CL spectra of synthetic minerals artificially activated by Pr (Blanc et al. 2000).

5.6.2.1 Magmatic Apatite

The presence of Pr in apatite samples, up to 424.4 ppm in the blue apatite sample, was confirmed by Induced-Coupled Plasma analysis (Table 1.3). Luminescence spectrum of apatite with broad gate width of 9 ms is shown in Fig. 4.3a where delay time of 500 ns is used in order to quench the short-lived luminescence of Ce^{3+} and Eu^{2+}. The broad yellow band is connected with Mn^{2+} luminescence, while the narrow lines at 485 and 579 nm are usually ascribed to Dy^{3+} and the lines at

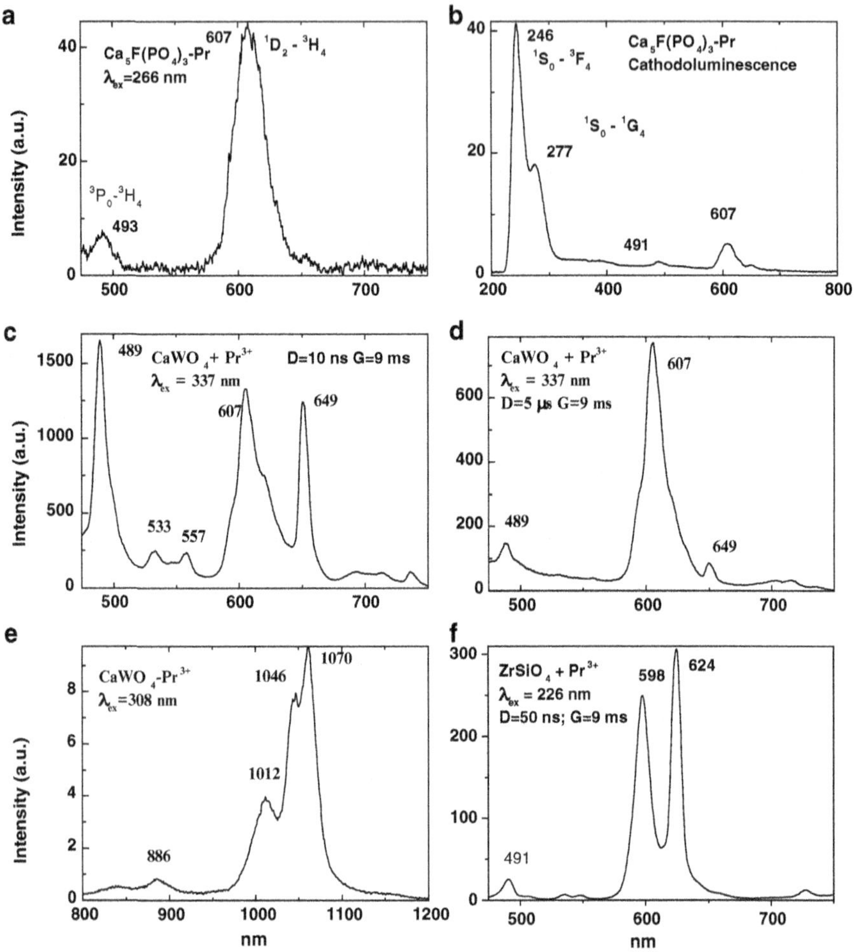

Fig. 5.7 **(a–f)** Laser-induced time-resolved luminescence spectra of synthesized apatite, zircon and scheelite artificially activated by Pr

604 and 652 nm, to Sm^{3+}. Only those luminescence centers are detected by steady-state spectroscopy. Nevertheless, with shorter gate width of 100 μs, when the relative contribution of the short lived centers is bigger, the characteristic lines of Sm^{3+} at 652 nm and Dy^{3+} at 579 nm disappear while the lines at 485 and 607 nm remain (Fig. 4.3b). It is known that such luminescence is characteristic of Pr^{3+} in apatite, which was proved by study of synthetic apatite artificially activated by Pr (Gaft et al. 1997a, 1999). The connection with Pr^{3+} was confirmed by strong emission at 485 and 607 nm under $\lambda_{ex} = 462$ nm, which is suitable for $^3H_4$-$^3P_0$ transition in Pr^{3+}. The lines at 485 and 607 nm have different decay times. After delay of 2 μs the first of the lines disappears indicating that they are connected with transitions from different levels. The line at 485 nm belongs to $^3P_0$-$^3H_4$ transition.

The short decay of the blue line is connected with nonradiative relaxation between ^3P_J and $^1D_2$ levels due to the presence of high frequency vibrations of (PO_4)-groups in the apatite lattice at 1040 cm^{-1}. The line at 607 nm belongs to $^1D_2$ level, which is confirmed by its effective excitation at 570 nm, the energy of which is much lower than the energy of $^3P_0$ level.

Luminescence of Pr^{3+} described above is also detected under excitations, which do not coincide with transitions inside this center. It may be ascribed to energy transfer from Eu^{2+} or Ce^{3+}, which have luminescence in the blue range.

5.6.2.2 Sedimentary Apatite

In sedimentary apatite, which are formed in marine conditions at ambient temperature, after high temperature treatment a set of very intensive and narrow luminescence lines appears with strongest maxims at 594, 619, 626, 634 and 645 nm (Fig. 4. 6a), these are related to REE^{3+} luminescence which was not observed in the unheated samples. The lines at 634 and 645 nm are characterized by a decay time of 0.5 ms and the line at 619 nm has a decay time of 0.1 ms. At 77 K a new lines appear with maxims at 603, 651, 653 and 656 nm and decay times of 0.6 ms, and at 615 nm with a decay time of 0.2 ms. The spectral form and long decay times of the sharp lines are sure indicators that this luminescence is connected with f-f transitions in REE^{3+}. For the correct interpretation of the luminescent lines, artificial activation of natural francolite by different REE was accomplished by heating the corresponding mixtures at 1200 K in air. It was found that all lines might be connected mainly with luminescence of Pr^{3+} (Gaft et al. 1996, 1997a). The strongest lines at 619, 634 and 645 nm may be ascribed to the $^3P_0$-$^3F_2$ and $^1D_2$-^3H_j transitions in Pr^{3+}. The luminescence of REE^{3+} is well known in magmatic apatites, but the spectral positions of lines are significantly different. In apatite the calcium ions are located in two crystallographically distinct sites, labeled Ca^{2+}(I) and Ca^{2+}(II). The luminescence of trivalent REE in magmatic apatites is due to their substitute in the Ca (I) position. The luminescence properties of magmatic apatites do not change after heating to temperatures up to 1200 K, but after thermal activation with REE, they enter the Ca^{2+}(II) position and the luminescence spectra change (Gorobets 1968; Tarashchan 1978). The spectral properties of the trivalent REE which substitute for Ca II in magmatic apatite are very similar to those obtained from the apatite of fossil fish teeth after heating, indicating that luminescence of Pr^{3+} on the Ca II position takes place in this case. Luminescence of trivalent REE is more intensive when the site symmetry is lower and the parity inhibition is partially lifted. The Ca^{2+}(II) site in apatite lattice has a lower symmetry then Ca^{2+}(I), and REE in site II may have parity forbidden electron transitions. This explains the great intensity of trivalent REE luminescence in heated sedimentary apatite.

Two possibilities exist which allow explaining the different luminescence of the REE in magmatic and sedimentary apatite. It is possible that in sedimentary ones REE are incorporated in Ca II position, but luminescence is not detected because of quenching by the components with high-energy phonons (water and organic matter). The latter are removed during heating with resulting luminescence. Nevertheless, francolite is different from the magmatic apatite due to crystallization at low temperature, while activation by REE results in luminescence in Ca^{2+}(I) position. Thus another explanation may be needed. It is possible that francolite accommodates, at least partly, REE in concentrated form (e.g. free minerals), which is usually not suitable for registration by luminescence due to concentration quenching. As a result of high temperature heating, thermal diffusion of REE in apatite lattice takes place. These changes initiate luminescence. The similarity with luminescence spectra of the francolite, which was artificially activated by the thermal diffusion of the REE, supports this interpretation.

It is interesting to note, that Pr^{3+} spectra obtained under $\lambda_{ex} = 308$ nm excitation do not resemble the shape and the lifetimes of those obtained under direct excitation to the Pr^{3+} levels. They exhibit the lines at 619, 634 and 645 nm with same long decay time components of ~750 μs (Fig. 4.6b). The excitation at 308 nm does not correspond for transitions inside Pr^{3+}. It was firstly supposed that an effective non-radiative energy transfer from the first excited state of uranyl to the levels ^3P_J (J = 0, 1, 2) and $^1I_6$ of Pr^{3+} may be responsible for the excitation, especially because the energy transfer between donor uranyl ion to Pr^{3+} was earlier demonstrated to take place in phosphate glasses (Reisfeld and Jörgensen 1977). Uranyl is present in natural sedimentary apatite and has strong green luminescence at 530 nm after heating (Fig. 4.6c). We studied artificial oxyapatite activated by Pr^{3+}, which does not contain uranyl impurity. In accordance with our model, this sample does not exhibit Pr^{3+} luminescence under excitation at 308 nm, while characteristic Pr^{3+} lines under 462 and 570 nm excitations exist. Nevertheless, the relatively short decay time of 100 μs for uranyl at negligible concentration of Pr^{3+} does not enable to explain the very long decay of Pr^{3+} of 650–750 μs. Besides that, excitation by UV longer than 308 nm with $\lambda = 337$–360 nm leads to strong uranyl luminescence but the lines of Pr^{3+} are absent (Fig. 4.6c).

The very long decay time of Pr^{3+} of 650–750 μs which is obtained only at 308 nm excitation in the presence of uranyl can be explained by energy transfer from excited uranyl ion to one of the ^3P states and followed by excited state absorption to $^1S_0$ level due to short UV laser excitation. The $^1S_0$ level serves as storage of an excited state of Pr^{3+}, which is slowly decaying to the lower lying $^3P_0$ level. As a result, a strong emission signal from ^3P_j is observed in the visible domain. Such kind of photon-cascade emission with high quantum efficiency under V-UV excitation is well known for Pr^{3+} in oxide matrix.

The active participation of $^1S_0$ level is indirectly accredited by the simultaneous observation of UV and visible emission in cathodoluminescence and synchrotron excited spectra (Fig. 5.8) Excitation into the 4f5d and higher lying bands evidently decays to the $^1S_0$ level located at 46,300 cm^{-1} which exhibits luminescence in wide band-gap hosts due to radiative de-excitation to the lower lying levels of Pr^{3+}. The $^1S_0$-$^3F_4$ transition at 246 nm is especially strong in oxyapatite. In F-apatite only the line at 269 nm is present. It may be explained by the relatively long-waved absorption edge in fluorapatite, which is at about 300 nm (Morozov et al. 1970).

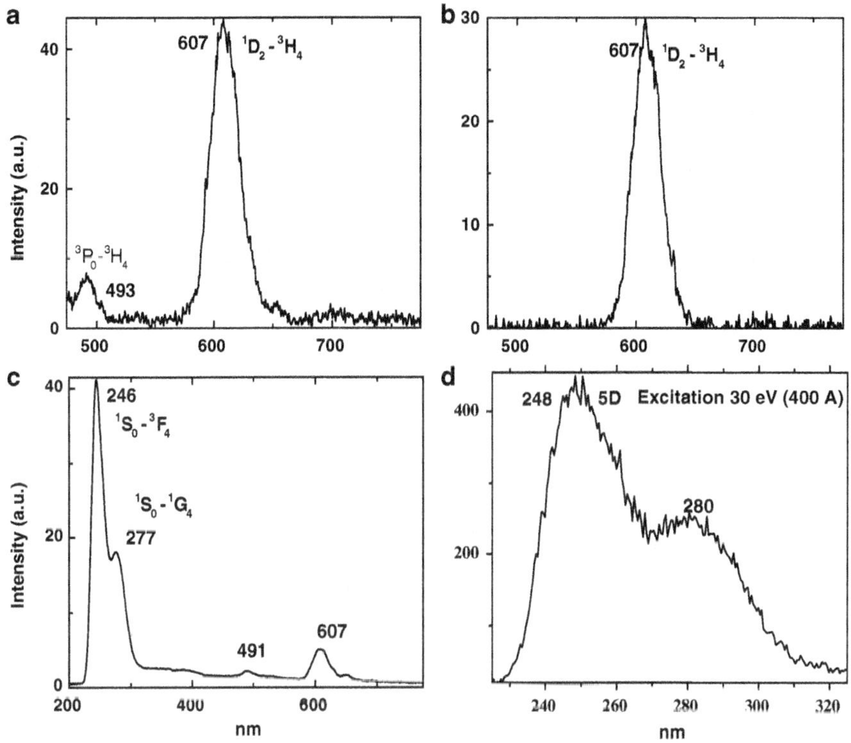

Fig. 5.8 (**a–d**) Cathodoluminescence and synchrotron excited spectra of Pr activated apatite

5.6.2.3 Scheelite

Luminescence spectrum of scheelite with broad gate width of 9 ms is shown in Fig. 4.11d. The narrow lines at 490 and 572 nm are usually ascribed to Dy^{3+} and the lines at 607 and 647 nm to Sm^{3+}. Nevertheless, the relative intensity of the line at 607 nm compared to the line at 647 nm is lower at longer delay times (Fig. 4.11e, f). Besides that with shorter gate width of 1 μs, when the relative contribution of the short lived centers is bigger, the characteristic lines of Sm^{3+} at 647 nm and Dy^{3+} at 575 nm disappear while the lines at 488 and 607 nm remain. Such luminescence is characteristic of Pr^{3+}, which was confirmed by time-resolved luminescence study of scheelite artificially activated by Pr^{3+} and Sm^{3+}. Unlike in apatite, the phenomenon exists not only under 308 nm, but also under 337 excitation due to higher energy of intrinsic WO_4 luminescence compared with UO_2. The active participation of blue intrinsic scheelite luminescence is confirmed by reabsorption line at 448 nm (Gaft et al. 1999), which corresponds to $^3H_4$-$^3P_2$ transition inside Pr^{3+}.

Luminescence of Pr^{3+} in the red and IR parts of the spectrum was found (Fig. 5.7). According to energy levels scheme of Pr^{3+} the following interpretation is possible: 840 nm – $^1D_2$-$^3F_2$, 886 nm – $^3P_1$-$^1G_4$ and 1070 nm – $^1G_4$-$^3H_4$.

5.6.2.4 Anhydrite

Ultra-violet X-ray excited luminescence of Pr^{3+} was firstly observed in minerals in anhydrite, but have been mistakenly ascribed to transitions from $^6I_{7/2}$ and $^6D_{5/2}$ levels in Gd^{3+} (Gaft et al. 1985). According to CL spectra of anhydrite artificially activated by Pr^{3+} (Baumer et al. 1997; Blanc et al. 2000) the emission lines at 228, 239, 258 and 268 nm (Fig. 4.30a) belong to Pr^{3+} and may be connected with $^1S_0$-$^3H_5$, $^1S_0$-$^3H_6$ and $^1S_0$-$^1G_4$ transitions. The lines of Pr^{3+} in the visible range are much weaker.

5.6.2.5 Zircon

Luminescence of Pr^{3+} in zircon is very difficult to detect under UV excitation even by time-resolved spectroscopy. The reason is that it has relatively short decay time similar to those of radiation-induced centers. Visible excitation, which is not effective for broadband luminescence, allows revealing Pr^{3+} luminescence lines, using high-resolution steady-state spectroscopy. Under such experimental conditions each element has individual lines, enabling confident identification of the spectrum possible (Gaft et al. 2000a). Only if radiation-induced luminescence in zircon is relatively weak, the lines of Pr^{3+} may be detected by UV excitation (Fig. 4.90c).

5.6.2.6 IR Emission

Under cw excitation with 514, 532, 780 and 785 nm Pr^{3+} lines are often detected but less than Nd^{3+}. They have been found in the IR part of the spectrum in several minerals, such as titanite (Jasinevicius 2009).

5.6.3 Nd^{3+}

Nd^{3+} energy levels arise from the $4f^3$ electronic configuration. The $4f^25d$ higher lying states are almost always completely in the UV range. Some transitions among $4f^3$ levels of Nd^{3+} ion are spin-allowed, such as $^4I_{9/2}-{}^4F_{3/2}$, $^4F_{5/2}$, $^4F_{7/2}$, $^4F_{9/2}$, $^4G_{7/2}$ or $^4F_{3/2}-{}^4G_{7/2}$, $^4D_{3/2}$ and so on, but this is not valid for the remaining ones. Neodymium has been recognized as one of the most efficient rare-earth luminescence centers in minerals (Lenz et al. 2013), while its emission has been found only in IR part of the spectrum. Nevertheless it is well known that Nd^{3+} may also generate UV-visible luminescence in certain matrixes, for example in YAG-Nd (Marech et al. 1989; Balda et al. 2001). The Nd^{3+} UV and visible luminescence spectra consists of many narrow lines whose half-widths reach only several cm^{-1}. It was found that the Nd^{3+} UV and visible luminescence depend on excitation wavelength. Most narrow luminescence lines together with slightly broader peaks were observed for $\lambda_{ex} = 355$ nm where there is strong Nd^{3+} absorption. Nearly 140 lines have been found under this excitation. The mostly important lines detected at room temperature are near 396, 401, 435, 461, 488, 500, 524, 550, 588, 609, 617, 620, 629, 638, 663, 712 nm. The measurements of luminescence decay times revealed that several various lifetimes are observed ranging from 3 to 300 μs. The observed Nd^{3+} UV and visible luminescence spectra were interpreted by the following way: the most intense lines originates from the UV lying Nd^{3+} level $^2F_{5/2}$: to $^4F_{5/2}$, $^2H_{9/2}$ (lines around 400 nm), to $^4F_{9/2}$ (lines around 435 nm), $^2H_{11/2}$ (lines in the range 450–464 nm), to $^4G_{5/2}$ (lines around 480 nm) and to $^2G_{7/2}$ (lines around 488 nm). In minerals such luminescence is mixed with other REE lines and not detected by steady-state spectroscopy.

5.6.3.1 Apatite

Neodymium in natural and artificial apatite is characterized by anomalous distribution of luminescence intensity in the groups at 1.06 and 1.3 μm. In each of these spectral groups there is one line whose intensity exceeds many times the intensity of the remaining lines of said group (Morozov et al. 1970). In laser-induced luminescence of natural apatites we also found somewhat different luminescence spectra (Fig. 4.4). Decay times of these lines are rather close and it is possible to suppose that all Nd^{3+} occupy the Ca(I) sites with different charge compensations.

5.6.3.2 Scheelite

Figure 4.10a presents scheelite luminescence in the near IR ranges of the spectrum. The usual characteristic lines of Nd^{3+} are detected in the spectral range up to 1.6 μm. Strong reabsorption lines of Nd^{3+} are characteristic for luminescence of many minerals, but were firstly detected in scheelite (Gorobets 1975).

Unusual behavior of the luminescence line at 417 nm has been detected by time-resolved spectroscopy. It is usually ascribed to Tb^{3+}, but sometimes in spectra with a narrow gate this line remains strong, while other lines of Tb^{3+} disappear (Fig. 4.10b, c). The supposition that those lines are connected with Nd^{3+} was confirmed by our study of $CaWO_4$:Nd, where, besides the known IR, the group of UV and violet lines with short decay times are detected, while in $CaWO_4$:Tb such lines are absent (Fig. 5.9) (Gaft et al. 2008b).

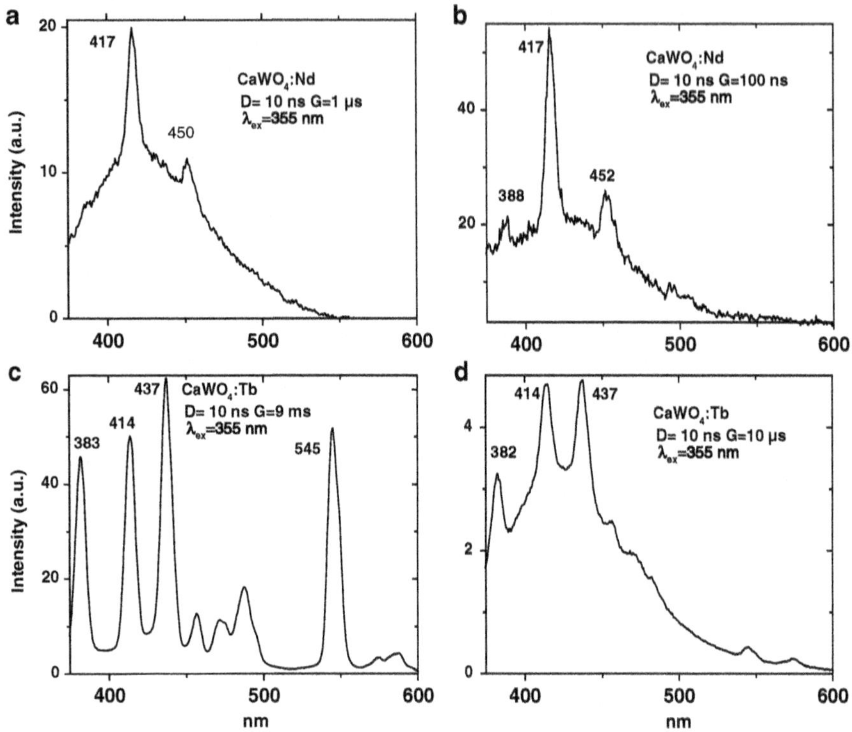

Fig. 5.9 (**a–d**) Laser-induced time-resolved luminescence spectra of synthetic $CaWO_4$ artificially activated by Nd and Tb

5.6.3.3 Fluorite

IR luminescence lines with relatively short decay times connected with Nd^{3+} are very strong in fluorite emission spectrum (Fig. 4.15c, d). Besides that, UV and violet lines with short decay time appear, which are ascribed to Nd^{3+} (Fig. 4.15a, b).

5.6.3.4 Barite

Figure 4.74a, b represent narrow luminescence lines detected in barite by time-resolved spectroscopy. Much weaker lines at 446 and 672 nm accompany the strongest one at 588 nm. They have a relatively short decay time of 5 µs and emphasized in the spectrum with short gate. Such combination of spectral and kinetic properties is not suitable for any trivalent REE besides $^2P_{1/2} \Rightarrow {}^4I_{9/2}$ (446 nm), $^4G_{7/2} \Rightarrow {}^4I_{11/2}$ (588 nm) and $^4G_{7/2} \Rightarrow {}^4I_{12/2}$ (672 nm) transitions in Nd^{3+}.

5.6.3.5 Titanite

Figures 4.80a, c demonstrate emission lines of titanite, which according to their spectral positions may be confidently connected with Nd^{3+}. Luminescence spectrum in the 860–940 nm spectral range, corresponding to the $^4F_{3/2}$-$^4I_{9/2}$ transition, contains six peaks at 860, 878, 888, 906, 930 and 942 nm, while around 1089 nm corresponding to $^4F_{3/2}$-$^4I_{11/2}$ transition it contains five peaks at 1047, 1071, 1089, 1115 and 1131 nm. The decay time of IR luminescence of Nd^{3+} equal to approximately 30 μs in titanite is evidently the shortest one in the known systems activated by Nd^{3+}. Typical radiative lifetime of this level depends on the properties of the solid matrix and varies from approximately 100–600 μs (Kaminskii 1996). To explain the fast decay time of Nd^{3+} in titanite, the energy level quenching by host matrix may be considered.

Another group of lines is detected in titanite luminescence spectrum, which may be considered as connected with Nd^{3+} emission. Those lines at 589, 658, 743 and 846 nm are especially strong in luminescence spectra with narrow gate excited by $\lambda_{ex} = 532$ nm (Fig. 4.80b). Such combination of emission lines with relatively short decay times is very unusual for minerals and may not be easily connected to any rare-earth element traditional for luminescence in the visible range. If to consider the possible connection with visible emission of Nd^{3+}, the detected lines correspond very well, for example, to electron transitions from $^2G_{7/2}$ level to $^4I_{9/2}$, $^4I_{11/2}$, $^4I_{13/2}$ and $^4I_{15/2}$ levels.

5.6.3.6 Zircon, Anhydrite, Calcite, Rhodonite, Feldspars

IR luminescence lines of Nd^{3+} are detected at 817 and 885 nm in zircon (Fig. 4.90e), 892 and 899 nm in anhydrite (Fig. 4.31c), 817 and 889 nm in calcite (Fig. 4.18b), 815, 875 and 890 nm in rhodonite (Fig. 4.167b), and 895 nm in feldspars (Fig. 4.99c).

5.6.3.7 CW Excitation

Under cw excitation with 514, 532, 780 and 785 nm Nd^{3+} lines dominate in the IR part of the spectrum in many minerals, such as apatite (Fig. 4.8), scheelite (Fig. 4.12), fluorite (Fig. 4.17), aragonite (Fig. 4.23), dolomite (Fig. 4.26), danburite (4.28), anhydrite (Fig. 4.32), hardystonite (Fig. 4.37), pectolite (Fig. 4.41), pyrochlore (Fig. 4.44), leucophane (Fig. 4.46), milarite (Fig. 4.48), walstromite (Fig. 4.50), diopside (Fig. 4.52), eudialite (Fig. 4.53), shortite (Fig. 4.54), burbankite (Fig. 4.55), svabite (Fig. 4.57), pyromorphite (Fig. 4.60d), vanadinite (Fig. 4.61), wulphenite (Fig. 4.62a, b), margarosanite (Fig. 4.66b), cerussite (Fig. 4.67a), malayaite (Fig. 4.70d), witherite (Fig. 4.75b), strontianite (Fig. 4.77b), titanite (Fig. 4.81), zircon (Fig. 4.92), baddeleyite (Fig. 4.94), wollastonite (Fig. 4.96a, c), albite (Fig. 4.101a), phenakite (Fig. 4.105b), grossular (Fig. 4.132), uvarovite (Fig. 4.133), zoisite (Fig. 4.139b), lazurite 4.162), delhayelite (Fig. 4.165), monazite and thorite (Figs. 4.178 and 4.180), xenotime (Fig. 4.181), bastnasite (Fig. 4.182), cerite (Fig. 4.183), parisite (Fig. 4.184), and shomiokite (Fig. 4.185).

5.6.4 *Sm^{3+}, Sm^{2+}*

Trivalent samarium activated minerals usually display an intense luminescence spectrum with a distinct line structure in the red-orange part of the spectrum. The radiating term $^4G_{5/2}$ is separated from the nearest lower level $^6F_{11/2}$ by an energy interval of ~7500 cm^{-1}. This distance is too large compared to the energy of phonons capable to accomplish an effective non-radiative relaxation of excited levels and these processes do not significantly affect the nature of their spectra in minerals. Thus all detected lines of Sm^{3+} luminescence take place from one excited level and usually characterized by long decay time.

An important feature of the electronic structure of Sm^{2+} is the low energy of the $4f^5$-$5d^1$ excited electronic configurations. As a result, the $4f^5$-$5d_1$ configurations interacts significantly with the ^5D_j levels of the 4f ground configuration and exerts a strong influence on the optical properties of Sm^{2+}. Sometimes the $4f^5$5d1 level of Sm^{2+} is located below its 4f levels, for example, in fluorite, resulting in band luminescence due to the 5d-4f transition with decay time of several μs. But in different hosts a line spectrum due to the 4f-$4f^5$ D_0-$^7F_1$ transitions has been observed (Tarashchan 1978).

5.6.4.1 Apatite

In apatite two kinds of Sm^{3+} luminescence are detected connected with substitution in different Ca sites (Fig. 4.3c), while the substitution in Ca(I) site is much more widespread. Artificial activation by Sm in vacuum is characterized by the lines at 465, 598 and 645 nm with decay of ~2 ms. After activation in air the lines at 607 and

654 nm appear with different excitation spectrum and shorter decay of ~1.6 ms (Tarashchan 1978; Gaft et al. 1997a). It is possible to suppose that the first group of lines appearing in vacuum may be ascribed to substitution on Ca(I) while the second one appearing in the air – on Ca(II) site. The splitting of the lines indicating the removal of the degeneracy is better seen in the case of activation in air. It is consistent with the lower symmetry of Ca(II) site.

Sm^{2+} luminescence is detected only at low temperatures, where the strongest line at 734 nm, connected with $^5D_0$-$^7F_2$ electron transition, is clearly seen (Fig. 4.2d). Such emission was found very rare, but the presence of Sm^{2+} in natural apatite may be seen by indirect way after oxidizing heating, where luminescence intensity of Sm^{3+} becomes enhanced evidently as result of Sm^{2+}-Sm^{3+} transformation.

5.6.4.2 Fluorite

The lines of Sm^{3+} connected with several types of centers are well studied in fluorite by steady-state luminescence spectroscopy (Tarashchan 1978; Krasilschikova et al. 1986). In time-resolved spectra it is mostly prominent after long delay times and is mainly characterized by the lines at 562, 595 and 651 nm (Fig. 4.14d).

Together with Sm^{3+} other group of lines is often detected with the main line at 685 nm, which also has very long decay time of several ms (Fig. 4.14d). It is very close to the known resonance line of Sm^{2+}. Under low power UV lamp excitation, luminescence of Sm^{2+} in fluorite is known only at low temperatures, stating from approximately 77 K, and is composed of narrow f-f transition lines and broad band of 4f-5d transitions (Tarashchan 1978; Krasilschikova et al. 1986). Evidently, under strong laser excitation luminescence of Sm^{2+} may be seen even at room temperature, where 4f-5d luminescence is usually quenched because of radiationless transition.

5.6.4.3 Anhydrite

The Sm^{2+} ion ($4f^6$) demonstrates 5d-4f broad emission together with intra-configurational $4f^6$ line emission (Fig. 4.31). It is worth noting that, despite their different origins, broad band and narrow lines have similar decay time. The possible reason is thermally stimulated electron exchange between lower 4f and higher 5d excited states. It ceases at 77 K and the 5d-4f broad emission is absent. Using different excitations several types of Sm^{2+} are detected in anhydrite.

5.6.4.4 Titanite

The line at approximately 600 nm has a long decay time of 1 ms. It is the strongest one in titanite luminescence spectrum under 266, 355 and 532 nm (Fig. 4.79b, c), but its relative intensity is much lower under 514 nm excitation (Gaft et al. 2003b). It appears that from all lines found in titanite luminescence spectra only two weaker ones at 563 and 646 nm have similar kinetic and excitation characteristics with the line at 600 nm. Such combination of luminescence lines is very typical for Sm^{3+}. Thus the emission spectrum of Sm^{3+} in titanite exhibits three peaks corresponding to $^4G_{5/2}$-$^6H_{5/2}$, $^6H_{7/2}$ and $^6H_{9/2}$. The intensity of $^4G_{5/2}$-$^6H_{7/2}$ transition is much stronger than the other emission transitions, which is in accordance with Sm^{3+} behavior in glasses (Jayasankar and Babu 2000). The little difference in spectral position and half-width of the line at 600 nm under different excitations evidences that several Sm^{3+} centers may present. The presence of relatively high Sm concentration in titanite is confirmed by absorption spectrum and ICP data (Table 4.7).

5.6.4.5 Calcite, Feldspars, Hardystonite, Pyromorphite, Scheelite, Zircon, Baddeleyite

Narrow lines with long decay time characteristic for Sm^{3+} are also detected at 601, 614, 650 and 710 nm in calcite (Fig. 4.18b), at 603 and 645 nm in feldspars (Fig. 4.99b), 566, 603, 652 and 714 nm in pyromorphite (Fig. 4.59d), 601, 609 and 647 nm in scheelite (Fig. 4.12f), 566, 604, 615 and 651 nm in zircon (Fig. 4.90b), 549 nm in and 603 nm in leucophane (Fig. 4.45c).

5.6.5 *Eu^{3+}, Eu^{2+}*

Laser-induced time-resolved luminescence spectroscopy presents a good opportunity to detected different valence states of europium and its different positions inside minerals lattice (Gaft et al. 2001b, 2005).

5.6.5.1 Eu^{3+}

Trivalent europium is an excellent ionic probe for materials and its luminescence properties are extensively studied. Eu is one of the mostly informative elements in mineralogy, especially when the ratio Eu^{2+}/Eu^{3+} may be assessed. Both oxidation states are luminescent, but the lines of Eu^{3+} in minerals are usually very weak and concealed by other centers. By steady state luminescence spectroscopy its luminescence has been confidently detected only in scheelite and anhydrite (Tarashchan 1978; Gorobets and Rogojine 2001).

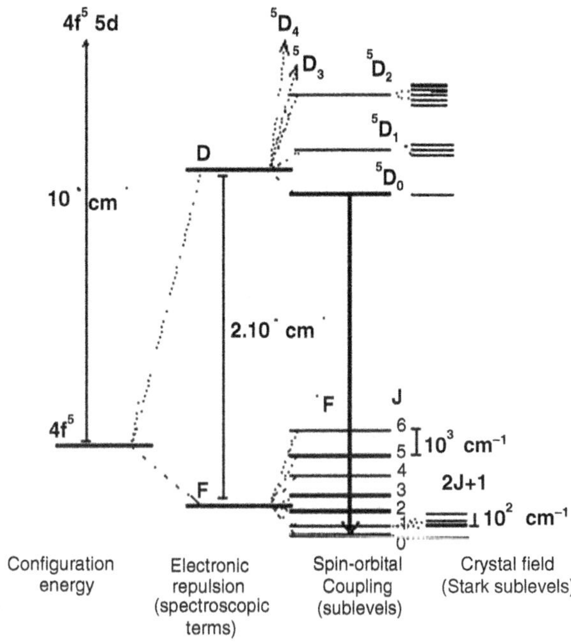

Fig. 5.10 Energy levels scheme of Eu^{3+}

The energy levels of Eu^{3+} arise from the $4f^n$ configuration. In a configuration coordinate diagram these levels appear as parallel parabolas ($\Delta R = 0$), because the 4f electrons are well shielded by the $5s^2 5p^6$ outer shells and the crystal field influence is weak. The emissions arising from f-f transitions yield sharp lines in the spectra. Because the transitions occur between states of the same parity, the lifetime of the excited state is long. The emission in Eu^{3+} corresponds to transitions from excited $^5D_0$ level to the ^7F_j ($J = 0, 1, 2, 3, 4, 5, 6$) levels of the $4f^6$ configuration. Because the $^5D_0$ level is not split by the crystal field (because $J = 0$), the splitting of the emission transition lines yields the crystal field splitting of the ^7F_j levels (Fig. 5.10). If Eu^{3+} occupies a crystal lattice site with inversion symmetry, optical transitions between levels of the $4f^n$ configuration are strictly forbidden as electric-dipole transitions (parity selection rule). They can only occur as the much weaker magnetic-dipole transitions which obey the selection rule $\Delta J = 0, \pm 1$ (but $J = 0$ is forbidden) or as vibronic electric-dipole transitions. If there is no inversion symmetry at the site of Eu^{3+}, the uneven crystal field components can mix opposite-parity states into $4f^n$-configurational levels. The electric dipole-transitions are now no longer strictly forbidden and appear as weak lines in the spectra, the so-called forced electric-dipole transitions. Some-transitions, viz. with $\Delta J = \pm 2, \pm 4$, are hypersensitive to this effect. Even for small deviations from inversion symmetry, they appear dominantly in the spectrum (Blasse and Grabmaier 1994; Reisfeld and Jörgensen 1977; Reisfeld 1973; Reisfeld et al. 2004). Intensity ratio analysis of the

$^5D_0 \rightarrow {}^7F_1$ and $^5D_0 \rightarrow {}^7F_2$ transition allows approaching the symmetry of the involved europium site. It is then possible to define the asymmetry ration R:

$$R = \frac{I(^5D_0 \rightarrow {}^7F_2)}{I(^5D_0 \rightarrow {}^7F_1)} \tag{5.28}$$

The asymmetry ratio R follows the evolution of the $^5D_0 \rightarrow {}^7F_2$ transition especially sensible to the crystal field. Parameter R increases when the covalent bonds with the neighbors are reinforced and when the symmetry of the site decreases. Crystal field theory enables to define certain parameters in order to characterize and distinguish the different europium site configuration. Crystal field parameter calculation involves some steps:

1. Site selective excitation in the $^7F_1 \rightarrow {}^5D_0$ absorption band;
2. Deconvolution of the $^5D_0 \rightarrow {}^7F_1$ transition for each excitation;
3. Determination of the three contributions, which are the three states of the crystal field for each environment. The contributions are labeled from the highest energy to the lowest one: ε_0, ε_- and ε_+, respectively. The ε_0 line is the thinnest and the most sensible to the excitation energy;
4. Determination of the energy of the Stark sub-level triplet (one triplet per site);
5. Representation of the Stark sub-level position in energy (cm^{-1}) for each site according to the excitation energy (cm^{-1});
6. Calculation of the B_{20} and B_{22} crystal field parameters using the three following equation.

$$E(\varepsilon_0) = E(^7F_1) + B_{20}/5 \tag{5.29}$$

$$E(\varepsilon_+) = E(^7F_1) - \left(B_{20} + 6^{(1/2)}.B_{22}\right)/10 \tag{5.30}$$

$$E(\varepsilon_-) = E(^7F_1) - \left(B_{20} - 6^{(1/2)}.B_{22}\right)/10 \tag{5.31}$$

Where $E(\varepsilon_0)$, $E(\varepsilon_+)$, $E(\varepsilon_-)$ represent the energy in wave number (cm^{-1}) of the three components and $E(^7F_1)$ the barycenter of the $^7F_1$ multiplet. The B_{20} and B_{22} parameters characterized the spitting of the $^7F_1$ level, which depends of the neighbor cation surrounding the luminescent ion and thus the environment of the rare earth element.

7. Calculation of the standardized B_2 parameter as followed:

$$B_2 = \left((B_{20})^2 + 2(B_{22})^2\right)^{(1/2)} \tag{5.32}$$

8. Representation of the evolution of the B_2 parameter in function of the excitation energy (cm^{-1}).

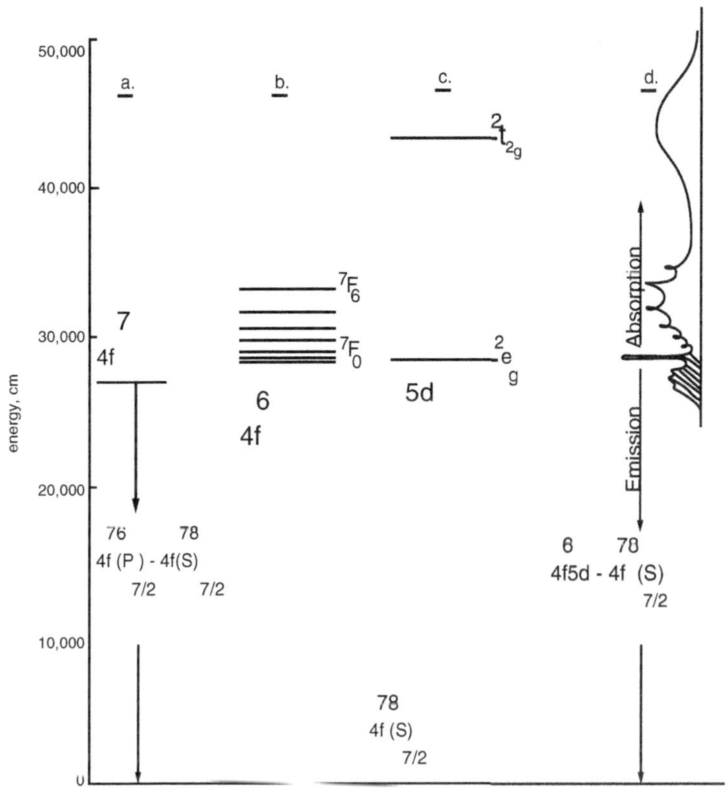

Fig. 5.11 Energy levels scheme of Eu^{2+}

5.6.5.2 Eu^{2+}

The absorption and emission spectra of divalent europium are due to electronic transitions between the $4f^{7+}$ and $4f^6 5d^1$ electronic configurations (Fig. 5.11). An approximate energy level scheme was proposed (Blasse et al. 1968) for the electronic transitions in Eu^{2+} by using strong field formalism to describe the 5d levels and the weak field formalism to describe the 4f orbitals. The ground state of Eu^{2+} is $^8S_{7/2}$ because of the $4f^7$ electronic configuration. In the configuration $4f^6 5d^1$, one electron occupies a 5d orbital which is split into two orbital sets, t_{2g} and e_g, by a cubic crystal field. Thus the energy terms are ^2E_g and $^2T_{2g}$ in full cubic symmetry.

In the group of divalent REE luminescence centers Eu^{2+} is the most well known in minerals. It shows a 5d-4f emission, which varies usually from UV to blue. The host lattice dependence of the emission color of the Eu^{2+} is mainly connected with covalence (the nephelauxetic effect), which will reduce the energy difference between the 4f and 5d configurations, crystal field splitting of the 5d configuration and the Stokes shift (Blasse and Grabmaier 1994). Nevertheless, in certain cases anomalous luminescence may be possible, identification of which may be based on the following aspects: an abnormally large Stokes shift and width of the emission band; a wavelength of emission that is not consistent with the wavelength anticipated from the properties of the compound; an anomalous decay time and thermal behavior (Dorenbos 2003). Such luminescence may be red, for example at 600 nm in BaF_2, with decay time is about 600–800 ns. This is due to the fact that the emitting level contains spin octets and sextets, whereas the ground state level is an octet, so that the optical transition rate is slower because of spin selection rule (Dorenbos 2003).

5.6.5.3 Apatite

Divalent europium is detected in apatite as a shoulder of Ce^{3+} luminescence when studied by steady-state spectroscopy (Gorobets 1968). The reason is that Eu^{2+} and Ce^{3+} emit in the same spectral range, but the Eu concentration in all investigated samples is very low compared to that of Ce. Besides that, since the 5d-4f transitions in Ce^{3+} is parity allowed and spin selection is not appropriate, the emission transition is fully allowed one and luminescence intensity is very strong. Thus Ce^{3} makes Eu^{2+} luminescence detection hardly possible (Fig. 4.2a). However, because Eu^{2+} has a much longer decay time, time-resolved spectroscopy enables the isolation of the Eu^{2+} luminescence in pure form. After a delay of 500 ns and with a gate width of 500 ns, the short-lived luminescence of Ce^{3+} is already quenched and the long-lived luminescence of other trivalent REE and Mn^{2+} is not detected yet. At such conditions the blue, relatively narrow band peaking at 450 nm appears with intermediate decay time of ~400 ns (Fig. 4.2b). Such an emission band is known as in natural and in synthetic artificially activated apatite and is connected with Eu^{2+}. On excitation with UV, two types of lowest excited states are possible, namely ^6P_j (f-f) or $4f^65d$ (f-d), depending on the host matrix. The nature of Eu^{2+} emission, namely either an intra-configurational (f-f) line or inter-configurational (f-d) band emission in a given host, is mainly decided by the effect of the ligand field on the Eu^{2+} energy levels. In the apatite case, the anions constituted by the $PO_4{}^{3-}$ network produce a strong nephelauxetic effect and hence the $4f^65d$ level is the lowest excited state, leading to band emission. It was found that in calcium fluor-apatite there is only one emission band. The emission maximum dependence on the halogen type confirms that this band can be assigned to the Ca(II) sites (Kottaisamy et al. 1994).

After a delay of several μs, luminescence of Eu^{2+} is already very weak, and narrow long-lived lines of trivalent REE dominate in the spectrum (Reisfeld et al. 1996). The lines at 589, 617, 651, and 695 nm (Fig. 4.2c) have never been detected in natural apatite by steady-state spectroscopy. According to their spectral position they may be ascribed to Eu^{3+}, but they are different from a known lines in synthetic apatites activated by Eu (Jagannathan and Kottaisamy 1995; Morozov et al. 1970; Piriou et al. 1987, 2001; Voronko et al. 1991). In order to clarify this problem we studied artificially activated samples by laser-induced time-resolved

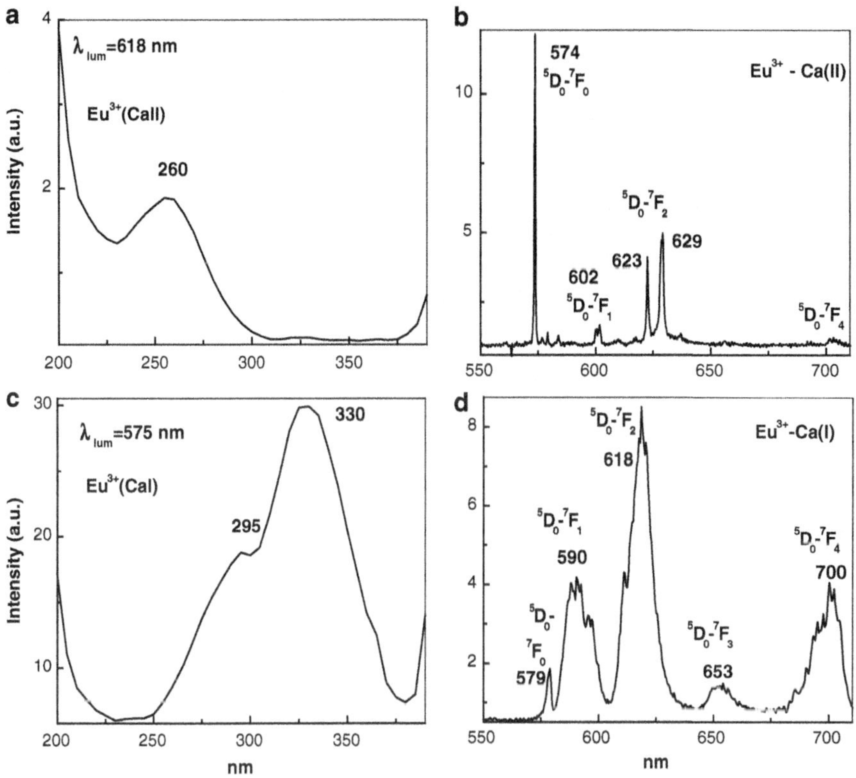

Fig. 5.12 (**a–d**) Laser-induced luminescence and excitation spectra of synthetic apatite artificially activated by Eu in vacuum (**a**) and in air (**b**)

luminescence spectroscopy (Gaft et al. 1997b). Two main centers appear after activation in air both characterized by the abnormal relative intensity of $^5D_0$-$^7F_0$ transition at 574 and 579 nm. Activation in vacuum leads to prominent changes. New lines at 590, 618 and 700 nm appear which dominate at $\lambda_{ex} = 384$ nm (Fig. 5.12).

Thus different luminescence takes place after activation in air and in vacuum. As it was already mentioned, in the apatite structure $Ca_5(PO_4)_3F$ there are two types of Ca site: Ca(I) with C_3 symmetry and Ca(II) with C_s symmetry. The $^5D_0$-$^7F_0$ transition has been reported to exist in cases where the site symmetry allows an electric dipole process: C_s, C_n, C_{nv}. This is consistent with the conclusion that the centers with the main line at 574 nm belongs to Ca(II) site with C_s symmetry, while the asymmetry of the crystal field is lower for the center with the main line at 618 nm. The ratio between $^5D_0$-$^7F_0$ and $^5D_0$-$^7F_2$ which are forced electric dipoles to $^5D_0$-$^7F_1$ which is magnetic dipole tells us about the symmetry of the site in which Eu^{3+} is situated (Blasse and Grabmaier 1994; Reisfeld 1973). In fluor-apatite activated in air the ratio is higher meaning a lower symmetry Ca(II) site, while in natural fluorapatite or synthetic one activated in vacuum the ratio is lower indicating high symmetry Ca(I) site. The decay of the corresponding transitions is longer for the center with the main line at 618 nm indicating the higher symmetry.

The excitation spectra of the line at 573 nm after activation in air and the line at 618 nm after activation in vacuum are totally different (Fig. 5.12). The excitation bands of the line at 618 nm have the higher energies. They are connected with charge transfer occurs by electron jumping from one of the highest filled molecular orbital of the ligand to the partly filled shell of the central atom (Reisfeld 1973). The higher energy of the charge transfer bands indicates the larger Eu^{3+}-O^{2-} distance corresponding to Ca(I) position. For the samples activated in vacuum by Eu and Na it was indeed found that Eu^{3+} luminescence in higher symmetry Ca(I) position is the only one and corresponds exactly to Eu^{3+} luminescence in natural fluorapatite (Gaft et al. 1997b).

Under laser excitation, Eu^{3+}(I) is especially prominent under $\lambda_{ex} = 266$ nm and it is the main form of Eu^{3+} in natural apatites. Extremely rare, 337 and 355 nm excitations lead to the new lines at 574, 623, 630 and 711 nm appearance (Fig. 4. 2d), which are similar with those received as result of activation in air. Thus they are connected with Eu^{3+} in the low symmetry Ca(II) site. Because of the lower symmetry its luminescence has a relatively short decay time and is more prominent in the spectrum with narrower gate width. This center is especially strong at liquid nitrogen temperature where the prominent $^5D_0$-$^7F_0$ line at 574 nm can be seen, unhidden by Dy^{3+} luminescence.

5.6.5.4 Fluorite

The luminescence center of divalent europium in fluorite is well known (Haberland et al. 1934; Tarashchan 1978; Krasilschikova et al. 1986; Barbin et al. 1996). It is clearly seen in laser-induced time-resolved luminescence spectra with decay time of 600–800 ns (Fig. 4.14a). In several samples the band with spectrum similar to those of Eu^{2+} has very long decay time and remains even after delay of several ms. Principally it may be connected with energy migration from UV emitting center with long decay time, for example, Gd^{3+}.

The luminescence of Eu^3 has not been detected in natural samples using steady-state spectroscopy, because it obscured by the strong emission band of Eu^{2+} and other centers. Because the decay time of Eu^{2+} is in the range of 600–800 ns, it is already quenched after a delay of 10 μs and only the long-lived luminescence of REE^{3+} is detected. Two groups of lines appear which may be ascribed to Eu^{3+}. The first is especially strong under excitation of 266 nm with the main line at 573 nm (Fig. 4.14b). The second type is also especially strong under excitation of 266 nm with the main lines at 595, 622, 645 and 700 nm (Fig. 4.14c). Two types of REE luminescence centers are well known in synthetic artificially activated fluorite (Stepanov and Feofilov 1956). Eu^{3+} may be located in the CaF_2 lattice in several non-equivalent positions with cubical, tetragonal, trigonal and rhombic symmetries. Decay times of the first group of luminescence lines are extremely long, possibly indicating high symmetry cubic symmetry. Another group of Eu^{3+} lines

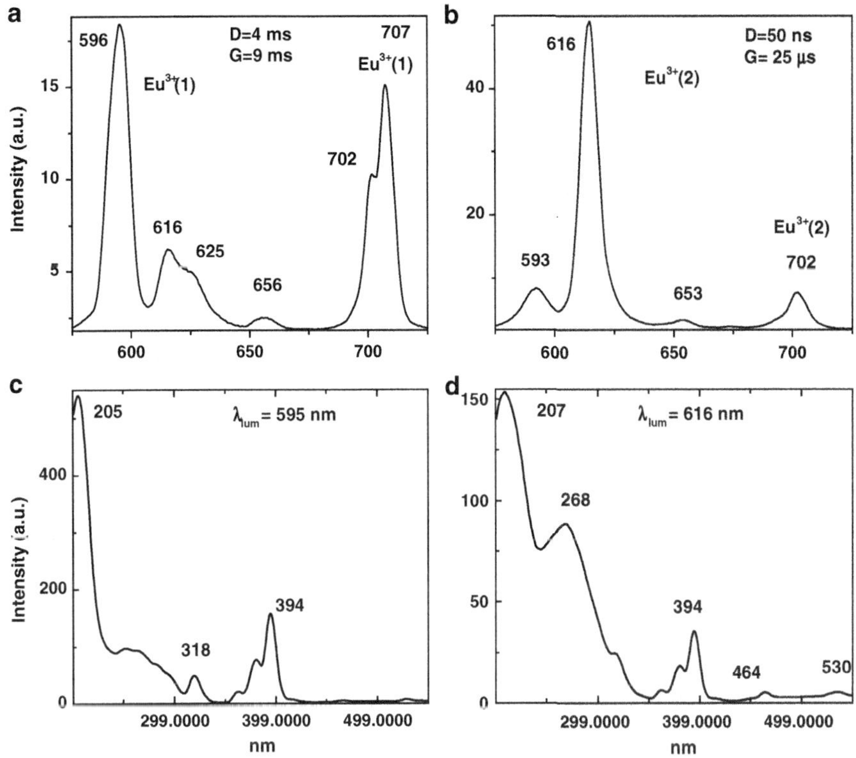

Fig. 5.13 (**a–d**) Laser-induced time-resolved luminescence spectra of synthetic zircon artificially activated by Eu (**a**, **b**) and excitation spectra of different Eu^{3+} centers (**c**, **d**)

with shorter decay time is connected with a Eu^{3+} site of lower symmetry in fluorite lattice. The Eu^{3+} line at 574 nm belongs to $^5D_0$-$^7F_0$ transition which exists only in cases where the site symmetry allows an electric dipole process: C_s, C_n, C_{nv} (Reisfeld 1973).

5.6.5.5 Zircon

Time-resolved luminescence spectroscopy of zircon revealed luminescence lines, which may be confidentially ascribed to Eu^{3+} center (Fig. 4.90d). Usually they are hidden by broadband yellow emission of zircon and may be detected only with long delay time using its much longer decay time compared to yellow luminescence. In order to interpret detected those lines time-resolved luminescence of synthetic $ZrSiO_4$ activated by Eu was studied (Gaft et al. 2000c) (Fig. 5.13). Time-resolved luminescence spectroscopy enabled to detect at least two Eu^{3+} centers with different decay times: main line 616 nm and much weaker lines at 593 and 702 nm with relatively short decay time, and main lines at 596 and 707 nm and weaker line at

616 nm with long decay time. The shorter decay and the higher ratio of forced electric dipole $^5D_0$-$^7F_2$ to magnetic dipole $^5D_0$-$^7F_1$ transition indicate the lower local symmetry of Eu-I center as it is normally observed. The relatively high intensity of $^5D_0$-$^7F_2$ transition in Eu-I may be connected with the absence of inversion symmetry on the site. This is consistent with the behavior of hypersensitive transitions with $|\Delta J| = 2$ (Blasse and Grabmaier 1994). In Eu-II center the magnetic dipole transition $^5D_0$-$^7F_1$ is dominant indicating higher symmetry site. The found difference in the excitation into charge transfer spectra of the two emissions at 596 and 616 nm confirms the existence of two sites with different symmetry, one associated with mainly single charge transfer band at 205 nm, another associated with two charge transfer bands at 207 and 268 nm. Charge transfer occurs by electron jumping from one of the highest filled molecular orbital of the ligand to the partly filled shell of the central atom. The higher energy of the charge transfer bands indicates the larger Eu^{3+}-O^{2-} distance. The line at 596 nm belongs to Eu-II center, which has only one type of the oxygen sites in coordination polyhedron having the highest symmetry and the longest decay up to 1.8 ms. The line at 615 nm belongs to Eu-I center, which has big difference between the largest and the shortest Eu-O distances having the lowest symmetry and the shortest decay of 125 μs.

Zircon belongs to the tetragonal system and is a positive uniaxial. The typical form shows the {111} and the {110} planes. The two orientations selected for luminescence polarisation study were the (110) plane, parallel to the basal section and the [100] row. In such case the axis perpendicular to the (110) plane will be called X. The orientation notation is made according to the so-called Porto notation (Porto et al. 1956). The $X_1(ZX_2)X_1$ orientation means that the laser light entered parallel to the X_1 axis of the crystal and is polarized in the Z direction, while the emission is collected along the X_1 axis with X_2 polarization. By polarization spectroscopy with high spectral resolution (less than 0.1 nm) six lines are observed for $^5D_0$-$^7F_1$ transition of Eu-II center instead of the maximum three allowed for an unique site (Fig. 5.14). In Z(XX)Z geometry which corresponds to observation of σ-polarized luminescence we see relatively strong $^5D_0$-$^7F_2$ line. It may be explained by splitting scheme where $^5D_0$-$^7F_2$ transition has four lines with σ-polarization while other three are forbidden both as electric and magnetic dipole transitions (Diecke 1968). In such case $^5D_0$-$^7F_1$ transition has two lines at 596 and 598 nm with π-polarization and one line at 597 nm with σ-polarization. The $^5D_0$-$^7F_4$ transition has four pure resolved lines at 708 nm with σ-polarization and three pure resolved lines at 701 nm with π-polarization while other lines are very weak. It is confirmed by observation in Y(ZZ)Y geometry which corresponds to observation of π-polarized luminescence where we see the strongest line at 597.1 nm in $^5D_0$-$^7F_1$ transition and at 707 nm in $^5D_0$-$^7F_4$ transition, while $^5D_0$-$^7F_2$ transition is practically absent. It is important to note that the $^5D_0$-$^7F_4$ transition is the strongest one in such case, which is firstly observed for Eu^{3+} luminescence. In the case of in Y(XX)Y geometry which also corresponds to observation of σ-polarized luminescence we see the three strongest lines of $^5D_0$-$^7F_1$ transition at 591, 592 and 593 nm and much weaker lines of $^5D_0$-$^7F_2$ transition at 614 and 627 nm and of $^5D_0$-$^7F_4$ at 707 nm.

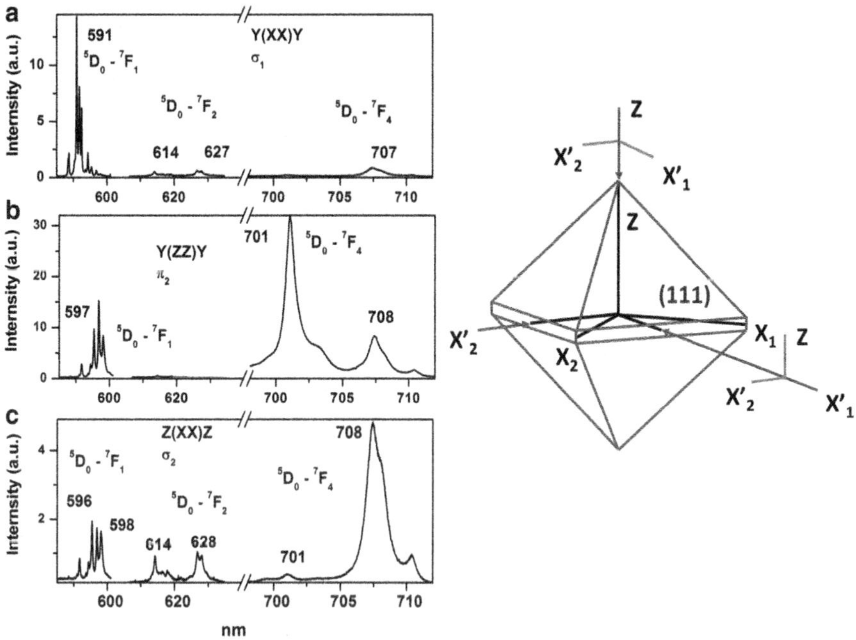

Fig. 5.14 (**a–c**) Laser-induced polarized spectra of synthetic zircon artificially activated by Eu

Thus at least three Eu^{3+} centers take place in zircon, which have been characterized in Table 5.3 by the main transitions, polarizations and decay times of each emitted level. These centers may be reasonably connected with Zr^{4+} cation substitution within tetragonal $ZrSiO_4$ structure with space group $I4_1/amd$ for two main reasons: the ionic radius of Eu^{3+} is 0.99 Å while of Zr^{4+} is 0.79 Å compared with 0.42 Å of Si^{4+}, and in addition, the eight oxygen atoms coordinated to Zr^{4+} cation, instead of four for Si^{4+}, which corresponds to usual one for Eu^{3+} ion. In fact $ZrSiO_4$ unit cell contains four ZrO_8 groups, which are equivalent in the sense that they have the same geometry but different orientation. The possible reason of several Eu^{3+} centers could be associated with local charge compensation by the way of electron-hole centers. The following schemes may be considered: (i) Substitution of Eu^{3+}-Zr^{4+} leads to the formation of the hole center on the oxygen, as in the case of substitution Y^{3+}-Zr^{4+} when O^- center SiO_4^{3-} forms (Marfunin 1979b); (ii) Zr^{3+} centers are known in zircon at the expense of adjacent oxygen vacancy or interstitial alkali metal ion (Marfunin 1979a). It may be substituted by Eu^{3+} for two reasons: the ionic radius of Zr^{3+} has to be between 0.79 Å for Zr^{4+} and 0.109 Å for Zr^+, so even closer to the radius of Eu^{3+}, and in addition, it has the same charge.

Table 5.3 Different Eu^{3+} luminescence centers in artificially activated zircon

Center	λ, nm I	τ, μs	Transition polarization
Eu^{3+} – I	587, 593 20	125	$^5D_0$-$^7F_1$
	616, 630 100	125	$^5D_0$-$^7F_2$
	653 5	125	$^5D_0$-$^7F_3$
	702 10	125	$^5D_0$-$^7F_4$
Eu^{3+} – II	596, 598 15	1600	$^5D_0$-$^7F_1$ π
	597 15	1600	$^5D_0$-$^7F_1$ σ
	614, 628 10	1600	$^5D_0$-$^7F_2$ σ
	653 1	1600	$^5D_0$-$^7F_3$ σ
	701 100	1600	$^5D_0$-$^7F_4$ σ
	708 100	1600	$^5D_0$-$^7F_4$ π
Eu^{3+} – III	591, 593 100	1800	$^5D_0$-$^7F_1$ π
	592 100	1800	$^5D_0$-$^7F_1$ σ
	614, 628 5	1800	$^5D_0$-$^7F_2$ σ
	653 1	1800	$^5D_0$-$^7F_3$ σ
	701 10	1800	$^5D_0$-$^7F_4$ σ
	707 10	1800	$^5D_0$-$^7F_4$ π

5.6.5.6 Baddeleyite

Luminescence of Eu^{3+} has in natural samples was found (Fig. 4.93b). It is well studied in artificially activated ZrO_2 (Gutzov et al. 1998; Gedanken et al. 2000; Reisfeld et al. 2000). The main emission occurs between the $^5D_0$ level to the ^7F_j multiplet with decay time of approximately 0.5 ms (Fig. 5.15). The luminescence intensity is relatively weak, but may be substantially increased by co-doping with nanoparticles of semiconductors, such as CdS. The origin of the intensification could be explained by the following reasoning. CdS nanoparticles have intensive absorption bands, while the formation of electron-hole pairs is not followed by radiation at room temperature. The energy from non-radiative recombination can be transferred to the high lying energy levels of the Eu^{3+}. These excited levels once again non-radiatively decaying to the long-lived $^5D_0$ level. Such a mechanism will increase the population of the emitting level, but not their emission probability to the ground level, which was confirmed by the unchanged decay time of Eu^{3+}.

5.6.5.7 Scheelite

Figure 5.16a, b represent luminescence spectra of scheelite enriched by Eu. Luminescence of Eu^{3+} is well known in steady-state spectra of scheelite (Tarashchan 1978; Gorobets and Kudrina 1976). In time-resolved spectroscopy its relative intensity is stronger after a long delay time, which is explained by longest decay time of Eu^{3+} in scheelite compared to other REE.

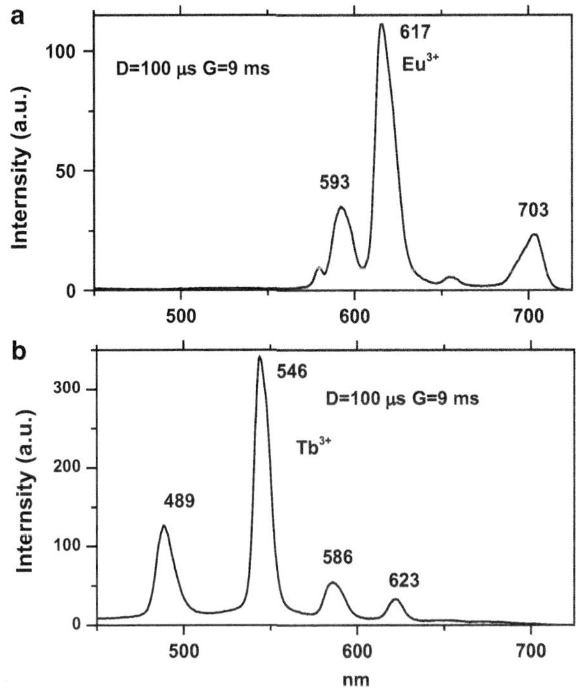

Fig. 5.15 (**a–b**) Laser-induced time-resolved luminescence spectra of ZrO_2 activated by Eu^{3+} (**a**) and Tb^{3+} (**b**)

The possible luminescence of Eu^{2+} in scheelite is very interesting problem. It was not detected by steady-state luminescence spectroscopy. The possible reason is that very strong intrinsic luminescence of scheelite is situated in the same spectral range, which covers the weaker emission of Eu^{2+}. We tried to solve this problem by time-resolved method using different decay times for intrinsic and Eu^{2+} bands. Time-resolved spectroscopy of synthetic $CaWO_4$:Eu (Fig. 5.16c, d) revealed the blue band peaking at the same spectral region, but with narrower band width compared to intrinsic scheelite emission. Thus it may be supposed, that blue emission consists of several luminescence bands and the participation of Eu^{2+} is principally possible.

5.6.5.8 Barite and Anhydrite

The spectral-kinetic parameters of the narrow band at 375 nm enable its confident identification as Eu^{2+} luminescence, which is confirmed by emission of synthetic $BaSO_4$ artificially activated by Eu (Fig. 5.17a). Such emission was also detected and interpreted by steady-state spectroscopy (Tarashchan 1978). It is interesting to note that very often such band is absent in natural barite and appears only

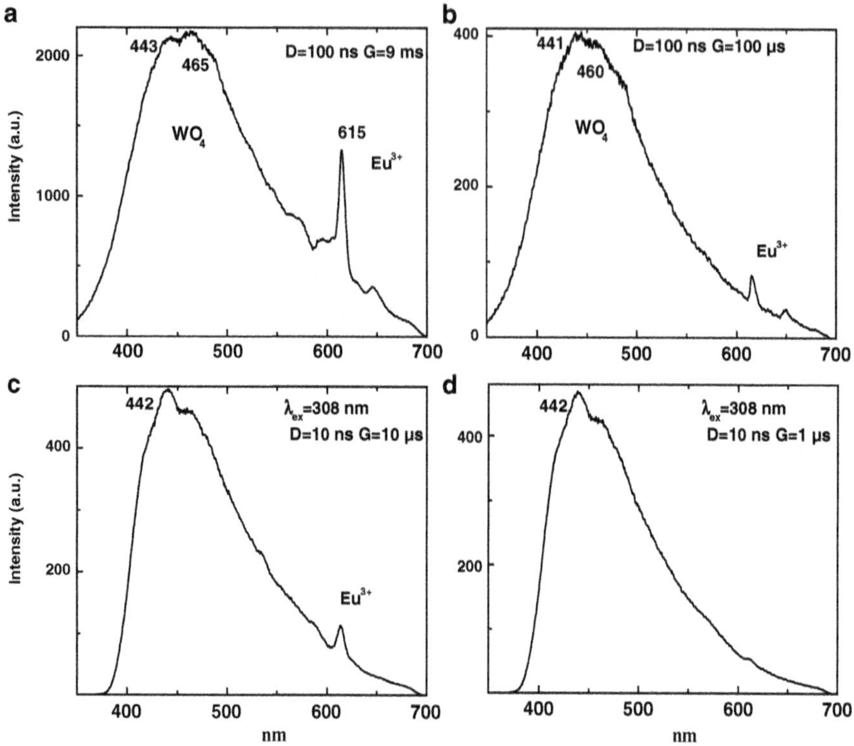

Fig. 5.16 (**a–d**) Laser-induced time-resolved luminescence spectra of scheelite with elevated concentration of Eu (**a**, **b**) and synthetic scheelite artificially activated by Eu (**c**, **d**)

after heating in air at 600–700 °C. Such a transformation is reversible, at least partly. Under X-ray excitation the intensity of UV band diminishes, and a new blue-green emission appears (Fig. 5.18). This shows some kind of transformation, which takes place in the barite lattice under these conditions. Several possibilities exist. It is possible that in barite the luminescence is quenched by the components with high-energy phonons. The water and organic matter may represent the latter. They are removed after heating and luminescence becomes visible. Another possibility is that barite accommodates REE in concentrated adsorbed form, which is not suitable for detection by luminescence due to concentration quenching. As a result of high temperature heating, thermal diffusion of Eu in lattice takes place with resulting luminescence. The isomorphic substitution in Eu^{2+} form is easy because its ionic radius (139 pm) is close to that of Ba^{2+} and Sr^{2+} (Gaft and Rudenkova 1993).

The other model is also principally possible. Europium initially enters the barite lattice as Eu^{+}, which oxidizes to Eu^{2+} at 700 °C. The relatively small difference

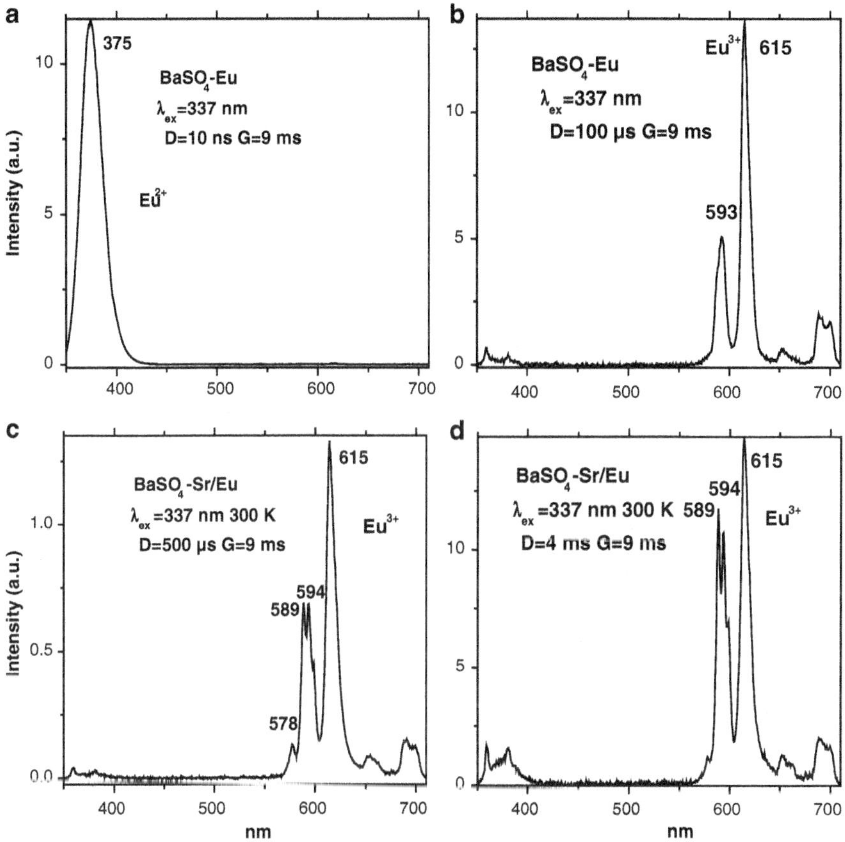

Fig. 5.17 (**a–d**) Laser-induced time-resolved luminescence spectra of Ba(Sr)SO_4:Eu

between the Ba^{2+} and the Eu^{+} ionic radii (1.5 and 1.7 Å) makes this substitution possible. Luminescence of Eu^{+} was still not observed in minerals, but is known in luminofors (Gorobets et al. 1968). Eu^{+} has $6s^1$ electron configuration and the mostly probable are electric-dipole electron transitions $6s^1$-$6p^1$, taking place between uneven $^7S_3{}^9S_4$ and even $^7P_{2,3,4}{}^9P_{3,4,5}$ terms. It usually generates the broadband luminescence in the UV-green part of the spectrum.

Luminescence of Eu^{2+} is common in barite, while trivalent Eu was only recently detected in natural barite (Dimova et al. 2006). After a delay of several μs, the Eu^{2+} emission becomes much weaker and very weak narrow lines of Eu^{3+} appear, peaking at 488, 544 and 615 nm (Fig. 4.72). Luminescence of Eu^{3+} and Eu^{2+} is well known in anhydrite (Tarashchan 1978; Gaft et al. 1985; Baumer et al. 1997). They are also found in time-resolved spectra under excitation at 266 nm (Fig. 4.30).

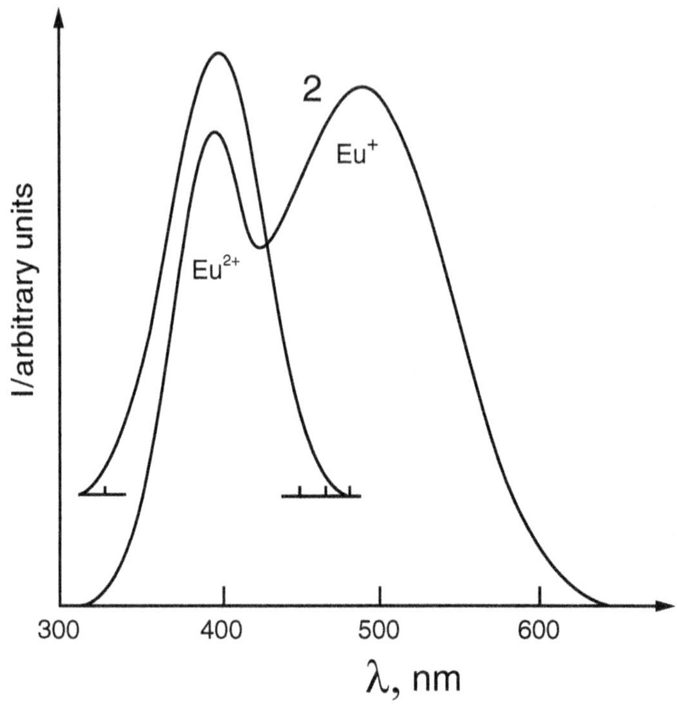

Fig. 5.18 Laser-induced luminescence of barite: *1*- after heating at 600 °C while violet luminescence of evidently Eu^{2+} appears; *2*-after X-ray irradiation of the heated sample where blue shoulder appears, supposedly connected with Eu^{+}

5.6.5.9 Calcite

Steady-state luminescence of Eu^{3+} was found in CL spectra (Haberman et al. 1996). Figure 4.18c represents laser-induced time-resolved luminescence of calcite under 266 nm. Luminescence lines of Eu^{3+} are seen. Hypersensitive $^5D_0$-$^7F_2$ (618 nm) emission dominates Eu^{3+} luminescence, showing local coordination without inversion symmetry. Another type of Eu^{3+} is characterized by strong $^5D_0$-$^7F_0$ line at 575 nm. Good luminescence of Eu^{3+} was found also in aragonite (Fig. 4.22b).

5.6.5.10 Danburite and Datolite

The narrow band at 437 nm with decay time of 650 ns in danburite luminescence spectrum belongs to Eu^{2+} luminescence (Fig. 4.27a), which was also detected by steady-state spectroscopy (Gaft et al. 1979). Besides that, under 308 nm excitation narrow lines appear at 580, 592, 611, 618, 655 and 692 nm (Fig. 4.27c) with long decay time, which confidently may be ascribed to Eu^{3+}. Under 266 nm excitation another group of lines appear with the main line at 575 nm connected with different type of Eu^{3+} (Fig. 4.27d).

The narrow band at 455 nm with decay time of 750 ns in datolite luminescence spectrum belongs to Eu^{2+} luminescence (Fig. 4.29b), which was also detected by steady-state spectroscopy (Gaft et al. 1979). Besides that, narrow lines appear in the orange part of the spectrum with the main one at 610 nm (Fig. 4.29c) with long decay time, which confidently maybe ascribed to Eu^{3+}.

5.6.5.11 Pyromorphite

Figure 4.59 demonstrates time-resolved luminescence of pyromorphite under 355 and 266 nm laser excitations. Luminescence lines of Eu^{3+} are clearly seen in both cases.

5.6.5.12 Feldspars

Most of the natural alkali feldspars and plagioclases rarely contain more than a few ppm of Eu and its emission is often very difficult to detect, especially when superposed by strong emission of other activators. In natural samples the blue-violet Eu^{2+} activated luminescence is known by steady-state spectroscopy (Haberland and Köhler 1939; Moroshkin et al. 1987). Time-resolved spectroscopy enables to detect separately narrow luminescence bands with decay times is the range of 600–800 ns, which confidently may be ascribed to Eu^{2+} (Fig. 4.98). In our study its emission maximum was changed from 403 nm in labradorite to 410 nm in plagioclase. It has to be mentioned that very often, additional blue bands are detected, but with much longer decay times, which may be mistakenly interpreted as Eu^{2+} in steady-state spectra. Eu^{3+} luminescence was firstly detected in feldspars by laser-induced time-resolved spectroscopy. It was excited by green laser light at 532 nm and is characterized by the narrow lines at 614 and 624 nm (Fig. 4.99c) with long decay time of 550 μs.

5.6.5.13 Charoite

Narrow band near 410 nm in time-resolved luminescence spectra of charoite (Fig. 4.39d) is connected with Eu^{2+} luminescence

5.6.5.14 Titanite

Based on similar behavior under different excitations, decay times and synchronous change of intensities in different samples, we detected two groups of lines in titanite luminescence spectra, which are evidently connected with Eu^{3+} emission

(Fig. 4.79d). The first grope contains the lines at 574, 589, 613 and 620 nm, while the second one the lines at 578, 590, 598, 608, 620, 652, 703, 705 and 711 nm (Gaft et al. 2003b). All lines have long decay times of approximately 840 μs. Transition $^5D_0$-$^7F_0$ presents in both centers, which exists only in cases where the site symmetry allows an electric dipole process (Reisfeld 1973). The most probable substitution for Eu^{3+} in titanite structure is instead of Ca^{2+} with corresponding charge compensation. The relative intensities of magnetic dipole transition $^5D_0$-$^7F_1$ and electric dipole transition $^5D_0$-$^7F_2$, together with strong $^5D_0$-$^7F_0$ transition in titanite is very similar to Eu^{3+} emission in low symmetry sites in apatite structure, where Ca^{2+} has seven nearest neighbors, six oxygens and one fluor. Thus Eu^{3+} in titanite may substitute for seven-fold coordinated Ca^{2+} where F initially substitutes one of the oxygen atoms. In such case charge compensation may be achieved by opposite substitution of F^- by O^{2-}.

5.6.5.15 Zoisite

The narrow band peaking at 440 nm with decay time of approximately 1 μs is evidently connected with Eu^{2+} center substituting for Ca^{2+} (Fig. 4.138d).

5.6.5.16 Leucophane

Luminescence of Eu^{2+} and Eu^{3+} has been found in X-ray excited luminescence spectra of synthetic leucophane activated by Eu (Prokofiev et al. 1982), while in natural samples only Eu^{2+} was detected. Laser-induced time resolved luminescence under 532 nm excitation enables to detect clear lines of Eu^{3+} with the strongest $^5D_0$-$^7F_0$ electron transition at 573 nm (Fig. 4.45e).

5.6.5.17 Sodalite and Tugtupite

Violet luminescence bands peaking at 405 and 430 nm with decay times of approximately several μs were detected in tugtupite and sodalite (Figs. 4.160 and 4.155). Such emission is typical for a Eu^{2+} impurity luminescence center. Such an interpretation is supported by the fact that impurities of Eu in quantities up to 500 ppm have been found in natural tugtupite (Gaft et al. 2009), which is quite enough for luminescence generation. The most probable position of Eu^{2+} is in Na site, because the ionic radius of Eu^{2+} is the closest to Na^+ among all elements in sodalite and tugtupite. The Na atoms are in the framework cavities on four-fold coordination in the sodalite and tugtupite structures. The ionic radius of Na^+ in four-fold coordination is 113 pm, while the ionic radius of Eu^{2+} is not known for such coordination. Nevertheless, ionic radii of Na^+ and Eu^{2+} are rather similar for octahedral and 8-coordination: ionic radii of Eu^{2+} are 131 pm in octahedral coordination and 139 pm in 8-coordination, respectively; while Na^+ are 116 and

132 pm in octahedral and 8-coordination, respectively. The possibility of such isomorphic substitution by Eu^{2+} with resulting luminescence was already proved for cryolite $Na_2[NaAlF]_6$ (Gorobets and Rogojine 2001).

5.6.5.18 Other Minerals

Luminescence of Eu^{3+} was found in chabazite (Kobayashi et al. 2011), baddeleyite (Fig. 4.93b), dolomite (Fig. 4.25c), pyrochlore (Fig. 4.43b), eudialite (Fig. 4.53a), burbankite (Fig. 4.55c) and Eu^{2+} in walstromite (Fig. 4.49d), charoite (Fig. 4.39d) and milarite (Fig. 4.47).

5.6.6 *Gd³⁺*

In all minerals the gadolinium luminescence spectra are completely located in the UV part of the spectrum and consist of several lines at 310–315 nm, corresponding to transition $^6P_{7/2}$-$^8S_{7/2}$. The main line is characterized by long decay time and is especially prominent in the spectra with long delay. Gd^{3+} is known as a good sensitizer of the other rare-earth ions luminescence. It is detected in spectra of fluorite (Fig. 4.16a), zircon (Fig. 4.90h), anhydrite (Fig. 4.30a), and hardystonite (Fig. 4.36b).

5.6.7 *Tb³⁺*

Luminescence spectra consisting of many lines due to ^5D_j-^7F_j transitions are observed for Tb^{3+}. The intensity of the emissions from $^5D_3$ decreases with increasing Tb^{3+} concentration due to cross-relaxation. Among the emission lines from the $^5D_4$ state, the $^5D_4$-$^7F_5$ emission line at approximately 550 nm is the strongest in nearly all host crystals when the Tb^{3+} concentration is high. The reason is that this transition has the largest probability for both electric-dipole and magnetic-dipole induced transitions. The Tb^{3+} emission has a broad excitation band in the region from 220 to 300 nm originating from the $4f^8$-$4f^75d^1$ transition. The intensity ratio of the emission from $^5D_3$ to that from $^5D_4$ depends not only on the Tb concentration, but also on the host material. Two additional factors have to be considered in order to determine the possible ratio of $^5D_3$ to $^5D_4$ intensity. One is the maximum energy of phonons that causes phonon-induced relaxation. If the maximum phonon energy is large, the ratio of $^5D_3$ to $^5D_4$ intensity becomes small. The other factor is the energy position of the $4f^75d^1$ level relative to $4f^8$ levels, which can be discussed in terms of the configurational coordinate model. If the minimum of the $4f^75d^1$ is low in energy and the Franc-Condon shift is large, there is a possibility that an electron

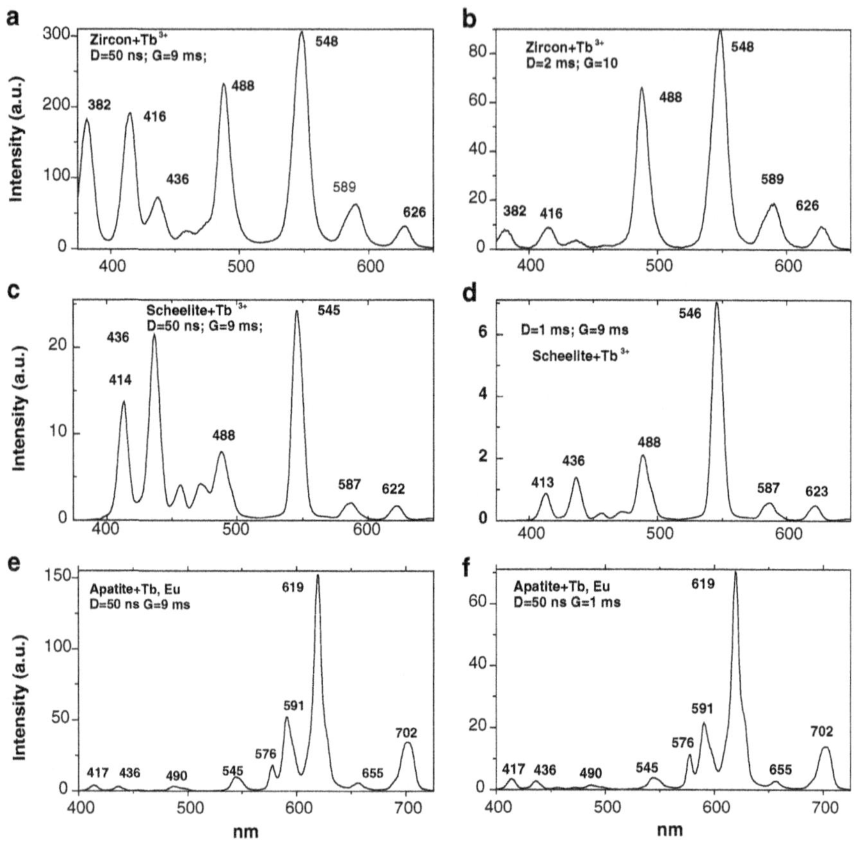

Fig. 5.19 a-f Laser-induced time-resolved luminescence spectra of synthetic zircon, apatite and scheelite artificially activated by Tb

excited to the $4f^7 5d^1$ level can relax directly to the $^5D_4$, bypassing the $^5D_3$ and thus producing only $^5D_4$ luminescence.

In order to correct identification of Tb^{3+} lines in minerals several of them were synthesized and artificially activated by Pr (Fig. 5.19). Besides that, comparison has been made with CL spectra of synthetic minerals artificially activated by Tb (Blanc et al. 2000).

5.6.7.1 Apatite

Usually the strongest lines at 380, 414 and 436 nm of $^5D_3$-^7F_j transitions (Fig. 4.3d) are known in natural apatite, which represent Tb^{3+} luminescence in the Ca (I) position. The intensity of the 545 nm line ($^5D_4$-$^7F_2$) of Tb^{3+} is strongest in red apatite, as compared to the others (Fig. 4.2c). Evidently Tb^{3+} is in the Ca(II) site.

The abnormal intensity is not connected with unusually high Tb concentration (Table 4.4), but with lower symmetry of this site. Simultaneous activation by Tb and Eu (Fig. 5.19e, f) demonstrates that blue emission lines of Tb^{3+} have shorter decay times compared with green ones in accordance with different symmetries of Ca(I) and Ca(II) sites.

5.6.7.2 Scheelite

In Tb^{3+} the luminescent lines are connected with electron transitions from different excited levels: $^5D_3$ and $^5D_4$. The lines at 436 and 414 nm have $\tau = 375$ μs while the lines at 546 and 489 nm are characterized by a very long $\tau = 2.4$ ms. In the time-resolved spectra of natural samples with 337 nm excitations, the previously hidden lines at 416 and 439 nm enable sure identification of Tb^{3+} (Fig. 4.11e, f).

5.6.7.3 Anhydrite, Feldspar, Fluorite, Zircon, Calcite, Dolomite, Zoisite

The strongest lines of emission of Tb^{3+} in these minerals are in the UV-violet and visible part of the spectrum (Figs. 4.30d, 4.99a, 4.14c, 4.19c, 4.25c, 4.90h, 4.138c).

5.6.8 *Dy*$^{3+}$

Dysprosium activated minerals have luminescence in the visible part of the spectrum. The spectra of Dy^{3+} in minerals are mainly characterized by narrow lines near 480 and 575 nm, accompanied by the weaker ones near 660 and 752 nm corresponding to transitions from level $^4F_{3/2}$ to the levels of multiplets ^6H_j and ^6F_j (Tarashchan 1978). Consequently, the spectra are not changed with delay time and excitation energy and all luminescence lines of Dy^{3+} are characterized by decay time. The best excitation at 350 nm is connected with 4f-4f $^6H_{15/2}$-$^4G_{7/2}$ transition.

Two main groups of narrow emission lines are detected in grossular spectra. The first one with two groups of lines around of 475 and 500 nm with long decay times of 600–630 μs (Fig. 5.20). The second group near 580 nm is difficult to detect on the background of broad emission band peaking at 590 nm, which evidently belongs to Mn^{2+}. They are more evident in spectra with narrower gate width, because the decay time of this broad band of 8.6 ms is longer than the decay time of narrow lines. In the clear form they may be seen in Tanzania tsavorite, where Mn^{2+} luminescence is actually absent. Those two groups of lines may be associated with Dy^{3+}, which presents in this specific sample in 2 ppm quantity. It has to be noted the presence of the relatively strong line peaking at 500 nm, which is not typical for Dy^{3+} luminescence. It is not known in emission spectra of other trivalent REE and has to be additionally studied.

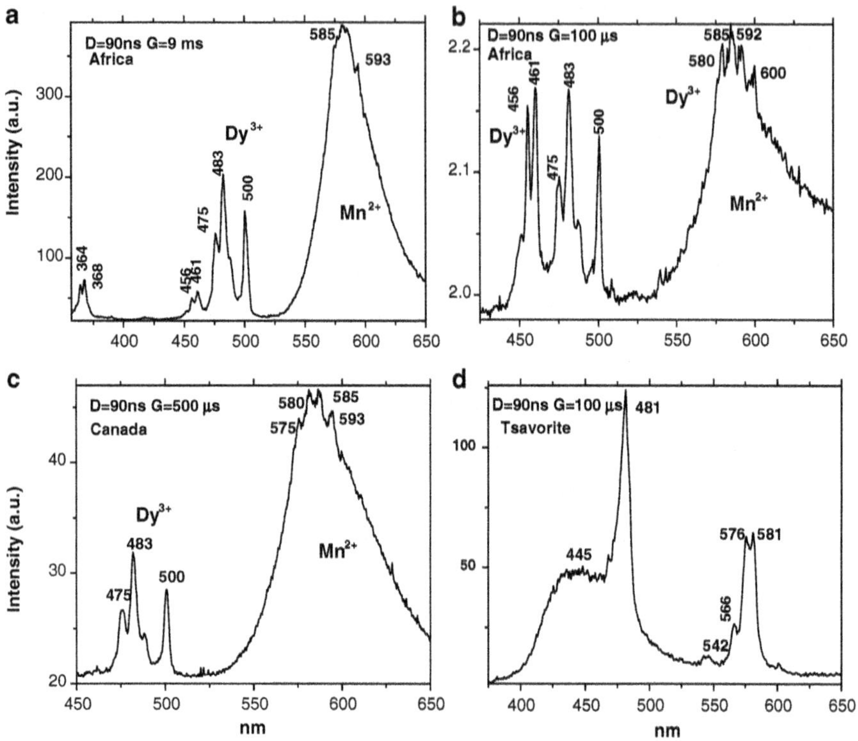

Fig. 5.20 (**a–d**) Time-resolved luminescence and excitation spectra of transparent Canada grossular at 300 K with zero delay and different gate and spectral resolution under excitation by 355 nm: 19 ms, low resolution (**a**), 500 μs, low resolution (**b**), 1 ms, high resolution in visible range (**c**), 500 μs, high resolution in UV range

Dy^{3+} luminescence is also detected in apatite (Fig. 4.3c), scheelite (Fig. 4.11e), zircon (Fig. 4.90a) where they usually dominate in the spectrum, anhydrite (Fig. 4.30c), calcite (Fig. 4.19d), zoisite (Fig. 4.138b), and feldspars (Fig. 4.99b). Several types of Dy^{3+} centers may be discerns in time-resolved luminescence spectra of fluorite. The main bands are at 478 and 573 nm (Fig. 4.16a). Under certain excitation and registration conditions it may be see that narrow lines at 588, 673 and 765 nm appear (Fig. 4.16b), which are typical for Dy^{3+} in cubic symmetry (Tarashchan 1978).

5.6.9 Er^{3+}

The main feature is green emission corresponding to transitions from state $^4S_{3/2}$ to the ground state $^4I_{15/2.}$ At the same time, an intense luminescence may be detected

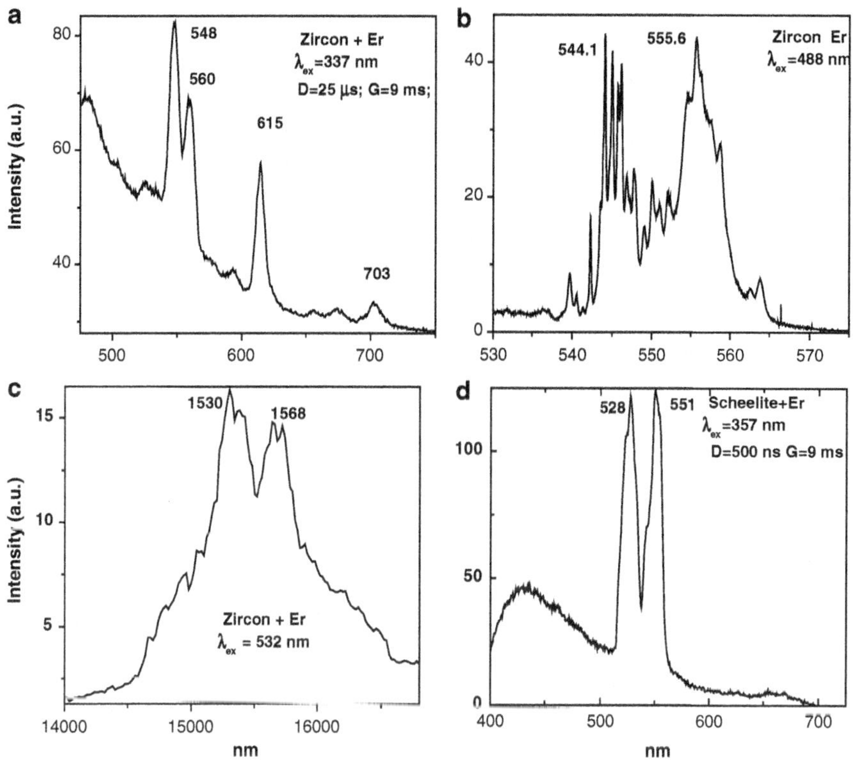

Fig. 5.21 (**a–d**) Laser-induced time-resolved (**a**, **d**) and steady-state (**b**, **c**) luminescence spectra of synthetic zircon and scheelite artificially activated by Er

at 1.5 μm, which is caused by resonance transitions $^4I_{3/2}$-$^4I_{15/2}$. The presence of green luminescence indicates that the de-activation of the high level accompanied by IR emission is not complete, but it results in relatively short decay time of Er^{3+} green emission. Thus the luminescence of Er^{3+} is better to detect in time-resolved spectra with narrow gate. In order to correct identification of Er^{3+} lines in minerals several of them were synthesized and artificially activated by Er (Fig. 5.21). Besides that, comparison has been made with CL spectra of synthetic minerals artificially activated by Er (Blanc et al. 2000).

5.6.9.1 Apatite

A relatively rare strong line appears in the luminescence spectrum of apatite near 544 nm (Fig. 4.3c). Spectrally it is similar to the line of Tb^{3+}, but is characterized by much shorter decay time and its intensity is not correlated with other lines of Tb^{3+}. By analogy with other hosts it is known that Er^{3+} and Ho^{3+} may generate luminescence in this spectral region. Artificial fluor-apatite activated by Er exhibits green and IR luminescence near 550 and 1500–1600 nm, respectively (Morozov et al. 1970). Cathodoluminescence spectra of artificial chlor-apatites activated by Er and Ho demonstrate green lines for both cases (Blanc et al. 2000). We investigate Er and Ho activated chlor-apatite samples under argon laser excitation with $\lambda = 488$ nm and it was found that both Er and Ho are characterized by luminescence lines at 545 nm, but for Ho the strong lines near 650–660 nm present which were not detected in natural fluor-apatites. IR luminescence allows detecting the band at 1540 nm the intensity of which is correlated with intensity of the lines at 545 nm (Gaft et al. 1997a). Thus the luminescence of Er^{3+} present in natural fluor-apatite corresponds to $^4S_{3/2}$-$^4I_{15/2}$ (545) and $^4I_{13/2}$-$^4I_{15/2}$ (1540) electron transitions.

5.6.9.2 Scheelite

The lines of Er^{3+} are difficult to point out because they have short decay times, which are comparable with the decay of broadband luminescence of scheelite at the same spectral region. Using a dye laser with 357 nm emission solved this problem. It is effective for Er^{3+} but not suitable for the excitation of the WO_4-center (Fig. 4. 10c). These lines of Er^{3+} are mostly clearly seen with narrower gate width when the luminescence of short-lived centers dominates. Investigation of time-resolved luminescence of synthetic $CaWO_4$:Er confirmed our interpretation (Fig. 5.21).

5.6.9.3 Zircon

Luminescence of Er^{3+} in emission spectra of zircon is very difficult to detect under UV excitation even by time-resolved spectroscopy. The reason is that it has relatively short decay time similar to those of zircon yellow luminescence, which usually is much stronger than Er^{3+} lines. Visible excitation, which is not effective for broadband luminescence, allows revealing Er^{3+} luminescence lines, using high-resolution steady-state spectroscopy (Fig. 4.90f).

5.6.9.4 Titanite

Three emission lines at 563, 853 and 978 nm (Fig. 4.80c) demonstrate correlated behavior and evidently connected with one emission center. Such emission is characteristic for Er^{3+}. The main feature of Er^{3+} luminescence in titanite is connected with $^4S_{3/2}$ excited state. It is responsible for green emission at approximately 550–560 nm corresponding to $^4S_{3/2}$-$^4I_{15/2}$ transition and for near IR emission at 850 nm corresponding to $^4S_{3/2}$-$^4I_{9/2}$ transition. The other important state is the $^4I_{11/2}$, which is upper level for the $^4I_{11/2}$-$^4I_{15/2}$ transition at about 980 nm.

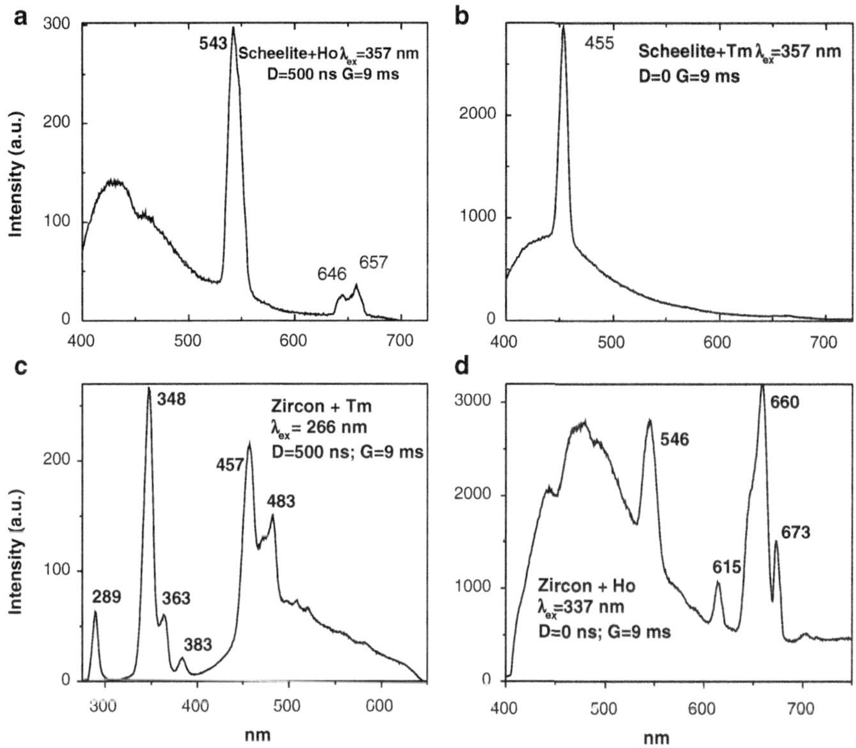

Fig. 5.22 (**a–d**) Laser-induced time resolved luminescence spectra of synthetic zircon and scheelite artificially activated by Ho and Tm

5.6.10 Ho^{3+}

Holmium activated minerals display green luminescence from a resonant transition from state $^5S_2$ to the ground state $^5I_8$. Because Ho^{3+} has luminescent lines connected with electron transitions from the same excited level, the spectra are not changed with delay time and excitation energy. The high intensity of the green luminescence indicates that the de-activation of the high level $^5S_2$ accompanied by IR emission from $^5I_7$ level is not complete, but it results in relatively short decay time of Ho^{3+} green emission. Thus the luminescence of Ho^{3+} is better to detect in narrow gate time-resolved spectra.

In order to correct identification of Ho^{3+} lines in minerals several of them were synthesized and artificially activated by Ho (Fig. 5.22). Besides that, comparison has been made with CL spectra of synthetic minerals artificially activated by Ho (Blanc et al. 2000). Luminescence of Ho^{3+} was detected in time-resolved luminescence spectra of scheelite (Fig. 4.10d), fluorite (Fig. 4.16c), and zircon (Fig. 4.90f).

Fig. 5.23 Absorption spectrum of titanite

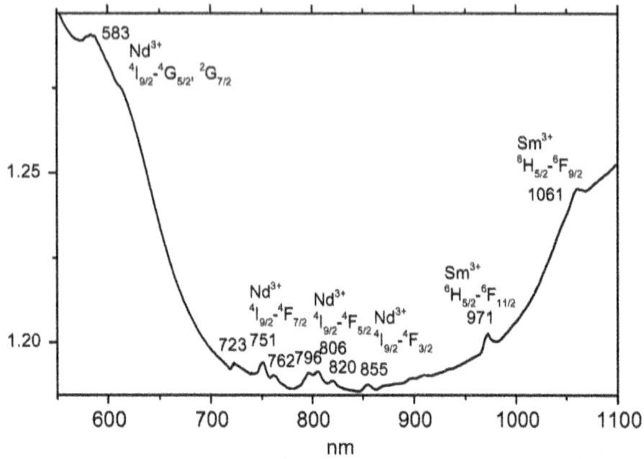

5.6.11 *Tm*$^{3+}$

Thulium displays in minerals an intense UV and blue visible luminescence with a line spectrum near 360 and 450 nm, correspondingly. They are connected with electron transitions from different excited levels $^1D_2$ and $^1D_4$ at 360–365 and 450–455 nm. The luminescence of Tm^{3+} is better to detect in time-resolved spectra with narrow gate, because it usually has relatively short decay time. The UV line usually has much shorter decay time compared with the blue one line. Different decay times from these levels are evidently connected with nonradiative relaxation due to the presence of high frequency vibrations in the lattice. The best excitation is at 355 nm, which is connected with transition $^3H_6$-$^1D_2$.

In order to correct identification of Tm^{3+} lines in minerals several of them were synthesized and artificially activated by Tm (Fig. 5.22). Besides that, comparison has been made with CL spectra of synthetic minerals artificially activated by Tm (Blanc et al. 2000).

The luminescence of Tm^{3+} was detected in apatite (Fig. 4.3c, d), scheelite (Fig. 4.10d), fluorite (Fig. 4.15a, b), zircon (Fig. 4.90h), anhydrite (Fig. 4.30c), calcite (Fig. 4.19d), and hardystonite (Fig. 4.36c). Group of several UV-blue lines with relatively short decay time of 30–35 μs presents in grossular spectra (Fig. 5.20). Such kind of emission is typical to Tm^{3+}.

Titanite has rather unusual luminescence of Tm^{3+} (Gaft et al. 2003b). The lines near 800 nm are very characteristic for its luminescence spectra under different excitations (Fig. 4.80c). It is interesting to note that the lines at 796, 806 and 820 nm present also in optical absorption spectra of titanite (Fig. 5.23). Such absorption lines are usually ascribed to electron transitions to $^4F_{5/2}$ level of Nd^{3+}. Thus in such case it is logical to suppose that emission lines at 800 nm have to be correlated with other Nd^{3+} luminescence, for example with IR emission from $^4F_{3/2}$ level.

Nevertheless, it is not the case. For example, with $\lambda_{ex} = 488$ nm excitation the lines at 800 nm are absent, while the IR emission of Nd^{3+} is very strong. The same situation is in ionoluminescence spectrum, where the lines near 800 nm are absent and luminescence of Nd^{3+} presents (Yang 1995). Thus it is possible to suppose that the lines near 800 nm with relatively short decay time of 60 μs belong to another luminescence center.

A possible candidate may be Tm^{3+}. For example, the doublets at 803 and 817 nm and at 796 and 813 nm are the strongest ones in cathodoluminescence spectra of fluorite and scheelite activated by Tm^{3+} (Blanc et al. 2000). It is possible to suppose that the strong lines at 805 and 820 nm with a relatively short decay time of 60 μs in titanite luminescence spectrum belong to Tm^{3+}. They appear under 532 nm excitation and evidently connected with $^1G_4$-$^3F_5$ electron transition.

Similar emission of Tm^{3+} was detected also in time-resolved luminescence spectra of leucophane (Fig. 4.45c, f) and walstromite (Fig. 4.49b).

5.6.12 Yb^{3+}, Yb^{2+}

IR luminescence of Yb^{3+} in minerals has been detected in steady-state luminescence of fluorite (Tarashchan 1978) and zircon (Gaft et al. 1986). Time-resolved luminescence spectra of apatite (Fig. 4.4b) and scheelite (Fig. 4.10a) also exhibit such emission. Emission lines in natural apatite are very similar to those in Yb^{3+} doped artificial apatite (DeLoach et al. 1994). The excitation of Yb^{3+} is strongly affected by energy transfer from Nd ions, because the direct excitation of Yb^{3+} is difficult, since it requires a comparatively short-wavelength UV radiation.

The luminescence of Yb^{2+} in the near UV and visible is due to $4f^{14}$-$4f^{13}5d$ transitions. The energy level scheme of the $4f^{13}5d$ configuration of Yb^{2+} in crystals is complicated due to the simultaneous action of the crystal field and spin-orbit coupling on the free ion states. The lowest energetic excited states, arising from $^3P_2$ free ion level, have $^3T_{2u}$ and ^3E_u symmetry. About 2000–3000 cm^{-1} above these levels there are $^3T_{1u}$ levels from the $^3H_5$ state. Only transitions between the ground state ($^1A_{1g}$) and states with T_{1u} symmetry are symmetry allowed.

It was found that, in the sequence: sulfates, phosphates, borates, silicates, aluminates, simple oxides the probability of observing normal df emission decreases and the probability of observing anomalous emission or no emission at all increases. Normal df emission is only observed in the sulfates MSO_4 (M = Ca, Sr, Ba) and in the condensed borate SrB_4O_7. Anomalous emission occurs in several phosphates and few silicates, but not in aluminates and CaO. In aluminates and simple oxides, even down to 4 K, no broadband Yb^{2+} emission has been reported. Most compounds for which normal df emission is observed contain large halogen ions. Anomalous emission is much more frequently observed when the small Yb^{2+} is the dopant than when Eu^{2+} is the dopant and normal df emission is never

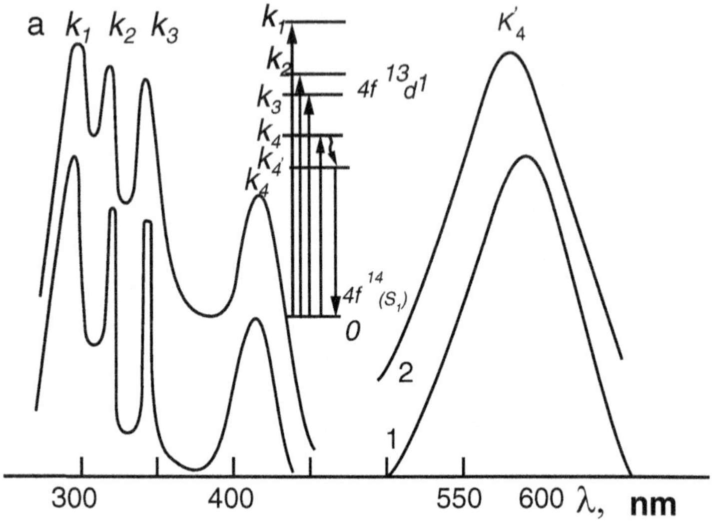

Fig. 5.24 Luminescence and excitation spectra of Yb^{2+} in danburite and datolite (Gaft and Gorobets 1979)

observed for Yb^{2+} on trivalent cation sites and always observed for Yb^{2+} on monovalent cation sites (Dorenbos 2003).

In minerals luminescence of Yb^{2+} was found in steady-state low temperature luminescence spectra of fluorite (Tarashchan 1978) and danburite and datolite (Gaft and Gorobets 1979; Fig. 5.24). It is characterized by broadband in the green part of the spectrum. The strong thermal quenching of Yb^{2+} is connected with its very small Stokes shift. It may be expected to find luminescence of Yb^{2+} in low temperature luminescence spectra of natural sulfates, because UV luminescence was detected in synthetic $CaSO_4$, $SrSO_4$ and $BaSO_4$ artificially activated by Yb^{2+}. The corresponding decay time was several ms at 20 K and the luminescence was quenched at approximately 200 K (Lizzo et al. 1994).

5.7 Transition Metals Elements

Transition metal ions have an incompletely filled d-shell, i.e. their electron configuration is d^n. The optically active electrons are thus bound to central potential as well as experiencing crystal field potential, and not shielded by outer electrons. Most transition metal ions are multi-valent. Mainly the number of 3d electrons and the crystal field determine their optical properties. Thus the groups below have similar optical behavior:

Configuration	$3d^1$	$3d^2$	$3d^3$	$3d^4$	$3d^5$	$3d^6$	$3d^7$	$3d^8$	$3d^9$	$3d^{10}$
Ions									Cu^{2+}	Cu^+
							Ni^{3+}	Ni^{2+}		
							Co^{2+}			
						Co^{3+}				
					Fe^{3+}	Fe^{2+}				
	Mn^{6+}	Mn^{5+}	Mn^{4+}	Mn^{3+}	Mn^{2+}	Mn^+				
	Cr^{5+}	Cr^{4+}	Cr^{3+}	Cr^{2+}	Cr^+					
	V^{4+}	V^{3+}	V^{2+}							
	Ti^{3+}	Ti^{2+}								

5.7.1 $3d^3$ Elements

The elements with such electronic configuration known as luminescent centers are Cr^{3+}, Mn^{4+} and V^{2+}. They are capable of substituting in a wide variety of metal oxide host systems. They are invariably oxygen-coordinated with six nearest neighbors, and may be in a pure octahedral or a distorted octahedral symmetry site. These luminescent centers exists in a d^3 configuration (Fig. 5.25), and the electronic repulsion, which results from placing three electrons in the same set of d-orbitals yields several states identified as "free ion" energy levels. The placement of such ions in a crystal lattice produces a further splitting of the free ion levels. The

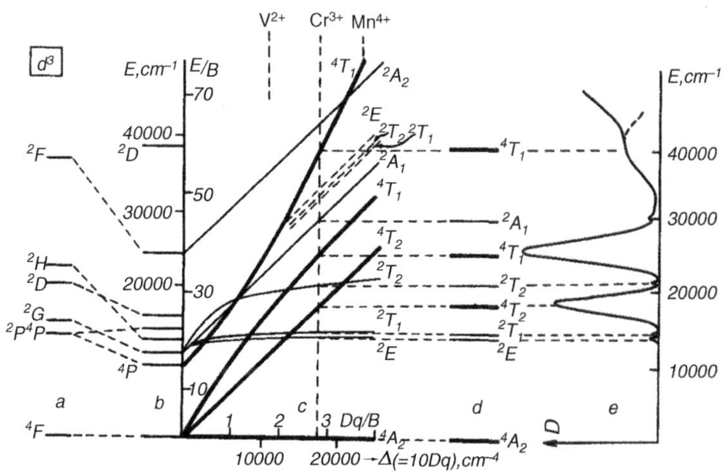

Fig. 5.25 Tanabe-Sugano diagram for d^3 elements

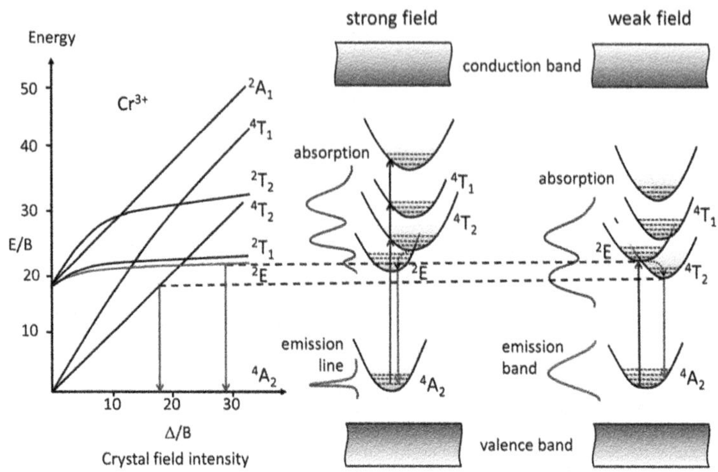

Fig. 5.26 Energy levels scheme of Cr^{3+} in different crystal fields

exact energy spacing between the energy levels is therefore determined by the strength of the crystal field experienced by luminescent center. Since different crystal field strengths are anticipated for each mineral host into which Cr^{3+} can be placed, it may be concluded that the luminescence may be a reliable indicator of the mineral structures properties.

The mostly important free ion states are ^4F ground level and excited ^2H level. In an octahedral field, the ^4F level splits into the $^4A_2$ ground state and the excited $^4T_2$ and $^4T_1$ states. The spin allowed transitions that could therefore be used to populate the excited states directly correspond to $^4A_2 \rightarrow {}^4T_2$ and $^4A_2 \rightarrow {}^4T_1$. Mostly Cr^{3+} appears in minerals as a trace constituent substituting for Al^{3+} in six-fold coordination. As such, the larger Cr^{3+} ion is placed in a site with small metal-ligand distances and the crystal field is large, typically 1650–1750 cm^{-1}. As may be seen on the Fig. 5.26, if the crystal field is sufficiently strong to make the doublet state the lowest energy excited state, there will be a red luminescence. Because the doublet state belongs to the same orbital configuration as the ground state, the Stokes shift and Condon offset will be small. As a result, the line width of emission is narrow and a considerable amount of phonon fine structure is observed even at room temperature. Such transition is forbidden by symmetry and spin selection rules and decay time is usually very long. Although the narrow line emission is essentially independent of crystal field and thus of site size and symmetry, the energy of the double levels does depend on Racah parameters B and C, which are dependent on the covalency of the metal-ligand bond and thus there is some variability in the line emission from mineral to mineral.

At weaker crystal fields, a strong broadband quartet-quartet $^4T_2 \rightarrow {}^4A_2$ transition appears, which is only symmetry forbidden and has much shorter decay time. Within the intermediate crystal field there is complicating mixing between doublet and quartet states with complicated spectra, non-radiative transfer and the temperature

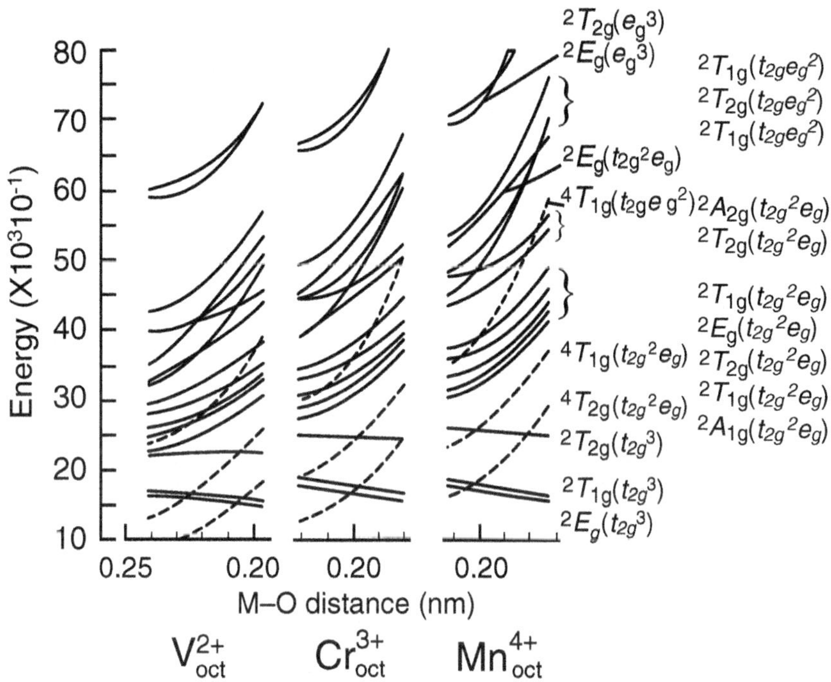

Fig. 5.27 Distance-dependent multiplet-energy diagram for Mn^{4+}, Cr^{3+} and V^{2+} (Ishii et al. 2002)

dependence of luminescence. It was suggested, that spin-orbit interaction, coupling ^2E and $^4T_2$ states, was responsible for breaking of the selection rule forbidding the ^2E-$^4A_2$ transition. The admixture of the $^4T_2$ wave function into the ^2E wave function, which depends on the doublet-quartet energy separation, determines the radiative lifetime of the doublet, which, therefore, may change significantly with crystal field. It has been empirically known that the energies of the lowest excited state of octahedrally coordinated $3d^3$ metals increase in the order $V^{2+} < Cr^{3+} < Mn^{4+}$. This tendency has been considered to originate from the difference in covalency, which reduced two-electron repulsion between the electrons occupying 3d orbitals. Recently this question was considered using first-principles electronic-structure calculation (Ishii et al. 2002). It was found the tendency that the 3d levels went toward lower energy as the atomic number increased. The effective ligand field splitting increased as the atomic number increased. Distance dependent multiplet-energy diagrams for these elements have been obtained (Fig. 5.27). It is similar to traditional Tanabe-Sugano diagram, but is different in that they are specific to every metal. The ground state is $^4A_{2g}$ and excited states concerning spin-allowed transition are expressed as broken lines, while the states concerning spin-forbidden transitions are expressed as solid lines. It is clearly seen that the ^2E_g and $^4T_{2g}$ states of these centers in octahedral coordination are located close to each other indicating a possible dependence of the shape of emission on the M-O distance.

5.7.1.1 Cr^{3+}

Cr^{3+} is very important ion because it is widely used in lasers (Reisfeld and Jörgensen 1977; Fabeni et al. 1991; Kaminskii 1996). The active ion lasers are based on the notion that a transition or rare-earth metal can be doped into inactive matrix, thus yielding isolated species where the metastable states are less prone to nonradiative decay. Some of the desired physical properties of matrices for active ion lasers are: hardness, low temperature dependent refractive index, high thermal conductivity and high transparencies. However, before it is possible to measure the above physical quantities, the crystal must be synthesized, and thus the matrix must fulfill some chemical requirements. Some of them are: that the matrix should melt congruently, that there should only be one ion in one geometry which has the same size as dopant ion, that the diffusion rate of the dopant should be high in order to obtain a homogenous crystal, and that the material should form large single crystal without defects. These qualities are all found in natural gemstones, that is why materials like sapphire, garnet, alexandrite, etc. are used in many laser systems. Thus luminescence of Cr^{3+} in minerals may yield extremely important information and was carefully investigated, where studied mineral include corundum, spinel, kyanite, topaz, beryl, emerald, diopside, spodumen and jadeite. Nevertheless, strongly different decay times of different transitions in Cr^{3+} centers make time-resolved spectroscopy useful for much better understanding of Cr^{3+}luminescence in minerals.

Topaz

Steady-state luminescence of topaz has been studied by Tarashchan (1978). At 77 K narrow lines at 680, 696, 712 and 730 nm have been ascribed to single Cr^{3+} (R-lines) and Cr^{3+}-paires (N-lines). We studied two transparent, four yellow and one orange natural topaz. Two transparent precious stones have not been subjected to destructive chemical analyses. In other samples their Cr, Mn and V concentrations have been measured by ICP-AS method (Table 4.19). Absorption spectrum typical for Cr^{3+} has been detected only in red topaz with highest Cr content of 500 ppm. Spectrum clearly revealed two broad bands typical of the octahedrally coordinated Cr^{3+} (Gaft et al. 2003a). The bands centered at 418 and 533 nm evidently correspond to the spin-allowed $^4A_{2g} \Rightarrow {}^4T_{2g}$ and $^4A_{2g} \Rightarrow {}^4T_{1g}$ transitions, respectively. The peak attributed to the spin-forbidden $^4A_{2g} \Rightarrow {}^2E_g$ transition is located near 685 nm, while other very weak lines at 696 and 712 nm also present.

Luminescence spectra at 300 K of all topaz samples contain strong narrow lines at approximately 680 and 684 nm and weaker line at 695 nm together with relatively broad band (half-width of 1100 cm^{-1}) peaking near 680 nm. In the purest form such luminescence was detected in precious transparent topaz (Ural) where other luminescence lines are not detected (Fig. 4.114a). The temporal behavior of the lines and the band in time-resolved spectra is somewhat different, where the band has shorter decay and dominates in the spectra with narrow gate starting from

100 μs. Nevertheless, the difference between decay times of the lines and the band is substantially less compared to the sites with strong and low crystal fields. Such behavior may be explained by Cr^{3+} in intermediate crystal field sites for which the crystal field parameters lie in the crossing region of the $^4T_2$ and ^2E states. Within the intermediate crystal field there is complicating mixing between doublet and quartet states with complicated spectra, non-radiative transfer and the temperature dependence of luminescence. In such case the emission from both $^4T_2$ and ^2E states may be expected. At 300 K the lower ^2E state acts as an energy reservoir for the $^4T_2$ state and thermal activation enables to emit from both levels. Because the $^4T_2$ and ^2E states are in thermal equilibrium emission from both levels should occur with similar decay. At 77 K the thermal activation is impossible and the broad band at 680 nm disappears (Gaft et al. 2003a). Besides that, the R_1 narrow line at 680 nm is quenched at 20 K, which indicates thermal equilibrium at 300 K between the populations of the two Kramers doublets of the ^2E_g crystal-field split state. At 300 K decay time of the R-line at 684 nm is changed from 160 to 270 μs in different samples, while at 20 K with the absence of phonon-electron interaction its exponential decay of approximately 2.4 ms is the same in all samples and becomes suitable for this spin-forbidden transition.

The time-resolved spectra of topaz samples with elevated Cr contents, starting from 180 ppm, contain additional luminescence lines with different excitation and temporal behaviors. Luminescence spectra at 300 K of red topaz under $\lambda_{ex} = 266$ nm contains only lines and band typical for single Cr^{3+} luminescence. Under $\lambda_{ex} = 355$ nm the lines at 697, 711 and 734 nm appear, while their intensities in the spectra with the narrower gates are higher compared with the lines at 684 nm, which may be explained by their shorter decay times (Fig. 4.114b, c). It is well known that with the increase in Cr^{3+} concentration, additional luminescence lines begin to appear at the longer wavelength side of the R-lines. Those lines are attributed to magnetically coupled Cr^{3+}-Cr^{3+} pairs and clusters. The shorter decay time compared with single Cr^{3+} luminescence and the strongest luminescence under 355 nm excitation, which is exactly twice of the energy of the line at 711 nm, are also compatible with pairs model.

Kyanite

Cr^{3+} in Strong Crystal Field

Figure 5.28 presents typical long-lived luminescence spectra of a natural transparent blue kyanite containing low Cr concentrations with a delay of 500 μs and a gate of 5 ms. These experimental conditions are optimal for detection of luminescence with a long decay time. At 300 K the luminescence spectrum consists of two narrow lines at 688.6 and 706.3 nm. Their FWHM, determined at high resolution, are both approximately $\Delta = 0.15$ nm. The spectral position of the lines is similar in different samples. Those lines are superimposed on a weak, broad band with a maximum around 700 nm. In addition, a series of weak lines are superimposed at both longer

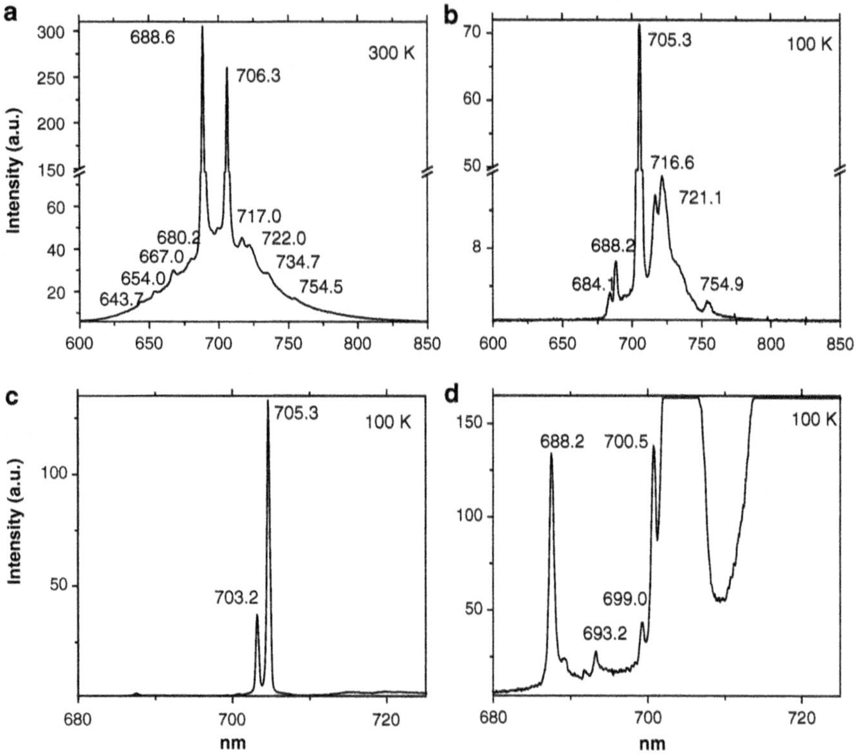

Fig. 5.28 (**a–d**) Long-lived luminescence spectra (D = 500 μs, G = 5 ms) of a transparent blue kyanite containing low Cr concentrations at 300 (**a**), and 100 K (**b**, **c**, **d**)

and shorter wavelengths than the two main lines peaking at 643.7, 654.0, 667.0, 680.2, 717.0, 722.0, 734.7, 754.5 nm (Fig. 5.28a). The intensity of the short wavelength line at 688.6 nm remains more or less constant down to approximately 225 K and after that becomes progressively weaker. It is very weak at 100 K peaking at 688.2 nm shifted 0.4 nm from the 300 K spectrum (Fig. 5.28b). The long wavelength line at 706.3 nm becomes more intense at lower temperatures, approximately 8–10 times more intense at 100 K, with the peak at 705.3 nm, a shift of 1.0 nm compared to the 300 K spectrum. Spectra taken with higher resolution resolves the λ_2 and λ_3 lines at 703.2 and 693.2 nm, respectively (Fig. 5.28c, d). The long wavelength line becomes the narrowest one with $\Delta = 0.05$ nm. At 100 K, the broad band disappears, but part of vibrational structure remains peaking at 684.1, 716.6, 721.1 and 754.9 nm (Fig. 5.28b).

As it was already mentioned, the narrow emission lines were previously ascribed to the so-called R-lines attributed to transitions from two sublevels of the ^2E_g level to the $^4A_2$ ground state of Cr^{3+} in a site with a strong CF. The low temperature behavior of the intensity was connected to thermal equilibrium of electrons between two sublevels of the ^2E_g level at 300 K, which is less efficient at lower temperatures.

As a result, at low temperature nearly all excited electrons emit from the lower energy sublevel.

The broad unstructured band peaking at 700 nm at 300 K is completely "frozen-out" at low temperature. Time-resolved luminescence spectroscopy enables the detection of this weak band that has a long decay time. This band is best seen in spectra that are obtained with a long delay time (500 μs) that removes the more intense band emissions that have much shorter decay times that prevent its detection in CW luminescence spectroscopy. The long decay time of this band is similar to that of the narrow line as would be expected in the case of three thermalized levels. At temperatures higher than 300 K, the narrow lines and side-bands broaden and weaken, so that at 400 K the luminescence spectrum consists mainly of single, broad, structureless band (Gaft et al. 2013c). The temperature-induced weakening of line luminescence accompanied by an intensification of the broad band luminescence may be explained by quick electronic transitions between the corresponding levels with increasing temperature. This suggests that the energy difference between those levels, ΔE, in kyanite is small enough to allow for mixing of the two excited states, even at 300 K. The origin of this broad band is uncertain. It may not be ascribed to the electronic $^4T_2 \rightarrow {}^4A_2$ transition in Cr^{3+}, which was postulated earlier (Wojtowicz 1991). Such an allowed transition would have a short decay time, which is not the case.

Vibrational structure for Cr^{3+} luminescence is well known in minerals (Tarashchan 1978). Because the Cr concentration in natural kyanite with strong, narrow line emission is very low, typically in the tenths of a ppb range, it is logical to suppose that the Cr ions do not interact with each other and those vibrations are not connected with so-called N-lines ascribed to Cr pairs. They are obviously phonon satellites of these narrow luminescence lines with long decay time. At low temperatures, only Stokes vibronic transitions appear, while at higher temperatures the anti-Stokes ones also appear. The lines at 684.1, 699.0, and 700.5 nm have not been detected previously. We managed to find them using a time-resolved luminescence technique that enabled the removal of broad emission bands with relatively short decay times. Those lines are seen only at temperatures below approximately 200 K (Fig. 5.28d). Based on the emission spectra, it appears that they are also connected to Cr^{3+} luminescence. This is supported by the observation of such narrow lines in the optical spectrum of Cr-containing kyanite where they are especially strong in β-polarization (http://minerals.gps.caltech.edu).

Luminescence spectra of kyanite obtained at low spectral resolution often contain a narrow line peaking at 694.3 nm with a very long decay time (Fig. 5.29a). Such a line is typical for ruby luminescence, which was confirmed by the ruby spectrum obtained under the same experimental conditions as for kyanite (Fig. 5.29b). Spectra obtained at higher resolution reveal that there actually is a doublet at 694.3 and 692.8 nm, which corresponds to the ruby spectrum (Fig. 5.29c). Besides, the excitation spectrum of the line at 694.3 nm (Fig. 5.29d) reveals two maxima peaking at 410 and 550 nm which correspond well to the natural corundum excitation spectrum (Tarashchan 1978). Thus, we conclude that those lines do not belong to kyanite, but rather to ruby micro-inclusions inside the

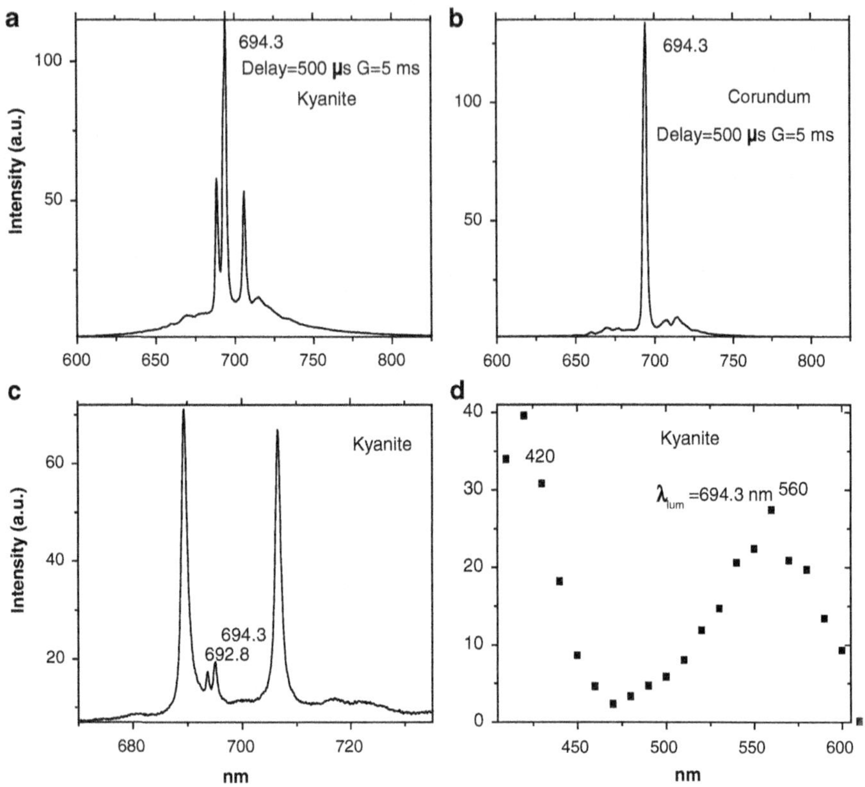

Fig. 5.29 (**a–d**) Low spectral resolution time-resolved luminescence spectra of Cr^{3+} in kyanite (**a**) and corundum (**b**) under the same experimental conditions (D = 500 μs, G = 5 ms) and high spectral resolution spectra of kyanite (**c**) with typical Cr^{3+} luminescence in corundum. Excitation spectrum of 694.3 nm line in kyanite (**d**) is also typical for corundum

kyanite matrix. The possibility that chromium substituting in octahedral aluminum sites in an Al-bearing oxide matrix may occur as a ruby precursor phase was proposed earlier by Platonov et al. (1998).

Figure 5.30 presents the excitation spectra of the λ_1 and λ_4 lines in the 410–715 nm spectral range at 300 and 100 K. At 300 K, the spectra for both lines are very similar and contain a broad band peaking at 600 nm (Fig. 5.30a) and two narrow lines at 687 and 704 nm (Fig. 5.30b). Another broad excitation band is evidently peaking at approximately 400 nm, which is out of spectral range presently studied. It is certain that there are differences in position of both broad bands between the absorption and excitation spectra (http://minerals.gps.caltech.edu). The differences were previously detected and explained by the contribution of the lower field sites to the absorption spectrum, while the excitation spectrum is due to the highest field site (Wojtowicz 1991; Platonov et al. 1998). The full-widths at half-magnitude (FWHM) of the broad excitation bands are approximately

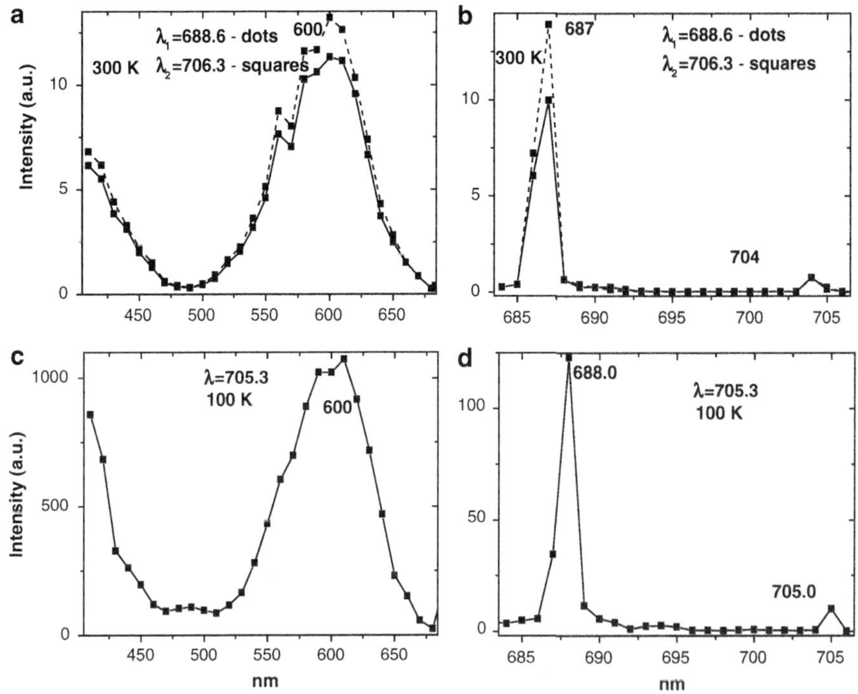

Fig. 5.30 (**a–d**) Excitation spectra of the lines at 688.6 (*dots*) and 706.3 (*squares*) nm at 300 K in 405–684 nm (**a**), 683.5–706.5 nm (**b**) and of the line at 705.3 nm at 100 K in 405–684 nm (**c**), 683.5–706.5 nm (**d**) spectral ranges

2220 cm^{-1}, which is small for a spin-allowed band. This width is comparable to the excitation spectrum of ruby where the FWHM is equal to 2000 cm^{-1}. At low temperature the excitation spectrum of the remaining line at 705.3 nm is essentially the same as at 300 K (Fig. 5.30c), but the broad excitation band is five times more intense compared to the narrow excitation line (Fig. 5.30d). Thus, based on these excitation spectra, it may be concluded that the λ_1 and λ_4 lines belong to the same luminescence center.

Under the assumption that the emission center is Cr^{3+}, the two broad bands in the excitation spectrum are due to transitions $^4A_2 \rightarrow {}^4T_1$ (410 nm) and $^4A_2 \rightarrow {}^4T_2$ (590 nm). These assignments enable the evaluation of the crystal field strength $D_q = 1700$ cm^{-1}, which is very close to previously published data (Platonov et al. 1998). Such a D_q corresponds well to a strong crystal field site for d^3 elements in octahedral coordination. The narrow lines observed in the excitation spectrum are characteristic for many chromium-bearing crystals. They are attributed to Cr^{3+} transitions from $^4A_2$ to the excited states ^2E, $^2T_1$ and $^2T_2$. Spin-orbit interactions admix $^4T_1$ and $^4T_2$ states having the same spin multiplicity as the ground state to the states ^2E, $^2T_1$, and $^2T_2$. In absorption spectra this causes an intensification of the respective absorption lines due to partial relaxation of the spin-multiplicity rule for the intercombinational ($\Delta S = 1$) electronic transitions $^4A_2 \rightarrow {}^2E$, $^4A_2 \rightarrow {}^2T_1$, and $^4A_2 \rightarrow {}^2T_2$. These lines in the kyanite spectrum are exceptionally intense compared to those in other Cr bearing minerals. It may be caused by more powerful admixing of the $^4T_1$ and $^4T_2$ states to the spin-forbidden ones.

All previous publications agree that the λ_1 and λ_4 lines have long decay times, but certain differences in values exist. For example, for the higher energy λ_4 line, which is not influenced by a spectrally close line with fast quenching, decay times at 300 K are 980 (Wojtowicz 1991) or 900 μs (Tolstoy and Shinfue 1960) and at 250 K are 1905 (Tolstoy and Shinfue 1960) or 1400 μs (Wojtowicz 1991). It was found that for high temperatures departures from a single exponential fit for decay time are evident for long-lived lines. The experimental data were approximated by a two exponential fitting procedure (Wojtowicz 1991). Our experimental data confirmed that at 300 K both lines under discussion have long decay times which are usually close to each other in each specific sample, while varying between different samples changing from 750 μs to 1.2 ms. Nevertheless for the lines with the longest decay times the emission decay curve is usually well described by a single exponential dependence. This dependence is observed in transparent blue kyanite samples which do not contain Fe-connected bands in their absorption spectra. On the other hand, decay curves of the samples with elevated Fe concentrations are non-exponential, as observed, for example, in greenish, gem quality kyanite samples. Thus we arrived at the conclusion that the non-exponential behavior can be explained by resonant energy transfer between a Cr^{3+} emission center and a quenching center, such as Fe^{3+} or Ti^{3+}.

In the case of an electric dipole-dipole interaction, the decay of the emission intensity is described by the Inokuti-Hirayama equation (Inokuti and Hirayama 1965):

$$I(t) = \exp\left\{-(t/\tau_0)-(Wt)^{1/2}\right\}, \tag{5.33}$$

where t is the time, τ_0 is the intrinsic decay time and W is the non-radiative decay rate. It was found that a very good approximation may be achieved if we take as τ_0 the single exponential decay time of the transparent blue sample. Using this approach, non-radiative decay rates were calculated which correlated well with iron levels in kyanite samples, namely the higher the Fe concentration the higher is the parameter W.

The decay time behavior with temperature lowering was studied for the transparent sample with a single exponential decay at 300 K. The decay times of both lines are equal at higher temperatures. They both become longer at lower temperatures up to approximately 5 ms at 100 K, when the decay of the short wavelength line becomes somewhat less compared to the long wavelength one. At temperatures down to 65 K where the 688.7 nm lines still may be detected, the difference between decay times becomes larger: 5.3 ms for 708.6 nm and 4.9 ms for 688.7 nm. At higher temperatures the decay times of both lines become shorter, ~380 μs at 450 K.

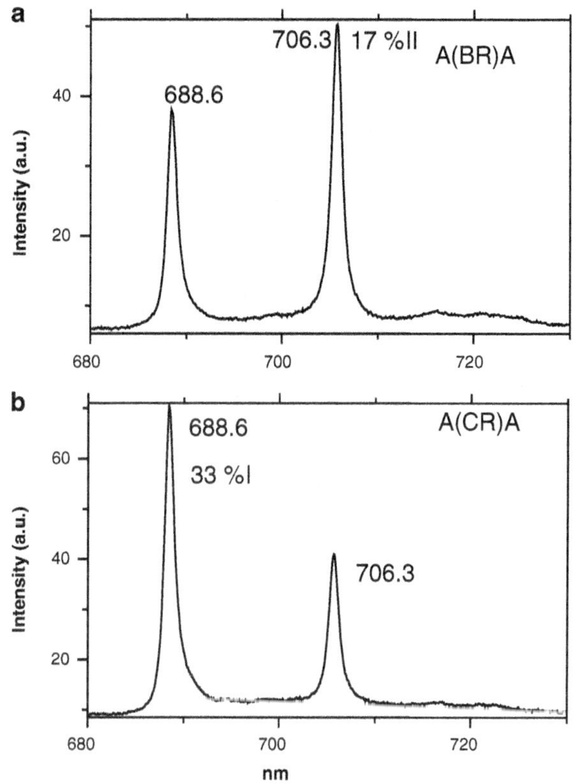

Fig. 5.31 (a–b) Luminescence spectra of gem quality kyanite for two different orientations

The polarization behavior of the luminescence lines of kyanite was not previously studied. We obtained such data using a greenish, transparent, gem quality sample in the following way. The excitation was oriented in the direction perpendicular the table of the stone, with the laser beam polarized in the B direction and the emission detected with zero angle geometry using a linear polarizing film. A comparison with known Raman spectra of differently oriented kyanite crystals revealed that the C(BR)C spectrum corresponds to 0° orientation, and the C(AR) C spectrum to 90° orientation (Gaft et al. 2013c). This study showed that the luminescence lines with long decay time are polarized in different directions (Fig. 5.31). As a result, the line intensity ratios are dependent on crystal orientation relative to the laser beam and on the direction of detection.

The main questionable point in the "natural" interpretation of those lines is their assignment to transitions from the split ^2E_g sublevels that have an extremely large splitting of approximately 350 cm^{-1}. This splitting is the largest reported for Cr^{3+} luminescence in minerals, where the lowest is 8 cm^{-1} in spinel and the former largest was 106 cm^{-1} in topaz. The large splitting value in topaz was explained by the lowered local symmetry of Al^{3+} that results from the substitution of two fluorine ions for two oxygen ions (Tarashchan 1978). Kyanite belongs to the P1 space group of the triclinic system. Cr^{3+} may substitute into any one of the four possible Al sites each having C_1 point symmetry and exhibiting a distorted octahedral configuration similar to the six-fold oxygen coordination in Al_2O_3. Thus the local symmetry of the Cr^{3+} sites in kyanite does not differ drastically from other minerals and does not explain the large splitting.

Besides, if we suppose that the luminescence is connected to the split ^2E_g levels in thermal equilibrium, another question appears. The theoretical intensity ratio of those long-lived lines should be equal to:

$$\frac{I_2}{I_1} = \frac{\tau_1}{\tau_2} e^{-\frac{\Delta E}{kT}} \tag{5.34}$$

because the degeneracies of those sublevels are equal. It is well known that the radiative decay times (τ) of both components of this doublet under low temperature conditions are always equal to each other (Tolstoy and Shinfue 1960). Thus, assuming equal decay times for both R-lines, this ratio at 300 K is approximately 0.18. As we already mentioned, one must be very careful in evaluation of the experimental data on the intensity ratio of those lines, because of their short life times and the influence of polarization. Nevertheless, using time-resolved spectroscopy to eliminate the short lived emission and powdered kyanite sample to remove the polarization influence, we found that this ratio at 300 K is in the 0.8–1.1 range, which is substantially higher than the calculated value. At 100 K the calculated ratio, 0.006, is closer to the experimental one, 0.15, but still is much lower.

A possible explanation is that emission is not from the split ^2E_g but rather is from ^2E_g and $^2T_1$ levels. The $^2T_1$ level is energetically very close to ^2E_g and is similar in its behavior, namely it does not depend on crystal field strength and its emission consists of narrow lines. The transitions $^2T_1 \rightarrow {}^4A_2$ and $^2E \rightarrow {}^4A_2$ belong to the same t_{2g} configuration and have both forbidden electronic dipole and spin forbidden character. Thus, the decay time for luminescence from $^2T_1$ also has to be long, but evidently different from the decay time from the ^2E_g level. Indeed, at low temperatures those decay times are different, 5.3 ms for 706.3 nm and 4.9 ms for 688.6 nm. Even though the difference is small it is consistently detected.

The possibility that those lines belong to different energy levels may be concluded from their absorption intensities at 77 K (Fig. 5.32). For optical transitions between the same pair of energy levels, the following relationship is used to convert the absorption to luminescence (Marfunin 1979)

$$\sigma_i = \tau_i \left(g_{gr}/g_i\right) C\left(\upsilon_{absi}/\upsilon^3{}_{emi}\right) = \tau_i \left(g_i/g_{gr}\right) C\left(1/\upsilon^2{}_{em}\right) \left(\upsilon_{abs} = \upsilon_{em}\right) \tag{5.35}$$

where σ_i – integrated absorption cross-section, τ_i – radiative decay time, g_i – degeneracy of the excited state, g_{gr} – degeneracy of the ground state, υ_{abs} – the mean absorption frequency of the absorption line and υ_{em} – the mean emission frequency of the emission line. Consequently, for the lines at 688.3 and 706.3 nm the following ratio applies:

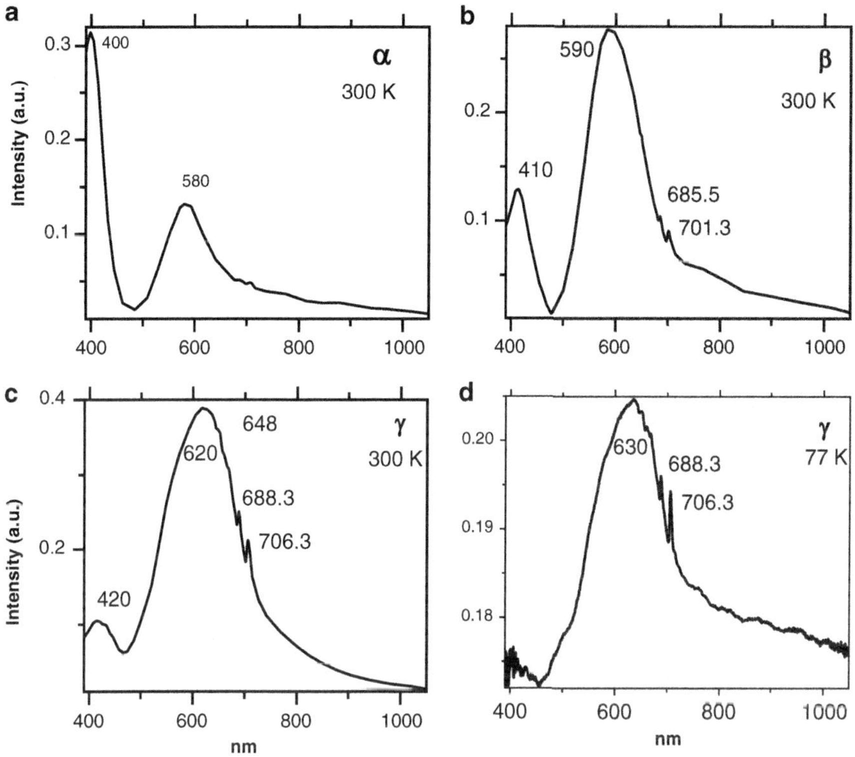

Fig. 5.32 (**a–d**) Absorption spectra of Cr containing kyanite at 300 K in α, β and γ (**a–c**) and at 77 K in γ (**d**)

$$(\sigma_{706.3}/\sigma_{688.3}) = (\tau_{706.3}/\tau_{688.3})(g_{688.3}/g_{706.3})(\upsilon^2{}_{688.3}/\upsilon^2{}_{706.3}) \sim (\tau_{706.3}/\tau_{686.3}) \times (g_{706.3}/g_{686.3}) \tag{5.36}$$

According to our experimental data, the $(\sigma_{706.3}/\sigma_{686.3}) = 3$ and $\tau_{708.6}/\tau_{688.8} = 1.1$. Thus $g_{686.3}/g_{706.3} \sim 3$. This is possible if the line at 686.3 nm belongs to $^2T_1$ level and the line at 706.3 nm to the lower sub-level of the split ^2E_g level.

In addition, the calculated intensity ratio in this case is closer to the experimental one, because the ^2E_g and $^2T_1$ levels have different degeneracies of 2 and 3, respectively. This makes the calculated ratio 1.5 times bigger. In such a case, the correlation between the calculated and experimental ratios at low temperatures up to approximately 125 K becomes very good, while at higher temperatures it is still big enough to be approximately correct. If one supposes that the 706.3 nm line is connected to the lower sub-level of the split ^2E_g level with a degeneracy of 1, than the calculated ratio becomes much closer to the experimental one, especially at high temperatures (Fig. 5.33).

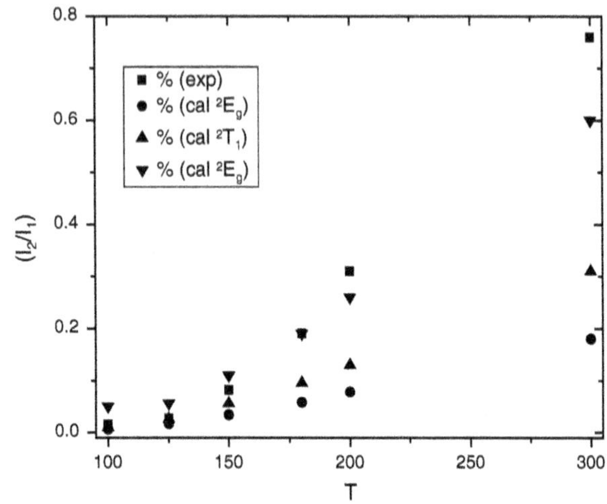

Fig. 5.33 Experimental and theoretical intensity ratios for different electronic transitions of the long-lived Cr^{3+} luminescence lines

The polarization of the ^2E_g and $^2T_1$ levels in different orientations is well known, thus the polarization behavior of the long-lived luminescence lines does not contradict this model of the ^2E_g and $^2T_1$ levels participation. It is not unequivocal evidence, because the R-lines luminescence also may be strongly polarized in different directions, as in alexandrite case (Bukin et al. 1981). Alternatively, polarization measurements show that both R lines in ruby are mainly polarized in σ polarization and that the π emission is an order of magnitude smaller (Margueron and Clarke 2007). Higher resolution polarized cathode luminescence spectra of corundum revealed that both R-lines are polarized in the same direction (Guguschev et al. 2010). According to a calculation of relative intensities of R-lines for trigonal symmetry assuming distortions of T_{1u} and T_{2u} symmetry, the σ and π polarized emission intensities are equal (Yamaga et al. 1992).

Thus it may be proposed that those narrow long-lived lines are connected to radiative transitions from ^2E_g and $^2T_1$ levels and not from the split components of the ^2E_g level. In such a case, in kyanite we do not have the largest splitting, but rather have zero splitting of the ^2E_g and $^2T_1$ levels. It is possible that the single lines are induced by a magnetic dipole mechanism and the ^2E_g and $^2T_1$ levels are not split by spin-orbit coupling. The participation of the $^2T_1$ level in Cr^{3+} luminescence was considered relatively rare, but it was proposed, for example, in case of spinel-type α-$ZnAl_2S_4$:Cr single crystals (Broussell et al. 1997).

Cr^{3+} in Intermediate Crystal Field

The decay times of the λ_2 and λ_3 lines (Fig. 5.34a) are relatively short at 300 K (approximately 60–80 μs), but become longer at low temperatures (465–500 μs at 100 K). Those lines are only weakly polarized. In the visible range, the luminescence is excited by two bands peaking at 485 and 610 nm (Fig. 5.34b). The

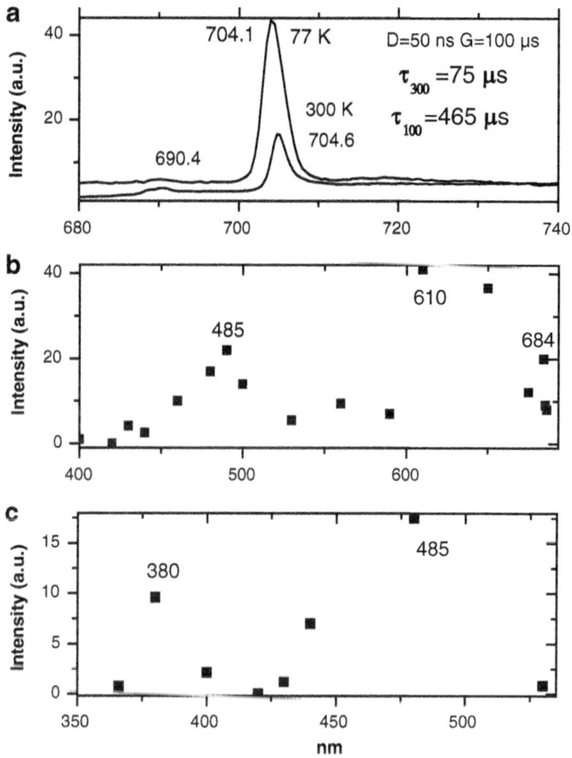

Fig. 5.34 (a–c) Luminescence at 300 and 77 K (**a**) and excitation spectra of short lived emission lines in blue kyanite in the visible (**b**) and UV-visible (**c**) parts of the spectrum

excitation study in the UV part of the spectrum revealed an additional excitation maximum of the short lived lines at approximately 380 nm (Fig. 5.34c), which is absent in excitation spectrum of the long-lived lines.

Those lines have been previously ascribed to $^2E_g \rightarrow {}^4A_2$ luminescence of Cr^{3+} in an intermediate crystal field site (Tolstoy and Shinfue 1960; Wojtowicz 1991). Nevertheless, several contradictions prevent us from accepting such an interpretation. Excitation spectra of two types of emission lines are very similar to those published earlier and determined as “a” and “b” environments (Platonov et al. 1998). Crystal field parameters calculated based on their excitation and polarized absorption spectra gave $D_q = 1720$ cm^{-1} and $B = 730$ cm^{-1} for “a” and $D_q = 1600$ cm^{-1} and $B = 570$ cm^{-1} for “b” environments. As was already mentioned, it was concluded that these great differences in the crystal field parameters cannot be explained by a distribution of Cr^{3+} between two or more of the four crystallographically different octahedral sites in the kyanite structure. The presence of a corundum precursor in kyanite was confirmed by our experiments. Nevertheless, those lines have long decay times typical for Cr^{3+} in strong crystal field.

Besides, such short decay times of the R-lines are usually associated with the thermal population of the closely spaced $^4T_2$ level with an allowed transition. Thus at low temperatures the broad band emission from $^4T_2$ dominates the spectrum, while at higher temperatures, the narrow R-lines become strong as a result of their thermal population from the lowest $^4T_2$ (Marfunin 1979). However, in our case the broad red band emission is absent in the low temperature spectrum.

Those considerations stimulated us to look for different interpretations, such as the possibility that a trace component different from Cr^{3+} is responsible for this emission, such as Mn^{4+} and V^{2+}. Nevertheless, in orange Mn-containing kyanite, such luminescence from Mn lines was not detected (Gaft et al. 2011). Another potential cause of luminescence is V, which is present in the studied samples in quantities sufficient for detectable luminescence. For a definite conclusion, synthetic kyanite samples artificially activated by Cr and V are needed.

Cr^{3+} in Weak Crystal Field

The strong luminescence band at 300 K has a maximum at approximately 775 nm and half-width of 75–85 nm with decay time of 15 μs (Fig. 5.35a). At 77 K, the intensity becomes much stronger while the decay time remains the same (Fig. 5.35b). Luminescence is excited in three bands peaking at 400, 575 and 860 nm (Fig. 5.35c).

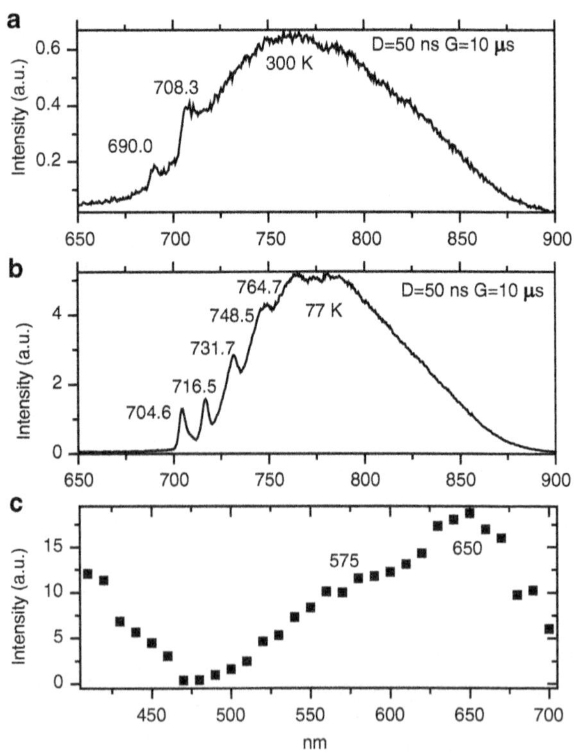

Fig. 5.35 (**a–c**) Luminescence at 300 (**a**) and 77 K (**b**) of the broad emission band and its excitation spectrum at 300 K (**c**) in blue kyanite

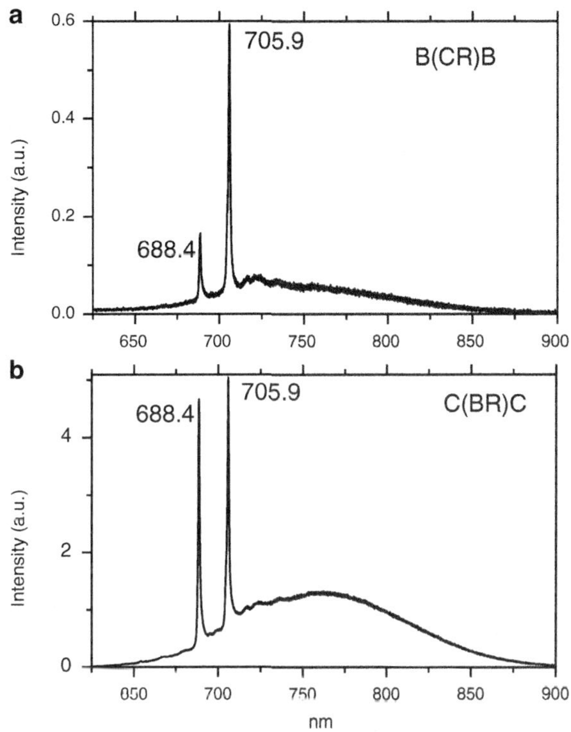

Fig. 5.36 (a–b) Luminescence of the broad red band in gem quality kyanite for two different orientations

This band emission is strongly polarized (Fig. 5.36). Such a band was previously ascribed to the $^4T_2 \rightarrow {}^4A_2$ emission for Cr^{3+} in a low crystal field site. The narrow, low temperature lines were interpreted as zero-phonon line (716.5 nm) and phonon replicas (731, 746, 765 and 779 nm), indicating a phonon energy of about 280 cm^{-1} (Wojtowicz 1991). But once again, it is very difficult to suppose that in the Al^{3+} positions which are structurally rather similar to each other there may be such different crystal fields enabling such narrow lines and band emissions of Cr^{3+}. Besides, the luminescence decay time of such Cr^{3+} centers is usually in the range of 100–150 μs (Boulon 2000). However, it actually does take place in sillimanite luminescence where artificial activation by Cr proved that the broad IR band peaking at approximately 790 nm at 300 K is connected to Cr^{3+} in a weak crystal field site which has a decay time of 160 μs that increases to 260 μs at 27 K.

Thus possibilities different from Cr^{3+} emission may be considered. A well known luminescence center in minerals is Fe^{3+} in tetrahedral coordination which is characterized by a broad red-IR band, but its transitions are spin forbidden and their decay times are usually in the milliseconds range. Ti^{3+} in octahedral sites with a broad spin-allowed luminescence band and a correspondingly short decay time of several microseconds may be much a more probable candidate. For example, Ti^{3+}

substitutes for Al^{3+} in the trigonally distorted octahedral Al^{3+} site of sapphire and is characterized by a strongly polarized band peaking at approximately 750 nm with a decay time of 3.1 μs (Fabeni et al. 1991).

Titanium, in quantities sufficient for luminescence, has been found in the kyanite samples that were studied (Chadwick and Rossman 2009). It is agreed that the oxidation state of Ti in kyanite is 4+ and that the blue kyanite color comes from Fe^{2+}-Ti^{4+} inter-valence charge transfer (Platonov et al. 1998). Ti^{3+} is not a generally stable oxidation state in terrestrial rocks (Waychunas 1987), however, instances of its occurrence are established, often due to the effects of naturally occurring ionizing radiation.

Sillimanite

Cr^{3+} in the Weak Crystal Field

Figure 5.37 presents luminescence (a) and (c) and excitation (b) and (d) spectra of green sillimanite (GRR1924) with zero delay and a narrow gate of 500 ms at 300 and 100 K. At 300 K luminescence is characterized by very broad band with

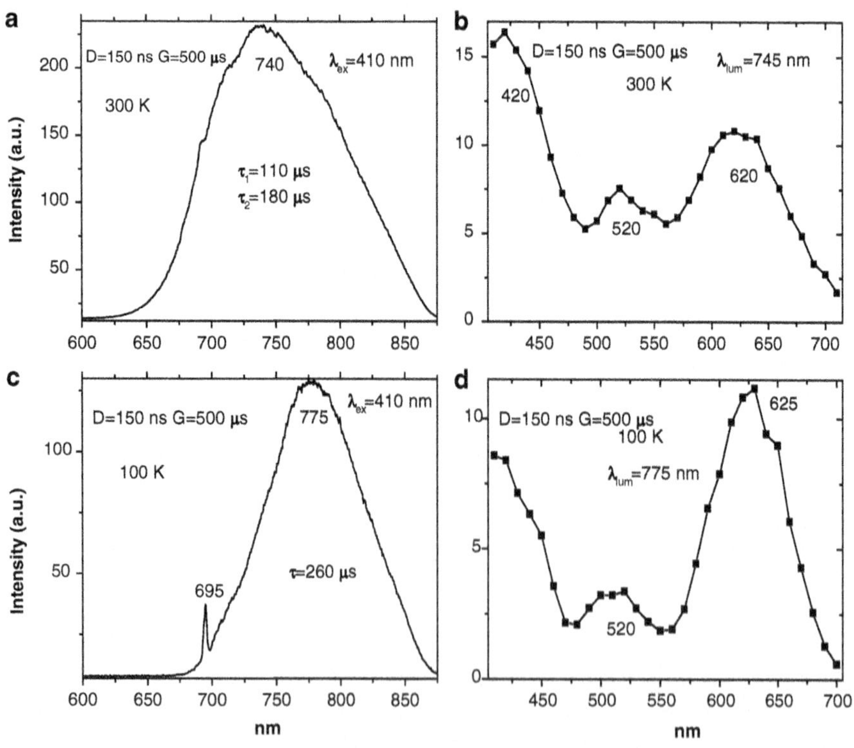

Fig. 5.37 (**a–d**) Luminescence at 300 and 77 K (**a**) and (**c**) and corresponding excitation (**b**) and (**d**) spectra of sillimanite GRR1924 connected to Cr^{3+} in weak crystal field structural positions

Δ of 150 nm peaking at approximately 740 nm with intermediate decay time τ of 150–200 μs and three excitation maxima at 420, 520 and 620 nm. At 100 K the band is substantially narrower with Δ of 100 nm peaking at approximately 775 nm with longer τ of 260 μs and approximately the same excitation maxima. Similar bands but with substantially lower intensity have been detected in greenish GRR518 sillimanite sample.

The broad red band with similar spectral and kinetic parameters was ascribed to Cr^{3+} luminescence in a weak crystal field (Wojtowicz and Lempicki 1988). Such an interpretation is confirmed by the fact that this emission has the greatest intensity in the spectrum of the sample GRR518 with substantially elevated Cr content of 0.15 %, which is detected by microprobe analyses and gives a green tint to yellow sillimanite (Rossman et al. 1982). In two other samples, where Cr luminescence is present, but with lower intensity, chromium was not found by electron microprobe analyses (Rossman et al. 1982), which may be explained by its quantity below microprobe sensitivity, but sufficient for detectable luminescence. In order to further prove this interpretation, we studied synthetic sillimanite activated by Cr (Gaft et al. 2012). Figure 5.37 presents its luminescence spectra with 532 nm excitation at 300 K with relatively narrow gate of 100 μs in order to selectively detect the luminescence with intermediate decay time. Decay time of this band is 150 μs, which corresponds well to that of natural sillimanite. At 77 K this band behaves similar to one in natural sillimanite, namely becomes narrower and moves in long wavelength direction. Thus it may be concluded that broad luminescence band peaking at 750–760 nm with decay time of 150 μs in natural sillimanite is connected to a Cr^{3+} emission in a weak crystal field. The mostly probable substitution is in an Al^{3+} octahedral position. Actually, sillimanite has two Al^{3+} sites and potentially two different Cr^{3+} luminescence types. Nevertheless, one of Al^{3+} structural positions is tetrahedral. It is well known that Cr^{3+} has a very high ligand field preference for octahedral sites and its presence in tetrahedral coordination is generally considered as impossible. The reason is the stability of the trivalent ion and its $3d^{3}{}^{4}A_2$ ground state with all t_{2g} spin-up orbitals filled (Fig. 5.38).

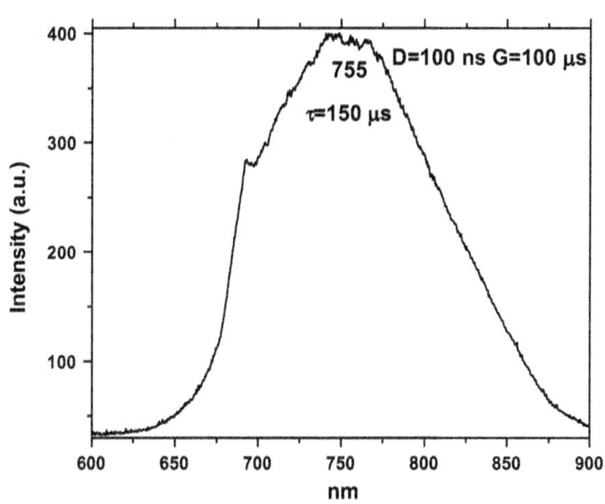

Fig. 5.38 Luminescence at 300 of synthetic sillimanite activated by Cr

Wollastonite

Spectra with narrow gates where the centers with short decay time are emphasized enables to detect broad bands at 794 nm with decay time of 5 μs and broad band at 840 nm with decay time of 190 μs (Fig. 4.95c). These bands may be ascribed to Cr^{3+} luminescence centers, in addition to Cr^{3+} with narrow R-lines at 720 nm, detected by steady-state spectroscopy (Min'ko et al. 1978). Wollastonite structure has three different types of six-coordinated calcium-oxygen groups, which enables formation of several types of Cr^{3+} luminescence centers. Nevertheless, luminescence of Cr^{3+} as result of Ca^{2+} substitution has not been confidently found yet and another interpretation is also possible. For example, ions of V^{2+} may be considered, which have similar luminescence properties with Cr^{3+} and may substitute in Ca^{2+} sites.

Corundum

Chromium activated ruby was the first laser material and its luminescence properties are carefully studied. It is a classical example of Cr^{3+} in octahedral crystal field. Here Cr^{3+} substitutes the Al^{3+} ions, while such possibility can be rationalized by an excellent chemical fit of Cr in place of Al. Ruby is a high crystal field material and thus the $^4T_{2g}$ state lies above the $^2E_{2g}$ level. Pumping is accomplished by a spin-allowed transition into the ^4T state, while emission occurs from ^2E level without vibrational broadening and almost all excited states can be converted into laser action. Both ^2E and $^4A_2$ states are split due to the trigonal crystal field and spin-orbit coupling. The splitting of the ^2E excited state is the reason for the 1.4 nm wavelength difference between the R_1 and R_2 lines. The splitting of the $^4A_2$ ground state is much less intensive and accounts for splitting of R_1 line into two lines less than 0.02 nm apart, which is only observed in low temperature measurements. The two main Cr^{3+} lines are well known for their marked frequency shift with increasing pressure (Mao et al. 1986).

Time-resolved spectra demonstrate doublet of R-lines with very long decay time of several ms, accompanied by sidebands and N-lines (Fig. 4.118).

Emerald

Emerald, Cr^{3+} doped beryl, has a beryl structure with the Cr^{3+} impurity ions in highly distorted octahedron sites. The discovery of lasing action in emerald stimulated investigation of its luminescence properties. It was established that its tuning range is approximately 730–810 nm, while luminescence consists of narrow line at 684 nm and band peaking at 715 nm with similar decay times of 62 μs. The relative intensities of those line and band are different in σ- and π-polarized spectra (Fabeni et al. 1991).

Laser-induced time-resolved luminescence spectra of natural emeralds also demonstrate R-lines of Cr^{3+} at 680 and 684 nm accompanied by narrow band peaking at 715 nm, which have similar decay times of approximately 55 μs (Fig. 4.123).

Garnet

Garnet activated by trivalent Cr^{3+} is promising system for tunable laser applications and those systems have been well studied. Cr^{3+} replaces Al^{3+} in octahedral sites with a weak crystal field. The transition involved in laser action is $^4T_2$-$^4A_2$, a vibrationally broadened band. At room temperature it has a maximum in the 715–825 nm range with decay time in the 100–250 μs range depending on ΔE between ^2E and $^4T_2$ levels. When the ΔE is maximal, narrow lines also appear from ^2E level. At low temperatures, when thermal activation of $^4T_2$ level is difficult, R-lines luminescence becomes dominant with the main line at 687 nm (Monteil et al. 1988). We studied pyrope artificially activated by Cr and also found two emission types described above (Fig. 5.39)

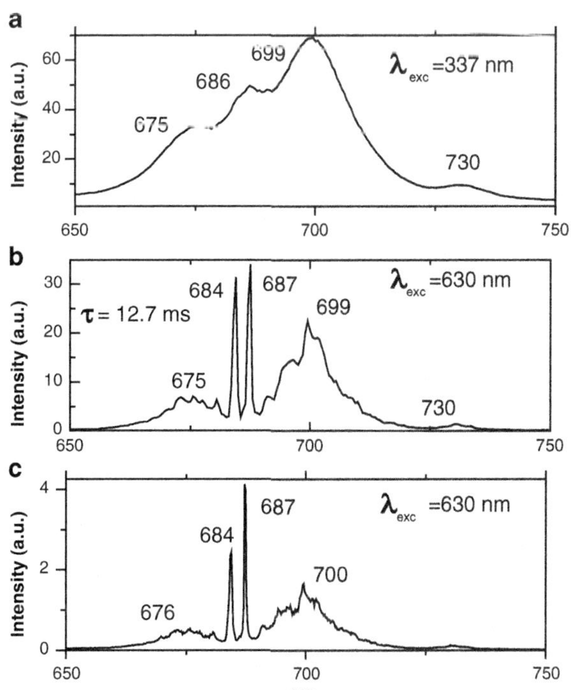

Fig. 5.39 (a–c) Laser-induced time-resolved luminescence spectra of pyrope activated by Cr^{3+}

Alexandrite

Alexandrite, the common name for Cr-doped chrysoberyl, is a laser material capable of continuously tunable laser output in the 700–800 nm region. It was established that alexandrite is an intermediate crystal field matrix, thus the non-phonon emitting ^{2}E state is coupled to $^{4}T_2$ relaxed state and behaves as a storage level for the latter. The laser emitted light is strongly polarized due to its biaxial structure and is characterized by decay time of 260 μs (Fabeni et al. 1991; Schepler 1984; Suchoki et al. 2002). Two pairs of sharp R-lines are detected connected with Cr^{3+} in two different structural positions: the first near 680 nm with decay time of approximately 330 μs is connected with mirror site fluorescence and the second at 690 nm with much longer decay of approximately 44 ms is connected with inversion symmetry sites (Powell et al. 1985). The group of narrow lines between 640 and 660 nm was connected with anti-Stokes vibronic sideband of the mirror site fluorescence.

Figure 5.40 presents time-resolved luminescence spectra of alexandrite (13.11) from Madagascar. At room temperature two lines peaking at 680.4 and 678.7 nm

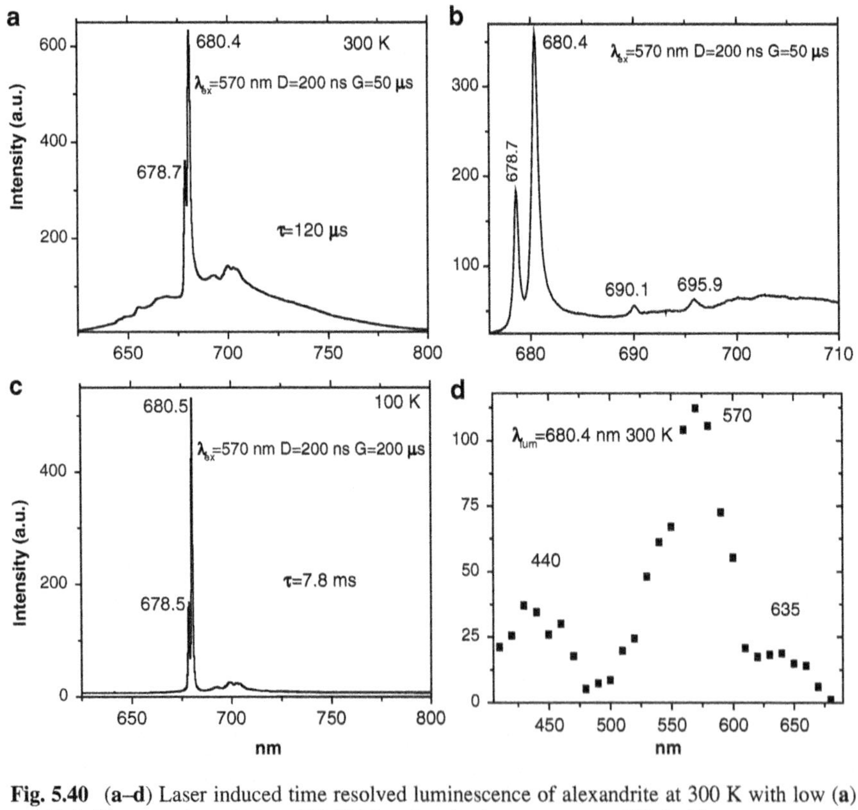

Fig. 5.40 (**a–d**) Laser induced time resolved luminescence of alexandrite at 300 K with low (**a**) and high (**b**) spectral resolution, 100 K (**c**) and excitation spectrum at 300 K (**d**)

Fig. 5.41 (a–c) Laser induced time resolved luminescence of alexandrite at 300 K (**a**), 100 K (**b**) and excitation spectrum at 300 K (**c**)

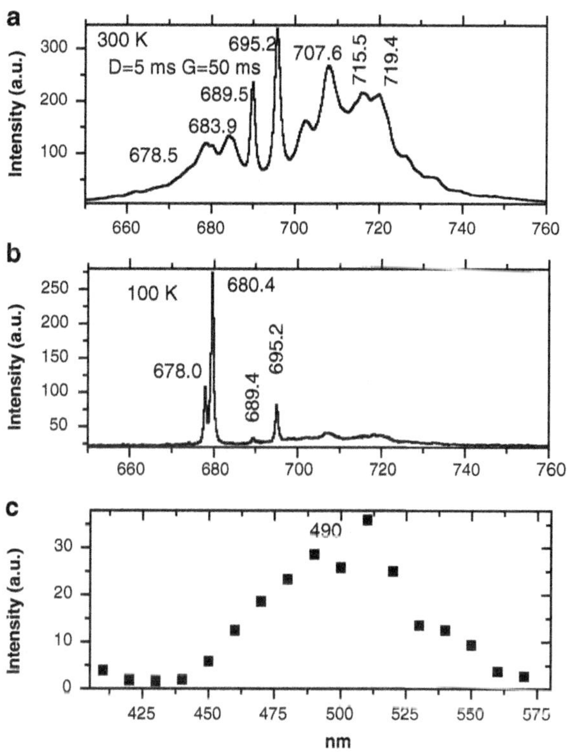

present accompanied by relatively weak band emission (Fig. 5.40a). Decay time of those lines is $\tau = 120$ μs. The spectra with high spectral resolution revealed also two weaker lines peaking at 695.9 and 690.1 nm (Fig. 5.40b). At the low temperature of 100 K the spectral position of the lines is practically unchanged (Fig. 5.40c) but the decay time becomes much longer reaching $\tau = 7.8$ ms. Excitation spectrum has two bands peaking at 420 and 580 nm (Fig. 5.40d) and is essentially the same as the absorption spectrum. Such spectral and kinetic behavior clearly indicates that it is Cr^{3+} luminescence center ascribed to mirror sites. Such luminescence remains unchanged under different excitations and with different delays and gates. Thus mainly $Cr^{3+}{}_{m}$ exists in this sample. It corresponds to the fact that Cr^{3+} is slightly larger than Al^{3+} and therefore, prefers to occupy the bigger mirror site.

Chemical data presented in Table 4.20 demonstrate that the alexandrite samples which exhibit only one type of chromium luminescence in its lattice, namely $Cr^{3+}{}_{m}$, are characterized by low concentrations of chromium impurity. Figure 5.41a presents time-resolved luminescence spectra at 300 K after very long delay of 5 ms of alexandrite from Siberia which is characterized by the highest chromium. It is structurally complicated emission composed of two narrow lines peaking at 690.3 and 695.9 nm superimposed by the broad emission band. All those narrow lines and structured broad band combination remains the same under different excitations,

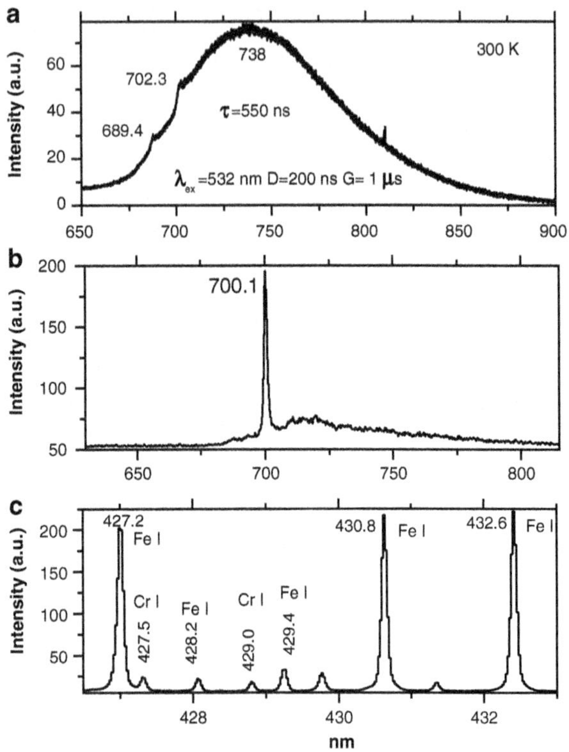

Fig. 5.42 (**a–c**) Laser induced time resolved luminescence of Cr^{3+} in forsterite at 300 K (**a**) and 100 K (**b**) and breakdown spectrum with characteristic Cr lines (**c**)

have the same long decay time of approximately 800 μs and evidently belong to one luminescence center. At low temperature of 100 K the broad band disappears (Fig. 5.41b). The excitation spectrum consists mainly from one band peaking at approximately 590 nm (Fig. 5.41c). Such kind of luminescence is similar to $Cr^{3+}{}_i$.

Forsterite

Cr^{3+}: Strong Crystal Field

Figure 5.42a presents the time resolved luminescence spectrum at 300 K of the Supat sample under 532 nm excitation with a very narrow gate of 1 μs. A broad band peaking at approximately 738 nm is detected with very short decay time of approximately 500 ns accompanied by weak narrow shoulders at 689 and 702 nm. At 100 K, the broad band was not detected, while the narrow line peaking at 700 nm dominates the spectrum (Fig. 5.42b). The 738 nm band and all the sharp lines correspond well with the emission of Cr activated forsterite where the lines with milliseconds lifetimes around 700 nm are ascribed to $'E + {}^4T_2$ transitions at low temperatures and a short-lived broad band peaking between about 690 and 750 nm

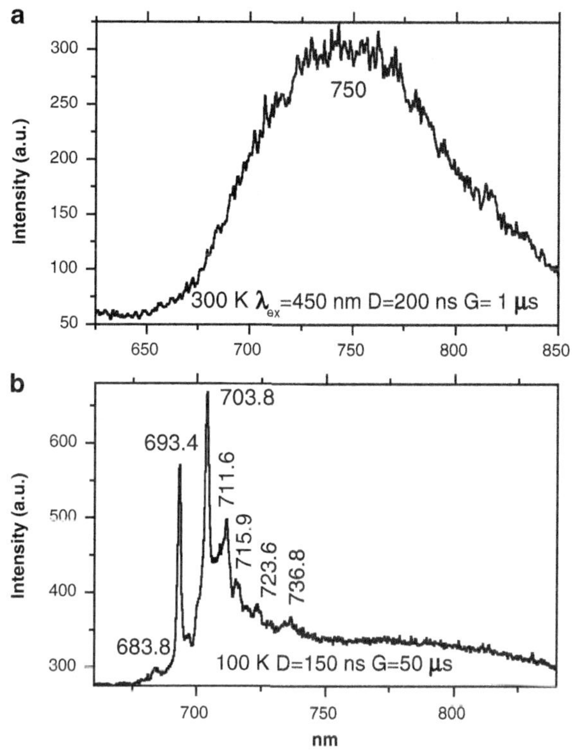

Fig. 5.43 (**a–b**) Laser induced time resolved luminescence of Cr^{3+} in forsterite at 300 K (**a**) and 100 K (**b**)

at high temperatures is ascribed to $^4T_2$, $'E + {}^4A_2$ thermalized emission (Moncorge et al. 1991). Such emission was ascribed to the M1 site (Glynn et al. 1991). Elevated chromium concentrations in the Supat sample is confirmed by its LIBS data, where characteristic lines of Cr I peaking at 427.5 and 429.0 nm are clearly detected (Fig. 5.42c). Chromium content in this sample is approximately 19 ppm (Table 4.22).

Another kind of Cr^{3+} luminescence in a strong crystal field site was found in the green sample from China. Figure 5.43a presents time-resolved luminescence spectra at 300 K under excitation by 450 nm where a weak, broad band peaking at 750 nm is detected with a very short decay time of 450 ns. At low temperature several intense narrow lines dominate the spectrum peaking at 693.4, 703.8, 711.6, 715.9, 723.6 and 736.8 nm (Fig. 5.43b). Such lines were previously ascribed in chromium activated forsterite to R-lines from Cr^{3+} in M1 positions associated with Mg vacancies needed for charge compensation (Glynn et al. 1991).

Such kind of Cr^{3+} luminescence is rather typical for natural forsterite. For example, Figure 5.44 presents luminescence features of several forsterite samples detected during Raman spectra study (http://rruff.info). Luminescence bands peaking at 740-750 nm under excitation by 532 nm at 300 K may be ascribed to Cr^{3+} in intermediate and strong crystal fields.

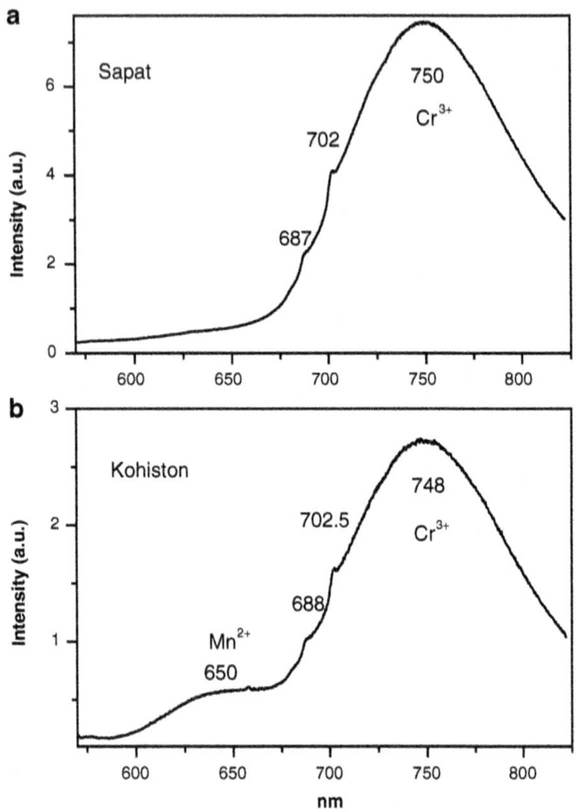

Fig. 5.44 (**a–b**) CW laser induced (532 nm) luminescence of Cr^{3+} in forsterite at 300 K

Cr^{3+}: Weak Crystal Field

Figure 5.45a presents time-resolved luminescence spectra at 300 K of the transparent sample from Russia but with a relatively narrower gate of 10 μs. The corresponding luminescence band peaking at approximately 805 nm is very broad ($\Delta = 140$ nm) and is characterized by a relatively short decay time of $\tau = 15$ μs. Its excitation spectrum in the visible range consists of one main band peaking at approximately 450 nm (Fig. 5.45b). It may be interpreted by comparison with synthetic forsterite luminescence activated by Cr. This band's spectral and kinetic properties correspond well to the low ligand field Cr^{3+} center which is characterized by a broadband emission peaking around 805 nm with a lifetime ranging between 12.5 and 18 μs (Moncorge et al. 1991). It may be supposed that such a low crystal field position may be associated with the M2 site. Because the average Mg-O band length of M2 sites is larger than that of Mg1, the crystal field strength of M2 sites is weaker.

Breakdown spectra of this sample do not reveal characteristic emission lines of Cr. It may be explained by the extreme sensitivity of laser-induced luminescence for the elements with allowed luminescence electronic transitions.

Fig. 5.45 (**a–b**) Laser induced time resolved luminescence (**a**) and excitation (**b**) spectra of Cr^{3+} in forsterite at 300 K

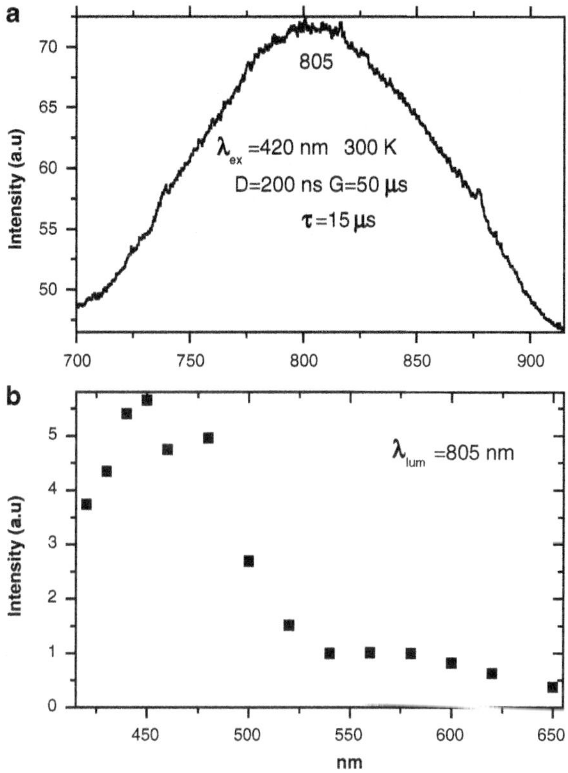

A forsterite near-IR band peaking at approximately 890 nm and detected under excitation by 780 nm (Fig. 5.46) was also ascribed to Cr^{3+} in a weak crystal field site in Cr-activated forsterite (Jia et al. 1991).

Boehmite, Diaspor

Time-resolved luminescence spectra of diaspore samples are demonstrated in Fig. 4.143. Two narrow lines at 694 and 692 nm, accompanied by line at 690 nm and phonon repetitions at 707 and 714 nm characterize the time-resolved luminescence spectra of this sample After long delay of several ms the relative intensity of the line at 692 nm becomes stronger, the evident reason is that this line has longer decay time compared to the line at 694 nm. After heating at 900 °C, the spectra are pronouncedly changed. Now they contained relatively broad line at 692 nm and a broad band peaking at 780 nm, which dominates the spectra with narrow gate (Fig. 4.143c, d). Such behavior evidences that the broad band has very short decay time compared with the line at 692 nm. After heating at 1000 °C, the spectra contain again two narrow lines at 694 and 692 nm in the time-resolved luminescence

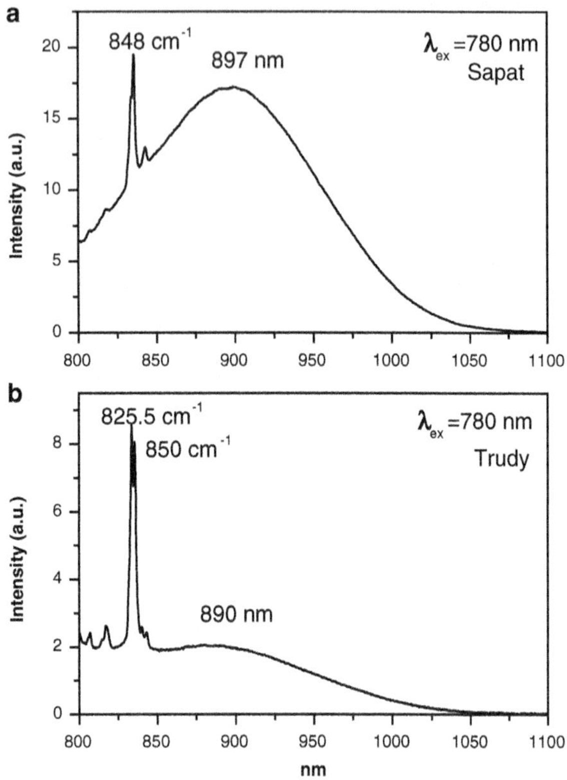

Fig. 5.46 (**a**–**b**) CW laser induced (780 nm) luminescence of Cr^{3+} in forsterite at 300 K (http://rruff.info). The Raman lines are indicated in wavenumber (cm^{-1})

spectra (Fig. 4.143e, f). However, the lines are strong and their relative intensities in time-resolved spectra not changed evidence that they have similar decay times. It is clear that luminescence of natural diaspor is connected with Cr^{3+} in strong crystal field position. The lines in the luminescence spectra of diaspore treated at 900 °C are related to Cr^{3+} luminescence in the meta-phase. The broad asymmetric line peaking at 692 nm with long decay time is evidently connected with inhomogeneously broadened R-lines of Cr^{3+} luminescence in octahedral sites with a strong crystal field and a very disordered environment. The broad band peaking at 780 nm with much shorter decay time is attributed to Cr^{3+} luminescence in the weaker crystal fields, where a strong broadband quartet-quartet ${}^4T_2 \rightarrow {}^4A_2$ transition appears, which is only symmetry forbidden. These spectral types are similar to those of Al_2O_3 powders heated to temperatures from 400 °C to 1000 °C (Pilonett et al. 2000). The lines in the luminescence spectra of diaspore treated at 1000 °C are related to Cr^{3+} luminescence in corundum. These lines are typical for ${}^2E \rightarrow {}^4A_2$ transitions of Cr^{3+} luminescence centers (R-lines) of Cr-corundum (ruby), which are characteristic for Cr^{3+} luminescence, which is substituting for Al^{3+} in octahedral symmetry. Both lines are long-lived and have the same decay times.

A narrow line at 692 nm with very long decay time accompanied by phonon repetitions characterizes the time-resolved luminescence spectra of boehmite (Fig. 4.144). After heating at 900 °C, this emission remains but additional broad band appears peaking at 725 nm with short decay time, which dominates the spectrum with narrow gate. After heating at 1000 °C, lines at 683, 686 and 690 nm accompanied by lines at 692 and 694 nm are observed in the steady-state luminescence spectra. These two groups of lines appear with different decay times time-resolved luminescence spectra: at 683, 686 and 690 nm with longer decay time and at 694 and 692 nm with shorter decay time. Luminescence of natural boehmite is connected with Cr^{3+} in strong crystal field position. The lines in the luminescence spectra of diaspore treated at 900 °C are related to Cr^{3+} luminescence in the meta-phase. The broad band peaking at 725 nm with short decay time is evidently connected with Cr^{3+} luminescence in weak crystal field site. The lines at 683, 686 and 690 nm in the luminescence spectra of boehmite treated at 1000 °C are related to Cr^{3+} luminescence in Al-spinel. The other lines at 692 and 694 nm are connected with presence of corundum. Durville et al. (1985) related the splitting and the inhomogeneous broadening of the ^2E level in the spectra of Mg-spinel formed during heating of magneso-alumino silicate glass to wide distribution of the Cr^{3+} sites in the microcrystallites. They related the wide distribution to the well-known disordered distribution of Mg^{2+} and Al^{3+} in the Mg-spinel.

Zircon

Luminescence of Cr^{3+} was firstly detected in synthesized $ZrSiO_4$ artificially activated by Cr (Gaft et al. 2000b). This host has been chosen, first of all, by the presence of tetrahedrally coordinated Si^{4+}, which has been assumed to accept Cr^{4+} in silicate crystals without any charge compensation. A second reason is the presence of Zr^{4+} in triangular dodecahedron with 4 close oxygen atoms arranged according to distorted tetrahedron. It was proposed that Cr^{4+} could replace either Si^{4+} or Zr^{4+} and the interpretation of absorption spectra led to the hypothesis of a replacement of Si^{4+} with Cr^{4+} (Beletti et al. 1995).

Luminescence spectra of $ZrSiO_4$: Cr at different temperatures under $\lambda_{ex} = 532$ nm demonstrate narrow lines and broad bands in the red-IR range. Cr^{3+} ions can be easily detected by distinct ^2E-$^4A_2$ transition (R-line) at 694 nm (Fig. 5.47b). It appears above 150 K whereas at lower temperatures only $^4T_2$-$^4A_2$ broadband emissions can be seen peaking at 775 nm. It indicates that $^4T_2$ level is lower than ^2E and consequently the relatively low crystal field site symmetry takes place. The ^2E excited level have the same decay time of 3–5 μs as the broad band, which means that it is thermally occupied at temperatures higher than 150 K. It is worth to mention that the $^4T_2$-$^4A_2$ band at low temperature demonstrates vibrational structure indicating tetragonally distorted excited state (Blasse and Grabmaier 1994). Besides that, the R-line of Cr^{3+} is strongly polarized and is detected with experimental geometry, which corresponds to observation of σ-polarized luminescence only. Under these observations and on the basic ideas that ionic radius of

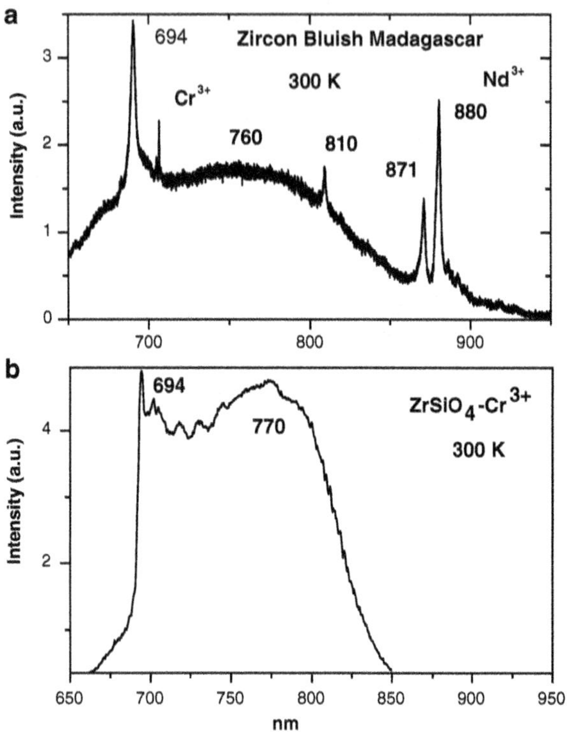

Fig. 5.47 (**a–b**) Laser-induced time-resolved luminescence spectra of Cr^{3+} in natural zircon (**a**) and Cr^{3+} in artificial $ZrSiO_4$-Cr (**b**)

Cr^{3+} in octahedral coordination (75.5 pm) is close to those of Zr^{4+} (73 pm), and that the Zr-O distances are close to Cr-O length in systems with Cr^{3+}, we can assume the presence of Cr^{3+} on the place of Zr^{4+} ion. Recently emission of Cr^{3+} has been found in precious variety of natural zircon (Fig. 5.47a).

Titanite

The lines at 686 and 693 nm with long decay time of approximately 1 ms in titanite emission spectrum are not correlated with any other lines and bands (Fig. 4.80c). Such lines are very typical for Cr^{3+} in high field coordination and may be connected with such center. The broad luminescence band appears peaking at 765, which may be ascribed to Cr^{3+} in weak field coordination. The band at 765 nm has distinct dips at 749, 762, 793, 798, 804 and 820 nm. Comparison with titanite absorption spectrum demonstrates that those lines exactly coincide with absorption spectrum of Nd^{3+} (Bakhtin and Gorobets 1992). Cr^{3+} is a good energy sensitizer, because it has broad, allowed absorption bands with broad emission spectrum, which overlaps the absorption bands of the lasing ion (Nd^{3+}, Ho^{3+}).

Thus radiative energy transfer to Nd^{3+} may explain "negative" lines. Such emission-reabsorption type of the energy transfer implies emission of light by a single ion and its absorption and emission by the other ion. Both ions should be activators and have sufficiently strong absorption bands, while the luminescence emission band of one of them must be overlapped by the absorption band of the other. Radiative energy transfer may increase the laser efficiency of Nd^{3+} substantially. For example, amplification of emission of Nd^{3+} as result of coating a laser rod by organic dye was about 50 % (Eyal 1988).

Decay time of Cr^{3+} band of approximately 150 ns is very short for such emission. Radiative energy transfer may not explain it because in such case the decay curves of each of the ions are independent of the presence of the other. Thus non-radiative energy transfer may also take part, probably via multipolar or exchange interactions. In such case the process of luminescence is of an additive nature and the lifetime of the sensitizer from which the energy is transferred is determined, apart from the probability of emission and radiationless transitions, by the probability of the energy transfer to the ion activator.

The low concentrations of Cr and Nd in natural samples (200 and 555 ppm, correspondingly) make it difficult to understand the strong energy transfer between them. Thus it may be supposed that the distribution of donors and acceptors are not random and that the presence of a dopant in a particular site in the crystal will affect the probability of a second dopant (either identical or different) locating in nearby sites. It is a known situation in minerals where exchange-connected impurity pares explain many peculiarities of minerals colors (Bakhtin and Gorobets 1992). In order to explain pair formation in laser materials the relative ionic radii of dopant and the host ion have been considered (Rotman and Hartman 1988). For a dopant to enter a lattice site, it is advantageous for the dopant to be approximately the same size as the host ion it is replacing. Failure to do so may severely limit its degree of substitution. To lower a crystal strain, the two dopants, which are considerably smaller and considerably larger than the ions they replace, have to collocate next to each other and form preferential pares with enhanced energy transfer. In titanite case such approach is not working because ionic radii of Nd^{3+} and Cr^{3+} are practically similar to ionic radii of Ca^{2+} and Ti^{4+}, correspondingly. In such case the charge compensation consideration are evidently the mostly important, namely $Nd^{3+} + Cr^{3+} - Ca^{2+} + Ti^{4+}$. Similar model was also proposed for titanite enriched by light rare-earth elements (up to 3.5 wt.%). It was concluded that the following mechanism of substitution consistent with compositional data: $[LREE]^{3+} + [Fe^{3+}] - Ca^{2+} + Ti^{4+}$.

Drastic change takes place in luminescence spectrum of titanite at low temperatures (Fig. 4.80d). At 77 K, Nd^{3+} luminescence intensity becomes lower and narrow line appears at 732 nm with long decay time of 2.5 ms accompanied by phonon repetitions. At even lower temperature of 20 K such emission totally dominates luminescence spectrum. Such behavior may be explained by Cr^{3+} in intermediate crystal field sites for which the crystal field parameters lie in the crossing region of the $^4T_2$ and ^2E states. Within the intermediate crystal field there is complicating mixing between doublet and quartet states with complicated spectra, non-radiative transfer and the temperature dependence of luminescence. In such case the emission from both $^4T_2$ and ^2E states may be expected. At 300 K the lower ^2E state acts as an energy reservoir for the $^4T_2$ state and thermal activation enables to emit only from $^4T_2$ level. At lower temperatures the thermal activation is impossible and ^2E level becomes luminescent. It is worth to note, that the energy of $^2E \rightarrow {}^4A_2$ bandwidth is rather low; at least in known minerals luminescence the spectral position of this so-called R-line is changing from 677 to 705 nm (Tarashchan 1978). From the other side, in LiSc(WO_4):Cr crystals the $^2E \rightarrow {}^4A_2$ transition in high field Cr^{3+} sites is situated near 760 nm (Grinberg et al. 2002) and 778 nm in $BaCaTiO_3$-Cr crystals (Jastrabik et al. 2002). Energy of the ^2E level in the d^3 electron system, according to the Tanabe-Sugano theory, increases a little with increase of the strength of the crystal field. The high-pressure low-temperature Cr^{3+} luminescence experiments show, contrary to the prediction of the Tanabe-Sugano theory, that the energy of the ^2E level of Cr^{3+} ions decreases slightly with increase of the crystal field strength (Suchoki et al. 2002). It was explained by nephelauxetic effect, i.e. decrease of interelectronic crystal field Racah's repulsion parameters B and C due to covalency of bonds between the central ion and ligands. Thus the red shift in Cr^{3+} luminescence may be attributed to greater covalency in titanite structure. Thus two different Cr^{3+} sites are detected in titanite structure in strong, and intermediate field sites. The energy transfer to Nd^{3+} evidently takes place only as result of allowed broad $^4T_2$ emission, because any long time tail in Nd^{3+} decay is absent as expected if transfer will come from strong crystal field sites.

Two different Cr^{3+} emission centers in strong and weak CF have been found in chromium-doped titanite samples peaking at 693 and approximately 800 nm have been found (Fig. 5.48). Emissions of centers related to traces of trivalent rare-earth elements are observed exclusively from chromium-free samples. Their Cr-doped analogues (containing the same REEs on the same concentration levels), in contrast, only show broad-band Cr^{3+} emission whereas all REE emissions are suppressed. This behavior is assigned to quenching of REE emissions by chromium centers (i.e., $REE^{3+} \rightarrow Cr^{3+}$ energy transfer) (Nasdalla et al. 2014).

Spinel

An identification of the nature of Cr^{3+} luminescence in synthetic spinel by line narrowing technique enabled to distinguish up to 25 different Cr^{3+} sites (Deren et al. 1996). Laser-induced time-resolved spectroscopy enables to see typical for spinel emission of Cr^{3+}, while the different broadness of the spectral lines at different time windows demonstrates that different Cr^{3+} sites present also in natural spinel (Fig. 4.119).

Fluorite

Luminescence of Cr^{3+} was not found in natural fluorite, but such emission has been found in fluoride hosts, including CaF_2, artificially activated by Cr (Payne 1988). It was found that despite a big difference in ionic radii of Ca^{2+} and Cr^{3+} the last one

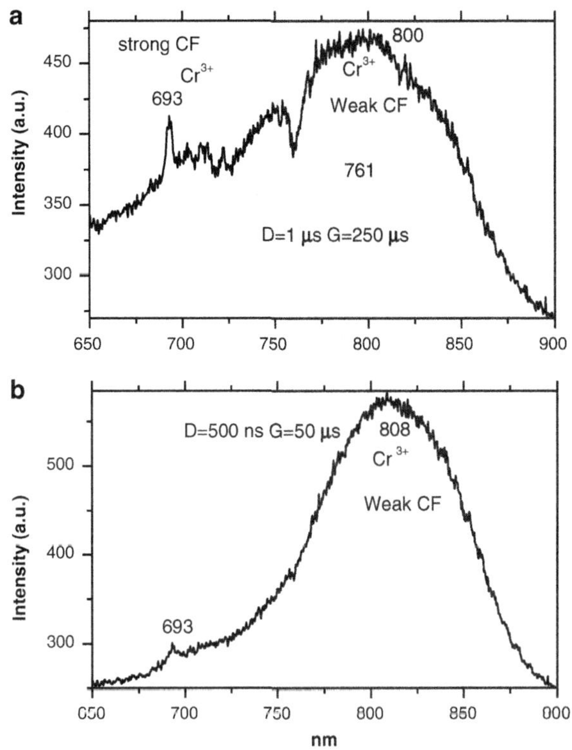

Fig. 5.48 (**a–b**) Laser-induced time-resolved luminescence spectra of Cr^{3+} in natural titanite activated by 0.2 % Cr

forms luminescence center, which is characterized at low temperature by broad emission band peaking at approximately 920 nm.

Other Minerals

Under different pulsed and cw excitations Cr^{3+} have been found in tourmaline (Fig. 4.134), clintonite (Fig. 4.146), phlogopite (Fig. 4.150), clinochlore (Fig. 4.151), euclase (Fig. 4.153) and dickite (Fig. 4.154).

Cr^{3+}-Pairs

The mostly known are Cr^{3+}-pair in ruby with elevated Cr contents (Powell et al. 1967). The satellite lines appear in absorption and luminescence spectra, which were interpreted by exchange pair coupling theory. Theoretical calculations based on DV-ME method were made for the first and fourth nearest neighbor ion pairs (Boulon 1987). The resulting scheme of ground and excited states well correspond to experimental data. The calculated results demonstrate that the first and fourth nearest neighbor ion pairs are anti-ferromagnetically and ferromagnetically coupled, correspondingly.

5.7.1.2 Mn^{4+}

In artificial phosphors luminescence bands due to Mn^{4+} exist from 620 to 715 nm. The spectrum has a structure consisting of several broad lines originating from transitions aided by lattice vibrations. Because the crystal field at the higher charged Mn^{4+} ion is stronger than for Cr^{3+} (Fig. 5.23), the emission is most probably connected with ^2E-$^4A_2$ transition, but in many cases the origin of the emitting and ground states is doubtful (Butler 1980; McNicol and Pott 1973; Stade et al. 1974; Bryknar et al. 2000).

Orange Kyanite

Figure 5.49 presents luminescence spectra of orange kyanite containing 0.1 % Mn with excitations by 532 and 355 nm at 77 K (Gaft et al. 2011). Under excitation by 532 nm, the spectra with long delay of 500 μs contain the strong lines at 679.3 and 705.2, accompanied by weaker lines at 677.0, 684.2, 689.6, 691.8, 694.7 and 697.3 nm (Fig. 5.49a). Spectra with narrow gate of 20 μs contain the strong lines at 679.3 and 697.0, accompanied by weaker lines at 683.8, 689.6, 702.2, 708.7, and 713.2 (Fig. 5.49b). Under excitation by 355 nm, the spectra with long delay of 500 μs contain the strong lines at 679.3 and 694.7 nm, accompanied by weak lines at 676.9, 689.3, 691.6, 697.5, and 703.7 (Fig. 5.49c). The spectra with narrow gate of 20 μs contain mainly the lines at 679.3, 696.7 and 712.9 nm (Fig. 5.49d). The lines at 679.3 and 677.0 nm have two decay components of 20–30 and 250–400 μs. Two lines at 694.7 and 691.7 nm have long decay time of 630–450 μs, while two short lived lines at 696.7 and 689.5 nm have short decay time of 60 μs. The line at 705.0 nm detected only under excitation by 532 nm has very long and evidently non exponential decay time. The new spectral feature in comparison to known in blue chromium containing kyanite at 300 K is the line peaking at 681.3 with a very short decay time of 2.3 μs. At 77 K its luminescence becomes much stronger; the maximum is moved to 679.3 nm and decay time of 290 μs is two orders of magnitude longer compared to 300 K. Many other lines appear at 77 K the strongest one with long decay time is at 694.7 nm, accompanied by weaker line at 691.8 nm and with short decay time is at 696.7 nm, accompanied by very weak line at 689.5 nm. It is logical to suppose that the narrow lines appearing in orange Mn bearing kyanite are connected with Mn^{4+}. It is supported by their excitation spectra, which contain one broad band peaking at 510–540 nm typical to Mn^{4+} absorption and luminescence excitation. The following main groups of lines may be separated based on similar decay times: 679.3 and 677.0 with long and short components in decay time; 689.9, 694.7, 691.6 and 697.4 with long decay time; 689.6 and 697.0 with short decay times. The first group is present at 300 and 77 K, while the last two

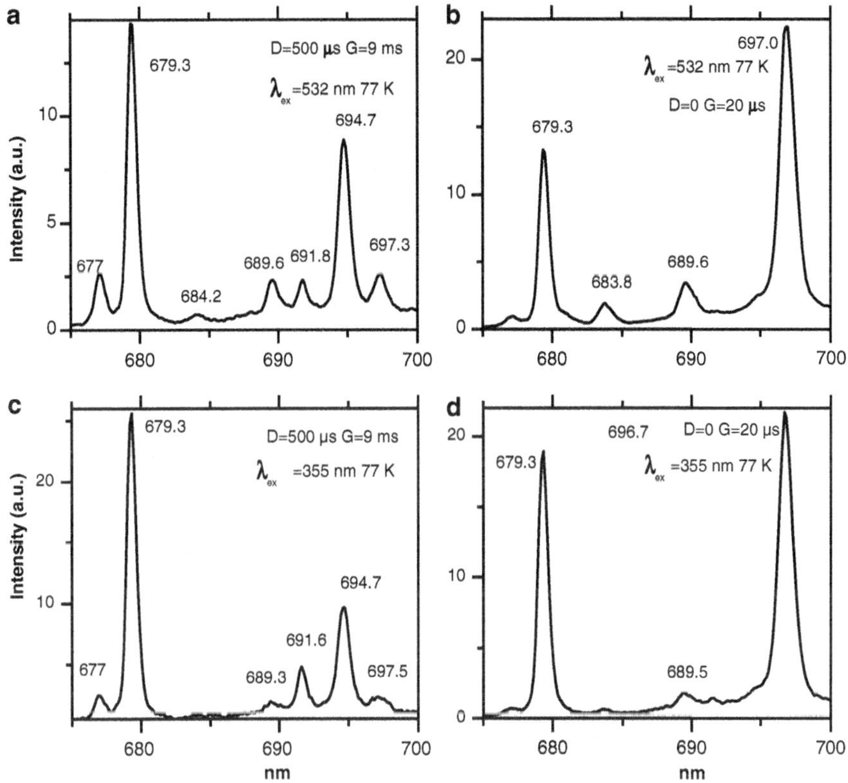

Fig. 5.49 (**a–d**) Laser-induced time-resolved luminescence spectra of Mn^{4+} in orange Mn-kyanite

are evident only at low temperature. They may be preliminary ascribed to Mn^{4+}-Al^{3+} substitutions in different sites inside the kyanite structure. For example, the two lines at 679.3 and 677.0 nm may be ascribed to radiative transitions from the split ^2E level in Mn^{4+} situated in strong crystal field position. The splitting in this case is 50 cm^{-1} which is quite reasonable for kyanite structure. The very short decay time at 300 K may be connected to effective radiationless energy migration from this center or to intrinsic temperature quenching. At low temperature both those processes are less effective and the decay time becomes substantially longer. It may be supposed that at 300 K the energy migrates to luminescence centers with broad emission band, which are absent at low temperatures. The four lines at 689.9, 691.8, 694.7 and 697.4 nm demonstrate similar behavior under different experimental conditions, such as excitation wavelength, temperature, delay and gate. They appear only at low temperatures and have long decay time typical for strong crystal field site. The presence of four lines and not usual two may be connected with radiative transitions not only from the split ^2E_g but also from the $^2T_1$ level. This level is energetically very close to ^2E_g and similar in its behavior, namely it does not depend on crystal field strength and its emission has to be characterized by narrow lines. The transitions $^2T_1 \rightarrow {}^4A_2$ and $^2E \rightarrow {}^4A_2$ belong to the same T_{2g} configuration and have forbidden electronic dipole and even spin forbidden character. Thus decay time for luminescence from $^2T_1$ has to also be long. The luminescence disappearance at 300 K may be connected to energy migration. The two lines at 697.0 and 689.6 nm may be ascribed to radiative transitions from the split ^2E_g level in Mn^{4+}. The splitting of the ^2E_g level of 154 cm^{-1} is rather big but reasonable. The data on the ^2E_g splitting for Mn^{4+} luminescence in minerals are absent, but for Cr^{3+} the smallest one is 8 cm^{-1} in spinel and the former "champion" is 106 cm^{-1} in topaz, where the big splitting value was explained by the lowering of the local symmetry of Al^{3+} as result of substitution by two fluorine ions for two oxygen ions (Tarashchan 1978). For the Mn^{4+}-Al^{3+} substitution the situation is different from Cr^{3+} because charge compensator is needed with corresponding lowering of the local symmetry, which may explain relatively big splitting. The short decay time may be explained by intermediate crystal field position. At weaker crystal fields, a strong broadband quartet-quartet $^4T_2 \rightarrow {}^4A_2$ transition is energetically very close to the ^2E_g level. It is only symmetry forbidden and has a much shorter decay time. Within the intermediate crystal field, there is complicating mixing between doublet and quartet states with complicated spectra, non-radiative transfer and the temperature dependent luminescence. The spin-orbit interaction, coupling the ^2E and $^4T_2$ states, was responsible for breaking the selection rule forbidding the $^2E \rightarrow {}^4A_2$ transition. The admixture of the $^4T_2$ wave function into the ^2E wave function, which depends on the doublet-quartet energy separation, determines the radiative lifetime of the doublet, which, therefore, may change significantly with crystal field.

Sillimanite

Emission lines of supposedly Mn^{4+} have been also found in sillimanite (Gaft et al. 2012). Figure 5.50 presents time-resolved luminescence spectra of sillimanite where those lines are exceptionally intensive at 300 and 100 K under excitation by 532 nm. At room temperature five narrow lines were detected peaking at 682.8, 691.3, 704.7, 709.7 and 717.0 nm (Fig. 5.50a). All those lines may be seen only after substantial delay time of approximately 1 ms which enables to quench the broad band emission of Cr^{3+} described above. At low temperatures at the same experimental conditions, namely delay and gate, the line at 682.8 nm disappears, the line at 691.3 nm is detected as a weak shoulder near the new line at 693.7 nm, while the last three lines become much more intensive (Fig. 5.50b). Low temperature spectra with higher resolution with zero delay and relatively narrow gate of 500 μs, which are optimal for line detection with relatively short decay time, exhibit that the line at 691.3 nm splits to two lines peaking at 691.5 and 692.9 nm (Fig. 5.50c). Low temperature spectra with higher resolution with big delay of 1 ms and very broad gate of 19 ms, which are optimal for line detection with relatively long decay time, exhibit that the line at 691.5 nm disappears, the line at 690.1 becomes visible, the line at 692.9 nm remains and the lines at 702.2, 707.5

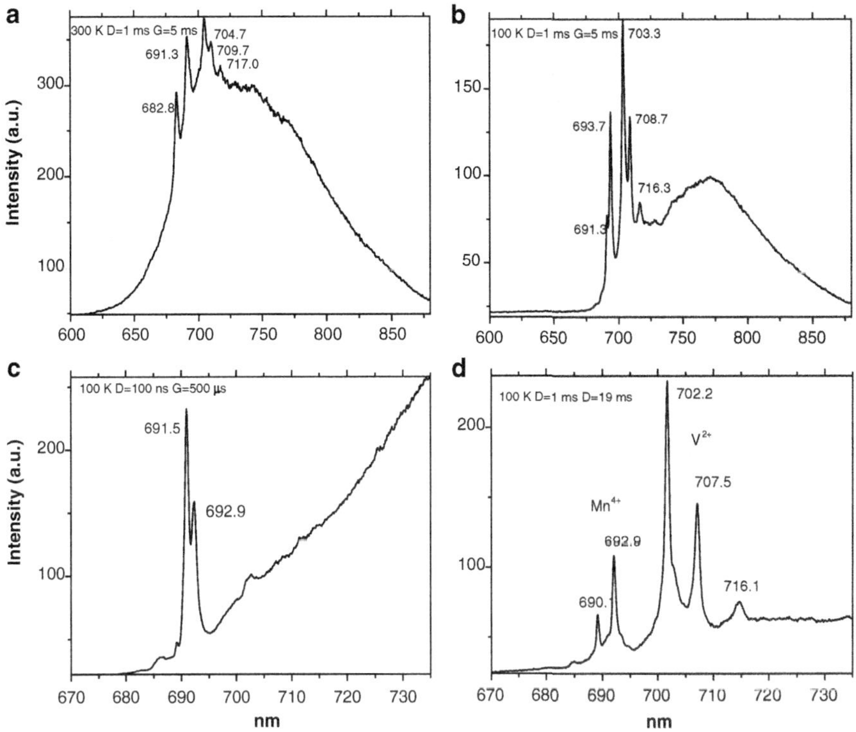

Fig. 5.50 (**a–d**) Laser-induced time-resolved luminescence spectra of Mn^{4+} and V^{2+} in sillimanite

and 716.1 nm are very strong (Fig. 5.50d). Excitation spectra of those narrow lines in visible range are mainly characterized by two broad bands, but peaking at different wavelengths.

Decay times of all those lines may be approximated by two components: 200–300 μs and several ms. Nevertheless it seems that only the lines at 692.9 and 691.5 nm really have short components of 350–400 μs, because for them the intensity of this component is nearly an order of magnitude more intensive than the long lived one. For all other lines the long lived component is substantially stronger and the short lived one is evidently connected to the background of the broad band emission. The lines at 690–693 nm range have long decay components in 1.9–2.3 ms range, while the lines at 704–750 nm range have the long decay components of 6–9 ms. The narrow lines luminescence in sillimanite is extremely complicated and presently we propose only the first approximation of their interpretation. Those long-lived narrow lines principally may be attributed to Cr^{3+} in a strong crystal field position. Nevertheless, they were not found in synthetic sillimanite activated by Cr. Besides, those lines were found in sillimanite samples with relatively low Cr contents. This motivated to look for another interpretation, such as the possibility that minor component different from Cr^{3+} is responsible for this emission, such as Mn^{4+} and V^{2+}. Many trace elements are present in sillimanite

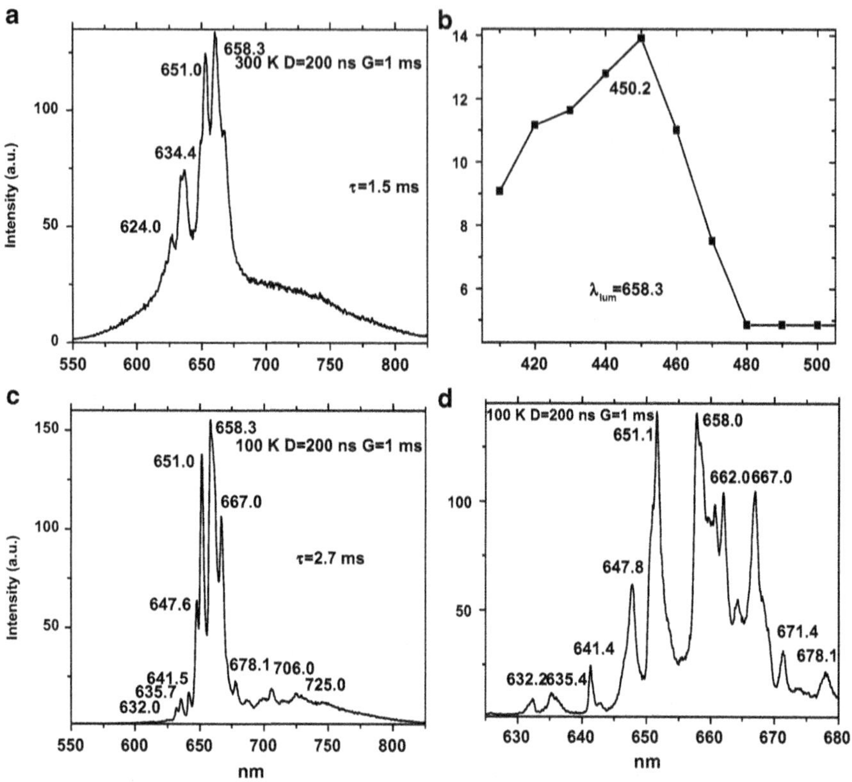

Fig. 5.51 (**a–d**) Laser-induced time-resolved luminescence and excitation spectra of supposedly Mn^{4+} in grossular

including Mn and V making such participation in luminescence possible. Thus we suppose that the lines peaking at 689.8, 691.5, 692.6 and 692.9 nm may be ascribed to several Mn^{4+} emission centers with different compensation mechanisms needed for Mn^{4+}-Al^{3+} substitution.

Grossular

Figure 5.51 presents time resolved luminescence and excitation spectra at 300 and 100 K of grossular from Franklin. One of the luminescence features are several narrow lines peaking at 624.0, 631.4, 634.8, 651.0, 658.4 and 664 nm, superimposed on relatively broad band peaking at 660 nm (Fig. 5.51a). All those lines and band have similar excitation spectra, consisting of a relatively narrow band peaking at 450 nm (Fig. 5.51b). The narrow line luminescence features have long decay times in the range of 1.5 ms. At lower temperature of 100 K, the broad band disappears, narrow lines become narrower, their relative intensities differ from 300 K spectrum, several weaker narrow lines become visible (Fig. 5.51c) and decay time becomes longer, approximately of 2.7 ms. More detailed information may be seen on spectrum with higher spectral resolution (Fig. 5.51d).

The luminescence spectrum of grossular might be interpreted to be connected to Sm^{3+} emission, which in many minerals is characterized by narrow lines in a similar spectral range with long decay times (Tarashchan 1978; Gaft et al. 2005). Nevertheless, its excitation spectrum is not typical for Sm^{3+} where several narrow lines present connected to f-f transitions (Shionoya and Yen 1999). We prefer another interpretation. Luminescence by Mn^{4+} is well known in artificial luminescent materials, including GGG garnets (Suchoki et al. 2004) where linear emission is similar to that found in natural grossular, taking into account the fact that different crystal fields lead to spectral shifts of this luminescence with strong crystal field influence.

Topaz

Time-resolved luminescence spectra of red topaz under $\lambda_{ex} = 532$ nm excitation are different from those under 355 nm excitation (Gaft et al. 2003a). The relative intensities of the lines at 697 and 711 nm are higher as in the spectra with narrower gates, and in the spectra with longer delays (Fig. 4.114d–f). The evident reason is that two different transitions with strongly different decay times take place both at 697 and 711 nm. At 20 K both those lines remain and the presence of at least two decay components for the line at 711 nm is evident (Fig. 5.52). The intensity of those "long lived" components is not correlated with Cr concentration. They are especially strong in the yellow sample with only 60 ppm of Cr, and even in transparent precious topaz, which was not analyzed on Cr, but evidently its content is low. In both samples the short-lived lines connected with Cr-pairs are absent, which may be seen on the spectrum with narrow gate.

Such luminescence may not be connected with Cr^{3+} pairs because of their longer decay time compared with single Cr^{3+}. Principally it may be connected with the following: energy migration from long-lived center on single Cr^{3+} centers; single Cr^{3+} in strong field position, luminescence center different from Cr^{3+}. To clarify this, we studied the influence of thermal treatment of topaz on its luminescence properties. After heating the yellow topaz with 180 ppm of Cr at relatively low temperature of 200 °C, the yellow color disappears and the sample becomes transparent. The intensity of the line at 711 nm with long decay strongly diminishes, which is clearly seen in the spectrum with long delay, while the line at 711 nm with short decay remains (Fig. 5.53a–d). Such behavior evidently excludes the possibility of single Cr^{3+} luminescence, because the trivalent state of Cr is the mostly stable in topaz structure and oxidation at such low temperature is improbable.

From the other side, such result enables us to consider the possible participation of other d^3 elements, such as Mn^{4+}, while such valence states are not stable and can appear and disappear as result of irradiation and thermal treatment. Unfortunately, the sensitivity for Mn of ICP-MS method is relatively low and it is confidently detected only in the red topaz. However, Mn was detected in several topaz samples,

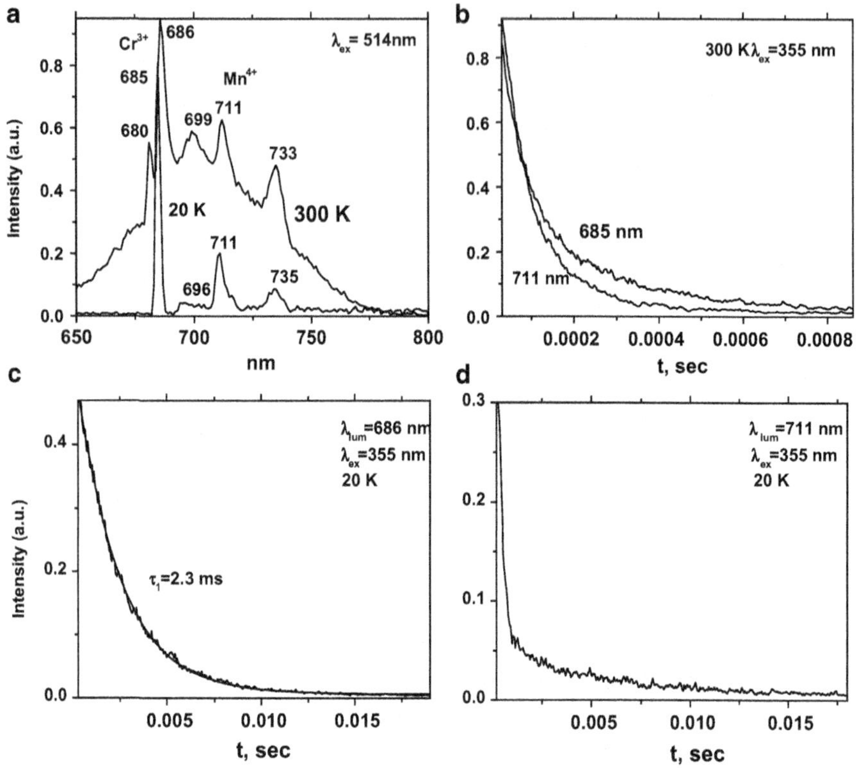

Fig. 5.52 (**a–d**) Laser-induced time-resolved luminescence spectra of topaz at different temperatures demonstrating different decays of Cr^{3+} and possibly Mn^{4+} centers

while it is not correlated with Mn^{2+}, found by EPR and another valence state was supposed (Schott et al. 2003). Ionic radius of Mn^{4+} of 0.67 Å is practically equal to ionic radius of Al^{3+} (0.675 Å). Thus from crystallochemical positions the presence of Mn^{4+} in topaz structure is quite possible. Strong absorption bands due to Mn^{4+} corresponding to the spin-allowed transitions to ^4T levels in the visible to near-UV region generate the yellow color of the phosphors, which corresponds to the yellow color of topaz. Thus we propose that connection may be considered of long-lived lines at 697 and 711 nm with Mn^{4+} centers, which may be formed as result of natural irradiation of topaz.

Eosphorite

Detected narrow red lines have been connected with Cr^{3+} (Fig. 4.140), but the sample contains 19.45 wt.% of Mn and less than 5 ppm of Cr and it may be supposed that Mn^{4+} centers also participate in emission.

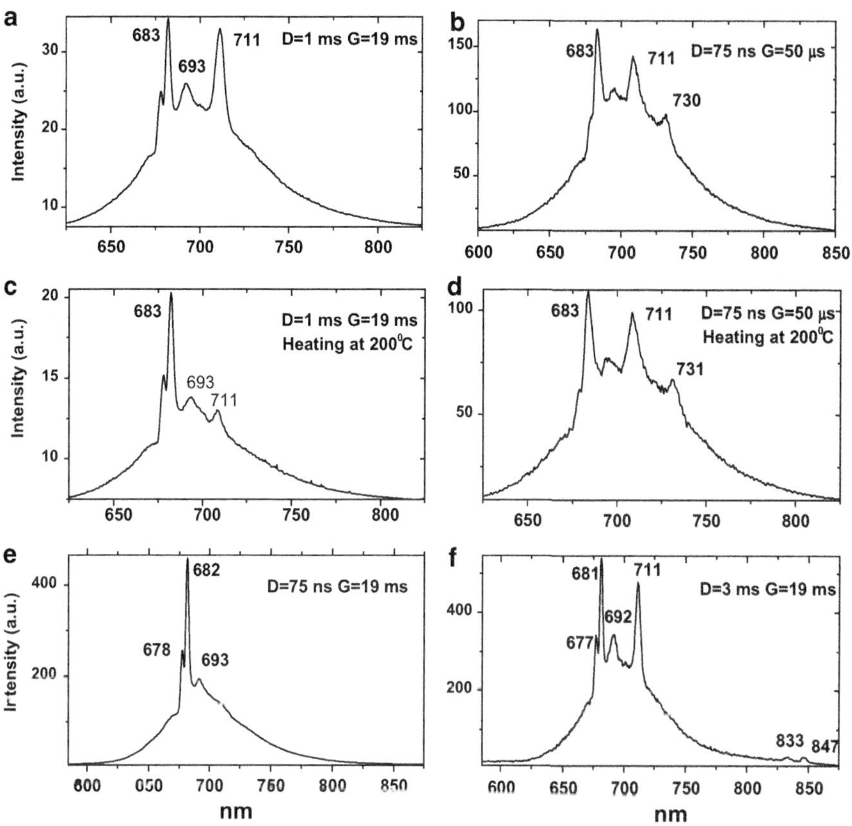

Fig. 5.53 (**a–f**) Laser-induced time-resolved luminescence spectra of topaz *before* (**a–c**) and *after* heating at 200 °C (**d–f**)

Beryl

Radio-luminescence of transition metals doped natural beryl has been studied (Chithambo et al. 1995). It was found that Mn containing samples gave intense red radio-luminescence with sharp emission lines, while the Mn activated beryl (morganite) emission is more than twice as bright as that from emerald. Such luminescence has been ascribed to Mn^{3+}, but it may be supposed that such emission is connected with Mn^{4+} luminescence. Laser-induced time-resolved luminescence spectra of natural morganite revealed band peaking at 730 nm, which may be preliminary ascribed to Mn^{4+} center (Fig. 4.122).

Corundum

Mn^{4+} fluorescence at 672 and 676 nm in artificial Al_2O_3-Mn^{4+} (Fig. 5.54) is known (Fergusson 1970; Geschwind et al. 1962), but was not detected in natural corundum.

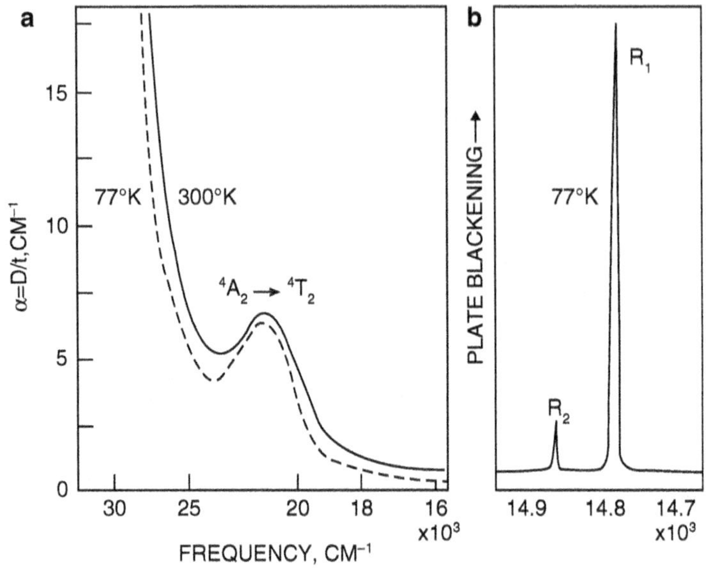

Fig. 5.54 (**a–b**) Luminescence of $Al_2O_3:Mn^{4+}$ (Fergusson 1970)

Benitoite

Under 355 and 532 nm excitations at 300 K the certain benitoite samples exhibit red band peaking at 720 nm with half-width of ~125 nm accompanied by narrow line at 680 nm (Fig. 4.82c). Time resolved spectroscopy with different decays and gates revealed that both band and line behave in a similar way, namely they have similar decay time of ~170 μs. It may be supposed that band and line belong to the same luminescence center. The broad band disappears at low temperatures. The line is much stronger at lower temperatures, where two additional weak lines appear at 695 and 702 nm. Decay time at low temperature is very long reaching $\tau = 1.1$ ms (Gaft et al. 2004).

Combination of broad emission band and narrow line is typical for elements with d^3 electronic configuration, such as Cr^{3+}, Mn^{4+} and V^{2+}. Manganese participation is supported by chemical analyses of benitoite, where chromium was never mentioned as micro-impurity, while Mn is known with concentrations changing from 0.03 % to 0.11 % (Laurs et al. 1997). Such concentrations are quite enough for luminescence generation. Substitution in Mn^{2+} form substituting for Ba^{2+} is doubtful because of the great difference in ionic radii, thus Mn^{4+} form is possible.

5.7.1.3 V^{2+}

Divalent vanadium belongs to $3d^3$ ions and is isoelectronic with Cr^{3+}. Hence all energy levels and spectral characteristics are similar. The difference between them

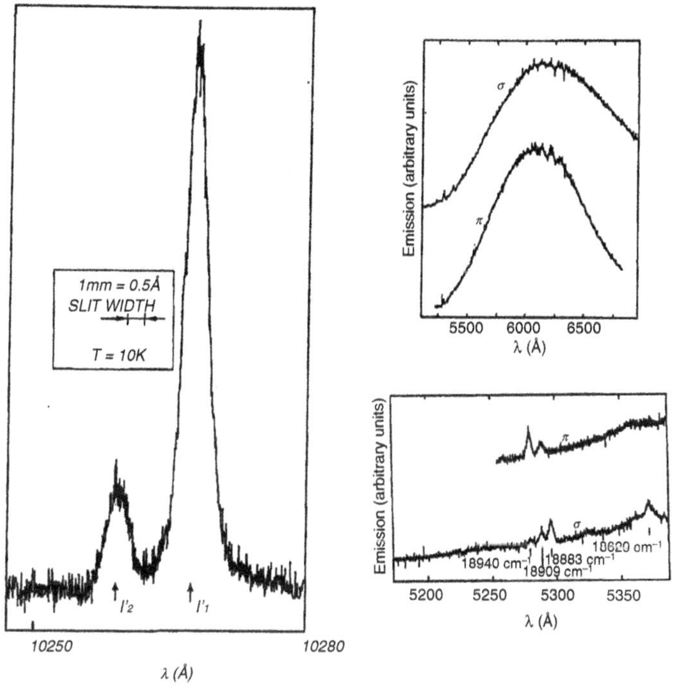

Fig. 5.55 Luminescence of Al_2O_3:V (Champagnon and Duval 1977)

lies in the lower crystal field strength for V^{2+} in comparison with Cr^{3+}. Luminescence of V^{2+} is known in Al_2O_3-V (Fig. 5.55) where it is characterized by linear luminescence from the ^2E levels as in Cr^{3+}, but shifted in the IR (855 and 856 nm) (Champagnon and Duval 1977).

Sillimanite

Similar to Mn^{4+}, V^{2+} participation in minerals luminescence was several times proposed on hypothetical level (Gaft et al. 2005; Koziarsca et al. 1994). We think that it is better grounded in a recently studied sillimanite luminescence (Gaft et al. 2012). It is well known that V^{2+} emission is always at longer wavelengths than Mn^{4+}. Thus the lines peaking at 703.1, 708.3, 740.8 and 747.9 nm (Fig. 5.50d) may be ascribed to V^{2+} luminescence center. Such an interpretation is further proved by the fact that those LIBS emission lines appear most strongly in sillimanite samples with very high vanadium concentration. They are much weaker in green sillimanite with elevated chromium content, but with lower vanadium (Fig. 5.56).

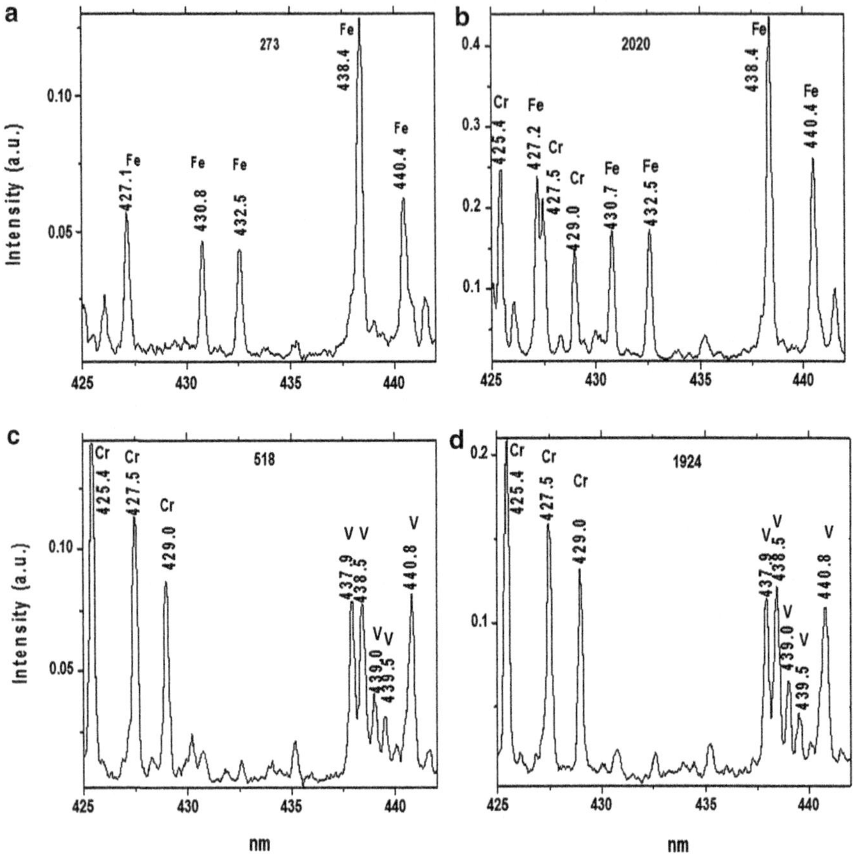

Fig. 5.56 (**a–d**) Breakdown spectra of different sillimanite samples demonstrating emission lines of Cr, Fe and V impurities

Grossular

Another example is grossular luminescence. At 300 K it is characterized by two broad bands peaking at approximately 690 and 716 nm accompanied by narrow lines at 698.7 and 702.0 nm (Fig. 5.57a). All spectral lines and bands have similar decay time of approximately 1 ms. At 100 K the long wavelength narrow line dominates the spectrum while its maximum shifts to 699.6 nm (Fig. 5.57b). It has very long decay time of 8.2 ms. Two broad bands are relatively much less intensive and exhibit many weak narrow line. Their decay time is also very long of 7.8 ms. Excitation spectrum has a strong maxima at 430 nm (Fig. 5.57c). It has to be noted that the excitation spectrum of V^{2+} and absorption spectrum of V in tsavorite (Fig. 5.57d) are rather similar. It may be supposed that the broad asymmetric band peaking at 715 nm may be ascribed to V^{2+}. Supporting evidence is similar spectral type of luminescence and its excitation in GGG activated by V^{2+} with Zr^{4+}

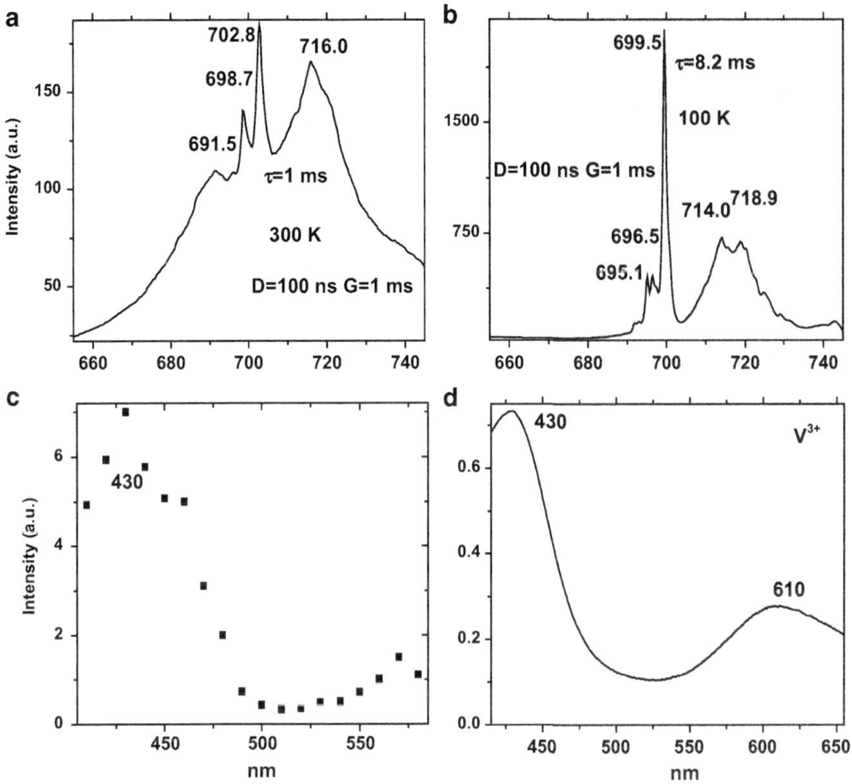

Fig. 5.57 (**a–d**) Time-resolved luminescence at 300 K (**a**), 100 K (**b**), luminescence excitation spectrum (**c**) and absorption spectrum (**d**) of green grossular (2.77)

co-activator (Fig. 5.58). An additional reason is that this emission dominates the tsavorite luminescence where V content is relatively high, which is confirmed by LIBS data (Fig. 5.59). Chemical connection with vanadium impurity is also confirmed by the fact that such kind of grossular luminescence dominates in samples with typical for vanadium absorption spectrum and elevated vanadium content (GIA Gem Database 2010).

Forsterite

Figure 5.60 presents time resolved luminescence spectra at 300 K of the green sample (Bulsam, North Carolina) under excitation by 532 nm with relatively narrow gate of 50 μs. The luminescence band has a maximum at approximately 700 nm accompanied by narrow lines at 680.4 and 685.3 nm (Fig. 5.60a). The decay time for the all spectral features is the same and equal to $\tau = 9.5$ μs. At lower temperature the broad band from the shorter wavelength range is quenched and the

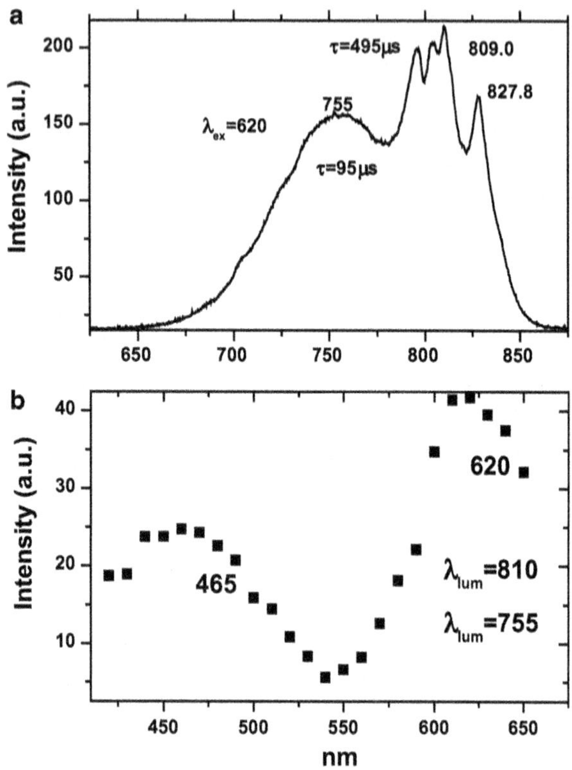

Fig. 5.58 (**a–b**) Laser-induced time-resolved luminescence (**a**) and excitation (**b**) spectra of V^{2+} in artificially activated by vanadium grossular

additional narrow line peaking at 689.9 nm appears (Fig. 5.60b). As at 300 K, all spectral features have the same decay time which is very similar to those at the room temperature. The spectrum remains the same under different excitations with excitation maxima peaking at 420 and 540 nm (Fig. 5.60c).

Such kind of emission, where the bands remain not quenched at low temperatures, was not detected in chromium activated forsterite, which was very carefully studied. Consequently its connection with other impurity may be supposed. Divalent magnesium in octahedron of the forsterite structure has ionic radius of 0.72 Å, which is suitable for substitution by V^{2+}, especially taking into account that charge compensation is not needed in this case. Similar emission was found also in magnesium bearing sinhalite $MgAlBO_4$ (Fig. 4.177).

Zoisite

The narrow lines at 692 and 710 nm in luminescence spectrum of zoisite have been connected with d^3 element emission, while Cr^{3+} and V^{2+} were considered as the possible candidates (Koziarsca et al. 1994). Laser-induced time-resolved

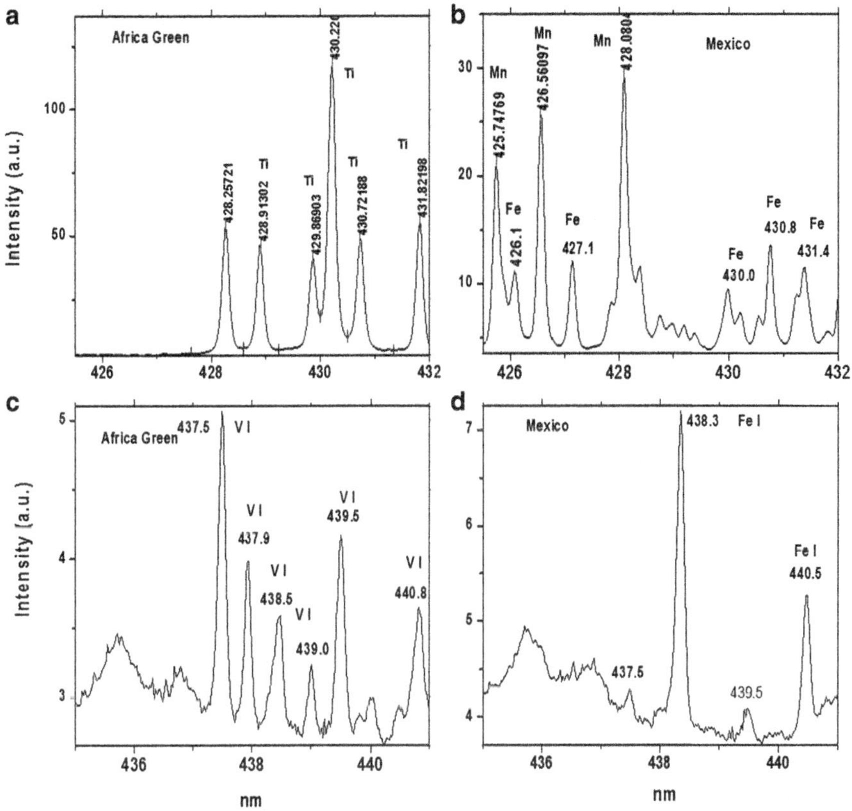

Fig. 5.59 (**a–d**) Breakdown spectra of different grossular samples demonstrating emission lines of Cr, Fe and V impurities

luminescence spectra of zoisite reveal the same lines (Fig. 4.138). We are inclined to connect these lines with V^{2+} for the reason that vanadium concentration in this sample is much higher than the chromium one.

5.7.2 3d² Elements

The elements with such electronic configuration known as luminescent centers are Cr^{4+}, Mn^{5+}, V^{3+} and Ti^{2+} (Fig. 5.61). Within this model, the ground state is a spin triplet $^3A_2$ (e^2 electron configuration) and the first excited state is triplet $^3T_2$ (t^12e^1 electron configuration). Above this state it is possible to see singlet ^1E ($t^2{}_2$ electron configuration) roughly independent of the crystal field strength and also another triplet state labeled $^3T_1$. It is the analogy with the evolution of the two first $^4T_2$ and ^2E energy levels of Cr^{3+} ion with a crossing point (Scott et al. 1997).

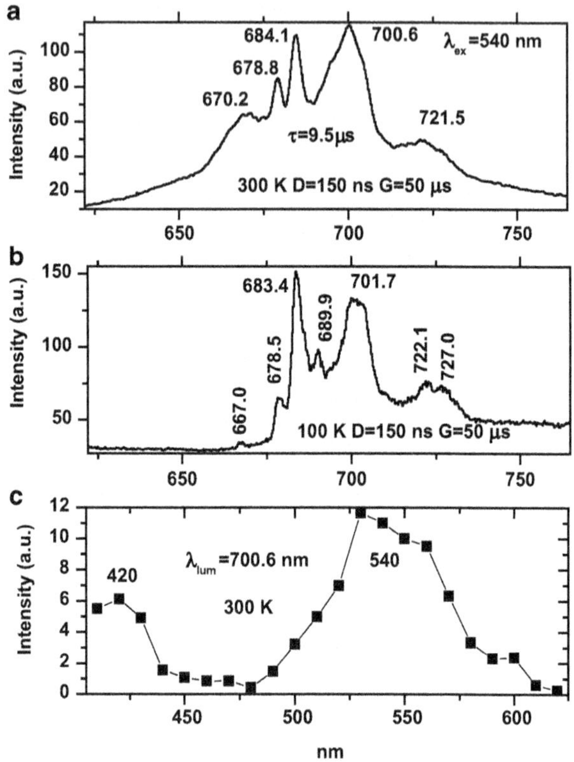

Fig. 5.60 (a–c) Laser-induced time resolved luminescence at 300 K (a), 100 K (b) and luminescence excitation (c) of green forsterite

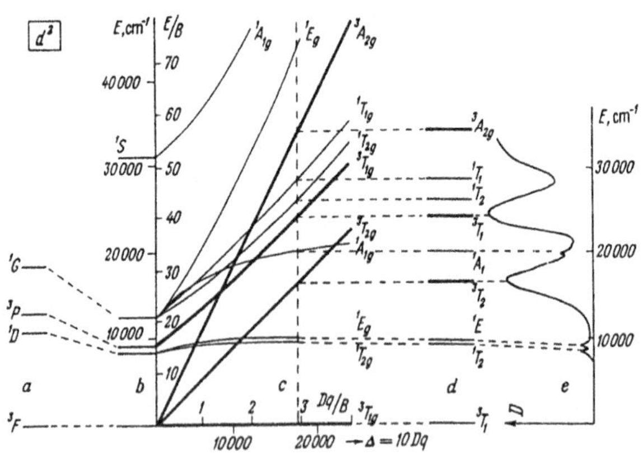

Fig. 5.61 Tanabe-Sugano diagram for d^2 ions

It has been empirically known that the energies of the lowest excited state of tetrahedrally coordinated $3d^2$ metals decrease in the order $Cr^{4+} < Mn^{5+} < Fe^{6+}$. As in the case of $3d^3$ elements, this tendency has been considered to originate from the difference in covalency, which reduced two-electron repulsion between the electrons occupying 3d orbitals. Recently this question was treated using first-principles

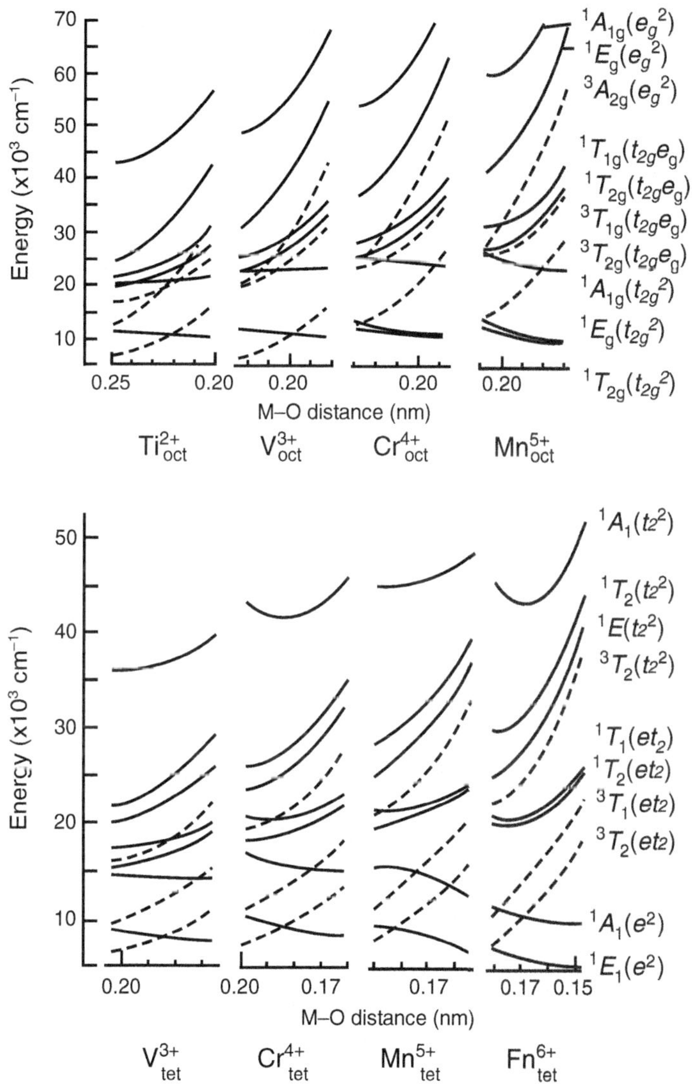

Fig. 5.62 Distance-dependent multiplet-energy diagram for Mn^{5+}, Cr^{4+}, Ti^{2+} and V^{3+} (Ishii et al. 2002)

electronic-structure calculation (Ishii et al. 2002). The same tendencies were found as for the $3d^3$ ions. Distance dependent multiplet-energy diagrams for these elements have been obtained (Fig. 5.62), which enable to envisage the typical shapes of the possible emissions. As in the Tanabe-Sugano diagram, in the tetrahedrally coordinated $3d^2$ system, the positions of the ^1E state, corresponding to narrow line emission, and the $^3T_2$ state, corresponding to broad band emission, are responsible for the shape of luminescence spectrum.

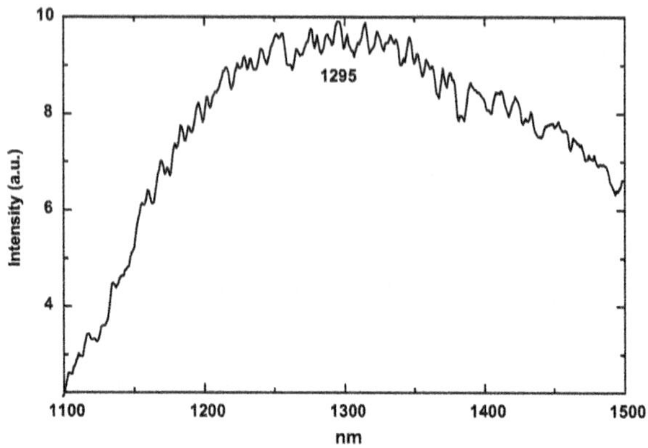

Fig. 5.63 Laser-induced luminescence of artificial apatite activated by Cr^{4+}

5.7.2.1 Cr^{4+}

The tetravalent chromium has attracted great interest in recent literature due to tenability over a large spectral range in both CW and pulsed operation of the forsterite ($Mg_2Si_2O_4$) laser. According to energy levels scheme (Fig. 5.61), ^1E and $^3T_2$ states in tetrahedral coordination of Cr^{4+} are located close to each other. This means that the shapes of the emission are sensitive to the M-O distances. In most cases of $Cr^{4+}{}_{tet}$, the lowest excited states have been considered as $^3T_2$, splitting under low symmetry, but some crystal fields led to ^1E state as the lowest (Boulon 1997). Many of the solid- state laser material with Cr^{4+} possess both tetrahedrally and octahedrally coordinated sites. Although the site for the dominant laser emission is the tetrahedrally coordinated one, the contribution of the octahedrally coordinated site is not negligible. The ^2E_g and $^4T_{2g}$ states of $Cr^{3+}{}_{oct}$ will be typically located in the energy region overlapping with the lowest states of $Cr^{4+}{}_{tet}$ (Fig. 5.62). Although the corresponding transitions are spin-forbidden ones with weak intensities, some confusing structures could be created in the near IR region.

Luminescence of Cr^{4+} was not detected in minerals yet, but the first candidates are evidently forsterite and apatite (Fig. 5.63).

5.7.2.2 Mn^{5+}

The experimental data recorded for Mn^{5+} are consistent with Tanabe-Sugano diagram by assuming the higher crystal field compared to Cr^{4+}, which leads to the higher energy of the emission with the occurrence of the singlet ^1E metastable state as a final state before emission and then longer lifetime from this level (Boulon 1987).

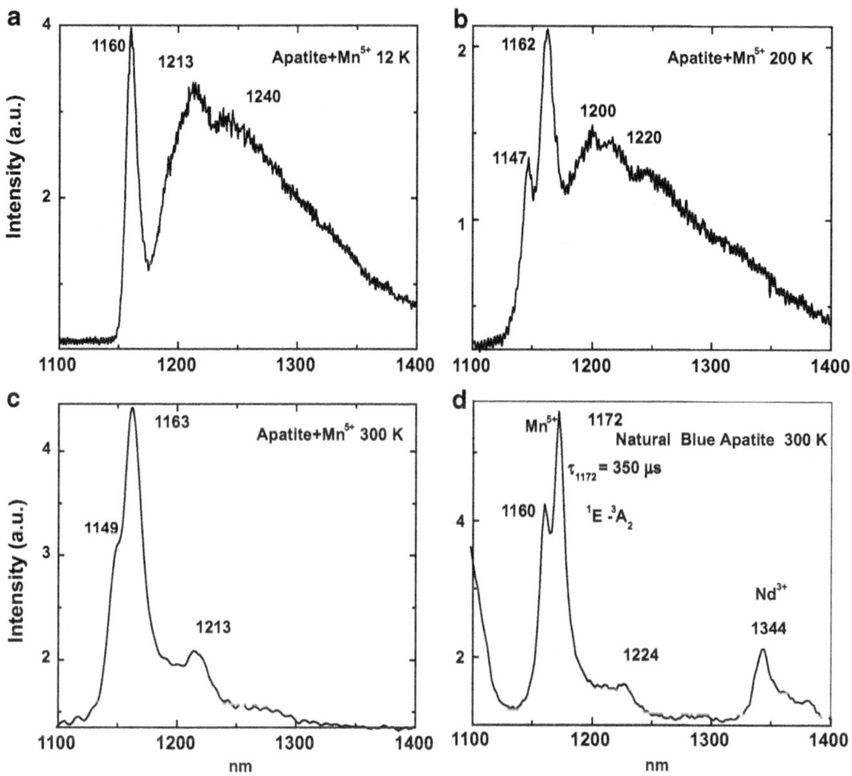

Fig. 5.64 (**a–d**) Laser-induced luminescence of Mn^{5+} in artificially activated (**a–c**) and natural blue (**d**) apatite

Magmatic Apatite

In the IR spectra of blue and green-blue magmatic apatites the strong narrow band was detected (Fig. 4.4b). It was associated with tetrahedrally coordinated Mn^{5+} on the site of P^{5+} (Gaft et al. 1997a). Such interpretation was firstly proposed by Kingsley et al. (1965) and confirmed by the study of artificial apatite activated by Mn^{5+} (Moncorge et al. 1994). According to Tanabe-Sugano diagram the ground state is spin triplet $^3A_2$ and the sequence of the energy levels of Mn^{5+} in natural apatite correspond to a strong field situation for which the first excited state is a singlet ^1E partially mixed with the higher $^3T_2$ state. The emitting state ^1E gives two sharp lines at 1149 and 1160 nm as result of splitting into two components. The situation in synthetic apatites corresponds to a lower field in which the first excited state is triplet $^3T_2$ giving rise to a broader emission bands (Fig. 5.64). Even with the existence of strong mixing between ^1E and $^3T_2$ excited states, the different origin of the narrow and broad bands is confirmed by their different decay times of 350 and 200 μs, correspondingly.

Sedimentary BioApatite

Certain sedimentary apatites have a blue color after oxidizing heating at 700–800 °C. It appears that before heating they are characterized by luminescence of Mn^{2+} while after heating luminescence of Mn^{5+} appears. For example, mastodon ivory, also called odontolite or bone turquoise is one of the most fascinating imitations of turquoise mineral used in the Middles Ages as a semi-precious stone. The understanding of the mysterious turquoise-blue colour formation induced by heat treatment in this material is the subject of much confusion. Former studies demonstrated that its chemical and structural composition corresponds to fluorapatite. Laser-induced time-resolved luminescence spectroscopy was used to determine the accommodation of the potential elements responsible for the colour (Reiche et al. 2001). It was proved that the colour of odontolite can be ascribed to Mn^{5+} traces in the fluorapatite host matrix. These manganese traces are situated on the tetrahedral phosphorus sites. These usually unstable cations are stabilized by the apatite host matrix. The characteristic luminescence revealed that there is a thermal diffusion from Mn^{2+} in the calcium sites to the phosphorus sites accompanied by an oxidation into Mn^{5+}, which produces an intense blue colour by an oxygen- Mn^{5+} charge transfer.

Besides apatite, luminescence of Mn^{5+} has been found in other phosphate related minerals, such as triplite (Ajo et al. 1997). It was proposed that the lattice distortion is smaller for the substituting Mn^{5+} ion than for the structural P^{5+}, which explains the occurrence of Mn^{5+} in several mineral phosphates, while the oxidizing state V is unusual for Mn in aqueous solutions.

5.7.2.3 V^{3+}

The energy levels diagram of V^{3+} is similar to this of Cr^{3+}, though these ions are not isoelectronic. Having the different number of electrons and hence different multiplicity, these ions have the same orbital ground state (^3F and ^4F) and one more state with the same multiplicity (^3P and ^4P) that leads to the three intense spin-allowed absorption bands with some weak narrow lines related to spin-forbidden transitions.

Luminescence of V^{3+} in minerals was proposed in tsavorite where the broad-band IR luminescence with short decay time less than 1 μs was associated with this center (Mazurak and Czaja 1996). The luminescent level has been identified as a component of the $^3B_2$ ($^3T_{2g}$) energy level. At least two different V^{3+} sites are detected.

Luminescence of V^{3+} has been found in α-Al_2O_3 under argon laser excitation at 514.5 nm, which consists of two narrow IR lines with the strongest one at 1026 nm, characterized by decay time of 1.6 μs at 77 K (Champagnon and Duval 1977). Connection of V^{3+} with 833 and 847 nm lines with long decay time in luminescence spectrum of topaz (Fig. 5.53e, f) is also possible (Gaft et al. 2003a).

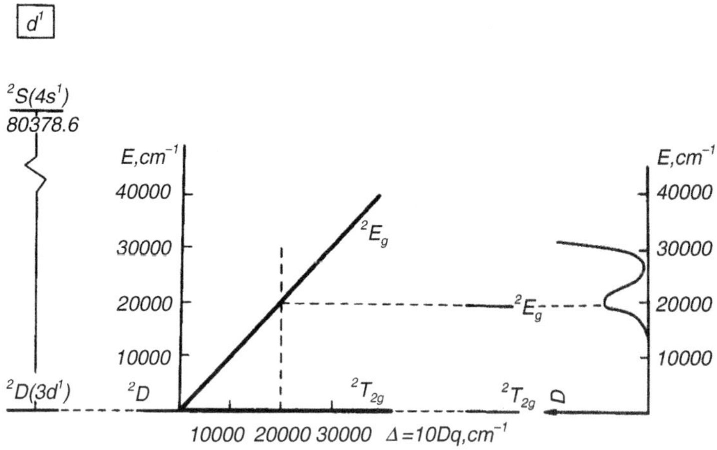

Fig. 5.65 Tanabe-Sugano diagram for d^1 ions

5.7.2.4 Ti^{2+}

Ti^{2+} can be stabilized in halide lattices with broad-band luminescence, characterized by considerable thermal quenching. Thus the potential minimum of the $^3T_2$ excited state lies below the $^1E(T_2)$ minimum and is thus the emitting state. If Ti^{2+} is forced into a very tight site, such as in $MgCl_2$, the crystal field gets larger and $^1E(T_2)$ becomes the emitting level with sharp lines IR emission. Upon warming the $^3T_2$ state is gradually populated and the sharp-line spectrum changes to a broad-band spectrum and the luminescence lifetime drops by more than two orders of magnitude (Güdel et al. 1998).

5.7.3 *3d¹ Elements*

The elements with such electronic configuration known as luminescent centers are Cr^{5+}, Mn^{6+}, V^{4+} and Ti^{3+} (Fig. 5.65). The ^2E ground state has a single unpaired electron occupying the π antibonding set of e-orbitals. In the first excited state this electron is promoted to the π and σ antibonding set of t_2-orbitals, giving rise to the $^2T_2$ ligand-field excited state. The ^2E-$^2T_2$ transition is usually appears as a weak broad absorption band in the near-infrared region.

5.7.3.1 Cr^{5+}: Zircon

IR luminescence detected in $ZrSiO_4$-Cr has excitation band peaking at 920 nm. Its luminescence spectrum at 300 K (Fig. 5.66) is characterized by relatively

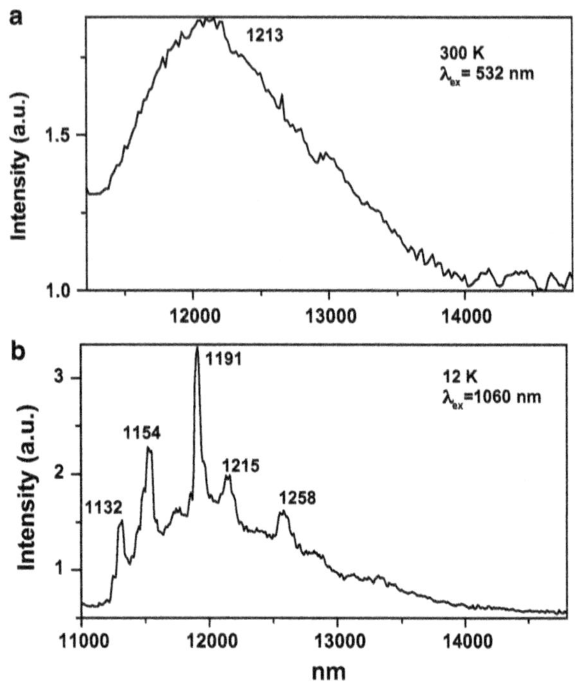

Fig. 5.66 (a–b) Laser-induced steady-state luminescence of $ZrSiO_4$: Cr^{5+}

unresolved broad band peaking at 1200 nm. It is very similar to Cr^{4+} luminescence in silicates, especially in forsterite except very short decay time shorter than the time resolution of our detection system about 200 ns. It is not suitable for Cr^{4+} with much longer decay in μs range (Boulon 1997). Luminescence at lower temperatures is much more intensive and spectra are characterized by several strong narrow lines with very short decay which appear already at 100 K. Once again, it is rather unusual for Cr^{4+}.

Luminescence spectra at low temperature, together with very short decay times are rather characteristic for Cr^{5+} luminescence, which was not known until recently, and which is especially similar to luminescence of $CrO_4{}^{3-}$ doped YPO_4 with zircon structure (Brunold et al. 1997). At 15 K it also consists of several narrow lines with very short decay time less than 20 ns. Thus we proposed to assign the luminescence spectrum to ^2E-$^2T_2$ allowed electric dipole transition or, more precisely, to $^2B_1$-$^2A_1$ components of D_{2d} symmetry in distorted tetrahedral $SiO_4{}^{4-}$ groups (Gaft et al. 2000b). The origin of the spectrum in $ZrSiO_4$: Cr is situated at 8881.00 cm^{-1} and built upon it progressions in two modes $\nu_1 = 170$–$180\ cm^{-1}$ and $\nu_2 = 440$–$445\ cm^{-1}$ are observed. The last one is the same as in Cr^{5+} doped YPO_4 where it has been connected with O-Cr-O bending mode (Hazenkamp and Gudel 1996). Nevertheless, IR and Raman spectra of pure zircon contain this band and thus it may be a lattice mode also. From the other side, mode at 170–180 cm^{-1} is absent in vibrational spectra of pure zircon and may be connected with vibrational frequencies of the $CrO_4{}^{3-}$ unit, which are between 270 and 830 cm^{-1}.

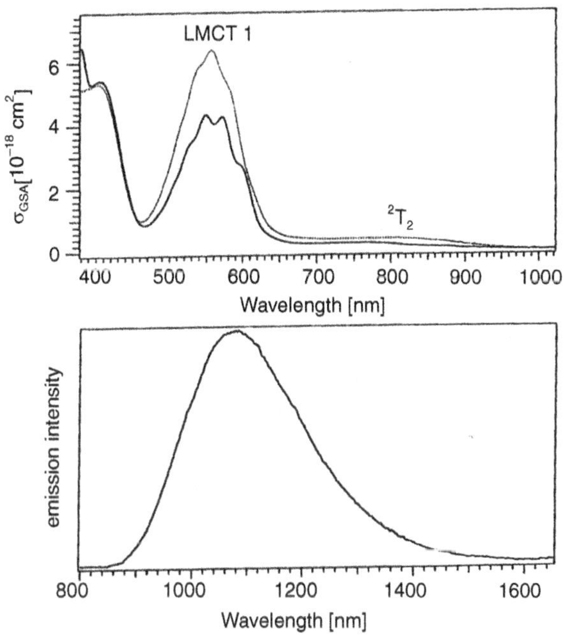

Fig. 5.67 Luminescence (**b**) and excitation (**a**) spectra of $BaSO_4$-Mn^{6+} (Brunold and Güdel 1997; Brunold et al. 1997)

EPR spectra of $ZrSiO_4$: Cr contain a single line with $g\| = 1.986$, $g_\perp = 1.956$ which is characterized by D_{2d} local symmetry ($K_m = 1$). Cr^{5+} ion has spin S – 1/2 and with any symmetry of the crystal field, splitting in the magnetic field always leads to the appearance of a single line in the EPR spectrum. The possible connection of the line with Cr^{5+} is confirmed by its intensity correlation with Cr concentration in the melt. The ratio of the main values of g-tensor is characteristic for Cr^{5+} ($3d^1$ configuration) in tetrahedral coordination (Krasnobayev et al. 1988).

5.7.3.2 Mn^{6+}: Barite

Luminescence of Mn^{6+} was not detected in minerals yet, but is well known in phosphors. The first possibility will be barite (Fig. 5.67), because $BaSO_4$-Mn^{6+} has IR luminescence and considered as a potential material for a tunable NIR laser (Brunold and Güdel 1997; Brunold et al. 1997).

5.7.3.3 Ti^{3+}: Benitoite and Titanite

Ti^{3+} belongs to d^1 configuration, which is the simplest one. The free ion has five-fold orbital degeneracy (^2D), which is split into two levels (^2E and $^2T_2$) in octahedral symmetry, which is quite common for transition metal ions. The only possible optical transition with excitation is from $^2T_2$ to ^2E. This transition is a forbidden

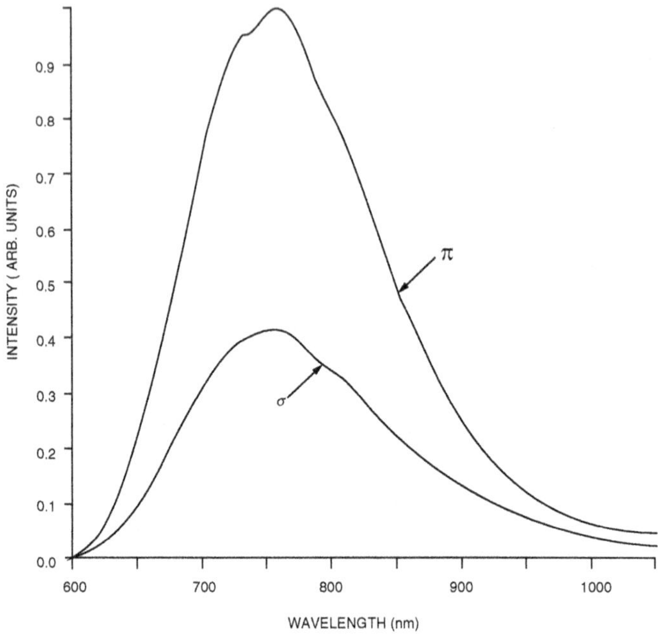

Fig. 5.68 Luminescence and absorbance of Ti^{3+} in artificial sapphire (Rappoport and Khattak 1988)

one, since it occurs between levels of the d-shell. Therefore the parity does not changed. The parity selection rule may be relaxed by coupling of the electronic transition with vibrations of suitable symmetry.

Ti^{3+} is able to create tunable laser in sapphire lattice (Rappoport and Khattak 1988; Fabeni et al. 1991). It substitutes the Al^{3+} ion of the host crystal in a trigonally distorted octahedral lattice site and is characterized by strongly polarized band peaking at approximately 750 nm with decay time of 3.1 μs (Fig. 5.68).

Luminescence of Ti^{3+} was not confidently detected in steady-state luminescence spectra of minerals. In Ti minerals studied by laser-induced time resolved spectroscopy broad read band have been found with decay time of several μs: at 660 nm in benitoite (Fig. 4.82) and 750 nm in titanite (Fig. 4.80f). At room temperature the benitoite band with a maximum at 660 nm has half-width of ~135 nm and may be approximated by one Gaussian (Gaft et al. 2004). One exponent with decay time of 1.1 μs approximates well its decay curve at room temperature in all spectral range of luminescence band. At lower temperatures up to 30 K this red luminescence intensity becomes approximately ten times higher and the spectrum undergoes certain changes, namely its maximum shifts in long wave direction to 668 nm and the band becomes a little narrower with half-width of ~105 nm. Such red emission is not excited by laser sources in the visible part of the spectrum, such as 488, 514 and 532 nm. Excitation spectrum at lower temperatures, when

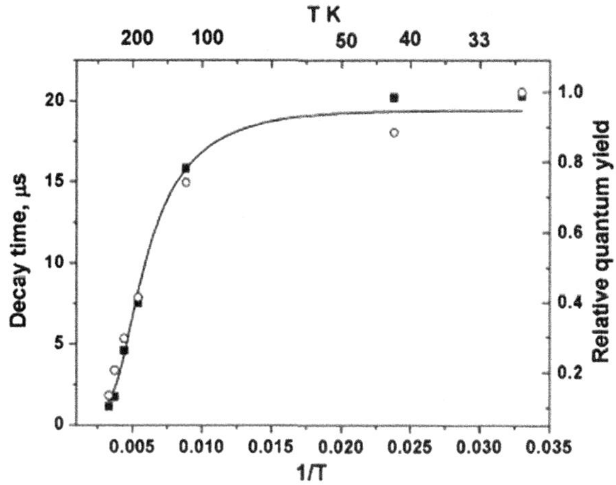

Fig. 5.69 Temperature dependence of quantum efficiency (Q) and decay time (τ) of red emission band of benitoite at 660 nm

luminescence intensity becomes strong enough for UV lamp source, contains one asymmetric band peaking at ~350 nm with half-width of ~75 nm and the corresponding Stokes shift is approximately 13,000 cm^{-1}. It may be approximated by to Gaussian curves peaking at 28,130 cm^{-1} (355 nm) and 30,279 cm^{-1} (330 nm). Decay curve at lower temperatures remains mono-exponential, but decay time becomes substantially longer changing to ~20 μs.

Comparison of this luminescence intensity in different samples reveals that any correlation is absent with any impurity concentration. Thus it was supposed that the mostly probable luminescence center is Ti^{3+}, which presence is quite natural in Ti bearing benitoite. The wide occurrence of Ti^{3+} minor impurities in minerals was detected by EPR. Like the other d^1 ions (V^{4+}, Mo^{5+}), Ti^{3+} ions occur often in minerals as electron center (Marfunin 1979). It may be realized in benitoite, which does have some natural exposure to gamma rays in its natural setting. There could be radiation centers, such as, for example, Ti^{4+} + gamma ray + electron donor ⇒ Ti^{3+} + electron hole. Benitoite color does not change with gamma irradiation to quite high doses (http://minerals.gps.caltech.edu) but luminescence is much more sensitive compared to optical absorption and can occur from centers at such low concentration that they do not impact the color of a benitoite.

Figure 5.69 summarizes the temperature behavior of decay time and quantum efficiency η of red benitoite luminescence at 660 nm in the forms of τ and η as a functions of 1/T. In such case the luminescence may be explained using simple scheme of two levels, namely excited and ground ones. The relative quantum yield at a specific temperature (T) was calculated as:

$$\eta(T) = I_0(T)/\ I_0(30K) \tag{5.37}$$

where I_0 is the maximum of the luminescence intensity at the corresponding temperature. The luminescence may be explained by using a simple scheme with two levels, excited and ground states. The relative quantum yield and decay time of the red emission may be described by simple Arenius equations:

$$\eta = \frac{1}{1 + A_d/A_r \exp(-U/kT)} \tag{5.38}$$

$$\tau = \frac{1}{A_r + A_d \exp(-W/kT)} \tag{5.39}$$

where A_r and A_d are radiation and non-radiation transition probabilities between excited state and ground level and $\boldsymbol{U}$ is energetic barrier between them. The best fit to the experimental results is received when $A_r = 5.4 \times 10^4\ s^{-1}$, $A_d = 2.66 \times 107\ s^{-1}$ and $U = 0.09$ eV. Two competing processes take place after excitation: emission with radiation probability A_r and non-radiative quenching due energetic barrier U with probability $A_d exp(-U/kT)$. At low temperatures the first one dominates and quantum yield is relatively higher with long decay time. At elevated temperatures non-radiative mechanism dominates with the following quenching and decay time shortening.

Such behavior is quite compatible with Ti^{3+} luminescence center. In such case, excitation peaking at 355 nm is connected with $^2T_2$-^2E transition, while the splitting of the ^2E state is a reason of the two shoulders presence in excitation spectrum. At lower temperature the opposite transition generates intensive luminescence band peaking at 660 nm with decay time of 20 μs. Thermal quenching and drastically reducing decay time with increasing temperature result from non-radiative transition from excited to ground state.

The main difference between considered benitoite luminescence and Ti^{3+} emission in well studied Al_2O_3 and $Y_3Al_5O_{12}$ is strong short-waved shift of benitoite emission and especially excitation bands. For example, excitation and absorption maximum of Ti^{3+} in sapphire is at 550 nm, while in benitoite at 355 nm. Nevertheless, it is not so uncommon for Ti^{3+} in different matrixes. The optical absorption spectrum of Ti^{3+} in $YAlO_3$, where Ti^{3+} substitutes for Al^{3+} in sites with orthorhombic symmetry, is composed of two broad bands with peak wavelengths of 434 and 492 nm, the energy separation of which is due to splitting of the excited ^2E state (Yamaga et al. 1992). Such luminescence parameters are closer to those found in benitoite.

5.7.3.4 V^{4+}

Diagrams of V^{4+} levels splitting in crystal fields of different symmetry are identical with those for isoelectronic Ti^{3+} ion. An emission spectrum attributed to V^{4+} was found in αAl_2O_3 in visible region at low temperatures (Champagnon and Duval 1979). It consists of broad band peaking at 625 nm and narrow lines at approximately 528 and 529 nm (Fig. 5.53b, c).

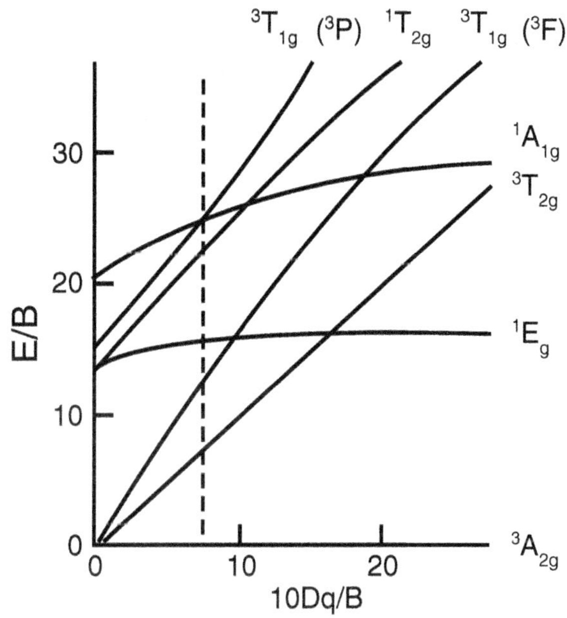

Fig. 5.70 Tanabe-Sugano diagram for the $3d^8$ coordination in an octahedral crystal field

Three slightly different octahedrally coordinated V^{4+} centers have been found in topaz by EPR method (Schott et al. 2003), thus its participation in topaz luminescence is possible.

5.7.4 $3d^8$ Elements: Ni^{2+}

Center Ni^{2+} has electronic configuration $3d^8$ (Fig. 5.70). Three luminescence bands in the near IR, red and green parts of the spectrum characterize Ni^{2+} in many synthetic luminofors (Blasse and Grabmier 1994). For example, broad IR band peaking at 1520 nm and connected with electron transition from the lowest excited state $^3T_{2g}$ has been found in enstatite artificially activated by Ni^{2+} (Moncorgé et al. 1999). Green luminescence is connected with $^1T_{2g}(D)$-$^3A_{2g}$ (F) while the red one with $^1T_{2g}(D)$-$^3T_{2g}$ (F) electron transition. The emissions from $^1T_{2g}$ state are not only observed at cryogenic temperatures but persist up to room temperature as, for example, in MgO (Mironova and Ulmanis 1988) (Fig. 5.71) and $KZnF_3$ (Grimm et al. 2003).

The possibility of Ni^{2+} participation in minerals luminescence has not been seriously considered yet, but we were confident that it has to be done. Ionic radius of Ni^{2+} is 69 pm in tetrahedral coordination and 83 pm in octahedral one. Thus it may substitute many cations with similar dimensions, such as Mg, Zn, Ca. When

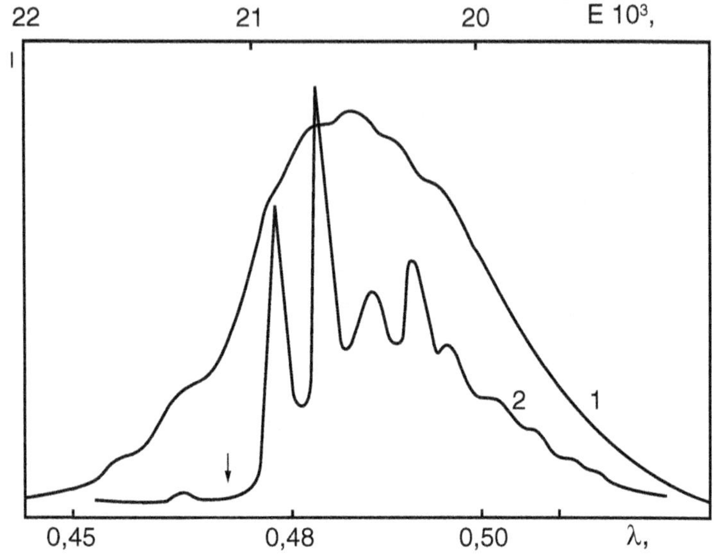

Fig. 5.71 Green luminescence of Ni^{2+} in MgO at room (*1*) and liquid helium (*2*) temperatures (Mironova and Ulmanis 1988)

Ni^{2+} ions are inserted at insulating host, their energy levels are strongly affected by the electrostatic potential (crystal field) generated by neighborhood ions. The d^8 free ion ground energy term 3F is transformed in a lower $^3A_{2g}$ energy level and in two excited levels $^3T_{2g}$ and $^3T_{1g}$. The next excited energy level ^1D split in ^1E_g and $^1T_{2g}$. The third level in increasing order is the 3P, which is transformed into a $^3T_{1g}$ level. For systems with weak crystalline field, the $^3T_{2g}(^3F)$ is the first excited level while for Ni^{2+} ions in stronger crystalline field the first excited is the ^1E (^1D) energy level. Three luminescence bands in the near IR, red and green parts of the spectrum characterize Ni^{2+} luminescence correspondingly connected with electron transition from the lowest excited state $^3T_{2g}$ or ^1E, with $^1T_{2g}(D)$-$^3T_{2g}$ (F) and $^1T_{2g}(D)$-$^3A_{2g}$ (F) electron transition. The emissions from $^1T_{2g}$ state are not only observed at cryogenic temperatures but persist up to room temperature. We studied GGG activated by Ni^{2+} and Zr^{4+} and effectively found red luminescence band which is very similar to those detected in natural grossular (Fig. 5.72). The relatively short decay time of such emission connected to this spin-forbidden radiative transition may be preliminary explained as being caused by the mixing of singlet and triplet states.

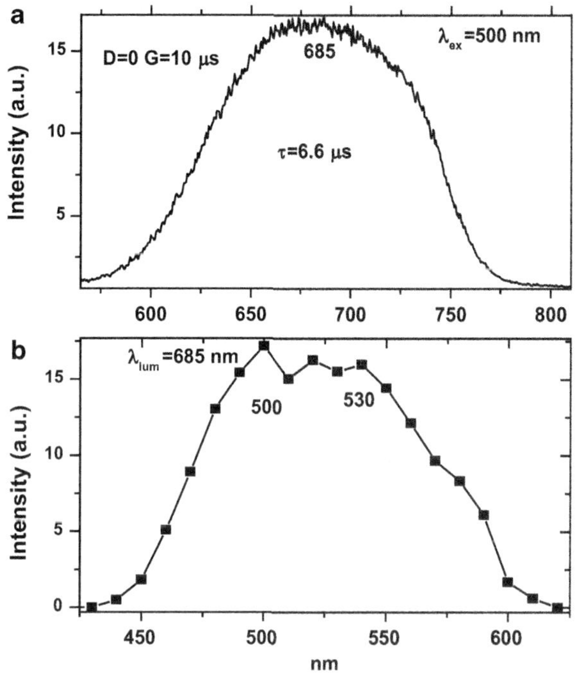

Fig. 5.72 (**a–b**) Luminescence (**a**) and excitation (**b**) spectra of supposedly Ni^{2+} in grossular

5.7.5 $3d^6$ Elements: Fe^{2+}

Ion Fe^{2+} has electronic configuration $3d^6$. It is characterized by extremely strong absorption and is the strongest visible luminescence quencher in minerals, but it has emission band in the IR part of the spectrum.

5.7.6 $3d^5$ Elements

5.7.6.1 Mn^{2+}

The $3d^5$ system manganese(II) with sextet (^6S) ground state is well-known activator in many minerals (Tarashchan 1978; Gorobets and Rogojine 2001). Emission bands due to Mn^{2+} have been observed in practically all classes of minerals. It is responsible for green, yellow and orange-red luminescence. For the case of octahedrally coordinated Mn^{2+}, as in calcite, the position of the minimum of the configurational curves occurs at different Mn^{2+}-O distances for states, which are dependent on the crystal field strength, D_q, such as $^4T_1$ and $^4T_2$. For states whose energies are independent of D_q, the minimum occurs at the same interatomic distance as in the ground state, $^6A_1$ (states ^4E_g and $^4A_1$). The consequence of this

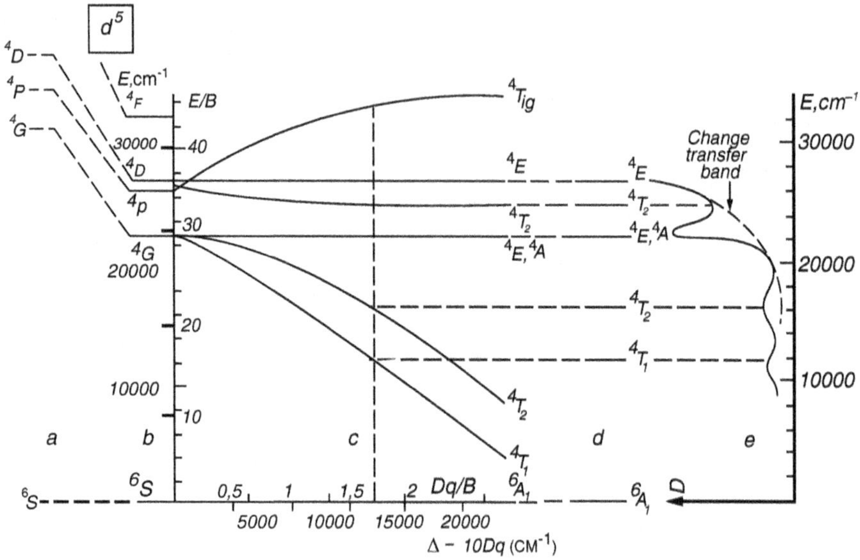

Fig. 5.73 Tanabe-Sugano diagram for d^5 elements

is that there is a progressive shift in the energy of the $^6A_1 \rightarrow {}^4T_1$ and $^6A_1 \rightarrow {}^4T_2$ transitions as D_q increases, producing band broadening and a shift toward longer wavelengths for both emission and absorption bands. It is easy to see (Fig. 5.73) that the curvature of the excited states leads to the width of the bands in the absorption spectrum. States not involved in bonding give rise to sharp absorption lines because they are only weakly coupled to host, while strongly coupled ion states vary considerably in energy with interatomic distance and yield broad bands.

Changes in coordination, such as from octahedral to tetrahedral, produce considerable change in D_q. The tetrahedral D_q value is smaller and the energy of all of the transitions in Mn^{2+} increases. This creates a shift toward short wavelengths. Octahedral Mn^{2+} coordination in calcite thus has red-orange emission at 620 nm (Fig. 4.19a), while the analogous transition for Mn^{2+} in willemite yields yellow-green emission at 525 nm (Fig. 4.85c, d). Prominent changes in the color of the emission may also occur from a change in the emission state. The $^4T_1$ state is responsible for emission in calcite and willemite, but the $^4T_2$ state is responsible for the green luminescence of fluorite (500 nm) and anhydrite (505 nm) even though Mn^{2+} has larger coordination sites.

It was originally stated that $CaCO_3$ activated with manganese cannot be excited by UV radiation, but $CaCO_3$ activated with lead, thallium or cerium and manganese shows an orange-red manganese luminescence under UV irradiation at room temperature (Blasse and Aguilar 1984; Sidike et al. 2002). Nevertheless, in many minerals luminescence of Mn^{2+} has been found with excitation spectra typical for this center without additional bands of Pb or Ce impurities.

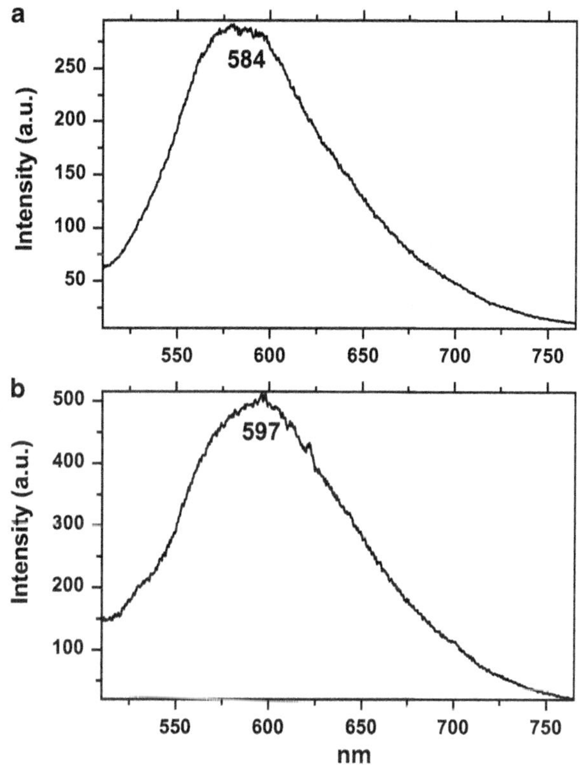

Fig. 5.74 (**a–b**) Laser induced time-resolved luminescence spectra of Mn^{2+} in Ca(I) and Ca (II) positions in artificially activated apatite

Apatite

The typical Mn^{2+} luminescence in natural apatite is connected with band peaking at 569 nm (Fig. 4.2e) with delay time of ~5 ms. Various laser lines were used to excite the luminescence but the spectra were all very similar. All portions of the curve have the same decay time and the spectra were found to be identical for different delay times and gate widths. All of this suggests that we a dealing with a simple single-ion fluorescence. Red apatite has unusual for natural sample luminescence of Eu^{3+} and Tb^{3+} in Ca(II) instead of Ca(I) sites. Luminescence of Mn^{2+} in this sample is also exceptional. Its spectrum is characterized by maximum at 583 nm (Fig. 4.2f) and extremely long decay time of ~12 ms. By analogy with artificially activated apatite (Ryan et al. 1970) it is possible to suppose, that luminescence of Mn^{2+} in Ca(I) position takes place in the red apatite. As in our case, Mn(I) band is 15 nm shifted to the red part of the spectrum compared to the Mn(II) one (Fig. 5.74). The longer decay time is consistent with the higher symmetry of the Ca(I) site. From the other side, according to Tanabe-Sugano diagram, the lower crystal field in Ca(I) position has to lead to the luminescence band shift in the opposite direction. Nevertheless excitation spectra of Mn^{2+}(II) in natural and artificially activated apatite contain only $^6A_1(S)$-$^4T_2(G)$ band, implying that the luminescence is connected with opposite $^4T_2(G)$-$^6A_1(S)$ transition after relaxation. Such

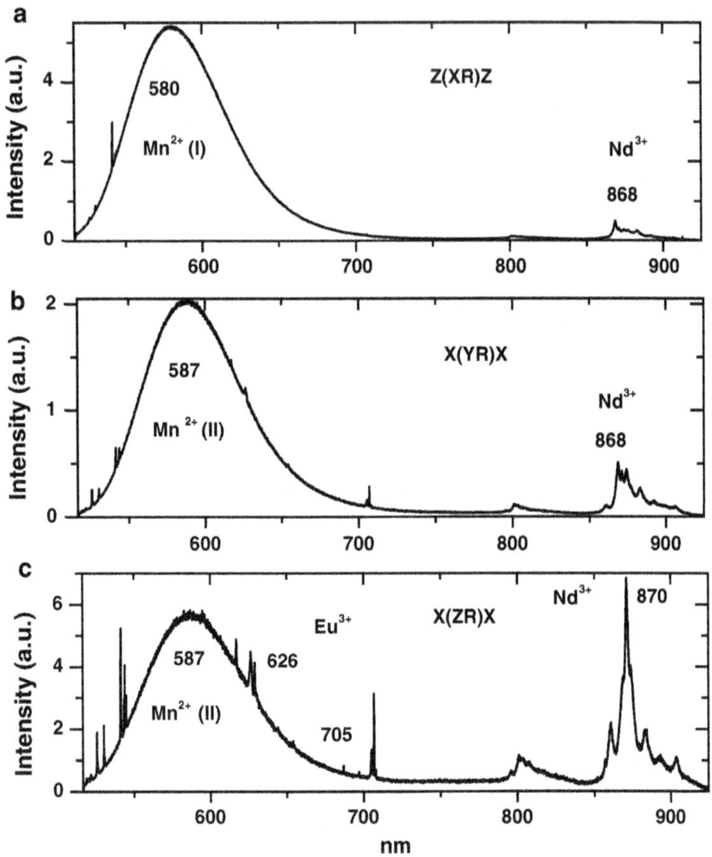

Fig. 5.75 (**a–c**) Polarized laser-induced luminescence of different Mn^{2+} centers in apatite

deactivation is due to electron-phonon interaction, which is very effective in apatite. The luminescence transition from the lower lying level leads to the lower photon energy even in the site with lower crystal field strength.

Situation with Mn^{2+} centers distribution between Ca(I) and Ca(II) positions in apatite lattice is opposite to this for REE^{3+}: in artificially activated apatite the Mn (I) center clearly dominates the fluorescence spectra (Ryan et al. 1970), while in natural one only Mn(II) centers have been detected (Tarashchan 1978). In order to clarify the distribution in different Ca positions, luminescence spectra have been measured with different polarizations from one section or from prismatic and basal sections with the same analytical conditions. As it was found earlier (Barbarand and Pagel 2001) the shapes of the spectra are usually the same for both crystallographic orientations, while the major difference in the spectra is their intensity, with the mean intensity for the basal section lower than for the prismatic face. Nevertheless, in certain cases polarization changes lead to different spectra (Fig. 5.75). In this case spectra are composed mainly of the Nd^{3+} and Mn^{2+} with relatively weak Eu^{3+} lines. The polarization change results in an inversion of the relative intensities of the luminescence bands: Mn^{2+}(I) emission dominates in one orientation and REE emission is practically not seen, but Mn^{2+}(II) in other orientations is much weaker compared to Mn^{2+}(I), while the REE are much stronger. The similar result was described for CL study of apatite, where Ce^{3+} band is significantly high for the zone analyzed in a section perpendicular to the c-axis, but the Mn^{2+} band is significantly higher for the zone analyzed parallel to the c-axis (Barbarand and Pagel 2001).

Feldspars

Luminescence of Mn^{2+} is well known in steady-state luminescence spectra of feldspars (Tarashchan 1978; Waychunas 1989; White 1990; Götze 2000; Gorobets and Rogojine 2001). Its green luminescence is predominantly detected in plagioclases. In K-feldspars the Mn^{2+} emission is less common because of the difficulty of the Mn^{2+}-K^{+} substitution. Its band is also very well detected in laser-induced time-resolved spectra peaking at 550–560 nm (Fig. 4.97a). It is characterized by extremely long decay time of 10–12 ms.

Apophyllite, Hardystonite and Esperite

Broad bands at 525 and 575 nm in time-resolved luminescence spectra of apophyllite under 355 nm excitation (Fig. 4.34c, d) with very long decay time of several ms may be ascribed to strongly forbidden d-d transitions in Mn^{2+} luminescence center. Two bands may be connected with isomorphous substitutions on Ca in Zn structural positions. The spectrum of a famous yellow-green esperite luminescence (Fig. 4.38a) consists of narrow band peaking at 545 nm with very long decay time of 9 ms. Such parameters together with typical excitation spectrum (Fig. 4.38b) enable confident identification of luminescence center as Mn^{2+}. Orange emission near 580 nm of hardystonite is also evidently connected with Mn^{2+} center (Fig. 4.36d).

Grossular

The luminescence band peaking at approximately 585–605 nm (Fig. 5.76) with long decay time is very typical for Mn^{2+} emission with its forbidden $^4T_1(G)$-^6S electronic transition. The mostly intensive excitation band in the range of 400–420 nm is connected to $^6A_1(S)$-$^4E^4A_1(G)$ electronic transition, the band at 450–475 nm to $^6A_1(S)$-$^4T_2(G)$, at 540 nm to $^6A_1(S)$-$^4T_1(G)$ electronic transitions. The optical absorption of Mn^{2+} in garnet (http://minerals.gps.caltech.edu) is clearly similar for luminescence excitation supporting the ascribing of this luminescence to Mn^{2+}. The most probable structural substitution is for Ca^{2+} position. It is worth to note the presence of several similar bands with slightly different maximum positions. The presence of two different luminescence bands peaking at 575 and 590 nm was also detected in thermo-stimulated luminescence emission spectrum of grossular (Yauri et al. 2008).

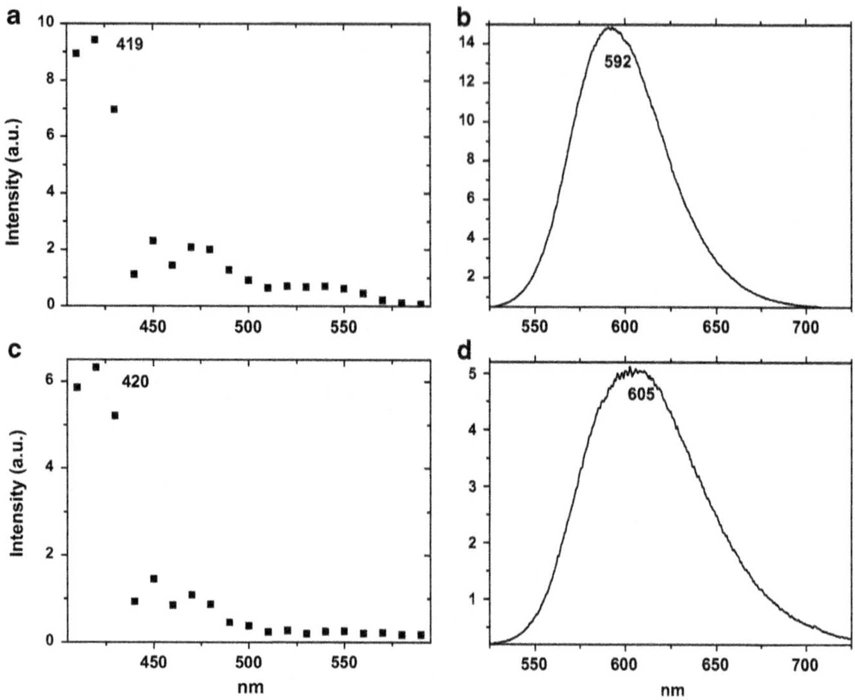

Fig. 5.76 (**a–d**) Laser induced time-resolved excitation (**a**, **c**) and luminescence (**b**, **d**) spectra of different Mn^{2+} in grossular

We do not know the possible candidate for such kind of luminescence besides Mn^{2+} (Gaft et.al. 2013b). The possible presence of several Mn^{2+} luminescence centers may be supposed while only one Ca^{2+} position takes place in grossular structure. The differences in the second coordination sphere are possible.

Spinel

Artificially activated spinels of composition $Mg_{1-x}Al_2O_{4-x}$:Mn^{2+} were studied (Mohler and White 1994). It was found that $MgAl_2O_4$:Mn^{2+} emits at 650 nm, while when $x = 0.05$ emission bands appear at 517 and 744 nm. They increase intensity and the 650 nm band disappears as x increases. The green emission is connected with Mn^{2+} in tetrahedral sites, while the red one evidently with Mn^{2+} in octahedral ones.

In natural spinel emission bands with maxims from 570 to 620 nm have been found interpreted as connected with Mn^{2+} in octahedral sites (Mironova et al. 1993). According such interpretation it is possible to ascribe the band at 612 nm to Mn^{2+} in octahedral positions (Fig. 4.119b).

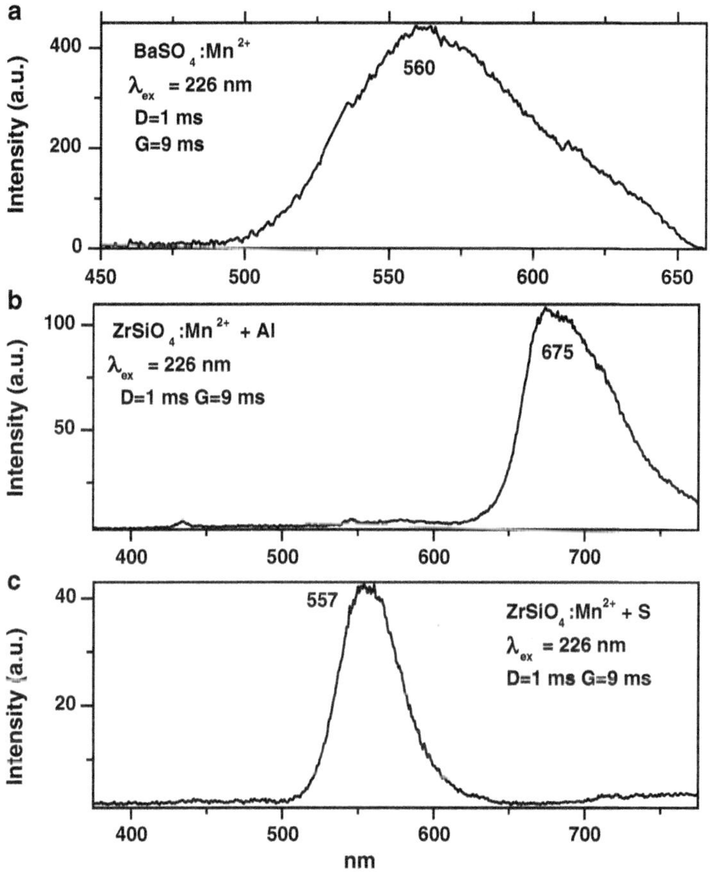

Fig. 5.77 (**a–c**) Laser-induced time-resolved luminescence of $BaSO_4$-Mn^{2+} (**a**) and $ZrSiO_4$-Mn^{2+} (**b**)

Zircon

Certain narrow luminescence bands with long decay time detected in time-resolved luminescence spectra of zircon may be principally connected with Mn^{2+}. luminescence. In order to clarify this, artificially activated $ZrSiO_4$-Mn has been studied (Gaft et al. 2002b). The ionic radius of Mn^{2+} of 0.97 Å is close to those of Zr^{4+} (0.88 Å) and the corresponding substitution is possible with proper charge compensation. The presence of Mn impurities in quantities in zircons under investigation was confirmed by ICP analyses (Table 4.15). Intensive narrow bands with very long decay times were found with spectral position strongly depended from co-activator (Fig. 5.77b, c). Principally, such difference may be connected with several kinds of Mn^{2+} centers with different local symmetry, for example, Mn^{2+}-Si^{4+} (tetrahedron) and Mn^{2+}-Zr^{4+} (octahedron) substitutions, but the small ionic radius of Si^{4+} of 0.4 Å makes such substitution very doubtful. At any case, such luminescence bands have never been detected in the luminescence spectra of natural zircon. It is possible to suppose that zircon contains Mn not in the form of Mn^{2+}, but as Mn^{4+}, which is confirmed by EPR data (Votyakov et al. 1993).

Barite

Manganese impurities are always found in barite samples and Mn^{2+} participation in barite luminescence was considered as possible. In order to check it we studied synthetic $BaSO_4$ artificially activated by Mn^{2+}. Relatively weak green luminescence has been found (Fig. 5.77a) with very long decay time of several ms, but such emission has not been found yet in natural barite.

Margarosanite

Two different types of orange luminescence may be detected in margarosanite (Gaft et al. 2013a). The first one is under excitation by 266 nm, where the emission band at 100 K becomes a little narrower compared to the 300 K, but its spectral position and decay time remain actually without big change (Fig. 5.78a, b). Such behavior is typical for Mn^{2+} luminescence with its forbidden d-d luminescent transition and correspondingly relatively long decay time. For the first glance, orange luminescence under excitation by 355 nm looks the same, but several details are different. The emission band at 100 K moves approximately 20 nm in long wavelength direction compared to the luminescence at 300 K. Decay time becomes nearly twice longer (Fig. 5.78c, d).

The possible explanation is that Mn^{2+} evidently substitutes for Ca^{2+} and calcium has two different positions in margarosanite crystal structure (Freed and Peacor 1969). In such case, luminescence under excitation by 355 nm may be ascribed to Mn^{2+} clusters known in many Mn bearing minerals (Gaft et al. 2005). It is quite natural for margarosanite with elevated Mn concentration of 0.8–1.0 %, which is far bigger than trace level.

Other Minerals

Characteristic bands of Mn^{2+} well studied by steady-state luminescent spectroscopy (Tarashchan 1978; Gorobets and Rogojine 2001) have been found in time-resolved luminescence spectra of calcite (Fig. 4.19a), dolomite (Fig. 4.25a), datolite

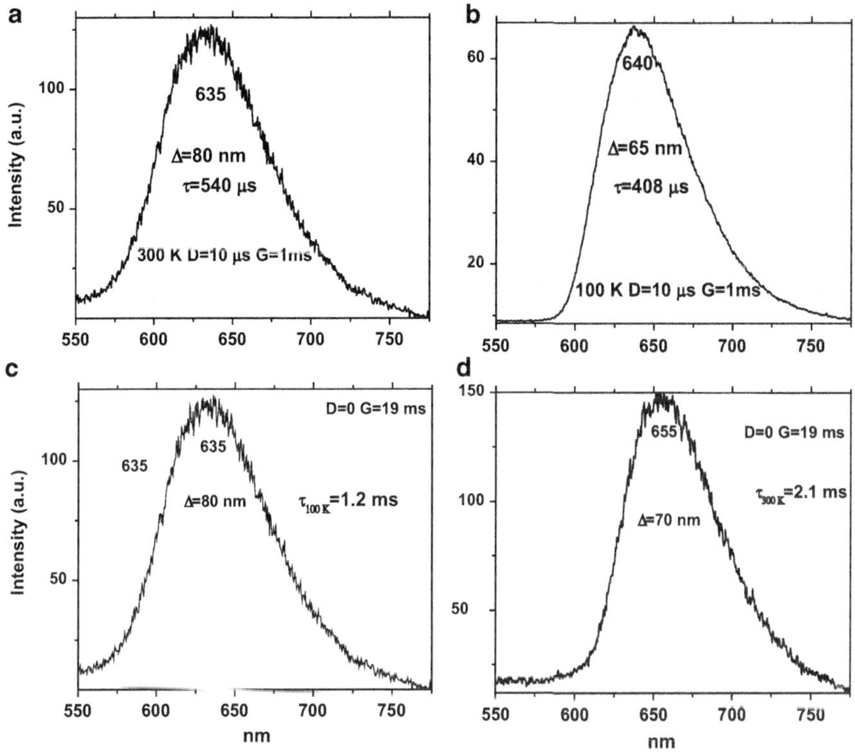

Fig. 5.78 (**a–d**) Laser induced time-resolved luminescence spectra of different Mn^{2+} in margarosanite

(Fig. 4.29d), fluorite (Fig. 4.14d), wollastonite (two bands at 555 and 603 nm; Fig. 4.95a, c), and spodumen (Fig. 4.142c).

Besides, the green structureless luminescence peaking at 510 and 494 nm in tugtupite (Fig. 4.160b, c) evidently belongs to an Mn^{2+} center for the following reasons (Gaft et al. 2009). Tugtupite structure has no octahedral sites, only tetrahedral, thus green Mn^{2+} luminescence is logical. The structural position of Mn^{2+} in tugtupite evidently may not be connected with substitution for Al^{3+}, Si^{4+} and Be^{2+}sites, because Mn^{2+} has significantly larger ionic radii (80 pm) as compared to those cations (53, 40 and 41 pm, correspondingly). Thus, substitution for a Na^{+} site in four-fold coordination is more probable, because the ionic radius of Na^{+} in four-fold coordination (113 pm) may be suitable for Na^{+} substitution and the luminescence green color corresponds to the tetrahedral coordination. Substitution of Mn^{2+} for Na^{+} with a luminescence center formation is known in many minerals (Gorobets and Rogojine 2001). The presence of two Mn^{2+} luminescence centers suggests that substitutions take place in different structural sites with different crystal field strengths. Such a possibility is confirmed by the fact that a very similar luminescence band, but peaking at 540 nm, has been detected in synthetic sodalite activated by Mn (Van Doorn and Schipper 1971).

The photoluminescence spectra, optical excitation spectra and decay curves of anthophyllite from Canada were obtained at 300 and 10 K (Sidike et al. 2010a). In the PL spectra obtained under 410-nm excitation, bright red bands with peaks at 651 and 659 nm were observed at 300 and 10 K, respectively. The origin of the red luminescence was ascribed to Mn^{2+} in anthophyllite from the analysis of the excitation spectra and PL decay times of 6.1–6.6 ms.

Natural calcite from Kuerle, Xinjiang, China, was studied (Sidike et al. 2006). It shows orange-red fluorescence when exposed to short-wave ultraviolet light (Hg 253.7 nm). The photoluminescence emission spectrum under 208 nm excitation consists of three bands: two UV bands at 325 and 355 nm and an orange-red band at 620 nm. The three bands are ascribed to Pb^{2+}, Ce^{3+} and Mn^{2+}, respectively, as activators. The Pb^{2+} excitation band is observed at 243 nm, and the Ce^{3+} excitation band at 295 nm. The Pb^{2+} excitation band is also observed by monitoring the Ce^{3+} fluorescence, and the Pb^{2+} and Ce^{3+} excitation bands, in addition to six Mn^{2+} excitation bands, are also observed by monitoring the Mn^{2+} fluorescence. These indicate that four types of the energy transfer can occur in calcite through the following processes: (1) $Pb^{2+} \rightarrow Ce^{3+}$, (2) $Pb^{2+} \rightarrow Mn^{2+}$, (3) $Ce^{3+} \rightarrow Mn^{2+}$ and (4) $Pb^{2+} \rightarrow Ce^{3+} \rightarrow Mn^{2+}$.

Photoluminescence (PL) properties of halite from Salton Sea, California, USA, are investigated at 10, 80, and 295 K (Sidike et al. 2006). The color of clear part of this halite is transparent under daylight, and the fluorescence under UV light at 253.7 nm is orange-red. The orange-red emission band at 640 nm is assigned to the electronic transitions within the $3d^5$ configuration of Mn^{2+} in halite (NaCl). The excitation spectrum of the orange-red fluorescence consists of inefficient excitation bands due to Mn^{2+} and efficient excitation bands due to Pb^{2+}. The existence of the Pb^{2+} bands in the excitation spectrum of the Mn^{2+} fluorescence suggests that there is an excitation process due to the energy transfer with the resonance type from Pb^{2+} to Mn^{2+}. Thermal quenching of the specimen is undertaken, and it is clarified that the clusters of the Pb^{2+}-vacancy pairs play an important part in the orange-red fluorescence from the halite. Analysis by inductively coupled plasma mass spectrometry (1CP-MS) reveals that this halite includes Mn (98 μg/g) and Pb (640 μg/g).

Mn^{2+}-Clusters

In such a case, the excitation does not stay on the same ion of Mn^{2+}, but can travel readily through the sub-lattice of in-resonance Mn^{2+} ions. The emission originates from Mn^{2+} ions associated with impurities and other defects. They occupy regular cation sites in the lattice and perturb the surrounding Mn^{2+} ions, lowering their energy levels relative to those of the unperturbed (intrinsic) Mn^{2+} ions. The diffusing excitons can now be trapped by the perturbed Mn^{2+} ions. At low temperatures the excitation cannot return to the exciton state; the excited, perturbed Mn^{2+} ions decay radiatively with a spectrum characteristic of the particular trap. Deeper traps are also present and are effective as quenching states, i.e. traps from which no emission occurs, but where the excitation is lost non-radiatively. At low temperatures the traps are effective, but at higher ones they begin to lose their trapped excitation energy by thermally activated back-transfer to the exciton level. From here deeper traps may trap the energy. Finally all the emitting traps are emptied and only the deep, non-emitting traps are operative. As a consequence the luminescence has been quenched. These quenching traps may be Ni or Fe ions. If the mineral contains two or three-dimensional chains Mn-O-Mn-O, excitation energy migrates over the lattice, which results in the following distinctions from luminescence properties of impurity Mn^{2+} in isostructural non-manganese minerals:

(a) sharp drop in luminescence quenching temperature due to an enhanced probability of excitation energy to hit quenching centers,
(b) long-wave shift of emission λ_{max}, because luminescence proceeds at local levels of Mn^{2+} near defects where the shift occurs either due to the electron-phonon interaction or due to a strong local crystalline field with the possible change of the radiation level of Mn^{2+}.

If there are only one-dimensional chains or Mn ions are linked through two oxygen ions, the energy migration does not take place and the luminescence peculiarities mentioned above are not observed. Dependence of luminescence features on peculiarities of Mn mutual arrangement and not only on there spacing makes clear indication that the main mechanism of the energy transfer is the exchange one (Powell and Blasse 1980). Steady-state luminescence of Mn-clusters in natural minerals of manganese has been found at liquid nitrogen temperatures (Gorobets et al. 1978; Gaft et al. 1981b) and interpreted in terms of minerals structure (Fig. 5.79).

Under powerful laser excitation with $\lambda = 532$ nm manganese minerals, such as rhodonite, rhodochrosite and axinite, exhibit orange-red luminescence even at room temperature. In rhodonite it is a band peaking at 620 nm with half-width of 85–95 nm and very long decay time of 5–6 ms (Fig. 4.167c). Such luminescence is typical for impurity Mn^{2+}. Similar picture takes place in axinite (Fig. 4.167d). In rhodochrosite luminescence is more complicated. The luminescence band has a maximum near 640 nm and a half-width of 80–90 nm, which is typical for impurity Mn^{2+}, but decay time of only 5–10 μs is very short for such center and indicates strong energy migration (Fig. 4.167a, b).

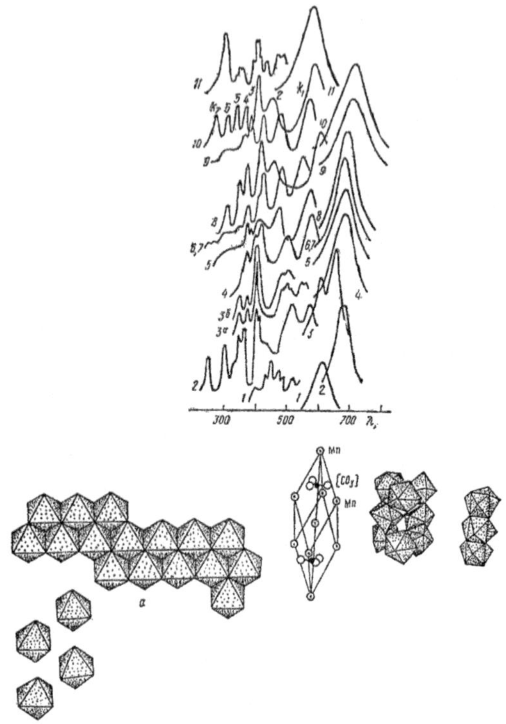

Fig. 5.79 Laser-induced luminescence of Mn bearing minerals (*1*-helvine; *2*-severgenite; *3*-Mn-milarite; *4*-triplite; *5*-rodonite; *6*-rhodochrosite; *7*-Mn-calcite; *8*-Mn-curchatovite; *9*-Mn-tantalite; *10*-dgimboite; *11*-Mn-apatite) and its connection with crystallochemical structure: a-rhodonite; b-rhodochrosite; c-Mn-apatite; d-carpholite $(MnAl_2(Si_2O_6)(OH)_4$ (Gaft et al. 1981a)

Excitation spectra of Mn^{2+} enable to calculate the crystal field force Δ and Raka parameters B and C. The relative position of the ground level $^6A_{1g}$ (S) and excited levels $^4A_{1g}$, 4Eg (G) and ^4E_g(D) does not depend on Δ. Using the frequencies of electronic transitions from the ground level to those excited levels (k_3 and k_5), parameters B and C may be calculated based on formulas:

$$k_3 = 10B + 5C, \; k_5 = 17B + 5C \tag{5.40}$$

Resulting B and C are substituted for:

$$k_1 = -\Delta + 10B + 6C - 26B^2/\Delta, \; k_2 = -\Delta + 18B + 6C - 26B^2/\Delta \tag{5.41}$$

And two independent values of Δ are received.

5.7.6.2 Fe^{3+}

Luminescence of Fe^{3+} may be considered from the same positions as Mn^{2+} emission because it also has d^5 configuration. Nevertheless, because of the large value of crystal field, the emission is shifted in the deep red when Fe^{3+} is present on tetrahedral sites. Fe^{3+} in octahedral coordination is predicted to emit in the near IR at about 900–1000 nm, but was not confidently observed in minerals. Steady-state luminescence of Fe^{3+} in minerals is mainly observed when it substitutes for Al^{3+} or Si^{4+} in tetrahedral sites in alumosilicates (Tarashchan 1978; White et al. 1986). It has luminescent transitions, which are forbidden and correspondingly the decay times are rather long.

Sillimanite

The luminescence spectrum of yellow sillimanite at 300 K is characterized by a broad band in the red with a half-width (Δ) of 100 nm peaking at approximately 765 nm with a very long decay time of 3.8 ms (Fig. 5.80a). At 77 K the luminescence shifts to 760 nm and narrows ($\Delta = 80$ nm) (Fig. 5.80b). This luminescence excitation has two maxima situated at 450 and 590 nm (Fig. 5.80c) (Gaft et al. 2012). Such broad band luminescence with long decay time found in yellow and brown samples is typical for forbidden d-d electron transitions in a Fe^{3+} center in tetrahedral coordination. The luminescence is connected to the forbidden $^4T_1(G)$ $^6A_1(S)$ electronic transition, whereas the excitation arises from the $^6A_1(S)$-$^4T_1(G)$, $^6A_1(S)$-$^4T_2(G)$, and $^6A_1(S)$-$^4A_1{}^4E_1(G)$ electronic transitions. Minor concentrations of iron in sillimanite are well-documented in the literature and were confirmed in our samples. This specific yellow sample contains 1.28 % of Fe_2O_3 and negligible contents of Cr and Ti. It is well known that Fe in the form of Fe^{2+} is an efficient luminescence quencher, while Fe^{3+} forms luminescent centers with concentrations up to approximately 1 % (Tarashchan 1978). The mostly probable tetrahedral position in the sillimanite structure is substitution into Si^{4+} tetrahedral sites. Besides broad band emission, narrow luminescence lines were found peaking at 694.0, 740.1 and 747.0 nm (Fig. 5.80b). Because they are detected even after very big delay time of several ms it may be concluded that they have very long decay time similar to those of Fe^{3+}. Similar lines were previously observed at low temperatures peaking at 694.0, 727.8, 739.7 and 746.9 nm. They were interpreted as "a false origin of the broadband Cr^{3+} $^4T_2$-$^4A_2$ emission" (Wojtowicz and Lempicki 1988). Nevertheless they were detected in sillimanite with negligible content of Cr and actually without broad emission band ascribed to $^4T_2$-$^4A_2$ emission of Cr^{3+} luminescence center. It may be supposed that those lines are connected to Fe^{3+} emission. In such case they may be interpreted as zero-phonon line at 694.0 nm and vibronic lines associated with $^4T_1(^4G)$-$^6A_1(^6S)$ transition. Similar luminescence was observed in $LiAl_5O_8$: Fe^{3+} (Abritta and de Souza Barros 1988).

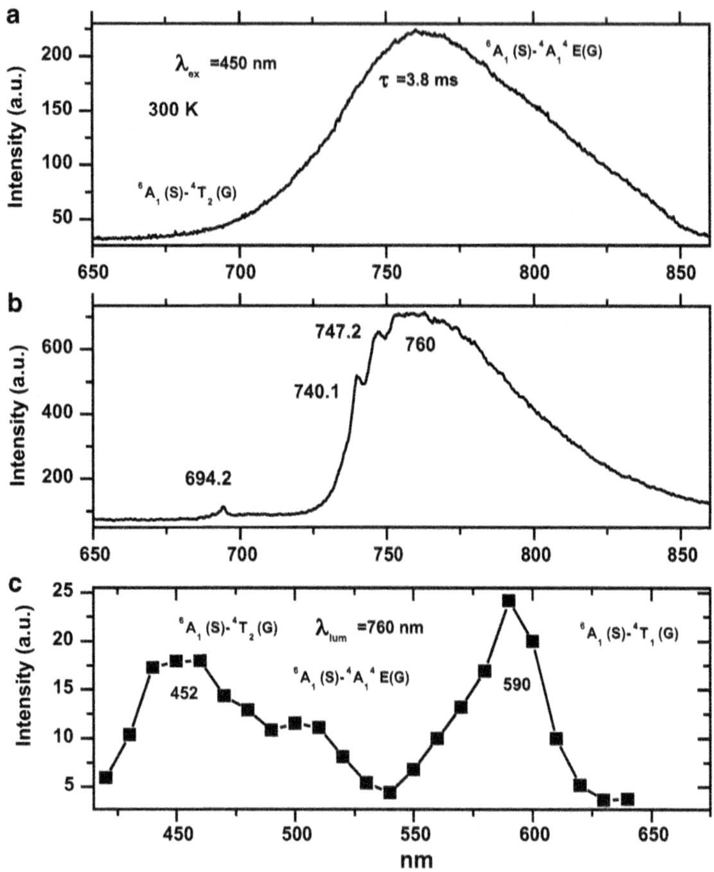

Fig. 5.80 (a–c) Laser Induced time-resolved excitation and luminescence spectra of Fe^{3+} in sillimanite

Other Minerals

In time-resolved luminescence, Fe^{3+} dominates spectra with long delay times. The examples may be seen in feldspars (Fig. 4.97), wollastonite (Fig. 4.95b), zircon (Fig. 4.91d), and beryl (Fig. 4.122b).

Fe^{3+} luminescence spectra were recently found in several additional minerals. Red luminescence in sodalite (Fig. 4.155), tugtupite (Fig. 4.160) and hackmanite (Fig. 4.158) with a long decay time of several milliseconds, evidently belongs to Fe^{3+} center for the following reasons (Gaft et al. 2009). The presence of Fe impurities with concentrations of several tens to hundreds ppm in sodalite and tugtupite was confirmed by ICP-AES data. The interpretation is strengthened by the fact that a very similar luminescence band has been detected in synthetic sodalite artificially activated by Fe (van Doorn and Schipper 1971). Excitation spectra of Fe^{3+} in both minerals are characterized by a very strong UV band, which is usually ascribed to sensitizer's presence, such as Pb^{2+} (Tarashchan 1978), because forbidden d-d excitation transitions are usually rather weak. Tugtupite structure consists of BeO_4, AlO_4 and SiO_4 tetrahedra. It is logical to suppose that the Fe^{3+} cation substitutes for the Al^{3+} cation, because their valences are the same and ionic radii in tetrahedral coordination are the closest (63 and 53 pm, correspondingly), whereas Be^{2+} has ionic radii of 41 pm and Si^{4+} 40 pm. The long wavelength shift of Fe^{3+} emission in hackmanite as compared to tugtupite could be connected with stronger crystal field strength inside AlO_4 tetrahedra in the sodalite structure as compared to tugtupite. This leads to stronger splitting of the emitting 4G level with a corresponding reduction of luminescence transition energy. Two types of emission bands connected with Fe^{3+} detected in tugtupite and sodalite, and peaking at different wavelengths, may be connected with several types of structural substitution, possibly for Al and Si tetrahedra. The interpretation of the red luminescence band in hackmanite, with relatively shorter decay time of less than 500 μs, is not clear. It may be supposed that it is also connected with Fe^{3+} center in different structural position, where the local symmetry is lower and correspondingly the decay time is shorter.

5.7.7 *3d⁴ Elements: Mn^{3+}*

Unlike Mn^{2+}, luminescence from Mn^{3+} is not known in minerals. However, the luminescence properties of Mn^{3+} are well studied in different artificial phosphors, including garnets. Specifically, gadolinium gallium garnet (GGG), activated by Mn^{3+}, was spectroscopically studied in the visible and near infrared spectral range (Kuck et al. 1998). In absorption studies a broad split band, peaking at 500 nm, was found connected to two spin-allowed transitions from the Jahn–Teller split ^5E ground state (in the following the two components are denoted as $^5E'$and $^5E''$) to the $^5T_2$ level (Fig. 5.81). The absorption spectrum is nearly temperature independent. In emission, however, temperature dependent spectra were measured. At low temperatures, only the $^1T_2$ level is thermally populated and the emission occurs from this level to the two Jahn–Teller components of the ^5E ground state and also to the $^3T_1$ state. Three emission bands were observed at 12 K, i.e. two broad bands ($^1T_2$-$^5E'$, $^5E''$) with maxima at 14,300 cm^{-1} (700 nm) and 11,700 cm^{-1} (850 nm), and a structured band around 9000 cm^{-1} (1100 nm, $^1T_2$-$^3T_2$). The population of the $^5T_2$ level increases with temperature, i.e. now the two spin-allowed transitions $^5T_2$ – 5E′, 5E″ were observed. At 300 K, these spin-allowed transitions dominate the emission spectrum and cover the double spin forbidden transitions completely. At 12 K, the decay times are a few milliseconds, while at higher temperatures lifetimes decrease and down to 0.5 μs.

Fig. 5.81 Tanabe-Sugano diagram for d^4 elements

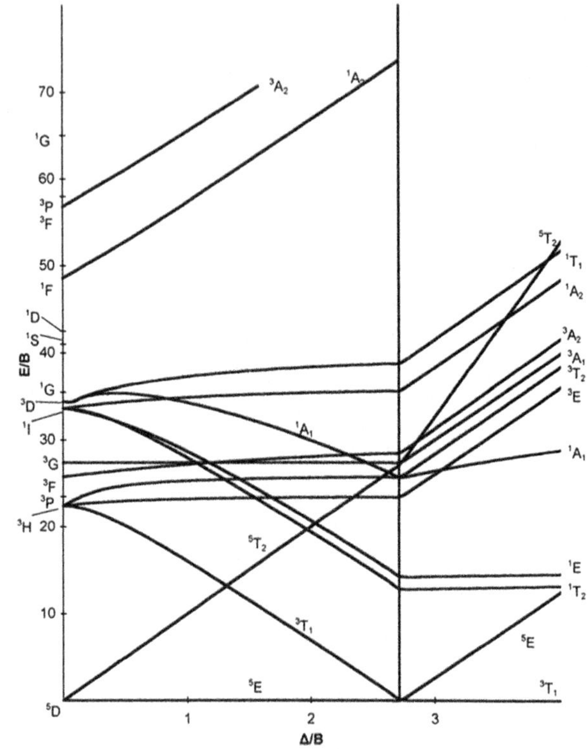

5.7.7.1 Grossular

Such behavior in the visible range is similar to that found for the short lived luminescence at 300 K in the Mexico grossular (Fig. 5.82b, c). Just as described above, such band disappears at low temperatures, and two bands with long decay times present at 100 K. The main difference is the short decay time compared to known Mn^{3+} emission and it has to be studied. This Mexican grossular contains 1.6 % of MnO, and Mn^{3+} presence was proposed earlier based on absorption spectroscopy (Geiger et al. 1999) with broad band peaking at 514 nm ascribed to Mn^{3+}. It is close enough to our excitation spectrum, keeping in mind that absorption and excitation spectra may differ (Gaft et al. 2013b)

5.7.7.2 Mn-kyanite

Another potential example of Mn^{3+} luminescence is orange Mn containing kyanite (Gaft et al. 2011). The broad luminescence band was detected under excitation by 355 nm at 300 K in orange kyanite with a maximum at 765 nm, relatively small

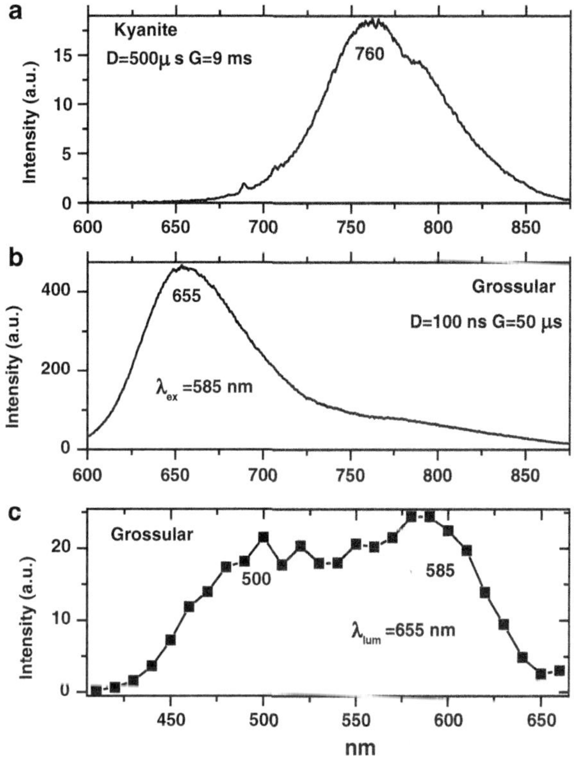

Fig. 5.82 (a–c) Luminescence and excitation spectra of supposedly Mn^{3+} in orange kyanite and grossular

half-width of 75 nm (Fig. 5.82a) and long decay time of 830 μs. The luminescence band with such spectral-kinetic characteristics was not detected in blue kyanite. Such a combination of spectral and decay parameters with temperature is different from any known luminescence centers in minerals, at least according to our knowledge. It has a rather long decay time for it to be connected to the $^4T_2 \rightarrow {}^4A_2$ luminescence of Cr^{3+} or to Ti^{3+} emission. Likewise, this decay time is rather short to be ascribed to Fe^{3+} or Mn^{2+}. The connection with $^4T_2 \rightarrow {}^4A_2$ luminescence transition in Mn^{4+} luminescence is doubtful because the crystal field at the higher charged Mn^{4+} ion is usually stronger than for Cr^{3+} and the luminescence of Mn^{4+} in weak crystal field positions was not detected yet. It may be natural to try to connect this luminescence band with Mn^{3+} which definitely presents in orange kyanite based on its absorption spectra. Spectroscopically, its behavior is consistent with what is typical for Mn^{3+}, namely the red band at 780 nm at 300 K with decay time of 0.83 ms excited in UV range and disappearing at 77 K. Thus it may be preliminary supposed that this band is connected to a Mn^{3+} luminescence.

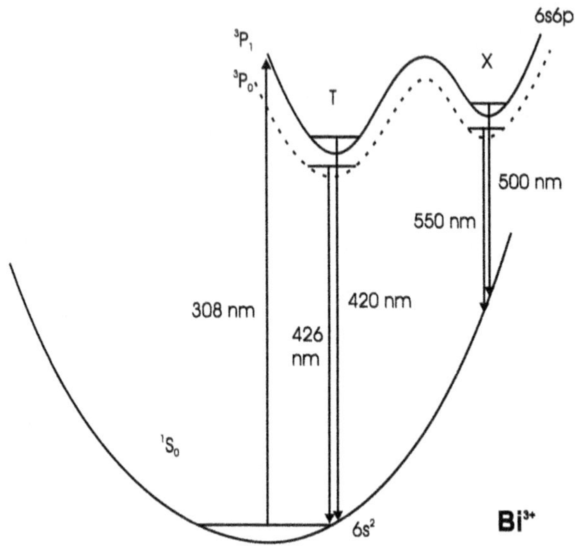

Fig. 5.83 Configurational coordinate diagram with emission and excitation transitions in Bi^{3+}

5.8 s^2 Ions

In this case filled s shell of a free ion gives rise to a $^1S_0$ ground state. The excited sp state gives a triplet $^3P_{0,1,2}$ for spins parallel and a singlet $^1P_1$ for spins antiparallel. In view of the selection rules, only the transitions between the singlets are allowed.

5.8.1 *Bi^{3+}, Bi^{2+}: Barite*

The luminescence of Bi^{3+} is quite diverse and depends strongly on the host lattice (Boulon 1987; Blasse and Grabmaier 1994; Blasse et al. 1994; Srivastava 1998). For the heavy $6s^2$ Bi^{3+} the transitions between the ground state and the $^3P_1$ state becomes additionally allowed by spin-orbit mixing of the $^3P_1$ and $^1P_1$ states. After excitation at low temperature, the system relaxes to the lowest excited state. Consequently, the emission at low temperatures can be ascribed to the forbidden transition $^3P_0$-$^1S_0$ and has a long decay time. Nevertheless, both $^3P_1$ and $^3P_0$ are emitting levels and they are very close so that at higher temperatures the luminescence from $^3P_1$ level may appear with similar spectrum, but shorter decay (Fig. 5.83).

Electron configuration of Bi^{2+} is $(6s)^2(6p)^1$, yielding a $^2P_{1/2}$ ground state and a crystal field split $^2P_{3/2}$ excited state (Hamstra et al. 1994). Because the emission is a 6p inter-configurational transition $^2P_{3/2}$-$^2P_{1/2}$, which is confirmed by the yellow excitation band presence, it is formally parity forbidden. Since the uneven crystal-field terms mix with the $(6s)^2(7s)^1$, $^2S_{1/2}$ and the $^2P_{3/2}$ and $^2P_{1/2}$ states, the parity

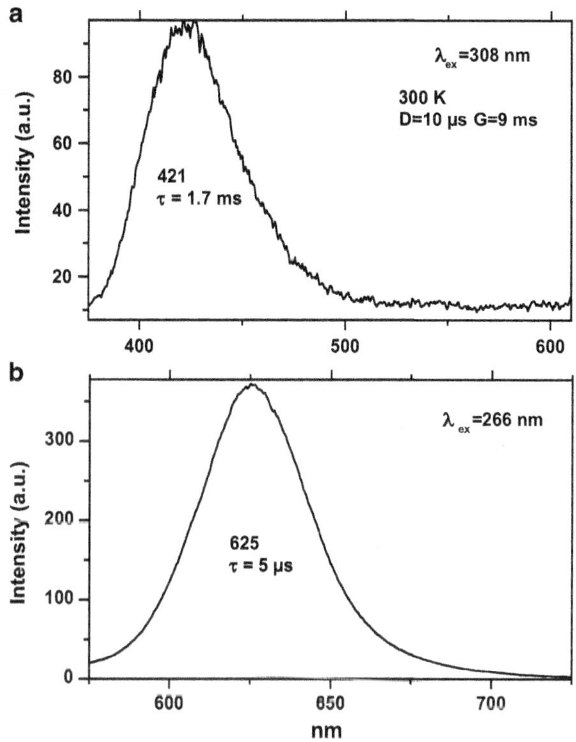

Fig. 5.84 (**a–b**) Laser-induced time-resolved luminescence spectra of $BaSO_4$:Bi

selection rule becomes partly lifted. The excitation transition $^2P_{1/2}$-$^2S_{1/2}$ is allowed one and it is demands photons with higher energy.

Under short wave 266 nm excitation of barite two bands are detected: the intensive blue one peaking at 460 nm with very short decay time of 100 ns and the relatively weak narrow orange one peaking at 625 nm with half-width of 30 nm and decay time of 5 μs (Fig. 4.73a). In order to present both bands in comparable magnitudes, the 1 μs delay is used to diminish the intensity of the short-lived blue emission. Such luminescence is not detected under longer UV laser lines. It is interesting to note that narrow orange band has been firstly reported for synthetic $BaSO_4$:Bi in 1886 (de Boisbaudran L 1885). Half a century later this was confirmed and connected with Bi^{3+} luminescence (Kroger et al. 1949). Another half a century later the new interpretation has been done and it was assumed that the luminescence center is Bi^{2+} (Hamstra et al. 1994). We also studied $BaSO_4$-Bi (Fig. 5.84) by laser-induced time-resolved technique and under 266 nm excitation and the narrow orange band, exactly similar to those in natural barite has been found. Thus connection of narrow orange band in luminescence spectrum of natural barite with Bi was confirmed. In attempt to clarify its connection with Bi^{3+} or Bi^{2+} we used step-like thermal treatment in air in order to stimulate Bi oxidation with possible change of valence state. In natural barite the narrow orange band totally disappeared after heating at 700 °C (Gaft et al. 2001a). Such behavior usually evidences that emission is connected with luminescence center in the lowest valence state, namely Bi^{2+}.

Figure 4.73b represents luminescence spectra of natural barite after long delay of nearly 1 ms, when the emissions with long decay time are emphasized and the narrow violet band at 426 appears. At liquid helium temperature this violet band becomes even narrower, shifts substantially in the short wavelength direction and its decay time of 3.1 ms becomes much longer (Gaft et al. 2001a). Such band is absolutely similar to those detected in $BaSO_4$-Bi and ascribed to Bi^{3+} luminescence center (Kroger et al. 1949).

The only possible position for Bi^{3+} and Bi^{2+} in barite lattice is in the Ba^{2+} site. The Ba ions in the structure of barite lie on the mirror planes and link the sulfate ions in such a way that 12 oxygen atoms coordinate each Ba. The exact ionic radius for Ba^{2+} in such coordination is unknown, but for coordination number 8 it is equal to 156 pm. Bi^{3+} in such coordination has ionic radius of 131 pm, which is very close. The exact ionic radius for Bi^{2+} is also unknown, but it has to be even bigger than for Bi^{3+}, which is compatible with big ion Ba^{2+}, especially for isomorphous substitution, when the charge compensation is not needed. The big half-width of Bi^{3+} luminescence band of 75 nm in barite is not surprising. It was observed that when Bi^{3+} is a substitute for big ion the Stokes shift increases and very broad luminescence bands take place (Boulon 1987; Blasse and Grabmaier 1994). The large variation of Bi^{3+} luminescence is ascribed to the amount of space available for the Bi^{3+} ion in the lattice. The Bi^{3+} ion in Ba position with coordination number 12 has enough space for relaxation after excitation with resulting large Stokes shift and half-width. The magnitude of this shift depends on the details of vibrational relaxation in the excited and ground states, which, in turn depend on the interaction of the emitting ion with the adjacent coordination ions in the structure. The lower Stokes shift and half-width for Bi^{2+} indicate the weak interaction where configuration curves are essentially flat.

5.8.2 *Pb^{2+}*

All considerations above for Bi^{3+} are also relevant for Pb^{2+}, which also belongs to s^2 elements (or mercury like).

5.8.2.1 Calcite

Figure 4.19c demonstrates time-resolved luminescence spectra of calcite, Franklin, under 266 nm excitation. A very intensive UV band at 312 nm with short decay time of 120 ns is detected. It may not be connected with Ce^{3+} emission, because its spectrum situated at longer wavelength near 400 nm (Fig. 4.19e). The excitation spectrum of the band at 312 nm consists of one band at 240 nm (Gaft et al. 2002a).

Tarashchan (1978) already ascribed the UV luminescence band at 325 nm with excitation band at 237 nm in pink calcite to Pb^{2+}. Decay time of 120 ns measured in our sample is consistent with such interpretation. Such UV band was unique in the calcite collection in our disposal and ICP-MS analyses of its impurities have been done (Table 4.7). It was found that Pb concentration in Franklin, NJ calcite of 450 ppm is approximately 50 times higher than in pink calcite, taken for comparison sake, while its Ce content is more than 10 times lower. Those data confirm the connection of UV band at 325 nm with Pb^{2+} luminescence center.

The only possible position for Pb^{2+} in calcite structure is instead of Ca^{2+} with coordination number of 6, where ionic radii of Ca^{2+} and Pb^{2+} are 114 and 133 pm.

5.8.2.2 Hardystonite

Luminescence spectra of hardystonite under 266 nm laser excitation reveal an extremely strong, rather narrow UV band at 355 nm, with a very short decay time of ~25 ns (Fig. 4.36b). Usually such bands in minerals are attributed to Ce^{3+} luminescence. However as another band was already confidently ascribed to this center (Fig. 4.36a) assignment appears problematic. In principle it is possible that several different Ce^{3+} centers occur in a structure, which are formed, for example, as a result of substitutions on Ca and Zn positions or because of different types of charge compensations. The first possibility may be excluded based on the large differences in ionic radii of Ce^{3+} (115 ppm) and Zn^{2+} in tetrahedral coordination (74 ppm), while the second possibility may be taken into consideration.

In order to check it, the excitation spectra of both UV bands have been determined (Gaft et al. 2002a). It was found that the UV band ascribed to Ce^{3+} according to its luminescence and decay, has an excitation spectrum which is typical for this center, while the excitation spectrum of the UV band at 355 nm is different, containing one excitation band with large Stokes shift of ~11,000 cm^{-1}. Thus it may be suspected that the UV band at 355 nm is connected not with Ce^{3+} in different structural position, but with a different luminescence center.

The luminescence of Pb^{2+} in synthetic $Ca_2ZnSi_2O_7$ has been reported (Butler 1980). The emission and excitation peaks occur at 251 and 347 nm, respectively with a Stokes shift of 10,000 cm^{-1}. It is very close to luminescence and excitation bands detected in natural samples. In order to prove the possible relation of the UV luminescence band at 355 nm to Pb^{2+} in natural hardystonite, its decay time as a function of temperature has been studied. These decay curves are very specific for mercury-like ions, where the emission at low temperatures is ascribed to the forbidden transition $^3P_0$-$^1S_0$ and has a long decay time. Nevertheless, both $^3P_1$ and $^3P_0$ are very close emitting levels so that at higher temperatures the luminescence from $^3P_1$ level may appear with a similar spectrum, but with shorter decay (Boulon 1987; Blasse and Grabmaier 1994). In contrast, the decay times of the broad emissions of Ce^{3+} or Eu^{2+} are not temperature-dependent. At room temperature the decay time of the UV band in hardystonite at 355 nm in hardystonite is very short (~25 ns). Nevertheless, with decreasing temperature a very long decay

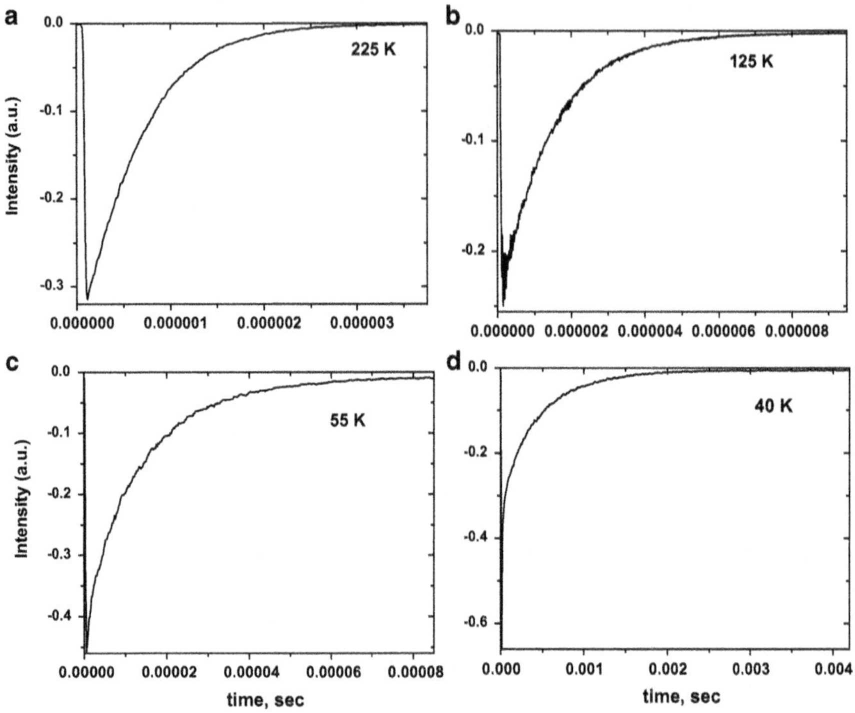

Fig. 5.85 (**a–d**) Decay time of Pb^{2+} in hardystonite as a function of temperature

component of approximately 150 μs appear while the luminescence spectrum remains practically the same (Fig. 5.85). This typical behavior is strong evidence that the emission band of hardystonite at 355 nm is caused by Pb^{2+}.

ICP analyses of the hardystonite sample in our study confirm a very high Pb content of 5000 ppm, while the Ce concentration of 55 ppm is two orders of magnitude lower. This elevated Pb concentration may be the reason of extremely short decay time at room temperature because of energy migration, with a corresponding decrease in decay time. The most likely position for Pb^{2+} in the hardystonite structure is at the Ca^{2+} site. The four-coordinated Zn site is less probable. Moreover, the difference between the ionic radii of Zn^{2+} and Pb^{2+} is quite large.

5.8.2.3 Hydrozincite

Figure 4.85a presents the time-resolved luminescence spectrum of a hydrozincite under 266 nm laser excitation. A relatively broad band is detected at 430 nm, which is responsible for well-known blue hydrozincite luminescence. Its spectral position and decay time of approximately 700 ns are typical for Eu^{2+} luminescence.

However, the excitation spectrum of this band consists of one narrow band at 240 nm (Fig. 4.85b), which does not correspond to a Eu^{2+} excitation spectrum. Two bands usually characterize the latter with relatively small Stokes shifts of 30–50 nm caused by crystal field splitting of the $4f^6 5d$-levels. Moreover, the measured Eu concentrations in the hydrozincite samples under investigation are very low (less than 0.5 ppm) and they do not correlate with the intensity of the blue luminescence, i.e. the band at 430 nm.

Three hydrozincite samples with different blue luminescence intensities were analyzed by ICP-MS. The first one with the strongest emission has 2000 ppm of Pb, the second with weaker luminescence has 600 ppm, while the last one, practically without emission, has only 35 ppm. Such strong correlation supports interpretation of a Pb^{2+} luminescence center as responsible for hydrozincite blue emission. In contrast, the Eu concentrations (another candidate for a luminescence center) are very low and do not correlate with the intensities of the blue luminescence.

In order to confirm the observed correlation of blue luminescence of hydrozincite with Pb^{2+}, artificial analogs of the mineral have been synthesized, activated by different amounts of Pb. It was found that the blue emission intensity increases with the Pb content, up to approximately 2000 ppm, while a decrease of luminescence at higher Pb concentrations is evidently connected with concentration quenching (Gaft et al. 2002a). The only possible position for Pb^{2+} in the hydrozincite structure is the site of Zn^{2+}. Evidently, Pb^{2+} substitutes for Zn^{2+} in octahedral coordination where the crystal field splitting Δ is especially strong, and this may be the reason for the long-wave Stokes shift of Pb^{2+} luminescence. Another reason may be the presence of hydroxyl groups, which strongly increase the crystal field splitting.

5.8.2.4 Barite, Anhydrite and Celestine

Luminescence of Pb^{2+} in synthetic alkaline earth sulfates is well known (Folkerts et al. 1995). In this study, $CaSO_4$:Pb^{2+} shows an emission band with a maximum at 235 nm at 300 K, while the excitation maximum is at 220 nm. The decay curve of the emission is single exponential with decay time of 570 μs at 4.2 K. Emission spectrum of $BaSO_4$:Pb^{2+} demonstrates a broad band peaking at 340 nm with excitation maximum at 220 nm, while in $SrSO_4$:Pb^{2+} the luminescence band has a maximum at 380 nm. In natural barite and anhydrite samples we detected several narrow UV bands, which may be connected with Pb^{2+}emission, but for confident conclusion additional study is needed. At any case, Pb^{2+} participation in natural sulfates luminescence has to be taken into consideration.

5.8.2.5 Sodalite and Tugtupite

Several UV-blue luminescence bands were detected in tugtupite and sodalite, which are characterized by two components with close spectra, but with strongly different decay times (Figs. 4.155 and 4.158). At room temperature the quenching is very fast, while at low temperature it is substantially longer (Gaft et al. 2009), which is typical for mercury like center luminescence. According to impurity analyses, the potential luminescence center may be Pb^{2+}, Tl^{+}, Sn^{2+} and Sb^{3+}, but specific identification demands additional study using artificially activated synthetic tugtupite and sodalite analogs.

5.8.2.6 Minerals of Pb

Pyromorphite

As it was described earlier, evidently Ce^{3+} emission in pyromorphite is characterized by very long decay component, which is not typical for this luminescence center. The possible reason may be energy migration from Pb ions, which are the major elements in these minerals. It is well known that Pb^{2+} is very effective sensitizer for many luminescence centers in minerals, including Mn^{2+} and REE (Marfunin 1979). Excitation spectrum of Ce^{3+} luminescence in esperite revealed UV band at 280 nm, which is not typical for Ce^{3+} luminescence and is possibly connected with Pb^{2+} excitation with energy migration.

Besides that, narrow luminescence bands with long decay components in pyromorphite luminescence spectra may be in principal connected with Pb^{2+} emission. The main problem with the luminescence of elements with high concentration is the concentration quenching. Nevertheless, it was found that the concentration quenching of Mn^{2+} luminescence in Mn-apatite is nearly absent (Gaft et al. 1981a). It was explained by the absence of energy migration by exchange mechanism in the apatite structure. Thus, the concentration quenching may be also negligible for Pb^{2+} in pyromorphite. The similar case is $PbSO_4$ where the luminescence of Pb^{2+} was found despite its high concentration. The simultaneous presence of two emission bands connected with Pb^{2+} is well known. For example, in $SrSO_4$:Pb^{2+} the emission spectrum shows an intense broad band with two maxima at 285 and 380 nm. It has been proposed that this luminescence originates from two different excited states: the shorter wavelength emission is assigned to ^3P-S_0 transition, while the longer wavelength band is attributed to emission from yet unidentified D-level (Folkerts et al. 1995).

For example, narrow bands in pyromorphite luminescence spectra at 380 and 439 nm (Fig. 4.59c) may be connected with Pb^{2+} emission. Two structurally different Pb sites in pyromorphite structure may cause the two bands in the luminescence spectrum. For comparison, it was already mentioned that the luminescence of all impurity centers is drastically different at Ca(I) and Ca(II) positions in the isostructural mineral apatite because the crystal field strength of the Ca (I) position of the apatite structure is unusually strong. As a consequence, in pyromorphite it may generate a strong splitting of the ^3P excited level for the Pb (I) site with a strong separation between the $^3P_0$ and $^3P_1$ levels. Thus, the forbidden luminescence transition $^3P_0$-$^1S_0$ takes place with long decay time. On the other

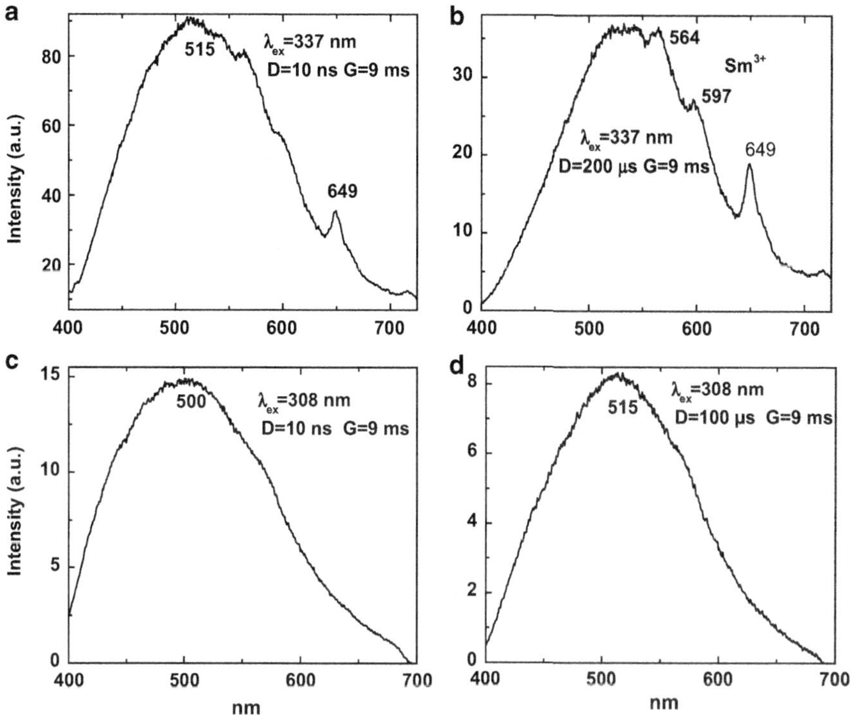

Fig. 5.86 (**a–d**) Laser-induced time-resolved luminescence spectra of cerussite

hand, in the Pb(II) position the splitting is less, and an allowed $^3P_1$-$^1S_0$ transition takes place.

Cerussite

Cerussite is characterized by several broad luminescence bands in the green-yellow part of the spectrum with different decay times (Fig. 5.86). Those bands are not confidently interpreted yet and only recently the idea of Ag^+ or Cu^+ participation has been proposed based on their close radii with Pb^{2+} (Gorobets and Rogojine 2001). We think that such interpretation is principally logical and has to be checked, but the possible participation of intrinsic lead also may be considered.

Margarosanite

Blue very broad luminescence band is found in margarosanite peaking at 450 nm with relatively short decay time of 10.6 μs. It becomes less intensive with temperature lowering and nearly disappears at approximately 150 K. Simultaneously a

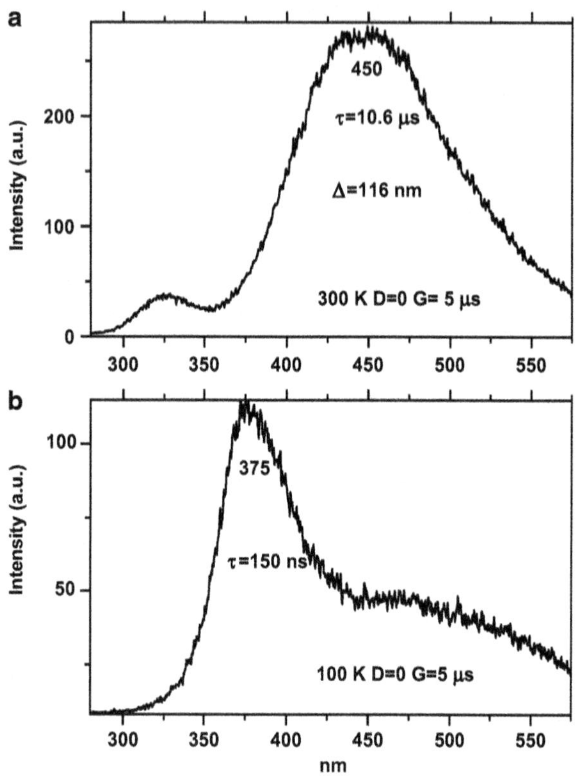

Fig. 5.87 (**a–b**) Laser-induced time-resolved luminescence spectra of supposedly Pb^{2+} in margarosanite

new UV band appears peaking at 375 nm with very short decay time of 150 ns (Fig. 5.87). It may be supposed that such luminescence is connected to another potential luminescent ion presents in margarosanite, namely to Pb^{2+} (Gaft et al. 2013a). In some cases two emission bands are typical for s^2 ions, where Pb^{2+} belongs, while their relative intensities change with temperature. The two emissions arise from different minima on the potential energy surface of the relaxed excited state. At low temperatures the high energy level which is populated by optical excitation emits with resulting UV luminescence. At elevated temperatures the barrier between the minima may be overcome and emission from minimum with lower energy occurs with resulting visible blue luminescence. In margarosanite, where Pb is a major element, concentration quenching and corresponding energy migration adds to the complexity of this phenomenon.

5.8.3 *Pb^+: Feldspars*

Luminescence of monovalent Pb was firstly proposed for explanation of IR luminescence band peaking at 860 nm in emission spectrum of feldspars (Kusnetsov and

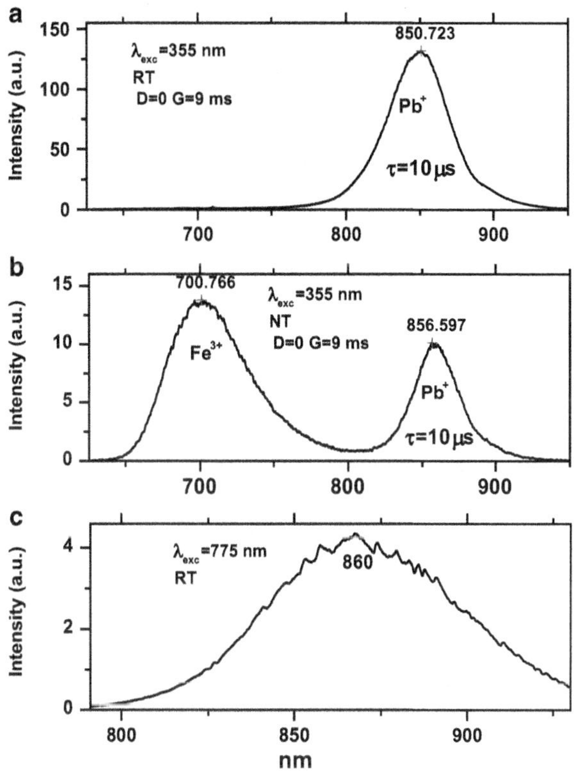

Fig. 5.88 (**a–c**) Laser-induced time-resolved luminescence spectra of Pb^+ in feldspar

Tarashchan 1978). Later new results have been reported on the characteristics of photo- and radioluminescence emissions at 865 and 910 nm emitted by potassium-rich feldspars (Erfurt 2003). The emissions were ascribed to electron transitions in Pb^+ from the $^72P_{1/2}$ and $^72S_{1/2}$ excited states to the $^62S_{1/2}$ ground state of Pb^+. Under laser excitation using 355 nm, similar band peaking at 850 nm appears at 300 K with relatively short decay time of 1 μs (Fig. 5.88a). At 77 K this band has decay time very similar to decay at 300 K (Fig. 5.88b). Such IR band was detected only in potassium feldspar with elevated Pb concentration. Our decay time results confirm the Pb^+ interpretation. Electron configuration of Pb^+ is $(6s)^2(6p)^1$, yielding a $^2P_{1/2}$ ground state and a crystal field split $^2P_{3/2}$ excited state. Because the emission is a 6p inter-configurational transition $^2P_{3/2}$-$^2P_{1/2}$, it is formally parity forbidden. Since the uneven crystal-field terms mix with the $(6s)^2(7s)^1$, $^2S_{1/2}$ and the $^2P_{3/2}$ and $^2P_{1/2}$ states, the parity selection rule becomes partly lifted and decay time is in μs range, which coincides with our data. I have to measure the excitation spectra. The easy excitation by 783 nm (Fig. 5.88c) corresponds to interpretation because for Pb^+ emission very small Stokes shift is typical (Kusnetsov and Tarashchan 1978).

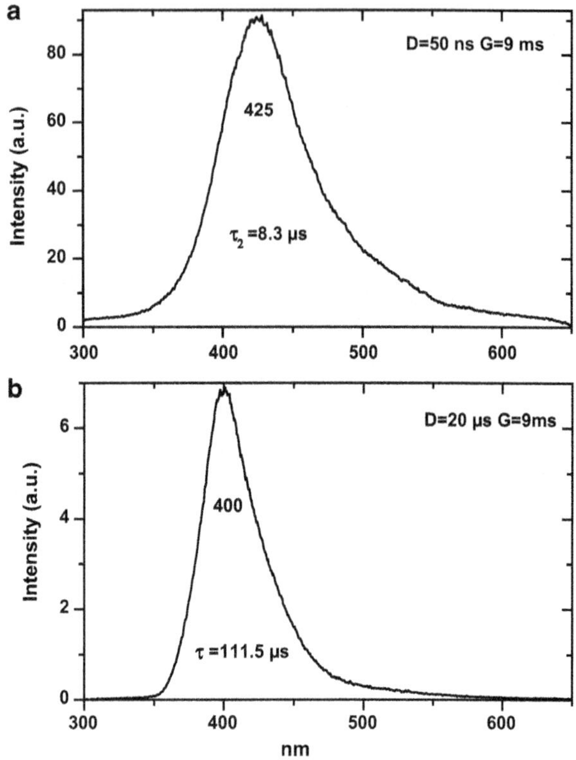

Fig. 5.89 (**a–b**) Laser-induced time-resolved luminescence spectra of Tl^+ in pezzattoite at 300 (**a**) and 77 (**b**) K

5.8.4 Tl^+: Pezzattoite

All considerations above for Bi^{3+} and Pb^{2+} are also relevant for Tl^+, which also belongs to s^2 elements (or mercury like). In minerals its UV luminescence peaking at 290 nm was found in microcline, where the isomorphous substitution of K^+ for Tl^+ takes place (Kusnetsov and Tarashchan 1978).

The blue band induced by UV excitation in pezzattoite has slightly different spectra at 300 and 77 K (Fig. 5.89a, b), while decay times are strongly different at these temperatures, namely short decay (8.3 μs) at room temperature and substantially longer decay (111.5 μs) at low temperature. It appears similar to what was described earlier in the case of mercury like ions, such as Pb^{2+} and Bi^{3+} (Gaft et al. 2005). For the heavy $6s^2$ ions the transitions between the ground state and the $^3P_1$ state becomes additionally allowed by spinorbit mixing of the $^3P_1$ and $^1P_1$ states. After excitation at low temperature, the system relaxes to the lowest excited state. The emission at low temperatures can be ascribed to the forbidden transition $^3P_0$-$^1S_0$ associated with a long decay time. Nevertheless, both $^3P_1$ and $^3P_0$ are emitting levels and they are very close so that at higher temperatures the luminescence from $^3P_1$ level may appear with slightly different spectrum, but shorter decay. Among these elements with potential luminescence, only traces of lead and thallium could be identified by ICP analysis in our sample (Panczer et al. 2010). The values are 1 ppm for Pb (or below detection limit for XRF) and 45 ppm (20–50 ppm by ICP MS) for Tl. The relatively high value leads us to consider the Tl^+ as potential candidate for the blue luminescence of pezzottaite. The possible isomorphous substitution is of Tl^+ for Cs^+ in the "beryl" channels. Pezzottaite is therefore one of the very rare mineral together with feldspar (K^+ substitution) and pollucite (Cs^+ substitution) demonstrating Tl^+ luminescence (Gorobetz and Rogojine 2001).

5.8.5 *Sn^{2+}*

Center Sn^{2+} belongs to $5s^2$ configuration. As impurity in artificial phosphors it is mainly responsible for narrow luminescence bands from UV to red part of the spectrum. Besides that intrinsic luminescence of tin compounds is also known (Donker et al. 1989).

5.8.5.1 Cassiterite

Based on considerable zonation, which was found in CL spectra of cassiterite with respect to W and Ti impurities it was concluded that W is responsible for blue luminescence, while Ti for the yellow one (Hall and Ribber 1971). Luminescence of synthetic nominally pure SnO_2 have been studied (Agekyan et al. 1974) and it was concluded that the green-yellow broadband emission is connected with electron- hole pare formation under UV excitation with the following recombination. Nevertheless, in natural monocrystals of cassiterite it was found that as luminescence and excitation bands are strongly polarized in direction perpendicular to C_4 axes (Gaft and Vorontsova 1982). Evidently that is not complied with recombination mechanism, because electron-hole pare formation does not depend on light polarization. It was also found that luminescence spectra under 250 nm excitation, where photons energy is large than energy gap of cassiterite, and 312 nm excitation, where photons energy is less than energy gap, are different, namely green and yellow-orange bands, correspondingly. It may be explained by different luminescence mechanisms, recombination in the first case and inner in the second.

In order to find impurity responsible for yellow emission, artificial SnO_2 have been studied activated by different impurities such as Ti, Nb, Ta, W, Fe (Gaft et al. 1982). From crystallochemical positions such ions are capable to substitute for Sn^{4+}. Nevertheless, the correlation of luminescence with any impurity was not found. From the other hand, in nominally pure super pure SnO_2 all luminescence bands have been found, which were detected in natural cassiterite (Fig. 5.90). Thus it was concluded that luminescence is connected to intrinsic, namely to Sn of diverse valence, possibly Sn^{2+}. Blasser et al. (1981) arrived at the same conclusion about intrinsic nature of SnO_2 luminescence.

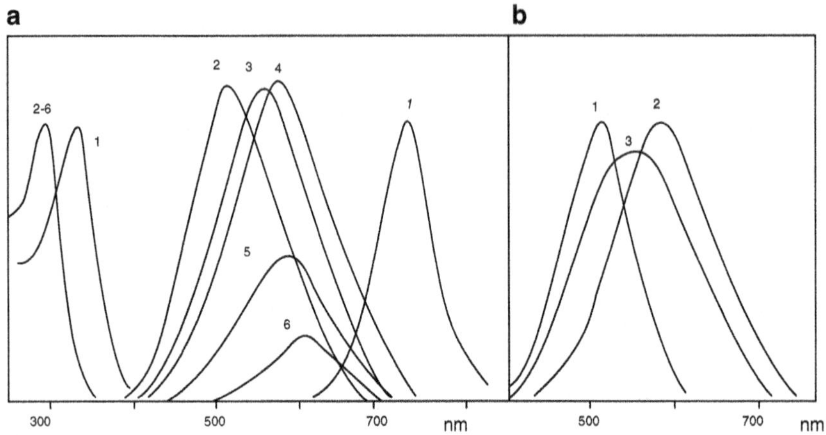

Fig. 5.90 Photoluminescence and excitation spectra of pure SnO_2: (**a**) at different temperatures: *1*–4.6 K; *2*–77 K; *3*–120 K, *4*–140 K; *5*–180 K; *6*–200 K; (**b**) after thermal treatment and ionizing irradiation (all spectra at 77 K): *1* – original SnO_2; *2*-heating in air at 1500 °C during 3 h; *3*-X-ray irradiation during 3 h of the sample heated at 1500 °C

5.8.5.2 Barite

We studies laser-induced time-resolved luminescence of synthetic $BaSO_4$ artificially activated by Sn and found several intensive UV and blue bands evidently connected with Sn^{2+} emission (Fig. 5.91). Similar bands have been also found in natural barite laser-induced luminescence spectra and we think that their connection with Sn center is quite possible.

5.8.6 *Sb^{3+}*

This luminescence center is well known in synthetic phosphors, for example in halophosphates, which are closely related to hydroxyapatite (Blasse and Grabmaier 1994). The Sb^{3+} doped calcium halophosphate is a very efficient blue-emitting phosphor under short wave 254 excitation (Fig. 5.92). When the halophosphate host lattice contains not only Sb^{3+}, but also Mn^{2+}, part of the energy absorbed by the Sb^{3+} ions is transferred to Mn^{2+}, which shows an orange emission. By carefully adjusting the ratio of the Sb^{3+} and Mn^{2+} concentrations, a white emitting phosphor can be obtained.

Another example is luminescence of Sb^{3+} in YPO_4 with tetragonal zircon structure (Oomen et al. 1988). The emission consists of two bands, one in the UV region and one in the visible part of the spectrum. The intensity ration of these bands is strongly temperature dependent (Fig. 5.93).

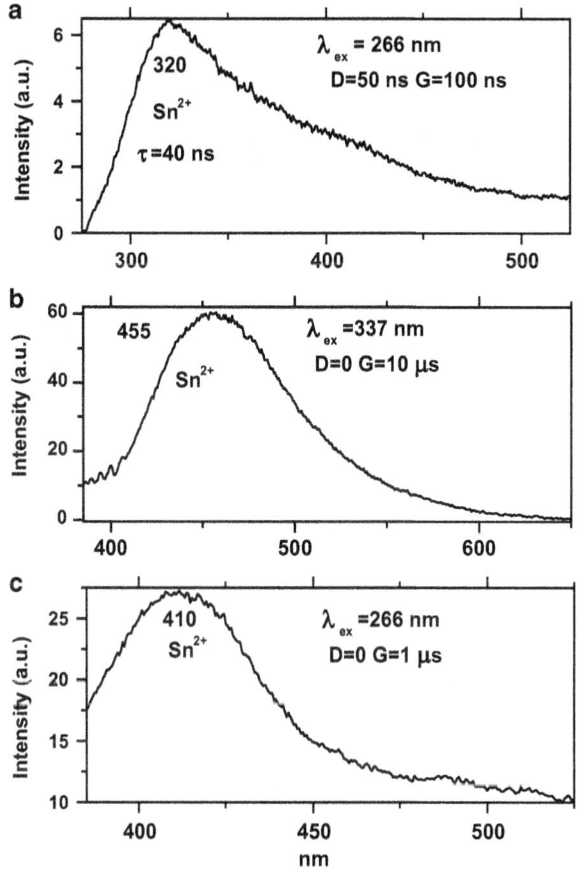

Fig. 5.91 (**a**–**c**) Laser-induced time-resolved luminescence spectra of $BaSO_4$-Sn

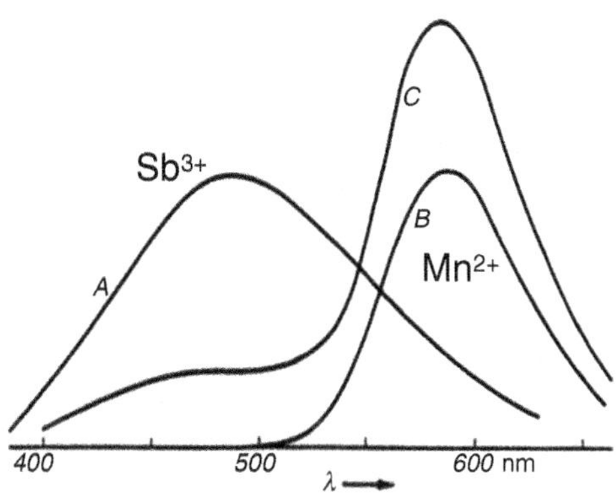

Fig. 5.92 Luminescence spectrum of calcium halophosphates: *A*-Sb^{3+} emission; *B*-Mn^{2+} emission; *C*-warm-white halophosphate (Blasse and Grabmaier 1994)

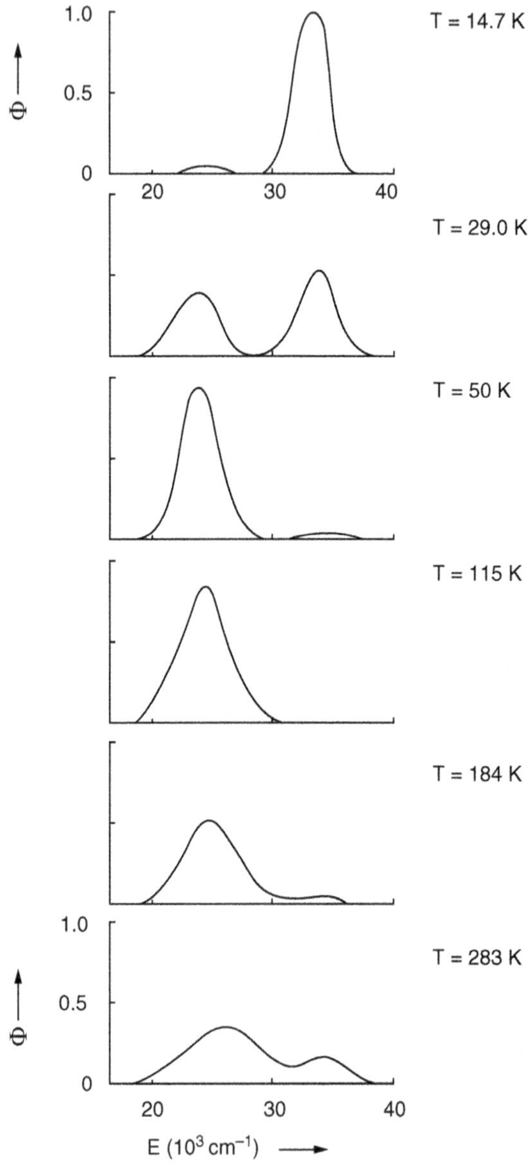

Fig. 5.93 Luminescence spectrum of YPO_4-Sb^{3+} (Oomen et al. 1988)

5.8.6.1 Swedenborgite

Swedenborgite blue luminescence at 300 K consists of broad band peaking at 410 nm with short decay time of 4.5 μs (Gaft et al. 2013a). At 100 K the spectrum is actually the same, but the decay time of 135 μs is much longer (Fig. 5.94). The problem with swedenborgite is that from classification of luminescent minerals position (Gorobets and Rogojine 2001) it is difficult to suppose the potential

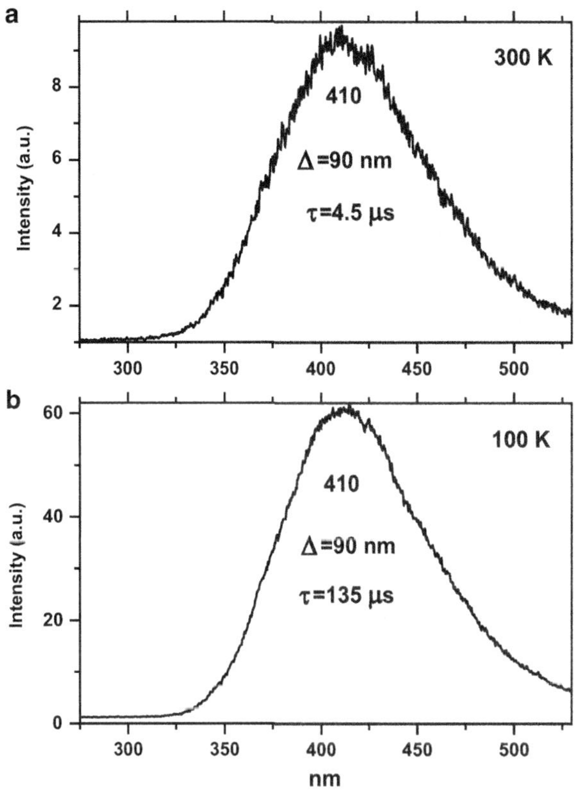

Fig. 5.94 (a–b) Luminescence of supposedly Sb^{3+} in swedenborgite

luminescent impurity. There are no known luminescence centers which may substitute for the Be^{2+} with its very small radius of 0.3–0.4 Å or for the Sb^{5+} with its very high valence state. The main luminescent cation substituting for Na^{+} is Mn^{2+}, but its luminescence properties are not consistent with found in swedenborgite. From the other side, such luminescence behavior is typical for s^2 ions, such as Sb^{3+}. In swedenborgite Sb presents in +5 state, which is not luminescent. If the minute presence of Sb^{3+} may be supposed, it is the mostly probable luminescence center.

5.9 d^{10} Ions

5.9.1 Ag^{+}

Complexes with monovalent d^{10} ions often show efficient luminescence at room temperature. The emission transition has been assigned to a d^9s–d^{10} transition, a ligand-to-metal charge-transfer transition, or a metal-to-ligand charge transfer transition, depending on the ligands (Boulon 1987). For the sulfur ligands a ligand-metal character is proposed as evident (Blasse and Grabmaier 1994). As in the case of Bi^3, the presence of several luminescence bands with strongly different decay times may be preliminary connected with excited state splitting.

5.9.1.1 Barite

Figure 4.71 presents typical laser-induced luminescence spectra recorded from natural barite with orange luminescence. At room temperature under 308, 337 and 355 nm excitations orange luminescence consists of very broad band peaking at 635 nm with half-width of approximately $\Delta = 150$ nm (Fig. 4.71a). At lower temperatures up to liquid helium the spectrum is very similar and only the half-width of $\Delta = 130$ nm becomes a little narrower (Fig. 4.71b). At all temperatures the spectra are not depending on delay times and gate widths. The possible conclusion is that only one luminescence center is responsible for this orange band. Decay time of luminescence is approximately 225 μs at 300 K and 275 μs at 4.2 K.

Let us first discuss why the orange luminescence cannot be assigned to some obvious reasons. The host itself may be excluded, because pure barite is not luminescent. The usual luminescence center in minerals, which are responsible for broad orange bands with relatively long decay time, is Mn^{2+}. This possibility may be excluded also for the following reasons. ICP data really reveals Mn impurity presence (Table 4.11), but the correlation between Mn concentration and orange luminescence intensity is absent. For example, two samples with strongest orange luminescence have 3.5 and 400 ppm of Mn. Nevertheless, we prepared and studied synthetic barite artificially activated by Mn. As result only extremely weak green band with very long decay time of 1 ms is seen, which is not detected yet in natural samples (Fig. 5.77a). According to EPR data ionic impurities of four-fold coordinated Mn^{5+} were detected in barite (Greenblat 1980), while luminescence of Mn^{6+} in $BaSO_4$-Mn has been found (Brunold and Güdel 1997). It is possible to suppose that they are the main form of Mn in natural barite.

The possibility of Cr^{3+} emission may be excluded because octahedral coordination is absent in barite structure. Luminescence of Fe^{3+} is crystallographically possible because tetrahedral surrounding presents in barite structure but Fe^{3+}-S^{6+} substitution is very difficult to suppose. The iron presence in ICP data is evidently connected with micro-impurities of iron minerals, which is usual for natural barite. Other ions, such as Ti^{3+} and Ni^{2+} with possible red luminescence, have ionic radii of 81 and 83 pm, correspondingly, which are small compared to 156 ppm of Ba^{2+}. ICP data confirm the absence of Ni in barite, while the minor quantities of Ti may be connected to Ti^{4+}.

In order to check the possibility of d^{10}-ions luminescence in barite lattice, the activation by Ag was accomplished (Gaft et al. 2001a). The main reason was that in all natural barite samples the Ag and Cu impurities have been determined by ICP analysis (Table 4.11). Besides that, weak orange luminescence on the tail of the strong UV band was detected in $BaSO_4$:Ag under X-ray excitation (Prokic 1979).

Fig. 5.95 (a–c) Laser-induced time-resolved luminescence of $BaSO_4$-Ag

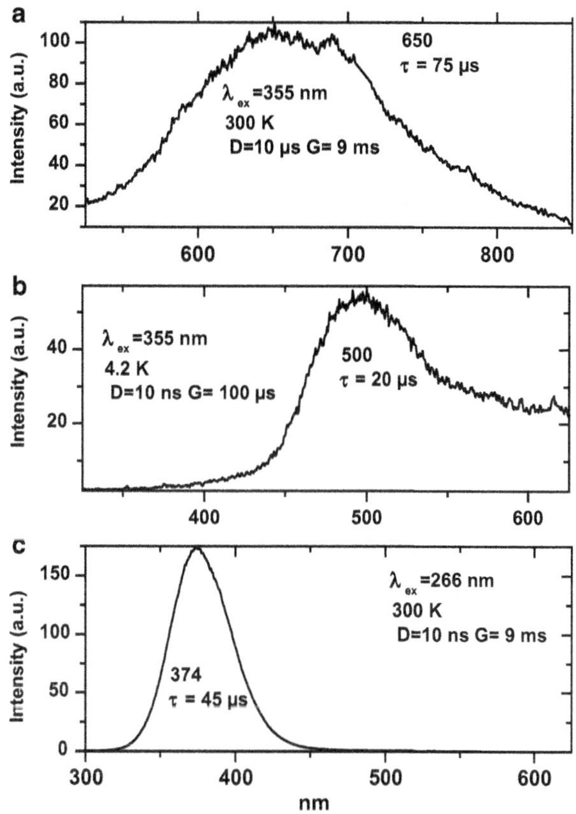

In $BaSO_4$:Ag (Fig. 5.95) three bands have been detected: red, green and UV. The red one peaking at 650 nm at room temperature has half-width of 150 nm and decay time of 75 μs and its spectrum is similar to broad orange band in natural barite. Decay time of this band in artificially activated barite is shorter compared to the natural one, which be connected with higher Ag content with resulting concentration quenching.

Ag^{+} is very big ion with ionic radius of 142 pm for coordination number 8, which is close to those of Ba^{2+}.

5.9.1.2 Ag Minerals -Chlorargyrite, Bromargyrite and Embolite

In this case emission is not connected with electron transitions inside Ag^{+} ion and may be interpreted in energy band scheme (Gaft et al. 1989). Under UV lamp excitation at 300 K those minerals do not exhibit luminescence but at lower temperature of 77 K very strong emission appear with blue-green color for chlorargyrite, yellow-green for embolite and yellow-orange for embolite.

Excitation spectra for all luminescence bands are similar and consist of one narrow UV band peaking at 350 nm. Under laser excitation luminescence appears even at 300 K (Fig. 4.166) with decay time of 200–250 ns. At 77 K the emission intensity becomes much stronger and decay time substantially longer, approximately of 3–6 μs. Time-resolved spectra demonstrate that minimum two luminescence bands present with different decay times.

An interpretation was done based on investigation of synthetic silver halogenides (Vacek 1971). The blue emission band in AgCl is intrinsic and connected with electron-hole recombination. At temperatures higher than 77 K holes start to migrate and may be recaptured on the deeper traps and once again recombine with electrons. This process explains luminescence color change at temperatures between 77 and 300 K. The orange emission of AgBr was connected with Mn^{2+} impurity, while the yellow-green one with I impurity. As for natural silver halogenides, under powerful laser excitation synthetic AgCl is also detected at 300 K, which is possibly connected with higher density of excited states and higher probability of luminescent transition.

5.9.2 *Cu^{+}: Barite*

Copper in minerals luminescence is usually considered only as an effective quencher. Nevertheless, it is well known that a bright blue luminescence is emitted from Cu^{+} ions in inorganic solids by UV light irradiation. It was found that these materials have potential application to tunable lasers. For example, in $CaO\text{-}P_2O_5$ glasses Cu^{+} is characterized by luminescence band at 440 nm with half-width of 100 nm and excitation maximum at 260 nm. Decay time of luminescence is approximately 25 μs (Tanaka et al. 1994). Red fluorescence possibly connected with Cu^{+} pair is also known (Moine et al. 1991).

In $BaSO_4$:Cu two bands are detected: red and violet (Fig. 5.96). The red one peaking at 740 nm at room temperature has half-width of 150 nm and decay time of 350 μs. The violet band peaking at 410 nm has at room temperature half-width of 35 nm and decay time of 7 μs. At lower temperatures both bands disappear and the new IR one is detected. Such deep red luminescence in natural barite has been described earlier (Gaft et al. 1985) while the possible connection with Cu was not considered.

5.10 d^0 Complex Ions

Complexes of transition metal ions with a formally empty d shell often show intense broadband emission with a large Stokes shift of 10,000–20,000 cm^{-1}. Examples mostly important for minerals are VO_4^{3-}, WO_4^{2-}, MoO_4^{2-} and

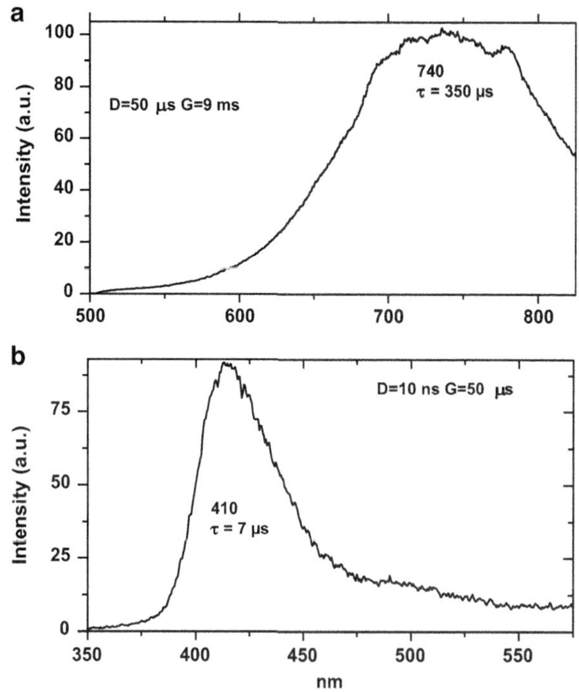

Fig. 5.96 (**a–b**) Laser-induced time-resolved luminescence spectra of $BaSO_4$:Cu

TiO_6^{8-}. Atomic orbitals s, p, d of the central atom and p orbitals of oxygen form molecular orbitals of the complexes (Fig. 5.97). The excited state is considered to be a charge-transfer state, i.e. electronic charge has moved from the oxygen ligands to the central metal ion (Blasse 1980; Blasse and Grabmaier 1994). The real amount of charge transfer is usually small, but a considerable amount of electronic reorganization occurs, in which electrons are promoted from bonding orbitals in the ground state to antibonding orbitals in the excited state. The value of ΔR is large, the Stokes shift is large, and the spectral bands broad. Especially the complexes with the lighter metal ions show long decay times of their emission. The spin-orbit interaction is strong in WO_4^{2-} because of high atomic number of W and the emission lifetimes decrease correspondingly. The lifetime of blue emission of WO_4^{2-} ion is as short as 10 μs, which is 10 times shorter than that of the MoO_4^{2-} and 100 times shorter than that of the VO_4^{3-} ion.

5.10.1 *WO$_4$: Scheelite*

The mostly important mineral example is natural scheelite. Scheelite emits a bright blue emission in a broad band centered at 425 nm (Fig. 4.11) with decay time of

Fig. 5.97 Energy levels scheme in tetrahedral molecular ions CrO_4, VO_4, MoO_4, WO_4 (Tarashchan 1978)

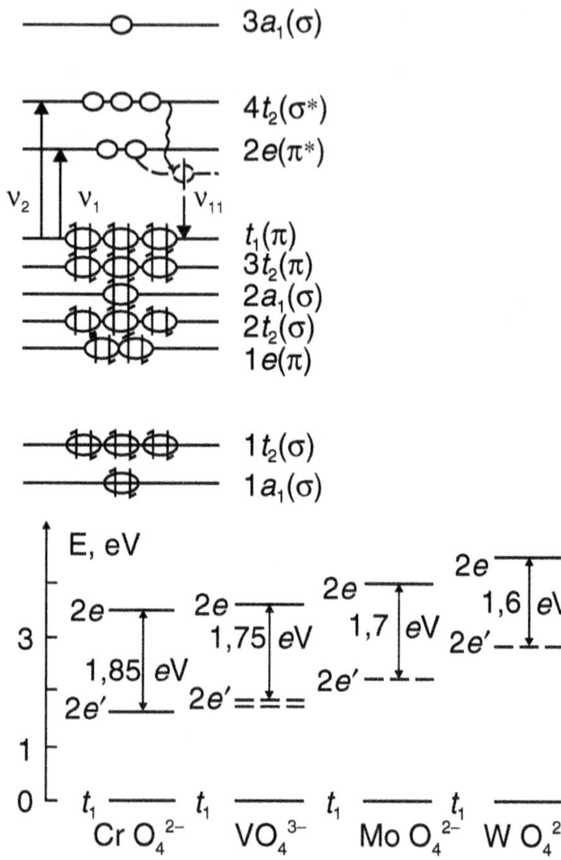

several μs. Calcium tungstate $CaWO_4$ has long been known as a practical phosphor, and has been carefully studied. The intrinsic blue luminescence center is the WO_4^{2-} complex ion in which the central W metal ion is coordinated by four O^{2-} ions in tetrahedral symmetry. It is widely accepted to use T_d symmetry to classify the terms and tentatively interpret the experimental data, while all of the modern work on tetroxo ions recognizes the fact that its symmetry is not maintained. Since the ground states $1t_2$, $1a_1$, 1e, $2t_2$, $2a_1$, $3t_2$ and t_1 one electron molecular orbitals are occupied, the WO_4^{2-} molecular complex can be excited via t_1-2e transition to the following molecular terms: $^3T_1$, $^3T_2$, $^1T_1$ and $^1T_2$ resulting from the (t^5_1, 2e) configuration. The lower lying triplet is believed to be responsible for luminescence. Other analogous T_d complex is molybdate MoO_4^{2-}, which is considered to be responsible for the green luminescence in scheelite (Tarashchan 1978).

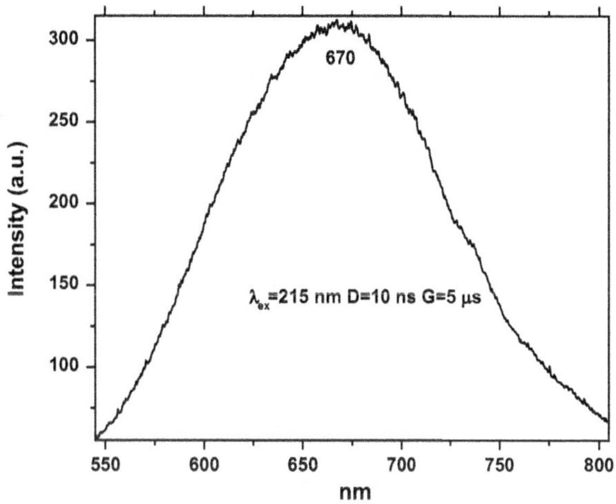

Fig. 5.98 Laser-induced time-resolved luminescence spectra of $BeAl_2O_4$:V

5.10.2 VO_6: Chrysoberyl, Pyromorphite and Beryl

Vanadium complex luminescence may be supposed in chrysoberyl where artificial chrysoberyl activated by V exhibits read broad band luminescence with decay time of approximately 5 μs under UV excitation (Fig. 5.98). Similarly, pyromorphite usually has orange to yellow UV excited luminescence characterized by broadband peaking at 580 nm (Fig. 4.60). It was earlier supposed that such luminescence is connected with $(VO_4)^{3-}$ emission center (Gaft 1984), but the possible role of Ag^+ also may be considered (Gorobets and Rogojine 2001). Beside, blue emission band in beryl luminescence spectrum with short decay time of 1 μs (Fig. 4.122a) can also be connected with $(VO_4)^{4-}$ complex, especially because the similar emission has been found in radioluminescence spectrum of beryl with elevated concentration of vanadium (Chithambo et al. 1995).

5.10.3 TiO_6: Benitoite, Chrysoberyl, Bahdadite, Baratovite and Aegirin

The luminescence of many titanium minerals was studied by steady-state luminescence spectroscopy and it was proposed that blue luminescence bands mutual for these minerals is connected with TiO_6 complex luminescence (Gaft et al. 1981a; White 1990). Figure 4.82a presents spectral properties of the blue emission from benitoite at 300 K. Under short and middle-wave UV laser excitation, such as at 266 and 308 nm, respectively, an intensive broad blue emission band peaking at approximately 420 nm with half-width of ~80 nm is detected. Spectra with different excitations, delays and gates revealed that this band consists of only one

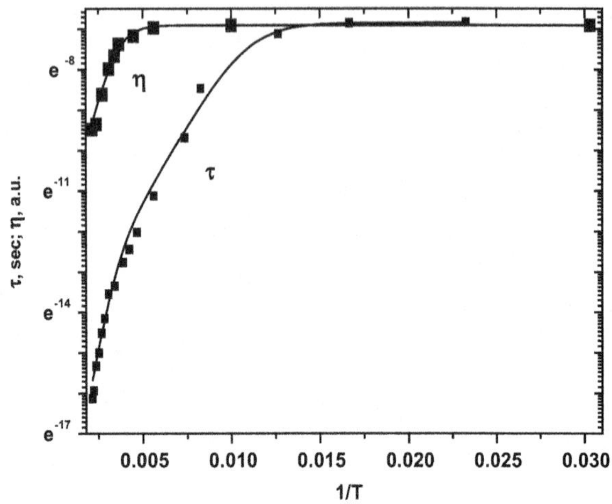

Fig. 5.99 Decay time and quantum yield of blue emission as a function of temperature: *dots* are experimental data and *solid lines* are best fit curves

component. The decay rate may be approximated by a mono-exponential curve with decay time of 2.6 μs. The excitation spectrum of the blue luminescence consists of a narrow band with half-width of 40 nm peaking at 290 nm and accompanied by the shoulder at 245 nm. At lower temperatures the luminescence efficiency becomes higher, while the spectral form is basically the same. At temperatures higher than room temperature the efficiency decreases and at 465 K is nearly quenched, while the band shape remains practically the same. Additional bands or shoulders do not appear at higher temperatures range. Decay curves at all temperatures remain mono-exponential, but the decay time becomes substantially longer with reduced temperature, from 1.1 ms at 40 K to 200 ns at 465 K (Gaft et al. 2004).

According to our experience all benitoite samples have intensive blue emissions under short wave and middle wave UV, and we have never seen or heard of any that did not. On the other hand, efficient luminescence has been observed from the TiO_4^{4-}, TiO_6^{8-} and even TiO_5^{6-} complexes. In addition, bazirite $BaZrSi_3O_9$ with the benitoite structure demonstrates luminescence of zirconate complexes ZrO_6^{8-} (Blasse 1980). The blue luminescence of impurity Ti^{4+} centers has been found in synthetic $BaSnSi_3O_9$-Ti^{4+} and $BaZrSi_3O_9$-Ti^{4+} with the benitoite structure (Konijenendijk 1981).

We thus have strong evidences in favor of an intrinsic model for the blue luminescence in benitoite, connected with the TiO_6 octahedra. Hence we have tried to determine if the spectroscopic experimental data may be explained based on such a model (Gaft et al. 2004).

Figure 5.99 summarizes the temperature behavior of decay time τ and quantum efficiency η of the blue luminescence from benitoite in the forms ln(τ) and ln(η) as a function of reciprocal temperature 1/T. Figure 5.100 demonstrates a suitable energy levels scheme. After excitation the metastable level 1 is populated due to nonradiative fast transition from excited level. Between levels 1 and 2 the

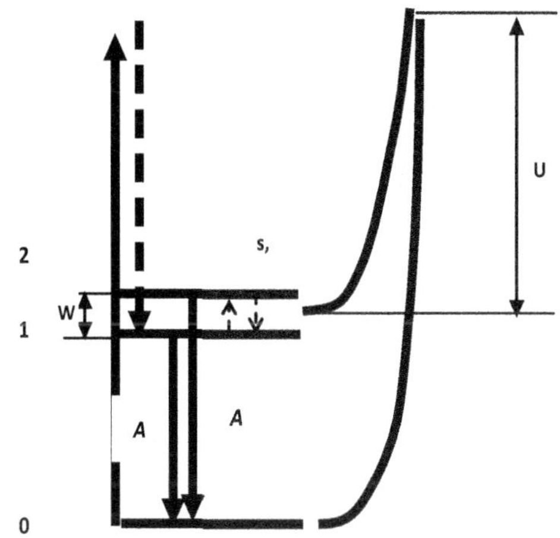

Fig. 5.100 Calculated energy levels scheme for blue benitoite luminescence

equilibrium population is established due to nonradiative transition. The relative quantum yield of the blue emission may be described by simple Arrhenius equation:

$$\eta = \frac{1}{1 + C\exp(-U/kT)} \tag{5.42}$$

where C is constant and U is energetic barrier between two closely spaced emitting excited states and excited level. The best fit to the experimental results is obtained when $U = 0.145$ eV or 1150 cm^{-1}.

The decay time is determined by the equation:

$$\tau = \eta \frac{1 + \exp(-W/kT)}{A_1 + A_2\exp(-W/kT)} \tag{5.43}$$

where A_1 and A_2 are radiation transition probabilities between the ground state and the first and second excited emitting states, respectively, and W is the energy difference between the first and the second excited emitting states. The best fit with experimental data is obtained for W = 0.06 eV or 460 cm^{-1}.

Such an energy levels scheme is compatible with the TiO_6 complex. Figure 5.101 presents a typical energy levels scheme for empty shell octahedral molecular complex centers. A molecular orbital calculation leads to e* and t_1 states for the lowest unoccupied molecular orbital and the highest occupied molecular orbital, respectively. By taking the e-t_1 transition into account, the excited electronic states of $t_1{}^5e$ electron configuration are found to consist of $^3T_{1u} \leq {}^3T_{2u} < {}^1T_{1u} < {}^1T_{2u}$ in order of increasing energy, with the ground state being a $^1A_{1g}$ state (Blasse 1980).

Thus the excitation band at 240 nm may be assigned to the allowed $^1A_{1g}$-$^1T_{2u}$ transition, and the weaker band at 290 nm assigned to the formally forbidden $^1A_{1g}$-$^1T_{1u}$ transition. Because of very long decay time, the emission at low temperatures must originate from the parent $^3T_{1u}$ triplet states, because the $^3T_{1u}$ state is the

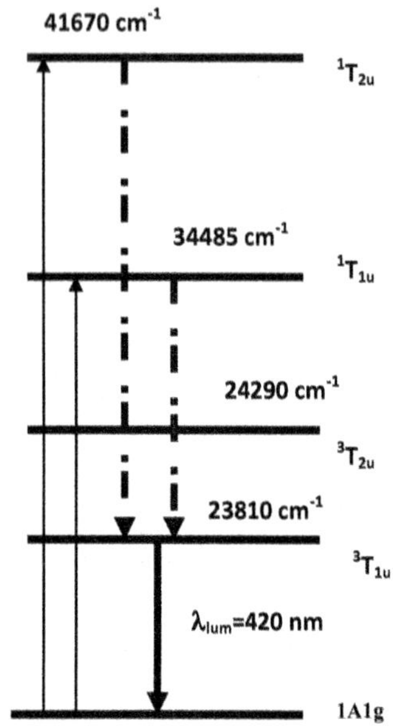

Fig. 5.101 Energy levels scheme for TiO_6 luminescence center in benitoite

only level which produces a sublevel from which a transition to the $^1A_{1g}$ ground state remains spin and orbitally forbidden. The presence of a higher level with higher radiation probability may be connected with the splitting of the emitting $^3T_{1u}$ levels into levels of A_{1u}, E_u, T_{1u} and T_{2u} symmetry due to spin orbit coupling. The T_{1u}-A_{1g} transition is allowed, while the others forbidden (Blasse 1980). One of the latter levels can act as an optical trap, which explains the temperature dependence of the decay time. In view of the small energy difference between the $^3T_{1u}$ and $^3T_{2u}$ levels, the $^3T_{2u}$ level may also play a role in the luminescence process at high temperatures.

Broad band emission was found in alexandrite from Brazil, where green luminescence presents peaking at 515 nm with decay time of 1.3 μs (Fig. 4.128b). Spectroscopically such emission may be preliminary connected to TiO_6 molecular complexes.

Baghdadite from Fuka, Okayama Prefecture, Japan shows a bright yellow fluorescence under UV (Hg 253.7 nm) excitation (Sidike et al. 2007). The photoluminescence (PL) spectrum at 300 K consists of one large band near 580 nm and two small UV bands at 318 and 397 nm. The optical excitation spectrum of the bright yellow fluorescence consists of two bands near 220 and 250 nm. The temperature dependence of the PL intensity exhibits linear thermal quenching. To reveal the origin of the bright yellow fluorescence from baghdadite, powder $Ca_3(Zr,Ti)Si_2O_9$ crystals are synthesized. Synthetic $Ca_3(Zr,Ti)Si_2O_9$ shows luminescence spectra similar to those of baghdadite, and the intensity of the yellow fluorescence is markedly increased by titanium addition. The origin of the bright yellow fluorescence from baghdadite is ascribed to the existence of titanium. Our study of decay time revealed that it is equal of approximately 10 μs, which corresponds to TiO_6 emission properties (Fig. 4.83).

The photoluminescence (PL) and optical excitation spectra of baratovite in aegirine syenite from Dara-i-Pioz, Tien Shan Mts., Tajikistan and katayamalite in aegirine syenite from Iwagi Islet, Ehime, Japan were obtained at 300 and 80 K (Sidike et al. 2008). Under short wave (253.7 nm) ultraviolet light, baratovite and katayamalite exhibited bright blue-white luminescence. The PL spectrum of baratovite at 300 K consisted of a wide band with a peak at approximately 406 nm and a full width at half maximum (FWHM) of approximately 6.32 kcm^{-1}. The excitation spectrum of the blue-white luminescence from baratovite at 300 K consisted of a prominent band with a peak at approximately 250 nm. The PL and excitation spectra of katayamalite were similar to those of baratovite. The luminescence from these minerals was attributed to the intrinsic luminescence.

5.11 Molecular Centers

5.11.1 Uranyl (UO_2)$^{2+}$

Uranyl is formed due to σ and π bonds of 5f, 6d, 7s electrons of uranium atom and 2p electrons of two oxygen atoms. Molecular orbitals of uranyl are formed by interaction of atomic orbitals of uranium and oxygen with the same type of symmetry and close energy levels. Bonding orbitals $1\sigma_u$, $1\sigma_g$, $1\pi_u$, $1\pi_g$ are filled with electrons with antiparallel spin directions. Nonbonding orbitals $1\sigma^*_g$, $1\sigma^*_u$, $1\pi^*_g$, $1\pi^*_u$ and so on are empty and form excitation levels. Ground state of uranyl is $\Sigma^+{}_{1g}$ (A_{1g}), which is singlet with summary spin 0 and center of symmetry. Excited state of uranyl is splitted. It has levels of different symmetry types A_{2g}, E_{2g}, A_{1g} and so on. These levels are splitted, in their turn, due to electron = phonon interaction forming a system of vibrational sublevels owing to vibration of oxygen atoms in uranyl. Absorption of photon by uranyl results in electron transition from the upper filled orbital $1\pi^4{}_g$ to one of excited levels of empty orbitals $1\delta_u$ and $1\varphi_u$ with vibrational sublevels. Electron transition in opposite direction results in emission of photon. Its energy is about $k_0 \approx 2 \times 10^{-4}$ cm^{-1}. This is zero-phonon electron transition. Its peak is accompanied by series of electron-phonon lines (Fig. 5.102). A feature distinguishing the molecular excitation and luminescence spectra of UO_2^{2+} is the fact that they arise from the transitions between electron-vibrational levels, i.e., the levels of electronic states split into sublevels whose spacing equals the vibrational frequency of the molecule (Marfunin 1979b). The color of luminescence of uranyl in different minerals is changing from a green-blue to the orange-red and the mean frequency interval alters from 840–880 to 680 cm^{-1}, correspondingly. The UO_2^{2+} exhibits always comparably long decay times in the range of 60–600 μs (deNeufille et al. 1981; Gaft 1989; Geipel 2006). Some of them fluoresce

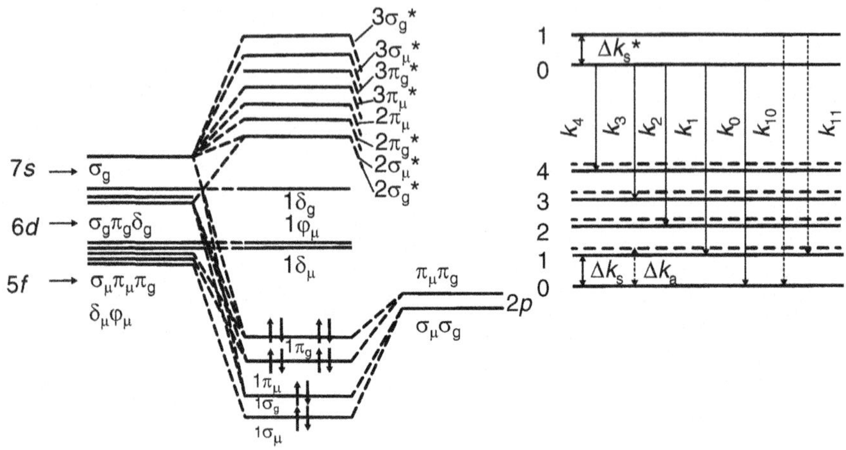

Fig. 5.102 Molecular orbitals scheme of UO_2 (Marfunin 1979b)

strongly at 300 K, some relatively weakly and others do not fluoresce at all. The emission of some of the weakly fluorescent uranyl minerals is dramatically increased by cooling to 77 K. The differences in the luminescence parameters are connected with the energy migration probability between the uranyls which, in turn, depends of the mineral structure, the connection type between various uranyl units and the chemical nature of ligands which interact with uranyl (Gorobets and Sidorenko 1974; Sidorenko et al. 1986; Gorobets and Rogojine 2001).

5.11.1.1 Apatite

Sedimentary carbonate-fluor-apatite $Ca_5(PO_4,CO_3)_3F$ is not luminescent under X-rays, and under UV lamp excitation it is characterized by broad structureless bands which are very similar to those encountered in many sedimentary minerals. It was concluded that this luminescence is due to different kindes of water-organic complexes (Tarashchan 1978). For this reason the luminescence properties of the sedimentary apatites is much less informative compared to magmatic apatite and attracted not much attention.

Carbonate-fluor-apatite accommodates large quantities of trace elements, mainly uranium, which are potential luminescence centers. It has been proposed that uranium may occur in phosphorites in the following forms: as a separate uraninite phase; as an adsorbed or structurally incorporated uranyl ion; as dominantly a U^{4+} replacement for Ca^{2+}; to be structurally incorporated appreciably as both U^{4+} and U^{6+} (Altschuler 1980). The steady-state luminescence of water-organic comlexes is strong and conceales the weaker characteristic luminescence of uranium containing centers, which can be detected by the difference in decay times only. The reason is that the decay time of water-organic complexes is characterized by two time intervales: less than 30 ns and more than 10 ms. Since the uranium centers have decay times in the microseconds range, it is possible to detect them by the time-resolved spectroscopy. In the time-delayed laser-induced spectroscopy, the luminescence spectra are recorded at a fixed moment after a laser pulse. These spectra may be different from the integrated steady-state ones since after a certain time short luminescence will be practically absent.

Spectra of the laser-induced luminescence of apatites with the long decay time (more than 1 μs) consist of several sets of relatively narrow bands in the green part of the spectrum, which are characterized by the clear vibrational structure (Panczer et al. 1998). The first type contains at 300 K the characteristic set of maxima at 484, 504, 526, and 546 nm. At 77 K the intensity only slightly increases, the spectrum is nearly the same, but the structure is more pronounced (Fig. 4.5c). The luminescent decay measured at 526 and 504 nm may be represented by the exponential fitting with the main decay component of 350 μs. The second type contains at 300 K the set of bands at 467, 486, 505, 526 and 547 nm (Fig. 4.5b). At 77 K the intensity is much stronger, the band at 467 nm disappears and the other bands are narrower. The quenching of the emission monitored at 526 nm at 300 K may be represented by the exponential fitting with the main decay component of 200 μs. The third type contains at 300 K the bands at 508 and 524 nm which are the strongest at 77 K (Fig. 4.5a). The quenching of the emission monitored at 524 nm at 300 K may be represented by the exponential fitting with the main decay component of 300 μs.

Thermal treatment has great influence on luminescence spectra. Certain cation may change valence and accommodation form during heating, with transformation from nonluminescent form to luminescent. Luminescent properties under UV excitation of the sedimentary phosphorites are strongly changed after the oxidizing thermal treatment. Thermal annealing at temperatures of up to 600 °C did not cause spectral changes, but above that temperature drastic alteration took place. The spectra at 300 K are characterized by very strong structuruless band with maximum at 527 nm. The decay curve has the main component of ~110 μs. At 77 K the spectrum is strongly changed with the intensiive narrow line at 522 nm (Fig. 4.5d). The decay is substantially slower with the main component of ~280 μs.

The spectral-kinetic parameters of the green laser-induced luminescence of the sedimentary apatites allow its association with UO_2^{2+} emission. The spectra presented in Fig. 4.5b, c are typical for uranyl minerals and it is possible to suppose that we are dealing with separate uranyl mineral phase on the apatite surface. Comparison with known uranyl minerals laser-induced luminescence shows that the most similar spectral-kinetic parameters have minerals andersonite $Na_2Ca(UO_2)(CO_3)_3 \times 6H_2O$ and liebigite $Na_2(UO_2)(CO_3)_3.10H_2O$, but this identification is ambiguous.

The spectra of the green laser-induced luminescence represented in Fig. 4.5a, together with their decay time, also allows its association with UO_2^{2+}. These luminescence spectra strongly differ from the spectral parameters of the all known uranyl minerals. For this reason this type of green luminescence is not possible to connect with finely dissipated uranyl phases. On the other hand, this luminescence is very similar in such different host minerals as sedimentary apatites, opalites, chalcedony, chert, quartz and barites. Luminescence independence from minerals structure evidences that it may be connected with uranyl adsorption on the minerals surface, supposedly in the form of $(UO_2 \times nH_2O)^{2+}$.

When the samples are heated the emission spectra are much less structured, more intensive and short-lived. It is known (Reisfeld and Jörgensen 1977) that uranyl which is a linear triatomic molecule can form complexes with ligands in the equatorial plane. When OH-group or F are the complexing species, the bond is very weak since there is no significant overlap of the molecular orbital of the ligands with those of the uranyl. As result, a well resolved spectrum is obtained, resembling gaseous molecules, resulting from the electron transitions from the oxygen to the 5f orbitals of the uranium. The spectra become much less resolved with stronger ligands occupying the equatorial plane. When water molecules are evolved at elevated temperatures, the OH groups disappear and, probably, P-ligands become strongly bounded with uranyl orbitals in the equatorial plane. The spectrum becomes less distinct, the intensity increases because of the less pronounced presence of a center of inversion which makes the parity-forbidden transitions in uranyl more allowed and correspondingly this is followed by shortening of the decay time.

5.11.1.2 Minerals of Uranyl

Luminescence of uranyl minerals is connected with uranyl emission. Energy migration plays very important role and its behavior was well analyzed by Gorobets and Rogojine (2001). In carbonates there is no energy transfer between $(UO_2)^{2+}$ and they therefore give bright blue-green emission at 300 K with maximum decay time of approximately 600 μs, narrow spectral lines and a large vibration constant of 820–840 cm^{-1}. In sulphates, phosphates and arsenates, uranyl ions are bond more strongly and their luminescence is green with decay times, vibration constants and temperatures of quenching a little lower. In silicates, vanadates, molybdates and hydroxides of uranyl strong exchange interaction and energy transfer take place. In this case, luminescence decay time falls to approximately 100 μs and vibration constant decreases to 750–680 cm^{-1}. Yellow, orange and red luminescence of these minerals can be observed at low temperatures.

5.11.1.3 U^{4+}

The reabsorption lines of U^{4+} have been found in zircon (Gaft et al. 1986) luminescence spectra. The strongest reabsorption lines at 592 and 656 nm (Fig. 4.91b) clearly seen, which are totally identical to absorption lines of U^{4+} in zircon (Platonov et al. 1984).

5.11.1.4 U^{6+}

Spectral parameters of the structured green luminescence which often appears in sedimentary apatite after oxidizing heating (Fig. 4.5d) are absolutely similar to those of U^{6+} luminescence in fluorite after thermal treatment (Tarashchan 1978). Principally, during the calcination of the sedimentary phosphates new mineralogical phases, including fluorite, may be formed. Taking these data into account, it is possible to conclude that after thermal treatment uranium is concentrated in fluorite lattice in the form of U^{6+}. It is interesting to note that in magmatic apatites the luminescence of uranium containing centres have not been discovered as before and after oxidizing heating. Thus it is reasonable to suppose that uranium is present mainly in U^{4+} form. The U^{4+} with ionic radius of 0.97 A may be located in the apatite structure instead of Ca^{2+} with ionic radius of 0.99 A. The mostly possible way of the excess charge compensation is the Na^+ for Ca^{2+} structural substitutions.

5.11.2 *O_2^-, S_2^-, M- and V_K-Centers*

A distinctive feature of the O_2^- and S_2^- luminescence spectra in minerals is a quasi-linear vibrational structure of the broad luminescence band (Tarashchan 1978). The O_2^- and S_2^- molecular ions are isoelectronic. From the molecular orbital diagram describing their electron structure the emission transition $^2\Sigma_g - {}^2\Pi_{3/2}$ is determined. When observing luminescence spectra at 77 K, a fine structure associated with the frequency of intra-molecular vibrations of O_2^- and S_2^- is detected. This frequency depends on the type of the molecular ion, on inter-nuclear distance and upon the particular position of the molecular ion in the structures. For S_2^- the maximum of the emission band lies within the range of 600–700 nm with a mean vibration frequency of 500–600 cm^{-1}, while for O_2^- the respective maximum is 450–550 nm with frequency in 800–1200 cm^{-1} range.

5.11.2.1 S_2^-: Sodalite, Tugtupite, Hackmanite and Lasurite

The luminescence properties of S_2^- centers correspond well to the spectral-kinetic characteristics of the structured yellow-orange spectra observed in time-resolved spectra of sodalite, tugtupite (Fig. 5.103), hackmanite (Fig. 5.104) (Gaft et al. 2009) and lasurite (Fig. 5.105). Molecules of S_2^- are unstable in the free-state, but become stabilized in a crystal lattice. Sulfur substitutes for Cl atoms in the tugtupite and sodalite structures, and are surrounded by four Na atoms (Tarashchan 1978). Luminescence spectra of S_2^- have vibrational structure due to symmetric vibrations of sulfur atoms in the molecules. Quite the contrary, the excitation spectrum of sodalite and tugtupite luminescence is structureless at room and low temperatures. Thus the vibrational sublevels of the ground state are not stable and do not actively

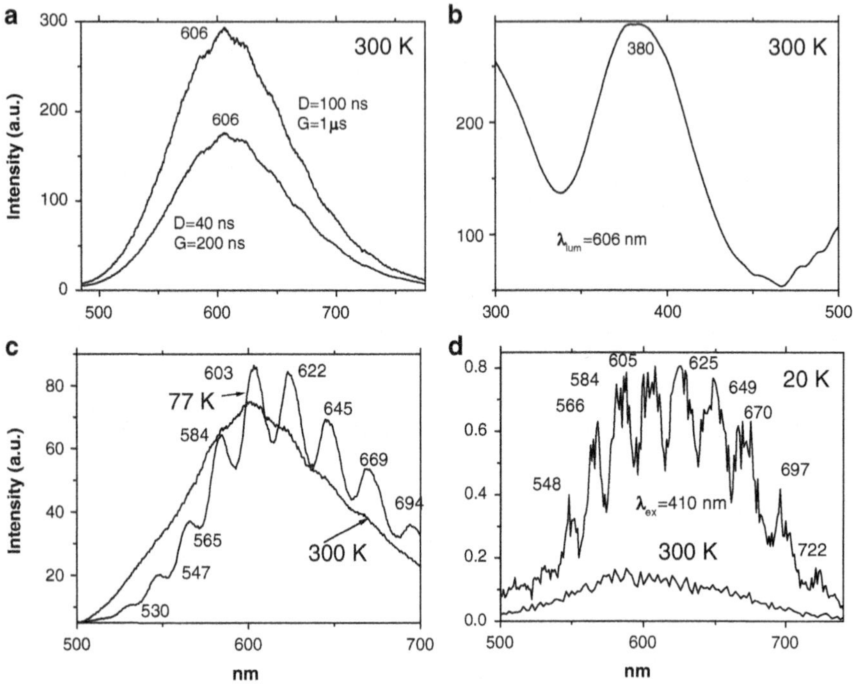

Fig. 5.103 (**a–d**) Laser induced time resolved luminescence of S_2^- in sodalite and tugtupite

participate in optical processes. Consequently the radiational transitions in the ion S_2^- take place from the vibrational sublevels of the excited state to the ground state. At room temperature the structure is very weak, which is evidently connected with fast deactivation from the excited vibrational sublevel. At low temperatures such deactivation is much less effective, and these spectra show up to ten equidistant lines related to phonon sublevels of the excited state of molecules S_2^-. The characteristic frequency of vibrational quantum is $D_k = 500–600\ cm^{-1}$ (Tarashchan 1978). The approximately 100 ns decay time of the structured band in tugtupite and sodalite at low temperature is shorter, compared to the decay time of approximately 135 ns at room temperature, while the luminescence intensity remains approximately the same. It corresponds well to the energy levels scheme of S_2^- centers where luminescence transitions from the excited vibrational sublevels are additionally quenched by the radiationless transitions to the bottom of the excited state. This explains the fact that after a relatively long delay time of 500 ns only structureless luminescence is detected in hackmanite at 300 K, while the weak vibrational structure is detected with shorter delay (Fig. 5.104).

It is interesting to note that S_2^- luminescence centers in alkali halides also have very short decay times of 60–75 ns at low temperatures (Hessman and Fiscima 1991). On the other hand, their decay time is strongly reduced as the temperature is raised above certain limits (80–90 K) which are accompanied by strong

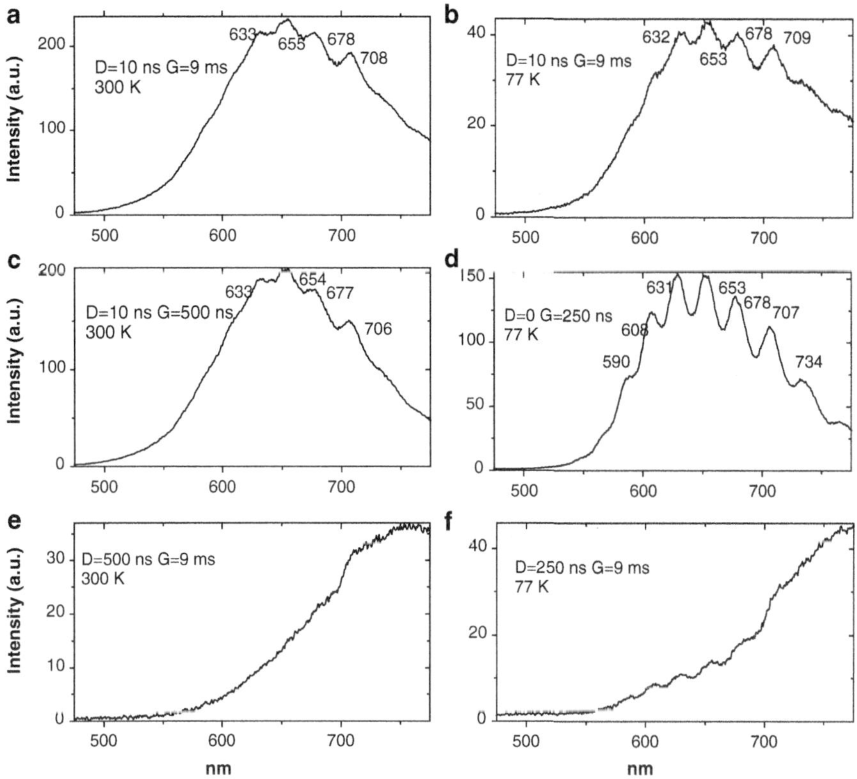

Fig. 5.104 (**a–f**) Laser induced time resolved luminescence of S_2^- in hackmanite

luminescence intensity quenching. This was ascribed to a non decay initiated by the intersection of the potential curves of electronic ground and excited states. This difference may be explained by the fact that the intersection of the potential curves of electronic ground and excited states in sodalite and tugtupite takes place at substantially higher temperatures as compared to the alkali halides. The traditional spectroscopy under steady excitation reveals just one type of S_2^- luminescence center in tugtupite, sodalite and hackmanite (Tarashchan 1978; Sidike et al. 2007). Laser-induced time-resolved luminescence spectroscopy with different excitations, delay times and gate widths for the first time enabled detection of several different types of S_2^- centers in these minerals, especially in sodalite. It was found that in certain samples the same spectra are detected with different excitations and gating, while sometimes they are quite different. For example, bands peaking at 618 and 644 nm, and not at the usual 603 nm, have been found. The most red-shifted luminescence of S_2^- was detected in a hackmanite sample peaking at 650 nm, with excitation by 355 nm (Gaft et al. 2009). The luminescence spectra diversity may be explained by the following. The energy of ^2P_g and ^2P_u electronic levels is influenced by local surroundings with resulting changes in luminescence spectra.

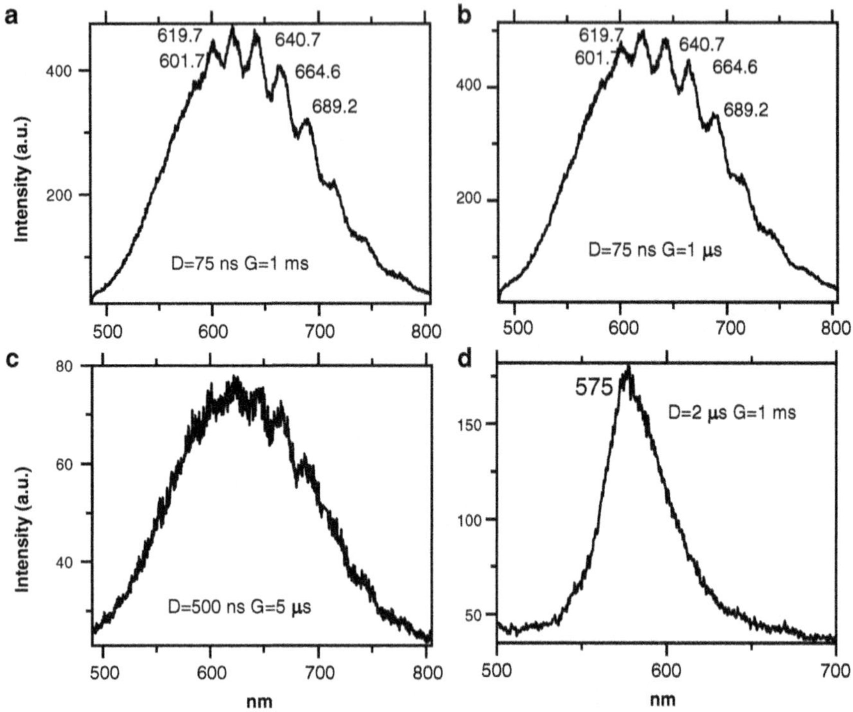

Fig. 5.105 (**a–d**) Laser induced time resolved luminescence of S_2^- in lasurite

According to Tarashchan (1978), the luminescence spectra of S_2^- in different minerals are rather similar, but changes in local symmetry lead to a shift in emission maxima of up strong yellow luminescence of S_2^- in the silicate mineral scapolite (Burgner et al. 1978), and it was found that the emission spectrum is dependent upon excitation wavelength, indicating the presence of S_2^- in sites with differences of the surrounding atoms. Another possibility is that luminescence centers different from S_2^- may be responsible for the detected luminescence. For example, Schlaich et al. (2000) obtained an emission spectrum of Se_2^- in synthetic sodalite at 300 K which contains a red band with strong vibrational structure, peaking at 14,000 cm^{-1} (715 nm). The electronic structure of the Se_2^- center is similar to that of S_2^-, and the luminescence band was also ascribed to a ^2P_u-^2P_g electronic transition. We did not find Se impurities in the studied samples, but its detection limit of 2 ppm by solution-ICP-MS is relatively high, and it is quite possible that laser-induced luminescence may be more sensitive. This question needs further investigation.

The photoluminescence and excitation spectra of sodalite from Greenland, Canada and Xinjiang (China) were observed at 300 and 10 K (Sidike et al. 2006). The features of the emission and excitation spectra of the orange-yellow fluorescence of these sodalite are independent of the locality. The emission spectra at 300 and 10 K consist of a broad band with a series of peaks and a maximum peak at

648 and 645.9 nm, respectively. The excitation spectra obtained by monitoring the orange-yellow fluorescence at 300 and 10 K consist of a main band with a peak at 392 nm. The luminescence efficiency of the heat-treated sodalite from Xinjiang is about seven times as high as that of untreated natural sodalite. The emission spectrum of the S_2^- center in sodalite at 10 K consists of a band with a clearly resolved structure with a series of maxima spaced about 560 cm^{-1} (20–25 nm) apart. Each narrow band at 10 K shows a fine structure consisting of a small peak due to the stretching vibration of the isotopic species of $^{32}S^{34}S^-$, a main peak due to that of the isotopic species of $^{32}S_2^-$ and five peaks due to phonon sidebands of the main peak.

The emission and excitation spectra of yellow luminescence due to S_2^- in scapolite were observed at 300, 80 and 10 K (Sidike et al. 2010b). Emission and excitation bands at 10 K showed vibronic structures with a series of maxima spaced 15–30 and 5–9 nm, respectively. The relative efficiency of yellow luminescence from one scapolite sample was increased up to 117 times by heat treatment at 1000 °C for 2 h in air. The enhancement of yellow luminescence by heat treatment was ascribed to the alteration of SO_3^{2-} and SO_4^{2-} to S_2^- in scapolite.

5.11.2.2 O_2^-

Relatively broad luminescence bands with vibrational structure even at 300 K peaking at blue-green spectral region in several minerals have been ascribed to O_2^- emission center (Tarshchan 1978).

5.11.2.3 M-center and V_K-center

M-center consists of two electrons captured by F anion vacancies in fluorite. It is characterized by red luminescence peaking at 725 nm and excited in 370 and 530 nm spectral ranges (Tarshchan 1978). This center is characterized by long decay time and is detected with relatively broad gate width. V_K-center is a hole center, which is formed by F^- ions which loses its electron. It is characterized by UV luminescence band peaking at approximately at 290 nm with very short decay time (Fig. 5.106).

5.12 Radiation-Induced Centers

Radiation induced luminescence centers and radiation influence on the luminescence properties of the other emission centers become more and more important both theoretically and for different applications (Nasdala et al. 2013).

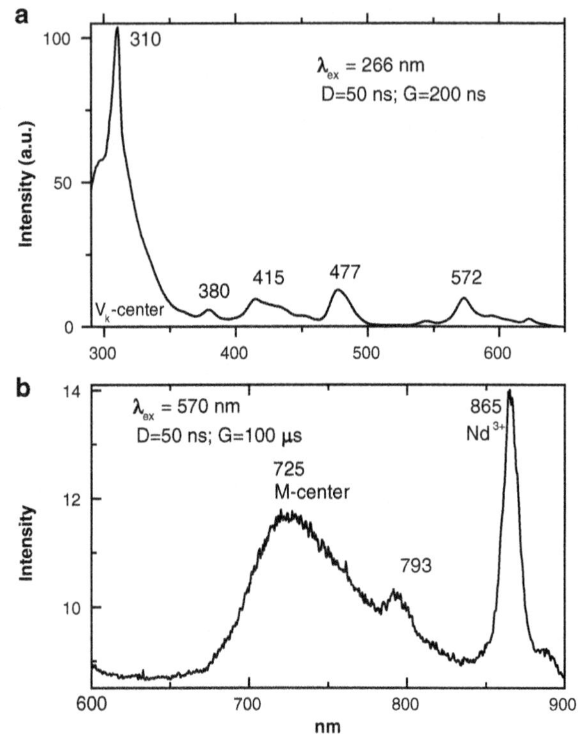

Fig. 5.106 (**a–b**) Laser-induced time resolved luminescence of Vk and M centers in fluorite

5.12.1 *Zircon*

Figure 4.91a represents luminescence band detected in all investigated natural zircons. It has excitation band peaking at 310 nm. The broad yellow band at 575 nm with decay time of 25–35 μs represents "classical" zircon luminescence. The picture is not principally changed with different delay time and gate width. After heating the sample up to 700 °C the luminescence intensity is nearly the same, but after that it is strongly reduced and at 800 °C the yellow luminescence disappears. Luminescence spectra at 77 and 4.2 K are not substantially different.

As it was already mentioned, the origin of this band was ambiguous. In order to clarify this we studied the irradiation influence on laser-induced time-resolved luminescence of the following samples (Gaft et al. 2003b):

- $ZrSiO_4$ activated by Eu with characteristic narrow Eu^{3+} lines and irradiated by different doses of thermal neutrons;
- $ZrSiO_4$ activated by Ca and P practically without luminescence and internally irradiated by alpha particles;
- Natural zircons heated at 800 °C during one hour when natural yellow broad-band luminescence nearly totally disappears and irradiated by different doses of alpha particles.

Fig. 5.107 (**a–c**) Laser-induced time-resolved luminescence spectra of $ZrSiO_4$-Eu^{3+} (**a**) irradiated by α-particles (**b**) and neutrons (**c**)

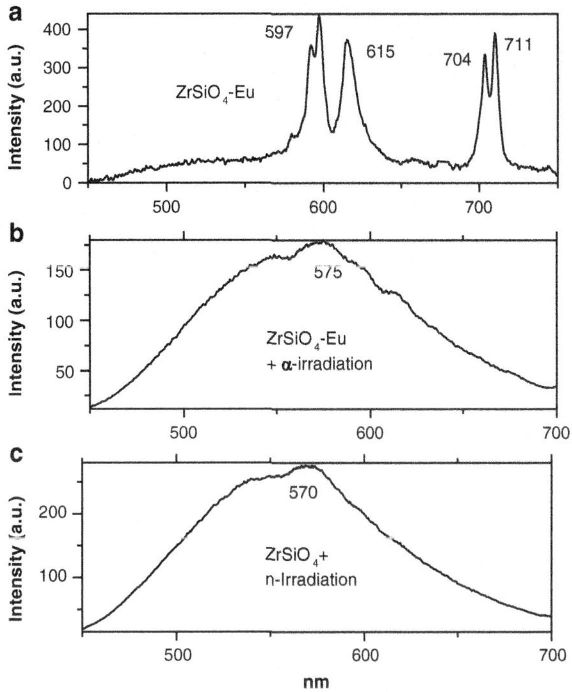

Non-irradiated $ZrSiO_4$-Eu is characterized by orange luminescence (Fig. 5.107a). After n-irradiation broad yellow luminescence band appears, which is much stronger then original Eu^{3+} emission. Such yellow luminescence is proportionally stronger with increasing irradiation time (Fig. 5.107b–c). Strong broad-band yellow luminescence appears also after alpha-irradiation in $ZrSiO_4$–Ca/P and in natural zircon after heating at 800 °C, which were not luminescence before irradiation. Investigation made at 300 and 77 K with different laser excitations, delays and gates proved that despite of very broad half-width and apparently asymmetric form, luminescence consists of only one band. Both for alpha and neutron irradiations, such induced luminescence is stable with time and in our study it has been detected several years after irradiation. The spectral shape of this radiation-induced luminescence is very close to those of natural zircon samples. Its decay time is of approximately 30–35 μs and the intensity remains stable under the heating up to 700 °C, which is characteristic for natural samples. Thus according to our knowledge it may be unequivocally concluded that "classical" yellow luminescence band in natural zircon with $\lambda_{max} = 575$ nm, half-width $\Delta = 160$ nm, decay time of $\tau = 30$–35 μs and thermally stable up to approximately 700 °C is connected with radiation induced centers. In minerals, the source of irradiation may be connected with radioactive decay of U and Th impurities.

For the interpretation of such centers it is important to note that yellow luminescence is connected only with neutron and alpha irradiations and not generated by higher doses of X-rays, beta and gamma irradiations. Thus corresponding centers have to be specific only for neutron and alpha treatment. According to EPR data (Hayashi et al. 1990) the neutron irradiation produces four typical signals in $ZrSiO_4$ with g = 2.000, 2.001, 2.003 and 2.008, while gamma-ray irradiation produces only the first two signals. The last one is completely extinguished already eight months after irradiation, while the signal with g = 2.003 remains. Laser-induced luminescence in our study is detected even 10 years after irradiation and it is possible to suppose that it is connected with radiation-induced center with g = 2.003. Such value of g-factor corresponds to $SiO^{n-}{}_m$ structural defects. Thermal stability of $SiO^{n-}{}_m$ defects have been studied by EPR and compared with those of integrated yellow zircon luminescence. It was found that any correlation is absent (Gaft et al. 1986). Thermal stability of radiation-induced part of yellow luminescence is correlated with those of $SiO^{n-}{}_m$ structural defects, but $SiO^{n-}{}_m$ defects are easily reproduced by X-ray irradiation after heating, while yellow luminescence not. Thus the exact type of the yellow luminescence center needs further clarification.

Recently the radiation induced nature of the yellow luminescence was confirmed by the study of helium irradiated zircon where at very low damage levels a broad-band yellow emission centre is generated whereas at elevated damage levels, this centre is suppressed and samples experience a general decrease in their emission intensity (Nasdala et al. 2011).

It has to be noted that the systematic changes in REE^{3+} emissions takes place depending on the accumulated radiation damage (Lenz and Nasdalla 2015). Structural reconstitution as caused by dry annealing resulted in intensity gains and decreases of half-widths of REE^{3+} emissions. The band half-widths of distinct luminescence Stark's levels of the $^4F_{9/2} \rightarrow {}^6H_{13/2}$ transition of Dy^{3+} (~17,250 cm^{-1}; ~580 nm wavelength) and the $^4F_{3/2} \rightarrow {}^4I_{9/2}$ transition of Nd^{3+} (~11,300 cm^{-1}; ~885 nm wavelength) were found to correlate sensitively with the degree of radiation damage accumulated. These REE^{3+} emissions are proposed as potential measure of the irradiation-induced structural disorder of zircon. The two emissions are considered particularly suitable because (1) they are commonly detected in PL spectra of natural zircon, and (2) they are hardly biased by other emissions or Stark's levels. Preliminary calibration curves that relate band-width increases to the a dose were established using a suite of well-characterized Sri Lankan zircon. Band broadening upon increasing corpuscular self-irradiation is assigned to increasing structural destruction, i.e., the increasing perturbation of REE^{3+} cationic lattice sites.

5.12.2 Calcite

5.12.2.1 Blue Emission

One specific variety of calcite is known as Terlingua-type (TT) because of its original discovery in large quantities at Terlingua, Texas, an old Hg-mining area.

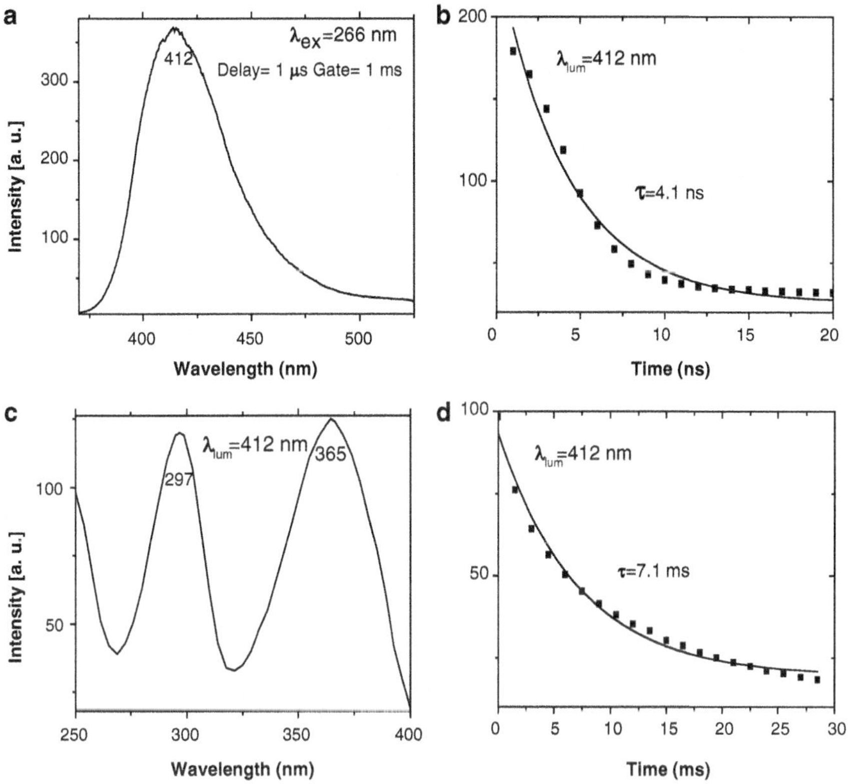

Fig. 5.108 (**a–d**) Luminescence (**a**), decays (**b**, **d**) and excitation spectra (**c**) of blue radiation-induced centers in calcite

It is characterized by unique intense blue emission under short-wave UV lamp excitation with an extremely long decay time, accompanied by pink-orange luminescence under long-wave UV excitation (Gaft et al. 2008a). In both cases the emission is especially strong in the more physically colored parts, which may be pinkish to red, and rarely green. Under laser excitations at 248, 266, 337 and 355 nm, and at 300 K, TT calcite has an intense UV to violet emission band peaking at 412 nm with a half-width of ~55 nm (Fig. 5.108a). The excitation spectrum of this band is composed of a short-wave tail in the spectral range less than 250 nm, and two UV bands peaking at 295 and 365 nm (Fig. 5.108c). The band intensity is much stronger in the pink calcite samples. The decay function contains two components, a very short one with $\tau_1 = 5$ ns and a very long one with $\tau_2 = 7.1$ ms (Fig. 5.108b, d). The total similarity of the violet luminescence spectra and its decay components under different excitations indicates that we are dealing with one luminescence center with two decay components and not with two different luminescence centers. At temperatures lower than 300 K the emission spectrum changes drastically. With decreasing temperature to 200 K, it has the same general form, but the emission intensity is constantly lower. A second band at 470 nm appears and dominates the spectrum at low temperatures. The intensity and maximum of this band is nearly the same from 120 to 40 K. However substantial changes are observed within the decay times. The main decay component of the band peaking at 412 nm in the ms range becomes longer with decreasing temperature, reaching $\tau_2 = 20$ ms at 265 K. The decay dependence of the band peaking at 472 nm is characterized by a prominent ns component starting from its appearance at approximately 220 K and extending down to 77 K. At still lower temperatures the luminescence spectrum, decay time and intensity remain approximately the same down to 4.2 K.

A violet emission band having two decay components, such as described in this paper, is very unusual for mineral luminescence. Traditional luminescence centers for minerals in this spectral range, such as Ce^{3+}, and Eu^{2+}, cannot explain it. Emission of Ce^{3+} in calcite is characterized by a rather broad UV luminescence band with two emission maxima and only one short decay lifetime component of 20 ns. Europium concentrations in our calcite samples are extremely low and Eu^{2+} decay behavior is usually characterized by one decay component of 600–800 ns. Thus direct connection between Eu^{2+} center and the violet luminescence band in our calcite specimens is very doubtful. Such luminescence decay properties may be explained by a configurational diagram scheme with one ground state and two very closely spaced excited states, where the radiative transition from the lowest excited state to the ground state is totally forbidden, while from the higher excited state to the ground state it is totally allowed. In such case, excited electrons exist in both lowest and higher excited states with n_l (lower) and n_h (higher) populations. At low temperatures n_l evidently is substantially higher than n_h, and luminescence intensity is low with a very short lifetime due to the few n_h electrons. With increased temperature, the higher excited state is more populated and the luminescence intensity is stronger with two decay components, one very short from the original n_h electrons and another very long due to electrons arriving from n_l.

Two types of luminescence centers have suitable energy level schemes to explain the violet band spectral kinetics: s^2 type elements such as Tl^+, Pb^{2+}, Bi^{3+}, Sn^{2+} and Sb^{3+}, described above, and radiation-induced F- centers. Nevertheless, TT calcite samples contain negligible quantities of those elements (Gaft et al. 2008a). The other possibility is a radiation-induced luminescence center. The best-known radiation-induced center is the F-center. It consists of an electron in a halide vacancy, for example in KCl. A good approximation of its absorption spectrum are the electronic states of the hydrogen atom. An F-center has a ground state of s-type and a split excited level, where the lower excited state is mainly of s-type and electron transitions to the ground state are forbidden, while the higher state is mainly of p-type and electron transitions to the ground state are allowed.

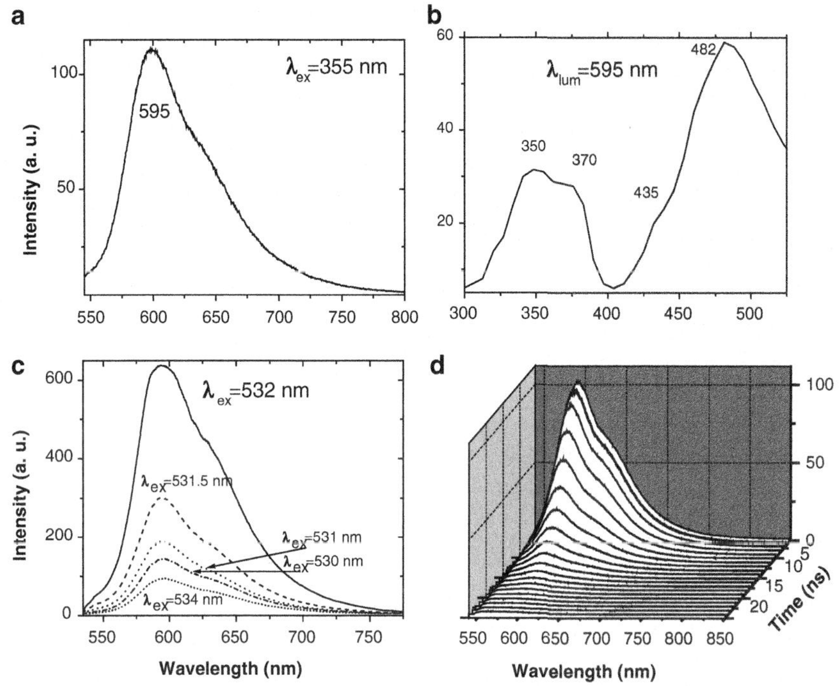

Fig. 5.109 (a–d) Luminescence (a, c) excitation (b) and decay (d) of orange radiation-induced centers in calcite

5.12.2.2 Orange Emission

Under long-wave UV (355 nm) and visible excitations a band peaking at 594 nm is observed with a half-width of ~95 nm (Fig. 5.109a). The excitation spectrum of this emission contains four maxima peaking at 350, 370, 435 and 480 nm (Fig. 5.109b). The OPO tunable source enabled excitation at energies very close to the luminescence band. This revealed a prominent excitation band at 532 nm, and thus a very small Stokes shift of 50–60 nm between the excitation and emission maxima (Fig. 5.109c). The luminescence band is not symmetrical, with a shoulder at 625 nm, but this feature remains in all time-resolved spectra with different delays and gates and does not resolve to separate emission bands. This band can be detected with an extremely narrow gate width, which is strong evidence that its decay time is very short, approximately 7–8 ns (Fig. 5.109d). This orange luminescence is especially strong in the most strongly colored parts of calcite (dark pink to red). At 40 K the band becomes very intense, while its spectrum and decay time remain practically the same. The orange band is thermally unstable, but in different samples it disappears after heating to different temperatures in a wide range from 230 °C to 500 °C. After this band disappears, another orange band becomes visible at 612 nm with a very long decay time of 8–10 ms. This band is consistent with a substitutional Mn^{2+} luminescence center.

The orange emission band with decay time of several ns is very unusual, because as a known broad orange-red luminescence bands in minerals have decay times from several μs (Ti^{3+}, Bi^{2+}, Cr^{3+}) to several ms (Mn^{2+}, Fe^{3+}, Cr^{3+}). It is interesting to note that the orange band peaking near 600 nm was detected in the thermally stimulated emission of Kazakhstan pink calcite, but without measurement of the decay time it was originally ascribed to a substitutional Mn^{2+} luminescence center. A possible exception from the usual connection of the orange emission in minerals with Mn^{2+} centers is the orange luminescence in rose-colored elbaite peaking at 570 nm, with emission intensity much stronger at lower temperatures. The decay time of such luminescence was not reported, but an allowed 2p-1s electron transition in color centers may have a lifetime in the ns range. The rose color of elbaite is thermally unstable and disappears after thermal treatment along with the orange luminescence. However, the thermal influence on the pink color and orange luminescence in TT calcite is different from elbaite. The color always disappears after heating at 230 °C, while the orange luminescence is sometimes stable up to 500 °C. In addition, X-ray irradiation of heated calcite samples does not restore the color and luminescence. Thus we must conclude again that the orange luminescence center in the TT calcites is probably due to a radiation-induced origin.

We have no geological evidence for the possible irradiation events leading to the formation of such radiation-induced centers in our calcite sample, but several indirect indications are present. The nature of pink coloration in calcite has been intensively studied and two possible models have been proposed. In Kazakhstan samples an elevated concentration of Pb was detected, from 0.1 % to 2 %. This color is thermally unstable and disappears after heating at 300 °C together with all absorption bands except one at 240 nm, which has been assigned to Pb^{2+} absorption. The thermally bleached calcite has its pink color restored after X-ray, gamma ray or electron irradiation. The pink color has been ascribed to absorption bands at 370 and 500 nm connected with Ca^{2+} vacancies and CO_3^{2-} units, respectively, which form as a result of ionizing irradiation. Usually such centers are not stable at room temperature, but it has been postulated that they can be stabilized by some kind of structural defects, such as Pb^{2+} substituting for Ca^{2+}. We studied a Kazakhstan calcite sample with pink color and high Pb concentration for comparison, but unusual luminescence was not detected. On the other hand, we did not find an elevated concentration of Pb in any pink calcite samples having the unique TT luminescence. In addition, the pink coloration in our samples is different from the Kazakhstan samples. It disappears after heating at ~200 °C and not 300 °C, and it is not restored after X-ray or beta irradiation. Thus it may be concluded that the pink color in our calcite is not connected with a Pb^{2+} impurity. Many TT pink calcites samples are enriched by rare-earth elements compared to the white or colorless areas in the same samples by a factor of 2–30, depending on the element. Thus our analytical data correspond with Rossman's interpretation, where the pink color is assigned to radiation damage centers, which often involve rare earths (http://minerals.gps.caltech.edu). The luminescence of the violet band is much more intense in the colored parts of the calcite compared to the colorless areas. It is well known that the sensitivity of luminescence spectroscopy is much greater than optical absorption spectroscopy, thus if any radiation-induced centers produce the pink calcite coloration, their concentration is definitely enough to generate luminescence activity.

Other evidence for natural irradiation is the presence of Natural Termoluminescence (NTL) in the pink calcite samples (Gaft et al. 2008a). NTL signal detected through violet-blue filters, corresponding to the violet-blue calcite emission spectrum, is very strong. Several peaks at 227 °C, 263 °C, 335 °C, 412 °C and 460 °C characterize it, although the first two peaks are sometimes absent. Kazakhstan samples also have NTL, but the spectra are different from our TT calcite results. In the Kazakhstan samples the most prominent peak is situated at 133 °C, while two weaker bands are present at 210 °C and 323 °C. The energy depths of corresponding traps have been calculated as 0.95 eV for the TL peak at 133 °C, and 1.83 eV for the TL peak at 323 °C. It has been calculated that at 30 °C the traps responsible for the first TL peak would be emptied after 800 years, which may be the reason for their absence in our TT samples. It is important to note that unusual violet luminescence or phosphorescence have never been mentioned for the Kazakhstan pink calcite samples. Moreover, such a band is absent in thermally stimulated emission spectra measured for different TL peaks. For temperatures below 133 °C the main emission bands are at 320 nm, evidently connected with Pb^{2+}, and a very broad band peaking at 450 nm of unknown nature. At temperatures higher than 300 °C these emission bands are absent. On the other hand, different types of violet afterglow were detected in our samples as a result of isothermal TL in all NTL maxima up to 400 °C. It may be seen that a long rise time is typical for several glow curves, especially at 335 °C, which explains the fact that the violet emission increases in intensity for a few seconds with continued short-wave UV excitation. It is important to note that all of our NTL studies were accomplished using a narrow range violet filter so that all other possible emission bands could not contribute to our signals.

The strong afterglow of the violet emission, which can still be observed for several seconds after the end of the excitation pulse, must be due to the trapping of electrons or holes. Figure 5.110 is a schematic illustration of a possible model for this process. As a result of natural ionizing irradiation, calcite contains both luminescence centers and also trapping centers. After UV excitation, intrinsic emission inside the violet luminescence center takes place as described above, with two decay components in the ns and ms ranges. In concert with this, light with energy larger than that needed to transfer the electrons from the traps to the conduction band yields free electrons and holes. One possibility is that the holes are trapped by the luminescent centers, whereas the electrons in the conduction band recombine with the holes, yielding emission. Another possibility is that a fraction of the electrons is trapped in the electron trap centers, from where they escape thermally or optically after some delay. Only then will they recombine with trapped holes, with recombination energy exciting the luminescence center. Hence, the corresponding emission occurs with considerable delay.

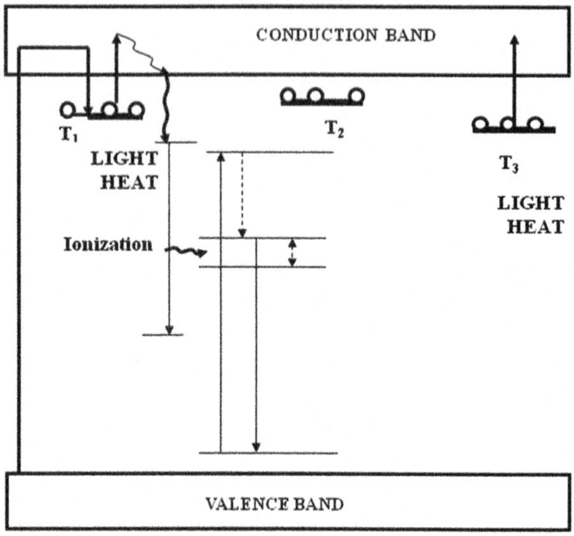

Fig. 5.110 Luminescence center and different traps in energy gap of calcite

5.12.2.3 Green Emission

Calcite grains taken from the same known location produced CL spectra with consistent peak positions for both main peaks (approximately 600 and 640 nm) and shoulder peak (515 nm). Upon exposure to neutron irradiation, the CL spectra of calcite grains exhibit a shoulder peak that increased in intensity linearly with neutron dose (Silletti et al. 2012). This correlation suggests that additional lattice defects are being introduced into the crystal lattice of the calcite by the radiation exposure. Although the CL signature resulting from these defects are permanent, they can be annealed away under extreme conditions through excessive heating and electron bombardment. The area of the shoulder peak is dependent on the orientation of the crystal to the electron beam, influencing the dose response curve greatly. Construction of a spindle stage for each grain sample in the vacuum chamber, would allow for rotation of the sample under the electron beam, allowing us to better account for this effect. The optimal direction with respect to the beam could be found within seconds, reducing the need for replicate measurements per sample. The ability to create a dose response curve for the effect of neutron irradiation on the CL of calcite could validate its use as a natural dosimeter. These results suggest that other CL active minerals and building materials can also act as passive detectors for ionizing radiation.

5.12.3 Apatite

Luminescence spectrum of the Canada apatite contains the yellow band, which is similar to Mn^{2+} emission in Ca(II) site (Fig. 5.111a). Nevertheless, this band has short decay time, which is not suitable for strictly forbidden d-d transitions in

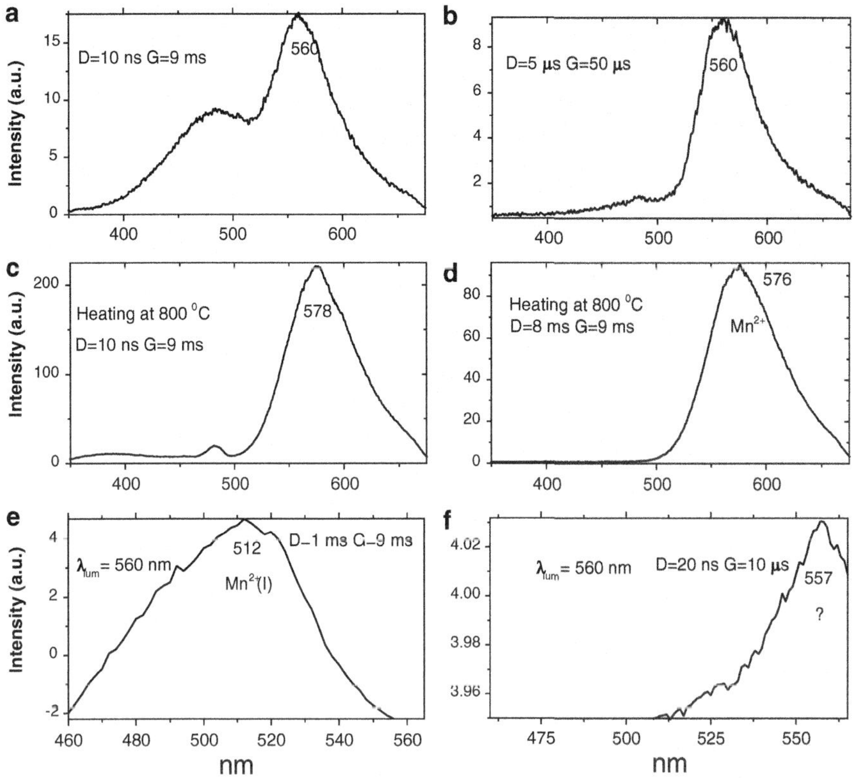

Fig. 5.111 (**a–f**) Unidentified emission center in laser-induced time-resolved luminescence spectra of apatite. (**a**) Time-resolved luminescence spectrum with broad gate; (**b**) Time-resolved luminescence with narrow gate where yellow band with short decay time dominates; (**c–d**) Time-resolved spectra after heating at 800 °C; (**e–f**) Excitation bands of Mn^{2+} and short-lived yellow band, correspondingly

Mn^{2+}. It dominates in the time-resolved spectrum with delay of 5 μs and gate of 50 μs when the shorter-lived centers are quenched, while the longer-lived ones are not detected (Fig. 5.111b). A change in the lifetime may be indicative of the energy transfer from Mn^{2+} by radiationless mechanism. A condition necessary for this mechanism is coincidence or a close distance between energy level pairs of the ion sensitizer and the ion activator. Here, the process of luminescence is of an additive nature and a longer duration and greater quantum yield of the activator luminescence accompany a reduced duration and decreased quantum yield of the luminescence of the sensitizer. The energy transfer explained the absence of Mn^{2+} luminescence in natural apatite, where its presence is ascertained from the EPR spectra, from levels $^4T_2$ and $^4T_1$ of Mn^{2+} to REE (Marfunin 1979b). In all samples where short-lived luminescence of Mn^{2+} is detected, the lines at 651 and 605 nm are unusually strong. They are well known in artificially activated apatite and belong to Sm^{3+} in Ca(II) site (Morozov et al. 1970). In natural apatite such luminescence is extremely rare and was previously detected in Canada and yellow apatites only. Thus it is principally possible to suppose that energy migration from Mn^{2+} to $^4G_{5/2}$ level of Sm^{3+} takes place. These lines excitation as result of energy migration from Mn^{2+} is confirmed by their unusually long decay time of 2 ms.

Nevertheless, such interpretation contradicts the fact, that after heating at 800 °C the short-lived yellow band disappeared and a usual long-lived Mn^{2+} luminescence becomes visible (Fig. 5.111c, d). Time-resolved excitation spectrum of short lived yellow band consists of one main broad band with extremely low Stocks shift and is absolutely different from those for Mn^{2+} (Fig. 5.111e–f). Certain similarity may be seen between this luminescence and short-lived orange emission in calcite, which has been ascribed to radiation-induced center. It is possible that natural irradiation may be a reason of orange luminescence in apatite also.

5.12.4 Topaz

Color centers in topaz can be produced by irradiation with fast neutrons and gamma-rays (Platonov et al. 1984). It was found that luminescence bands are connected with such color centers, which may be detected by steady state and time-resolved spectroscopy (Marques et al. 2000). The detected luminescence bands have been connected with both impurities and structural changes

5.12.5 Radiation-Induced Luminescence in Sand

Naturally occurring quartz is exposed to ionizing radiation. In the environment, this radiation comes from the radioactive decay of elements contained within the sediment (plus a component from cosmic ray flux). The effect of incident ionizing radiation is to redistribute electronic charge within the crystal lattice. A portion of the re-distributed electrons becomes trapped on charged defects within the crystal. When the quartz sand is buried, it absorbs environmental radiation and the 'trapping' of redistributed electrons begins. When the sediment is at the surface, the absorption of incident sunlight detaches electrons from light-sensitive electron traps, thus 'zeroing', 'bleaching' or 're-setting' the grain. This property is widely used for so called optical dating of sediments (Huntley et al. 1985). Detection of these trapped electrons involves illuminating the sand with light of a specific wavelength. As the sample is illuminated, electrons trapped in light sensitive traps are freed and diffuse through the lattice. A fraction of the free electrons go to 'recombination sites' ('holes' left behind due to the absence of an electron). As the electrons recombine, light is emitted mainly in the ultraviolet part of the spectrum. The presence of the emitted light evidences the presence of the sand, which was brought from the depth on the surface during this specific night. Two types of optical stimulation are possible. In continuous wave (CW) the excitation is continuous and the emitted luminescence is detected during stimulation. This arrangement requires the use of filters to discriminate between the stimulation light and the emitted light. In the initial studies of the quartz, the green light at 514.5 nm from an argon laser demonstrated that this energy is sufficient to empty the electron traps directly. Excitation spectra of such luminescence demonstrated that blue light at 470 nm is the mostly effective. The use of CW excitation presents the use of blue excitation because it is spectrally very close to emission. Pulse stimulation technique enables to overcome such problem. Here stimulation source is pulsed and the luminescence is monitored only between the pulses. Since the emission is not detected while the pulse is on, this arrangement can extend the range of stimulation wavelengths that may be used. In order to use the pulsed method, the data on decay time of quartz luminescence are crucial. It was found that for natural quartz emission responsible for our task is characterized by decay time $\tau = 27 \pm 5$ μs.

5.13 Reabsorption Lines of Oxygen and Water

5.13.1 Apatite: Luminescence Spectroscopy

The laser-induced time-delayed (more than 1 μs) luminescence spectra of magmatic and sedimentary apatites contain several "negative" lines at the red part of the visible spectrum (Fig. 4.7). The correlation analyses reveal that this group is subdivided into two (Gaft et al. 1997c): the strongest line at 760 nm accompanied by the weaker line at 687 nm; doublet at 720 nm accompanied by triplet at 823 nm. The following results allow us to conclude that these lines are not connected with noise or artifacts: the spectral features are "negative", thus they may not be connected with second order lines or incidental source of light; the spectrum is presented without any smoothing or other mathematical treatment.

It is clearly seen that negative lines are much stronger then the noise. Besides that, the negative lines are always situated at the same places. The invariability of the spectral positions provides the evidence that they are not connected with fluctuations of the laser pulses and detection system. Thus it may be concluded that we a dealing with reabsorption mechanism. The optical absorption spectra of natural apatites in the range 600–900 nm contain several lines and bands connected with Nd^{3+}, Pr^{3+}, Mn^{5+}, SO_3^- (Gorobets 1975; Gilinskaya and Mashkovtsev 1995) but they do not coincide with the negative lines detected in our study. The optical spectroscopy data connected with other minerals and solids have been also checked, but all attempts were unsuccessful (Platonov 1979).

Nevertheless, these absorption lines though not previously mentioned in solids are well known to absorb the visible light in the atmosphere (Measures 1985; Peixoto and Oort 1992). The strongest absorption lines of molecular oxygen are named as A-band, or 760 nm, and B-band, or 687 nm. The 760 nm band is very

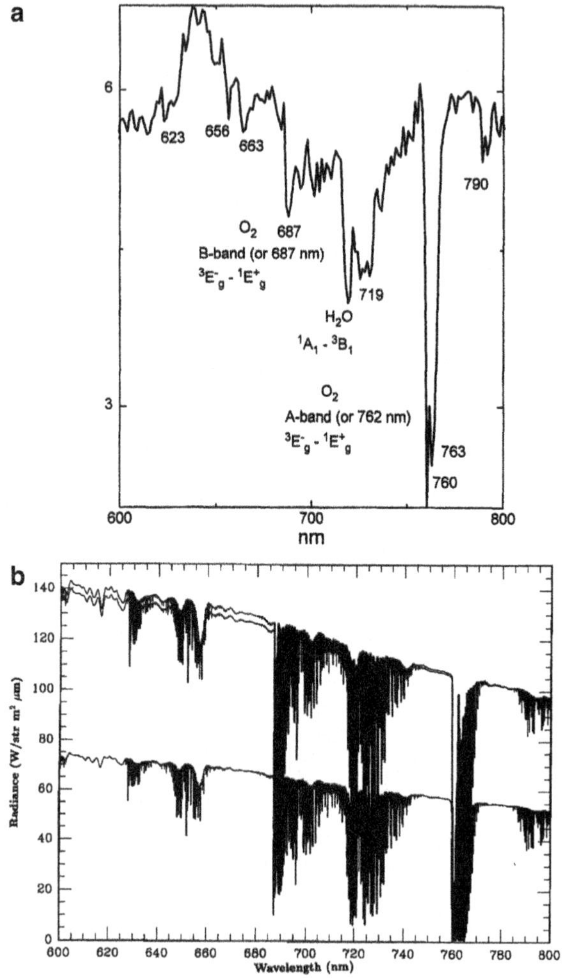

Fig. 5.112 (a–b) Reabsorption spectrum of apatite (**a**) and water and oxygen absorption in the atmosphere (Kuze and Chance 1994)

famous in astronomy, because its presence in the atmosphere is used as a test of photosynthetic activity on a distant planet. The strongest absorption lines of water in the visible range consist of a doublet at 718 and 729 nm and a triplet at 818, 823 and 829 nm, which exactly coincide with reabsorption lines in apatite. These is a striking identity between reabsorption spectrum of the apatite and absorption spectrum of the atmosphere determined with high spectral resolution and signal to noise ratio (Kuze and Chance 1994). Not only the strong lines, but also the weak ones are the same. Even the splitting of A-band is similar in both cases. It consists of the narrower and more intensive line situated at 760 nm and the broader and less intensive one situated at 763 nm (Fig. 5.112).

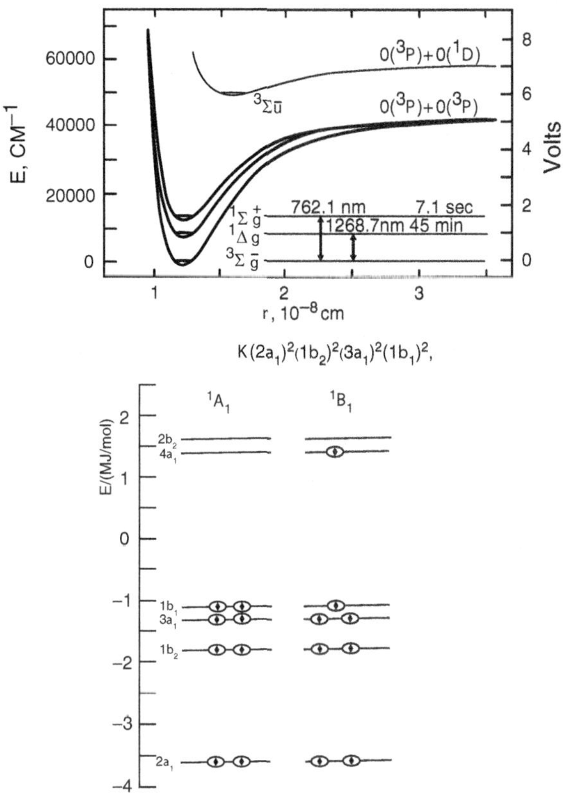

Fig. 5.113 Energy levels scheme of molecular oxygen and water (Ogryzlo 1965)

The absorption of molecular oxygen and water in the air is very weak. It is explained by the energy levels schemes (Fig. 5.113) of molecular oxygen and water (Wallace and Hunten 1968). The corresponding electron transitions to excited states, which give rise to absorption bands of molecular oxygen and water, are forbidden on the bases of spin and symmetry. The low transition probability is reflected in the long radiative lifetime of 7 s from $^1E^+{}_g$ state of molecular oxygen. Why in such case the absorption in apatite is relatively strong? The possible explanation is that this forbiddenness, strictly observed in the spectra of free molecules, is less stringent inside crystals where forbidden transitions often occur owing to interaction with heavy metal impurities, such as U, Fe, Mn, which are responsible for spin-orbit coupling relaxing to some extent the forbiddenness of the spin-forbidden transitions. It is worthwhile to note that the long lifetimes are in the limit of the zero pressure and may be considerably modified even by perturbations from surrounding gases.

The next interesting problem is why reabsorption lines disappear at liquid nitrogen temperature? In emission-reabsorption process both centers behave as independent systems, and do not interact directly. Thus energy migration is not temperature dependent. The possible explanation is that at liquid nitrogen temperature (77 K) the oxygen and water exist not as gas and liquid, but as liquid and solid, correspondingly. It is known that in the spectra of condensed or compressed oxygen there are bands that are not due to transitions of isolated O_2 molecules. They are attributed to transitions in molecular complexes, perhaps short-lived collision complexes. The strongest band near 633 nm is attributed to the (0–0) transition associated with simultaneous excitation of two $3E^{-}{}_{g}$ oxygen molecules to the $1\Delta_g$ state (Ogryzlo 1965). We do not see such lines at 77 K. Another possibility that the interaction between molecular oxygen and water may be decreased at lower temperature and electron transitions remain strictly forbidden.

Thus it is possible to conclude that the optical active centers in apatite, which are responsible for the reabsorption lines, are molecular oxygen and water. Apatite structure is characterized by the existence of channels, running along the C-axis of the hexagonal structure. It has been proposed that the space existing in the channels of non-stoichiometric apatites may be available for molecular trapping. The most evident example is given by oxygen containing apatites. It has been shown, in the case of phosphor-calcium apatites, that they occupy all the available space of the channels and their maximum amount is approximately 0.5 molecular percents. Water has been suggested to exist in vacancies of the channels. However quantitative data are scarce because of the difficulty to distinguish molecules inside the crystal and molecules adsorbed on its surface (Rey et al. 2009).

Besides that, investigation of Eu^{3+} luminescence in different kinds of artificial apatites leads to analogous conclusion. It is apparent that the charge compensating species for the trivalent rare-earth elements occupying the divalent calcium may appear in the form of halogen being replaced by oxygen. However, the profound dependence of the Eu^{3+} spectra on the type of halogen ion involved in apatites makes to conclude that the halogen in the second type of calcium position remains intact. Hence, it is suggested that the charge compensating species, namely free oxygen, should occur as an interstitial in the vicinity of the Eu^{3+} center (Jagannathan and Kottaisamy 1995). Thus the spectroscopic conclusion is in accordance with crystallochemistry of apatite, namely with possible accommodation of molecular oxygen and water in different ways: by structural incorporation and by adsorption.

5.13.2 Apatite: Photoacoustic Spectrophotometry

Such unusual phenomena as reabsorption lines of molecular oxygen and water require a confirmation by an additional independent technique. The best way is to try to detect the corresponding absorption lines by UV-visible spectroscopy. The problem is that sedimentary apatite samples are in the form of non-transparent powder, which is not suitable for optical spectroscopy. The photoacoustic spectroscopy (PAS), which allows measurement of absorption spectra of powdered opaque samples, has been chosen for this purpose (Gaft et al. 1998). PAS works as follows:

light from a broad wavelength source is chopped by a rotating sector. The light, after monochromator, is directed into a gas tight cell, which contains the sample. A miniature microphone is connected to the cell. In the absence of photophysical or photochemical reactions, the light degrades to heat. This heat flows to the surface, and causes a heating and cooling of the gas in immediate contact with the sample. The acoustic wave thus produced is then detected with a miniature microphone. The microphone is an exceptionally sensitive instrument and can detect acoustic waves of very little power. The signal from the microphone is detected using a phase sensitive detector. This form of detection can detect a signal at a specific frequency 100 dB below the noise level. The resulting signal is that of the emission of the light modified by the transfer characteristics of the spectrometer and the absorption of the sample. In order to obtain the true spectrum one obtains the ratio of this signal to that of carbon black powder, which is assumed to be uniformly absorbing (Rosenzweig 1990).

It was found that the absorption spectrum of the apatite is characterized by the bands at 775 and 850 nm, which are close to reabsorption bands in the same sample and according to absorption in atmosphere may be connected with molecular oxygen and water, correspondingly (Fig. 5.114).

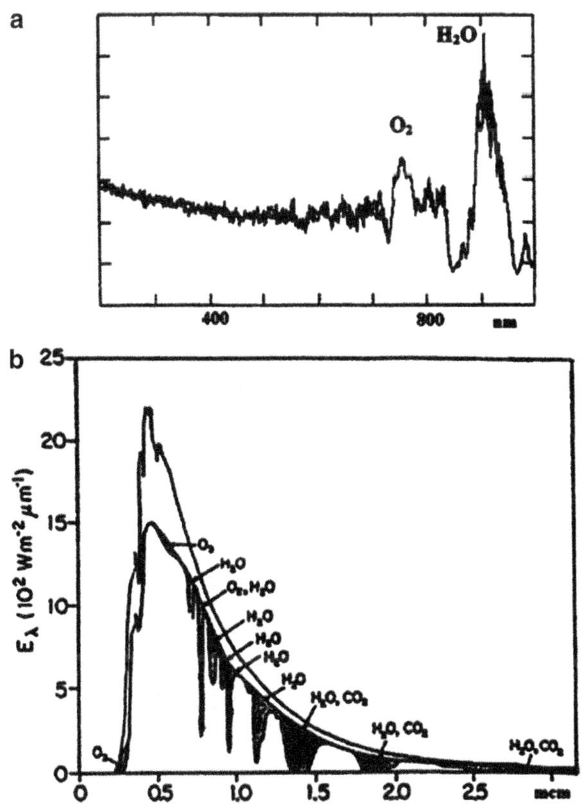

Fig. 5.114 (a–b) Absorption lines of molecular oxygen and water in the air (**a**) and photoacoustic spectra of apatite (**b**)

5.13.3 Diamond

Reabsorption lines of molecular oxygen and water have been detected in time – resolved luminescence spectra of certain diamonds.

5.14 Luminescence Centers in Diamonds

The perfect diamond crystal consists of perfectly regular array of carbon atoms, every atom being symmetrically surrounded by four others, and forming an electron-pair bond with each one. Energy of about 5.5 eV is required to remove an electron from one of these bonds and the crystal is therefore transparent for ultra-violet wavelengths greater than 225 nm. For shorter wavelengths it will absorb strongly. No crystal is perfect and the deviation from perfection results in the optically active defects so common in diamonds. Several defects have a strong absorption in the UV and visible parts of the spectrum and are responsible for different kinds of luminescence. The defects may be chemical or structural, or a combination of both. It has to be noted that in the diamonds luminescence spectra very narrow intensive lines of purely electronic zero-phonon transitions are distinguished together with broad bands with maxima, which extend into the long wavelengths direction and represent a superimposition of the phonon vibrational replicas of the electron transitions. Because of the fact that the vibration energy at different points of the Brillouin zone is dissimilar, and also owing to the formation of di- and tri-phonon repetitions upon addition of similar and different phonons, 10–15 overlapping lines appear, which form a broad band perceived as blue, green, yellow, orange and red emission. At lower temperature the sharp lines dominate in the spectrum, but at room temperature conditions they are drastically quenched and the broad relatively structureless bands prevail. We concentrated on the room temperature luminescence, which is especially important for diamonds sorting. Interpretation of luminescence centers in diamonds is very complicated issue. The main models and spectral-kinetic properties may be found in Zaitsev 2005 and references inside.

5.14.1 N_3

The N_3 optical center is one of the best known in steady-state luminescence spectra diamond. It is connected with three substitutional nitrogen atoms bounded to a common carbon atom or a vacancy, the ground state being a $^2A_1$ level and the excited state where luminescence originates a ^2E state (C_{3v} point group). The zero-phonon line occurs at 2.985 eV and absorption and emission spectra show very closely a mirror relationship (Bokii et al. 1986). The N_3 prompt luminescence decay is exponential and equal to 40 ns. Time-resolved luminescence spectroscopy enables to detect that N_3 center has some metastable levels between the emitting and ground state. One of the decay paths of these metastable levels is delayed N_3 luminescence, which occurs

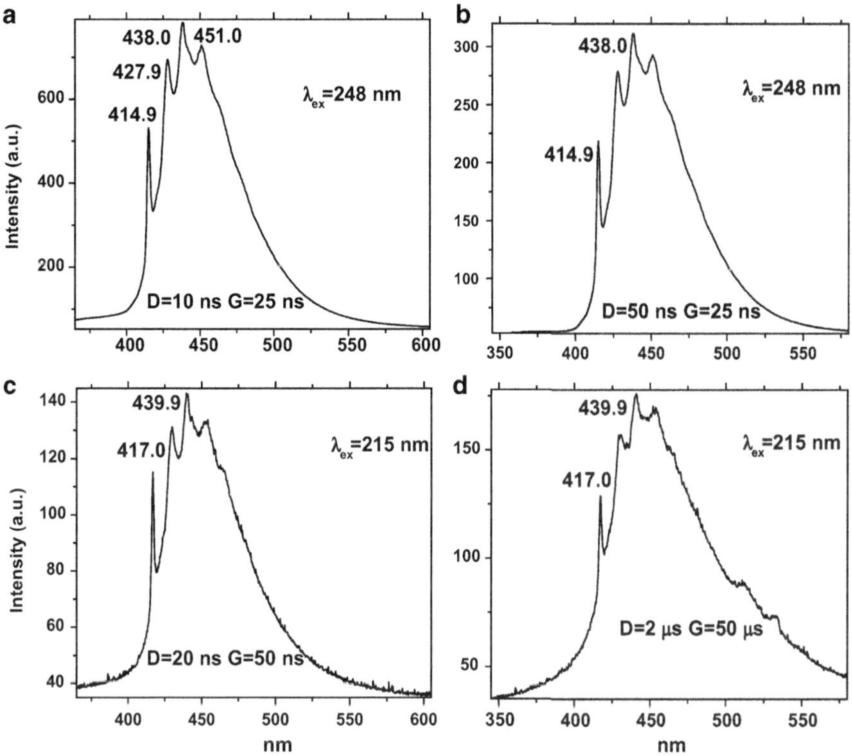

Fig. 5.115 (**a**–**d**) Laser-induced time resolved luminescence of N_3 center in diamond with different decay times

above 170 K. Besides this, time-resolved spectroscopy enables to detect another center with zero-phonon lines with energy very similar to N_3 center (Pereira and Santos 1993; Pereira and Santos 1994). Under steady-state conditions there is some superposition of luminescence and excitation spectra of N_3 and this center and they cannot be spectroscopically separated. By time-resolved spectroscopy the centers were separated and the second one, named 2.96 eV was connected with Al impurity.

Our study of time-resolved luminescence of diamonds revealed the similar behavior (Panczer et al. 2000). Short-decay spectra usually contain N_3 luminescence centers (Fig. 5.115a, b) with decay time of $\tau = 30$–40 ns. Despite such extremely short decay, sometimes the long-delay spectra of the same samples are characterized by zero-phonon lines, which are very close in energy to those in N_3 centers. At 77 K $\lambda_{ex} = 308$ nm excitation decay curve may be adjusted to a sum of two exponents of $\tau_1 = 4{,}2$ μs and $\tau_2 = 38.7$ μs (Fig. 5.115c), while at 300 K only shorter component remains. Under $\lambda_{ex} = 384$ nm excitation even longer decay component of $\tau_3 = 870$ μs may appear (Fig. 5.115d). The first type of long leaved luminescence may be ascribed to 2.96 eV center, while the second type of delayed N_3 luminescence is ascribed to the presence of two metastable states identified as quartet levels at the N_3 center.

N_3 is the only luminescence center with an appreciable fine structure at 300, while its zero-phonon line at 415 nm is characterized by different intensities down to practically total disappearance. The decrease of zero-phonon line intensity is accompanied by an increase of the 360 nm line in the excitation spectrum (Gaft et al. 2005). The cause may be the N_3 clusters formation of N_3 centers with the ensuing concentration quenching resulting from energy migration. The transfer is of the emission-reabsorption type, namely, emission by one luminescence center and its reabsorption and emission by the other center. This is explained by the similarity of the energies of emission and reabsorption. The following model is possible: a feature distinguishing the N_3 center is that its absorption is similar to its excitation and presents a mirror reflection of its photoluminescence. The line at 415 nm is present in the emission and in the absorption spectra being subjected to the concentration quenching. This is confirmed by the fact that this process is particularly prominent in yellowish diamonds, which are characterized by elevated N_3 contents. A strong reabsorption line at 415 nm in these diamonds is clearly seen in the spectrum, which was received with a time delay of 1 microsecond, when N_3 luminescence is no longer detected because of its short decay time.

5.14.2 H3, H4, S2, S1 and S3

The H_3 center is connected with pare of nitrogen atoms, which captures one vacancy. It is well known in steady-state luminescence spectra of diamonds. It belongs to the C_{2v} point group, the ground state being $^1A_1$ level and the excited state from which luminescence takes place a $^1B_1$. Both emission and absorption spectra present a zero-phonon line at 2.463 eV. The decay time of the H_3 luminescence is of 15–20 ns (Bokii et al. 1986; Davies 1994). Time-resolved luminescence spectroscopy revealed also a delayed luminescence spectrally identical to the fast one, but with a temperature dependent decay time of the order of a few ms. Results were interpreted as due to reversible non-radiative transitions from the emitting $^1B_1$ excited level to triplet levels (Pereira and Monteiro 1990; Pereira and Monteiro 1991). The H4 one is principally the same as H3 but with slightly different configuration.

Defect S2 is believed to be due to combination of two vacancies and one nitrogen atom, while S3 one is connected with combination of several nitrogen atoms and vacancy and S1 with vacancy-nitrogen pare. In our study we found that H3, H4, S2 and S3 centers are characterized by relatively broad bands with λ_{max} at 520–545 nm, sometimes accompanied by very weak zero-phonon lines at 489 and 523 nm (S2), 498 (S3) and 503 (H3) nm. It is very difficult to distinguish between the centers of this group, especially when they present together. Under pulse laser excitation the decay time differences enable more definite recognizing. Different decay components in the green part of the spectrum allow establishing the presence of H3 (12 μs) and S3 (126 and 213 μs) centers. These broad bands are sometimes accompanied by narrow lines of GR1 center at 794 nm and by system at 700 and 788 nm (Davies 1994; Bokii et al. 1986).

5.14.3 A-band

It appears that there are A-bands of different nature. One of the A-band model is radiative recombination at dislocations. This model concerns a relatively narrow A-band peaked at 440 nm. This band is usually observed in low-nitrogen type II diamonds. There is a dislocation related model of the A-band considering luminescence on pure non-decorated dislocations excluding D-A recombination. This model is used for the A-band with a maximum at 415 nm. The second model of the A-band is intra center transitions at the B1(N9) centers (platelets). This model relates to the broad A-band with a maximum at 480 nm observed in natural type I diamonds.

In natural diamonds the spectral position of the maximum of the dislocation related A-band may range from 445 to 415 nm. The maximum measured in delayed CL on natural diamonds exhibits a jump from 435 to 470 nm when the delay exceeds 27 μs. The A band luminescence exhibits a non-exponential decay. The decay time may change from less than 1 ms to greater than 50 ms (Zaitsev 2005).

In some CVD diamond films the decay of the A-band exhibits very fast components in the range 0.1–1 ns attributed to the intrinsic radiative decay of directly excited D-A pairs. In some CVD diamond films (both undoped and nitrogen doped) the A-band shows a single exponential decay, This single decay time has the largest value at RT ranging from 4.8 to 5.3 ns.

References

Abritta T, de Souza B (1988) Luminescence and photoacoustic measurements of $LiAl_5O_8:Fe^{3+}$. J Lumin 40&41:187–188

Agekyan V, Shiryapov I, Oreshnikova L (1974) Exiton-phonon transitions in tin dioxide. FTT 16:2473-2475 (in Russian)

Ajo D, Favaro M, Pozza G et al (1997) Near-Infrared photoluminescence in phosphate minerals and related glass-ceramics. J Mater Sci 32:4217–4220

Altschuler Z (1980) The bearing of geochemistry on the recovery of uranium and rare earth in phosphorites. Proc 2nd Internat Cong Phosphorus Compounds, Boston, p 605–625

Bakhtin A, Gorobets B (1992) Optical spectroscopy of minerals and ores and its use in geology. Kazan University Publishing, Kazan (in Russian)

Balda R, Sanz M, Mendioroz A et al (2001) Infrared-to-visible upconversion in Nd^{3+}–doped chalcohalide glasses. Phys Rev B 64:144101–144102

Barbarand J, Pagel M (2001) Cathodoluminescence study of apatite crystals. Am Mineral 86:473–485

Barbin V, Jouart J-P, D'Almedia T (1996) Cathodoluminescence and laser-excited luminescence spectroscopy of Eu^{3+} and Eu^{2+} in synthetic CaF_2: a comparative study. Chem Geol 130:77–86

Baumer A, Blanc P, Cesbron F, Ohnenstetter D (1997) Cathodoluminescence of synthetic (doped with rare-earth elements) and natural anhydrites. Chem Geol 138:73–80

Beletti A, Borromei R, Oleari L (1995) Absorption spectra of zircon crystals doped with Cr(IV): $ZrSiO_4:Cr^{4+}$. Inorg Chim Acta 235:349–362

Blanc P, Baumer A, Cesbron F, Ohnenstetter D, Panczer G, Remond G (2000) Systematic cathodoluminescence spectral analysis of synthetic doped minerals: anhydrite, apatite, calcite,

fluorite scheelite and zircon. In: Pagel M, Barbin V, Blanc P, Ohnenstetter D (eds) Cathodoluminescence in geosciences. Springer, Berlin/Heidelberg/New-York, pp 127–160
Blasse G (1980) The luminescence of closed-shell transition-metal-complexes. New developments Struct Bond 42:1–41
Blasse G, Aguilar M (1984) Luminescence of natural calcite ($CaCO_3$). J Lumin 29:239–241
Blasse G, Grabmaier B (1994) Luminescent materials. Springer, Berlin/Heidelberg/New York
Blasse G, Wanmaker J, Vrigt J, Bril A (1968) Fluorescence of Eu^{2+} activated alkaline-earth aluminates. Philips Res Rpts 23:201–209
Blasse G, Dalhoeven G, Choisnet J, Studer F (1981) On the luminescence of titanium activated stannates. J Solid State Chem 39:195–198
Blasse G, Meijerink A, Nomes M, Zuidema J (1994) Unusual Bismuth luminescence in Strontium Tetraborate (SrB_4O_7:Bi). J Phys Chem Solid 55:171–174
Bokii G, Bezrykov G, Kliyev Yu et al (1986) Natural and synthetic diamonds. Nauka, Moscow (in Russian)
Boulon G (1987) Spectroscopy of post transition metal ions. In: Di Bartolo B (ed) Spectroscopy of solid-state laser-type materials. Plenum Press, New York, pp 223–266
Boulon G (1997) Broad band centers applied for laser materials: example of tetrahedrally coordinated centers. In: Di Bartolo B (ed) Spectroscopy and dynamics of collective excitations in solids. Plenum Press, London, New York, pp 561–573
Boulon G (2000) Transition metal ion lasers – Cr^{3+}. In: Webb C, Jones J (eds) Handbook of laser technology and application. Volume II. Laser design and laser systems. CRC Press, London, pp 307–337
Broussell I, Fortin E, Kulyuk L, Popov S, Tezlevan V (1997) Luminescence of the Cr^{3+} ions in spinel-type α-$ZnAl_2S_4$: Cr single crystals. J Lumin 72&74:640–644
Brunold T, Güdel H (1997) Excited-state absorption and laser potential of Mn^{6+}-doped $BaSO_4$ crystals. J Opt Soc Am B 14:2373–2377
Brunold T, Hazenkamp M, Güdel H (1997) Luminescence of CrO_4^{3-} and MnO_4^{2-} in various hosts. J Lumin 72–74:164–165
Bryknar Z, Trepakov V, Potůček Z, Jastrabík L (2000) Luminescence spectra of $SrTiO_3$:Mn^{4+}. J Lumin 87–89:605–607
Bukin G, Matrosov V, Orekhova V et al (1981) Growth of alexandrite crystals and investigation of their properties. J Cryst Grouse 52:537–541
Burgner R, Scheetz B, White W (1978) Vibrational structure of the S_2^- luminescence in scapolite. Phys Chem Miner 2:317–324
Butler K (1980) Fluorescent lamp phosphors. Pennsylvania State University Press, University Park, p 117
Chadwick K, Rossman GR (2009) Orange kyanite from Tanzania. Gems Gemol 45:146–153
Champagnon B, Duval E (1977) Emission spectrum of V^{3+} -αAl_2O_3. J De Phys 38:L-299–L-301
Champagnon B, Duval E (1979) Emission spectrum of V^{4+}- alpha Al_2O_3: Jahn-Teller effect in the fundamental $^2T_{2g}$ and excited ^2E_g states. J Phys C Solid State Phys 12:L-425–L-429
Chithambo M, Raymond S, Calderon T, Townsend P (1995) Low temperature luminescence of transition metal-doped beryls. J Afr Earth Sci 20:53–60
Czaja M, Bodył-Gajowska S, Mazurak Z (2013) Steady-state luminescence measurement for qualitative identification of rare earth ions in minerals. J Miner Petrol Sci 108:47–54
de Boisbaudran L (1885) Sur la fluorescence des terres rares. C R Acad Sci Paris 101:552–588
Davies G (1994) Properties and growth of diamonds. Gordon Davies, King's Colledge, London
DeLoach L, Payne S, Kway W et al (1994) Vibrational structure in the emission spectra of Yb^{3+} -doped apatite crystals. J Lumin 62:85–94
DeNeufville J, Kasdan A, Chimenti R (1981) Selective detection of luminescent characteristics of uranyl geologic targets. Appl Opt 20:1279–1296
Deren P, Malinowski M, Strek W (1996) Site selection spectroscopy of Cr^{3+} in $MgAl_2O_4$ green spinel. J Lumin 68:91–103
Dieke G (1968) Spectra and energy levels of rare earth ions in crystals. Interscience, New York

Dimova M, Panczer G, Gaft M (2006) Spectroscopic study of barite (Kremikovtsi deposit, Bulgaria) with implication for its origin. Ann Geol de la Peninsule Balkanique 67:101–108
Donker H, Smit W, Blasse G (1989) On the luminescence of some tin-activated alkaline-earth orthophosphates. J Electrochem Soc 136:3130–3135
Dorenbos P (2003) Anomalous luminescence of Eu^{2+} and Yb^{2+} in inorganic compounds. J Phys Condens Matter 15:2645–2665
Durville F, Champagnon B, Duval E, Boulon G (1985) Lase spectroscopy of spinel microcrystallite in a Cr^{3+}-doped silicate glass. J Phys Chem Solid 46:701–707
Erfurt G (2003) Infrared luminescence of Pb^{+} centres in potassium-rich feldspars. Phys Stat Sol (a) 200:429–438
Eyal M (1988) Energy transfer for better efficiency of Nd3+ lasers. Proc SPIE 1182:140–150
Fabeni P, Pazzi G, Salvini L (1991) Impurity centers for tunable lasers in the ultraviolet and visible regions. J Phys Chem Solid 52:299–317
Fergusson J (1970) Spectroscopy of 3d complexes. Prog Inorg Chem 12:159–294
Folkerts H, Hamstra M, Blasse G (1995) The luminescence of Pb^{2+} in alkaline-earth sulfates. Chem Phys Lett 246:135–138
Freed R, Peacor D (1969) Determination and refinement of the crystal structure of margarosanite $PbCa_2Si_30_9$. Zeitschrift fur Kristallographie, Bd 128:213–228
Gaft M (1984) Luminescence of vanadium containing centers in natural phosphates. Miner J 4:83–86 (in Russian)
Gaft M (1989) Luminescence of minerals under laser excitation. Ministry of Geology, Moscow (in Russian)
Gaft M, Gorobets B (1979) Photoluminescence of manganese minerals. J Prikl Spectrosc 6 (31):987–990
Gaft M, Rudenkova I (1993) Laser-induced luminescence of barite after thermal treatment. J Therm Anal 42:187–195
Gaft M, Vorontsova L (1982) Luminescence of cassiterite and the possibilities of its practical use. Miner J 4(5):75–78
Gaft M, Gorobets B, Malinko S (1979) Studies of luminescence of boron minerals. Docl Acad Nauk SSSR 244(5):171–174
Gaft M, Gorobets B, Homyakov A (1981a) On luminescence nature of titanium and zirconium minerals. Docl Acad Nauk SSSR 263(3):1234–1237
Gaft M, Gorobets B, Naumova I (1981b) Relation between luminescent properties and crystallochemistry of manganese minerals. Miner J 3:80–90 (in Russian)
Gaft M, Gorobets B, Marshukova N, Pavlovskii A (1982) Tin minerals luminescence and its use in the prospecting of tin ore deposits. Docl Acad Nauk SSSR 266(1):217–220
Gaft M, Bershov L, Krasnaya A, Yaskolko V (1985) Luminescence centers in anhydrite, barite, celestite and their synthesized analogs. Phys Chem Miner 11:255–260
Gaft M, Rassulov V, Zukova V, Rakov L (1986) Nature of zircon photoluminescence. Miner J 8 (3):74–78 (in Russian)
Gaft M, Scorobogatova N, Rassulov V, Moroshkin V (1989) The use of natural silver halogens luminescence for mineral prospecting. Miner J 11:58–64 (in Russian)
Gaft M, Reisfeld R, Panczer G et al (1996) Luminescence of Eu^{3+}, Pr^{3+} and Sm^{3+} in carbonate-fluor-apatite. Acta Phys Pol A 90:267–274
Gaft M, Reisfeld R, Panczer G et al (1997a) Accommodation of REE and Mn by apatite. Opt Mater 8:149–156
Gaft M, Reisfeld R, Panczer G et al (1997b) Luminescence of Eu^{3+} in high symmetry Ca position in apatite structure. J Lumin 72–74:572–574
Gaft M, Reisfeld R, Panczer G, Champagnon B (1997c) Reabsorption lines of molecular oxygen and water in natural apatite. Opt Mater 8(1–2):143–149
Gaft M, Trabjerg I, Reisfeld R, Panczer G (1998) Absorption of molecular oxygen and water in apatite by photoacoustic spectroscopy. Spectrochim Acta A 54:1721–1724

Gaft M, Reisfeld R, Panczer G et al (1999) Luminescence of Pr^{3+} in minerals luminescence of Pr^{3+} in minerals. Opt Mater 13:71–79

Gaft M, Panczer G, Reisfeld R, Shinno I (2000a) Laser-induced luminescence of rare-earth elements in natural zircon. J Alloys Comp 300–301:267–274

Gaft M, Boulon G, Panczer G et al (2000b) Unexpected luminescence of Cr-doped zircon crystals $ZrSiO_4$. J Lumin 87–88:1118–1121

Gaft M, Panczer G, Reisfeld R, Shinno I (2000c) Three centers of Eu^{3+} luminescence in zircon $ZrSiO_4$. J Lumin 87–89:1032–1035

Gaft M, Reisfeld R, Panczer G et al (2001a) The nature of orange luminescence of mineral barite. Opt Mater 16(1–2):279–290

Gaft M, Reisfeld R, Panczer G et al (2001b) Laser-induced time-resolved luminescence as a means for discrimination of oxidation states of Eu in minerals. J Alloys Comp 323–324:842–848

Gaft M, Panczer G, Reisfeld R, Uspensky E (2001c) Laser-induced time-resolved luminescence as a tool for rare-earth elements identification in minerals. Phys Chem Miner 28:347–363

Gaft M, Seigel H, Panczer G, Reisfeld R (2002a) Laser-induced time-resolved luminescence of lead Pb^{2+} in Franklin, NJ, minerals. Eur J Miner 14:1041–1048

Gaft M, Panczer G, Reisfeld R et al (2002b) Laser-induced time-resolved spectroscopy of broad luminescence bands in zircon ZrSiO4. Miner Petrol 76:235–246

Gaft M, Nagli L, Reisfeld R, Panczer G, Brestel M (2003a) Time-resolved luminescence of Cr^{3+} in topaz $Al_2SiO_4(OH, F)_2$. J Lumin 102–103:349–356

Gaft M, Nagli L, Reisfeld R, Panczer G (2003b) Laser-induced time-resolved luminescence of titanite. Opt Mater 24:231–241

Gaft M, Nagli L, Waychunas G (2004) The nature of blue luminescence of natural benitoite $BaTiSi_3O_9$. Phys Chem Miner 31:365–373

Gaft M, Reisfeld R, Panczer G (2005) Modern luminescence spectroscopy of minerals and materials. Springer, Berlin/New York

Gaft M, Nagli L, Panczer G et al (2008a) The nature of unusual luminescence of natural calcite $CaCO_3$. Am Miner 93:158–167

Gaft M, Reisfeld R, Panczer G, Dimova M (2008b) UV-Visible luminescence of Nd^{3+} in minerals. Alloys Comp 451:56–61

Gaft M, Nagli L, Panczer G, Yeates H (2009) Laser-induced time resolved luminescence of tugtupite, sodalite and hackmanite. Phys Chem Miner 36:127–141

Gaft M, Nagli L, Panczer G et al (2011) Laser-induced time-resolved luminescence of orange kyanite Al_2SiO_5. Opt Mater 23:1476–1480

Gaft M, Strek W, Nagli L et al (2012) Laser-induced time-resolved luminescence of natural sillimanite Al_2SiO_5 and artificial Al_2SiO_5 activated by Cr. J Lumin 132:2855–2862

Gaft M, Nagli L, Panczer G, Yeates H (2013a) Laser-induced time resolved luminescence of natural margarosanite $Pb(Ca, Mn)_2Si_3O_9$, swedenborgite $NaBe_4SbO_7$ and walstromite $BaCa_2Si_3O_9$. Eur J Miner 25:71–77

Gaft M, Yeates H, Nagli L, Panczer G (2013b) Laser-induced time resolved luminescence of natural grossular $Ca_3Al_2(SiO_4)_3$. J Lumin 137:43–53

Gaft M, Nagli L, Panczer G, Rossman G (2013c) Long-lived laser induced time-resolved luminescence of Cr^{3+} in kyanite Al_2SiO_5. J Spectrosc Dyn 3:22–29

Gedanken A, Reisfeld R, Sominski E et al (2000) Sonochemical preparation and characterization of europium oxide doped in and coated on ZrO_2 and ISZ. J Phys Chem B 104:7057–7065

Geiger C, Stahl A, Rossman GR (1999) Raspberry red grossular from Mexico. Eur J Miner 11:1109–1113

Geipel G (2006) Laser-induced fluorescence spectroscopy. In: Vij D (ed) Handbook of applied solid state spectroscopy. Springer, New York, pp 577–593

GIA Gem Database (2010) Edward J. Gübelin Collection, 34991

Geschwind S, Kisluk P, Klein M et al (1962) Sharp-line fluorescence, electron paramagnetic resonance, and thermoluminescence of Mn^{4+} in α-Al_2O_3. Phys Rev 126:1684–1686

Gilinskaya L, Mashkovtsev R (1995) Blue and green centers in natural apatites by ERS and optical spectroscopy data. J Struct Chem 36:89–101 (in Russian)
Glynn T, Imbusch C, Walker G (1991) Luminescence from Cr^{3+} centers in forsterite Mg_2SiO_4. J Lumin 48&49:541–544
Gorobets B (1968) On the luminescence of fluorapatite doped with rare earth elements. Opt Spectrosc 25:292–294 (in Russian)
Gorobets B (1975) Reabsorption of luminescence emission in minerals because of neodymium impurities. Zap Vses Mineral Ob-va 3:357–359 (in Russian)
Gorobets B, Kudrina M (1976) Typomorphic features of scheelite as revealed by their rare-earth elements luminescence spectra. Const Svoy Miner, Kiev, Naukova Dumka 10:82–88 (in Russian)
Gorobets B, Gaft M, Laverova L (1978) Photoluminescence of manganese minerals. J Prickl Spectr 28:1100–1102
Gorobets B, Rogojine A (2001) Luminescent spectra of minerals. Handbook. RPC VIMS, Moscow
Gorobets B, Sidorenko G (1974) Luminescence of secondary uranium minerals at low temperature. At Energy 36:6–13
Gorobets B, Novozhilov A, Samoilovich M (1968) Monovalent europium in NaCl-Eu and KCl-Eu. DAN SSSR 180:1351–1354
Götze J (2000) Cathodoluminescence microscopy and spectroscopy in applied mineralogy. Technische Universität Bergakademie Freiberg, Freiberg
Grimm J, Wenger O, Güdel H (2003) Broadband green upconversion luminescence of Ni^{2+} in $KZnF_3$. J Lumin 102–103:380–385
Grinberg M, Barzovska J, Shen Y et al (2002) High pressure photoluminescence characterization of the $LiSc(WO_4)$ crystal doped with chromium. International conference on luminescence and optical spectroscopy of condensed matter, Budapest, Abstracts, p 31
Güdel H, Brunold T, Hazenkamp M et al (1998) Luminescence of transition metal ions in unusual oxidation states. Electrochem Soc Proc 29:225–234
Guguschev C, Götze J, Göbbels M (2010) Cathodoluminescence microscopy and spectroscopy of synthetic ruby crystals grown by the optical floating zone technique. Am Miner 95:449–455
Gutzov S, Bredol M, Wasgestian F (1998) Cathodoluminescence study of europium doped zirconia and cassiterite powders. J Phys Chem Solid 59:69–74
Greenblatt M (1980) Electron spin resonance of tetrahedral transition oxyanions (MO_4^{n-}) in solids. J Chem Educ 57:546–551
Haberland H, Köhler A (1939) Uber die blaue Fluoreszen: von naturlichen Silikaten in UV lichte und iib er syntetischen verzeiche an silikatschmelzen mit eingebautem zweiwertigem. Naturwiss 27:275–281 (In German)
Haberland H, Karlik B, Przibram K (1934) Zur Fluoreszen: des Fluoriten II. Sitzber Akad Wiss Wien, Abt Iia 143:151–161 (in German)
Haberman D, Neuser R, Richter D (1996) Low limit of Mn^{2+} activated cathodoluminescence of calcite: state of the art. Sed Geol 101:1–7
Hall M, Ribber P (1971) An electron microprobe study of luminescence centers in cassiterire. Am Miner 56:31–45
Hamstra M, Follkerts H, Blasse G (1994) Materials chemistry communications. Red bismuth emission in alkaline-earth-metal sulfates. J Mater Chem 4:1349–1350
Hayashi M, Shinno I, Taguchi S, Sugihara S (1990) ESR signals of zircon irradiated with thermal neutrons and gamma-rays. Min Pet Econ Geol 85:27–33
Hazenkamp M, Güdel H (1996) Luminescence properties of chromium(V) doped into various host lattices. J Lumin 69:235–244
Hessman W, Fiscima F (1991) The lifetime of the excited O_2^-, S_2^-, Se_2^- and Te_2^- centers. Phys Stat Sol 167:679–685
Huntley D, Gofrey-Smith D, Thewalt M (1985) Optical dating of sediments. Nature 313:105–107
Inokuti M, Hirayama F (1965) Influence of energy transfer by the exchange mechanism on donor luminescence. J Chem Phys 43:1978–1983

Ishii T, Ogasawara K, Adachi H (2002) First-principles analysis for the multiplet structures of tetrahedrally and octahedrally oxo-coordinated 3d2 and 3d3 transition metals. J Chem Phys 116:471–479

Jagannathan R, Kottaisamy M (1995) Eu^{3+} luminescence: a spectral probe in $M_5(PO_4)_3X$ apatites (M = Ca or Sr; X = F^-, Cl^-, Br^- or OH^-). J Phys Cond Matt 7:8453–8460

Jastrablk L, Kudyk B, Kapphan S et al (2002) Luminescence of $Ba_{0.77}Ca_{0.23}TiO_3$:Cr. International conference on luminescence and optical spectroscopy of condensed matter, Budapest, Abstracts, p 84

Jajasankar C, Babu P (2000) Optical properties of Sm^{3+} ions in lithium borate abd kithium glasses. J Alloy Comp 307:82–95

Jasinevicius R (2009) Characterization of vibrational and electronic features in the Raman spectra of gemstones. MSc thesis, Department of Geosciences, University of Arizona, Tucson, 147 p

Jia W, Liu H, Jaffe S, Yen W (1991) Spectroscopy of Cr^{3+} and Cr^{4+} ions in forsterite. Phys Rev B 43:5234–5242

Jørgensen C, Judd B (1964) Hypersensitive pseudo-quadrupole transitions in lanthanides. Mol Phys 8:281–290

Kaminskii A (1996) Crystalline lasers: physical processes and operating schemes. CRS Press, Boca Raton

Kasyanenko E, Matveeva O (1987) Ультрафиолетовое поглощение и люминесценция исландского шпата. J Prikladn Spectrosc 46:943–949

Kingsley J, Prener J, Segall B (1965) Spectroscopy of $MnO_4{}^{3-}$ in calcium halophosphates. Phys Rev 137:A189–A202

Kobayashi S, Sidike A, Yamashita N (2011) Luminescence spectra of chabazite-Ca, a zeolite mineral. Phys Chem Miner 39:465–470

Konijnendijk W (1981) Luminescence of BaSnSi3O9:Ti^{4+} compared to BaZrSi3O9:Ti^{4+}. Inorg Nucl Chem Lett 17:129–132

Kottaisamy M, Jagannathan R, Jeyagopal P et al (1994) Eu^{2+} luminescence in $M_5(PO_4)_3X$ apatites, where M is Ca^{2+}, Sr^{2+}, and Ba^{2+}, and X is F^-, Cl^-,Br^- and OH^-. J Phys (D) Appl Phys 27:2210–2215

Koziarsca B, Godlewski M, Suchoki A et al (1994) Optical properties of zoisite. Phys Rev B 50:12297–12300

Krasilschikova O, Tarashchan A, Platonov A (1986) Color and luminescence of natural fluorite. Naukova Dumka, Kiev (in Russian)

Krasnobaev A, Votyakov S, Krohalev V (1988) Problems of the applied spectroscopy of minerals. Nauka, Ekaterinburg (in Russian)

Kroger F, Overbeek J, Goorissen J (1949) Bismuth as activator in fluorescent solids. J Electrochem Soc 96:132–138

Kuck S, Hartung S, Hurling S et al (1998) Emission of octahedrally coordinated Mn^{3+} in garnets. Spectrochim Acta A 54:1741–1749

Kuze A, Chance K (1994) Analysis of cloud top height and cloud coverage from satellites using the O_2 A and B bands. J Geophys Res 99:481–496

Laurs B, Rohtert W, Gray M (1997) Benitoite from the New Idria District, San Benito County, California. Gems Gemol 33:166–187

Lenz C, Nasdala L (2015) A photoluminescence study of REE^{3+} emissions in radiation-damaged zircon. Am Miner 100:1123–1133

Lenz C, Talla D, Ruschel K et al (2013) Factors affecting the Nd^{3+} (REE^{3+}) luminescence of minerals. Miner Petrol 107:415–428

Lizzo S, Mejernic A, Blasse G (1994) Luminescence of divalent ytterbium in alkaline earth sulphates. J Lumin 59:185–194

Mao H, Xu J, Bell P (1986) Calibration of the ruby pressure gauge to 800 kbar under quazi-hydrostatic conditions. J Geophys Res 91:4673–4676

Marech J, Jacquier B, Pedrini C, Boulon G (1989) Selective one-photon and two stepwise excitations of Nd^{3+} -YAG and visible fluorescence in Nd-YAG. Mater Chem Phys 21:237–259

Marfunin A (1979a) Physics of minerals and inorganic materials. Springer, Berlin/Heidelberg/New York
Marfunin A (1979b) Spectroscopy, luminescence and radiation centers in minerals. Springer, Berlin/Heidelberg/New York
Margueron S, Clarke D (2007) Effect of polarization and uniaxial stress on R-line luminescence of single crystal sapphire. J Appl Phys 101:7–11
Marques C, Santos L, Falcão A et al (2000) Luminescence studies in colour centers produced in natural topaz. J Lumin 87–89:583–585
Mazurak Z, Czaja M (1996) Optical properties of tsavorite $Ca_3Al_2(SiO_4)_3$:Cr^{3+}, V^{3+} from Kenya. J Lumin 65:335–340
McNicol B, Pott G (1973) Luminescence of Mn ions in ordered and disordered $LiAl_5O_8$. J Lumin 6:320–334
Measures R (1985) Laser remote sensing: fundamentals and applications. Wiley, New York
Min'ko O, Bahtin A, Shepelev Yu (1978) Luminescence spectra of wollastonite as typomorphic feature of wollastonite rocks origin. Abstracts of USSR conference, Tallinn, pp 81–83 (in Russian)
Mironova N, Ulmanis U (1988) Radiation defects and iron group ions in oxides. Zinatne, Riga (in Rusian)
Mironova N, Skvortsova V, Smirnov A et al (1993) Optical spectra and electron spin resonance of 3d-ions in magnesium aluminium spinel. Miner Zhurn 15:36–45 (in Russian)
Mohler R, White W (1994) Influence of structural order on the luminescence of oxide spinels: manganese activated spinels. Mater Res Bull 29:1109–1116
Moine B, Pedrini C, Duloisy E et al (1991) Fluorescence properties of Cu + ion in borate and phosphate glasses. J De Phys IV 1:C7-289–C7-292
Moncorge R, Cormier G, Simkon D, Capobianco J (1991) Fluorescence analysis of chromium doped forsterite (Mg_2SiO_4). IEEE J Quantum Electron 27:114–120
Moncorgé R, Manaa H, Boulon G (1994) Cr^{4+} and Mn^{5+} active centers for new solid state laser materials. Opt Mater 4:139–144
Moncorgé R, Bettinelli M, Gyot Y et al (1999) Luminescence of Ni^{2+} and Cr^{3+} centers in $MgSiO_3$ enstatite crystals. J Phys Condens Matter 11:6831–6841
Monteil A, Boulon G, Garapon C (1988) Optical spectroscopy of a Cr^{3+} doped gadolinium calcium magnesium zirconium oxide (Gd, Ca)$_3$(Ga, Mg, Zr)$_2$$Ga_3O_{12}$ garnet. J Lumin 39:1367–1373
Moroshkin V, Gorobets B, Bushev A (1987) Luminescence of microcline and plagioclases from mica p egmatites as their exploration and genetic guides. Izvestiya Acad Nauk SSSR (Geol) 10:87–91 (in Russian)
Morozov A, Morozova L, Trefilov A, Feofilov P (1970) Spectral and luminescent characteristics of fluorapatite single crystals activated by rare earth ions. Opt Spectrosc 29:590–596
Nasdala L, Grambole D, Götze J et al (2011) Helium irradiation study on zircon. Contrib Miner Petrol 161:777–789
Nasdala L, Grambole D, Ruschel K (2013) Review of effects of radiation damage on the luminescence emission of minerals, and the example of He irradiated $CePO_4$. Miner Petrol 107:441–454
Nasdala L, Stoyanova-Lyubenova T, Gaft M et al (2014) Photoluminescence of synthetic titanite-group pigments: a novel quenching effect. Chem Erde-Geochem 74:419–424
Ogryzlo E (1965) Why liquid oxygen is blue. J Chem Educ 42:647–652
Oomen E, Smit W, Blasse G (1988) Jahn-Teller effect in the Sb^{3+} emission in zircon-structured phosphates. Phys Rev 37:18–26
Panczer G, Gaft M, Reisfeld R et al (1998) Luminescence of uranium in natural apatites. J Alloys Comp 275–277:269–272
Panczer G, Gaft M, Marfunin A (2000) Systems of interacting luminescence centers in natural diamonds: laser-induced time-resolved luminescence and cathodoluminescence spectroscopy.

In: Pagel M, Barbin V, Blanc P, Ohnenstetter D (eds) Cathodoluminescence in geosciences. Springer, Berlin/Heidenberg/New York, pp 359–373

Panczer G, Gaft M, de Ligny D, Champagnon B (2010) Luminescence characteristics of the Cs bearing "beryl", pezzottaite Cs(Be$_2$, Li)Al$_2$(SiO$_3$)$_6$H$_2$O. Eur Miner J 22:605–612

Payne S (1988) Optical properties of Cr^{3+} in fluoride hosts. J Lumin 40–41:305–306

Peixoto J, Oort A (1992) Physics of climate. Springer, Berlin

Pereira E, Monteiro T (1990) Non-radiative processes in deep centers of diamonds. J Lumin 45:443–446

Pereira E, Monteiro T (1991) Delayed luminescence of the H3 center in diamond. J Lumin 48–49:814–818

Pereira E, Santos L (1993) The 2.96 eV centre in diamond. Phys B 185:222–227

Pereira E, Santos L (1994) Dynamical processes of the 2,818 eV centre in diamond. Diamond Relat Mater 4:26–32

Pillonnet A, Garapon C, Champeaux C et al (2000) Fluorescence of Cr^{3+} doped alumina optical waveguides prepared by pulsed laser deposition and sol–gel method. J Lumin 87–89:1087–1089

Piriou B, Fahmi D, Dexter J (1987) Unusual fluorescence properties of Eu^{3+} in oxyapatites. J Lumin 39:97–102

Piriou B, Elfakir A, Quarton M (2001) Site-selective spectroscopy of Eu^{3+-}doped sodium lead phosphate apatite. J Lumin 93:17–26

Platonov A (1979) Color of minerals. Naukova Dumka, Kiev (in Russian)

Platonov A, Taran M, Balizkii B (1984) The natures of semi-precious stones color. Nedra, Moscow (in Russian)

Platonov A, Tarashchan A, Langer K et al (1998) Electronic absorption and luminescence spectroscopic studies of kyanite single crystals: differentiation between excitation of FeTi charge transfer and Cr^{3+} dd transitions. Phys Chem Miner 25:203–212

Porto S, Giordmaine J, Damen T (1956) Depolarization of Raman scattering in calcite. Phys Rev 147:607–614

Powell R, Blasse G (1980) Energy transfer in concentrated systems. Struct Bond 42:43–85

Powell R, DiBartolo B, Birang B, Naiman C (1967) Fluorescence studies of energy transfer between single and pair Cr^{3+} systems in Al_2O_3. Phys Rev 155:296–308

Powell R, Xi L, Gang Xu, Quarles G (1985) Spectroscopic properties of alexandrite crystals. Phys Rev B 32:2788–2797

Prokic M (1979) Mechanics of thermoluminescence in natural barites. J Phys Chem Sol 40:405–412

Prokofiev I, Gorobets B, Gaft M, Lurie Yu (1982) Rare earth luminescence centers in leucophane, shortite, meionite and monacite. Miner Sborn Lvov State Univer Vyshchaya Shkola 36:76–79 (in Russian)

Rappoport W, Khattak C (1988) Titanium sapphire laser characteristics. Appl Opt 27:2677–2684

Reiche I, Vignand C, Champagnon B, Panczer G et al (2001) From mastodon ivory to gemstone: the origin of turquoise color in odontolite. Amer Miner 86:1519–1524

Reisfeld R (1973) Spectra and energy transfer of rare earths in inorganic glasses. Struct Bond 13:53–98

Reisfeld R (2015) Optical properties of rare earths in condensed phase, theory and applications. AIMS Materials Science 2, 2, 37–60

Reisfeld R, Jörgensen K (1977) Lasers and excited states of rare earths. Springer, Berlin/Heidelberg/New York

Reisfeld R, Gaft M, Boulon G et al (1996) Laser-induced luminescence of REE in natural fluor-apatite. J Lumin 9:343–351

Reisfeld R, Gaft M, Saridarov TS et al (2000) Nanoparticles of CdS with Eu and Tb in zirconia films having intensified luminescence. Mater Lett 45:154–156

Reisfeld R, Zigansky E, Gaft M (2004) Europium probe for determination of site symmetry in glass films, glasses and crystals. Mol Phys 102:1319–1330

Rey C, Combes C, Drouet C, Glimcher MJ (2009) Bone mineral: an update on chemical composition and structure. Osteoporos Int 20:1013–1021
Rosenzweig A (1990) Photoacoustic and photoacoustic spectroscopy. Kriger, Florida, 309 pp
Rossman GR, Grew E, Dollase W (1982) The colors of sillimanite. Am Miner 67:749–757
Rotman S, Hartman F (1988) Non-radiative energy transfer in doubly-doped solid-state laser materials. SPIE 1182:151–161
Ryan F, Ohlmann R, Murphy J et al (1970) Optical properties of divalent manganese in calcium fluorophosphate. Phys Rev B 2:2341–2352
Schepler K (1984) "Fluorescence of inversion site Cr" ions in alexandrite. J Appl Phys 56:1314–1318
Schlaich H, Lindner G, Feldman J et al (2000) Optical properties of Se_2^- and Se_2 color centers in red selenium ultramarine with the sodalite structure. Inorg Chem 39:2740–2746
Schott S, Rager H, Schurmann K, Taran M (2003) Spectroscopic study of natural gem quality "Imperial" – topazes from Ouro Preto, Brasil. Eur J Miner 15:701–706
Scott M, Henderon B, Gallagher H, Han T (1997) Optical spectroscopy of $(MnO_4)^{3-}$ and $(VO_4)^{5-}$ in $Sr_{10}(VO_4)F_2$. J Phys Condens Matter 9:9893–9897
Shionoya S, Yen W (1999) Phosphor handbook. CRC Press, Boca Raton
Sidike A, Kusachi I, Yamashita N (2002) Energy transfer from Pb^{2+} to Mn^{2+} in fluorescent halite from Salton Sea, California. J Miner Petrol Sci 6:278–284
Sidike A, Kusachi I, Yamashita N (2006) Yellow fluorescence from baghdadite and synthetic $Ca_3(Zr, Ti)Si_2O_9$. Phys Chem Miner 32:665–669
Sidike A, Sawuti A, Wang X-M et al (2007) Fine structure in photoluminescence spectrum of S_2^- center in sodalite. Phys Chem Miner 34:477–484
Sidike A, Kusachi I, Kobayashi I et al (2008) Photoluminescence spectra of S_2^- center in natural and heat treated scapolites. Phys Chem Miner 35:137–149
Sidike A, Jilili N, Kobayashi S et al (2010a) Photoluminescence properties of anthophyllite. Phys Chem Miner 37:83–89
Sidike A, Kobayashi S, Zhu H-J, Yamashita N (2010b) Photoluminescence of baratovite and katayamalite. Phys Chem Miner 10:705–710
Sidorenko G, Gorobets B, Dubinchuk V (1986) Contemporary methods of mineralogical analysis of uranium bering rocks. Energoatomizdat, Moscow (in Russian)
Siletti D, Brokus S, Earlywine E et al (2012) Radiation-induced cathodoluminescent signatures in calcite. Radiat Meas 47:195–200
Srivastava A (1998) Luminescence of divalent bismuth in M^{2+} BPO_5 ($M^{2+} = Ba^{2+}$, Sr^{2+} and Ca^{2+}). J Lumin 78:239–243
Stade J, Hahn D, Dittmann R (1974) New aspects of the luminescence of magnesium titanate part II: activation with manganese. J Lumin 8:318–325
Stepanov I, Feofilov P (1956) On two types of luminescence spectra in artificial fluorite crystals. Docl Acad Nauk SSSR 108:615–618
Suchoki A, Biernacki A, Arizmendi L (2002) Nephelauxetic effect in luminescence of Cr^{3+} doped lithium niobate and garnet. International conference on luminescence and optical spectroscopy of condensed matter, Budapest, Abstracts, p 35
Suchoki A, Biernacki S, Boulon G et al (2004) Enhanced Zeeman effect in GGG:Mn $^{4+}$, Ca crystals. Chem Phys 298:267–272
Tanaka K, Yano T, Shibata S et al (1994) Cu^+ -doped CaO-P_2O_5 glasses for lasers. J Non-Cryst Solid 178:9–14
Tarashchan A (1978) Luminescence of minerals. Naukova Dumka, Kiev (in Russian)
Tolstoy N, Shinfue L (1960) Kinetick of emission of chromium luminophores. Opt Spectrosc 9:415–419
Vacek K (1971) Some remarks on photoluminescence of AgCl crystals excited by laser or after deformation. Czech J Phys B21:303–308
Van Doorn C, Schipper D (1971) Luminescence of O_2^-, Mn^{2+} and Fe^{3+} in sodalite. Phys Let A 34:139–140

Voronko K, Maksimova G, Sobol A (1991) Anisotropic luminescence centers of TR^{3+} ions in fluorapatite crystals. Opt Spectrosc 70:203–206 (in Russian)

Votyakov S, Krasnobaev A, Krohalev V (1993) Problems of applied spectroscopy of minerals. Nauka, Ekaterinburg, 233 p (in Russian)

Wallace L, Hunten D (1968) Dayglow of the oxygen A band. J Geophys Res 3:4813–4833

Waychunas G (1987) Synchrotron radiation XANES spectroscopy of Ti in minerals: effects of Ti bonding distances, Ti valence and site geometry on absorption edge structure. Am Miner 72:89–95

Waychunas G (1989) Luminescence, X-Ray emission and new spectroscopies. In: Hawthorne FC (ed) Spectroscopic methods in mineralogy and geology. Rev Mineral. Mineralogical Society of America, Washington, DC, pp 638–698

White W (1990) Photoluminescence, candoluminescence, and radical recombination luminescence of minerals. In: Coyne L, McKeever W, Blake D (eds) Spectroscopic characterization of minerals and their surfaces. American Chemical Society, Washington, DC, pp 118–134

White W, Masako M, Linnehan et al (1986) Absorption and luminescence of Fe^{3+} in single-crystal orthoclase. Am Miner 71:1415–1419

Wojtowicz A (1991) Luminescence of Cr^{3+} in kyanite. J Lumin 50:221–230

Wojtowicz A, Lempicki (1988) Luminescence of Cr3+ in sillimanite. Phys Rev B 39:8695–8701

Yamaga M, Yosida T, Henderson B et al (1992) Electro paramagnetic resonance and optical spectra of Ti^{3+}-doped $YAlO_3$. J Phys Condens Matter 4:7285–7294

Yang C (1995) Ionoluminescence techniques for geological applications. Department of Nuclear Physics, Lund Institute of Technology, Lund

Yauri J, Cano N, Watanabe S (2008) TL, EPR and optical absorption in natural grossular crystal. J Lumin 128:1731–1737

Zaitsev A (2005) Optical properties of diamond. Springer, Berlin

Chapter 6
Complementary Laser Based Spectroscopies

In any practical application in the mineralogy field the main task is composition analyses. There is no single detection technique that can by itself provide a 100 % probability of different minerals detection combined with a low false alarm rate. We suggest system approach, which combines different laser-based technologies having orthogonal detection and identification capabilities. In such a way the strength of one technique may compensate for the weaknesses of the others, and the vulnerability of one detection device could be compensated for by another detection device. Clever combination of the detection techniques may achieve detection probabilities and false alarm rates that are more acceptable than those of systems based on one method only.

The same equipment, which is used for time-resolved luminescence application is suitable for other laser-based spectroscopies. Thus several spectroscopic methods may be applied simultaneously. The mostly important technique, which may be used together with time-resolved luminescence, is laser-induced breakdown spectroscopy. Several books have been recently published devoted to Laser Induced Breakdown Spectroscopy (LIBS) (Cremers and Radziemski 2013; Miziolek et al. 2006; Singh and Thakur 2007; Noll 2012; Hahn and Omenetto 2010; Hahn and Omenetto 2012). LIBS aspects were considered applied to the analysis of minerals, rocks and related materials (Senesi 2014). Thus only the theoretical aspects which are the mostly relevant to our research devoted to the real time online quality control of minerals will be considered.

M. Gaft et al., *Modern Luminescence Spectroscopy of Minerals and Materials*,
Springer Mineralogy, DOI 10.1007/978-3-319-24765-6_6

6.1 Laser Induced Breakdown Spectroscopy

6.1.1 *Plasma Creation*

As most commonly used, the technique employs a pulsed laser and a focusing lens to generate a plasma that vaporizes a small amount of a sample. A portion of the plasma light is collected and directed to a spectrometer. The spectrometer disperses the light emitted by excited atoms, ions, and simple molecules in the plasma, a detector records the emission signals, and electronics take over to digitize and display the results.

A general and simplified LIBS process may be presented as follows. When a short duration laser pulse of sufficient energy density is focused onto the surface of a target, the surface temperature instantly increases to above the vaporization temperature. This violent photon-material interaction could include single- or multi-photon absorption, dielectric breakdown, and others still unknown mechanisms. Compared with the rate of energy delivery from the laser pulse, the energy dissipation through vaporization is relatively slow. As a result, before the surface layer vaporizes, the underlying layer of material reaches critical temperatures and pressures, and forces the surface to explode. Within the same laser pulse, usually a few nanoseconds, the ablated material continues to absorb energy from the laser beam and forms luminous plasma. In the laser-induced plasma atoms, ions, molecules, and clusters exist, in their different energy states. The temperature of the plasma is in the range of 10^4–10^5 K, and the electron number density is in the order of 10^{15}–10^{19} cm^{-3}. The plasma finally starts to decay including radiative processes, where each element has its own characteristic lines and the spectra emitted are used to determine the sample's elemental constituents.

The laboratory experimental setup is actually the same as for laser induced time resolved luminescence. It includes laser source, spectrometer, detector, beam focusing and guiding optics and computer. The important difference from luminescence is the typical laser source. Various laser types are applied for LIBS studies. The most wide spread in practical applications are flashlamp-pumped solid-state lasers with Nd: YAG as laser medium operated in the Q-switch mode to generate high-energy laser pulses with pulse durations in the nanosecond range. For pulsed Nd: YAG lasers typically applied for LIBS, Xe flash lamps are used with typical life times of 20–60 million flashes depending on the total energy dissipated during the flash lamp discharge. Diode-pumped systems offer a better efficiency. The specific features of solid-state lasers in comparison with gas, liquid, or semiconductor lasers are high power, short pulses, efficient frequency conversion, and the capability to transmit the laser radiation via fiber optics.

Double pulse approach became recently very popular. One of its advantages is that it may be easily introduced in industrial analyzers. The main difference from the single pulse approach can be summarized as follows: (i) the first laser pulse creates a plasma and the second pulse propagates at high velocity in the rarified medium created by the first pulse. (ii) the plume size is wider in the double-pulse case when compared to that observed with a single pulse. The region of the plume characterized by a high temperature is also wider in the dual pulse case; (iii) the enhancement in the intensity of the spectral lines is associated with a larger mass removal from the ablation pulse, attributed to the lower shielding effect of the laser beam from the buffer gas, as shown by the increase in the crater volume ablated on the sample surfaces. This enhancing effect allows improving the detection limits and consequently extending the analysis also to elements present as traces in the sample; (iv) a higher proportion of ionic line emission is observed in some double-pulse configurations, the behavior of the intensity of ionic lines changes as a function of inter-pulse separation and observation delay times, while the enhancement is different for ions and neutral atom transitions; (v) re-heating seems to be effective at short inter-pulse separation used and transitions with much higher excitation energy are observed with two pulse operation; and (vi) there is a correlation between the excitation energy of the transition investigated and the enhancement observed in double-pulse operation.

The differences with SP LIBS are the mostly interesting during an earliest phase (0–500 ns) of the lifetime of the plasma: (i) the intensity of continuous radiation is nearly a factor of two lower and the lines are substantially narrower, implying a lower density of electrons; (ii) the relative line-intensity of singly and, especially, doubly-ionized species is higher than for neutral species; (iii) the plasma lifetime for both neutral and ionized species is longer, but still different for different ionization states. The lifetime of doubly-ionized species is very short and after a delay of 200–300 ns they are not detected. This lifetime dependence of differently ionized species leads to a strong dependence of the experimentally determined amplification of line-intensities on the experimental time conditions and valence states of the species under interest. A mechanism is proposed enabling qualitative explanation of the experimental data collected in a dual-pulse plasma experiment. It is based on the assumption that, when the second pulse arrives, there is a net increase in the amount of material in the plume (because of increased ablation by the laser pulse), and at the same time an increase in the plasma plume volume (at a much larger rate than the increase of the overall mass of the species). This increase of plasma volume leads to an effective reduction of the overall density of the plasma, and therefore to a reduction of the line broadening that is estimated from the Stark model, and to increasing plasma lifetime. The remaining second pulse energy is absorbed by free electrons (inverse Bremsstrahlung), resulting in an increase of their energy compared with SP excitation. These high-energy electrons increase the concentrations of singly and doubly-ionized species in the plasma plume, via collisions with atoms and singly-ionized species, thereby losing their own energy. As a result, the Bremsstrahlung radiation significantly diminishes in the UV–visible spectral range compared with longer wavelengths under DP laser excitation, while the relative intensity of singly and doubly-ionized species is more pronounced. The increase of plasma lifetime is evidently connected with an increased volume of the plasma plume (reduction of its density). This leads to a diminishing number of collisions with free-electrons, because of the lower electron density (Nagli et al. 2011).

6.1.2 Plasma Analysis

From the analytical point of view, plasma emission can be divided into several time domains. The simplified approach to plasma time history is that soon after plasma initiation, continuum and ionic spectra are seen. The continuum is the "white light" from the plasma that contains little spectroscopic information and the ions result from electrons ejected by neutral atoms. As the plasma decays, these are followed by spectra from neutral atoms, and eventually simple molecules formed from the recombination of atoms. Throughout the temporal history, one observes a diminishing continuum spectral background due to recombination of free electrons with ions. Actually the situation is more complicated and rich of information starting from delay of 1 ns and after the actual plasma termination.

6.1.2.1 Fraunhofer Type Absorption

The first stage is actually characterized by a broad emission originating from the *Bremsstrahlung* of the free electrons and electron-ion recombination and it has duration of a few hundred nanoseconds. Weak lines show up on the strong continuum and they are mostly identified as ionic lines of the plume constituents. This time domain was considered as not suitable for analytical applications and minimum delay time of 100–500 ns was recommended to remove the *Bremsstrahlung*.

Nevertheless it was found that absorption phenomena take place during the earliest time of plasma life, which may be used for analytical purposes. Absorption measurements in LIBS plasmas can be envisaged in different ways. One way is to use a fixed frequency laser (for example, the same laser that created the plasma) and directly measure the plasma transmission. Another way is to use an external, tunable laser traversing the plasma at different delays with respect to plasma formation. When the spectral profile of the laser is much narrower than the absorption profile of the atomic (ionic) transition investigated, the spectral shape of the transition can be directly obtained by scanning the laser wavelength across the entire profile and the number density of the absorbing species can be evaluated. A final way is to use the plasma itself as a continuum radiation source for the absorbing species. This can be done using the LIBS plasma as an external excitation source and focusing its radiation on the atom reservoir; in this way, the laser-induced plasma plays the role of the primary excitation source in a conventional atomic absorption experiment. It was found that by analyzing the plasma emission at early times (less than ~100 ns), when the spectral continuum dominates the spectrum, one notices atomic absorption features corresponding to strong transitions involving the ground state or low lying levels. The last approach has been recently exploited by Ribie're and Che'ron (2010) by using two lasers to produce two closely spaced, independent plasmas on the same sample and investigating the resulting absorption features at different inter-pulse delays. Principally using this technique no calibration of the detection system is needed,

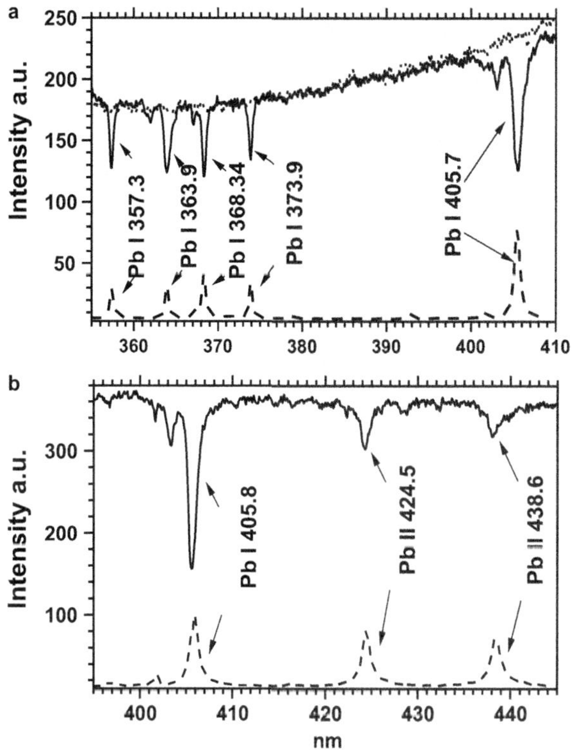

Fig. 6.1 (**a–b**) Fraunhofer type absorption of Pb in laser induced plasma

the only requirement being that the interaction between the laser radiation and the atoms must be linear.

The approach with only one laser source was proposed, where the collinear double-pulse LIP during its earliest stages of evolution, namely first 5–10 ns, is very like to solar radiation (Nagli and Gaft 2013). It consists of a hot compact plasma core created by the second pulse and a larger volume of ablated material left from the first pulse. The core irradiates the broadband continuum spectrum, which is partly absorbed by the much cooler outside plume containing atoms and ions in their ground or close to ground states. Fraunhofer-type absorption lines were experimentally detected and theoretically calculated for Pb (Fig. 6.1) and Si. Besides, similar Fraunhofer lines of Ti, Al, Ca, Mg, and Cu were found in pure metals, alloys, and minerals. Those lines exist only during 0–15 ns of plasma life after the second laser pulse. Experimental conditions, such as the laser pulse energy densities, delay time between the laser shots, and data acquisition time optimal for Fraunhofer lines detections, were found. New unidentified lines were detected in Fraunhofer spectra of Pb, Si, and Ti that were not seen in the emission spectra of these elements. These new lines are explained by the motional stark effect that split certain atom energy levels in collinearly DP LIP plasma. It was demonstrated that Fraunhofer lines may be used for analytical purposes, for

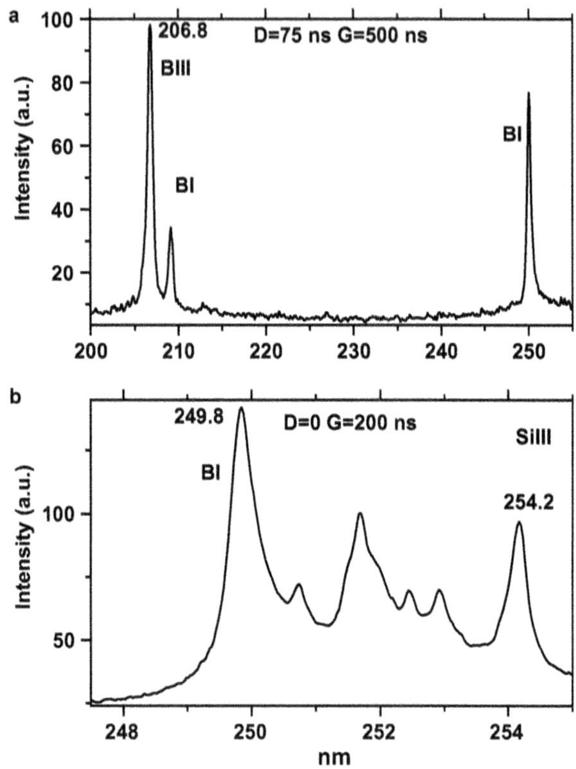

Fig. 6.2 (**a–b**) Emission of B III and Si III ions in laser induced plasma

example, in the case of high concentrations where loss of sensitivity takes place because of self-absorption. They provide the possibility for the new direct and calibration-free LIBS analytical applications, both in laboratory and industrial conditions.

6.1.2.2 High Ionized Ions

Next time window is from 10 to 100 ns where the emission from doubly ionized species in laser induced plasmas has not been properly investigated before since most analytical measurements were made at relatively long delays. It was found (Gaft et al. 2011c) that doubly-ionized species, such as B III, Al III, Si III and Fe III, can exist during the first 150–200 ns of the plasma lifetime in plasmas produced in air by typical lasers with irradiances of 109–1011 W/cm^2 (Fig. 6.2). The emission from these ions was detected using both the double- and single-pulse excitations. The sum of the second ionization potential and the energy of corresponding excited states is approximately 30 eV. The presence of doubly charged ions in the early plasma was additionally confirmed by computer simulations using a collision-

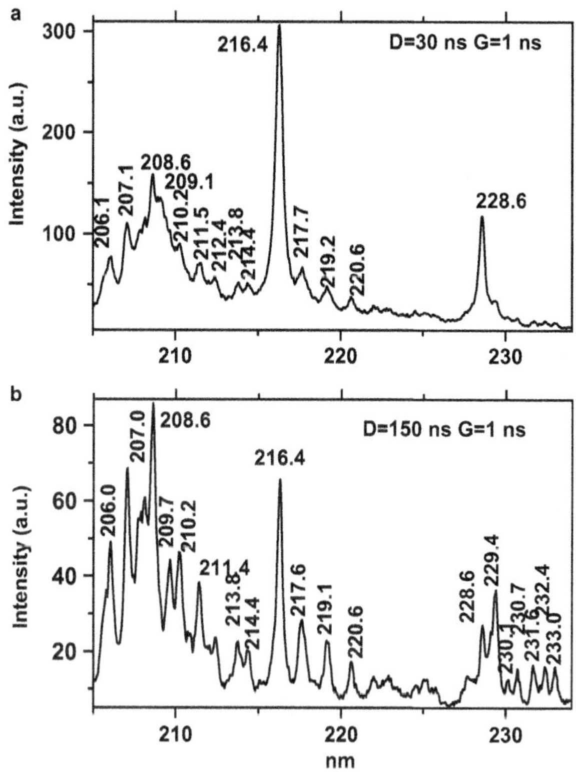

Fig. 6.3 (**a–b**) Emission of Zr IV and Zr III in laser induced plasma

dominated plasma model. The emission from doubly ionized species may be used for analytical purpose.

Besides, triple ionization was found in emission spectra of a laser-induced zirconium plasma at early times after plasma formation in the short UV spectral range from 190 to 240 nm (Gaft et al. 2013). Many lines from highly ionized Zr ions, such as Zr IV at 216.4 and 228.7 nm and Zr III at 194.1, 194.7, 193.1, 196.3, 196.6, 197.5, 199.0, 200.3, 200.8, 202.8, 203.6, 205.7, 206.1, 207.1, 208.6, 210.2, 211.4, 217.6, and 219.1 nm have been found in the plasma under ambient and vacuum conditions (Fig. 6.3). These lines could be detected in both single-pulse and double-pulse modes.

6.1.2.3 Atoms and Low Ionized Ions

The third time domain, from 100 ns to 30 μs, is characterized by an emission spectrum, where narrow atomic lines dominate corresponding to the elements present in the plume and the line strength is proportional to the atomic concentration. This time regime lasts for tens microseconds, where in its early stage ionic

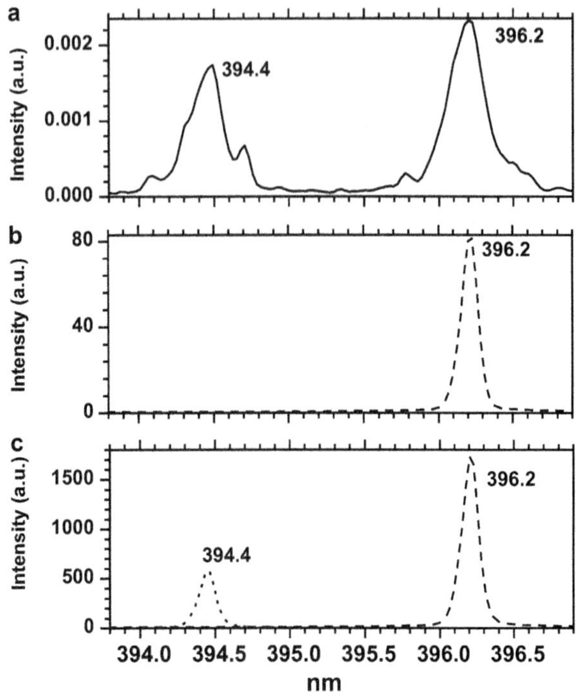

Fig. 6.4 (**a–c**) Al plasma emission measured at backward direction: LIBS (**a**), stimulated emission excited at 256.8 nm (**b**) and lasing spectra (**c**) under pumping by 256.8 nm (*dash*) and 257.5 nm (*dot*) 4 μs after plasma plume creation by 1.06 μm laser pulse

lines are strong, while the atomic lines dominate later on. This stage is well presented in literature.

The recent development is that under appropriate resonant optical pumping of a pre-formed laser plasma plume, stimulated emission and lasing occur (Nagli and Gaft 2015a). They manifest themselves as emission of intense, polarized and collimated beams in the forward and backward directions that exist only during a pumping pulse.

Figure 6.4 presents examples of the Al plasma emission spectra measured in backward direction, such as regular plasma emission (Fig. 6.4a), plasma plume emission pumped at 257.5 nm (Fig. 6.4b) and plasma plume emission placed in optical resonator (Fig. 6.4c). Solid curve is regular LIBS, dash curves presents the excitation at 257.5 nm and dot curve presents the excitation at 256.8. All results are measured 4 μs after plasma plume creation, 0 ns acquisition delay and with gate width of 1 μs. Plasma emission lines Full Width at Half Maxima (FWHM) is about 0.25 nm and FWHM of optically pumped emission lines is about 0.13 nm that evidently is determined by our optical system spectral resolution (0.1 nm). This strong collimated emission lasts only during existence of a pumping OPO pulses, i.e. about 4 ns.

Stimulated emission (SE) and lasing effects are found in plasma of atoms of elements from the 13th and 14th groups of the periodic table. Atoms of Al and In in laser-induced plasma both possess a three-level lasing system. In Tl plasma both three- and four-level laser systems are present. Laser-induced plasma of elements from the 14th group (Ge, Sn, Pb) also have strong SE and lasing effect, exhibiting four-level or quasi-four-level laser systems. Intense, collimated backward SE in a LIP plume may strongly enhance the selectivity and sensitivity of inductively-coupled plasma(ICP-OES) and laser-induced breakdown spectroscopy (LIBS) used for elemental analyses of different substances.

6.1.2.4 Molecular Emission

The fourth time domain is characterized by an emission spectrum, where molecular narrow bands dominate. It starts from approximately 15–25 μs and lasts for tenths of microseconds. Molecular analysis by LIBS has been much less investigated than elemental LIBS. Molecular LIBS refers to the spectra emitted by different molecules and diatomic radicals present in the plasma and resulting from the ablation of the target or from subsequent chemical reactions with the ambient air. Molecular spectroscopy with LIBS could be considered an contradictive term, since the high temperatures provided by the plasma would dissociate completely any molecular species present. Nevertheless, since the plasma evolves in time and cools down, an appropriate selection of the experimental parameters (laser energy, delay time) allows the observation of molecular emission. The well known examples are spectra given by CN violet system, CH, the C_2 Swan system, CaO, AlO, TiO and OH emission spectra in air (Parigger 2013). Several new molecular emissions have been found during minerals research which are as scientifically and practically important.

Halogens and Alkali-Earth Molecules

It was found that molecular LIBS is one of the potential ways to improve plasma analysis sensitivity for several elements which are difficult to detect by their atom or ion emission, especially in ambient conditions. The famous example is the case of halogens. The difficulty of detecting halogens is attributed to, among other factors, their energy level distribution. For example, the strongest emission lines for fluorine and chlorine are in the VUV spectral range (95.5 and 134.7 nm for F and Cl, respectively). Detection capability in this region is limited by atmospheric and materials absorption, as well as by detector sensitivity. Optical transitions applicable for LIBS detection but less strong than those in the VUV exist in the upper visible and in the near IR (500–850 nm). These are transitions between the excited states (n + 1)s ^4P–(n + 1)p ^4D manifold (n = 2 for F and n = 3 for Cl), corresponding to wavelengths centered around 685.6 nm for F and 837.6 nm for Cl. Detection in this spectral region is more practical than in the VUV region, but these lines have upper levels of 10.40 and 14.50 eV, respectively, above the ground state. The detection limits for these lines are less than satisfactory for demanding applications.

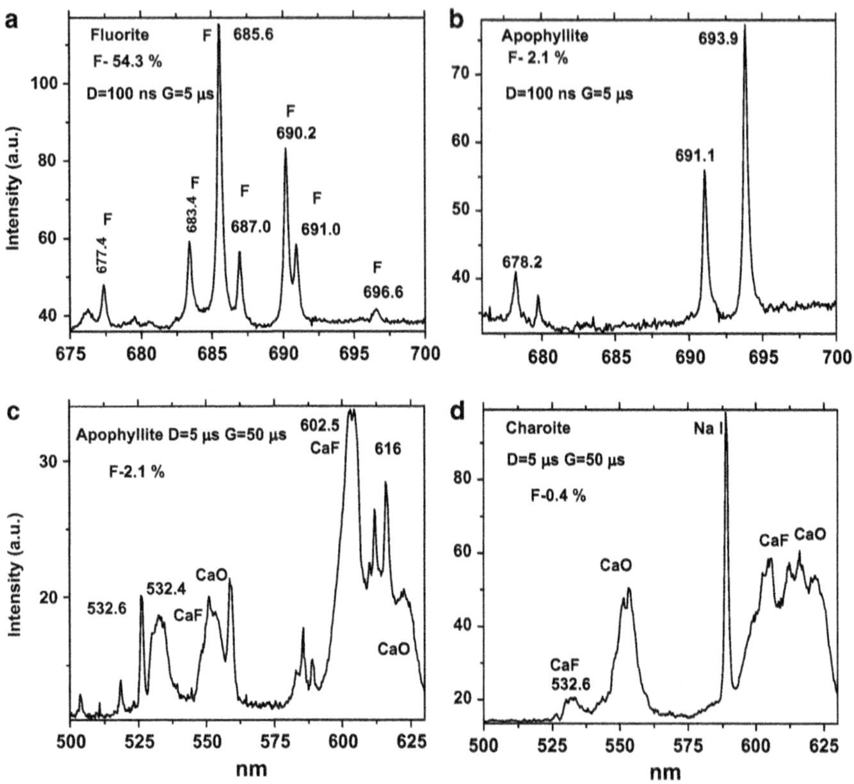

Fig. 6.5 (**a–d**) Comparison of ionic F detection sensitivity in single-pulse breakdown spectra of fluorite CaF_2 (**a**) and apophyllite $KCa_4Si_8O_{20}(F,OH)\times 8H_2O$ (**b**) with molecules CaF detection in apophyllite $KCa_4Si_8O_{20}(F,OH)\times 8H_2O$ and charoite $(Ca,Na)_2Si_4O_{10}(OH,F)\times H_2O$ (**c, d**)

At the same time in arc-induced plasma, where similar to LIBS halogens do not provide atomic or line spectra of sufficient intensity to permit their detection by normal spectroscopic procedures, it was found that halogens form molecules whose spectra may be easily identified. For example, CaF molecular bands have been recommended for use in analytical work as a test for the presence of fluorine. Similarly, chlorine may be detected by observation of the CaCl bands (Pearse and Gaydon 1965).

Similar to this, it was found that the same situation takes place in Laser Induced Plasma (Gaft et al. 2014b). For example, Fig. 6.5 compares molecular CaF and ionic F I detection sensitivity in SP LIBS. Two minerals were used: fluorite with approximately 50 % fluorine content and apophyllite with approximately 2 %. The strongest F I emission lines at 685.6 and 690.2 nm appeared in a fluorite emission spectra with a minimal delay time of 100 ns (Fig. 6.5a). Mineral apophyllite, with 25 times lower fluorine content than fluorite, did not exhibit F I emission lines (Fig. 6.5b). CaF bands were clearly visible not only in apophyllite (Fig. 6.5c), but also in charoite with

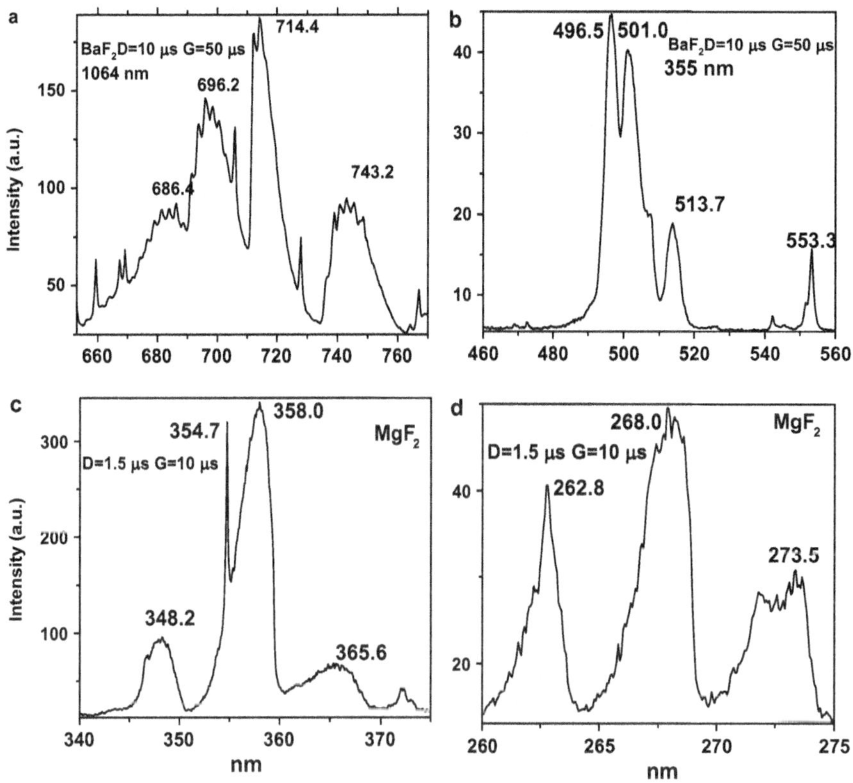

Fig. 6.6 (**a**–**d**) Single-pulse breakdown spectra of BaF_2 excited by 1064 nm (**a**) and 355 nm (**b**) containing BaF molecules emission and of MgF_2 excited by 1064 nm containing different MgF emission bands (**c, d**)

approximately 0.4 % fluorine content (Fig. 6.5d). Under DP excitation, ionic emissions were stronger compared to SP-LIBS by 4 times while molecular bands were stronger by 10 times, indicating that DP-LIBS enhances molecular sensitivity over SP-LIBS more than ionic sensitivity. Similarly, Mg and Ba forms MgF and BaF molecules with strong characteristic band emission (Fig. 6.6).

Chlorine behavior is similar to fluorine. Figure 6.7 compares the LIBS sensitivities of molecular CaCl and ionic Cl I. Using the SP mode, a weak line of Cl I peaking at 837.6 nm was detected in $CaCl_2$, whose chlorine content is approximately 64 % (Fig. 6.7a). It is not detected in the mixture of $CaCO_3$ with 10 % $CaCl_2$. Using molecular CaCl emission, Cl was detected not only in this sample (Fig. 6.7b), but also in the mixture of $CaCO_3$ with 1 % of $CaCl_2$ (Fig. 6.7c). In DP plasma, the Cl I ion emission intensity is approximately 3 times higher than in SP plasma and its very weak line may be detected in the mixture of $CaCO_3$ with 10 % $CaCl_2$. Using CaCl molecules, Cl is detected in the mixture with 0.4 % of $CaCl_2$ (Fig. 6.7d).

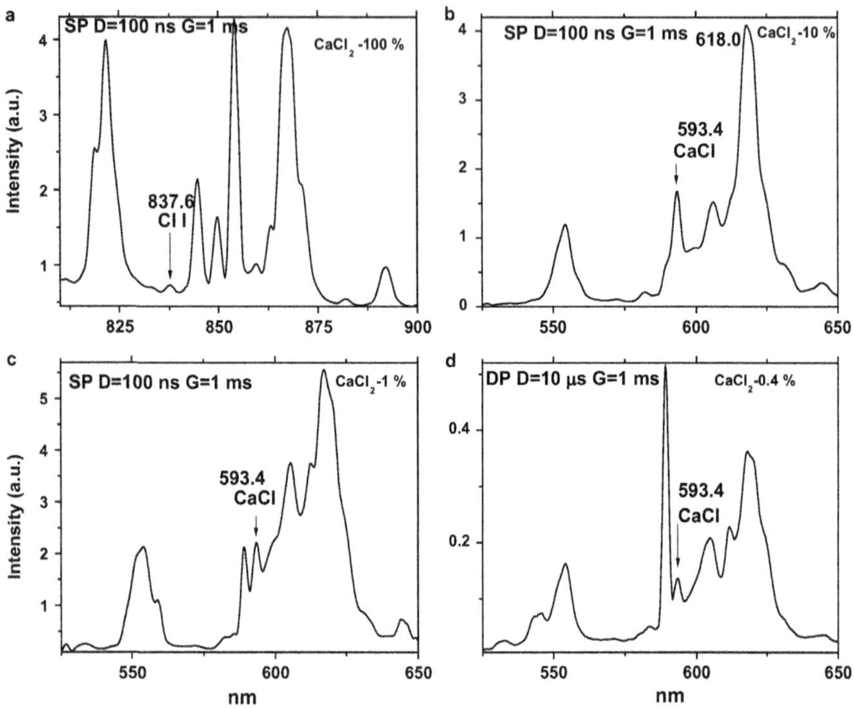

Fig. 6.7 (**a–d**) Comparison of ionic Cl detection sensitivity in single-pulse breakdown spectra of $CaCl_2$ (**a**) and molecular CaCl detection sensitivity in the mixture of $CaCO_3$ with 10 % $CaCl_2$ (**b**) and 1 % $CaCl_2$ (**c**), where molecular Cl emission were not detected. In double-pulse breakdown spectra CaCl molecules may be detected in the mixture with 0.4 % of $CaCl_2$ (**d**)

Boron Bearing Molecules

While studying LIP of several boron bearing minerals, several unusual bands have been found in their LIP emissions (Gaft et al. 2015). Figure 6.8 presents plasma emission spectra of several boron bearing minerals with $CaBO_2$, $MgBO_2$, BO and BO_2 molecules emissions. All those bands dominate the spectra after long delay time of approximately more than 30 μs. It may be seen that BO and BO_2 molecules is detected for any boron bearing material, including metallic boron, while complex molecules needs simultaneous presence of several elements. It was proved that emission of complex molecules, such as $CaBO_2$, $MgBO_2$ and $SrBO_2$, was for the first time found and interpreted in laser-induced plasma. They dominate the plasma emission after a very long delay time of more than 30 μs. Such molecules have been identified by mass spectroscopy, but have not until now been found in any kind of optical emission and absorption analysis.

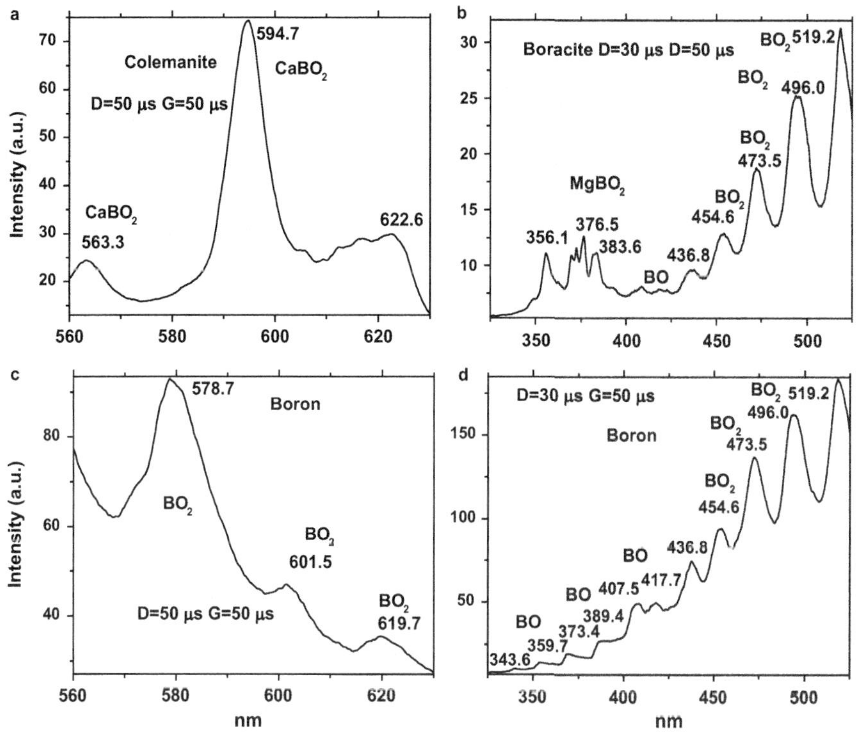

Fig. 6.8 (**a–d**) Single pulse breakdown spectra of Ca-B bearing colemanite (**a**) in visible range, Mg-B bearing boracite in UV range (**c**) and pure metallic boron in the same visible (**b**) and UV (**d**) spectral ranges

LIBS-Molecular Laser Induced Fluorescence (MLIF)

LIBS in its simplest form suffers in some cases from relatively low detection limits. Several methods for improvement have been proposed. One such method is combining LIBS with laser-induced fluorescence spectroscopy (LIBS-LIF), in which LIBS-created atoms and ions in a plume of plasma are analyzed using LIF, first proposed by Measurres and Kwong (1979). This method is applicable to atoms and ions that have resonant optical transitions between their ground and excited energy states in the accessible UV-VIS spectral range. The analytical potential of LIF for molecular excitation formed in laser plasma was recently investigated (Nagli and Gaft 2015b).

LIBS-LIF spectra are very different from usual LIBS spectra (Fig. 6.9). Estimated LIF decay time under VIS excitation is as rapid as the excitation OPO pulse width of 4 ns. The cause of such emission decay time shortening may be collisional quenching of the molecular excited states in LIP or thermal quenching of the excited states at high plasma temperature. Due to such short emission lifetime, MLIF measurements were done with a short gate width of W = 10 ns; a delay of

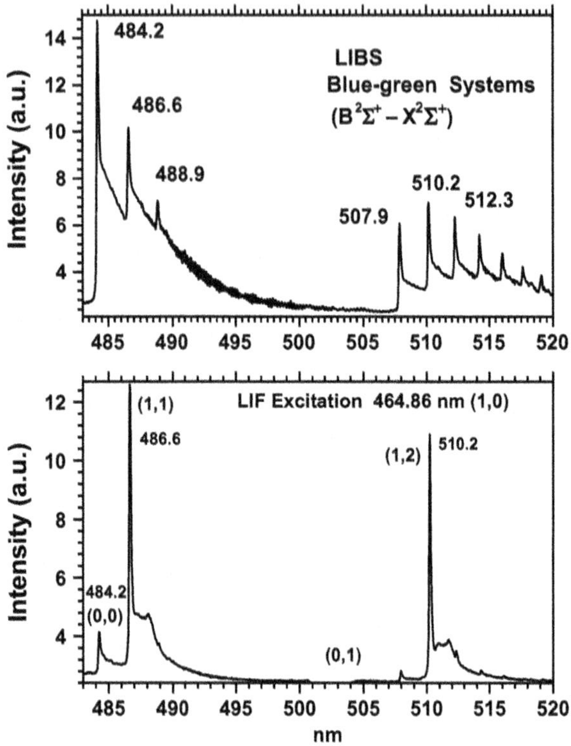

Fig. 6.9 (**a–b**) LIBS (**a**) and LIBS-MLIF (**b**) of AlO molecular emission

D = 10 μs was used, as in LIBS experiments. With such a short gate width, ordinary LIBS is almost undetectable, but optical excitation of the AlO molecules leads to strong AlO emission. An excitation wavelength corresponding to a definite molecular energy level *e' ν' J'* leads to the appearance of a series of strong emission bands directly from this energy level. Figure 6.9b shows typical AlO MLIF spectra excited by resonant (1,0) transition.

Such an approach, termed LIBS-MLIF, may be very effective in certain cases, both for analytical purposes and for isotopic analysis. For example, recently laser-ablation molecular isotopic spectrometry (LAMIS) was proposed for boron isotope analysis (Russo et al. 2011; Mao et al. 2011). LIBS-MLIF was tested for this task and proved that an easy and confident detection of different boron isotopes using well known molecular emissions (Mulliken 1925) is possible. Spectra of metallic boron with a natural concentration ratio of 4 between ^{11}B and ^{10}B isotopes were compared with corresponding spectra of an enriched boron sample with ^{11}B isotope concentration more than 96 %. MLIF (0,n = 1,2...) transitions in the $B^2\Sigma^+$-$X^2\Sigma^+$ system for both samples were excited at the (0,0) transition at 233.13 and at 233.19 nm for ^{11}BO molecules and ^{10}BO molecules, respectively. Both isotopes have been found in boron bearing mineral, while under both excitations, the ^{11}B isotope-enriched sample exhibited typical molecular emission bands characteristic for ^{11}BO molecules and ^{10}B molecular emission was not found (Fig. 6.10)

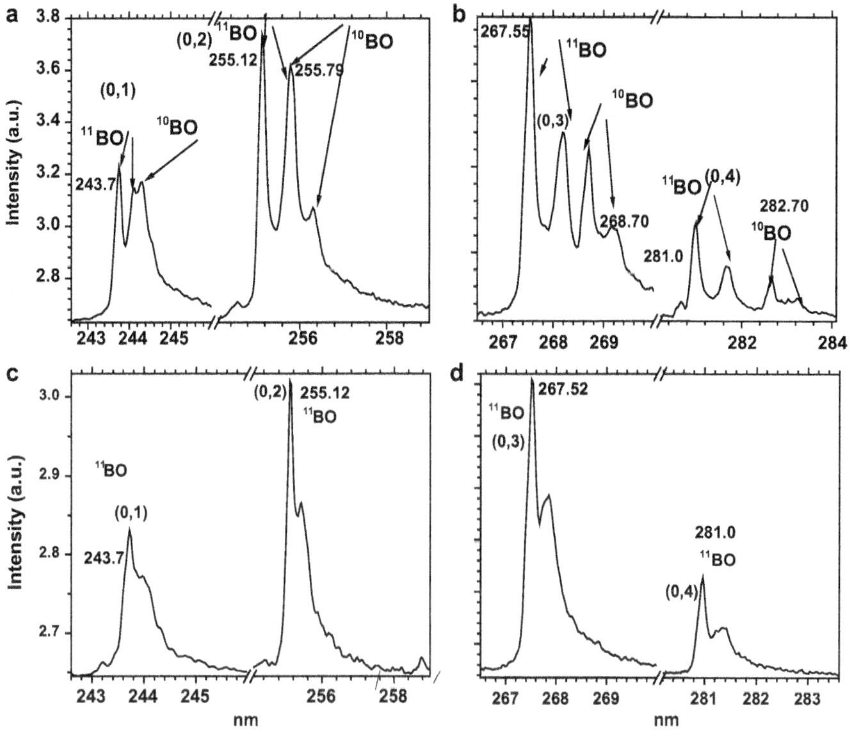

Fig. 6.10 (**a–d**) LIBS-MLIF B $^2\Sigma^+$-$X^2\Sigma^+$system spectra of a boron-bearing mineral with natural isotope ratio excited at the (0,0) transition in excitation for the ^{10}BO (**a, b**); (**c–d**) LIBS-MLIF B $^2\Sigma^+$-$X^2\Sigma^+$ system spectra of an enriched ^{11}BO (96 %) isotope sample excited at (0,0) transitions of ^{11}BO and ^{10}BO molecules (233.13 or 233.19 nm, respectively) (**c-d**)

Lanthanum and Yttrium Bearing Molecules

It was known that certain rare-earth elements (REE), such as Y and La, form strongly luminescent simple molecules with oxygen in arc induced plasma when the ambient air typically supplies the O atom of the molecule (Pearse and Gaydon 1965). It interesting to note that the mostly effective are REE without luminescence ability in elemental form. Three LaO yellow, blue and red systems (Fig. 6.11a–c) were detected in bastneasite (Ce,La)CO_3(F,OH) which are approximately 7–10 times more sensitive than the strongest La I emission lines at 492.2 and 500.0 nm. Orange luminescence of YO molecule (Fig. 6.11d) has been found in mineral xenotime YPO_4. The comparison between the strongest Y II emission lines at 360.0, 370.9, 377.4 and 490.2 nm and YO emission bands is favorable for the molecular emission.

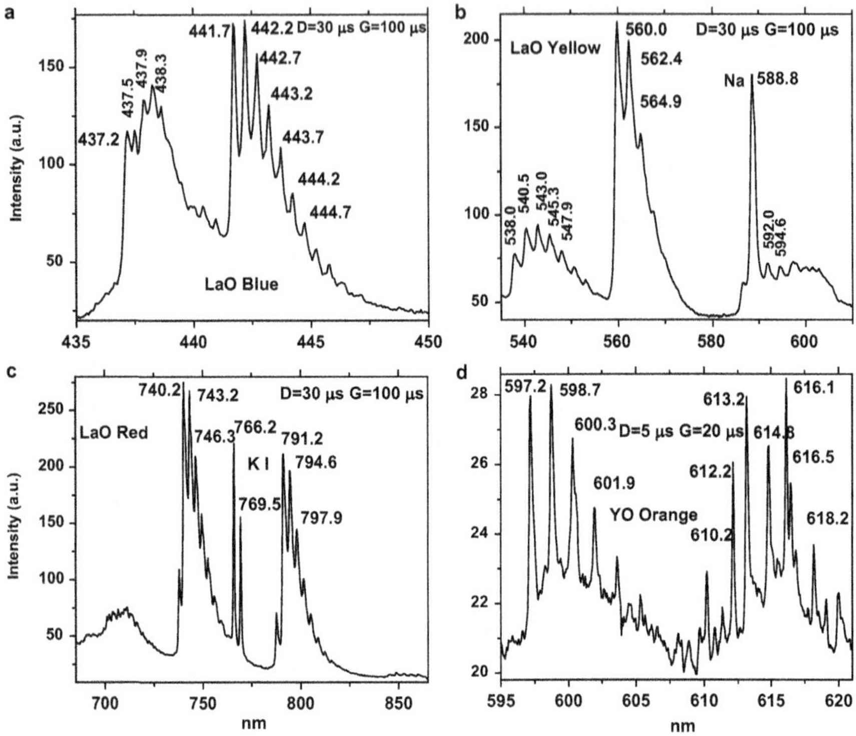

Fig. 6.11 (**a–d**) Single pulsed breakdown spectra of La-bearing bastneasite (**a–c**) and Y-bearing xenotime

6.1.2.5 Plasma Induced Luminescence (PIL)

The fifth time domain represents a new type of luminescence. Previously it was considered, that after approximately 100 μs any plasma emission is non-existent and consequently any useful information may not be received. Recently it was found that in certain cases emission still exists even after several milliseconds following the plasma excitation (Gaft et al. 2011b). It takes place when the matrix, where plasma is excited by laser, is capable to luminescence and contains luminescence centers. According to our knowledge, it is a new luminescence excitation mechanism and, correspondingly, new luminescence type, named as Plasma Induced Luminescence (PIL). It is excited in the matrix by laser-induced plasma and exists independently and simultaneously with atomic, ionic and molecular plasma emission.

Figure 6.12 presents typical breakdown spectra of natural magmatic apatite $Ca_5(PO_4)_3F$ in 385–660 nm spectral range. It is well known that in selected diapason of 385–660 nm after delay time of approximately 500 ns the spectrum is mainly characterized by Ca I and Ca II emission lines, while the Ca II emission

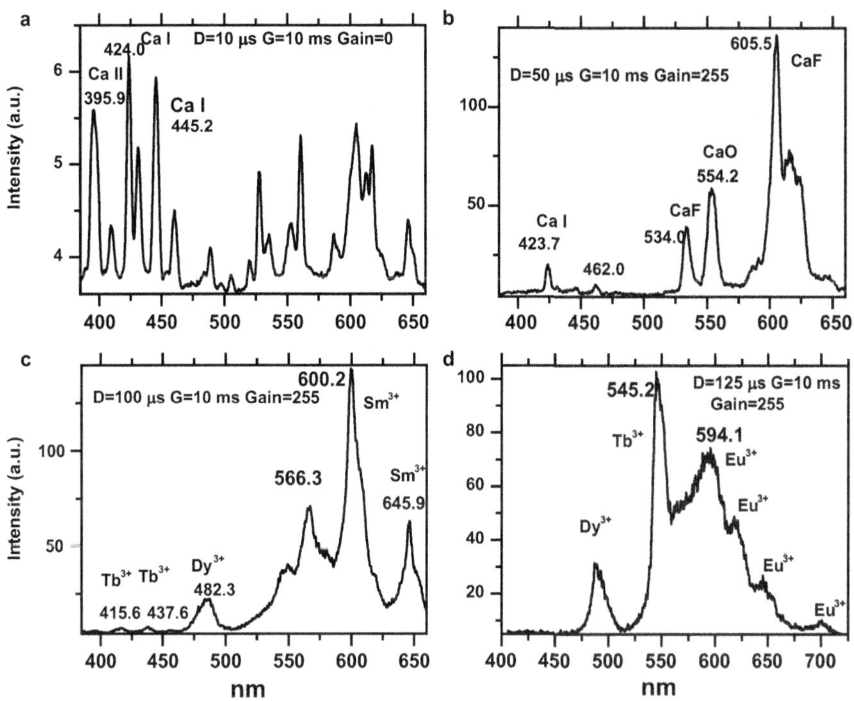

Fig. 6.12 (**a–d**) Plasma induced luminescence of trivalent REE in apatite

dominates (NIST database). With low spectral resolution typical for the present experiments, the strongest Ca II doublet at 393.4 and 396.8 are not spectrally resolved and appear as a relatively broad line peaking at 395 nm. After delay of 10 μs those Ca II lines which have relatively short decay times compared to Ca I are substantially less intensive compared to smaller delay, while Ca I lines at 422.7, 429.8, and 444.6 ones persist (Fig. 6.12a). After longer delay of 50 μs, even the strongest Ca I line at 422.7 nm is extremely weak, while the relatively broad lines at 532.8, 552.7, 604.4 and 615.0 nm dominate the spectrum (Fig. 6.12b). Those narrow bands are approximately 100 times weaker compared to the lines with delay of 10 μs and have been measured with gain of 255, which is nearly three orders of magnitude more sensitive compared to gain of 0 used for 10 μs delay. They are connected to emission of CaO and CaF molecules. After even longer delay of 100 μs new lines appear peaking at 415.6, 437.2, 484.2, 566.3, 598.4, 645.4 (Fig. 6.12c). Those broad lines with very long decay time look totally different compared to the known plasma lines and molecular bands. It is important to note, that mineral apatite is known for its luminescence properties under UV excitation, mainly connected to trivalent rare earth elements (REE^{3+}) which are characterized by emission with relatively narrow (for luminescence) lines. While emission lines found with delay time of 100 μs are unusual for breakdown spectra and may not be interpreted for apatite based on any known atomic or molecular spectra database, such spectrum is quite typical for laser-induced time-resolved luminescence of this mineral. The lines which become visible after 100 μs delay may be ascribed to trivalent rare-earth elements luminescence, namely to Tb^{3+} and Sm^{3+} which have longer decay times compared to trivalent Ho, Er and Pr. The line at 487 nm is usually ascribed to Dy^{3+}, but other typical line at 575 nm is absent. After even longer delay time of 125 μs, the Tb^{3+} lines at 415.6 and 437.2 disappear, the line at 485 nm remains, the line at 543.7 nm dominates the spectrum, very broad band peaking at 585 nm becomes more prominent and weaker lines on its background appear at 594.6, 617.3 and 644.5 nm appear (Fig. 6.12d). Those changes can be explained based on luminescence approach. The dominating lines at 488.0 and 543.7 nm may be connected with Tb^{3+} luminescence from $^5D_4$ level, which has longer decay time compared to the luminescence lines at 415.6 and 437.2 nm from $^5D_3$ level. The broad orange band is connected with Mn^{2+} luminescence, which has very long decay time of 4–5 ms. The lines at 594.6, 617.3, 644.5 and 698.6 nm are typical for Eu^{3+} (I) luminescence inside apatite structure, which is also characterized by very long decay time.

Figure 6.13 presents PIL spectra of other luminescence centers different from trivalent RRE which present in well known luminescent minerals: green luminescence centers of uranyl adsorbed by quartz (Fig. 6.13a), Mn^{2+} in willemite (Fig. 6.13b) and calcite (Fig. 6.13c) with green and orange luminescence, correspondingly, blue emission of radiation-induced center in calcite (Fig. 6.13d) and red luminescence of Cr^{3+} in corundum (Fig. 6.13e) and Fe^{3+} in feldspar (Fig. 6.13f). All luminescence centers discussed above, trivalent REE, Mn^{2+}, Cr^{3+}, Fe^{3+} and radiation-induced center with blue luminescence in calcite, are characterized by long decay times, from tens of microseconds to several milliseconds, which is explained by forbidden character of the corresponding electronic transitions. Thus it is understandable that their luminescence persists and becomes dominant after relatively long time following the plasma creation when the plasma emission is quenched. The duration of this stage depends on relative intensities of plasma and luminescence emissions and usually changes from 5 to 100 μs. Nevertheless, sometimes emission with short decay time also presents in PIL spectra.

Figure 6.14 presents PIL of several centers which are characterized by short and very short intrinsic decay times, such as 1300 ns for Eu^{2+} in fluorite CaF_2 (Fig. 6.14a), 20–30 ns for orange luminescence in calcite (Fig. 6.14b) and 25–30 μs for radiation induced yellow luminescence of zircon $ZrSiO_4$ (Fig. 6.14c). Such luminescence detection after delay time of 125–150 μs, where it has to be totally quenched, implies that the life time of luminescence is actually much longer than decay time determined by intrinsic transition probabilities. Actually, it is three orders of magnitude longer.

It is well known that laser-induced plasma is a powerful source of the emitted energy from X-rays to IR. With such powerful and diverse excitation both direct and recombination excitations are possible. First, UV–visible photons with different energies may directly excite luminescence centers which exist in matrix. Usually accompanying luminescence is much less intensive than emission of atoms and ions in plasma and is not detected during first stage of plasma life.

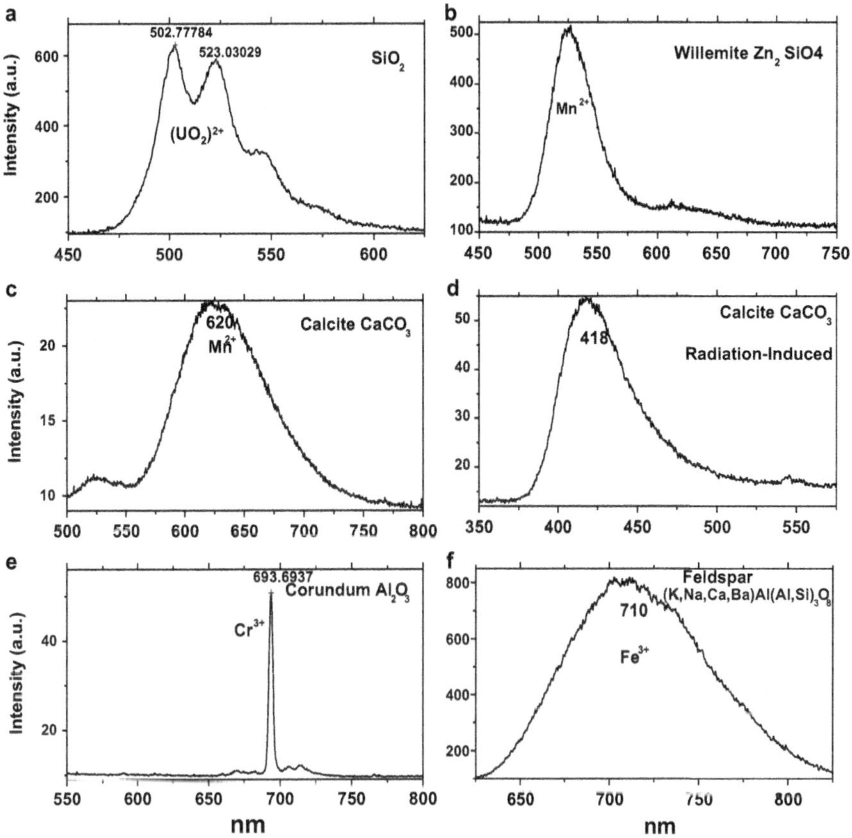

Fig. 6.13 (a–f) Plasma induced luminescence of different centers with long decay time

Nevertheless, the plasma decay time is relatively short and after certain span of time its emission is substantially quenched. From the other side, luminescence centers with long intrinsic decay time from tens of microseconds to several milliseconds become dominant after certain time following the plasma creation. The duration of this time depends on relative intensities of both emissions and usually changes from 5 to 100 μs. The luminescence centers with short intrinsic decay time detection after very long delay evidences that they emit not through direct excitation on their excited states, but by recombination mechanism. It results of free carriers formation by the energy bigger that the corresponding energy gap, which is equal to 5.5 eV (225.7 nm) for calcite, 7.1 eV (174.8 nm) for zircon, 7.4 (167.7) and 11.8 eV (105.2 nm) for fluorite. It may be done by deep UV photons and also X-rays, which are emitted from all parts of the absorption, interaction and transport plasma regime with energy ranging from 50 eV to 1 keV or above depending on the temperature of the plasma. Those free carries are trapped inside the matrix and consequently are thermally liberated in accordance to phosphorescence mechanism. The transfer of

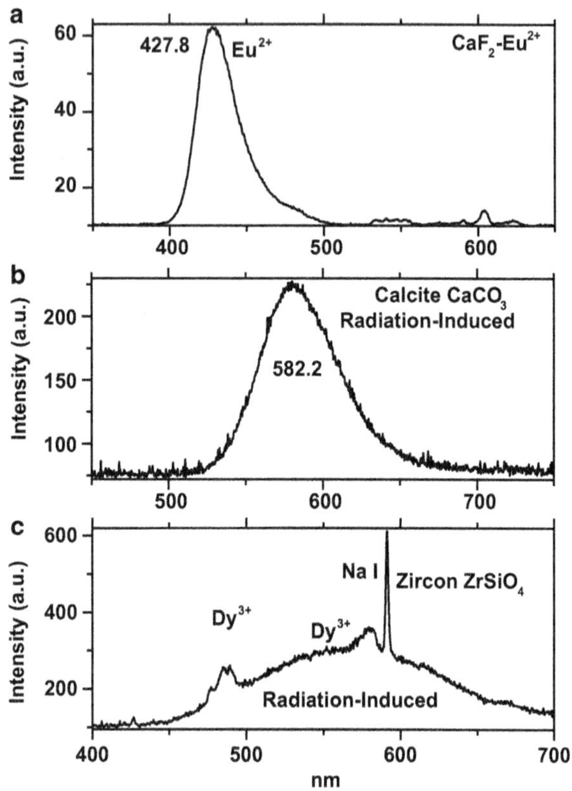

Fig. 6.14 (a–c) Plasma induced luminescence of different centers with short decay time

the excitation energy may be completed through electron–hole recombination or through an intermediate stage, storage of the light sum by the electron and hole-trapping levels.

6.1.3 *Quantitative Analysis by LIBS*

The quantitative aspects of LIBS even for laboratory applications may be considered its Achilles' heel, first due to the complex nature of the laser–sample interaction processes, which depends upon both the laser characteristics and the sample material properties, and second due to the plasma–particle interaction processes, which are space and time dependent. All those parameters influence strongly on plasma conditions and, consequently, on emission lines intensities. Consequently it may sound correct, that LIBS is a mix of great potential and severe limitation, that one of the major problems that precludes its more quantitative use is a lack of reproducibility of spectra at a given wavelength on a shot to shot basis, and that the sensitivity remains modest, precision mediocre and matrix dependence strong (Hahn and Omenetto 2012).

Many models and techniques have been developed in order to overcome such problems. Usually, calibration curves are needed for the quantitative determination of the elements. They emerge from a number of reference samples with known elemental composition, but a limit is set to the applicability of the method, because a similar composition for the unknown sample is also required. This can be good for laboratory measurements, where reference and unknown samples can be embedded in the same matrix, but is a severe restriction for field experiments. Thus, it was concluded that field experiments are limited to a semi-quantitative analysis in case of highly variable or unknown rock compositions.

It sounds very pessimistic but LIBS has been shown to be capable of performing quantitative analysis at a level considered adequate for the specific analytical task investigated. Therefore, the problem is not that LIBS cannot perform quantitative analysis, but rather how to make LIBS become as accurate and reliable as the other spectroscopic methods. It will be demonstrated, that in many industrial online applications LIBS is capable to give the quantitative analytical data which are accurate enough for process control tasks (Gaft et al. 2014a).

6.1.3.1 Laboratory Analysis

In allochromatic minerals, minor and trace chemical components are commonly associated with the presence of luminescence and color, and, from the gemological point of view, they are also responsible for the difference between a common mineral and a gemstone. Metal ions from the first row of transition elements in the periodic table, especially Ti, V, Cr, Mn, Fe, and Cu, are the most important causes of luminescence and color in oxides and silicate gemstones. Thus such elements analysis may be very useful for luminescence and color interpretation. Gem-quality alexandrite, hiddenite and kunzite, elbaite and topaz minerals were characterized through a multi-methodological investigation based on EMPA-WDS, LA-ICP-MS, and laser-induced breakdown spectroscopy (Rossi et al. 2014). With respect to the others, the latter technique enables a simultaneous multi-elemental composition without any sample preparation and the detection of light elements, such as Li, Be and B. The criteria for the choice of minerals were: (a) the presence of chromophore elements in minor contents and/or as traces; (b) the presence of light lithophile elements (Li, Be and B); (c) different crystal chemistry complexity. The results show that LIBS can be employed in mineralogical studies for the identification and characterization of minerals, and as a fast screening method to determine the chemical composition, including the chromophore and light lithophile elements. We proved such application for sillimanite, forsterite and grossular luminescence study. The surface multi-elemental analysis of a section of a stalagmite using the LIBS technique was done. The relative variations of elemental concentration were evaluated by the relative variations of emission intensity from the analyzed elements. A certain number of minor or trace elements, Na, Mg, Al, Si, K, Fe, Sr, detected in measurements showed fluctuations of relative variations of concentration significantly greater than that of the main element of the matrix Ca when a profile measurement was made across the sample. Such fluctuations were attributed to concentration changes of these elements in the sample. The relative variations of concentration for different elements were further measured over a delimited area through a two-dimensional scanning. Multi-elemental concentration variation mappings were performed for detected elements using the data from such scans. The multielemental concentration mappings showed a close correlation with colored laminar structure observed under natural light of the analyzed surface, demonstrating the trace element origin of the colored structure in the studied speleothem (Ma et al. 2010).

6.1.3.2 Analytical Application in Mining

One of the key goals of mining industry is to develop technologies for superior exploration and resource characterization through real time measurements of ore quality. This enables reduction of out-of-seam dilution, significantly reducing mineral processing and energy costs. Presently, most surface mining operations must rely on laboratory-based analysis to monitor the extraction process. This requires samples to be collected, transported to the instrument, analyzed, recorded, and data transmitted back to the pit location. This process is slow, subject to numerous errors, and does not support real time management of the extraction process.

Generally, LIBS as an analytical technique, has many advantages for such kind of applications. LIBS, in principle, can directly address and simultaneously detect all neutral and ion spectral features of all atomic and molecular species of all elements present in any type of sample and its surrounding environment using a single laser shot. The analyses process is fast, for no sample preparation is needed. Furthermore, LIBS is good for analyses of both non-conducting and conducting samples, regardless of their physical states, i.e. solids, liquids or gases. The capability for simultaneous multi-element determination, localized microanalysis, and surface analysis is also appealing. Finally, with the advent of miniaturized lasers, field-portable instrumentation is possible. Remote analysis is also possible by transporting the laser beam through the fiber optic cable or by focusing the laser beam on the sample.

Nevertheless, strong limitations are also typical for LIBS as an analytical technique. The major drawback of LIBS in industrial quantitative applications is related to the stability of the spectroscopic plasma signals, which undergo considerable fluctuations that originate from the very nature of the method. The following factors are the mostly important in real life scenarios.

Matrix Effects

Real industrial application usually means that the uncontrolled mixture of several minerals present. The meaning is that a significantly different signal response is observed at a given spectroscopic transition of an element present at the same concentration level in several different samples. Such different response can be attributed to differences in the laser–sample interaction, resulting from changes in the ablation mechanism and efficiency. Thus the nature of the sample has a

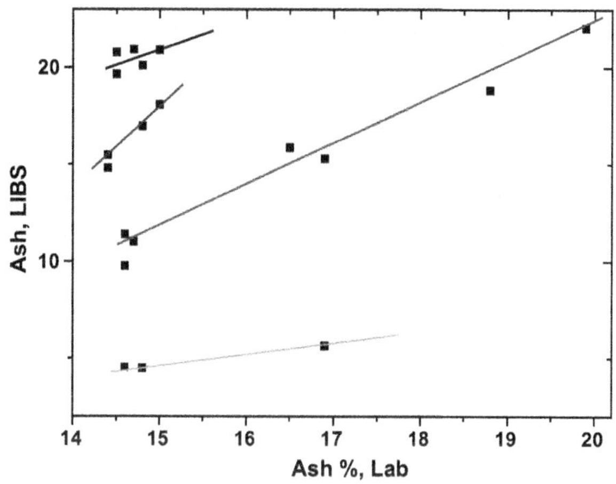

Fig. 6.15 Correlation between LIBS data and ash contents in four different coal types

powerful influence on the excitation conditions, leading to what is generally known as matrix effects. Matrix effects can be divided in two broad categories, namely those associated with the entry of the material into the plasma and those taking place in the plasma.

For example, Fig. 6.15 demonstrates the correlation between online LIBS and laboratory determined ash concentrations in 21 coal samples from Australia. The results have been unusual, because not one but evidently four different trends may be seen. The possible explanation was that we have different types of coal, supposedly four, while different matrix demand to use different parameters in the formulas for intensity-concentration transformations. Afterward it was confirmed that the coals actually include samples from different mined areas: main seam, underground mine, Houston Creek and Sutter Creek.

In order to reduce matrix effects in online mode, the only realistic way is correction of the effect by directly measuring some property of the matrix and establishing a correction function. In certain cases, the presence of characteristic elements in the corresponding matrix, which may be determined by LIBS, may be used.

Self-Absorption

The loss of sensitivity at high concentrations is most often due to self-absorption. It may be detected by "flat-top" behavior of the emission line, while in less extreme situations the effects are often not so immediately recognizable while they considerably affect the line intensity values. Self-absorption typically is observed for emission lines in which the lower level of the transition is the ground state or close to the ground state because they have the highest probability of reabsorbing by a member of the same species. Because of the high density of atoms in the microplasma and its characteristically high temperature and electron density gradients, the outer layer of the plasma will be populated by 'cool' atoms, residing mostly in

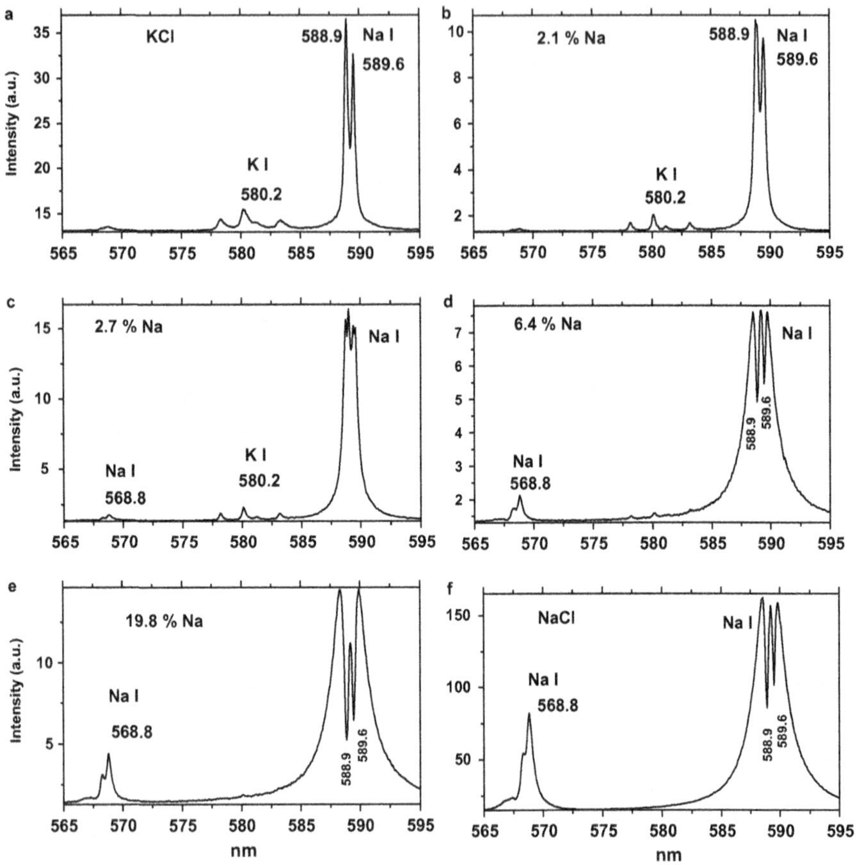

Fig. 6.16 (**a–f**) Breakdowns spectra of industrial potash fertilizers with different quantities of Na impurities

the ground state. The central core of the plasma will contain a higher density of excited atoms. As these atoms decay to the ground state, the emitted photons corresponding to resonance transitions will have a high probability of being absorbed by the 'cooler' atoms in the outer layers, thereby reducing the observed intensity of the emission line. As the concentration of the atoms in the target sample increases, the number of 'cooler' atoms in the outer layer increases and self-absorption becomes evident.

Thus for quantitative analyses it is better to use the analytical lines which are connected to the transitions on not ground state. For example, in industrial KCl fertilizers the self-absorption of the Na impurity becomes visible at approximately 2.0 % level (Fig. 6.16). It is clearly seen that non resonant line of Na peaking at 568.8 nm is much better correlated with Na concentration than famous resonant doublet at 589.9 and 589.6 nm. The same approach may be used for Mg analysis, where non-resonant Mg II lines peaking at 292.9 and 293.7 nm are less subjected to

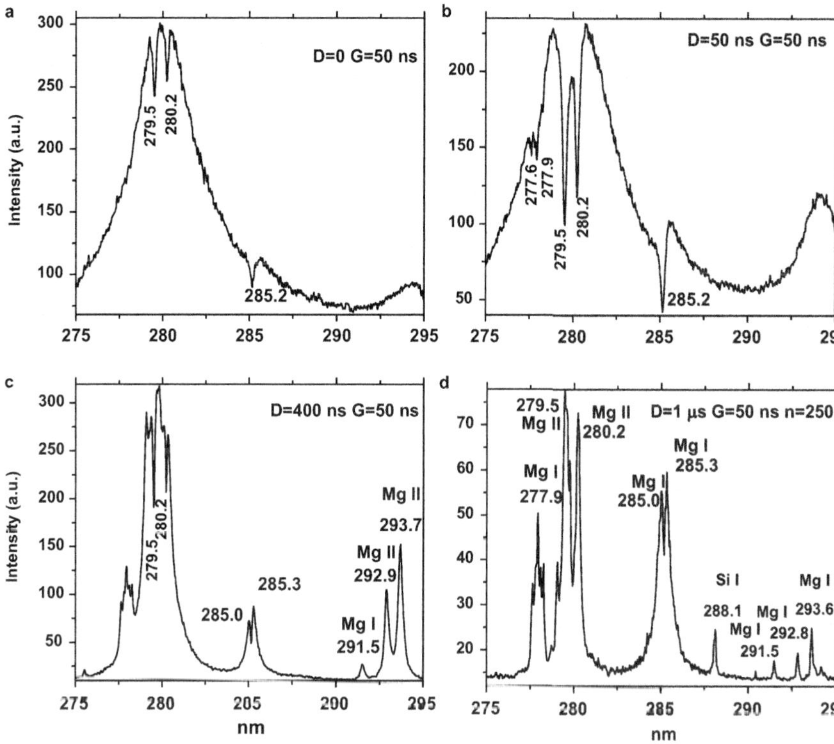

Fig. 6.17 (**a–d**) Breakdown spectra of magnesite during different time windows

self-absorption than more intensive resonant lines at 285 (Mg I) and 280 (Mg II) nm (Fig. 6.17). Besides, it seems that reabsorption is less pronounced after certain time from the plasma creation compared to the beginning of the plasma life.

Moisture

For the most of field application the water content is constantly changed. The humidity of the material reduces the emission intensities from its plasma. There are several reasons that contribute to this effect: The main reason is that some laser energy is wasted for water heating rather than for plasma formation. Also the aerosol production mechanism is suppressed by the water contents, thus reducing the signals at high repetition rates. On the microscopic level, water increases the energy requirement for the ablation of the substrate. Besides that, these effects act in different ways for various matrices. The reason is that certain matrixes adsorb the water much better than the others. If the water does not penetrate into the material, much more water is present at the surface than in the opposite case. Thus changeable moisture content negatively affect the analytical results. From the other side, LIBS can be an alternative for moisture quantification, since atomic emission from hydrogen and oxygen as well as molecular emission from OH radicals can be used as proxy indicators of water. For example, it was proved that that the LIBS spectrum could be well correlated to the water content in cheese samples. After optimization of the laser energy to minimize the oxygen influence from the air, the oxygen emission lines normalized by the CN emission as an internal standard were used as an indicator of the cheese sample moisture (Liu et al. 2012).

Such approach appears to be also useful for online LIBS analyses. For example, it was found that the intensity of Hα line in breakdown spectra can be represented as a linear combination of atomic percentage of hydrogen atoms in organics and in water, whereas the coefficient of proportionality for hydrogen in water is about three times higher than that for hydrogen in the organics. The great difference of intensity ratio of Hα/Mg II 292.86 nm for oil shale and for limestone enables reliably discriminate between oil shale and limestone independently of their moisture content. The measurements demonstrated that the moisture content cannot seriously affect the LIBS capability for the on-line monitoring of oil shale quality (Paris et al. 2015).

The idea to measure the sample moisture using LIBS data was checked on certain industrial materials, such as iron agglomerate and iron concentrate. It was based on measuring the ratio of plasma line of hydrogen and the line from the material constituent with more or less constant concentration, such as Fe. It was found that H/Fe intensities ratios usually exponentially depends on moisture content (Fig. 6.18). It is clear that this idea may work good enough only in non-organic material which does not contain the high concentration of H atoms and all detected H is from the moisture. In case of coal, for example, additional devices for online moisture content control may be recommended.

During online comparison of LIBS and neutron based analyzers of ash in coal it was found that strong moisture changes lead to substantial alterations in breakdown spectra resulting in errors in ash evaluation. Careful study of such influence enabled to relate the moisture with spectral changes, and consequently to evaluate the coal moisture based on LIBS data. The correlation enables to make corrections for ash evaluation. Big periodic errors for the LIBS ash analyses were correlated with strong moisture changes and removed after correction for moisture have been made (Gaft et al. 2008).

Evaluation of moisture influence on different industrial online control applications revealed, that usually the plasma intensity becomes substantially lower with moisture content increasing. Figure 6.19 illustrates this for iron agglomerate and iron concentrate cases. Nevertheless, certain lines intensities ratios remain nearly unchanged and may be used for online analysis.

Geometrical Parameters

For field applications the laser-sample distance is constantly changing and the physical form of the rock is different (pebbles of different size from tenth of centimeters to tenth of microns). All those parameters influence strongly on plasma conditions and, consequently, on emission lines intensities. Laser-target distance corrections may be achieved by constantly measuring it with corresponding changes in the optical

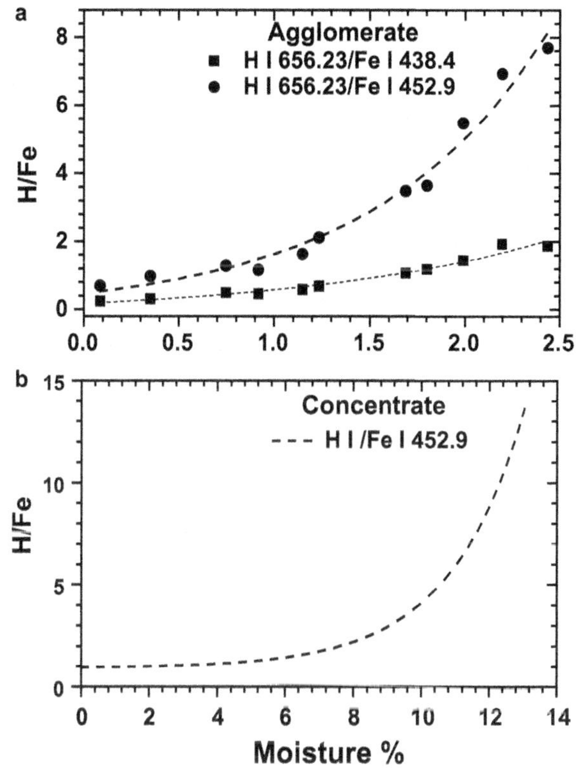

Fig. 6.18 (a–b) Dependence between moisture content and H/Fe emission lines ratio for iron agglomerate (**a**) and concentrate (**b**)

module. The aggregate state of the analyzed material also influences strongly on laser created plasma properties. For example, the comparison between coal pieces and coal dust breakdown spectra demonstrate striking differences (Fig. 6.20). It may be seen that all ions emissions are much weaker in the dust compared to the stones up to the nearly total disappearance of Ca II emissions. Evidently it may be connected to the relatively low plasma temperature of the coal dust.

6.2 Raman Spectroscopy

Another laser based technique, which may be used together with time-resolved luminescence, is Raman spectroscopy. It is very well known technique, which is widely used in mineralogy. Several reviews have been recently published mostly devoted to minerals (Dubessy et al. 1994; Nasdala et al. 2004; Smith and Dent 2005; Jasinevicius 2009; Dubessy et al. 2012; Panczer et al. 2012). Thus only the theoretical aspects which are the mostly relevant to our research devoted to the real time online quality control of minerals will be considered.

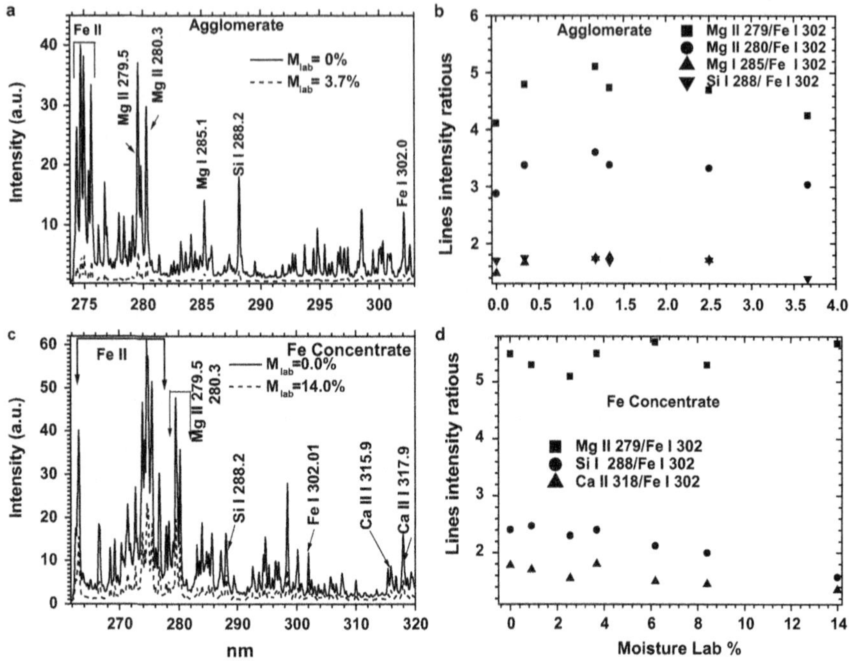

Fig. 6.19 (**a–d**) Breakdown spectra of Fe agglomerate (**a**) and concentrate (**c**) with different moisture contents and different emission lines ratios for agglomerate (**b**) and concentrate (**d**)

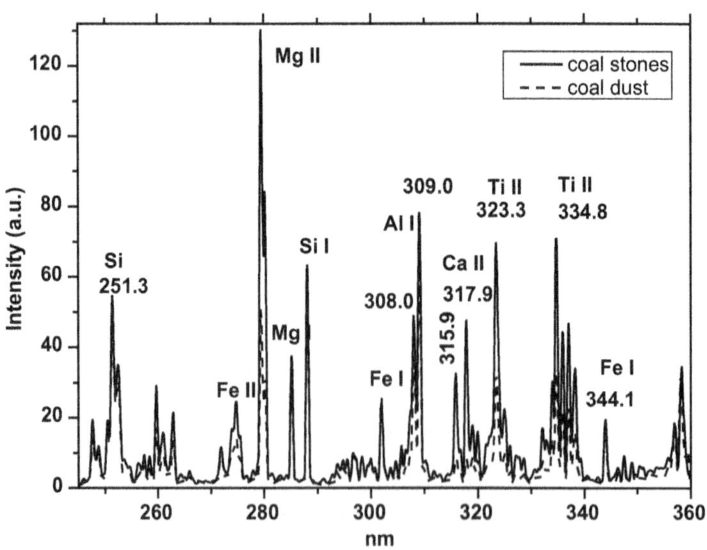

Fig. 6.20 Breakdown spectra of coal lumps and coal dust in the similar experimental conditions

6.2.1 Introduction

When radiation passes through a transparent medium, the species present scatter a fraction of the beam in all directions. In 1928, the Indian physicist C. Raman discovered that the wavelength of a small fraction of the radiation scattered by certain molecules differs from that of the incident beam. Furthermore, the shifts in wavelength depend upon the chemical structure of the molecules responsible for scattering. Raman spectra are obtained by irradiating a sample with a powerful laser source of monochromatic radiation. During irradiation, the spectrum of the scattered radiation is measured. At the very most, the intensities of Raman lines are 10^{-6} of the intensity of the Rayleigh diffusion line and as a consequence, their detection and measurement are difficult. The scattered radiation is of three types: namely Stokes, anti-Stokes, and Rayleigh. The last, whose wavelength is exactly that of the excitation source, is significantly more intense than either of the two types. The Raman peaks appearing at lower energies relative to the excitation energy resembles the Stokes shift found in luminescence and it is the reason for their name. Shifts toward higher energies are termed anti-Stokes and those lines are generally much weaker than the corresponding Stokes lines, but fluorescence may interfere with the observation of Stokes shift but not always with that of anti-Stokes, which therefore may be more useful, despite their lower intensity. It is important to note that the patterns of the shifts on the two sides are identical. The magnitude of Raman shifts is independent of the wavelength of excitation and shift pattern is independent on the type of laser.

The Raman effect can be described in terms of transitions between vibrational energy levels. When an energetic photon strikes a molecule in its ground state, it may raise the molecule to a higher virtual state. Since this is not a stable energy state for the molecule, two things can occur. Most probably the molecule returns to its ground vibrational state and emits a photon with the same energy and frequency as the exciting photon. This is called Rayleigh scattering. However, some of the excited molecules will not return to the ground state, but to some excited vibrational state. Such a molecule emits a photon, which has a lower energy than the exciting photon, the energy difference being equal to the difference between the initial and final vibrational states. This is Raman scattering, Stokes type. If a molecule, which is in the first excited vibrational state, absorbs the photon then the molecule is again raised to some high, non-stable energy state. Most probably, this molecule then return to the ground state, and doing so, emits a photon, which has a higher energy than the exciting photon. The difference in energy between the exciting photon and the emitted photon is equal to the energy difference between the two excited vibrational states of the molecule. This is Raman scattering, anti-Stokes type. The relative populations of the two energy states are such that Stokes Raman is much favored over anti-Stokes. Rayleigh scattering has considerably higher probability of occurring than does Raman scattering because the most probable event is reemission by the return of these molecules to the ground state.

The physical origin of Raman scattering may be viewed from a simple classical perspective in which the electric field associated with the incident light interacts with the vibrating crystal. In particular, this interaction occurs through the polarizability of the material: that is, its ability to produce an induced dipole in an electric field. It is the oscillations in the polarizability produced by vibrations, which cause Raman scattering: if a vibrational motion produces a change in the polarizability of a material, then the vibration is Raman active. Notably, while the induced dipole is a vector, the polarizability is a second rank tensor, and it is the symmetric properties of this tensor, which allow the prediction of which families of modes will be active when observed along different crystallographic directions.

Wave model of Raman scattering may be shortly presented by the following way. A beam of radiation having the frequency ν_{ex} is incident upon a mineral. The electric field E of this radiation can be described by the equation:

$$E = E_0 \cos(\omega t) = E_0 \cos(2\pi v_{ex} t). \tag{6.1}$$

When the electric field of the radiation interacts with the electron cloud of the mineral bond, it induces a dipole moment *m* in that bond that is given by

$$m = \alpha E = \alpha E_0 \cos(2\pi v_{ex} t), \tag{6.2}$$

where α is the polarizability tensor of the bond. In order to be Raman active, the polarizability of the bond, which is a measure of the deformability in an electric field, must vary as a function of the distance between nuclei according to the equation:

$$\alpha = \alpha_0 + (r - r_{eq})(\partial\alpha/\partial r), \tag{6.3}$$

where α_0 is polarizability of the bond at the equilibrium distance r_{eq}. The change in internuclear separation varies with the frequency of the vibration ν_v is given by:

$$r\text{-}r_{eq} = r_m \cos(2\pi v_{ex} t), \tag{6.4}$$

where r_m is maximum internuclear separation relative to the equilibrium. Thus

$$\alpha = \alpha_0 + r_m \cos(2\pi v_{ex} t)(\partial\alpha/\partial r). \tag{6.5}$$

Extending the equation of polarizability with that of the internuclear separation, and by substituting the initial equation of the dipole moment, one obtains an expression for the induced dipole moment *m*:

$$m = \alpha\, E_0 \cos(2\pi v_{ex} t) + E_0 r_m \cos(2\pi v_{ex} t)\, \cos(2\pi v_{ex} t)(\partial\alpha/\partial r). \tag{6.6}$$

Recalling from trigonometry that $\cos x \cos y = [\cos(x+y) + \cos(x-y)]/2$ and applying it to the previous equation, gives:

$$\begin{aligned} m = \; & \alpha\, E_0 \cos(2\pi v_{ex} t) + (E_0/2)\, r_m(\partial\alpha/\partial r) \cos[2\pi t(v_{ex} - v_v)] \\ & + (E_0/2)\, r_m(\partial\alpha/\partial r) \cos[2\pi t(v_{ex} + v_v)]. \end{aligned} \tag{6.7}$$

The first term of the last equation represents Rayleigh scattering, which occurs at the excitation frequency ν_{ex}. The second and the third terms correspond to the

Stokes and anti-Stokes frequencies of $(\nu_{ex} - \nu_{v})$ and $(\nu_{ex} + \nu_{v})$. Here, the excitation frequency has been modulated by the vibration frequency of the bond. It is important to note that Raman scattering requires that the polarizability of a bond varies as a function of distance, that is $(\partial\alpha/\partial r)$ must be greater than zero if a Raman line is to appear.

The theory of the Raman scattering shows that the phenomenon results from the same type of vibrational changes that are associated with infrared (IR) absorption. Thus, the difference in wavelength between the incident and scattered radiation corresponds to wavelengths in the mid–infrared region. Indeed, Raman scattering spectrum and IR spectrum for given specie often resemble one another quite closely. There are, however, enough differences between the kinds of groups that are IR active and those that are Raman active. It is not surprising when it is considered that the basic mechanisms, although dependent upon the same vibrational modes, arise from processes that are mechanically different. IR absorption requires that a vibrational; mode of the molecule have a change in dipole or charge distribution associated with it. Only then can radiation of the same frequency interact with the molecule and promote it to an excited vibrational state. In contrast, scattering involves a momentary distortion of the electron distribution around a bond in a molecule, follows by reemission of the radiation as the bond returns to it ground state. In its distorted form, the molecule is temporarily polarized, namely develops momentarily an induced dipole, which disappears upon relaxation and reemission.

A Raman spectrum is a plot of light intensity versus photon energy. In the vibrational spectroscopy it is usual to express the photon energy by the wavenumbers, defined as $\nu = 1/\lambda$. In the Raman spectroscopy the use of absolute wavenumbers would be impractical, because the wavelength and, with that, the absolute wavenumber of each obtained Raman bands must always depend on the wavenumber of the incident light ν_0. However, only the wavenumber shift, which corresponds to the atomic vibration in the sample, is of analytical interest. It has therefore, become usual to express Raman shifts using relative wavenumber, i.e., the wavenumber difference between incident and scattered light, $\nu_1 = \nu_0 - \nu$. By definition, the Rayleigh line is set at zero Raman shift and Stokes Raman bands have positive and anti-Stokes Raman bands have negative relative wavenumbers.

Thus each band in a Raman spectrum represents the interaction of the incident light with a certain atomic vibrations. Atomic vibrations, in turn, are controlled by the sizes, valences and masses of the atomic species of which the sample is composed, the bond forces between these atoms, and the symmetry of their arrangement in the crystal structure. These factors affect not only the frequencies of atomic vibrations and the observed Raman shifts, respectively, but also the number of observed Raman bands, their relative intensities, their widths and their polarization. Therefore, Raman spectra are highly specific for a certain type of sample and can be used for the identification and structural characterization of unknowns.

For many minerals, the types of vibrational modes may be divided into two categories: internal and lattice modes. Internal modes are vibrations, which can be associated with those of a molecular unit, shifted and split by interaction with the crystalline environment in which the molecular unit is bonded. Such internal modes are typically associated with the most strongly bonded units in a crystal, and thus with the highest frequency vibrations. Lattice modes comprise both a range of comparatively low-frequency vibrations not readily describable in terms of molecular units, and so called external modes. External modes are those, which involve motions of a molecular unit against its surrounding lattice.

6.2.2 *Raman and Luminescence*

It is well known that a molecular process closely related to Raman spectroscopy is luminescence. Its quantum yield can be as high as unity, depending on the system. For minerals that exhibit even weak luminescence, the Raman scattering will be completely obscured. The usual approach to avoid luminescence is to change the laser wavelength to avoid electronic excitation, but it is not always possible to find an effective wavelength. Sometimes this necessitates using a near-IR laser and an interferometer detector (Fourier Transform Raman Spectroscopy). However, NIR excited Raman spectroscopy is inherently insensitive, giving relatively low signal/noise ratio. The reason is that the intensity of the Raman signal is nearly proportional to the intensity of the incident laser light and inversely proportional to the fourth power of the laser wavelength, namely $I_{Raman} \sim 1/\lambda^4$. Besides that, a number of minerals still fluoresce strongly under NIR laser excitation (Jasinevicius 2009). Specifically, narrow luminescence lines of trivalent REE may be very similar to Raman signals (Table 6.1). An alternative approach is Shifted-Subtracted Raman Spectroscopy (SSRS). The simple way is to subtract broad luminescence background using fitting procedure and to leave sharp Raman bands behind. Unfortunately random variations in the pixel-to-pixel sensitivity may be larger than the actual Raman signals. However, every spectrum recorded with a given detector under the same experimental conditions will have the same irregularity in response. Thus, sets of spectra are recorded at two slightly shifted, usually for 1 nm, excitations (it may be done using dye or OPO laser). In such a way luminescence spectrum remains the same, while the Raman lines are shifted. The pairs of shifted spectra are then subtracted and Raman spectrum becomes visible (Bell et al. 2000).

Another approach is Shifted Excitation Raman Difference Spectroscopy (SERDS) implemented with two wavelength-stabilized laser diodes with fixed wavelength separation. It is an effective method for dealing with the effects of fluorescence in Raman spectroscopic analysis. Both qualitative and quantitative SERDS analysis of a variety of strongly fluorescing samples is possible, including binary liquid mixtures. This application is enabled by the Volume Bragg Grating technology, which allows manufacturing of compact low-cost high-power laser sources, suitable for extending the SERDS methodology to portable Raman spectrometers (Volodin et al. 2013).

Table 6.1 Trivalent REE emission in IR part of the spectrum

Center	Emission, nm	Transition	Excitation, nm
Nd^{3+}	1060	$^4F_{3/2}$–$^4I_{11/2}$	532 and 780
	890	$^4F_{3/2}$–$^4I_{9/2}$	
Er^{3+}	800	$^4I_{9/2}$–$4I_{15/2}$	780
	980	$^4I_{11/2}$–$4I_{15/2}$	
	545	$^4S_{3/2}$–$^4I_{15/2}$	532
	850	$^4S_{3/2}$–$^4I_{13/2}$	
	650	$^4F_{9/2}$–$4I_{15/2}$	
	800	$^4I_{9/2}$–$4I_{15/2}$	
	980	$^4I_{11/2}$–$4I_{15/2}$	
Ho^{3+}	543	$^5S_2$–$^5I_8$	532
	750	$^5S_2$–$^5I_7$	
	1015	$^5S_2$–$^5I_6$	
Pr^{3+}	950	$^1D_2$–$^3F_4$	532 nm
	810	$^1D_2$–$^3F_2$	
	770	$^1D_2$–$^3H_6$	
	580	$^1D_2$–$^3H_4$	
	1040	$^1G_4$–$^3H_4$	532 and 780
Tm^{3+}	710	$^3F_2$–$^3H_6$	532
	770	$^3F_3$–$^3H_6$	
	800	$^3H_4$–$^3H_6$	532 and 780
Yb^{3+}	1050	$^2F_{5/2}$–$^2F_{7/2}$	532 and 780

6.2.2.1 Visible and UV Raman

One of the effective ways to exclude luminescence influence on the Raman is UV Raman approach (Asher 2002). The total vibrational Raman scattering cross section σ_{mn} (in cm^2/molecule sr) for a vibrational transition m-n for an isolated molecule averaged over all its orientations is (Steinfeld and Wormhoudt 1998):

$$\sigma_{mn}(\upsilon_0) = I_{mn}/I_0 = \left(C\,\upsilon_0(\upsilon_0 - \upsilon_{mn})^3\right)/\alpha_{mn}(\upsilon_0)^2 \tag{6.8}$$

where I_{mn} is the scattered light intensity (in photons/srcm^{2}s), I_0 and υ_0 are the incident excitation beam flux (photons/scm^2) and frequency (cm^{-1}), $C = 1/NK$ is the constant where N is concentration (mol/cm^3) and K (cm) depends on sample geometry, υ_{mn} is the vibrational frequency of the Raman mode (cm^{-1}) and $\alpha_{mn}(\upsilon_0)$ is the polarizability tensor for the m-n transition for the excitation frequency υ_0. The fourth power dependence of $\sigma_{mn}(\upsilon_0)$ on incident frequency means that the Raman intensity increases by about 20 times when using UV laser radiation at 248 nm instead of the second harmonic of a Nd–YAG laser (532 nm). An additional increase in Raman intensity may be achieved due to the dependence of $\sigma_{mn}(\upsilon_0)$ on the excitation frequency. When the excitation frequency is close to an electronic transition of the molecule, strong enhancement of the particular Raman vibration mode may occur, and as a result strong enhancement of particular Raman lines will be observed. The pre-resonance Raman

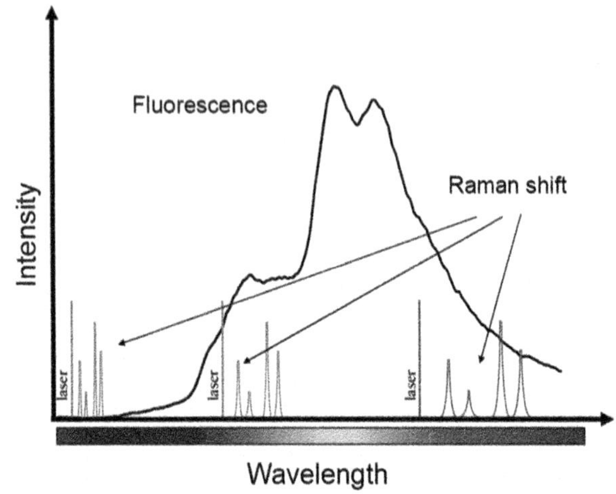

Fig. 6.21 Raman lines movement inside luminescence Stokes shift in UV Raman

cross-section dependence on excitation frequency is described by Albrecht's A-term expression (Albrecht and Hutley 1971):

$$\sigma_{mn}(\upsilon_0) = K\,\upsilon_0(\upsilon_0 - \upsilon_{mn})^3 \times \left(\upsilon_e^2 + \upsilon_0^2\right)/\left(\upsilon_e^2 - \upsilon_0^2\right), \tag{6.9}$$

where K is a constant and υ_e is the frequency of the transition to the resonant excited state. The factor $\upsilon_0(\upsilon_0 - \upsilon_{mn})^3$ is the inherent frequency dependence of the Raman cross section and the term in brackets describes the frequency dependence of the polarizability $\alpha_{mn}(\upsilon_0)$. When the difference between the laser excitation frequency, υ_0, and the frequency of the transition to the excited electronic state, υ_e, approaches zero, the Raman scatter cross section in Eq. (6.9), $(\upsilon_e^2 + \upsilon_0^2)/(\upsilon_e^2 - \upsilon_0^2)$, approaches infinity. In practice, the cross section reaches a very large limit, but not infinity. Resonance Raman is especially important for organic and biological molecules, since strong absorption typically occurs in the deep UV, and resonance enhancements of millions of times have been reported (Smith and Dent 2005). Many minerals also have fundamental absorption bands in the UV part of the spectrum, thus the possibility of resonance Raman exists.

Another advantage of excitation in the UV range is the possibility to move the Raman information out of the luminescence region. The Raman frequency shifts depend mainly on the identity of the molecule, not on the frequency of the excitation laser (except for some semiconductors and the D and D' bands in disordered graphite). Hence, infrared, visible and ultraviolet lasers produce the same set of Raman shifts for a specific molecule. In the UV range, with very energetic photons, the Raman lines are very close to the incident exciting frequency, while in this spectral range, luminescence is still very weak because of its inherently large Stokes shift (Fig. 6.21). According to literature data (Asher and Johnson 1984), the full Raman range, up to 4000 cm^{-1}, is normally free from luminescence when the excitation wavelength is shorter than about 250 nm.

6.2.2.2 Gated Raman Spectroscopy

Using CW lasers, Raman scattering is collected together with luminescence. In fact, the interaction time for Raman scattering is virtually instantaneous (less than 1 ps), whereas luminescence emission is statistically relatively slow, with a minimum time of hundreds of picoseconds lapsing between electronic excitation and radiative decay.

Thus, if we illuminate a sample with a short laser pulse, all of the Raman photons will be generated during the pulse, whereas most of the luminescence photons will be emitted at much longer times after the pulse. If the detection system is gated so as to detect only those photons scattered or emitted during the laser pulse, we will collect all of the Raman photons but reject the majority of the luminescence. Ideally, such a system should achieve the highest possible rejection ratio while having high throughput and a time resolution or gating time, short enough to match the duration of the laser pulse and correspondingly the Raman flux. Existing gated intensified charge coupled device (ICCD) cameras and photomultipliers typically operate on nanosecond timescales, with the fastest devices reaching hundreds or tens of picoseconds. Two rather different time-resolved approaches have been most effective in yielding high-quality Raman data from highly luminescent materials. The first uses a streak camera, which performs with a gating time of 10 ps. The second method uses a Kerr gate with a 3-ps full-width at half-maximum, which achieved three orders of magnitude suppression of the background from a fluorophore with a decay time of 2 ns. It produced excellent Raman signals, which were completely obscured in non-gated spectra. For practical field applications, picoseconds lasers are still very exotic, but even nanosecond lasers may be effective in many cases. The rejection ratio, $F_{R/L}$, between the luminescence emission and Raman scattering can be estimated by the following equation (Everall et al. 2001):

$$F_{R/L} = 1/(1 - e^{-\tau_g/\tau_l}) \tag{6.10}$$

where τ_g is the gate width and τ_l is the luminescence lifetime. The equation has proven to be a good approximation of the actual rejection ratio in most cases, especially when $\tau_g << \tau_l$ For example, in case of a decay time t = 1 ms, which is typical for luminescence centers with forbidden transitions, such as Mn^{2+}, Fe^{3+} and Cr^{3+} in strong crystal field (forbidden R–R lines), the luminescence signals will be diminished by 105 times. In the case of uranyl $(UO_2)^{2+}$ and REE^{3+} (besides Ce^{3+}) with $\tau = 100$–800 μs, the luminescence signals will be diminished by approximately 104 times. In the case of luminescence centers with t in the range of 1 μs, such as Eu^{2+}, Cr^{3+} in weak crystal field (allowed transition) and $(MeO_n)^{m-}$ complexes, the quenching will be ~ 100 times. In the case of luminescence centers with a very short decay time of 10–100 ns, such as water–organic complexes absorbed on minerals surface, Ce^{3+} and Pb^{2+}, the luminescence quenching may be not sufficient and gating in hundreds picoseconds range will be needed.

6.2.2.3 Visible Gated Raman Spectroscopy

Raman and Broad Luminescence Bands

Calcite is a mineral with strong luminescence, and in many cases it is impossible to detect its CW Raman spectra using blue or green excitations. Nevertheless, the fact that the Raman shift is more or less independent of the excitation wavelength while the luminescence has the certain excitation spectrum suggests using a wavelength that excites the Raman bands without luminescence. The usual way to avoid luminescence in CW Raman spectroscopy is to select the laser wavelength in the red or NIR field to avoid electronic excitation. Using this approach, it is usually possible to find such an excitation wavelength where the luminescence interference is negligible. The disadvantage is that the intensity of the Raman signal is inversely proportional to the fourth power of the laser wavelength, making NIR-excited Raman inherently insensitive, giving relatively low Signal to Noise Ratio. It demands either prolonged collection times or significantly increased power delivered to the sample. It may be successfully employed in academic fundamental research and routine mineral identification where the time factor is not important. Figure 6.22 presents how Raman and luminescence may be separated using the green second harmonic of a pulsed Nd–YAG laser at 532 nm (Gaft and Nagli 2009). This specific calcite sample under UV-visible excitation is characterized by very strong broad-band luminescence of Mn^{2+} with extremely long decay time $\tau = 11$ ms (Fig. 6.22a). Thus, in the CW Raman experiments, luminescence is mostly detected while the strongest Raman line at 1085 cm^{-1} may be hardly seen (Fig. 6.22b). One opportunity to see only Raman lines is to use NIR excitation by 785 nm. This calcite is non-luminescent under such excitation, and thus its Raman lines are clearly visible (Fig. 6.22c). Another way is to use the gated Raman technique with a gate width of 10 ns, which enables suppression of the long-lived luminescence of Mn^{2+} and reveals the hidden Raman lines (Fig. 6.22d).

Raman and Narrow Luminescence Lines

In the case described above, luminescence was composed of a broad band, corresponding to several thousands of cm^{-1}. In the case of narrow luminescence lines, it may be even more difficult to distinguish between Raman and luminescence signals. For example, luminescence of 4f electron elements, such as trivalent rare-earth elements (REEs), can have line widths equivalent to several cm^{-1}, which is the typical value of Raman lines. The REE-related emissions are therefore easily mistaken as Raman bands, and vice versa; publications describing the Raman spectrum of REE-bearing minerals should hence be referred to with appropriate caution. Reliable distinction between narrow line luminescence and Raman bands is possible by obtaining multiple Raman spectra with different excitation wavelengths. Raman lines will appear with equal Raman shifts in all spectra, while luminescence lines are characterized by certain fixed wavelengths that consequently mimic different Raman shifts in spectra obtained with different excitation

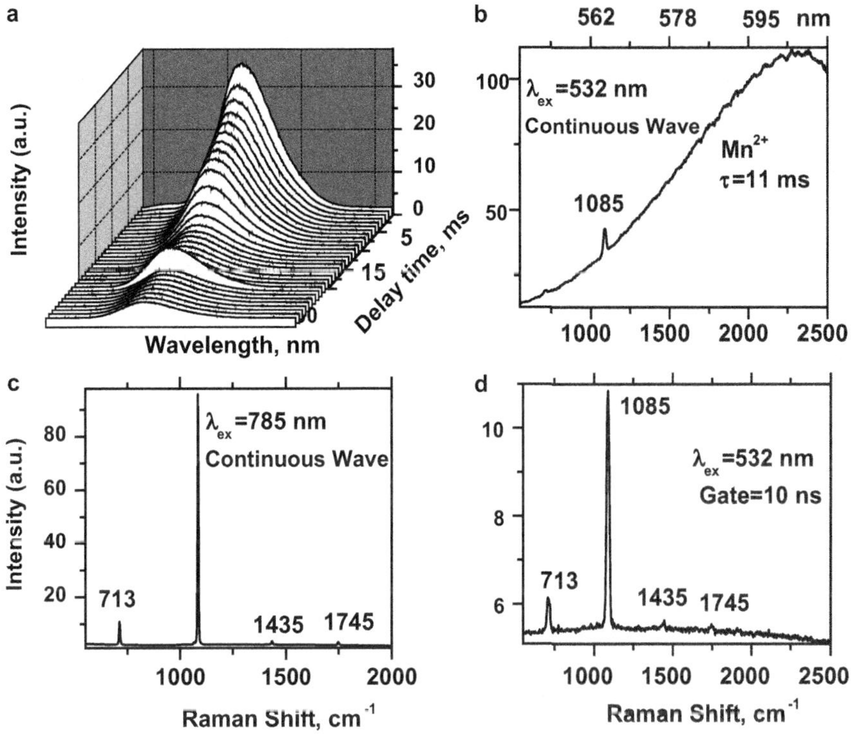

Fig. 6.22 (**a–d**) Long-lived orange luminescence spectrum ($\lambda_{ex} = 532$ nm) of Mn^{2+} in calcite andits decay as a function of delay time (**a**) CW Raman spectrum: $\lambda_{ex} = 532$ nm (**b**) and $\lambda_{ex} = 785$ nm (**c**). Gated Raman spectrum ($\lambda_{ex} = 532$ nm) with gate width of 10 ns (**d**)

wavelengths (Nasdala and Hanchar 2005). For example (Lenz et al. 2015 and references within), a green laser excitation (Ar + 514.5 nm) spectrum of a sample consisting of crystalline ZrO_2 and amorphous SiO_2 showed Raman bands of ZrO_2 and additional bands in the Raman shift range 1000–1100 cm^{-1}. The latter were assigned to Raman bands of crystalline $ZrSiO_4$ and SiO_2. The red-laser excitation (He-Ne 632.8 nm) Raman spectrum of the very same sample (Nasdala et al. 2004) however only showed ZrO_2 Raman bands. This indicates that the additional bands were green (wavelength 541–546 nm) PL emissions (most likely caused by trace of Er^{3+}). They were recorded unintentionally in the same spectral range as the Raman bands just because green laser-excitation was used in that Raman experiment. Similarly, the Raman spectra of zircon (obtained also with green laser excitation) were obscured by Er^{3+} related PL-emissions, and they were treated as Raman bands. "Raman spectra" of fluorite in the Raman-shift range 400–2400 cm^{-1} were presented and REE-related PL emissions were interpreted as Raman modes. Possible confusion of REE^{3+} emissions with Raman bands has been discussed for fluorapatite case obtained with 532 nm excitation, where Sm^{3+} emissions (wavelengths about 650 nm) are recorded with apparent Raman shifts of 3300–3500 cm^{-1}

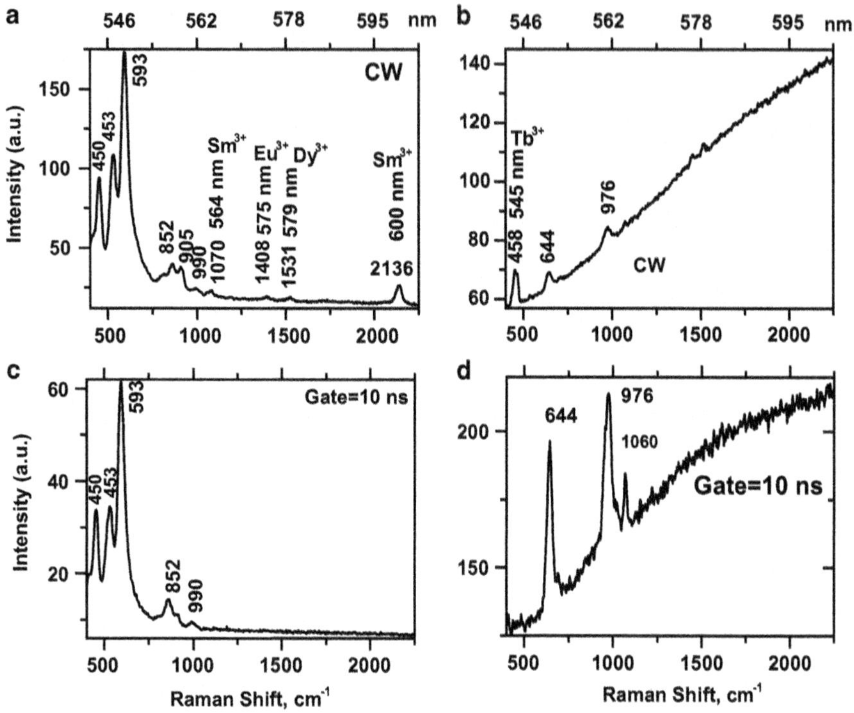

Fig. 6.23 (**a–d**) CW (**a, b**) and gated (**c, d**) Raman spectra ($\lambda_{ex} = 532$ nm) of titanite and rhodonite, respectively. The titanite lines at 600, 579, 575, 564 and 588 nm visible only in CW spectra belong to Sm^{3+}, Dy^{3+} and Eu^{3+} luminescence. Rhodonite lines visible only with a broad gate occur at 577, 579 and 545 nm belong to Dy^{3+} and Tb^{3+} luminescence

and therefore may be mistaken easily as O–H stretching bands of structurally bound hydroxyl groups.

Another method is to use the gated Raman technique, as decay times of f–f transitions of trivalent REEs are usually very long, i.e. in the range of tens and hundreds of microseconds (Gaft and Nagli 2009). Thus, though they may be present in CW spectra, they would disappear in gated spectra with narrow gate width. For example, a CW Raman spectrum of titanite under excitation of 532 nm contains several narrow lines that may be interpreted as Raman signals (Fig. 6.23a). In Raman spectra with narrow gating of 10 ns, some of these lines disappear. Those that remain definitely belong to Raman lines because trivalent REE lines with such short decay time do not exist (Fig. 6.23c). The lines at 600, 579, 575, 564 and 558 nm visible only in CW spectra belong to Sm^{3+}, Dy^{3+} and Eu^{3+} luminescence centers. A similar situation takes place in the mineral rhodonite (Fig. 6.23b, d). Lines visible only with a broad gate occur at 577, 579 and 545 nm and are connected with Dy^{3+} and Tb^{3+} luminescence centers luminescence centers. Similar situation is in zircon where Ho^{3+}, Er^{3+}, Sm^{3+} and Eu^{3+} luminescence lines present together with Raman lines (Fig. 6.24).

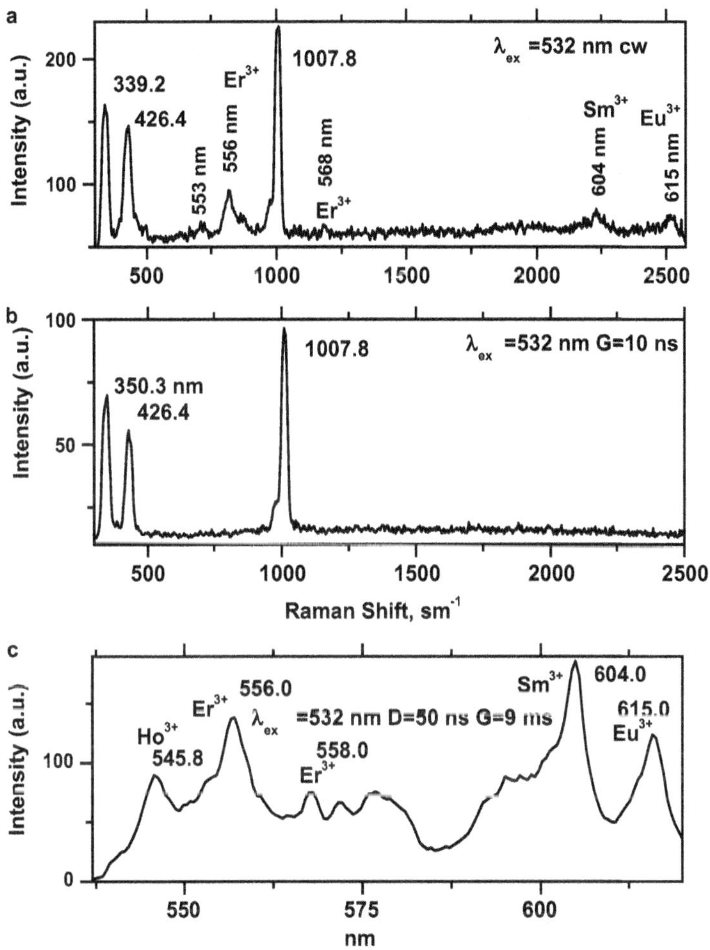

Fig. 6.24 (**a–c**) CW (**a**) and gated (**b**) Raman spectra ($\lambda_{ex} = 532$ nm) of zircon. The lines at 547, 556, 558, 604 and 615 nm visible only in CW spectra belong to Ho^{3+}, Er^{3+}, Sm^{3+} and Eu^{3+} luminescence (**c**)

Sometimes luminescence lines appear in the midst of the "regular" Raman lines in the 200–900 cm^{-1} spectral range. For example, titanite time resolved spectra under excitation by 532 nm reveal pure Raman spectrum with zero delay and 10 ns gate (Fig. 6.25a) while after delay of 500 ns when all Raman signals are definitely quenched, trivalent REE emission clearly dominates the spectrum (Fig. 6.25b).

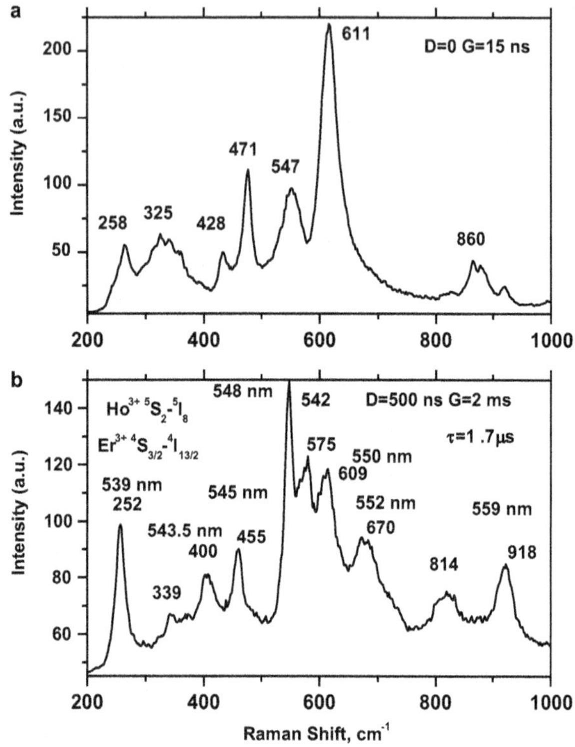

Fig. 6.25 (**a–b**) Raman (**a**) and luminescence (**b**) of titanite under excitation by 532 nm in different time windows

6.2.2.4 UV Gated Raman Spectroscopy

Terlingua-type white and especially pink calcites are characterized by a radiation-induced orange luminescence band, which is spectrally very similar to the emission from Mn^{2+} but has a very short decay time of 3–5 ns (Fig. 6.26a). Luminescence from these calcites is so strong that it is present in the CW Raman spectrum, even using NIR excitation of 785 nm (Fig. 6.26c). Because the decay time of this emission is very short, even under excitation at 532 nm with 10 ns gating, the Raman lines are not detected above the luminescence background (Fig. 6.26b); only extremely narrow gating of 0.5 ns using a pulsed laser with 50 ps pulse duration reveals the strongest Raman line more or less clearly above the intense luminescence (Fig. 6.26d). For such cases, the UV gated Raman technique may be more effective, since Raman emission under excitation at 355 nm is at least five times more intense than at 532 nm and 24 times more than at 785 nm excitation (Gaft and Nagli 2009).

Nevertheless, Terlingua-type calcites are characterized by extremely strong violet luminescence of radiation origin (Fig. 6.27a) with very short decay times of 3–5 ns. For this reason, 10 ns gating enables only limited detection in the pink variety (Fig. 6.27b); it is more effective for the white variety (Fig. 6.27c), because violet emission is less intensive here. Because time domain separation alone is not effective for 355 nm excitation, we used a deeper UV excitation at 266 nm (Fig. 6.27d). Here,

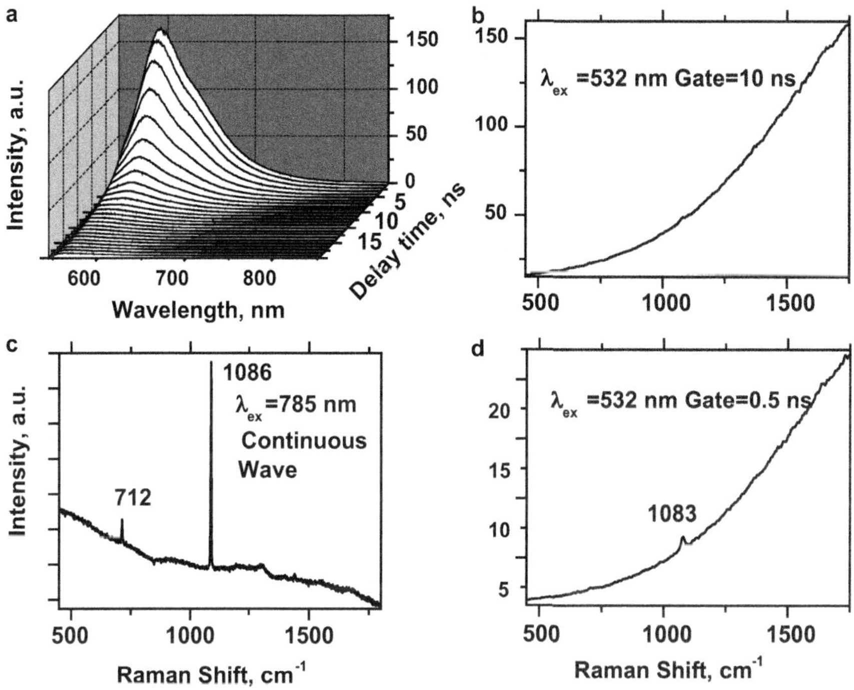

Fig. 6.26 (**a–d**) Short-lived orange luminescence spectrum ($\lambda_{ex} = 532$ nm) of radiation-induced center in calcite and its decay as a function of delay time (**a**). Gated Raman spectra with excitation at 532 nm and gate widths of 10 ns (**b**) and 0.5 ns (**d**). CW Raman spectrum with excitation at 785 nm (**c**)

UV Raman signals are only 3.2 times stronger compared to excitation at 355 nm, but most important is the fact that the violet band luminescence spectrum remains the same under 355 and 266 nm excitations, thus the luminescence Stokes shift is substantially larger for the excitation at 266 nm compared to 355 nm (145 and 55 nm, respectively). On the other hand, with excitation at 266 nm, a Raman shift of, for example, 1000 cm^{-1} corresponds to 7 nm displacement from the laser line, and therefore the Raman lines appear separated from the violet luminescence region. For comparison, with 532 nm excitation 1000 cm^{-1} corresponds to approximately 28 nm displacements and with 785 nm excitation to approximately 67 nm. Thus, the Raman lines interfere with the corresponding luminescence region. It is important to note that under CW UV excitation at 248 nm of both studied calcite samples, the Raman lines are not detected because of strong luminescence background. This contradicts the conclusion that the Raman range up to 4000 cm^{-1} is normally free from luminescence when an excitation wavelength shorter than about 250 nm is used (Asher and Johnson 1984). It may be correct for organic materials, but it is evidently not the case for many minerals, where intense UV emissions with very small Stokes shift often exist under short UV excitation. It was found that the most problematic luminescence center for UV gated Raman is Pb^{2+}, which has relatively narrow emission bands in the UV part of the spectrum with short decay time $\tau = 100$–120 ns.

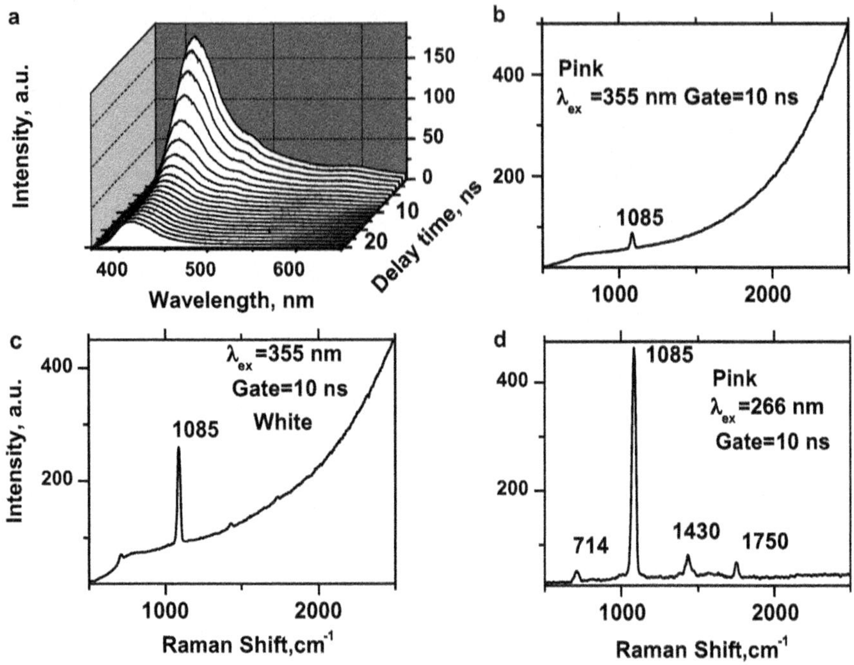

Fig. 6.27 (**a–d**) Short-lived violet luminescence spectrum ($\lambda_{ex} = 355$ nm) of radiation-induced center in Terlingua-type calcite and its decay as a function of delay time (**a**), gated Raman spectra ($\lambda_{ex} = 355$ nm) in pink (**b**) and white (**c**) varieties, and gated Raman spectra ($\lambda_{ex} = 266$ nm) in pink variety (**d**)

Table 6.2 Ratios of different intensities in calcite Raman spectra obtained with different excitation wavelengths, λ_{ex} (nm)

λ_{ex}	ν_4/ν_1	ν_3/ν_1	ν_4/ν_3
785	0.11	0.02	6.7
532	0.18	0.02	10
266	0.05	0.09	0.5
248	0.03	0.2	0.3
224	0	0	0

6.2.2.5 Resonance Raman in Minerals

Comparing the Raman spectra with different excitations, it may be seen that the relative intensities of the different Raman lines change with excitation wavelength (Gaft and Nagli 2009). For example, the relative intensity of υ_3 and $2\upsilon_2$ vibrations is much higher in spectra with excitation at 266 nm compared to excitation at 785 nm. This selective enhancement of Raman lines becomes even more pronounced under excitation at 248 nm (Table 6.2). It is evidently connected with resonance or pre-resonance Raman effects, which are known to provide an additional increase in Raman cross section above the inverse fourth power dependence on excitation

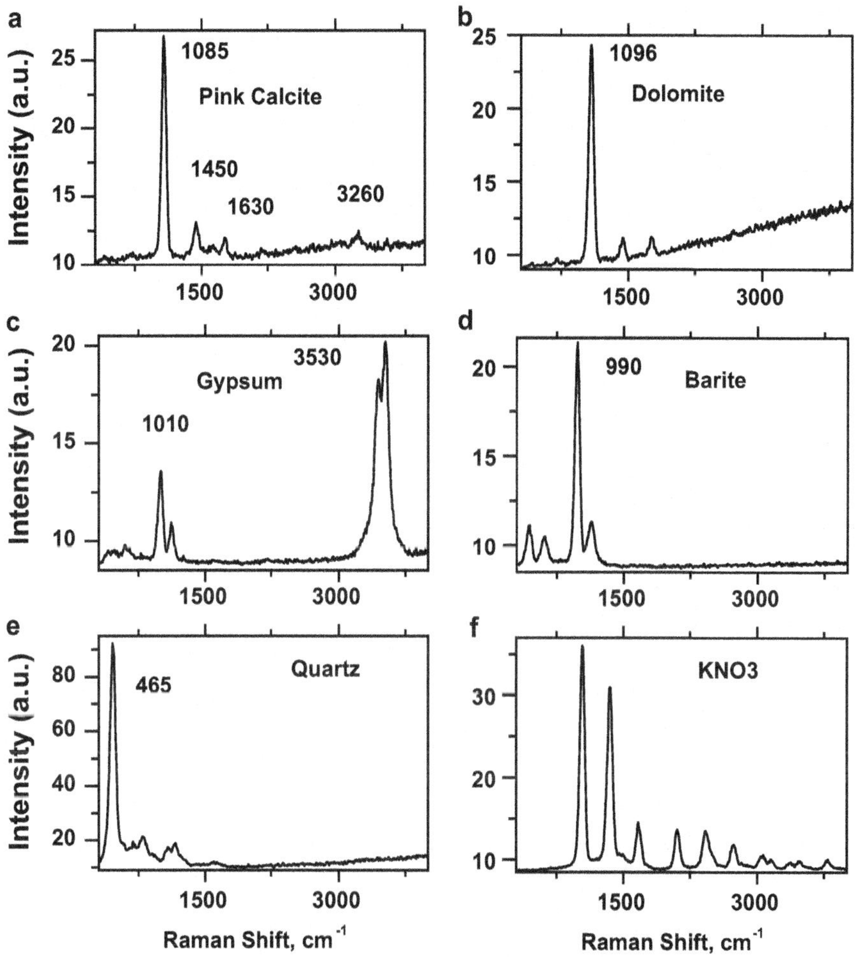

Fig. 6.28 (**a–f**) Gated Raman spectra of different minerals ($\lambda_{ex} = 248$ nm, gate = 10 ns) at controlled experimental conditions of Terlingua-type pink calcite (**a**), dolomite (**b**), gypsum (**c**), barite (**d**) and quartz (**e**) in comparison with KNO_3 (**f**). Data were used for absolute Raman cross-section evaluation using KNO_3 as a reference

wavelength. We evaluated the absolute cross sections of several minerals with 248 nm excitation, where KNO_3 was used as a standard material (Fig. 6.28 and Table 6.2). It was found that the Raman lines in the 710–730 cm^{-1} region, which are due to the υ_4 vibration, behave unexpectedly with changes in excitation wavelength. They are relatively strong under NIR and green excitations, become much weaker under 266 and 248 nm (Table 6.3) and disappear under 224 nm excitation (Storrie-Lombardi et al. 2001). It contradicts the classical λ^{-4} dependence of the Raman cross section and needs further study.

Table 6.3 Raman cross-section of several minerals

Mineral	Raman line	Cross-section cm^{-2} $molecule^{-1}sr^{-1}$
Calcite	1085	3.1×10^{-27}
Dolomite	1096	3.1×10^{-27}
Gypsum	1010	1.7×10^{-27}
	3530	2.4×10^{-27}
Barite	990	2.7×10^{-27}
Quartz	465	1.0×10^{-26}
KNO_3	1050	4.5×10^{-27}

6.2.3 Raman for Minerals Remote Detection

The transition of Raman spectroscopy from a time consuming, experts-only method to a standard analytical technique was propelled by several technological advances in optics and electronics. The key contribution was improvements in lasers and notch filters which efficiently eliminate the contribution of Rayleigh scattered light. Raman spectroscopy is a powerful technique for mineralogical analysis, where the sharpness of spectral features allows for unambiguous detection of minerals, especially in the presence of mixtures. Ten years ago, the possibility of CW Raman technique for remote detection of minerals was recognized. Later, it was found that the intense ambient light background during the day and luminescence in general are significant problems with this method. To overcome the first problem, the system may be used at low ambient light conditions, such as during the night or under cloud cover. To solve the luminescence interference, an NIR laser may in theory be used, but it demands either prolonged collection times or significantly increased power delivered to the sample, which is not possible in real-life scenarios. Utilizing a pulsed laser system and a gated receiver totally surmounts limitations caused by ambient light. A system using a pulsed, frequency-doubled Nd–YAG excitation laser (532 nm) was developed and used for measuring remote Raman spectra of rocks and minerals for Mars and Venus missions (Sharma et al. 2003; Angel et al. 2012). High-quality Raman spectra of carbonates, silicates, hydrous silicates and sulfate minerals were obtained at distances from 10 to 66 m.

Over the past decade, Raman spectroscopy has continued to develop as a prime candidate for the next generation of in situ planetary instruments, as it provides definitive structural and compositional information of minerals in their natural geological context. A time resolved Raman spectrometer have been developed that uses a streak camera and pulsed miniature microchip laser to provide picosecond time resolution (Blacksberg et al. 2010). The ability to observe the complete time evolution of Raman and fluorescence spectra in minerals makes this technique ideal for exploration of diverse planetary environments, some of which are expected to contain strong, if not overwhelming, fluorescence signatures. In particular, it was found that conventional Raman spectra from fine grained clays, sulfates, and phosphates exhibited large fluorescent signatures, but high quality spectra could be obtained using the time-resolved approach.

Time-resolved Raman has not been used commercially in large part due to the cost, size, and complexity of detectors with adequate time response, namely, traditional photocathode based imagers such as intensified CCDs and streak cameras. Until recently, a solid-state alternative was deemed out of reach due to limitations in combined detector format, bandwidth, and sensitivity. But the recent advent of compact and low-power silicon single-photon-avalanchediode-(SPAD)-based sensors with the potential for large scale manufacturability can enable a new generation of high temporal resolution imaging Raman spectrometers. It offers a significant reduction in size, weight, power, and overall complexity, putting it on par with instruments that do not have time resolution, but providing enhanced science return. Time-resolved Raman became recently achievable using an all-solid-state detector array – a 128 × 128 SPAD – as an alternative to the streak camera (Blacksberg et al. 2011).

According to our experience, gated detection with pulsed excitation at 532 nm is not optimal. UV gated Raman spectroscopy could be much more effective for remote detection of minerals, as has already been proven for standoff chemical (Sedlachek et al. 2004) and explosives detection (Gaft and Nagli 2008). Compared with 532 nm gated, application of UV gated Raman spectroscopy may enable remote detection of many additional minerals, and extend the detection distance to hundreds of meters. As demonstrated above, UV gated Raman spectroscopy has several advantages over visible or NIR Raman spectroscopy. The main advantages are Raman enhancement because of λ^{-4} dependence, resonance Raman and essentially luminescence-free spectral range. UV gating of a few nanoseconds substantially quenches all luminescence signals with relatively long decay time. One of the additional advantages of UV Raman is that because of the higher energy of UV photons, a relatively broad spectral range of Raman lines, from 400 to 5500 cm^{-1}, corresponds to only 40 nm and may be detected simultaneously on an ICCD detector with spectral resolution sufficient to Raman spectroscopy. Thus, water and OH Raman lines in the high-frequency region may be detected together with vibrations of minerals' structural units in the low-frequency region, which makes identification ability more definite (Fig. 6.29). Raman cross-section of water at 248 nm is 120 times larger than at 532 nm and 120 times larger than at 785 nm because of pre-resonance enhancement. Detection of water and water-bearing minerals is important to support existence of life on other planets (Misra et al. 2006).

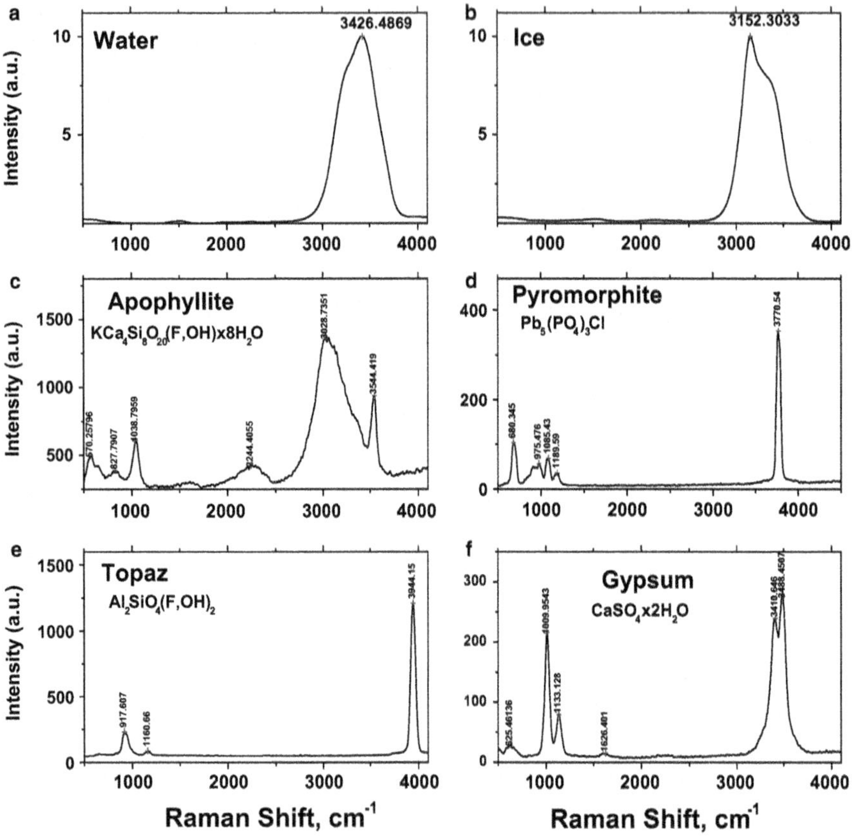

Fig. 6.29 (**a–f**) Gated Raman spectra of water and OH in different minerals ($\lambda_{ex} = 226$ nm, gate = 10 ns)

6.2.4 *Raman and Luminescence*

6.2.4.1 Rock Drill Core

A capable, rapid and online measurement technique was developed for analyzing the mineral composition of rock drill core samples (Kauppinen et al. 2014). Raman and laser-induced fluorescence (LIF) techniques have been combined. The sample set under study was segments of rock drill core collected from a Ni–Cu multi-metal mine situated in Kevitsa, Finland. Scanning drill core samples using LIF technique produces images which present a map of colors based on emissions of fluorescent minerals. Some parts of the sample set, based on the exposed colors, were measured through Raman spectroscopy and the minerals are identified using reference spectra from a public source. The combination of LIF and Raman is then used to produce mineral maps of the drill core samples and finally an estimation of the mineral abundance per units of drill core area was developed.

6.2.4.2 Diamonds ID

In the jewels industry there is needed to be able to quickly identify any particular diamond and to be able to distinguish it from the others. For example, since the appearance of diamonds living the polishing factory have no resemblance to diamonds that entered it, the opportunity exists for the polished gems to be switched for less valuable ones (Yifrach and Neta 1992). The combination of luminescence (spectra of luminescence and excitation, decay time and intensity of luminescence) and Raman properties at 300 K together with suitable program for data treatment allows for a reliable identification of specific diamonds and distinguishing them from others.

Time-resolved spectroscopy is especially effective, because 50–70 % of diamonds have strong blue luminescence, which hinders the discrimination, and to distinguish between N3 and N3-like centers is extremely important. Besides that, after 1 μs delay the N3 is quenched and more individual bands with longer decay appear. For example, Fig. 6.30 presents diamond spectra under excitation by 355 nm with different delay times and gate widths. The main luminescence center in this case is N3 and only it is detected in steady state spectrum. After delay time of 200 ns the N3 emission is partially quenched and with gate width of 500 ns the narrow line appears, which according to our knowledge was not previously detected in diamonds (Fig. 6.30a). It is characterized by narrow zero-phonon line at 529.5 nm (2.342 eV), only the second one after N3 center, which may be clearly detectable at room temperature. It has decay time of approximately 200 ns and may be seen after delay of 50–100 ns while the short-lived N3 emission is mainly quenched. After delay time of 400 ns the N3 emission is totally quenched and with gate width of 1 μs the broad green band peaking at 517 nm is detected (Fig. 6.30b). With broader gate with of 100 μs the luminescence remains mainly the same, but its maximum is slightly moved to 515 nm (Fig. 6.30c). After bigger delay of 4 μs and gate width of 400 μs narrower band peaking at 520 nm is the main in the spectrum (Fig. 6.30d). With the same delay, but with broader gate of 500 μs the broad emission band peaking at 540 nm appears (Fig. 6.30e). Broad band ($\Delta = 125$ nm) peaking at 585 nm (2.12 eV) is sometimes detected (Fig. 6.30f) with decay time of 10–20 μs, which appears after delay time of several μs and broad gate width of 1–2 ms. Figure 6.31 presents luminescence of diamond with dominant cw N3 luminescence, received with excitation by 248 nm. It may be seen that N3 and zero-phonon line at 529.5 nm are also detected with such excitation (Fig. 6.31a). After N3 luminescence quenching by delay time, broad band ($\Delta = 125$ nm) peaking at approximately 450 nm with long decay time of 1 ms usually appears. Very often such band dominates luminescence spectra with excitation by 248 nm even with narrow gate (Fig. 6.31b), while prominent reabsorption lines at 415.2 and 402.3 nm are detected. The broad band is evidently connected

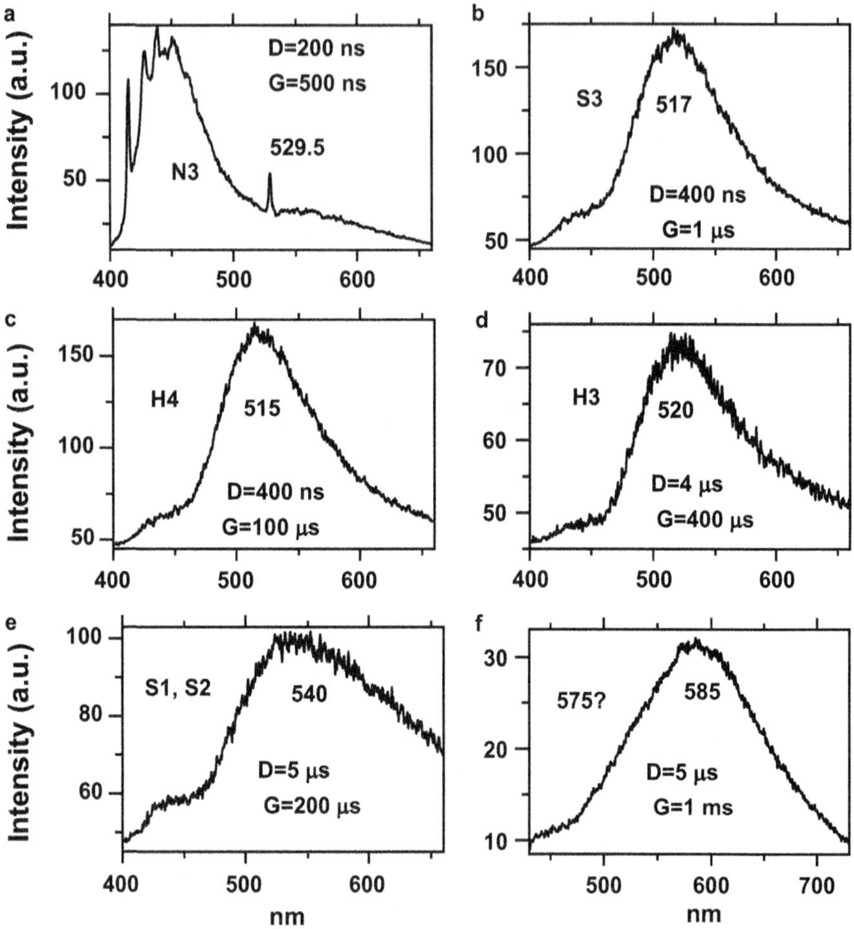

Fig. 6.30 (a–f) Laser-induced time-resolved luminescence spectra of diamond ($\lambda_{ex} = 355$ nm). Only N3 luminescence center is detected by CW luminescence

with A-center, which is one of the most characteristic luminescence features of natural diamonds. Figure 6.32 presents luminescence of diamond with dominant cw N3 luminescence, received with excitation by 215 nm. It may be seen that short lived A band emission dominates the spectrum with actually zero delay time together with reabsorption lines peaking at 414.5 and 402.5 nm (Fig. 6.32a). After delay times starting from 30 ns N3 center is the main one, which is also characterized by short decay time (Fig. 6.32b, c). When this emission is totally quenched, long lived A band luminescence may be seen (Fig. 6.32d).

It demonstrates that time-resolved luminescence of diamond is much more informative for ID task than the steady-state one. Vast amount of luminescence centers are known in natural diamonds (Zaitsev 2005) and if the specific diamond

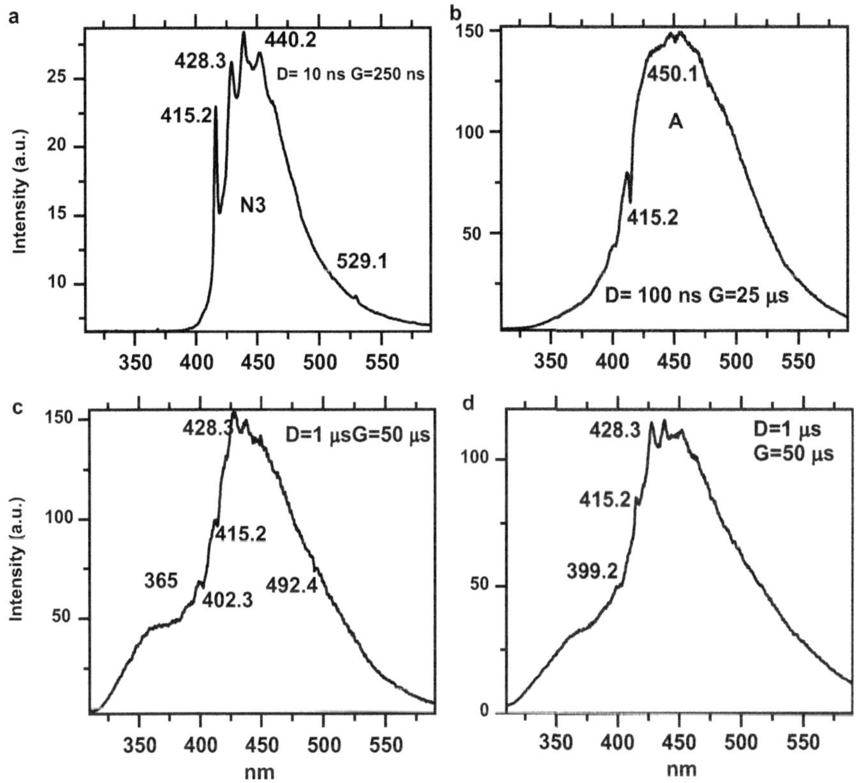

Fig. 6.31 (**a–d**) Laser-induced time-resolved luminescence spectra of diamond ($\lambda_{ex} = 248$ nm). Only N3 emission center is detected by CW luminescence

contains several luminescence centers the definite identification is relatively easy. Nevertheless, according to our experience approximately many natural diamonds contain only N3 blue luminescence centers, which make luminescence based identification problematic. In such case together with luminescence, strong Raman line of diamonds may be used. The spectral position of the diamond Raman line may change considerably depending on the diamond sample. In natural diamonds of different origin the line can be found in the range from 1331 to 1346 cm^{-1} (Zaitsev 2005). The spectral width of the diamond Raman line in perfect natural diamond may be as low as 1.5 cm^{-1}, while in disordered manmade diamonds the width of the line can attain a value of 40 cm^{-1}. Thus Raman line also may be a potential distinctive feature of specific diamonds. Our aim was to evaluate if gated Raman and time-resolved luminescence spectroscopies combination is effective for identification of diamonds with N3 dominant luminescence.

The spectra of studied diamonds may be sub-divided on two main groups according to the Raman/Luminescence intensities ratio. Figure 6.33 presents the first group where this ratio changes from 3.6 to 1.7 and down to 1. It was found that

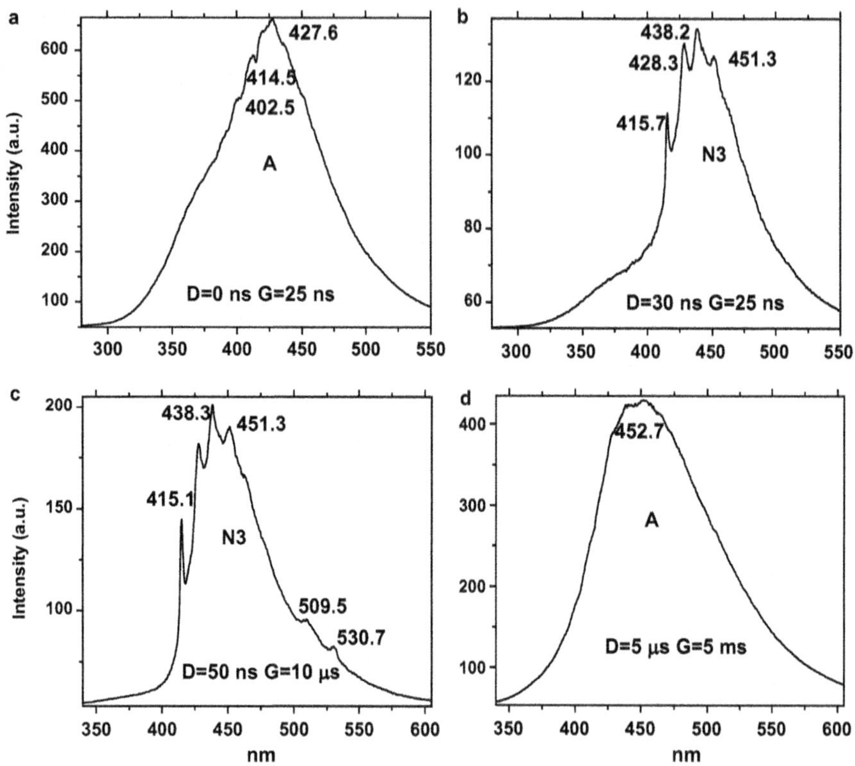

Fig. 6.32 (**a–d**) Laser-induced time-resolved luminescence spectra of diamond ($\lambda_{ex} = 215$ nm). Only N3 emission center is detected by CW luminescence

this ratio is stable parameter independent on diamond positioning inside experimental device and may be used for definite identification. Similar results have been received for Raman/Luminescence ratio of diamonds using different excitation wavelengths (Lamprecht et al. 2007). It was found that the Raman and luminescence intensities showed a variation of a factor of 10 for 4 different orientations, but the orientation of a diamond in the beam did not change the Raman/Luminescence ratio.

Diamonds belonging for the second grope has low and very similar Raman/Luminescence ratios, namely between 0.13 and 0.2. To differentiate between those diamonds several additional spectral features may be used. N_3 is the only luminescence center with an appreciable fine structure at 300 K, while its zero-phonon line at 415 nm is characterized by different intensities down to practically total disappearance. For example, the ratio between the maximum intensity of N_3 band at 440 nm and the 415 nm line intensity is changed in studied samples between 1.7 and 1.2. Figure 6.34 demonstrates the comparison between two samples with similar Raman/Luminescence ratio and even similar absolute intensity, while the ratio of

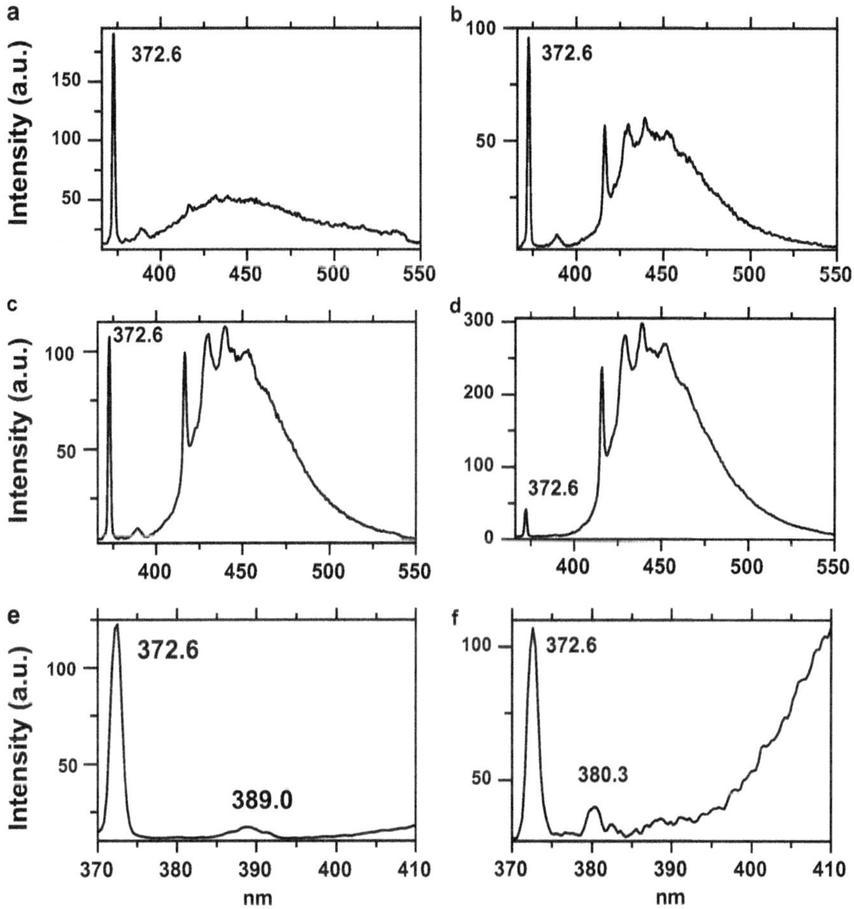

Fig. 6.33 (a–f) Combination of Raman and luminescence signals of the first grope of diamonds ($\lambda_{ex} = 355$ nm)

I_{440}/I_{415} is substantially different. Another parameter is decay time of diamonds fast blue luminescence, which is different in different samples and may be used for identification purpose. The main reason is that several luminescence centers exist with emission spectrum very similar to those of N3, but with longer decay time of 500–700 ns. Under steady-state conditions there is some superposition of luminescence and excitation spectra of N_3 and this center and they cannot be spectroscopically separated. By time-resolved spectroscopy the centers were separated and the second one, named 2.96 eV was connected with Al impurity. Figure 6.34b, c demonstrate that the ratio between the second and the third components in kinetic series with delay difference between them of 10 ns is different, which evidently connected with the presence of N3 and 2.96 eV centers together, but in different ratios.

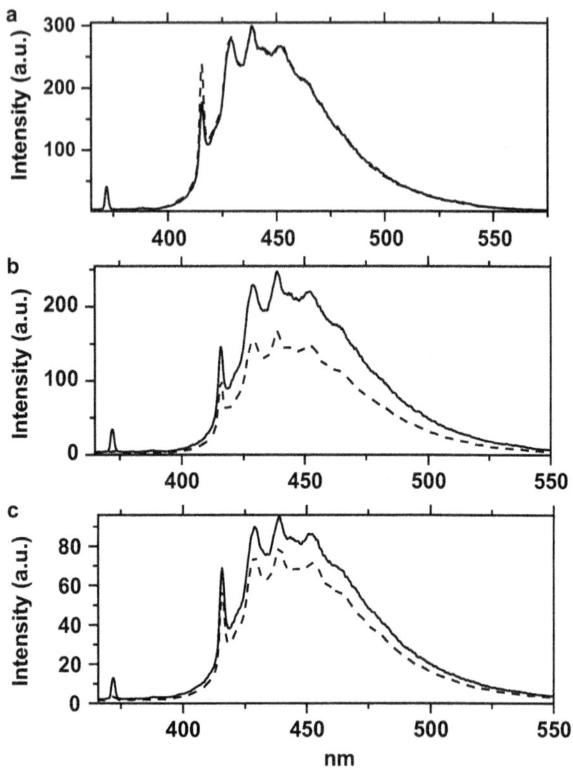

Fig. 6.34 (a–c) Differentiation between diamonds with similar Raman/Luminescence ratios: different I_{440}/I_{415} ratios in (**a**) and different ratios between the second and the third components in kinetic series ($\Delta = 10$ ns)

For the following Raman-luminescence "individualization", where the gating time was very narrow (Fig. 6.35a) we used "delayed" time resolved spectroscopy with excitation by 355 nm and delay time of several μs (Fig. 6.35b), which is usually quite enough. In certain cases the "fast" (Fig. 6.35c) and "slow" (Fig. 6.35d) components of the luminescence with excitation by 248 nm give additional identification features.

The specific advantage of 248 nm excitation is its ability to distinguish between types I and II diamonds. The reason is that the perfect diamond crystal consists of a perfectly regular array of carbon atoms, every atom being symmetrically surrounded by four others, and forming an electron-pair bond with each one. Energy of about 5.5 eV is required to remove an electron from one of these bonds, and the crystal is therefore transparent for ultra-violet wavelengths greater than 225 nm. For shorter wavelengths it will absorb strongly. No crystal is perfect and the deviation from perfection results in optically active defects, which have a strong absorption in the UV and visible parts of the spectrum. Most type IIa (and rare type IaB) diamonds transmit short-wave ultraviolet light while most of the diamonds (type Ia) do not. Thus with 248 nm excitation, UV Raman signal is only detected in type II diamonds, while in type I diamonds it is extremely weak or absent (Fig. 6.36).

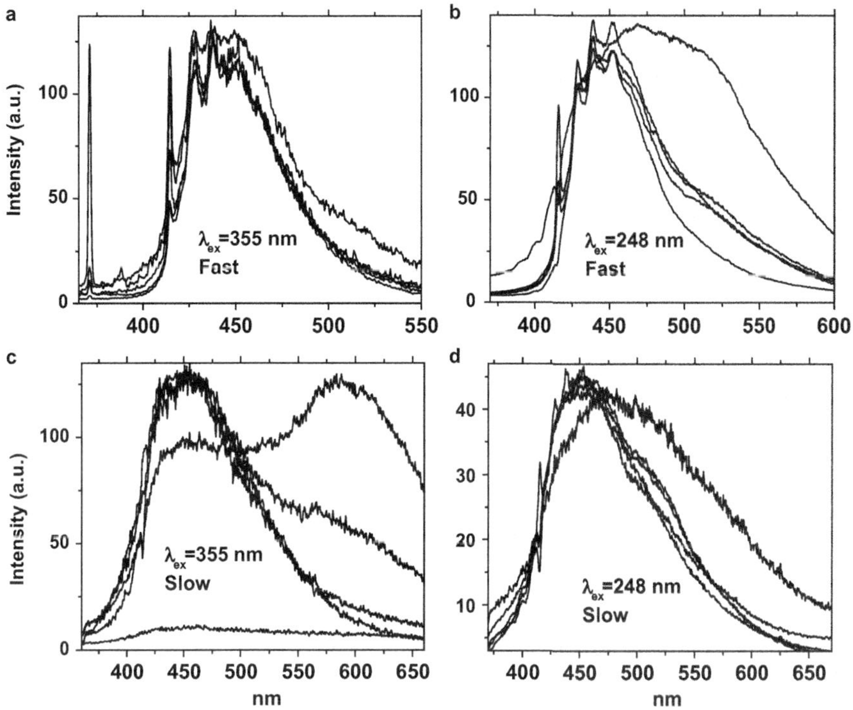

Fig. 6.35 (**a–d**) Time-resolved luminescence spectra of diamond at room temperature ($\lambda_{ex} = 355$ and 248 nm)

Fig. 6.36 Raman signal with excitation by 248 nm for I and II type diamonds

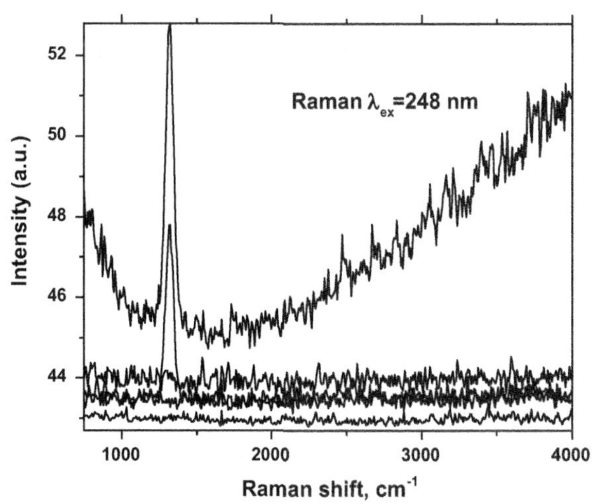

References

Albrecht A, Hutley M (1971) On the dependence of vibrational Raman intensity on the wavelength of the incident light. J Chem Phys 55:4438–4443

Angel M, Gomer N, Sharma S, McKay C (2012) Remote Raman spectroscopy for planetary exploration: a review. Focal point review. Appl Spectrosc 66:137–150

Asher S (2002) Ultraviolet Raman spectrometry. In: Chalmers J, Griffiths P (eds) Handbook of vibrational spectroscopy. Wiley, New York, pp 557–571

Asher S, Johnson C (1984) Raman spectroscopy of a coal liquid shows that fluorescence interference is minimized with ultraviolet excitation. Science 225:311–313

Bell S, Bourguignon E, Grady A et al (2000) Identification of dyes on ancient Chinese paper samples using the subtracted shifted Raman spectroscopy method. Anal Chem 72:234–239

Blacksberg J, Rossman GR, Gleckler A (2010) Time-resolved Raman spectroscopy for in-situ planetary mineralogy. Appl Opt 49:4951–4962

Blacksberg J, Maruyama Y, Charbon E, Rossman GR (2011) Fast single-photon avalanche diode arrays for laser Raman spectroscopy. Opt Lett 36:3672–3676

Cremerz D, Radziemski L (2013) Handbook of laser-induced breakdown spectroscopy, 2nd edn. Wiley, New York

Dubessy J, Orlov R, McMillan P (1994) Raman spectroscopy in earth science. In: Marfunin A (ed) Advanced mineralogy 2, methods and instrumentation. Springer, Berlin/Heidelberg/New York, pp 138–146

Dubessy J, Caumon M-C, Rull F, Sharma S (2012) Instrumentation in Raman spectroscopy: elementary theory and practice. In: Dubessy J, Caumon M-C, Rull F (eds) Raman spectroscopy applied to earth sciences and cultural heritage, vol 12. EMU Notes in Mineralogy, London, pp 83–172

Everall N, Hahn T, Matuosek P, Parker A, Towrie M (2001) Picosecond time-resolved Raman spectroscopy of solids: capabilities and limitations for fluorescence rejection and the influence of diffuse reflectance. Appl Spectrosc 55:1701–1708

Gaft M, Nagli L (2008) Laser-based spectroscopy for standoff detection of explosives. Opt Mater 30:1739–1746

Gaft M, Nagli L (2009) UV gated Raman spectroscopy of minerals. Eur J Miner 21:33–42

Gaft M, Dvir E, Modiano H, Schone U (2008) Laser induced breakdown spectroscopy machine for online ash analyses in coal. Spectrochim Acta B 63:1177–1182

Gaft M, Nagli L, Groisman Y (2011b) Luminescence excited by laser induced plasma. Opt Mater 34:368–375

Gaft M, Nagli L, Gornushkin I, Groisman Y (2011c) Emission of double ionized atoms in laser-induced breakdown spectroscopy. Anal Bioanal Chem 400:3229–3237

Gaft M, Nagli L, Gornushkin I (2013) Laser-induced breakdown spectroscopy of Zr in short ultraviolet wavelength range. Spectrochim Acta B 85:93–99

Gaft M, Nagli L, Groisman Y, Barishnikov A (2014a) Industrial online raw materials analyzer based on laser-induced breakdown spectroscopy. Appl Spectrosc 68:7–18

Gaft M, Nagli L, Eliezer N et al (2014b) Elemental analysis of halogens using molecular emission by laser-induced breakdown spectroscopy in air. Spectrochim Acta 95:39–47

Gaft M, Nagli L, Eleizer N, Groisman Y (2015) Boron- and iron-bearing molecules in Laser-Induced Plasma (LIP). Spectrochim Acta B 110:56–61

Hahn D, Omenetto N (2010) Laser-Induced Breakdown Spectroscopy (LIBS), Part I: review of basic diagnostics and plasma-particle interactions: still-challenging issues within the analytical plasma community. Appl Spectrosc 64:335–366

Hahn D, Omenetto N (2012) Laser-Induced Breakdown Spectroscopy (LIBS), Part II: review of instrumental and methodological approaches to material analysis and applications to different fields. Appl Spectrosc 66:347–419

Jasinevicius R (2009) Characterization of vibrational and electronic features in the Raman spectra of gemstones. MSc thesis, Department of Geosciences, University of Arizona, Tucson

Kauppinen T, Khajehzadeh N, Haavisto O (2014) Laser-induced fluorescence images and Raman spectroscopy studies on rapid scanning of rock drill core samples. Int J Miner Proc 132:26–33
Lamprecht G, Himan H, Snyman L (2007) Detection of diamond in ore using pulsed laser Raman spectroscopy. J Miner Proc 84:262–273
Lenz C, Nasdala L, Talla D et al (2015) Laser-induced REE^{3+} photoluminescence of selected accessory minerals – an "advantageous artefact" in Raman spectroscopy. Chem Geol
Liu Y, Gigant L, Matthieu Baudelet M, Richardson M (2012) Correlation between laser-induced breakdown spectroscopy signal and moisture content. Spectrochim Acta B 73:71–74
Ma Q, Motto-Ros V, Lei W et al (2010) Multi-elemental mapping of a speleothem using laser-induced breakdown spectroscopy. Spectrochim Acta B 65:707–714
Mao X, Bol'shakov A, Choi I et al (2011) Laser ablation molecular isotopic spectrometry: strontium and its isotopes. Spectrochim Acta B 66:767–775
Measures R, Kwong H (1979) TABLASER: trace (element) analyzer based on laser ablation and selectively excited radiation. Appl Opt 18:281–286
Misra A, Sharma S, Lucey P (2006) Remote Raman spectroscopic detection of minerals and organics under illuminated conditions from a distance of 10 m using a single 532 nm laser pulse. Appl Spectrosc 60(2):223–228
Miziolek A, Palleschi V, Schechter I (eds) (2006) Laser induced breakdown spectroscopy (LIBS): fundamentals and applications. Cambridge University, Cambridge
Mulliken R (1925) The isotope effects in band spectra II: the spectrum of boron monoxide. Phys Rev 25:259–296
Nagli L, Gaft M (2013) Fraunhofer type absorption line splitting and polarization in confocal double pulse laser induced plasma. Spectrochim Acta B 88:127–135
Nagli L, Gaft M (2015a) Lasing effect in laser-induced plasma plume. Opt Commun 354:330–332
Nagli L, Gaft M (2015b) Combining laser-induced breakdown spectroscopy with molecular laser-induced fluorescence. Appl Spectr (Published online)
Nagli L, Gaft M, Gornushkin I (2011) Comparison of single and double-pulse excitation during the earliest stage of laser induced plasma. Anal Bioanal Chem 10:3207–3216
Nasdala L, Hanchar G (2005) Comment on: application of Raman spectroscopy to distinguish metamorphic and igneous zircon. Anal Lett 38:727–734
Nasdala L, Smith D, Kaindl R, Ziemann M (2004) Raman spectroscopy: analytical perspectives in mineralogical research. In: Beran A, Libowitzky E (eds) EMU notes in mineralogy, spectroscopic methods in mineralogy. Vol. 6/7. Mineralogical Society of America, Chantilly, pp 281–343
Noll R (2012) Laser induced breakdown spectroscopy: fundamentals and applications. Springer, Berlin
Panczer G, De Ligny D, Mendoza C et al (2012) Raman and fluorescence in earth sciences. In: Dubessy J, Caumon M, Rull F (eds) Raman spectroscopy applied to earth sciences and cultural heritage. Chapter 2, vol 12. EMU Notes in Mineralogy, London, pp 61–80
Parigger C (2013) Atomic and molecular emissions in laser-induced breakdown spectroscopy. Spectrochim Acta B 79–80:4–16
Paris P, Piip K, Lepp A et al (2015) Discrimination of moist oil shale and limestone using laser induced breakdown spectroscopy. Spectrochim Acta B 107:61–66
Pearse R, Gaydon A (1965) The identification of the molecular spectra. Chapman & Hall, London
Ribiere M, Cheron B (2010) Analysis of relaxing laser-induced plasmas by absorption spectroscopy: toward a new quantitative diagnostic technique. Spectrochim Acta B 65:524–532
Rossi M, Dell'Aglio M, De Giacomo A et al (2014) Multi-methodological investigation of kunzite, hiddenite, alexandrite, elbaite and topaz, based on laser-induced breakdown spectroscopy and conventional analytical techniques for supporting mineralogical characterization. Phys Chem Miner 41:127–140
Russo R, Bol'shakov A, Mao X et al (2011) Laser ablation molecular isotopic spectrometry. Spectrochim Acta B 66:99–104

Sedlachek A, Ray M, Wu M (2004) Application of UV Raman scattering to non-traditional stand-off chemical detection. Trends Appl Spectrosc 5:19–38

Senesi G (2014) Laser-Induced Breakdown Spectroscopy (LIBS) applied to terrestrial and extra-terrestrial analogue geomaterials with emphasis to minerals and rocks. Earth Sci Rev 139:231–267

Sharma S, Lucey P, Ghosh M et al (2003) Stand-off Raman spectroscopic detection of minerals on planetary surfaces. Spectrochim Acta A 59:2391–2407

Singh J, Thakur S (2007) Laser-induced breakdown spectroscopy. Elsevier, Amsterdam

Smith E, Dent G (2005) Modern Raman spectroscopy: a practical approach. Wiley, Hoboken

Steinfeld J, Wormhoudt J (1998) Explosives detection: a challenge for physical chemistry. Ann Rev Phys Chem 49:203–232

Storrie-Lombardi M, Hug W, McDonald G et al (2001) Hollow cathode ion lasers for deep ultraviolet Raman spectroscopy and fluorescence imaging. Rev Sci Instrum 72:4452–4459

Volodin B, Dolgy B, Lieber C et al (2013) Quantitative and qualitative analysis of fluorescent substances and binary mixtures by use of shifted excitation Raman difference spectroscopy. In: Mahadevan-Jansen A, Vo-Dinh T, Grundfest W (eds) Advanced biomedical and clinical diagnostic systems XI progress in biomedical optics and imaging, vol 14. SPIE, Bellingham, pp 8572 11–8572 12

Yifrach A, Neta U (1992) Method and apparatus for identifying gemstones, particularly diamonds. US Patent 5,118,181

Zaitsev A (2005) Optical properties of diamond. Springer, Berlin

Chapter 7
Laser Based Spectroscopies for Minerals Prospecting

7.1 Laser Induced Luminescence

Powerful lasers as excitation source make possible to use luminescence for remote sensing. The basis of electromagnetic remote sensing is the recording by sensors placed on board aircraft or satellite of analogical or digital data, proportional to the intensity of an electromagnetic beam, reflected, emitted, or backscattered from the surface of the Earth. Where the source of illuminating beam is the Sun, remote sensing is sad to be passive. Remote-detection systems of Solar-Stimulated Luminescence are also known using Fraunhofer line discriminator. Fraunhofer lines (423, 486, 518, 589, 656, 687 and 760 nm) are bands of reduced intensity in the solar spectrum caused by the selective absorption of light by gaseous elements in the solar atmosphere. The sampling of these gaps in the smooth continuum, when compared with the direct solar emission, allows the determination of luminescence response.

Luminescence Lidar (Light Detection and Ranging) is active instrument, which sends out coherent waves to the object concerned. A fraction of the transmitted energy is transformed by the objects and sends back to the sensor. Lidar instruments measure both the traveling time interval between sensor/object/sensor as well as the difference between emitted and returning energy, providing information on the exact position of the objects and on the material the objects are made of. Spectral selectivity was achieved usually with the aid of narrow band interference filters.

Within the past two decades environmental issues have drawn to attention the ease with which the biosphere can be perturbed. Lidar is playing an ever increasing role in studying and solving the different kinds of environmental problems. This method is ideally suited for detecting, identification and mapping the oil spills and another kinds of pollutants, part of them according to their own luminescence properties and part because of "quenching" the pure water luminescence. The last method may be used for the surveying the total organic loading of natural bodies of water, which is very important because high contents of organic materials in the water can render the water harmful to man and the aquatic ecosystems (Measures 1985). Totally

M. Gaft et al., *Modern Luminescence Spectroscopy of Minerals and Materials*,
Springer Mineralogy, DOI 10.1007/978-3-319-24765-6_7

ecologically pure luminescent tracers may be used to study water movement and drifts balance for natural bodies of water. Based on this, it is possible to recommend the placements of new hydrotechnical industrial enterprises and recreation zones and the protection measures for already existing (Shteinman et al. 1997).

If the purpose of the given mission is to simply detect the presence of a specific target, such as mineral, then the fluorescence return signal recorded within some pre-selected spectral window may be satisfactory. If, however, the mineral has to be distinguished from a similar background, then detection of fluorescence within one or even two spectral intervals may be inadequate. Indeed, from luminescence point of view, emission intensity is very difficult to use, requiring a thorough knowledge of the overlap between the laser beam and the area of observation, the transmission losses, and the instrument calibration. In addition, the nature of the medium under investigation has to be well known, including the laser penetration depth and the scattering coefficient, which may be controlled by the background. It is clear that the amplitude of the return signal is only relative parameter possessing rather limited information.

In a remote sensing method operating under daylight conditions, the first principal limitations to detection of the faint fluorescence emission is imposed by the effects of solar radiation reflected from the mineral surfaces. The steady components of the solar radiation can be blocked electronically, but any fast-changing variations in reflected light and the electronic noise intrinsic to the PMP generated by the solar radiation will determine the minimum detectable change in intensity that can be ascribed to fluorescence. The second principal limitation is the presence of other unwanted luminescent components on the surface. Two main fields of laser-induced luminescence background of different nature can be revealed on the earth surface during minerals prospecting: organic field consists of vegetation and inanimate organic substances disseminated in rocks (Fig. 7.1) and barren host rocks. This luminescence background is very strong, covers all visible range and in many cases prevents targets detection and identification by steady-state laser induced luminescence method.

In the first attempts to overcome background problem using decay time, the variation of the fluorescence decay time as a function of wavelength across the entire emission profile for a variety of materials have been used (Measures et al. 1974; Measures 1985). For a variety of rocks and minerals, it was proved that this information represents a new kind of signature, the so called fluorescence decay spectrum, that possesses considerable discrimination power, being able to characterize the irradiated material with far superior precision than the normal luminescence spectrum (Fig. 7.2).

In the following steps, decay times of different backgrounds, such as organic matter and host rocks, have been compared with decay times of industrial minerals (Seigel and Robbins 1982, 1985). It was found that decay time of complex organic molecules is characterized by two components of $\tau_1 \leq 20$ ns and $\tau_2 \approx 8$–10 ms, while barren host rocks with luminescence mainly connected with long-lived Fe^{3+} and Mn^{2+} have decay time of several ms. Thus the intermediate vacant decay time range exists, which is a good "window" for targets detection and identification.

According to accumulated experience, the discovered luminescence in this selective window may be related to certain ore bodies, such as uranium, wolfram,

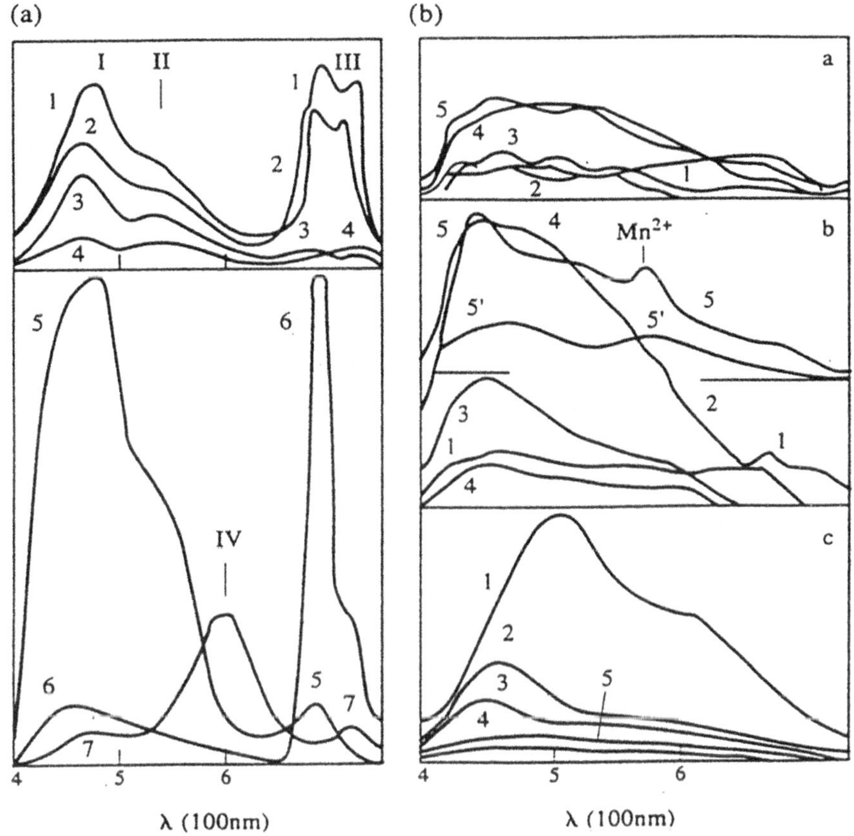

Fig. 7.1 Laser induced luminescence spectra of different natural objects: (a)-grass (1), green foliage (2), whortleberries (3), green needle (4), moss Sphagnum (5), -algae (6), and lichen Umblicaria sp (7); (b) - a-magmatic rocks: pyroxenite (1), gabbro (2), andesite (3), gneissodiarite (4), gneissogranite (5); b-metamorphic rocks: shale (actinolite-chlorite) (1), shale (quartz-sericite) (2), shale (quartz-mica) (3), marble (dark colored) (4), horn with wollastonite (fresh split) (5), horn (weathered) (5'); c- sedimentary rocks: salt (1), limestone (2), carbonate soils (3), snow (4), clay, gravel, sand, non-carbonate soils having almost no luminescence (Gorobets and Rogojine 2001)

molybdenum, zinc, boron, and phosphorus bearing minerals. Another case is so-called pathfinder minerals such as calcite, apatite, barite, and fluorite. These minerals presence may be related to the processes of differentiation of some ore forming elements and their accumulation in local structural and geomorphological traps. The accumulation of luminescent substances in separate sites of the Earth crust is the result of concentration of rare and ore elements. The ore elements give mineralogical and geochemical anomalies that can be used in searching for ore bodies. Such anomalies may form luminescent haloes in and around ore bodies, which have been mapped in the several minerals deposits (Kupriyanova and Moroshkin 1987; Gorobets and Walker 1994; Gorobets et al. 1995; Bushev and Portnov 2000). For example, the luminescent characteristics of calcite, apatite,

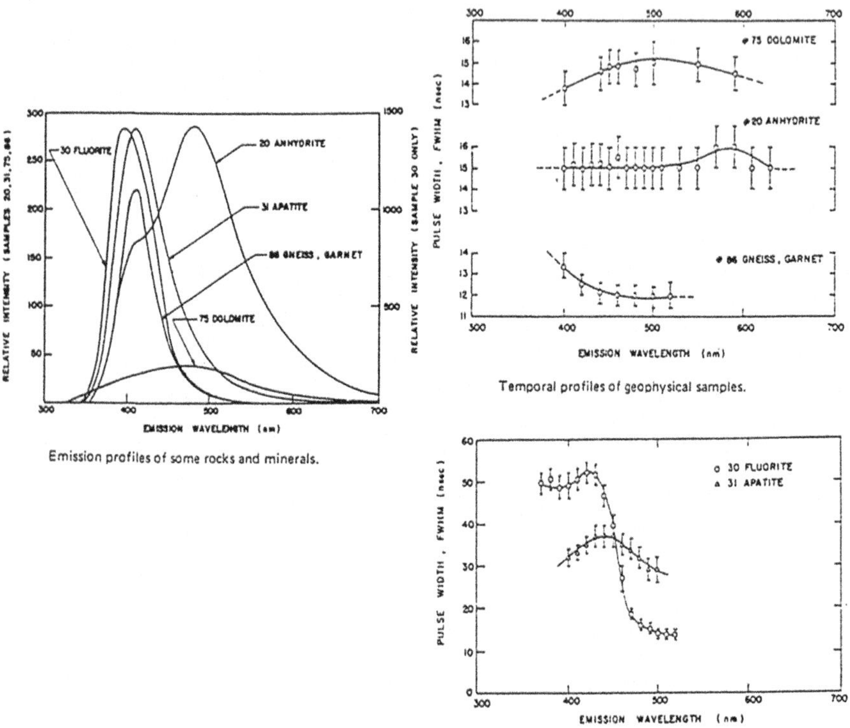

Fig. 7.2 Temporal profile of luminescence of minerals (Measures 1985)

zircon, barite and fluorite show the considerable influence of kimberlite on their country rocks. The luminescent haloes stretch up to some hundred meters and even to 1–1.5 km outside kimberlite bodies. Haloes around kimberlite-like pipes due to mechanical scattering of minerals were also detected with the aid of sand analysis. The most informative was the non-magnetic fraction smaller than 0.25 mm. Occurrence of apatite of "mantle" type with bluish-violet glow of Ce^{3+} and Eu^{2+} within haloes surrounding some magnetic anomalies significantly increase the probability that a source of mantle fluids is located under them. Besides, luminescence haloes of mica pegmatite veins, connected with albite-oligoclase with violet-blue glow and microcline with violet glow of Eu^{2+}, haloes in ruby-bearing skarn, connected mainly with calcite with red emission of Mn^{2+}, haloes of emerald-bearing bodies, mainly connected with apatite of "mantle" type and plagioclase with blue emission of Eu^{2+}, haloes of mountain crystal veins, related to calcite with red emission of Mn^{2+}, apatite, microcline and albite, and haloes of gold-bearing ore deposits, mainly related to outcrops of apatite, fluorite, calcite, scheelite and molybdoscheelite, have been described (Gorobets and Rogojine 2001). The luminescence of Eu^{3+} in scheelite may be served as an ore guide. It is particularly strong in luminescence spectra of scheelite from gold-bearing quartz veins (Gorobets and Kudrina 1976; Uspensky et al. 1998).

An air-bone method of exploration of ore deposits has been realized with the help of luminescence Lidar. Lidar embodies high power UV laser and a coaxial telescope detection system installed aboard a helicopter. The laser yields pulses having peak power 0.4–0.5 mJ, time delay 1 μs, gate width 1 ms, repetition rate 30–50 Hz, wavelength varying from 275 up to 400 nm. The laser "footprint" is about 400–500 cm^2 and the distance between the closest footprints about 1 m. Air-bone survey is carried out from an altitude of 50–75 m. Sensitivity is 0.2–0.5 % by exposed area for most of minerals (Seigel and Robbins 1982, 1985).

AIS Sommer GmbH of Germany delivered laser-induced fluorescence (LIF) analyzer for quality control in minerals and mineral processing (Broicher 2000). The LIF analyzer includes two light detector systems with three photomultipliers each, which evaluate three spectral bands in two time windows each. It was done in Kiruna phosphorous iron ore mine, Sweden. The limitation of LIF analysis is that its accuracy depends on the complexity of the composition of the ore and the concentration and fluorescence properties of the critical minerals in relation to all the other minerals present. The phosphorous iron ore in Kiruna is ideal for LIF analyzes, because its iron minerals are practically non-luminescent, while magmatic apatite is strongly fluorescent with intensive emissions of Ce^{3+} and Eu^{2+}.

7.2 Laser-Induced Breakdown Spectroscopy: Agronomical Application

In recent years there has been an increased focus on the development of reliable methods for elemental analysis techniques, particularly in contaminated soils. Environmental, as well as human health concerns have become the subject of numerous studies aimed at examining the effects of long-term inputs of potentially harmful trace elements in plant and soil samples, due to natural and anthropogenic origins. Current analytical methods for the detection of these elements mainly include the use of atomic absorption (AA), inductively-coupled plasma atomic emission spectroscopy (ICP-AES) and x-ray fluorescence spectroscopy (XRF). While these provide accurate quantitative elemental analysis, their use generally involves large amounts of time spent in laboratory preparation and testing procedures for sample analysis. LIBS has received much attention in research, due to its potential for in-situ, rapid, multi-elemental analysis (Ferreira et al. 2011; Lasheras et al. 2011). Results from this technique could be utilized in decision making for time-sensitive contamination events or other remedial operations (Ismael et al. 2011).

LIBS demonstrated a great capacity for accurate analysis of samples of various compositions, even for in-situ sampling of soils that may exhibit surface dust accumulation. LIBS remote sensing capabilities also make this method an attractive option for a variety of soils analysis and agronomic applications. The ability to obtain immediate results from analyses allows for on-site evaluation of contaminants or pollutants, which can be particularly useful for environmental and remediation monitoring after such events as chemical leaks, oil spills, and related chemical exposure to soils. Semi-quantitative analyses could be used to map contaminated areas for site-specific clean-up efforts. Applied to agricultural lands, the system should be able to evaluate plant essential nutrient concentrations and concentrations of specific compounds within the soil. This would only enhance precision agriculture applications focused on precise placement of agricultural fertilizers, pesticides, and herbicides.

7.3 Laser-Induced Breakdown Spectroscopy: Curiosity Mission on Mars

ChemCam is an active remote sensing instrument suite running successfully on MSL since 18 months (Maurice et al. 2012). It uses laser pulses to remove dust and to profile through weathering coatings of rocks up to 7 m away. As it was presented above, LIBS detects molecular CaF and CaCl emission which is much brighter than the associated atomic lines. It appears that these molecular emissions have been detected in some of the ChemCam spectra and enabled to report the first detection of chlorine and fluorine with ChemCam. It is also the first time ever that fluorine has been detected on the surface of Mars (Gaft et al. 2014b; Forni et al. 2014, 2015).

In light of the laboratory data, the CaCl band at 593.7 nm were found in two ChemCam spectra that match this feature. The feature can unambiguously be attributed to CaCl since no other emission lines are present between 593.0 and 594.0 nm. The two spectra are the ninth point of the Crest target acquired on Sol 125 and the ninth point of Measles Point target acquired on Sol 305. These two observations are characterized by a very high calcium concentration and show evidence for sulphur lines. No other high calcium spectra with sulphur lines show this molecular feature. The presence of the Cl I emission line at 837.8 nm is perhaps visible on the Crest-9 spectrum. Finally, these two observations, in contrast to the other sulphur-rich samples show evidence for high sodium content, making NaCl a possible phase for the detected chlorine that recombine in the plasma with the calcium to form the observed molecular line.

At least 20 ChemCam observations exhibit molecular emissions identical to those ascribed to CaF molecular emission have been observed. How-ever, the magnitude of the feature at 603 nm is variable and among these 20 observations many do not exhibit the corresponding feature at 533.0 nm because either it is too weak to be visible or it is absent. However, the shape of the 603 nm band makes the identification of fluorine unambiguous. The strongest signature is found in the fifth point on Epworth soil that was observed on Sol72. Epworth-5 contains a relative large amount of fluorine since the main atomic lines of F I are also observed. This spectrum also exhibits phosphorous emission lines at 253.7, 255.4 and 255.5 nm which, on the single shot observation, are correlated with calcium and fluorine making fluor-apatite ($Ca_5(PO_4)_3F$) the best candidate. However, there is too much fluorine in Epworth 5 for fluor-apatite only. Indeed fluor-apatite contains 3.8 wt%

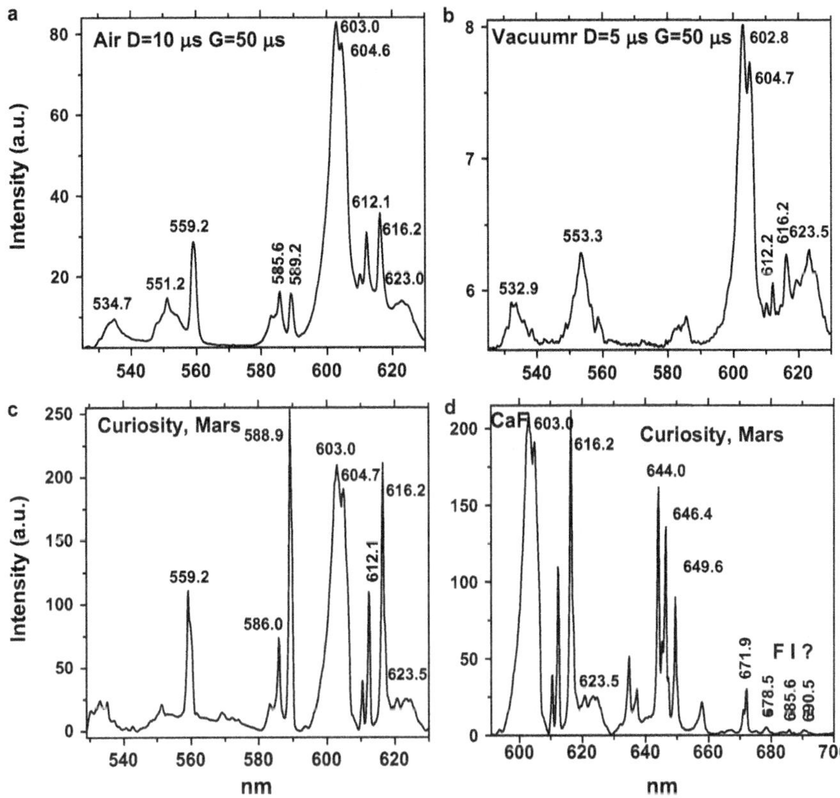

Fig. 7.3 (**a–d**) A comparison of CaF spectrum in fluor-apatite in air (**a**), in a vacuum chamber imitating Mars' atmospheric pressure (**b**) and the bands detected by Curiosity ChemCam device (**c**). Very weak emission IR lines evidently belong to F I ions (**d**)

fluorine for 55.6 wt% CaO. The CaO content of Epworth-5 is of the order 14 wt%, 4 times lower. For this CaO content we would expect only 0.5 wt% fluorine if all the calcium were in the fluor-apatite. When compared to a pure fluorite spectrum (Fig. 7.3), preliminary estimate of the fluorine content is about 10 wt%.

Consequently the fluorine must be contained in another phase and fluorite (CaF_2) is a good candidate. The other observations can be divided into two sets, the first is dominated by relatively high calcium content like the observation points located in the Rocknest unit. Some of the observations belonging to this set also exhibit P I emission lines making fluor-apatite the preferred phase. In many other cases phosphorous is not visible, but fluor-apatite remains the most plausible host phase though the CheMin soil analysis at Rock-nest did not indicate the presence of any apatite phases suggesting that the Epworth and Pearson ($\leq$1 wt% F) fluorine observations may be something other than apatite. The second set of spectra is dominated by Al rich and Ca poor points which are essentially represented by the conglomerates Goulburn and Link. In these spectra the Ca is so low that the observed fluorine cannot be borne by a calcium rich phase and it is probably more likely included in an Al rich phase like topaz or muscovite. Both of these are good candidates since, in Link at least, the hydrogen signature is relatively prominent.

7.4 Hyphenation of Laser Based Spectroscopies

Laser-induced time-resolved luminescence spectroscopy, Raman and LIBS are complementary techniques, the first one providing information on specific luminescence centers, the second one on molecular and structural information, and the third one on elemental composition. In all techniques, a laser is used for exciting the sample, light is collected from the sample, the light is dispersed using a visible or UV grating monochromator and its spectrum is measured with an array detector, usually an ICCD. For all techniques, a pulsed laser is definitely preferential. Although both techniques are very similar, there are many instrumental and measurement parameters which are quite different, including laser and detector timing, detector gating, collection optics and coatings and spectral range and resolution. These differences lead to trade-off issues that will be considered in the proper design of a combined system.

7.4.1 LIBS-Raman

A combined Raman/LIBS system would be a valuable tool for mineral analysis. Raman spectrometry would be used to provide molecular information and detect organic molecules, while LIBS can determine elemental composition, including minor components in the sample. The first standoff Raman/LIBS system was reported by Sharma et al. (2007) for planetary measurements. The combined system was capable of 10-m standoff distances, where various silicates and carbonates were studied. The results showed that combining the cations data from LIBS with Raman anion data would allow remote identification and characterization of samples on planetary surfaces. Wiens et al. (2005) also developed a combined system for geological analysis and stated that a difficulty for combining the two techniques while sharing components is how to cover the larger spectral range for LIBS while maintaining the sensitivity for Raman. Combined system standoff measurements have also been made under Martian conditions.

The main problem of measuring simultaneously Raman and LIBS signals in field application is that LIBS signals intensity are many orders of magnitude higher than intensity of the Raman signals and therefore in ordinary LIBS measurements Raman is not measurable together with LIBS. To avoid these difficulties, Misra et al. (2010) proposed to measure signals from different points of irradiated sample. Due to Gaussian lateral distribution of the focused laser spot intensity, outer part of laser spot does not create LIBS on sample surface but is sufficient intense to create measurable Raman signal from this part of the surface. The proposed idea is to create optics and spectrometer that separately looks on outer part of the irradiated sample to measure Raman signal and at the same time looks on central part of the irradiated sample measuring LIBS signal. For these purpose, laser pulse must be sufficiently intense and not tightly focused – the trade-off between getting irradiated surfaces without LIBS and getting strong enough LIBS signal from central part of the plasma plume.

Raman scattering and LIBS differs not only by intensity, but also in timing scale: Raman signal appears and presents only in time of existing laser pulse (~4–6 ns for Nd:YAG laser), while LIBS becomes measurable after 300 ns $\div$ 2 μs from laser pulse completing and exists tenths of μs. It is possible to separate Raman and LIBS signals not laterally but by applying specific ICCD camera timing. ICCD timing must consists of two time settings: (i) short delay or even negative delay D_1 ($-3 \div 2$ ns) relatively to laser pulse intensity maximum and short gate width W_1 (data acquisition time) that must be not longer than FWHM of the laser pulse ($4 \div 6$ ns) for Raman signal measuring, and (ii) long delay D_2 ($0.5 \div 2$ μs) and broad gate width W_2 ($5 \div 10$ μs) for LIBS measurements.

In another possible timing scheme Raman and LIBS can be measured separately by electronically switching between these two timing sets. In field applications where the data consist of many accumulated laser shots, it is easily realized. Still another setup may include two spectrometers, first with time gated ICCD camera for Raman measurements and second with gated CCD camera for LIBS measurements. Figure 7.4 presents Raman and LIBS signals excited by second harmonic Nd:YAG laser (532 nm, 15 mJ focused on spot about 50 μm) using short delay-narrow gate for Raman measurements and long delay – broad gate for LIBS measurements.

7.4.2 Raman-Laser Induced Timed Resolved Luminescence

Laser-induced fluorescence is complementary to Raman spectroscopy. Time-resolved measurements allow for the separation of Raman and long-lived luminescence, allowing each to be measured simultaneously with the same excitation source. Sharma et al. (2009) developed the first Raman/Luminescence standoff system, which was capable of standoff measurements up to 100 m. The luminescence portion of the system was capable of measurements up to 5 km, since fluorescence intensity is orders of magnitude higher than Raman scattering intensity. Bozlee et al. (2005) developed a combined system that measured LIF of minerals using excitation wavelengths in the UV and near-UV and used a gated 532 nm Raman system to measure various fluorescent samples. The optimal way is evidently to use UV laser as an excitation source both for Raman and luminescence. Although both techniques are very similar, there are many instrumental and measurement parameters which are quite different, including laser and detector timing, detector gating, collection optics and coatings and spectral range and resolution.

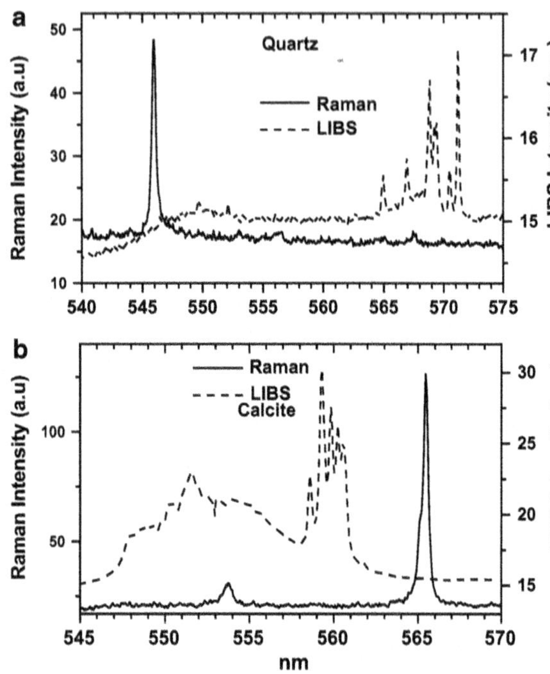

Fig. 7.4 (**a–b**) Simultaneously detected LIBS and Raman signals of quartz (**a**) and calcite (**b**) received with short timing set for Raman and long timing set for LIBS

These differences lead to trade-off issues that will be considered in the proper design of a combined system.

Recently the compact time-resolved (TR) Raman, and fluorescence spectrometers suitable for planetary exploration under NASA's Mars Instrument Development Program have been developed (Sharma et al. 2012). The compact Raman and fluorescence spectrometers consist of custom miniature spectrographs based on volume holographic gratings, and custom miniature intensified CCD cameras. Using a small frequency-doubled Nd:YAG pulsed laser (35 mJ/pulse, 20 Hz) and 50 mm camera lens, TR Raman and laser-induced fluorescence spectra of minerals from rare-earths and transition-metal ions in time domain and in the spectral range of 400–800 nm can be measured.

7.5 Anti-Terror Applications

Another aspect of minerals, rocks and foliage laser-induced spectroscopic properties became important recently all around the world, especially in Israel. Different types of camouflage simulating foliage and local rocks have been intensively used by terrorists to cover the roadside bombs and other explosive devices. The spectroscopic properties of different explosive materials and covers were studied in comparison with natural objects, which are imitated. The task was to evaluate the possibility to use the standoff non-contact laser-based techniques in order to greatly reduce the risk associated with trace sampling, hopefully reduce inspection time, and improve detection. The request is for any technology that meets the standoff requirement and detects explosives amounts in the range of micro-grams (or less) to tens of micro-grams per cm^2. The main question was is the standoff detection feasible in real life scenarios against background interferences? Two main groups have been studied: explosives and camouflage.

7.5.1 Explosives

7.5.1.1 Luminescence, LIBS and Second Harmonic Generation (SHG)

Despite of the fact that several explosives as military grade RDX, other RDX-based materials, such as C4, Urea Nitrate (UN) and especially Semtex generate strong emissions (Fig. 7.5), they are not characteristic enough to use them for their identification emissions on the background of naturally occurring objects. The similar conclusion, that luminescence properties provide little hope for the use of UV-excited fluorescence as a technique to perform safe standoff detection of adsorbed explosive particulates under real-world conditions with a useful degree of reliability was reached by Phifer et al. (2006).

A lot of research has been devoted to LIBS remote detection of explosives. Actually LIBS was the first technique proposed for remote detection of explosives (Haley and Thekkadath 1998). The patent describes performing detection without sampling by applying laser radiation, causing micro-denotation of explosive particulates that are found on a suitcase surface. The explosive is detected by measuring the characteristic emission of the explosive micro-detonation. The analysis of the LIBS spectra of explosive simulants on the open air proved that the intensity ratios between C_2 and CN and between O and N correlate to the molecular structure (Portnov et al. 2003). It indicates the possibility of distinguishing one chemical class from another and in optimal cases even identifying specific compounds by LIBS. Furthermore, the detection and characterization of explosives at distances up to 45 m using standoff LIBS has been demonstrated (López-Moreno et al. 2006). The C_2 Swan system, as well as the H, O and N emission intensity ratios were necessary parameters to identify the material such as an organic explosive, organic non-explosive and non-organic samples. One difficulty detecting energetic materials is that the contribution to the oxygen and nitrogen signals from air can impede the identification of the explosive. Techniques for reducing the air entrainment into the plasma, such as using argon buffer gas, generating the LIBS plasma with a femtosecond laser pulse and using a collinear double-pulse configuration, were studied for this application. A new technique has been developed and demonstrated using microwaves to enhance LIBS detection sensitivity. The microwaves enlarge and sustain the laser generated LIBS micro plasma to produce more spectral emission. Laboratory studies have demonstrated LIBS sensitivity enhancements

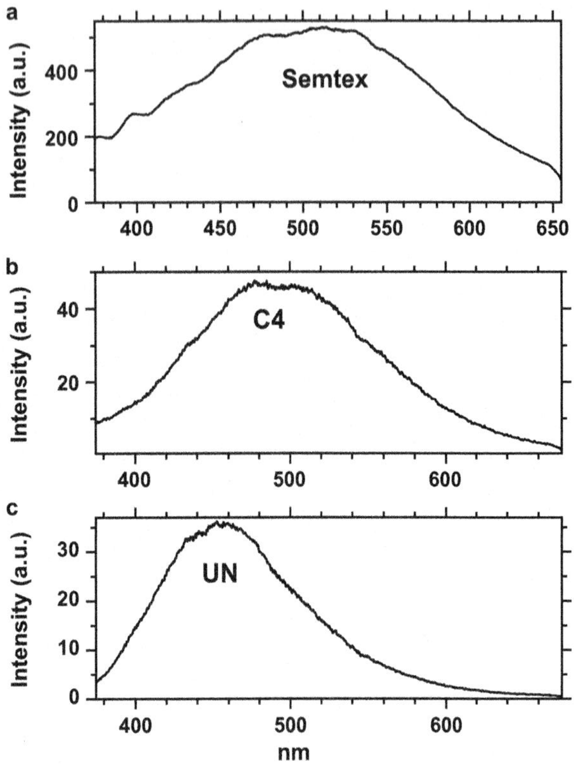

Fig. 7.5 (**a–c**) Laser-induced luminescence of Semtex (**a**), C4 (**b**) and Urea Nitrate under excitation by 355 nm

in excess of a factor of 500. The microwave technology can be beamed over long distances to allow implementation at distances in the range of 10–400 m (Gottfried et al. 2009). Despite of certain potential of LIBS approach, after many experiments we arrived to conclusion that it is not the optimal technique in most of real life scenario, specifically because of potential damage for tested humans and property.

Second Harmonic Generation (SHG) is a "three wave mixing" process in which a photon is scattered at a wavelength that is half of the original radiation wavelength, i.e. two photons combining into a single photon carrying the energy of both photons. It was found that solid trinitrotoluene TNT is characterized by a strong SHG response (Fig. 7.6)

7.5.1.2 Raman

According to our experience, Raman spectroscopy is the best tool for explosive materials identification. All known industrial and homemade explosive are characterized by specific Raman "fingerprints" under excitation by 532 nm (Fig. 7.7). The spontaneous Raman signal of explosives was measured at a distance of 30 m with a Nd:YAG laser operating at 532 nm with an 18 mJ/pulse and 8-ns pulse duration

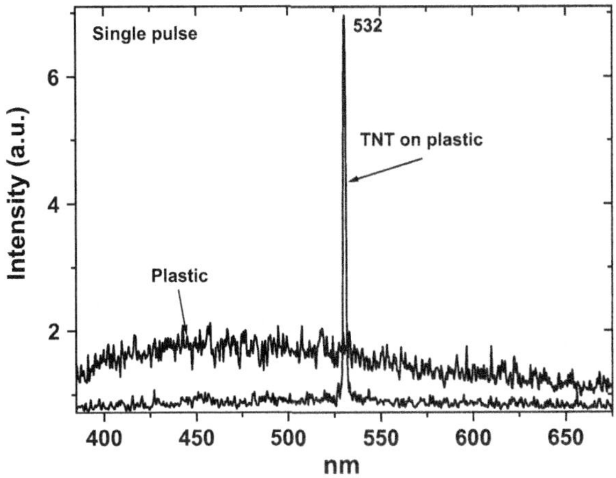

Fig. 7.6 Second harmonic generation by TNT

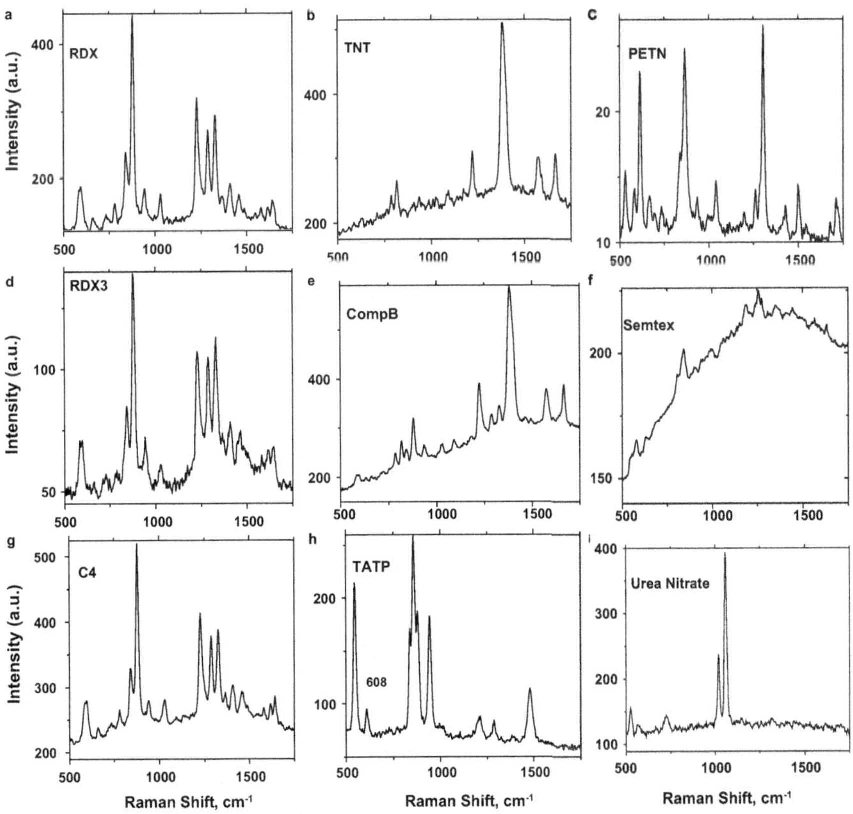

Fig. 7.7 (**a–i**) Raman spectra of explosive materials (RDX, TNT, PETN, Semtex, C4, TATP, Urea Nitrate...) under excitation by 532 nm

(Gaft and Nagli 2008). UN had the strongest signal with S/N of approximately 3.5 for 500 laser pulses. TNT and TATP explosives were confidently identified only after 2000 pulses. The Raman lines of RDX in C4 are obscured by the strong luminescence from the binder. These spectra suggest that the minimal detection quantity using 532-nm excitation is approximately 150 μg for UN, 500 μg for TNT and TATP and 750 μg for C4 (due to strong interfering luminescence). Although the spectra are somewhat noisy, the principle of remote detection is clearly established. Lawrence Livermore National Laboratory (LLNL) has published similar results with 532 nm excitation to detect explosive materials mixed in a silica matrix at a concentration of 8 % (Carter et al. 2005).

Certain conclusions have been made related to green excitation for standoff Raman detection of explosive materials: Raman spectroscopy is principally suitable for remote detection of trace amounts of explosive materials; one of the main problems limiting the use of visible light excited Raman spectroscopy is strong luminescence that generally is at the same spectral range as expected Raman spectra. Such luminescence may be subdivided on two types: internal luminescence of explosive materials itself, and external luminescence of relevant backgrounds. This luminescence signal may in some cases completely swamp the weak Raman spectra. Laser used in our equipment has the following disadvantages: its intensity suitable for Raman excitation is not eye safe; it excites in many cases very strong interfering internal and external luminescence with decay times much less than 10 ns, preventing luminescence extinguishing with gating of 10 ns. Thus we have to solve simultaneously two main problems: to find eye safe laser and to eradicate background luminescence. Our final task is to find such conditions when all relevant background materials will behave as "aluminum" in experiments with green excitation.

It was found that UV Raman may be perspective tool for explosives remote detection. On the first stage UV Raman experiments with explosive materials and the mostly typical backgrounds using 355 and 266 nm excitations were accomplished. As it was expected, the Raman lines are very close to incident exciting frequency, while in this spectral range the luminescence is still very weak, because of Stokes shift. Thus by UV excitation, separation in spectral domain was achieved. Besides, Raman signals under UV excitations are 100–200 times stronger compared to green (532 nm) excitation with similar experimental conditions, namely the same laser pulse energy, laser-target distance, detection efficiency and gate width. It may be explained by combination of high-energy UV exciting photons and UV fundamental absorption bands existence with resulting resonance Raman. Investigation of Raman signals of explosives and luminescence signals of different backgrounds proved that UV excitation using 266 nm is the mostly effective. The main reason is that practically always the relevant backgrounds luminescence is much weaker and situated far away from Raman signals relatively to 355 nm. Besides, Raman scattering is about three times stronger at 266 nm compared to 355 nm. The very important exception is RDX and RDX based explosives, which have relatively weak Raman signals, while they are much stronger with excitation by 355 nm.

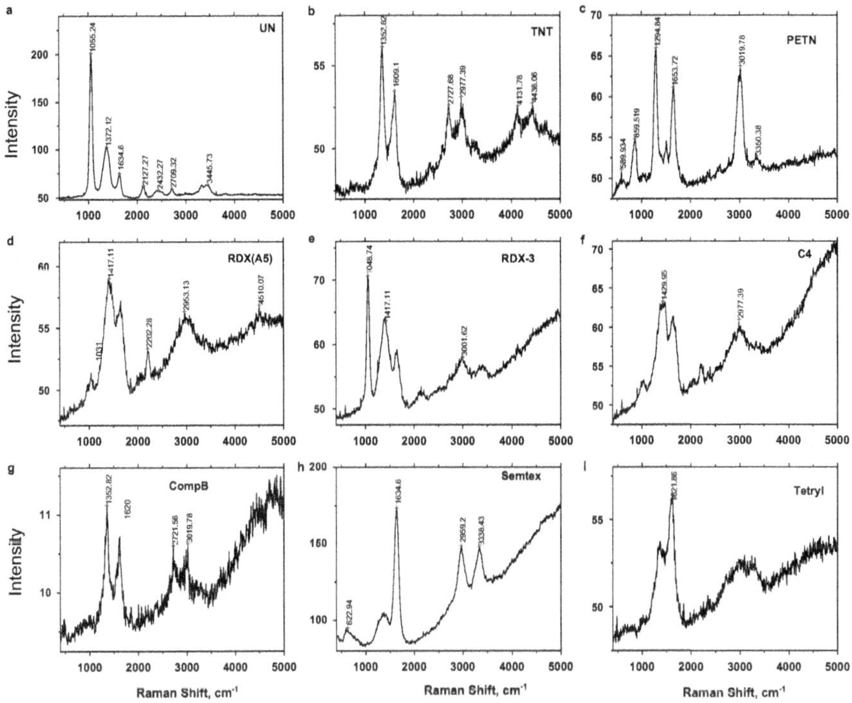

Fig. 7.8 (a–i) Raman spectra of explosive materials (Urea Nitrate, TNT, PETN, RDX(A5), RDX3, C4, CompB, Semtex, Tetryl) under excitation by 248 nm

Very promising results with 355 and 266 nm excitations stimulated our research in even deeper UV, namely using excitation by 248 nm, received from KrF excimer laser. This step was stimulated by the following considerations: Based on Raman theory, 1.3-fold increase in signal is predicted compared to excitation by 266 nm, and 4.2-fold increase in signal is predicted compared to excitation by 355 nm; Moving from 355 to 266 nm excited Raman spectra demonstrated that we are dealing with pre-resonance Raman and deeper UV may enable substantial signal increasing; according to the Asher and Johnson (1984), the full Raman range, up to 4000 cm^{-1}, is normally free from luminescence when excitation wavelength shorter than about 250 nm is used. The experimental results may be summarized as following (Fig. 7.8): the major signal increasing takes place for all RDX based explosives, such as military grade RDX, C4 and Semtex (Table 7.1) (Nagli et al. 2008). The Semtex fluorescence background with 532, 355 or 266 nm excitations completely masks any Raman bands, making it impossible to identify the sample as Semtex. The Raman spectrum with excitation by 248 nm has very little fluorescence background and clear Raman bands;

Very clear Raman signal of PETN was received, whose lines are also detected in Semtex spectrum; Strong Raman signal exists from TNT and TNT related compounds. Extremely good signals exist from homemade explosives, such as TATP

Table 7.1 Raman cross section dependence of explosives on excitation wavelength

	Cross section σ [cm²/molecule × sr]						
λ_{ex}	UN 1050 cm^{-1}	UN 1010 cm^{-1}	TATP 2992 cm^{-1}	TATP 889 cm^{-1}	PETN 1290 cm^{-1}	RDX 887 cm^{-1}	TNT 1340 cm^{-1}
248	1.4×10^{-27}	–	1.8×10^{-28}	4.3×10^{-29}	3.7×10^{-28}		9.8×10^{-29}
266	3.5×10^{-28}	–	9.5×10^{-29}	3.4×10^{-29}	1.3×10^{-28}	–	4.3×10^{-29}
355	2.6×10^{-29}	5.8×10^{-30}	1.8×10^{-29}	7.0×10^{-30}	3.2×10^{-29}	1.8×10^{-30}	–
500	–	–	–	–	–	3.6×10^{-31}	–
520	8.3×10^{-30}	1.3×10^{-30}	3.6×10^{-30}	1.1×10^{-30}	1.3×10^{-29}	–	6.0×10^{-31}
532	6.2×10^{-30}	1.1×10^{-30}	1.5×10^{-29}	6.2×10^{-30}	6.3×10^{-30}	2.4×10^{-31}	4.9×10^{-31}
625	1.1×10^{-30}	5.0×10^{-31}	6.8×10^{-31}	6.1×10^{-31}	3.0×10^{-30}	1.3×10^{-31}	2.0×10^{-31}

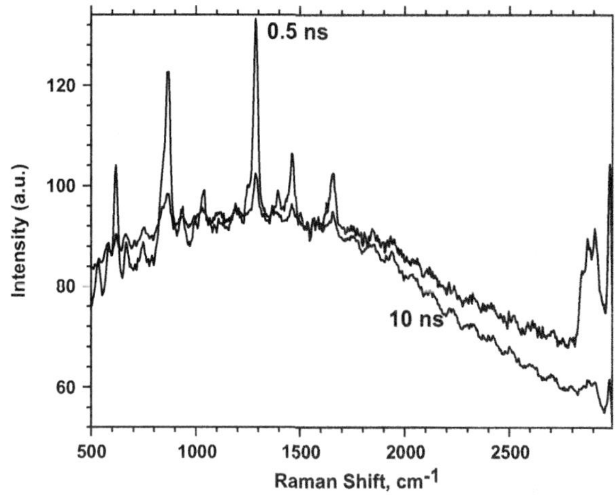

Fig. 7.9 Raman spectra of Semtex under excitation by 532 nm (50 ps) and gate width of 10 and 0.5 ns

and UN; Besides fluorescence absence in Raman relevant spectral range, the additional major difference is that only a part of Raman spectrum at visible excitation is apparent in the spectrum with excitation by 248 nm. It may be inferred that it is impossible to differentiate between compounds because of simplification of the Raman spectra caused by resonant enhancement. This is not true, because even the spectra of similar compounds such as 4MNT, 34DNT and TNT have different UV Raman spectra (Sands et al. 1998).

Ultra-fast gated Raman spectroscopy may give additional advantages (Fleger et al. 2009). Using excitation source with 50 ps pulse duration (532 nm) and gating with 500 ps, it was possible, for example, to detect characteristic Raman lines of RDX in Semtex, which were not seen using gate of 10 ns because of strong Semtex luminescence with decay time of 5–6 ns (Fig. 7.9). Such approach enables to detect effectively luminescence signals with decay time longer than 10 ns. In such case the Raman/luminescence ratio is one order of magnitude higher with 0.5 ns gate compared to 5 ns gate. This advantage becomes less effective for luminescence signals with decay time of 1 ns and less, where gate of 0.5 ns enables only 2–3 times improvement in comparison with 5 ns gate.

Another techniques tested for this task is Coherent anti-Stokes Raman Spectroscopy (CARS) where a four-wave parametric process in which three waves, two at a "pump" frequency, ω_p, and one at Stokes frequency, ω_s, are mixed in a sample to produce a new coherent beam at the anti-Stokes frequency $\omega_{as} = 2\omega_p - \omega_s$. As the frequency difference $\omega_p - \omega_s$ approaches a Raman-active vibrational mode in the sample, the efficiency of the wave mixing is greatly enhanced due to nonlinear polarization of the medium. By varying ω_s, it is possible to scan the vibrational frequency range and generate an anti-Stokes spectrum with all of the information contained in a conventional spontaneous Raman spectrum, but with much higher intensity. The CARS signal intensity was compared to that of spontaneous Raman scattering and its dependence on the distance between the sample and the detection system was found. It was found that CARS allows favorable detection as compared to Raman. It is estimated that for pulse energies of 10 mJ of the pump and Stokes laser beams, detection of trace amounts of samples from distances of ~10–200 m might be possible applying CARS, depending on the species (Portnov et al. 2010).

7.5.2 Camouflage

Discovering of camouflage which imitates the color of surroundings is a potential task for laser based spectroscopies. The camouflage and background usually have different luminescence parameters despite of the fact that by color they are practically the same. The reason is the ultimate individuality and complexity of luminescence phenomenon which is very difficult to imitate. For example, most of contemporary camouflage nets imitate foliage color very well, but their luminescence under, for example, excitation by 532 nm is practically zero. At the same time, foliage is always characterized by strong red chlorophyll emission (Fig. 7.10).

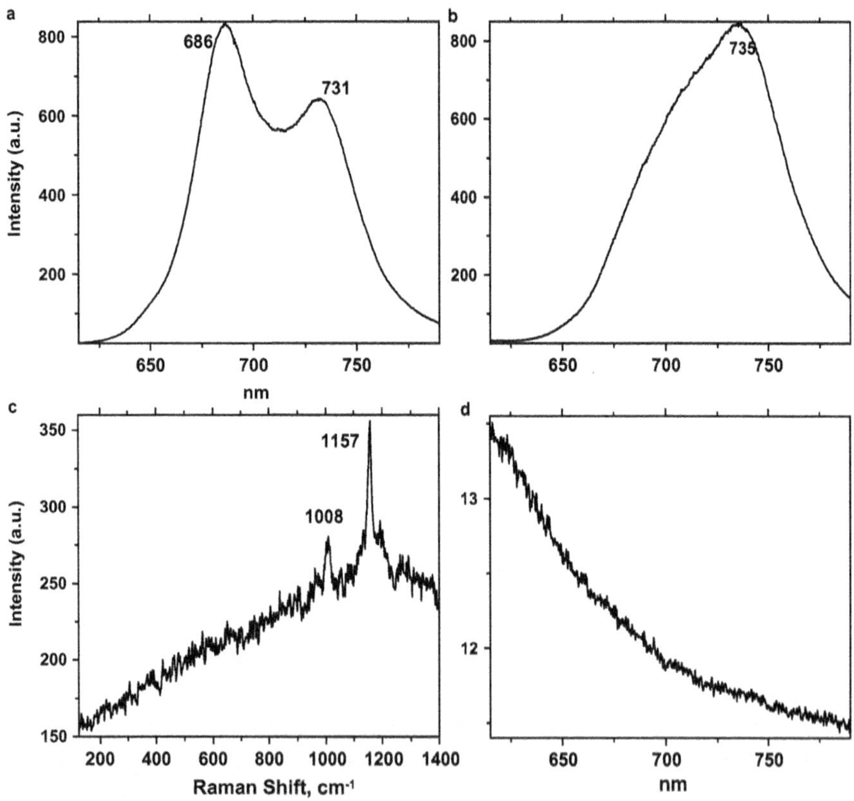

Fig. 7.10 (**a–d**) Luminescence (**a, b**) and Raman (**c**) spectra of foliage in comparison with actually zero signal from most of camouflage nets ($\lambda_{ex} = 532$ nm)

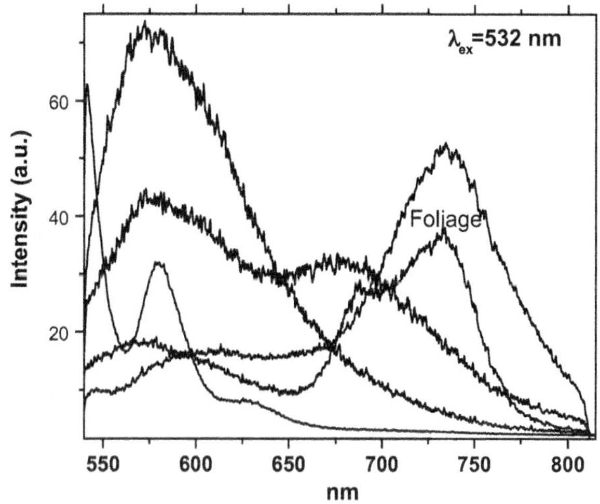

Fig. 7.11 Luminescence spectra of foliage and several sophisticated camouflage nets imitating foliage luminescence

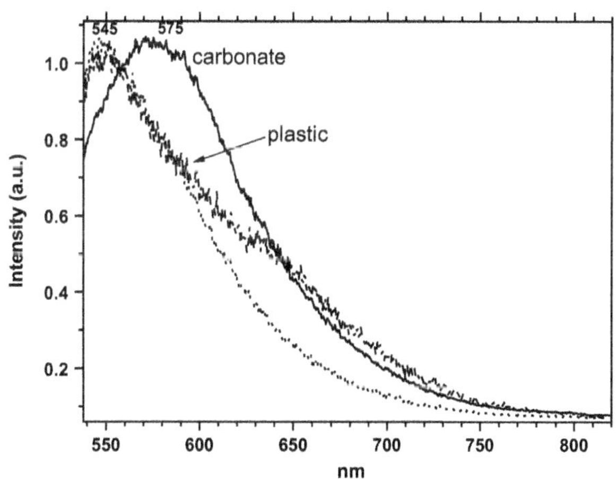

Fig. 7.12 Laser-induced time-resolved luminescence of carbonate rocks and plastic camouflage imitating those rocks

Nevertheless, the mostly sophisticated nets imitate not only the color, but also the luminescence spectrum of the plants (Fig. 7.11), but spectra and decay time differs from the chlorophyll ones.

Another issue is plastic materials used to cover the road side bombs, which imitate the color of local rocks. Time-resolved luminescence spectra under excitation by 532 nm may be used for definite differentiation between those materials for definite differentiation between those materials differentiation between those materials (Fig. 7.12).

References

Asher S, Johnson C (1984) Raman spectroscopy of a coal liquid shows that fluorescence interference is minimized with ultraviolet excitation. Science 225:311–313

Bozlee B, Misra A, Sharma S, Ingram M (2005) Remote Raman and fluorescence studies of minerals samples. Spectrochim Acta A 61:2342–2348

Broicher H (2000) Bulk sorting by LIF: quality control of ores for bulk sorting and blending – by laser-induced fluorescence analysis. Min Eng 52:73–77

Bushev A, Portnov A (2000) Photoluminescent haloes of gold-bearing deposits. Izvestia VUZov, Geologia I Razvedka 2:40–42 (in Russian)

Carter J, Angel S, Lawrence-Snyder M et al (2005) Standoff detection of high explosive materials at 50 meters in ambient conditions using small Raman instrument. Appl Spectrosc 59:769–775

Ferreira EC, Milori D, Ferreira EJ et al (2011) Evaluation of laser induced breakdown spectroscopy for multielemental determination in soils under sewage sludge conditions. Talanta 85:435–440

Fleger Y, Nagli L, Gaft M, Rosenbluh M (2009) Narrow gated Raman and luminescence of explosives. J Lumin 129:979–983

Forni O, Gaft M, Toplis M et al (2014) First fluorine and chlorine detection with CHEMCAM on MSL. 45th Lunar and Planetary Science Conference 1328

Forni O, Gaft M, Toplis M, Clegg S et al (2015) First detection of fluorine on Mars: implications for Gale crater's geochemistry. Geophys Res Lett 42:1020–1028

Gaft M, Nagli L (2008) Laser-based spectroscopy for standoff detection of explosives. Opt Mater 30:1739–1746

Gaft M, Nagli L, Eliezer N, Groisman Y, Forni O (2014b) Elemental analysis of halogens using molecular emission by laser-induced breakdown spectroscopy in air. Spectroch Acta B 95:39–47

Gorobets B, Kudrina M (1976) Typomorphic features of scheelite as revealed by their rare-earth elements luminescence spectra. Const Svoy Miner, Kiev, Naukova Dumka 10:82–88 (in Russian)

Gorobets B, Rogojine A (2001) Luminescent spectra of minerals. Handbook. RPC VIMS, Moscow

Gorobets B, Walker G (1994) Origins of luminescence in minerals: a summary of fundamental studies and applications. In: Marfunin A (ed) Advanced mineralogy 2, methods and instrumentation. Springer, Berlin/Heidelberg/New York, pp 138–146

Gorobets B, Portnov A, Rogojine A (1995) Luminescence spectroscopy of the earth. Radiat Meas 24:485–491

Gottfried J, De Lucia F, Munson C, Miziolek A (2009) Laser-induced breakdown spectroscopy for detection of explosives residues: a review of recent advances, challenges, and future prospects. Anal Bioanal Chem 395:283–300

Haley L, Thekkadath G (1998) Laser detection of explosive residues. US Patent 5,760,898

Ismael A, Bousquet B, Michel-le Pierres K et al (2011) In situ semi-quantitative analysis of polluted soils by laser-induced breakdown spectroscopy (LIBS). Soc Appl Spectrosc 65:467–473

Kupriyanova I, Moroshkin V (1987) On the possibilities of use of luminescence properties of plagioclase and apatite as emerald indicators. Izvestia AN SSSR (Ser Geol) 9:84–90 (in Russian)

Lasheras R, Bello-Galvez C, Rodriguez-Celis E, Anzano J (2011) Discrimination of organic solid materials by LIBS using methods of correlation and normalized coordinates. J Hazard Mater 192:704–713

López-Moreno C, Palanco S, Laserna et al (2006) Test of a stand-off laser-induced luminescence detection of explosive residues on solid surfaces. J Anal Atom Spectrosc 21:55–60

Maurice S, Wiens R, Saccoccio M et al (2012) The ChemCam instrument suite on the Mars Science Laboratory (MSL) rover: science objectives and mast unit description. Space Sci Rev. doi:10.1007/s11214-012-9912-2

Measures R (1985) Laser remote sensing: fundamentals and applications. Wiley, New York
Measures R, Houston W, Stephenson D (1974) Laser induced fluorescent decay spectra – a new form of environmental signature. Opt Eng 13:494–501
Misra A, Sharma S, Lucey P (2010) Method and apparatus for remote Raman and laser induced breakdown spectrometry. US Patent 2010/0171951 A1
Nagli M, Gaft M, Fleger Y, Rozenbluh M (2008) Absolute Raman cross sections of some explosives: trend to UV. Opt Mater 30:1747–1754
Phifer C, Schmitt R, Hargis P (2006) Studeis of the laser induced fluorescence of explosives and explosive compositions, SAND2006-6697. Sandia National Laboratories, Albuquerque
Portnov A, Rosenwaks S, Bar I (2003) Identification of organic compounds in ambient air via characteristic emission following laser ablation. J Lumin 102–103:408–413
Portnov A, Bar I, Rosenwaks S (2010) Highly sensitive standoff detection of explosives via backward coherent anti-Stokes Raman scattering. Appl Phys B 98:529–535
Sands H, Hayward I, Kirkbride T, Bannett R, Lacey R, Batchelder D (1998) UV-excited resonance Raman spectroscopy of narcotics and explosives. J Forensic Sci 43:509–513
Seigel H, Robbins J (1982) Method and apparatus for the remote detection of certain minerals of uranium, zinc, lead and other metals. UK Patent Application, GB 2089029
Seigel H, Robbins J (1985) Luminescence method-new method of air and ground exploration of ore deposits. ITC J 3:162–168
Sharma S, Misra A, Lucey P et al (2007) Combined remote LIBS and Raman spectroscopy of sulfur-containing minerals, and minerals coated with hematite and covered with basaltic dust at 8.6 m. Spectrochim Acta A 68:1036–1045
Sharma S, Misra A, Lucey P, Lentz R (2009) A combined remote Raman and LIBS Instrument for characterizing minerals with 532 nm laser excitation. Spectrochim Acta A 73:468–476
Sharma S, Misra A, Acosta T, Lucey P (2012) Time-resolved remote Raman and fluorescence spectrometers for planetary exploration. In: Turner M, Kamerman G (eds) Laser Radar technology and applications XVII, vol 8379. SPIE, Bellingham, pp 83790J/1–83790J/12
Shteinman B, Berman T, Inbar M, Gaft M (1997) A modified fluorescent tracer approach for studies of sediment dynamics. Israel J Earth Sci 46:107–113
Uspensky E, Brugger J, Graeseer S (1998) REE geochemistry systematic of sheelite from the Alps using luminescence spectroscopy: from global regularities to local control. Schweiz Miner Petrogr Mitt 78:31–54
Wiens R, Sharma S, Thompson J, Misra A, Lucey P (2005) Joint analyses by laser-induced breakdown spectroscopy (LIBS) and Raman spectroscopy at stand-off distances. Spectrochim Acta A 10:2324–2334

Chapter 8
Minerals Radiometric Sorting and Online Process Control

In many mineral-processing installations it is of critical importance to provide an efficient mineral purification. Complicated processing steps involving grinding to very fine particle sizes carry it out. It is necessary to grind the particles in order to liberate the impurities from a mineral. However, purification of the coarse feed particles before grinding can provide significant benefit in overall energy efficiency and the final product purity. When the particles representing impurities can be separated before grinding they do not contaminate the feed stream for further processing and additionally they are not unnecessarily ground. This can be achieved by radiometric sorting carried out before the grinding stage when the particles have size in a range of several centimeters.

Mineral sorting is a physical process in which particles are separated on the basis of their mineralogical composition. None of the minerals in the material treated undergoes any chemical change. The objective of any mineral-beneficiation process is to split the material being treated into two or more process streams. The intention is that valuable minerals are diverted into one stream, usually called the concentrate stream, while non-valuable minerals present are diverted into a second stream, which is called the tailings stream. At the same time as little as possible of the valuable minerals should be lost to the tailing stream. Sometimes when the mineral separations are difficult, a third stream, known as a middling stream, is produced.

Because the basis for mineral beneficiation is the physical separation of particles of different mineralogical composition, the success of such a particle separation as a means of separating one type of mineral from another depends on two prerequisites:

(1) Most fundamental prerequisite is that the mineral to be beneficiated should, for the most part, be physically disengaged or liberated from the gangue, and
(2) There should be sufficient differences in the physical properties of particles of different mineralogical compositions at the attained degree of liberation.

Provided that the physical properties of the minerals contained in the particulate ore are significantly different, the particles of distinct mineralogical compositions can be separated physically to achieve a mineral separation. Particles that consist

M. Gaft et al., *Modern Luminescence Spectroscopy of Minerals and Materials*,
Springer Mineralogy, DOI 10.1007/978-3-319-24765-6_8

predominantly of valuable mineral(s) will report to the concentrate under the influence of the exploited physical force, but will dilute that concentrate with the valueless mineral(s) with which they are physically locked. Particles that consist predominantly of gangue minerals will report to the tailings, and the contained value-bearing minerals will be lost. What is needed to make good mineral separations is good mineral liberation, i.e. the proportion of multi-mineral particles should be minimized. If the degree of liberation in the run-of-mine material is insufficient, a higher degree of liberation is achieved by a size-reduction process (crushing and grinding), which reduces the particles into smaller fragments. In addition, each mineral beneficiation process requires an optimum size range for efficient processing; therefore, sizing operations are also required prior to mineral beneficiation operations. Two types of sizing operations are recognized, namely screening and classification. The former uses rigid surfaces that are uniformly perforated with apertures, and the surface acts as a go-no-go gauge so that particles are separated into oversize and undersize streams. Classification is based on the fact that particles of the same shape and density but of different sizes settle in a fluid (water, air) at different rates.

Ore sorting has long been used in mineral processing operations, and was first instituted as hand sorting according to a humanly perceived difference in the ore. Minerals applications today generally require automated sorting methods to cope with a high tonnage throughput in an economic fashion, and large mechanical sorters have been developed to meet this need. There are significant opportunities for the use of such technology in many minerals processing operations, but to date, industry applications have been limited to a few niche processing areas such as for diamonds, magnesite, talc and uranium ores.

The modern automated sorters (Fig. 8.1) are machines which examine feed lumps on an individual basis, compare the measured properties of each lump with predetermined criteria, and then separate the lumps, using an applied force, into different products according to the measured properties. A sorter treating $-120+60$ mm material typically treats 85–100 t/h and a unit treating $-60+30$ mm material -30–40 t/h. Ore types treated by sorters around the world include different types of rocks (Mokrousov and Lileev 1979; Salter and Wyatt 1991; Cutmore and Eberhard 2002; Wotruba and Harbeck 2010). These sorters were installed for one of the following duties: pre-concentration of plant feed, intermediate and finished products production. The main benefits are as follows:

- A pre-concentration plant reduces the main treatment plant's head-feed and can remove waste rocks for a fraction of the cost of treating the material through the complete process operation. In addition, the main plant may be fed at its optimum feed rate with a consistent grade of material thereby increasing overall plant economic performance. Besides that, very often the reject grade from a sorting plant is lower than the final tailings grade from the overall process without sorting thereby increasing overall economic recovery.
- If some distance separates the plants and the pre-concentration plant is situated nearer to the ore hoisting shaft or open pit then ore transport costs can be reduced.

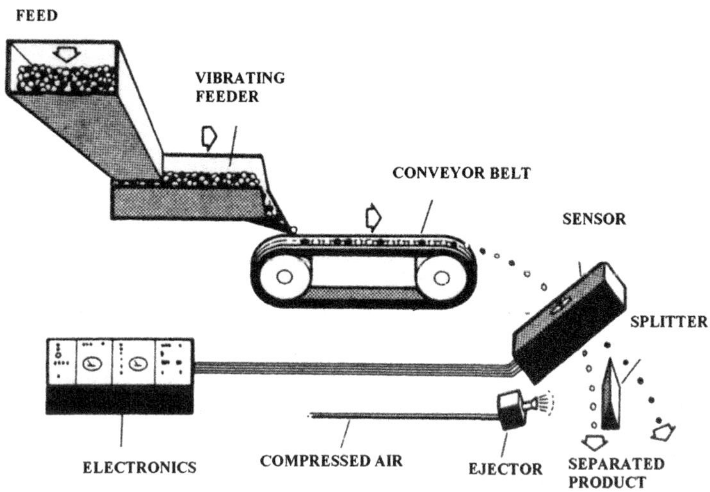

Fig. 8.1 Schematic representation of luminescent sorter with vibrational feeder, conveyer belt, sensor, splitter and ejector

- The application of sorting can increase the life of the mine by increasing reserves to include in-situ and stockpiled previously below cut-off grade material. It also decreases the need for selective mining and often allows the previously stockpiled low grade or waste rock to be processed economically.
- The application of sorting can remove lumps with elevated concentrations of harmful components.
- Because of rich deposits exhaustion and necessity to draw the low grade ores into production the volume of excavated rocks grows sharply and the velocity of waste accumulation constantly increases. Huge quantities of the mining industry waste not only fill the vaster territories of the Earth surface but also influence very badly on the ecological situation. But the ecological damage from this waste may be reduced due to additional products, which could be extracted here, because approximately 20–30 % of useful minerals now are throwing away to the tailing dumps.

At present time luminescent sorters are mainly used for the processing of diamonds, but also for the scheelite, fluorite and others types of ores (Mokrousov and Lileev 1979; Salter and Wyatt 1991; Gorobets et al. 1997). The sources of luminescence excitation in these sorters are X-ray tubes and UV lamps, when the first is more powerful and the second is more selective. UV impulse lasers employment as excitation sources enables to combine the very high power and extreme selectivity with the opportunity to use for the discrimination not only spectral, but also kinetic parameters of luminescence. Besides, other laser based spectroscopic techniques may be used with the same equipment. In the case of sorting the method must be much more effective than in remote sensing, since the distance from the excitation source to the sample is less, background from the daylight is absent and lumps of rock are washed and uniformly oriented. Despite of all this, such kind of sorting is not yet used. The main reason is the extreme complexity of natural raw materials and the absence of systematic investigations of minerals laser-induced spectroscopies.

8.1 Luminescence Sorting

According to spectral-kinetic parameters, the optimal conditions of luminescence excitation and detection, so called selection window (SW) parameters, were calculated by the following way. At optimal for the useful component excitation, the luminescence spectra, decay time and intensity were determined for this mineral and for the host rock. After that, on the personal computer was calculated the proportion between useful and background signals for the full spectral region after each 50 ns after laser impulse. For calculation the spectral band was simulated by the normal distribution low and the decay curve by mono-exponential function. The useful intensity was multiplied by the weight coefficient, which corresponds to the concentration at which this component must be detected.

Relations between luminescence intensity in the corresponding SW and useful components concentration in the lumps were examined statically on an individual basis. For this task pilot equipment was developed, working in polychromator regime with decay time registration in each spectral channel. Part of the channels monitors the luminescence of useful component and remaining part detects the background. Personal computer analyses all information and calculates the contribution of background to the information about mineral of impotence. This pilot allows to make one-source or two-sources excitation, to detect luminescence immediately after laser pulse or after definite time and to account for the matrix effect, when the optical parameters of host rock influence on the luminescence intensity (Gaft et al. 2005).

The main problem for luminescent sorting is the absence of natural luminescence in the case of very many industrially important minerals. In such case artificially induced luminescence may be used. White (1984), for example, described such approach to differentiation of samples. Here the natural luminescence of accept and reject fractions is similar. The basis of separation of magnesium-bearing ore particles is obtained by first conditioning the exposed magnesium-rich mineral with a surface coupling agent of hydroxquinoline, then irradiating the conditioned ore to excite and induce fluorescence and then effecting separation the magnesium-rich minerals from the lean ore particles by detection of the difference of the intensity of the fluorescence. This method detects ores having high magnesium contents on the surfaces of the pebbles. Such approach was also proposed for beneficiation of the precious metals ores (Gaft et al. 2005).

8.2 LIBS Sorting

LIBS is a very promising potential tool for radiometric sorting. In many cases even qualitative analysis may be effectively used in radiometric sorting of minerals. In this sense we do not need to determine a detailed chemical composition of the minerals, but rather the aim is to find the specific lines, which enable to identify the corresponding mineral. It was proved that practically in all cases it is possible to find the emission lines, which enable to differentiate between accepted and rejected elements. Based on minerals LIBS library, a method has been developed for real time detection and content evaluation of minerals or trace concentrations of elements in minerals as they are conveyed on a moving belt, using the intensity ratios of the emission lines characteristic for specific elements or minerals. Because associated minerals always have different chemical compositions, namely major or minor elements, the relative intensities, defined by their characteristic spectral lines, enables all phases to be consistently identifiable and assessed within a short time span which is consistent with both LIBS and the moving belt system (Gaft et al. 2002).

Three main methods of pebbles irradiation and survey are used in radiometric sorting: (a) integral where irradiation and survey zone is bigger than the pebble with maximal dimensions, namely simultaneous analyses of the whole sample; (b) scanning with narrow aperture where irradiation and survey zone have a strip form perpendicular to pebbles movement trajectory, namely passive scanning using pebbles movement; (c) point scanning where irradiation and survey zone is much smaller than pebbles dimensions, where at each time interval a small part of pebbles surface is analyzed, while an active scanning is accomplished in direction perpendicular to pebbles movement.

It is evident that LIBS sorting may not be accomplished by integral irradiation, because it is based upon plasma analyses generated on relatively small area by focused laser beam. Nevertheless, if each pebble is composed of only one mineral, one laser pulse will be representative enough in order to determine the pebble composition. For uniformly embedded ores the situation is different and 2–5 % of the pebble area has to be covered for the confident identification of useful mineral. Non-uniformly disseminated ores demand even more detailed scanning, which has to be determined individually for each specific case. Let consider parameters of laser and detector, which may be suitable for such cases. The following sorting conditions are selected as the mostly representative: the pebbles size of $-20+40$ mm and movement velocity in irradiation zone of 1 m/s. In such case the maximum sample area is 1600 mm^2 and 30–80 mm^2 has to be scanned for uniformly disseminated ores during 40 ms when the pebble will be under laser beam. According to our experience, one laser pulse at 1064 nm with energy of 5 mJ and duration of 8 ns generates plasma from the area of 0.2 mm^2. Contemporary Nd-YAG lasers, for example Quanta-Ray PRO-290, may generate 1000 mJ/pulse with frequency rate of 50 Hz. In such case approximately $0.2 \times 200 = 40$ mm^2 may be sampled during one pulse. Working with 50 Hz frequencies, one pebble will be analyzed twice and 80 mm^2 sampling may be achieved, which is compatible with practical demands.

8.2.1 *LIBS Online Control*

In any industrial production process, there are variability in the raw materials. Periodic laboratory measurements miss the variation of batch parameters and do not allow the on-line control of the production process. Thus any mineral production process is optimized to neutralize variations and to be on the "safe side". In real life scenario very frequently process control is based on experience, visual evaluation, and gut feeling. Detecting material variations in real time enables refinement of the technological process for improvements in efficiency and profitability. Contemporary on-line analysis instruments are improving productivity and adding value to mineral resources throughout the world by providing real-time process data that is automatic, continuous, rapid and accurate. The previously available sensing technologies include nuclear, optical, electrical, magnetic, acoustic, thermal, and various others specific to the ore. Mostly the properties are used, which are indirectly connected with mineralogical composition of accepted and rejected minerals. The two methods which directly determine the elements of minerals and may compete with LIBS in such applications are X-ray Fluorescence (XRF) and Prompt Gamma Neutron Activation Analyses (PGNAA).

8.2.1.1 Competing Techniques: XRF

XRF is produced by bombarding a sample with X-rays. As a result of the interaction, X-rays that are characteristic to the elements inside the sample are emitted and analyzed to determine the elements present within the sample. Low energy X-rays enable relatively simple shielding and easy licensing. The main analytical disadvantage of online on-belt XRF analyzer is that it is not suitable for detection of light elements. If the energy of characteristic emission is less than 5 keV, the interference is very strong because of air absorption and water on the minerals surface. It is claimed that elements starting from an atomic number *more than thirteen* and even eleven may be measured in online mode. Nevertheless, actual industrial applications are known for elements starting from S with $Z = 16$ (Gesing 2007). In contrast, LIBS technology does not suffer the inherent restriction of detecting elements because of their atomic weight and has sufficient analytical ability for online control of all elements on the Periodic Table. Besides, in many cases XRF method is not selective enough, because many elements present with characteristic emission similar to those of valuable components (Table 8.1). Finally, a laser unlike XRF does not create any ionizing radiation.

Comparison between online XRF and LIBS analyzers was conducted on quartz sand with different impurities. The main one is Fe, which is well detected by both

Table 8.1 Characteristic association of metals in rocks (Ostapenko 1990)

Range of XRF spectrum	Element	Interfering element I	II	III
High energy	W	Fe, Cu, Zn, Mo, Sn, As	Pb, Bi	
	Pb	Fe, Cu, Zn, Sb, Ba, Sn	W	Bi
	Hg	Fe, Sb, Ba, As	W	Pb
	Bi	Fe, Mo, Sn, Zn	W	Pb
Average energy	Ba	Fe, Pb	Sb, Sn	Cs
	Cs	Fe		Ba
	Sb	Fe, Cu, Zn, Hg, Pb, As	Ba	Sn
	Zn	Fe, Cu, As, Zn, W, Pb	Mo, Ba	Sb
	Mo	Fe, Cu, W, Pb, Zn, As	Sn, Cs, Ba	
Low energy	Zr	Fe, Ti		
	Rb	Li, Be, Cs, Ta		
	As	Fe, Sn, Sb	Cu, Zn	Hg, Pb
	Zn	Fe, Mo, Sb, Sn, Ba	Fe, As, Pb	Cu
	Cu	Mo, Sb, Sn, Ba	Fe, As, Pb	Ni, Zn
	N		Fe	Co, Cu
	Fe		Ti	
	Cr		Fe	
	Pb	Fe, Cu, Sn, Sb	W, Hg, As	Bi
	W	Mo, Sn	Pb, As, Fe	Cu, Zn

techniques. The comparison demonstrates good correlation with linearity coefficient of $R^2 = 0.87$ (Fig. 8.2). The results of both analyzers have been compared with laboratory XRF device. It was found that the average online LIBS error is approximately 13 ppm, while for online XRF it is approximately 21 ppm. It may be concluded that both analyzers are suitable for this task. But LIBS analyzer was capable to control simultaneously several additional important impurities, such as Ca, Al and K, which may not be detected by XRF in online mode.

8.2.1.2 Competing Techniques: PGNAA

The physical principle of PGNAA systems is that neutron activation of the sample stimulates prompt gamma emission, which is analyzed to determine what elements are present and their concentrations. PGNAA methodology has proven to be effective from analytical and technological standpoints, as it enables all relevant elements' online determination in a real-life scenario. The main problem is its potential health and environmental hazard and the necessity of meeting strict regulatory requirements. The real radiation dose rate is rather low, but not low enough to avoid the need for protective areas around the analyzer imposed by radiation protection authorities. For example, in Germany a distance of at least 4 m

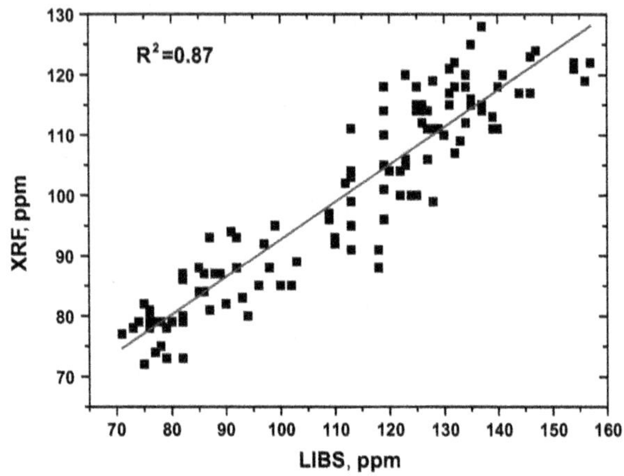

Fig. 8.2 Correlation between XRF and LIBS online analyzers used for real time Fe impurity control in quartz sand

to the analyzer is required to reduce the irradiation rate, because at the neutron source door the dose rate is 4–26 times higher than Europe's acceptable secure dose rate. There are additional problems with PGNAA: the machine is very bulky (minimum weight is 3200 kg); the radioactive source, californium-252, is a highly toxic material and can be used for preparing dirty bombs, so very strict measures to avoid Cf_{252} theft must be taken; there is high potential of personnel exposure during source replacement, thus it can only be done by the analyzer supplier's staff; the average service life of the source is 2.54 years and it has to be replaced every 2 years; continuous exposure to neutrons and gamma rays by the material on the stopped conveyor belt risks residual activation of the mineral or belt (metal lining); the system has to be transported under extreme safety conditions with high transportation cost; paperwork and import procedures (up to 1 year) for reprocessing a worn source; there is a gap under the belt and over the radioactive source, thus continuously irradiated dust will be end up as highly activated nuclear waste whose cleaning may be a safety issue; and difficult licensing, installation, and maintenance because of extensive radiation problems. LIBS technology provides absolute safety for personnel and the environment. It simplifies the implementation of equipment by eliminating the need to obtain a license, maintain strict safety measures, and undergo inspections, which are required for neutron technologies. A mobile mode, where one system is used for several conveyers, is possible for LIBS but not for a PGNAA unit.

From an analytical point of view, PGNAA is characterized by relatively low sensitivity. According to our knowledge, a minimum thickness of material on the conveyer belt is needed, typically 100–150 mm, to make the effective analysis possible. Unlike this, for LIBS a minimum thickness is the monolayer of the particles. Besides, the minimal signal accumulation time for PGNAA is typically 1–2 min, making fast process control impossible. For LIBS technology, the full spectral information is received in every single laser pulse, making short control time intervals possible; for example, 20–30 s in one of our applications. There are several additional disadvantages of PGNAA when compared with LIBS: PGNAA results depend on the weight of irradiated material, and industrial analytical scales have an accuracy of ±10 %, while the LIBS technology does not demand any additional measurement besides the spectroscopic ones; the PGNAA measurement is strongly influenced by Fe and Cl thermal neutron capture from the analyzed material and the conveyer belt itself. For example, approximately 1 % of chlorine saturates the detector. LIBS enables the measurement of any element in the presence of any other elements.

LIBS additional advantage is that it is a laser-based technique, therefore easily combined with other laser spectroscopy techniques, such as time-resolved luminescence and gated Raman. Specifically for the mining industry, such spectroscopic combination would enable analyses of both elements and minerals with characteristic luminescence or Raman signals, while PGNAA and XRF can only analyze elements.

A potential advantage of PGNAA in comparison with XRF and LIBS is its ability to measure the volume and not only the surface. The penetration depth of a laser beam is relatively shallow, no more than approximately 100–120 μm with a single laser pulse burst depending on laser pulse energy and material properties, which is not-comparable with material depth on conveyer belt changing from tenths of centimeters to several meters. So, on a moving belt conveyer only the surface of the material is analyzed while we need to know the impurity concentrations of the volume. Moreover, only tiny spots on the particles' surface are analyzed. Thus the greatest challenge was to prove in ambient conditions of material transported on a conveyor that it is possible to transform LIBS data to quantitative elemental content evaluations, and that high-frequency sampling of surface chemistry is commensurate with the average of the total cross-section and length. It is clear that the LIBS analyzer has to be employed on a surface which statistically represents the whole volume of the material. For industrial control, you need the data for certain time periods, from 30 s to 5 min (approximately 60–600 m of conveyor belt run). For such large quantities of sample, the surface in most cases is statistically relevant, since bulk materials typically get to the conveyor belt after explosive/crushing processes that are random in nature. Its efficacy has been proven by many industrial installations of XRF, which is also of the surface analyzer type, and by our LIBS installations. According to our accumulated experience with powder materials such as iron concentrate and sintering batch, to get the surface representative of the volume you need the following: get material to the conveyor belt via averaging bunker; choose the installation point after the mixing drums; mix the material by inexpensive mechanical tools. The applicability of LIBS surface analysis has been practically confirmed by all current industrial installations.

In practice such comparison was done between LIBS and PGNAA online devices analyzing ash content in coal on the product conveyor at coal Colliery, South Africa (Gaft et al. 2008). It was in line with a Coalscan 9500X PGNAA analyzer, which had been in operation on the same conveyer for several years with a stated standard deviation of $\sigma = 0.5$ %. The conveyer was equipped with an automatic sampler and hourly composite samples were analyzed by the Optimum laboratory. The data from the Coalscan and the laboratory analysis were used to provide an accuracy comparison with the readings obtained from the LIBS analyzer.

The test was accomplished according to standard procedure. The first stage of the final test was 2 weeks long. It was established that the mean of the LIBS unit readings of 13.2 % ash is compared favorably with 13.4 % for the laboratory results and 13.3 % for the Coalscan. The standard deviation for the Coalscan of 0.67 is the lowest of the three determinations. This is evidently due to the high number of values reported by Coalscan and the fact that the unit derives its values from the bulk of the material on the conveyer. The laboratory value of 1.1 exhibits the highest standard deviation and that of the LIBS unit of 0.83 is intermediate to the other two (Table 8.2). A special test was conducted after the 2 week period to compare the performance of the LIBS unit over shorter time interval. The test comprised taking 30 samples from the conveyer, using automatic sampling, at 5 min intervals. To do this, the complete primary sample was redirected to the by-pass chute in the sampling circuit, where it could be collected in a plastic bag. The sample was split, using a riffle chute, into two equal portions. One portion was prepared and analyzed at the Optimum laboratory and the second portion by an independent external control laboratory. The data presented show that the sample means and standard deviations derived from the laboratory results and LIBS unit determinations are quite close. Student's *t*-test was employed to determine whether the means of the distributions are statistically similar. The outcome of the test was positive and confirms that there is a 95 % probability that the two sample distributions were derived from the same population. Thus the LIBS analyzer has proven to be able to accurately determine the quality of coal on a moving conveyor belt in real time. Thus it was concluded that the LIBS analyzer has proven to be able to accurately determine the quality of coal on a moving conveyor belt in real time. The performance of the unit was evaluated and the results were found to be statistically valid.

8.2.2 Challenges in Laboratory to Mining

Many challenges exist in transferring a LIBS system from the laboratory to the moving belt conveyer at a mine. The difference between these environments are numerous and consequential: uniform pellets versus variable sample forms; a fixed laser-sample distance versus a variable distance; stationary samples versus moving samples; averaging many pulses over the surface of a single sample to improve signal-to-noise ratio versus the possibility of only one pulse on a sample; uniform mineral composition versus variable compositions; dry or uniformly wet samples versus varyingly wet samples; a hygienic clean room versus a dusty, harsh and corrosive atmosphere; indoor air conditioning versus heat and humidity; and highly trained technicians versus field operators. Moreover, many means of improving of

Table 8.2 Comparison of ash values determined by online PGNAA and LIBS and control Laboratory data

	PGNAA	Laboratory	LIBS
Number of observations	2866	250	1015
Mean	13.35	13.43	13.17
Max	30.0	17.8	16.5
Min	10.14	10.0	0.827
Standard deviation	0.672	1.064	0.827

LIBS analytical abilities, such as vacuum, buffer gases, picosecond and femtosecond lasers and sensitive ICCD cameras are not practical in a real-life scenario.

8.2.2.1 Analytical Lines Selection

At the first step, the temporal history of the corresponding plasma has to be studied by recording the emission features at predetermined delay times for the gated intensifier, while the intensities of the lines were obtained by integration of the peak area. Such information is very helpful in order to obtain a good compromise between the signal/noise ratio and the emission intensity of the selected lines. It is clear that the proper time delay is related to the energy of the laser, its wavelength, the target characteristics and the surrounding atmosphere (Sabsabi and Cielo 1995). Our results are mainly agreed with known temporal behavior of emissions from different species in the air: ions appear at early times (<2 μs), followed by neutral atoms (2–10 μs), and then simple molecules (>10 μs) and plasma induced luminescence (>50 μs).

In laboratory applications, the intensity of LIBS emission depends on the concentration of the element, because other parameters remain more or less the same. In industrial applications, intensity is influenced by factors such as changes in laser-to-sample distance, different water content and different matrixes. These factors strongly influence plasma parameters, and intensities of specific spectral lines and ratios between them usually vary for the same element concentration. Good understanding of laser-induced plasma properties is a vital factor for any industrial application. Thus, when evaluating any potential industrial task, much effort is invested in determining the best analytical lines to use. Dealing with a very broad range of minerals and rocks, in many cases finding an analytical line is not a simple task. For example, the traditional approach for analytical LIBS is to take a delay time of 100–500 ns in order to quench the Bremsstrahlung emission. Working with low concentrations of certain elements, such as boron, in the presence of much higher levels of iron, the usual analytical lines of B I at 249.6 and B II at 345.1 nm, detected using such a delay time, were nearly totally obscured by Fe emission lines. Additional research revealed that a B III emission line that exists during first 50–100 ns of plasma life may be used for analysis. Besides boron, many other elements, such as Fe, Pb, Ti, Al, Zr and others, also have highly ionized emitting ions. Moreover, during the first 50 ns of plasma life, we found Fraunhofer-type absorption lines that have the potential for direct and calibration-free analytical applications.

At the later end of temporal plasma history, after long delay times very important analytical information was revealed. For example, spectra from diatomic molecules of halogens and alkali-earth elements were detected after a 25 μs delay; these spectra enable fluorine and chlorine detection much more easily than the traditional F and Cl ionic lines. After a delay of more than 100 μs, sometimes plasma-induced luminescence was detected; it is very effective for rare-earth elements identification.

Table 8.3 presents several analytical lines which turned out to be the mostly important for several industrial installations. It was found that practically in all cases it is possible to find the emission lines, which enable to differentiate between accepted and rejected elements. Based on minerals LIBS library, a method has been developed for real time detection and content evaluation of minerals or trace concentrations of elements in minerals as they are conveyed on a moving belt, using the intensity ratios of the emission lines characteristic for specific elements or minerals.

8.2.2.2 Analytical Algorithm Development and Accuracy Evaluation

In order to evaluate a LIBS' potential for on-line analysis, samples with known concentrations of relevant elements are analyzed using a laboratory-scale conveyer and a LIBS analytical setup, imitating the industrial machine. Once the customer provides a detailed description of the technological task for the online stream analyzer, and we decide that it is feasible, he should procure and supply the representative samples of the material from the planned point of installation of the analyzer. This is essential for the initial calibration and evaluation of its accuracy, development of the analytical algorithm, and modification of the instrumentation for the specific material according to technical tasks. This requires selecting samples of each material (every mineralogical type) so that they represent at least 20 different values of each measured parameter, for each prospective point of installation. These values should represent as uniformly as possible the whole variation range of the elements concentrations studied between minimum and maximum values and should differ by not less than the laboratory error.

The influence of laser-to-sample distance, sample humidity and matrix changes on the lines relative intensities are also studied. In order to compensate, at least partially, for these parameters' influence, the laser-to-sample distance is controlled by an distance sensor, and a corresponding adjustment in optics height is made. The moisture content is controlled by LIBS data using H I emission line and an additional device traditionally used for online humidity control. The specific mineral matrix also affects the analytical line intensity, and usually several different minerals are present on the conveyer. Matrix changes may in certain cases be compensated by adjustments in LIBS analysis that is dependent on mineral type. Thus each mineral present in the specific rock is studied separately in order to find its individual breakdown spectrum for the sake of identification.

It is quite clear that the absolute intensity of an emission line is inadequate; two different kinds of line ratios must be found: (1) those that are least insensitive to

Table 8.3 LIBS lines of different elements important for minerals sorting

Element	λ_{emis}, nm
Al I	308.2, 309.3, 394.4, 396.2
AlO	485.0
B I	249.6
B II	345.1
B III	206.6
$CaBO_2$	596
C I	193.1, 247.8
C III	229.7
Ca I	422.7, 430.0, 445.0, 527.0, 558.0, 617.0
Ca II	316.0, 318.0
CaO	606
Cl I	837.6
CaCl	596
Cr I	425.5, 427.5, 429.0
Cu I	324.8; 327.4, 510.6, 515.3, 521.8
F I	685.6
CaF	532.0; 603.8; 604.5
Fe I	302.0, 344.0
Fe II	274.8
FeO	566, 590, 624
K I	766.5; 769.9
La I	492.2, 500.0
LaO	450.0, 560.0, 740.0, 795.0
Mg I	285.2, 383.8, 518.3
Mg II	279.5, 280.2
Mn I	403.1, 403.3, 404.1
NaI	589.00, 589.6
P I	213.5, 214.9, 215.4, 253.3, 255.3
Pb I	257.8, 261.4, 265.1, 266.4, 280.2, 357.2, 405.8
S I	180.7, 182.0, 182.6, 921.3, 922.8, 923.8
S II	542.9, 543.3, 545.4, 547.4
Si I	250.7, 288.1
Ti I	430.6
Ti II	323.4, 334.9
V I	437.9, 438.4, 440.8
Y II	360.0, 370.9, 377.4, 490.2
YO	600.0, 615.0
Zn I	213.9, 330.2, 334.5
Zn II	209.9, 250.2, 255.8
Zr I	360.1
Zr II	339.2, 343.8
Zr III	194.7, 196.3, 197.5, 199.0, 200.3, 203.6, 205.7, 206.1

changes in plasma parameters, to provide a stable and accurate indication of element concentration levels; and (2) those that are relatively sensitive to plasma parameters, to use as plasma parameter indicators. Laboratory-scale LIBS analysis of samples enables determination of the optimal analytical lines and their radiometric combinations. Analytical algorithms are developed to transform the spectra to concentration measurements. This is done to minimize the absolute deviation between spectroscopically predicted and actual (laboratory) analysis of the samples.

The next development task is algorithm improvement as result of industrial tests made with the LIBS analytical unit above the conveyer belt. The measuring device may be used for pure analytical task where the main parameter is the highest possible accuracy. It is important to emphasize that an online analyzer is not a laboratory piece of equipment. Its value is not the specific evaluation of a specific sample, but in the overall use of the analyzed properties to monitor quality levels, trends, changes and so on. The rock composition changes dramatically even during short periods of time, thus even low accuracy continuous measurements are substantially better than more accurate data, that are much less representative. Thus the immediate continuous nature of online analysis represents decisive advantage for technological process control and the accuracy demand is determined by its usefulness for the resulting technology improvement.

Pure analytical accuracy may be determined by comparison with another analytical device measuring the same set of samples. To define the accuracy for online applications is not a trivial task, because the real determination has to be the comparison with another online analyzer, measuring the same material. Nevertheless, this opportunity very rarely exists, so comparison with the existing offline technique – routine analyses of rock samples in the analytical laboratory – is important. It is clear that established laboratory analyses of specific samples are more accurate, while on-line analyses have a substantial advantage because of their continuous nature. Rock composition can change dramatically and quite often there is a large change even over a few feet of deposit depth. Thus even less accurate continuous measurements may be substantially better than more accurate data taken less frequently. In the data described below LIBS average errors (differences between chemical analyses and LIBS values) are calculated for all evaluated elements and compared with the corresponding laboratory errors. However there are many variables that could lead to distinctive errors in laboratory data, mainly sampling errors due to ineffective sampling stations, sample preparation errors and sample analysis errors in laboratory. For example, minimal quantity (Q_{min}) of sample in kg for laboratory control is usually calculated according to Richards-Chechott formula:

$$Q_{min} = k \times {d_{max}}^2$$

where d_{max} – top size in mm and k is coefficient of proportionality which depends on material characteristics: variation in the valuable component content, mineral size, and the component content in the raw material (presently in practice 0,1-0,5-1,0).

For example, for one of our applications in phosphate industry $Q_{min} = 125$ kg, from which representative 50 g have to be separated for analysis. The total error is usually composed of the sampling (15 %), sample preparation (80 %) and pure analytical (5 %) errors. The total relative analytical error in this case was approximately 20 %, while the pure analytical is only 3–6 %.

Therefore a number of tests were conducted to determine the sampling, analytical, and sample-splitting errors in LIBS and laboratory measurements of the same materials. In the first set of tests, the sample of phosphate rock collected was divided at the rotating splitter repeatedly and the total sample was used in making up samples for the laboratory. Up to 10 smaller samples were prepared from the sample taken. These were submitted to the laboratory over several weeks and were not identified as duplicate samples. A second set of tests consisted of two people taking samples at the same time. These samples were submitted to the laboratory at the same time, but were not identified as duplicate samples. It was usually found that the errors of the LIBS analyses are not significantly worse than the laboratory ones. Thus the immediate continuous nature of LIBS on-line control represents a decisive advantage (Gaft et al. 2007).

Another issue is how to determine whether or not the accuracy is sufficient. Because the analyzer is intended for process control, the demand on the accuracy has to be optimized by technologists according to how valuable the result is for the task. Non-technologically justified demand on the accuracy of the online analyzer may prevent its implementation, with corresponding loss of process control and economy.

8.3 Selected Industrial Applications

8.3.1 *Diamonds*

Diamond bearing host rocks are crushed, then the diamonds are separated from the lighter minerals by gravimetric methods such as rotating pans. This produces a heavy mineral concentrate. One of two things is done with the concentrate. The concentrate is blended with water and passed over a grease table to catch the diamonds. The diamonds will adhere to the grease whereas most of the remaining minerals do not. In another option, the concentrate is passed in a dry state through an x-ray sorting machine, which identifies the diamonds by the X-ray induced fluorescence.

8.3.1.1 Luminescence Sorting

Diamond luminescence was studied mainly with the two following aims: to carry out a fundamental investigation of its physical properties and to determine the optimal conditions for luminescent sorting of diamond bearing rocks. For the first task, diamond photoluminescence was studied at liquid nitrogen temperature at which luminescence centers are marked by characteristic zero-phonon lines and are much more informative then at room temperature. For the second task, were diamond is one of the first minerals for which luminescence sorting was used, luminescence properties should be studied at 300 K. On the first stages it was established that X-ray luminescence of A-band always characterizes natural diamonds, while certain diamonds are without photoluminescence under UV lamps excitation. Thus it was concluded that sorting under X-ray excitation is better and such separators are used now in diamonds industry. Such conclusion was not changed after first experiments with nitrogen lasers, which were relatively weak in order to achieve the needed recovery.

It is important to note that the modern X-ray diamond sorters use pulsed excitation. According to statistical analyses, 2 kg of rock contains one diamond crystal and 10^8 particles of other minerals including 10^6 particles of minerals with luminescence under X-ray excitation. The mostly abundant luminescent minerals are plagioclase, zircon and calcite. Luminescence of diamond is selected, first, by spectroscopic filtration, which is very good to "kill" the yellow zircon luminescence, and, second, using temporal behavior of diamond emission under X-ray pulse excitation. The diamond is characterized by two decay components of A-band, short of 100 ns and long of 3–10 ms, while plagioclase has decay time of 10 μs and calcite of 40 ms. During X-ray pulse with 10 ms duration, plagioclase reaches its excited state on 100 %, while for calcite this parameter is only of 1–2 % and for diamond of 10–15 %. Detection is accomplished in time-delayed mode, where it is started 10 ms later after the end of X-ray pulse. During this time the plagioclase emission is totally quenched, while the calcite signal remains on the 1–2 % level. Diamond signal becomes twice lower, but it is well above the noise and may be electronically treated.

At present time new deposits have been discovered which contain diamonds without luminescence under X-ray excitation. Sometimes the portion, which is unrecoverable in X-ray luminescent sorters, is big enough to warrant the development of a new sorting method. The luminescence spectra were investigated under excimer (193, 222, 248 and 308 nm), nitrogen (337 nm) and dye laser (340–360 nm) excitations (Fig. 8.3). It was found that for such diamonds the UV excitation by powerful UV lasers is very effective. The best way is to use the luminescence excitation by two lasers with different wavelength – the excimer laser with the line at 222 nm which excites the A-band, and the nitrogen laser which excites mainly S3 and "578 nm" centers. A-band is present in approximately 95 % of investigated diamonds, but with the big intensity variation. Practically all the diamonds with the weak A-luminescence are characterized by strong luminescence under nitrogen laser excitation. Detection is carrying out at two spectral-kinetic selective windows – between 420 and 480 nm with time delay of 3–9 ms and between 450 and 650 nm with time delay of 1–20 μs. Comparative investigation yielded the following results: in 12 % of the sampling laser excitation is much more effective and may be used even for recovering diamonds from the wastes of X-ray sorters; in 85 % of

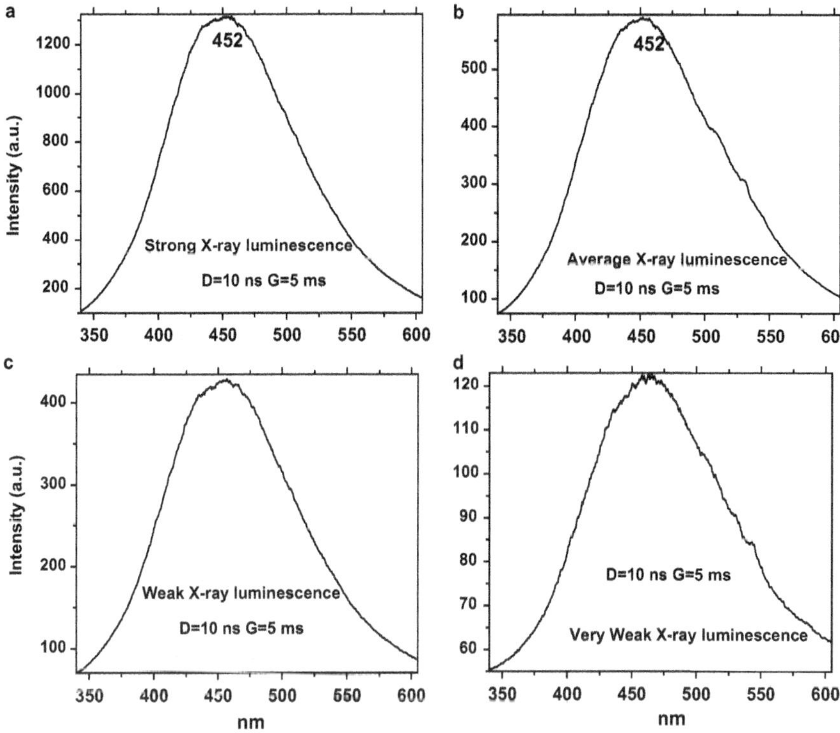

Fig. 8.3 (**a**–**d**) Laser-induced luminescence spectra of diamonds with different X-Rays excited luminescence intensity under excitation by 215 nm

the sampling laser excitation is better and allows for a higher feeding rates and recovery efficiency; in 3 % of the sampling the diamonds were not luminescent under both kinds of excitation (Gaft et al. 2005).

8.3.1.2 Raman Sorting

Raman spectroscopy has great potential for such applications but has been practically used before only for diamonds. Diamond is cubic (O_h^7) with tetrahedral coordination of C—C bonds around each carbon atom. The Raman spectrum of diamond is characterized by only one, very strong fundamental vibration at 1332 cm^{-1}. Gudaev et al. (1999) propose Raman spectroscopy using a CW He–Ne laser at 632.8 nm for detecting a wide variety of diamonds from a selection of minerals which are typically present in diamondiferous ore. However, laser-induced luminescence of diamonds and other minerals often cover the region examined for detecting the characteristic Raman shift of diamond. For this reason, the viability of using pulsed laser-excited Raman spectroscopy, as a method for diamond detection from ore, has been investigated (Lamprecht et al. 2007). Various pulsed laser wavelengths from 266 to 1064 nm were used, as well as CW lasers for comparison. It was found that pulsed lasers offer a decided advantage above CW resulting in superior Raman to luminescence signal ratios.

8.3.2 Industrial Minerals

Industrial minerals are natural minerals and rocks used as raw materials or functional additives in a wide range of manufacturing and other industries. The characteristics of the minerals sold depend upon the physico-chemical characteristics of the ore body from which they are extracted. Processing of the minerals before sale can be simple (mainly crushing, grinding and classifying), but may also be very sophisticated for the most expensive mineral types.

8.3.2.1 Calcium Carbonates

Industrial Calcium Carbonate ($CaCO_3$) is mainly produced by extraction/milling of natural ore. The suitable ore-bodies include chalk, limestone, marble, and travertine. Sufficiently pure ore-bodies are selected to allow direct exploitation. The main mining applications include cement and metallurgical industries where the main task is impurities, such as Si, Mg, Fe and Al control. LIBS is successfully used for this tasks (Fig. 8.4).

Limestone Control

LIBS technology was tested for limestone processing (Frick-Begemann et al. 2010). Of primary importance here is the magnesium content, so as to differentiate pure limestone from dolomitic rock. In the analysis of production samples, a MgO content in the range from 0 to over 20 % was covered. A value of 5 % MgO can be taken as a typical maximum value for the usability of the aggregate for cement and steel production. The measurement process was calibrated with production samples whose MgO content had been determined with X-ray fluorescence analysis. For measurements on moving specimens, the process standard deviation was around 1.3 % MgO, since here measurements are taken at other automatically selected points of the specimens, which do not represent the same spatial averaging as the XRF analyses. The scattering of the individual measured values is not critical for the characterization of the composition of a material stream as here based on temporal averaging, a conclusion can be made with regard to the mean content. How far localized laser direct analysis is sufficiently representative for effective single particle sorting can only be established in sorting tests in realistic conditions. For this purpose, a demonstration rig was set up according to the concept described above, which enables the sorting of production

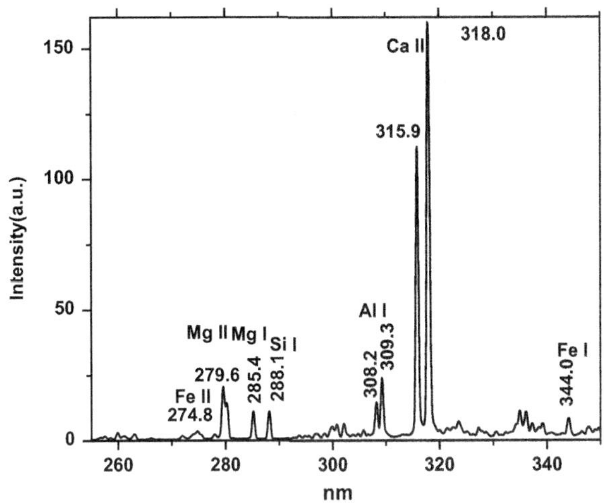

Fig. 8.4 Breakdown spectrum of carbonate

batches in the testing centre. Without further pretreatment, a batch of 435 kg of unknown composition from an Austrian limestone quarry was separated into two fractions on the basis of their MgO content; these can be classed as limestone and dolomite. As a threshold value for the classification, in the following example, a MgO content of 10 % was chosen. The sorted fractions were then comminuted as a whole and homogenized so that based on a representative specimen the average composition of the fraction could be determined with XRF. Most rocks have around 2 % MgO. The distribution soon drops to higher contents, extending to around 20 % MgO. Classification according to the defined limit leads to a classification of 79 % (weight) as limestone and 13 % as dolomite. A content of 8 % could not be classified, mainly because the stabilization zone was too short, so that the particles did not lie still on the belt conveyor. The XRF measurements resulted in a MgO content of 3.6 % for the limestone and 10.5 % for dolomite compared to 4.5 % for the entire batch. If the unclassified content is included in the limestone fraction instead of the return flow, the MgO content does not change significantly. The averaged analysis results of the laser measurements result with practically the same average value for the two classes in contents of 3.1 % and 13.3 %. The somewhat wider spread of the values can in turn be attributed to the inhomogeneities of the samples. Summing up, with production-oriented conditions (3 m/s belt speed, untreated production samples), effective single particle sorting based on laser direct analysis was achieved. With a discharge of only a good eighth of the material, the MgO content of the useful rock could be lowered significantly by one fifth from 4.5 % to 3.6 %. In further tests with other batches and sorting thresholds MgO-depletions were also realized, for example from 5.1 % to 3.7 %.

Drill Dust Control

Carbonates have been analyzed by LIBS in the drill dust (Noll et al. 2014). During drilling in mineral quarries, it is desired to measure inline the composition of the rock being drilled. Whereas analyses inside the drill hole are not EASY, the dust which is generated from the rock can be analyzed. Its composition was measured continuously by a LIBS analyzer which is attached to the dust extraction hose. For this task a transportable LIBS system was designed, which is easily installed at the drill rig and operates automatically. A sealed housing is used which is temperature controlled by air-cooling only. A high stability with respect to vibrations is essential to achieve reliable operation while mounted at a running drill rig. A compact DPSS laser source is employed, operating at up to 50 mJ and 100 Hz. A CCD spectrometer allows adapting the system to a large variety of rock compositions. The measurement results can be directly related to the depth in the deposit using the progress monitor of the drill rig as both dust transportation and analysis are carried out without significant delay. Thereby a depth resolution of a few centimeters is achieved. For evaluation of the measurement results, dust collection samples are taken and processed offline. During a dedicated evaluation campaign, samples were collected every 2 m during the drilling process. A comparison of LIBS and XRF data representing the $CaCO_3$ variation as a function of the drill depth in a limestone quarry was satisfactory. From the depth-resolved data of a series of drill holes a spatial model of the deposit can be calculated. This method can provide a detailed picture of a mineral deposit and thus assist the assessment and optimized excavation.

8.3.2.2 Boron Bearing Minerals

Borates and borosilicates are used in hundreds of products and processes in ways that touch the lives of almost everyone. Major uses include glass production, detergents, agriculture, enamels, and ceramic glazes. Apart from the industrial aspect, boron is an essential micronutrient, integral to a plant's life cycle and, as it becomes increasingly obvious, essential to animal life as well.

Luminescence Sorting

Datolite and danburite have been sorted by neutron separation (Mokrousov and Lileev 1979). These minerals are characterized by strong UV-violet luminescence under laser excitation, which may enable to sort them from quartz host rock. Because decay time of organic matter on quartz and Ce^{3+} in datolite are very close, the differences in excitation and emission spectra may be used. Pilot unit was constructed and transported to the plant (Dalnegorskoe deposit, Russia) and sorting in static regime was tested for the ore preconcentration and for the reprocessing of the coarse waste from heavy media separation (Gaft 1989). All

Table 8.4 Results of datolite $CaB(SiO_4)(OH)$ ore laser luminescent sorting

		Concentration			Recovering
Products before and after sorting	Yield	B_2O_3	$CaCO_3$	Fe_2O_3	B_2O_3
1. **Before**					
Ore (−120 + 70) mm	100	8.4	17.6	4.1	
After					
Middling product	69.6	11.1	14.3	3.0	92.0
Coarse waste	30.4	2.4	25.0	6.5	8.0
2. **Before**					
Ore (−50 + 30) mm	100	7.64	9.9	4.6	
After					
Concentrate	43.0	14.0	7.4	2.2	78.8
Middling product	9.0	5.2	12.0	7.0	6.1
Coarse waste	48.0	2.4	11.8	6.3	15.1
3. Products of liquid separation (−50 + 30)					
Before					
Waste	100	4.3	32.0	2.04	
After					
Middling product	30	10.6	22.2	2.25	74
Coarse waste	70	1.6	36.0	1.95	26

sorting products were analyzed on useful (B) and harmful (Ca, Fe) elements. It was found (Table 8.4) that for datolite ore with lumps size −50 + 30 mm it is possible to receive technological concentrate and waste product. The technological products after luminescent sorting are better then after heavy media separation, which is used on the factory. Besides, 30 % from the old waste may be extracted with the average concentration of useful component more then in excavated ore.

LIBS Sorting

A technological task exists where the lumps of Fe-B ore with low, medium and high B concentrations have to be identified in an online regime. The traditional technology for boron real time detection involves the online determination of B10 concentration via neutron absorption. However, it is not easy-to-implement technology because of safety issues. The potential of LIBS sorting was evaluated using laboratory scale equipment.

Traditional analytical lines for B identification are 345.1 and 249. 8 nm of B II and B I, correspondingly. Nevertheless, if high level of Fe presents in the rock, those lines are difficult to detect because of strong Fe emission interference. The first one may be detected with high boron content, but at average and low levels the Fe emission interference is very strong (Fig. 8.5a, b). From the other side, B III emission is situated in the spectral range which is free from Fe emission (Fig. 8.5c). Thus this line was selected as the analytical one for this task. The representative

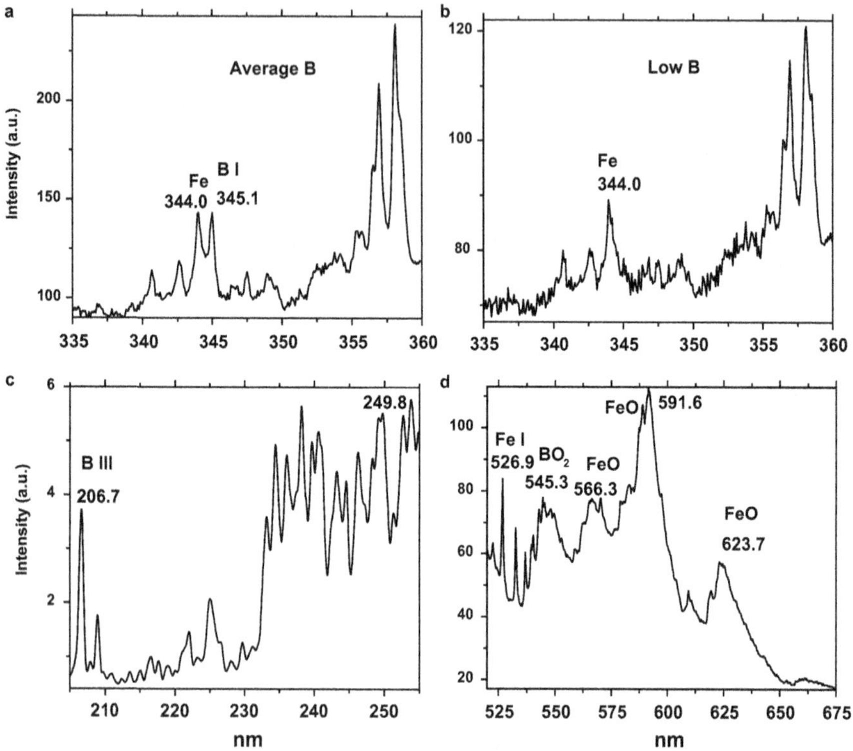

Fig. 8.5 (**a–d**) Breakdown spectra of B-Fe bearing rock with B II emission at 345.1 nm (**a**, **b**), B III and B I emissions and 206.7 and 249.8 nm, correspondingly (**c**) and B and Fe molecules (**d**)

sample containing 18 black-grey lumps with various mineralogical compositions and B_2O_3% levels was studied in a static regime by the laboratory scale LIBS analyzer. A single pulse LIBS set-up was used including a Big Sky Nd-YAG laser (1064 nm, 50 mJ/pulse, 7 ns pulse duration) and industrial gated spectrometer which was set at zero delay and 1 ms gate. It was found that for B-Fe ore lumps five laser shots accumulated from different points of each sample is sufficient for the positive characterization of a lump.

The results displayed in Fig. 8.6 demonstrate the correlation between laboratory data on B_2O_3 concentrations in each pebble determined by neutron absorption technique and B_2O_3 concentrations evaluated by laboratory LIBS set-up. It is clearly seen that products with low, medium and high B_2O_3 concentrations can be separated by LIBS. Only one sample out of 18 with the medium boron concentration (according to the laboratory data) was erroneously identified as belonging to the low level group. From our experience, such the results obtained during the laboratory-scale test look quite encouraging and promising for high separation efficiency required by industrial applications. The weight of each lump is approximately 100 g. For example, using a 1 kHz laser frequency and analyzing 5 points

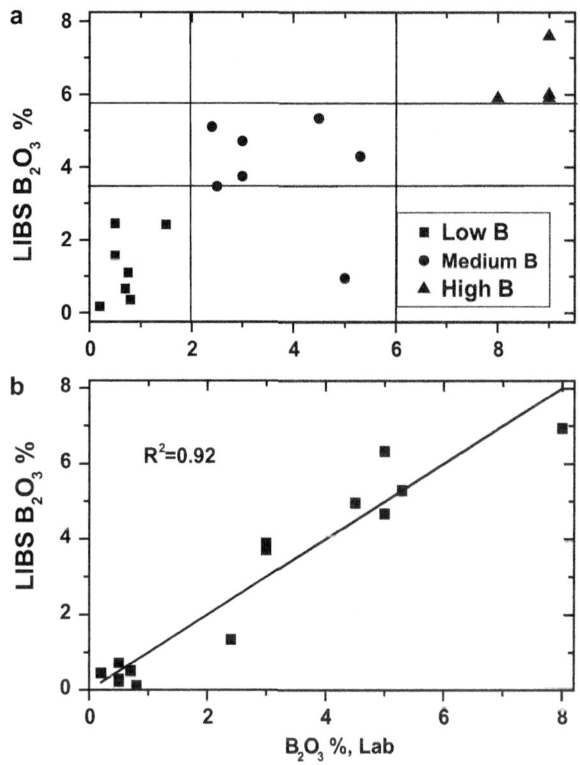

Fig. 8.6 (**a**–**b**) Correlation between online LIBS data based on B III line (**a**) and BO2 band (**b**) with B concentrations

from each sample, it may be possible to sort 200 lumps and, correspondingly, 20 kg of rock per second. The resulting productivity is 72 t/h which is quite reasonable for this specific product.

Another LIBS feature may be also used for sorting B-Fe ore. Figure 8.5d demonstrates breakdown spectra of such samples after very long delay time where molecular emission dominates the spectrum. It was found that BO_2 molecular orange emission peaking at 545 nm is clearly detected after a very long delay time of 75 μs. It is especially effective in DP mode and is easily detected in boron-iron ore. Figure 8.6b presents the comparison between LIBS and laboratory data for 16 B-Fe pebbles with different B concentrations. It is clearly seen that products with low, medium and high B_2O_3 concentrations can be separated by LIBS. Only two samples out of 16 with medium boron concentration (according to the laboratory data) was erroneously identified as belonging to the low-level group. From our experience, such results obtained during the laboratory-scale test are quite encouraging and promising for achieving high separation efficiency required in industrial applications.

8.3.2.3 Microcline and Plagioclase

Feldspar is by far the most abundant group of minerals in the earth's crust, forming about 60 % of terrestrial rocks. The classification of a wide range of rocks is based on their feldspar content. Most frequently, commercial feldspar is mined from pegmatite, feldspatic sands or aplite deposits. For some applications, the ore has to undergo processing in order to remove some accessory minerals, such as quartz, mica, while in other cases their presence is advantageous. The two main properties that make feldspar useful for industry are its alkali and aluminum content. Main applications include the ceramic and glass industries.

For separation between microcline and plagioclase X-ray excited luminescence is used (Gorobets et al. 1997). The main problem is that their emission spectra are very close. Under laser excitation these minerals also demonstrate blue emissions connected with Eu^{2+}, which have close spectral and kinetic parameters. Nevertheless, the differences in excitation spectra (Fig. 8.7) enable effective sorting.

Because the main reason of microcline-plagioclase sorting is to control the Na/K ratio, LIBS may be excellent separation technology. The analytical lines of Na and K are very strong and specific in order to promise confident identification (Fig. 8.8).

8.3.2.4 Silica

LIBS Control

Crystalline silica is the name of a group of minerals composed of silicon and oxygen. The term crystalline refers to the fact that the silicon and oxygen atoms are arranged in a three-dimensional repeating pattern. The main minerals included in this group are: quartz, cristobalite and tridymite. Quartz is the most common of them and, after feldspar, is the second most abundant mineral on the Earth's surface. Crystalline silica, basically in the form of quartz, has been mined for thousands of years. It is really one of the building blocks of ancient and current civilizations, being also vital for modern technologies. Main applications include glass production, foundry, ceramics, building materials, etc. Crystalline silica has also acquired a fundamental place in the information society, as the source of silicon.

One of the mostly difficult problems in quartz sorting is online impurities evaluation in optically pure quartz and feldspar content determination. The main task was to analyze in online mode Fe, Mg, Al, Na, Ca and K minor elements in optical quartz. Figure 8.9 demonstrates that all relevant elements are easily detected by online LIBS combining UV and visible spectral ranges. Sometimes important task is feldspar grains detection containing in optical quartz batch. Such kind of the phase analysis is very difficult for the regular LIBS, but in this case may be done using PIL. The specific feldspar was characterized by intensive Fe^{3+} and Eu^{2+} luminescence (Fig. 8.10). The first one has very long decay time and such emission was effectively detected in laser-induced plasma after long delay time.

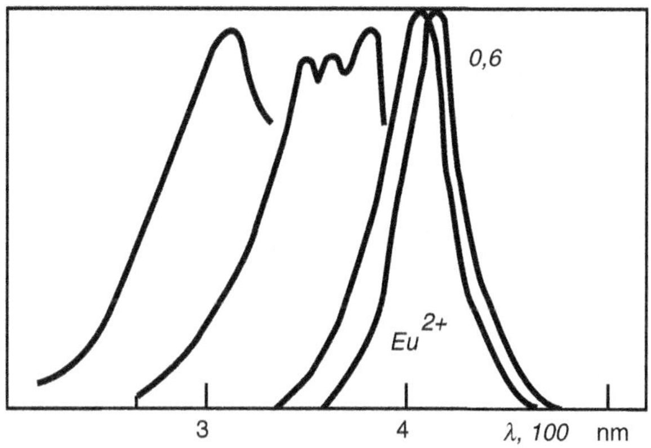

Fig. 8.7 Excitation and luminescence spectra of microcline and plagioclase

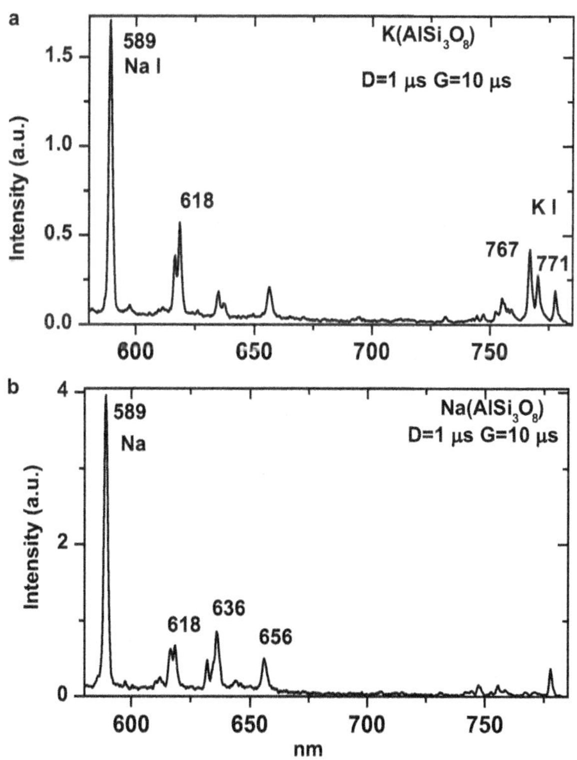

Fig. 8.8 Time-resolved LIBS of (**a**) microcline and (**b**) plagioclase

Second Harmonic Generation Control

Another property of quartz which may be used for its analysis and control, is its ability for Second Harmonic Generation (SHG) (Meisner 1994). The important property of short-pulse lasers is their production of high peak powers. If one

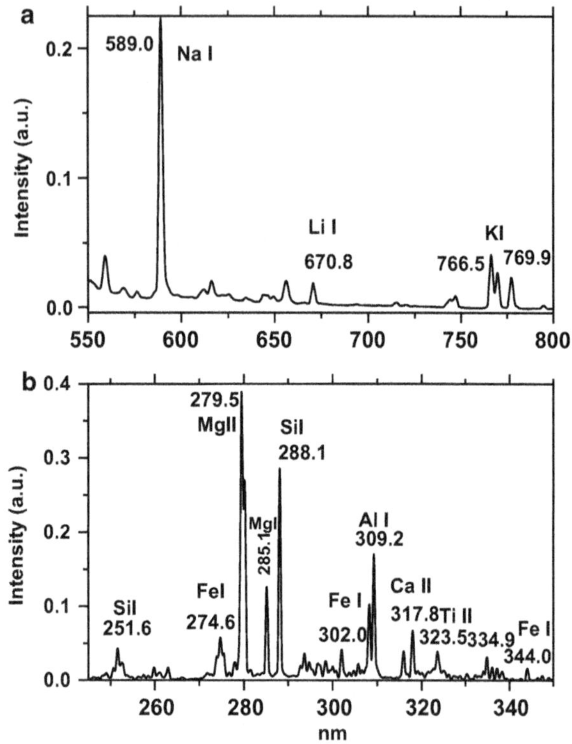

Fig. 8.9 (**a–b**) Breakdown spectra of optical quartz in visible (**a**) and UV (**b**) spectral ranges

considers two independent light beams, which intersect at some point in space, it is common experience that they pass through each other as if the other did not exist. This is a linear process and is observed in everyday life. However, if the light intensity is sufficiently great then non-linear coupling will occur, altering the properties of each beam. Second harmonic generation is one of many effects resulting from non-linear coupling between optical beams. The term "non-linear optics" is used to describe optical processes in which materials react nonlinearly to the electric field of the light incident on them. The result is that new frequencies can be produced and light beams can interact with each other when mixed together.

One common use of SHG is to convert the output of a fixed-frequency laser into a different spectral region. For example, the Nd-YAG laser operates in the near IR at a wavelength of 1064 nm, while SHG is routinely used to convert the wavelength of the radiation to 532 nm. For $\chi^{(2)}$ processes, the conversion efficiency can be up to 30 % for phase-matched case with a nanosecond laser pulse.

The study of nonlinear optical characteristics of minerals began in 1961 when Franken and his collaborators first observed how the red beam of the ruby laser with the $\lambda_1 = 694$ nm partly transformed to UV with the $\lambda_2 = \lambda_1/2$ after its passage through quartz crystal. Certain minerals besides quartz may be a non-linear medium. Prior to the invention of the lasers, the field of experimental nonlinear optics was virtually nonexistent, but growth in this area of spectroscopy has been

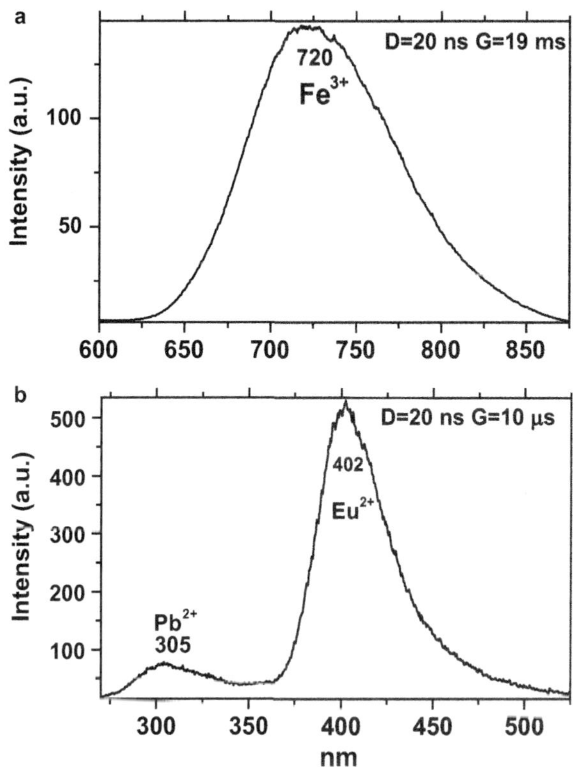

Fig. 8.10 (**a–b**) Laser induced luminescence of Fe3+ (**a**) and Eu2+ (**b**) in feldspar

dramatic, with a variety of nonlinear optical phenomena finding use daily. Using coherent spectroscopy and the method of Kurtz the space symmetry groups for 50 mineral species have been refined. A laser method for the rapid determination of the concentration of quartz and other non-centrosymmetric minerals in powder samples and thin sections was developed. The property of the quartz very effectively to convert IR laser irradiation at 1064 nm to green light at 532 nm (Fig. 8.11) was used in the area of mineral separation, for example, in order to remove minerals of feldspar group, which do not double the frequency of the laser radiation, from quartz-feldspar raw material. Quartz, which contains gas-liquid inclusions and quartz, which does not, differ both in the intensity of the SH and in the angular distribution of the intensity. This phenomenon nay be used in quartz separation. Certain perspectives were gained for the gold-containing ores, i.e. those in which gold associates with quartz (Meisner and Kuz'min 1986; Meisner 1994).

8.3.2.5 Fluorite

Fluorine for industrial use is almost exclusively obtained from fluorite. There are many possible fluorite-ore types. It occurs as fissure-fill and manto-replaced types,

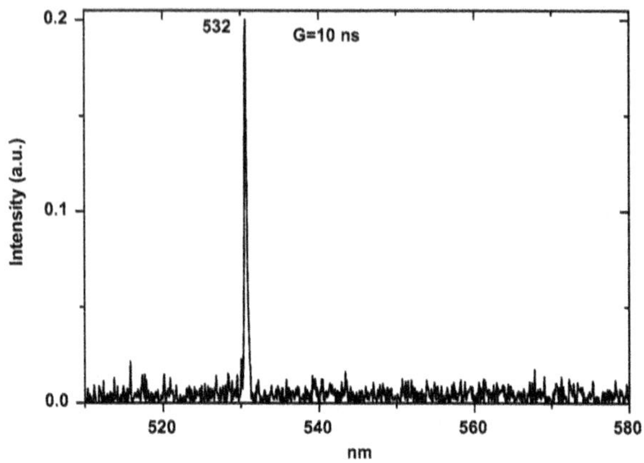

Fig. 8.11 Second harmonic generation by natural quartz under 1064 nm irradiation

and in pegmatites, residual/eluvial deposits, carbonates and phosphates deposits. It is a gangue mineral in many base metal deposits, but due to stringent trace metal requirements often can't be economically beneficiated to a saleable product. Most fluorite ores are selectively mined, and ore from some minor vein deposits is then upgraded to a saleable product by sorting.

Luminescence sorting of fluorite is well known technique, which is based on strong blue luminescence of Eu^{2+}. LIBS technique may be useful for fluorite ores sorting. Long lived emission of CaF molecule is a very effective distinctive feature of this mineral.

8.3.2.6 Barite and Gypsum

LIBS and Raman Control

Barite belongs to bulk minerals and is very rarely used as a source of metallic barium, but it is used in drilling fluids, paints, glass and as filler in paper products. Conventional open pit methods and processes usually selectively extract barite by crushing and pulverizing. High-grade ores may not require any further concentration prior to sale. For lower grade ores luminescence sorting based on UV band at 350 nm under X-ray excitation may produce a coarse-grained concentrate (Gorobets et al. 1997). As the drilling industry will not accept barite containing galena and other sulphides, or siderite, a saleable concentrate cannot usually be made from the barite gangue in base metal deposits. It may be proposed that LIBS and Raman methods may be effectively used for radiometric sorting of barite because of strong and characteristic emission lines of Ba and strong Raman bands of barite (Fig. 8.12a, b).

Most gypsum $CaSO_4 \times 2H_2O$ is used in building construction, to make plasterboard or as a setting retardant in Portland cement, but there is also substantial usage

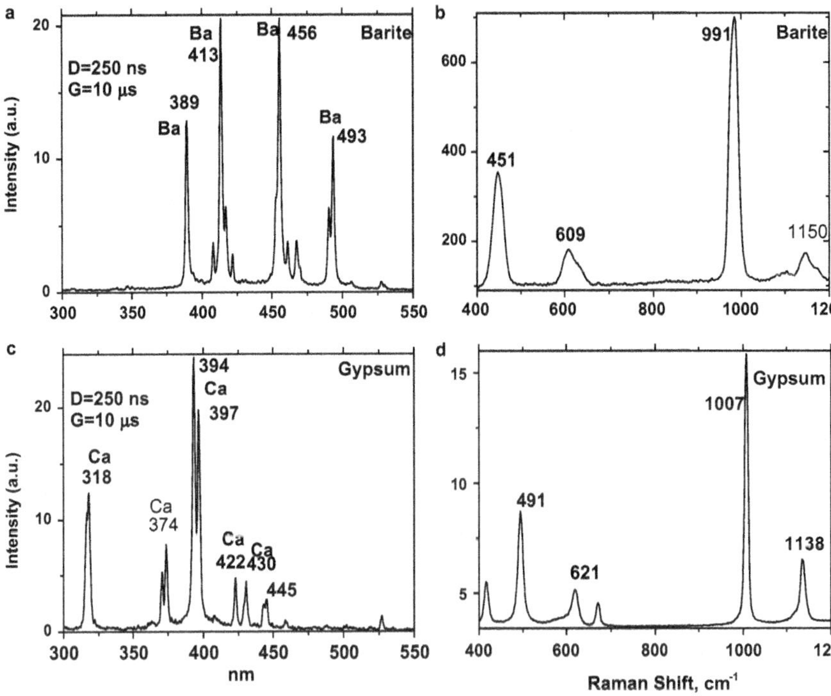

Fig. 8.12 (**a–d**) Breakdown (**a**, **c**) and Raman (**b**, **d**) spectra of barite and gypsum

in developed countries as a soil conditioner. Gypsum sorting may be made using its strong and characteristic Raman signal (Fig. 8.12c, d).

LIBS of Sulfur

Another spectroscopic feature which may be used for those minerals identification is sulfur emission. Nevertheless, there are significant challenges to detect sulfur with LIBS. Only a few spectral lines of sulfur are suitable for its detection and quantification. Strong emission lines occur either in the vacuum ultra-violet (VUV) or in the near infra-red (NIR) spectral region. For sulfur included in sulfide minerals most publications have reported the recording of the 180.7, 182.0 and 182.6 nm VUV spectral lines. The NIR spectral lines at 921.3, 922.8 and 923.8 nm were used for S detection in building materials. The demands for the on-line mineral analysis exclude measurements in vacuum, because it has to be accomplished in ambient conditions (Gaft et al. 2009b). This is a critical challenge for LIBS technology, especially because its main competitors in the on-line mineral analysis field, such as Prompt Gamma Neutron Activation Analysis and X-Ray Fluorescence, are effective in on-line sulfur evaluation. It was found that sulfur bearing minerals, such as

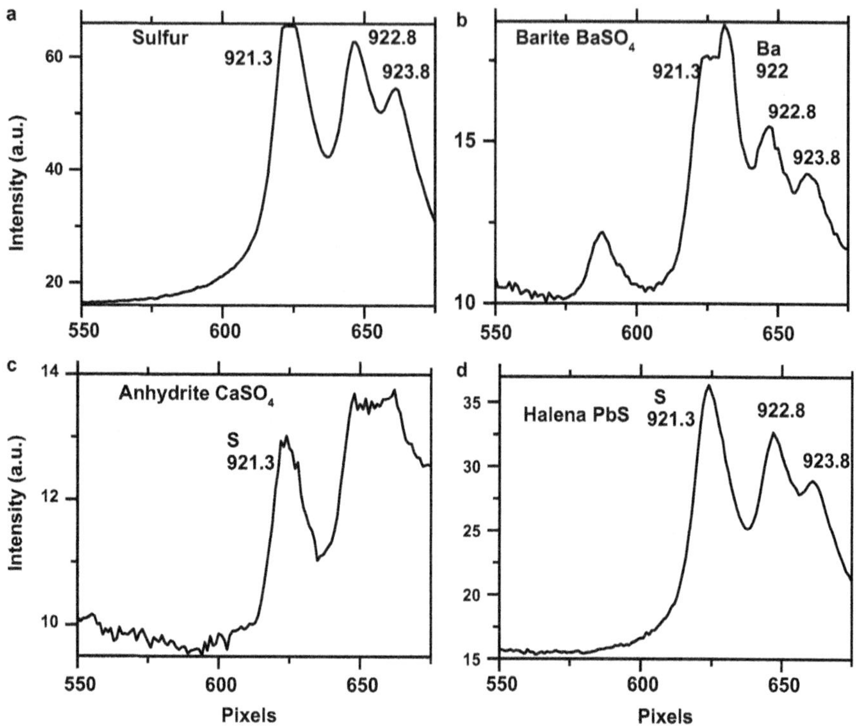

Fig. 8.13 (**a–d**) Breakdown spectra of sulfur bearing minerals in NIR spectral range

barite, anhydrite and galena has relatively strong emission lines in IR which may be used for those minerals online control in real life scenario (Fig. 8.13).

8.3.2.7 Bauxite and Alumina: Al

As the Al industry continues to grow the concentrations of impurities in raw materials are increasing with significant implications to meet customer chemical specifications. The mostly important impurities are Cr, Ti, Ni and V of the coke and Fe, Si, Ga, Zn of the alumina. These impurities must be removed in order to get good quality aluminum, because they make it brittle and liable to corrosion. In some smelters the concentration of these impurities is beginning to exceed customer cast product specifications. Rapid and automatic analysis allows industrial minerals producers to promptly detect changes in incoming raw materials, allowing them to take appropriate action in the process streams.

For example in Russia, the absence of high quality local bauxites, with silicon module (A/S) up to 10 with corresponding 50 % Al_2O_3 content, and strategic reasons led to creation of unique technologies to process domestic low grade bauxites with A/S in 3–6 range. The term bauxite is used for naturally occurring mixtures of aluminum monohydrate (boehmite or diaspore) and trihydrate gibbsite $Al(OH)_3$, including impurities which are typically clay minerals, free silica, iron hydroxides and titania. Thus online bauxite control enables to detect bauxite parts with low aluminum contents and to remove them from the following beneficiation process.

The applicability of radiometric sorting by XRF technique of Russia bauxite rocks has been extensively studied (Gorobets and Rogojine 2001). It was proved that for $-100+50$ and $-50+25$ fractions of North Ural deposit it is possible to remove carbonate from bauxite using intensity lines of Ca and Fe. As a result, the CO_2 content was lowered from 18.0 % to 4.1 %, while the A/S was increased from 10.2 % to 14.1 %. For Timanskoe deposit, the main task was to improve the A/S of the low grade hematite-boehmite bauxite. It may be done by separation of minerals with high Fe and Si contents, mainly chlorites, using characteristic Fe emission. It was proved that A/S may be raised by 1.6.

Another source is non-bauxitic raw materials such as nepheline $Na_3K(AlSiO_4)_4$ and alunite $KAl_3(SO_4)_2(OH)_6$. At the moment 4 major bauxite mines and 2 nepheline sources/mines are in operation in Russia. The Al_2O_3 content is substantially lower than in bauxite ores and in order to produce 1 t of Al, usually 3–4 t of bauxite is enough, while 7–8 t of nepheline and alunite are needed. Nevertheless, nepheline ore is a part of apatite-nepheline, nepheline sienite and urtite deposits. It enables complex process with nepheline concentrate used for alumina and apatite concentrate for fertilizers production. Alunite is also complex source, containing alumina, sulfur anhydrite and alkali Na and K metals. Once again, all those elements may be definitely detected and controlled by online LIBS technique.

Good XRF radiometric sorting results prove that the corresponding rocks are very contrast and online process control may be very effective. The main potential advantages of LIBS technique compared with XRF are the following. XRF it is not suitable for the light elements detection, usually with atomic number less than 20. If the energy of characteristic emission is less than 5 keV, the interference is very strong because of air absorption and water on the minerals surface. Thus very relevant elements, such as Al and Si, may not be detected directly and all control may be accomplished using accessory elements analysis, such as Ca and Fe. LIBS can detect all elements relevant for online control, including Al, Si, Ca, Fe and Ti (Fig. 8.14).

Aluminum belongs to elements with strong molecular emission which may be effectively used for analytical purposes (Fig. 8.15).

Bauxite is used for refractory materials production which is used mainly in the steel industry for transport ladles and electric arc furnace roofs, and in the cement and lime industry for rotary kiln linings. Besides, there are three aluminium polymorphs used in refractory industry, namely sillimanite, andalusite and kyanite, which share the same simple chemical formula (Al_2SiO_5). Those minerals differ in their crystallographic structures and stabilities with respect to both pressure and temperature and sillimanite is most desirable as a refractory raw material, as it fires with little or no change in volume and requires no pre-firing. LIBS is very helpful in

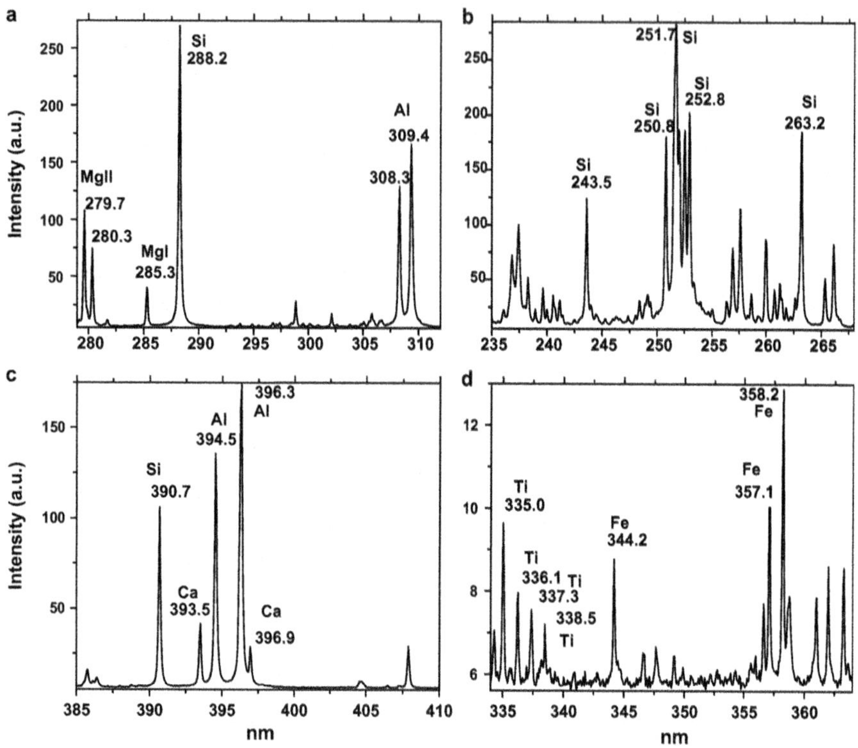

Fig. 8.14 (a–d) Breakdown spectra of Al bearing minerals

such case, because not only elemental, but also mineralogical composition may be evaluated (Fig. 8.16). It may be done using plasma induced luminescence (Gaft et al. 2011).

8.3.2.8 Magnesium Ores: Magnesite

Figure 8.17 presents high resolution breakdown spectra of the magnesium bearing minerals which are the mostly important for refractory industry. Usually LIBS is considered as an elemental composition analyzer, but in this specific case even mineralogical information may be received. Magnesium plasma emission contains both atomic (Mg I) and ionic (Mg II) lines, while their intensities ratio depends on mineral thermo-mechanical properties. Besides, at certain Mg level the reabsorption process begin with "negative" lines appearance. It is especially pronounced in magnesite with highest Mg concentration. Thus LIBS gave opportunity not only to analyze the Mg content, but also to evaluate the mineralogical composition, which is a very important advantage compared to XRF and PGNAA. Figure 8.18 presents magnesite spectrum with larger spectral range where the

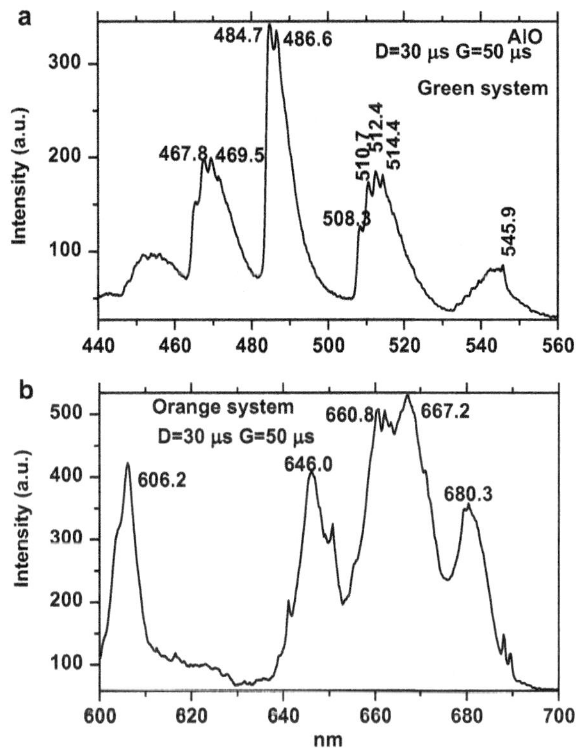

Fig. 8.15 (**a–b**) Breakdown spectra of molecular AlO emission: green (**a**) and orange (**b**) systems

mostly important "poisonous" impurities silicon and calcium are definitely detected.

At a crushing and concentration plant, a LIBS analyzer is used to grade the crushed ore fractions of up to 150 mm based on Ca and Si online analysis. Concentrations of Ca and silicon determines the material quality and depending on CaO and SiO_2 content in the portion of material passing along the conveyor during 60 s, the SCADA system, realizing control of the automatic gate on the basis of the operator's order and data received from LIBS analyzer, determines where (to what bunker) this portion of the material should be directed. The accuracy and reliability of the analyzer's readings were well confirmed in an application tasked with determining whether silicon oxide content exceeded the threshold value of 1.45 % (corresponding to 2.9 % silicon after losses of 50.6 % during burning for raw magnesite). The raw magnesite was graded into a bunker, and burning it yielded a batch of fired periclase (fireproof powder) with a silicon oxide content of less than 3 %. The relative deviation between the analyzer's readings and laboratory data was found to be less than 5 %. Such deviation is not worse than the accuracy of the laboratory analyses themselves. By implementing MAYA, the capabilities to reject the 10–15 % of material with non-corresponding parameters from the flow of qualitative raw magnesite and to extract an additional 10–30 % volume of

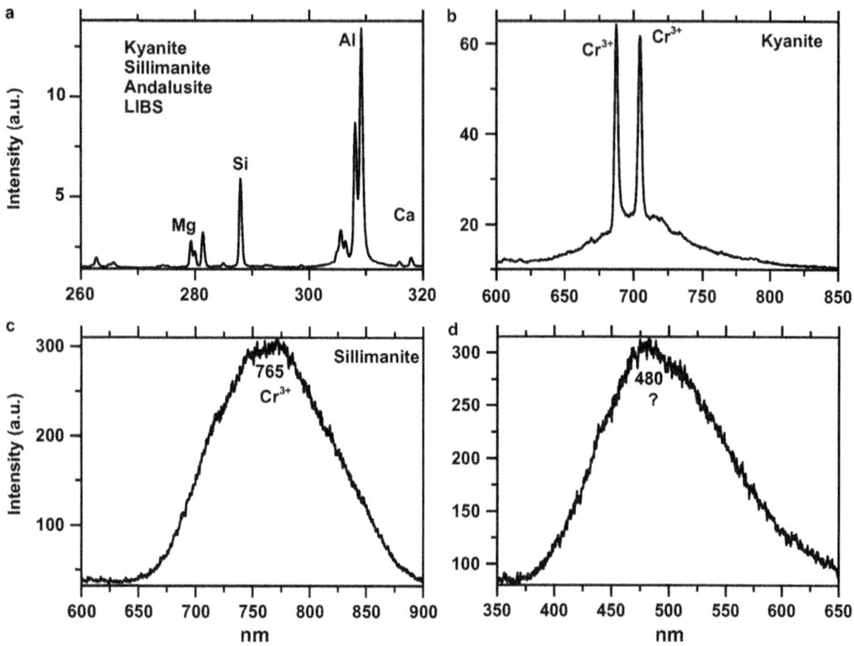

Fig. 8.16 (**a–d**) Breakdown spectra kyanite, sillimanite and andalusite (**a**) and their plasma induced luminescence (**b**, **c**, **d** correspondingly)

high-quality raw material from the flow of the medium grade crushed magnesite were realized (Gaft et al. 2014a).

8.3.2.9 Manganese Ore

One example we studied is an Mn-Fe ore which has to be sorted piece by piece based on Mn concentration. Online LIBS analyzer enables to definitely detect the characteristic Mn and Fe emission lines with 50 mJ Nd-YAG laser (1064 nm) excitation (Fig. 8.19a). Correlation between online-determined and laboratory data of Mn and Fe concentrations (Fig. 8.19b, c) enables to divide the individual lumps on waste (0–10 % Mn), intermediate product (10–30 % Mn) and concentrate (more than 40 % of Mn). To evaluate LIBS' lump sorting industrial potential, we determined how many analytical laser pulses are needed for proper Mn content evaluation and sorting according to predefined categories. The results demonstrated that an analysis made by five laser pulses had approximately 75 % accuracy to identify the correct grade for high and low grades of Mn-bearing lumps of 3–5 cm size. Three laser pulses give the correct result with 70–73 % of correct identifications. Five laser pulses were selected as a reasonable number for each analysis of a lump sorting task as a compromise between sorting speed and analytical precision. For

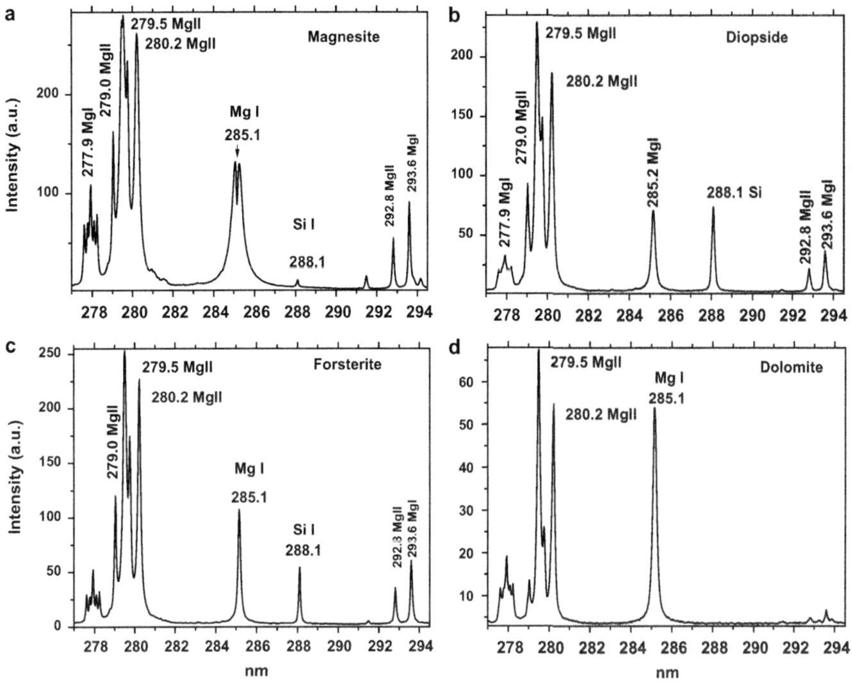

Fig. 8.17 (**a–d**) High resolution breakdown spectra of different Mg bearing minerals: magnesite (**a**), diopside (**b**), forsterite (**c**) and dolomite (**d**)

Fig. 8.18 Breakdown spectra of magnesite ore

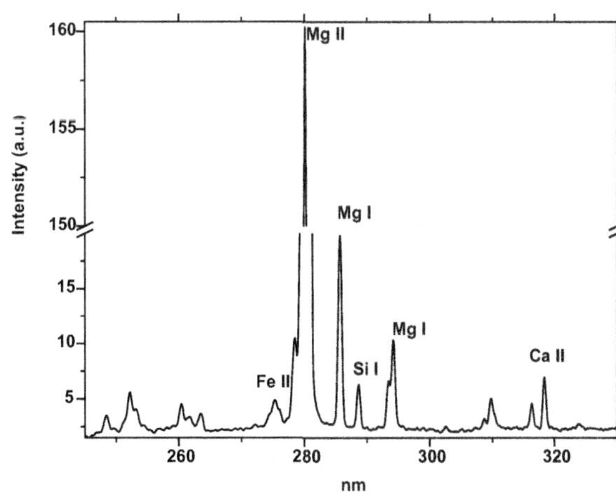

example, with 200 Hz repetition rate, 40 lumps of 3–5 cm size may be analyzed per second – 144,000 per hour, which translates to 43.2 tph. Using two such lasers the throughput may be increased up to approximately 90 tph, which is quite acceptable for such applications.

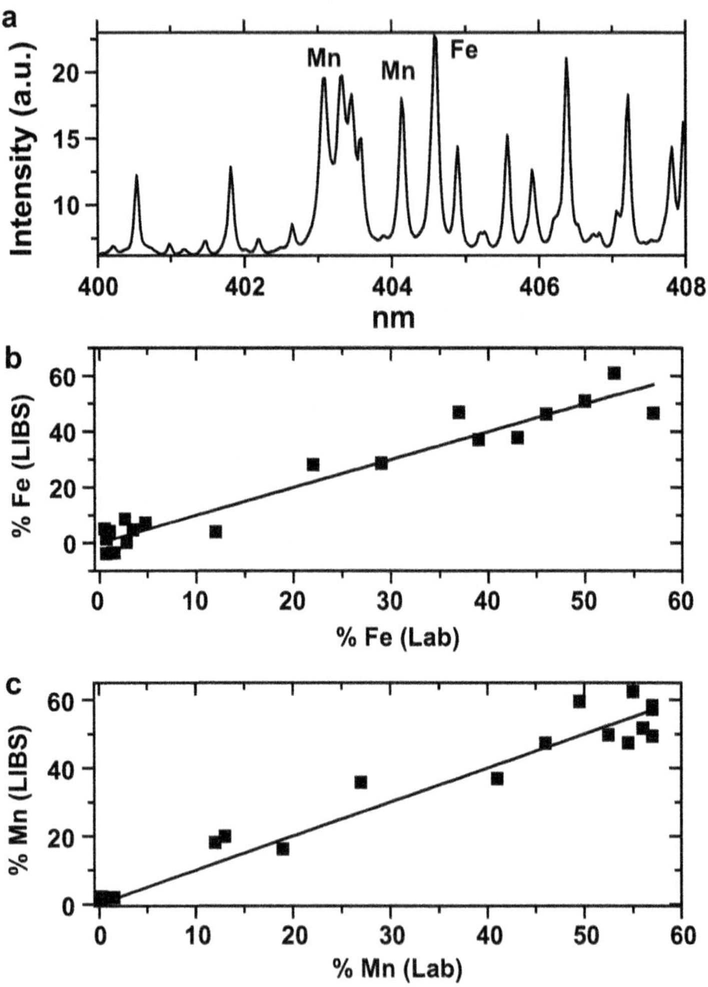

Fig. 8.19 (**a–c**) Breakdown spectra of manganese-iron ore (**a**) and correlation between online LIBS and traditional laboratory data on Fe (**b**) and Mn (**c**)

8.3.2.10 Phosphates

Among the deleterious materials (Fe_2O_3, Al_2O_3, CaO, MgO) in phosphate rock, dolomite is the most troublesome. Dolomite causes higher consumption of sulfuric acids, reduces filtration capacity, and lowers the P_2O_5 recovery in fertilizer manufacturing. Therefore the MgO content is a very important index in evaluating the quality of the phosphate concentrate. In Florida, the MgO in final phosphate concentrate is usually required to be less than 1 %.

Phosphate rock is one of the basic materials for fertilizers. USA phosphate production represents approximately one third of the world's total, while about

70 % of the US production comes from Florida. With the depletion of the low-dolomite, easy-to-process Bone Valley siliceous phosphate deposits, phosphate mining moves further south and southeast, and the phosphate matrix becomes leaner in grade and high in dolomite. The phosphate deposits in Southern Extension will be low in P_2O_5 and high in dolomite impurity, and may be divided into two zones. The upper zone may be processed using current technology, while the lower one is highly contaminated by dolomite. Geological and mineralogical data show that about 50 % of the phosphate resource would be wasted if the lower zone is bypassed in mining, and about 13 % of the resource would be lost if the dolomite pebble in the lower zone is discarded (Gao et al. 2002).

The high dolomite zones have been bypassed during mining operation because there is no technology available to process the high dolomite pebble economically. It has been well recognized that the development of technically and economically feasible beneficiation technology to process dolomite pebbles is very important in order to improve the recovery of phosphate value, extend Florida phosphate reserves and prolong the life of Florida phosphate industry.

This makes it more critical to control the quality of the pebble product, particularly the MgO content. Current practices require either stockpiling of pebble product until quality control data from sampling become available, or making the shipping or discarding decision based on visual observations of rock as it is being produced. Obviously, this practice can result in the shipping of undesirable products to the chemical plant or the discarding of acceptable pebbles. A reliable instantaneous analytical method is the ultimate solution to this problem.

Previously, gravitation separation and photometric sorting are mainly used for the treatment of phosphates with high dolomite concentration (Salter and Wyatt 1991). However, their effectiveness in many cases is limited due to minimal density and reflectance differences between dolomite containing lumps and phosphate rock.

Luminescence Sorting

X-ray excited steady-state luminescence has been checked as distinctive feature for the phosphorites, Karatau (Gorobets et al. 1997). Sedimentary apatite (francolite) is not luminescent under X-ray excitation and red Mn^{2+} luminescence of dolomite was employed. As a result, concentration of MgO in accepted fraction was lowered from 5.0 % to 2.4 %. The use of this method is limited because in many cases X-ray luminescence of dolomite is very low or non-existent. Besides that, other minerals without luminescence come together with accepted phosphate fraction, including siliceous shale, which is harmful for the following deep beneficiation. Thus new methods have to be developed for detection and content evaluation of dolomite in Florida phosphates on moving belt conveyors.

Our study of sedimentary apatite from Israel proved that laser-induced time-resolved luminescence is a perspective tool for evaluation of sedimentary phosphate ores with high dolomite content (Gaft et al. 1993b). The idea was based on the fact that natural apatite contains several characteristic luminescence centers, which enables to differentiate it from dolomite. The mostly widespread characteristic luminescence center in sedimentary apatite is uranyl $(UO_2)^{2+}$ with typical vibrational green band luminescence under nitrogen laser excitation (Fig. 8.20a, b). Nevertheless, it appears that such luminescence is absent in phosphate rock samples from Florida, evidently because of extremely low uranium concentration (Fig. 8.20c, d). In order to find potential luminescence centers, ICP-MS analyses of Florida phosphates was accomplished. From discovered REE, theoretically Dy^{3+} is the best candidate because of favorable combination of concentration, quantum efficiency, and spectral and temporal characteristics. Narrow lines of Dy^{3+} with long decay time of 500–600 μs appear under $\lambda_{ex} = 355$ nm in Florida apatite, while they have never been detected in Florida dolomite (Fig. 8.20e, f). The reason is that dolomite is very bad matrix for luminescence centers formation, even if he has suitable concentrations of corresponding impurities.

In the second stage, more than 200 samples of randomly chosen Florida phosphate rock samples have been analyzed for Dy^{3+} luminescence. The results prove the assertion, that Dy^{3+} lines are a good distinguishing feature between apatite and dolomite. They are not detected in a single dolomite and it allows 100 % detection of dolomite pebbles. The problem is that Dy^{3+} luminescence intensity is strongly varying from apatite to apatite and in certain samples it was not confidently detected under existing experimental conditions. The average content of apatite with confident Dy^{3+} luminescence is slightly less than 70 %. To improve the distinguishing feature efficiency, the optimal excitation has to be chosen. It is well known that Dy^{3+} has very narrow excitation bands. In order to determine the best excitation source of Dy^{3+}, the excitation time-resolved spectroscopy has been accomplished. For this task the tunable dye laser pumped by excimer XeCl (308 nm) was used. The selected dye enables the tuning in the 345–365 nm spectral range, which corresponds to the $^4F_{9/2}$ excited state of Dy^{3+} responsible for luminescence lines at 480 and 573 nm. It was found that the optimal excitation for Dy^{3+} in apatite is at 351 nm, which enables to increase Dy^{3+} luminescence by 25–30 % compared to $\lambda_{ex} = 355$ nm. Besides to be the optimal excitation for Dy^{3+}, the excitation at 351 nm exactly corresponds to the emission wavelength of XeF laser.

LIBS Online Sorting

This methodology was previously successfully used for express on-line analyses of phosphate ores. The commercial TRACER™ 2100 Laser Element Analyzer, a laser-induced breakdown spectroscopy instrument, was utilized for rapid analyses of phosphate ores at the mine site (Rosenwasser et al. 2001). Excellent calibrations were achieved for P and Mg with correlation coefficients significantly above 0.98. The instrument demonstrated strong potential of the LIBS for use in on-site, real-time or grading. LIBS apparatus was developed for applications in the phosphate industry (Smith et al. 2000). This includes rapid elemental analysis of P, Mg, Si and Fe in materials taken from various stages of phosphate mining and processing using a rapid, off-line measurement approach. A compact LIBS spectrometer was

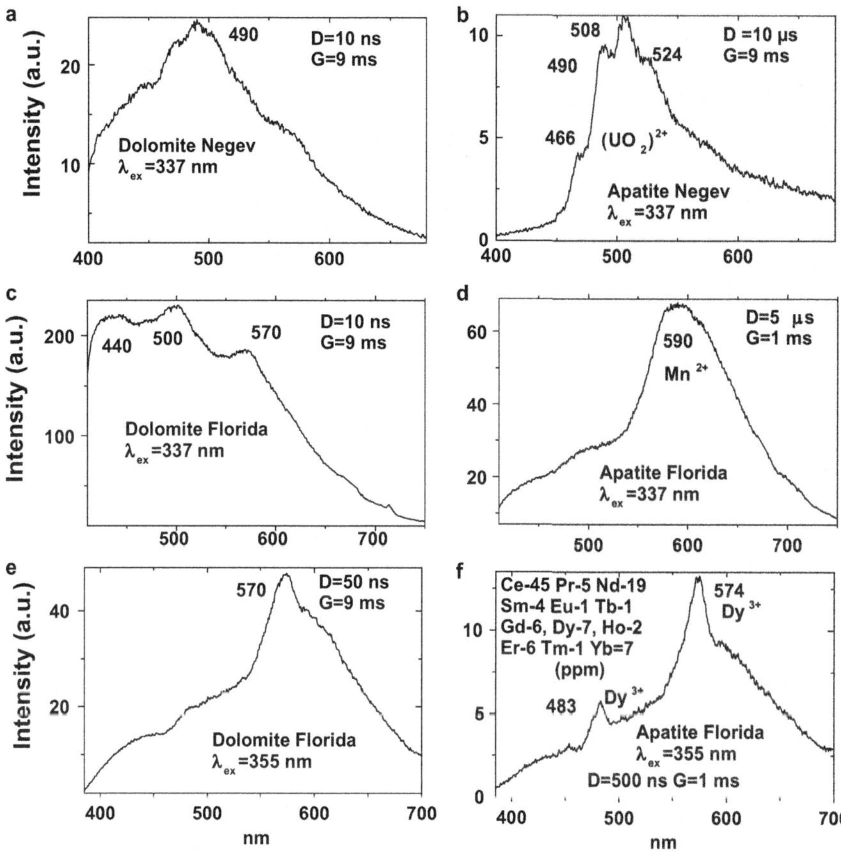

Fig. 8.20 (a–f) Laser-induced time-resolved luminescence spectra of dolomite and apatite

developed and evaluated in both the laboratory and in an industrial setting. The samples were dry fine powders deposited on an adhesive tape. Nevertheless, such equipment is developed for chemical analysis with analysis time of several minutes per sample and is not suitable for online control. The attempt was made to develop an LIBS analyzer for routine, rapid pseudo-on-line and real-on-line performance with the desire to apply LIBS to on-line slurry characterization. Nevertheless it was concluded that on-line quantitative chemical analysis in phosphate slurries does not appear to be practical at this time using LIBS. The reasons are the amount of water in the slurry and the large sizes of the particles analyzed (Winefordner 2000).

In order to use LIBS for real-time evaluation of dolomite content on the moving belt apatite and dolomite have to be confidently identified by LIBS with one laser pulse. Figure 8.21 demonstrates typical breakdown spectra of Florida apatite and dolomite under 355 nm Nd-YAG excitation with energy pulse of 15 mJ. Dolomite (Ca, Mg)$(CO_3)_2$ contains Mg as major element and does not contain P and F, and phosphate $Ca_5(PO_4)_3(F,O)$:$Ca_5(PO_4)_3CO_3$ contains as major elements P and F, and

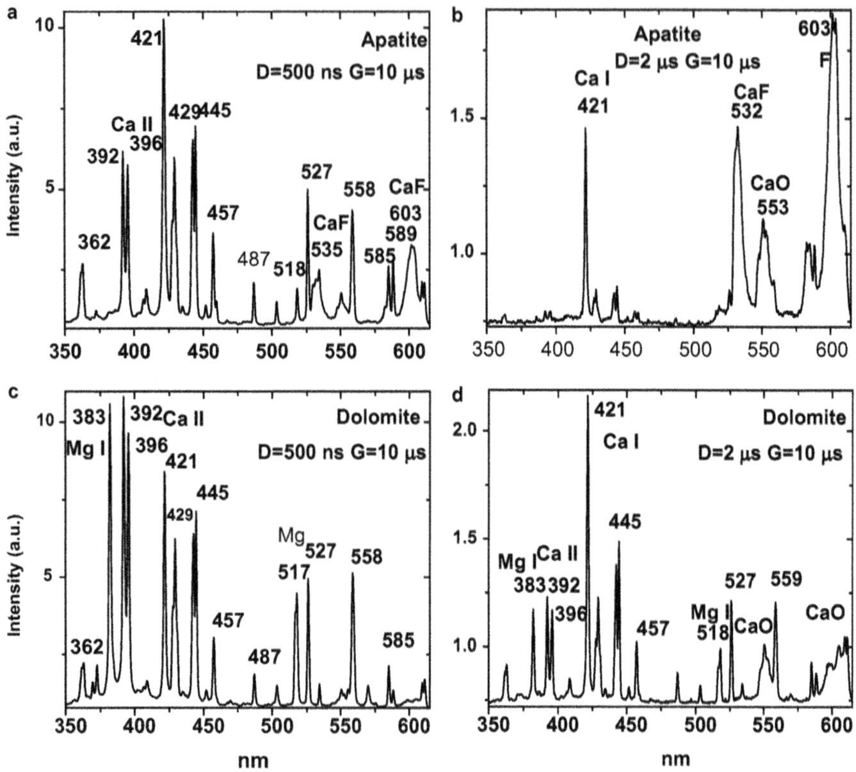

Fig. 8.21 (**a–d**) Time-resolved LIBS of Florida dolomite and apatite under $\lambda_{ex} = 355$ nm

does not contain Mg, which gives a good opportunity to differentiate between them by LIBS. According to expectation, it is clearly seen that the narrow band at 603 nm, which is connected with CaF, belongs only to apatite, while the lines at 383 and 518 nm, which are connected with Mg, belong only to dolomite. Thus LIBS enables confident dolomite and apatite identification.

To check the possibility of LIBS on preliminary unknown samples, 160 pebbles were randomly selected on the moving belt of the Four Corners Mine. They have been analyzed in conditions, similar to those on the moving belt, namely from the distance of 1 m using only one laser pulse (351 nm of excimer XeF laser). Three samples taken from the belt with time intervals of approximately 10 min were divided for two products each, named as "bad" (dolomite) and "good" (apatite). All products were subjected to chemical analyses in IMS-Phosphates analytical laboratory. The results are presented in Table 8.5. It is clearly seen, that in the "bad" fraction contains mainly dolomite with the mean $MgO/P_2O_5 = 0.7$, while the "good" fraction is mainly apatite with the mean $MgO/P_2O_5 = 0.015$, namely nearly 50 times lower. Besides that, the dolomite products contain elevated concentrations of Fe_2O_3, Al_2O_3 and insoluble residue compared to apatite products (by 1.5, 2.5 and

Table 8.5 Chemical analyses of apatite and dolomite fractions after LIBS separation

	No pebbles	MgO	P_2O_5	BPL	F	Fe_2O_3	Al_2O_3	Insoluble
Reject	20	2.2	3.5	7.6	0.4	1.1	2.2	70.4
Accept	31	0.5	28.6	62.5	3.6	0.7	1.1	12.9
Reject	18	2.5	5.9	12.9	0.5	1.1	3.8	63.9
Accept	33	0.4	28.1	61.42	3.3	0.8	0.8	13.3
Reject	13	4.6	4.5	9.81	0.3	1.3	2.0	56.5
Accept	45	0.3	27.4	59.80	3.4	0.8	1.4	14.2

4.7 times, correspondingly). Consequently, removing of the rock with elevated dolomite content enables to lower together with Mg concentration, other harmful impurities also.

Industrial applications for online process control in the phosphate industry may be divided into two groups: control of poisonous impurities (mainly Mg) and content control of P_2O_5 in raw materials and post-flotation concentrate (Gaft et al. 2008; Gaft et al. 2014a). For practical applications the first harmonic of Nd-YAG (1064 nm) is the best laser source, because it is powerful and exists in ruggedized industrial versions. Figure 8.22 presents typical LIBS spectra of phosphorus-bearing apatite $Ca_5(PO_4)_3F$ (Fig. 8.22a) and magnesium-bearing dolomite (Ca,Mg)CO_3 (Fig. 8.22b). Using LIBS, it was a relatively simple application to detect online which portions of raw materials or concentrates were Mg and P-enriched, and to make changes to the mining technology accordingly. As an example, pebble portions with elevated Mg were moved aside and as a result its concentration in final pebble product was substantially lower and the corresponding price per ton higher. Subsequent laboratory control of the removed portions of the rock confirmed that nearly 93.0 % of decision-making based on the LIBS online analyzer were correct. Another example is P_2O_5 control, where its high concentrations (20–30 %) may be analyzed using P I emission lines near 255 nm (Fig. 8.23a). It is a more difficult task when relatively low P_2O_5 concentrations are accompanied by very high Fe content. The lines near 255 nm are not easy to use because they are nearly obscured by Fe I and Fe II emission lines. In such a case, spectral detection near 215 nm is preferred, because the emission line of P I is present while the emissions from F and Si ions are relatively weak (Fig. 8.23b–c). In one specific case, the main interfering factor was the presence of copper minerals with strong emission near 215.0 nm (Fig. 8.23d).

Fluorine Online Analysis

The technological task was LIBS online control on laboratory scale conveyer of fluorine content in phosphate rock. Figure 8.24a, b demonstrate breakdown spectra of this phosphate rock with well-defined analytical lines of elemental P, Ca, Si, Mg, Al, Fe and molecules of CaF and CaO. The ionic F analytical lines are not visible in the samples with its low contents. The analytical algorithm was developed which on

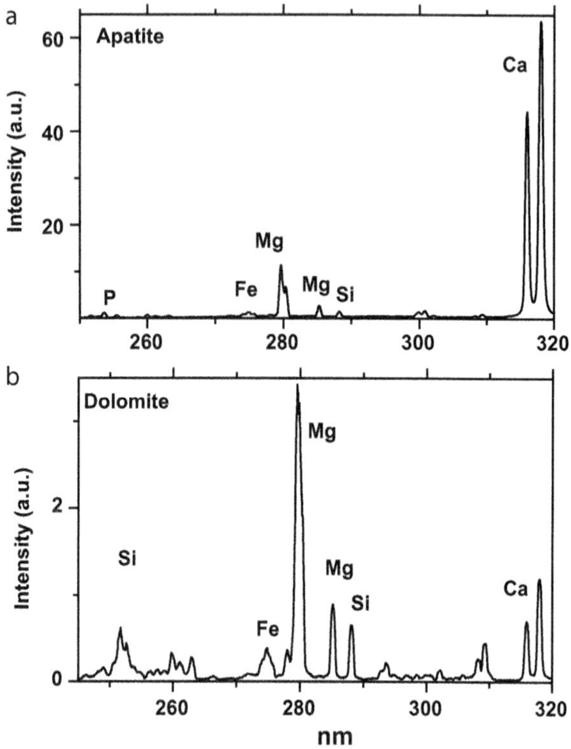

Fig. 8.22 (**a–b**) Online breakdown spectra of apatite (**a**) and dolomite (**b**) which are the main minerals in Florida phosphate deposits

the first stage classifies the sample as belonging to high or low fluorine concentration, where the border value is 0.2 % of F. On the second stage the samples with low F contents are analyzed by more sophisticated algorithm. Vertical lines represent content deviation inside each sample. As a result, all samples with F content above 0.2 % are successfully screened from the rest of the samples (Fig. 8.24c). No false negative or false positive cases have been seen. Samples with F content below 0.2 % F after second analytical algorithm (Fig. 8.24d) demonstrate good linearity and low analytical error of ±0.01 % (Abs) and ±13.9 % (Rel) with standard deviation inside each sample of ±0.013 %. Those data prove that online real time LIBS process control of fluorine is feasible.

Remote Sensing

Online control of minerals and rocks moving on a conveyer belt allows for sorting of phosphate rock after mining pumping and separation. It would be even more effective to situate the online analytical system in the first stages of the mining extraction process even before the conveyer. It is a usual situation when ore blocks include gangue or material below cut-off grade, reducing the grade and value of the

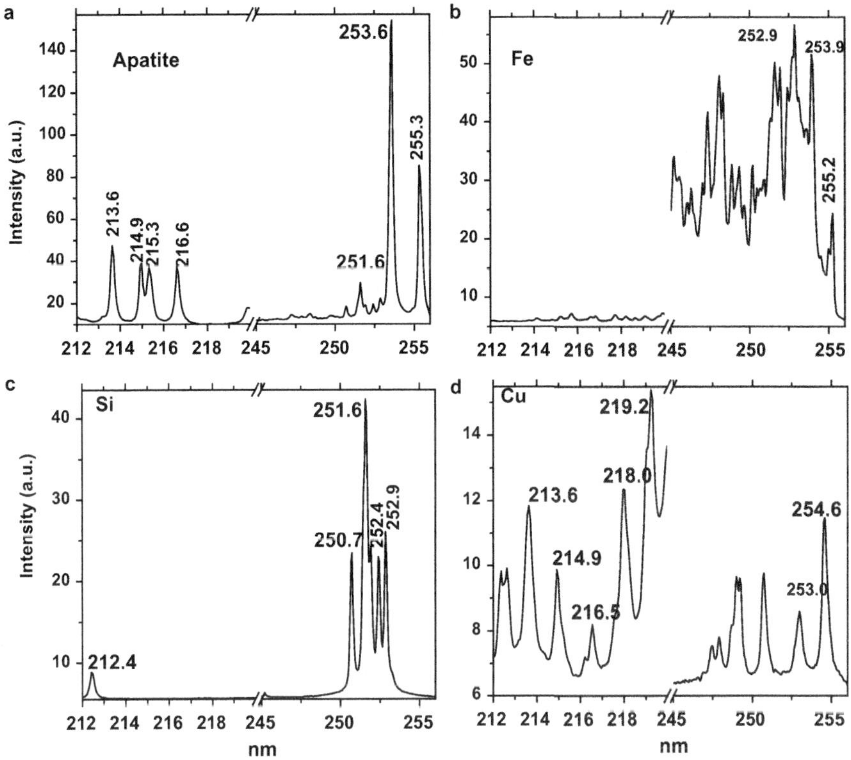

Fig. 8.23 (**a**–**d**) Online breakdown spectra of phosphorus (**a**), iron (**b**), silicon (**c**) and copper (**d**) in the spectral range of 212–255 nm

block. Similarly, waste blocks often include valuable ore which will be discarded and wasted. Thus express rock characterization during the digging stage not only eliminates the cost of mining, pumping and separation, but also reduces the possibility of discarding good rock rather than mining it. Presently the samples are excavated, transported to the laboratory, analyzed, and then the data are transmitted back to the mine. This process is slow, error prone, and does not allow real-time management of mining. Mining costs can be significantly reduced by in situ element analysis, enabling real-time selection of the highest grade ore. Practically it may be used for rock analysis at the mine face, rock in the dragline bucket or rock prior to sluicing it into the pumping pit. This required the construction of a remote online analyzer capable of identifying rock quality from distances from several to tenth of meters.

LIBS standoff mineralogical applications have been previously considered and it was proved that effective information may be obtained remotely (Cremers and Radziemski 2006; Sharma et al. 2003; Palanco and Laserna 2004; Salle et al. 2007). To prove such possibility in real life scenario, Laser Distance Spectrometry (LDS) constructed and successfully tested in real field conditions a remote LIBS (ReLIBS)

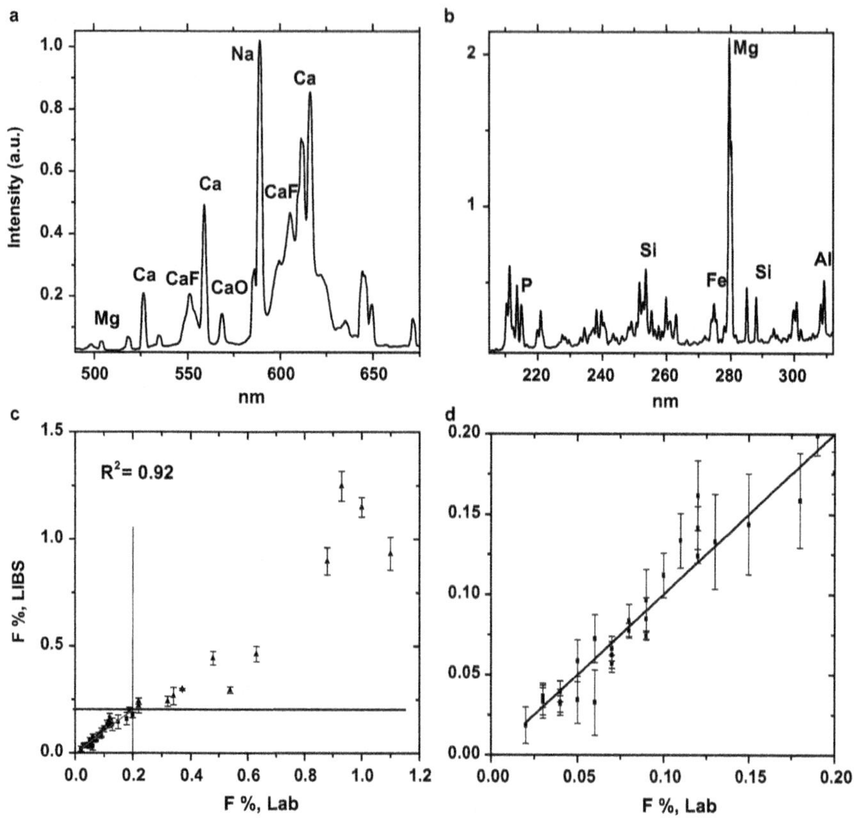

Fig. 8.24 (**a–d**) Breakdown spectra in visible (**a**) and UV (**b**) spectral ranges and correlation between online F analysis and laboratory data in whole range of F concentrations (**c**) and in low concentrations range (**d**)

unit capable of performing phosphate rock analysis at 5–25 m distances. The ReLIBS unit contains two pulsed Nd-YAG lasers (100 mJ, 8 ns) capable to work in single pulse and double pulse modes, two gated spectrometers with changeable delay time (100–100 μs) and gate width (10 μs–1 ms), optical telescopic focusing and collecting assemblages, autofocus and scanning mechanisms, laser distance meter, video camera, heater, chiller, fans, blowers and industrial PC. During the sampling process, telescope automatically scans the defined area and at each sampling point an auto focus process is performed according to the measured target distance. When focusing is accomplished, the exciting laser is being activated and at each laser pulse the spectrometer receives spectra which are transferred for analysis by the corresponding software.

The main task of ReLIBS is to use this analyzer for "horizon control" in open mining. Phosphate rock is usually found 15–50 ft beneath the ground in a mixture of phosphate pebbles, sand and clay known as phosphate "matrix." The sandy layer above the matrix, called the overburden, is removed using electrically operated draglines. Equipped with large buckets, these draglines remove the overburden, placing it in the previously mined voids, and excavate the matrix, depositing it into a shallow containment area or slurry pit. There, high-pressure water guns turn the material into a watery mixture called slurry, which is sent through pipelines to a processing facility, where phosphate rock is physically separated from the sand and clay in the matrix. Presently the dragline operators are advised by field geologists on what to mine based on information from drill cores and visual observation. Nevertheless, such information may be erroneous. This generally results in at least some inclusion of overburden or base material in with the phosphate matrix or matrix is cast aside with the overburden or left it in the mine cut. For example, in one of the tests, 56 samples of "Overburden", "Matrix" and "Bed" according to the dragline operator practice were collected from seven mines and analyzed. Out of the 51 samples which were clearly "Mine" or "No Mine", only 75 % were correct. Compared to what was actually being mined, "correct" mining would have increased production by 9 % and reduced MgO by 50 %. But actual mining was being controlled by the dragline operators. What if a geologist had been present? The same 56 samples were viewed by 5 geologists and characterized as to "Mine" or "No Mine." Average accuracy was 79 % (73–84 %). All 5 got the right answer on 31 of the samples and the wrong answer on 3 of the samples. Both the dragline operators and geologists were mining "Matrix" containing as low as 2 % BPL or with an MgO over 8 %. Both the dragline operators and geologists were leaving overburden or bed containing over 30 % BPL and less than 0.2 % MgO. It's not always possible to "see" the difference between matrix and non-matrix and chemical analysis is required to make the correct determination. At one mine, samples of overburden, matrix and bed at each dragline were taken each day. Several hundred samples were taken. Depending on the criteria used for "good rock", accuracy of what was being mined (compared to should) was 70–85 %.

The impact of correct mining will be to increase both production and improve mined rock quality. But where is the best place for ReLIBS unit? As result of the field experience, the washing pit was selected which is where the drag line operator places the excavated matrix for its transformation to slurry and the following pumping to the beneficiation plant. The advantages of this position are the following: the information for drag line operator is practically real time, only one bucket later compared to information from a unit mounted on drag line machine being the fastest; not every bucket has to be controlled, but only those in the unclear excavating zones; the distance from the ReLIBS to the pit can be approximately 15–20 m, which is in the range of existing capabilities; the relief is substantially less changeable compared to the actual surface; the safety issue is substantially less severe because there are no personnel present in this area.

For safety reasons, it was not possible to take actual field samples for laboratory analysis to correlate with the specific ReLIBS samples. Thus the spectral types based on different emission lines relative intensities as an indicator of the rock types were determined. In order to do this, the spectra of the well characterized rock samples, were measured in laboratory conditions. Based on calibration curves

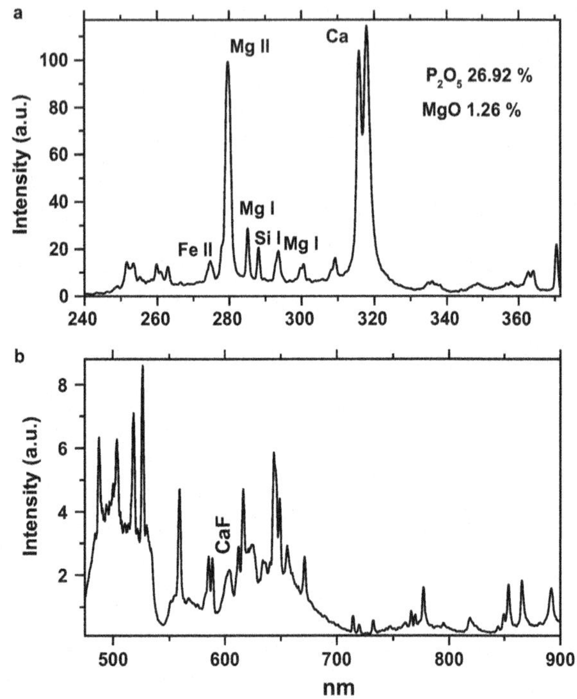

Fig. 8.25 (**a–b**) Typical UV (**a**) and visible (**b**) remotely detected (15–18 m) breakdown spectra of the rock with high P2O5 and relatively low MgO content

developed using the calibration samples, spectroscopic data have been used to give the chemical compositions. For example, Fig. 8.25 presents the breakdown spectrum of the rock with high P_2O_5 and relatively low MgO content.

This ReLIBS analyzer proved the feasibility of distant real time chemical analysis of phosphate rocks excavated by drag line machine in that it:

A. Differentiated between overburden, matrix and bottom materials;
B. Determined the P_2O_5 content;
C. Determined the MgO and Iron content in the matrix samples.
D. The optimum position (considering safety and mechanical feasibility) of the ReLIBS was determined to be at washing pit before the matrix is slurried to pump to the beneficiation plant.

Based on all the information, it looks like approximately 20 % of analyzed different rock portions represent material that should not have been mined as it would give rock of low value to the chemical plant. This is similar to the results reported previously in 1997 when 20–25 % of the matrix mined was found to not be worth mining. With such improvement in "not mining" the matrix that should not have been mined, there is significant value. At an annual rock production rate of 4 million tons, and 20 % of the matrix being mined and subsequently thrown away when it is caught by the MAYA on the pebble belt, the "wasted cost" can be significant. At a probable cost of $5/ton for digging, transporting, washing and screening the rock, the annual cost could be over \$4,000,000. Another possible source of the wrong mining, is how much overburden or bottom material was not mined. This also will be studied in the nearest future.

8.3.2.11 Potash Quality Control

The main task for LIBS online analyzer in potash quality control is sodium impurity evaluation (Groisman and Gaft 2010). It is the mostly important for high quality product where the highest acceptable Na concentration is 2 %. Figure 8.26 presents visible and UV spectra of KCl material with several lines belonging to K and Na ions. Those elements, especially with their high concentrations, are difficult to analyze because of their strong self-absorption process. Figure 8.27 with high spectral resolution demonstrates self-absorption influence on K and Na emission spectra form and spectral position. The self-absorption is the strongest for ground state transitions and can cause a deviation of the calibration curves from linearity. Non linearity in emission calibration curves may be also due to different kinds of energy transfer between excited-non excited atoms (ions) resulting in corresponding line broadening. The problem of calibration non-linearity is usually solved by using optically thin (non-resonant) emission lines. Different K and Na lines and their ratios have been checked in SP and DP modes with different delay and gates and the optimal algorithm enables good correlation with real analytical data.

Another possible way to overcome the self-absorption phenomenon is to use LIP absorption technique (Nagli et al. 2012). Absorption measurements in the proposed DP LIBS configuration provide the direct method for concentration determination that may be free from the above mentioned drawback. Opposite to emission, where light originates from a compact plasma region with high Optical Density (OD), the OD in absorption is not too large even for strong atomic and ionic transitions. This is because (i) the absorption layer of "cold" atoms in plasma periphery is thin and (ii) the analyte concentration is reduced by plasma expansion and scaled as R^{-3} with plasma radius R. Note that concentrations of elements in the first plasma are determined by absorption. To verify this, we prepared eight samples by mixing NaCl and KCl powders and measuring Na I Fraunhofer lines (Fig. 8.28). OD(λ) (the same as absorbance) as a function of the wavelength for the measured Fraunhofer lines of Na was determined as $\log (I_0(\lambda)/I(\lambda))$, where $I_0(\lambda)$ is the background continuum radiation in the vicinity of the spectral line and $I(\lambda)$ is the peak intensity of this line. OD(λ) is related to the absorption coefficient $\kappa(\lambda)$ by $\mathrm{OD}(\lambda) = 2.3\kappa(\lambda)l$ where $\kappa(\lambda) = CN$ where N is the densities of atoms in cm^{-3}, C is the constant, and l is the optical path in cm. So, the measured optical density of Na lines is proportional to the concentration of Na in the sample. Figure 8.28c presents the correlation between OD (that is not higher than 0.45) of Na and Na concentration. It can be seen that for all measured concentrations the dependence is linear.

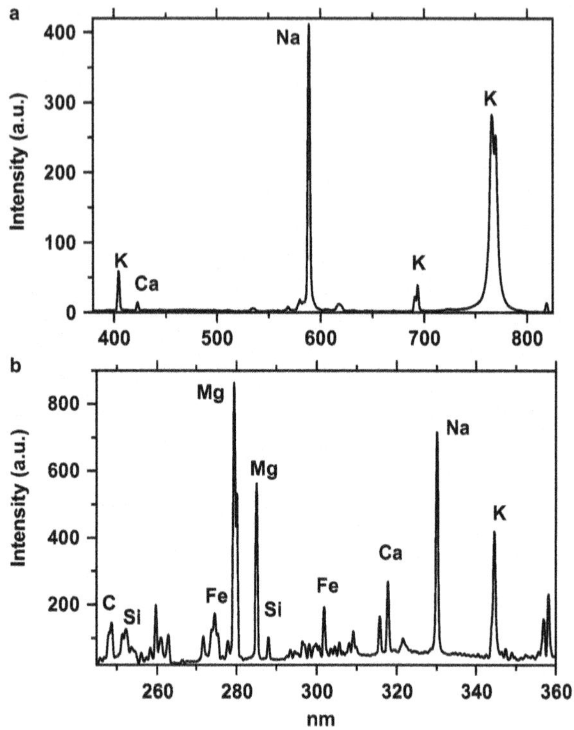

Fig. 8.26 (**a–b**) Breakdown spectra of KCl in visible (**a**) and UV (**b**) spectral ranges

8.3.3 Ferrous Metals: Si and Ca in Iron Sintering Mix

Iron sintering mix control and composition stabilization. For an efficient sintering process, a constant and optimized basicity of raw mix without short and long term fluctuations is a must. Achieving real-time automatic process control without human factor influence requires on-line elemental composition data. Figure 8.29a presents typical breakdown spectra of the sintering mix and the results of industrial LIBS unit test data, where laboratory CaO control data are compared with online analyzer readings. One hundred and forty samples have been taken from conveyer belt and send to laboratory for control analysis. It was found that the correlation of $R^2 = 0.88$ is very high, the average standard error being 0.31 % mean absolute error or 7.3 % mean relative error. It may be seen that online LIBS data follow all trends in sintering mix chemical composition (Fig. 8.29b). The accuracy achieved by LIBS was considered as relevant for online process control and the analyzer was incorporated in a factory's automatic system to supervisory control and data acquisition (SCADA), which provides real-time access to data on the elemental composition of the iron ore mix fed from storage to the sintering machine before dosing fluxes (limestone and dolomite) (Gaft et al. 2014a).

The production capacity of the sinter plant is 14.5 million tons/year. The environment is extremely challenging: dusty, with wide temperature variations

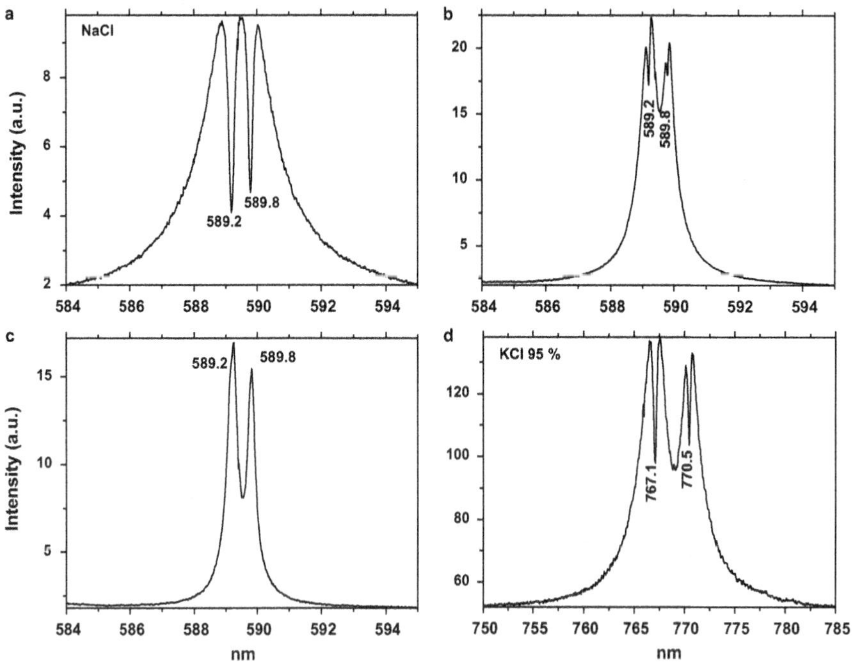

Fig. 8.27 (**a–d**) Breakdown spectra with high resolution of KCl with strong self-absorption

from −10 to +40 °C. The basicity level of sinter is 1,2–1,6 units and the task is to keep the set parameter on the ±0.05 level. It is adjusted based on the chemical analysis of the iron ore mix coming from the storage area by adding fluxes. CaO content in the iron ore mix fluctuates significantly, due to poor blending and material segregation in the pile. The inconsistency is greatly increased by addition of 25–35 kg/t of lime into the sinter pile. After successful pilot tests with the LIBS analyzer, the sintering technologists developed a system of automatic flux dozing based on communication of CaO content in the iron ore mix from the online LIBS analyzer and making continuous automatic adjustment of the flux flow rate during subsequent proportioning, thereby reducing the fluctuations of the sinter basicity. The usual time interval for flux adjustment is 20–60 s. Implementation of the LIBS analyzer enabled several improvements: receiving and transmitting the data on chemical composition of iron ore mix in automatic mode to the process control systems, in order to use them for on-line control of technological process; complete elimination of the human factor causing analysis errors, primarily related to sample taking and sample preparation procedures; decreased deviation of sinter basicity beyond the ±0.05 limits from 8–10 % to 2–4 %; decreased coke consumption in the blast furnace process by 0.6–1.0 kg/ton of iron.

Another potential tool for silicon analysis in iron mix is LIBS-MLIF spectroscopy. Figure 8.30a presents the corresponding SiO molecular emission in Si matrix where strong molecular characteristic lines are clearly detected. When silicon

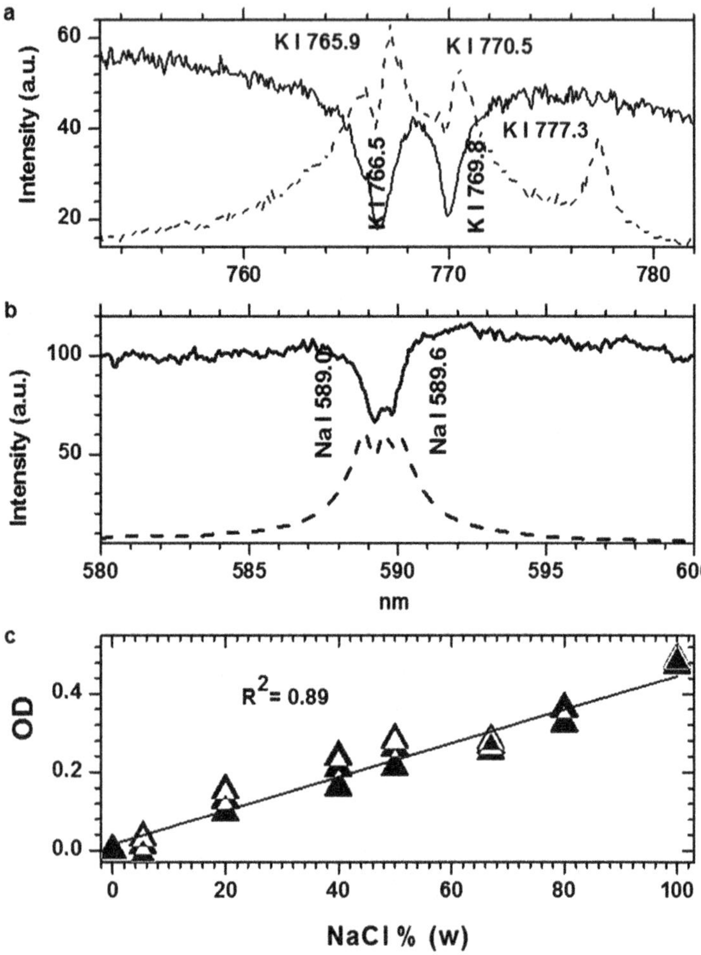

Fig. 8.28 (**a–c**) KCl (**a**) and NCl (**b**) plasma. *Solid curves* are absorption spectra *Dashed curves* are emission spectra measured. *Calibration curve* for K contents in NaCl measured using absorption spectroscopy (**c**)

concentration in iron mix is relatively low, namely less than 1.0 %, those lines are substantially weaker, but nevertheless may be definitely detected (Fig. 8.30b). Traditional LIBS technique on the same material is much less satisfactory, because even the mostly intense Si atomic line at 288.16 nm can hardly be discerned on the background of interfering Fe emission lines (Fig. 8.30c). It is important to note that combination of LIBS with Si atomic LIF on this material proved to be at least 10 times less sensitive.

Similar iron mix material was analyzed but with additional task for online control of carbon. Figure 8.31 presents corresponding breakdown spectrum and it may be seen that classical C I emission line at 247.8 nm may not be detected

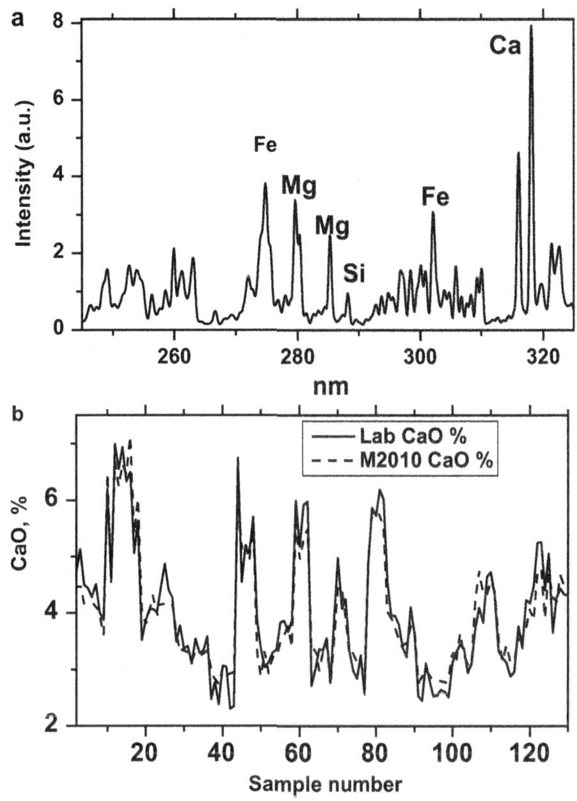

Fig. 8.29 (**a–b**) Typical breakdown spectra of the sintering mix (**a**) and correlation between online LIBS and laboratory data trends following (**b**) on CaO concentrations

because of strong Fe emission lines, while the C III line at 229.7 nm is easily seen (Fig. 8.31a). Correlation with laboratory data on carbon content is pretty good (Fig. 8.31b).

8.3.4 Base Metals

The base metals are those that are either found with copper or alloyed with copper and now include all the copper, lead, zinc, arsenic and tin ores. Tin works as a much better metal to alloy with copper than arsenic in making bronze – but it is much rarer. Zinc actually wasn't known to the ancients as it either was completely oxidized and went right up the furnace chimneys or reacted with copper to make brass directly in the furnace. The very large "grey area" in minerals applications would appear to be in sorting many base metal ores that require less proven or completely new sensing techniques for sorting. The perceived industry benefits have not been sufficient to support well-tested developments in this area. There are a large number of published articles on the development of sensing techniques for sorting base metals, a lower number on industry demonstrations of developed

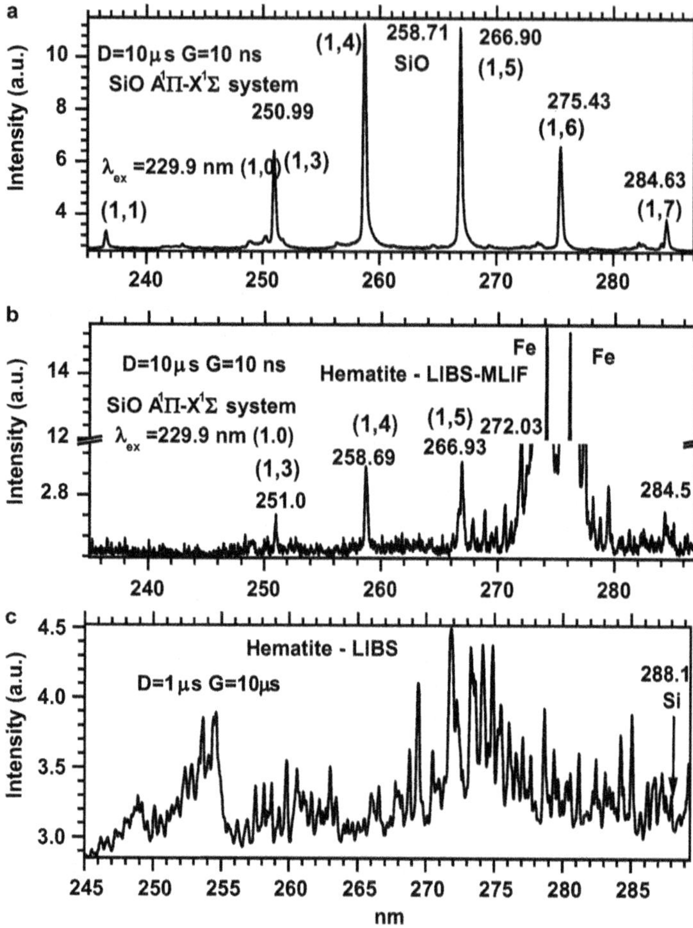

Fig. 8.30 (**a–c**) LIBS-MLIF spectra of SiO in pure Si (**a**) and iron mix with less than 1.0 % of Si impurity (**b**). For comparison, LIBS of the same iron mix is presented (**c**)

technologies (particularly in Russia), and very few on commercial operations using these technologies. One explanation for this rather poor record over a period of at least 20 years is that there are entrenched misconceptions in regard to the economic viability of sorting. A recent renewal of interest in new mineral sorting applications would seem to indicate that this view may be changing on the basis of demonstrated operational experience with modern sorters over the past decade.

8.3.4.1 Copper Bearing Minerals

Copper ores quality control is usually needed for several tasks. Breakdown UV spectra of copper-bearing ore in online mode detect clear analytical lines of major and minor elements such as Cu, Ca, Si, Fe, Mg and Zn (Fig. 8.32a). Sulfur analysis

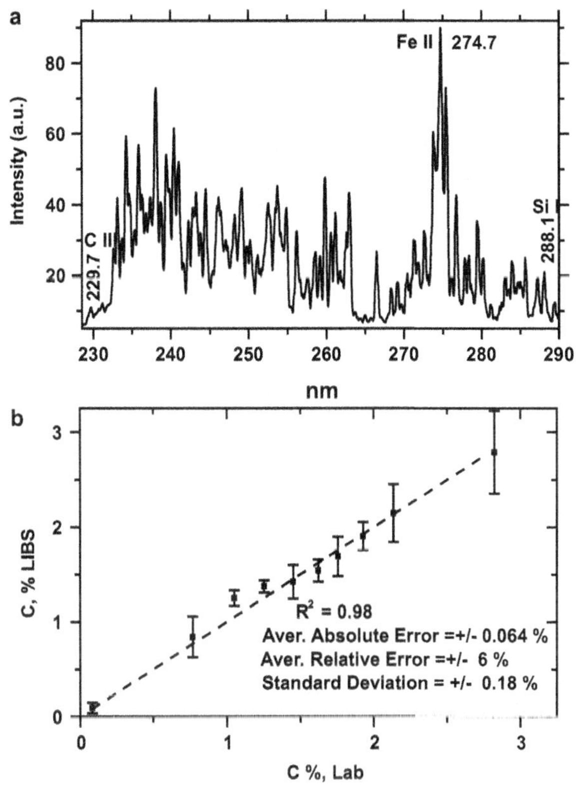

Fig. 8.31 (**a**–**b**) Typical breakdown spectra of the sintering mix (**a**) and correlation between online LIBS and laboratory data trends following (**b**) on C concentrations

in online mode is a challenging task, but for concentrations in the 15–30 % range it may be done using S I analytical lines in the 920–925 nm spectral range (Fig. 8.32b). In this specific case, online control of Cu, Ca Si and S was the task and laboratory scale tests gave very correlations coefficients R^2 in 0.9–0.96 range for all these elements. At such sulfur levels, it may be detected in air using IR lines in 920–925 nm range (Gaft et al. 2009c).

Another important task is online control of Mo impurity in copper-bearing ore. Online LIBS analyzer enables to detect definitely even trace Mo concentrations in the hundreds of ppm range (Fig. 8.33). Correlation coefficients with laboratory data R^2 of Cu, Mo and Fe are in 0.92–0.98 range, promising the possibility of process control in industrial online LIBS application (Gaft et al. 2014a).

8.3.4.2 Nickel Bearing Minerals

The first example is Ni in Cu-bearing ore requiring online control of both elements' contents. Both Cu and Ni are definitely identified by online LIBS analyzer (Fig. 8.34a). Correlation between laboratory data on Cu and Ni and laboratory-scale LIBS tests for two different fractions with R^2 in 0.89–0.94 range promise the

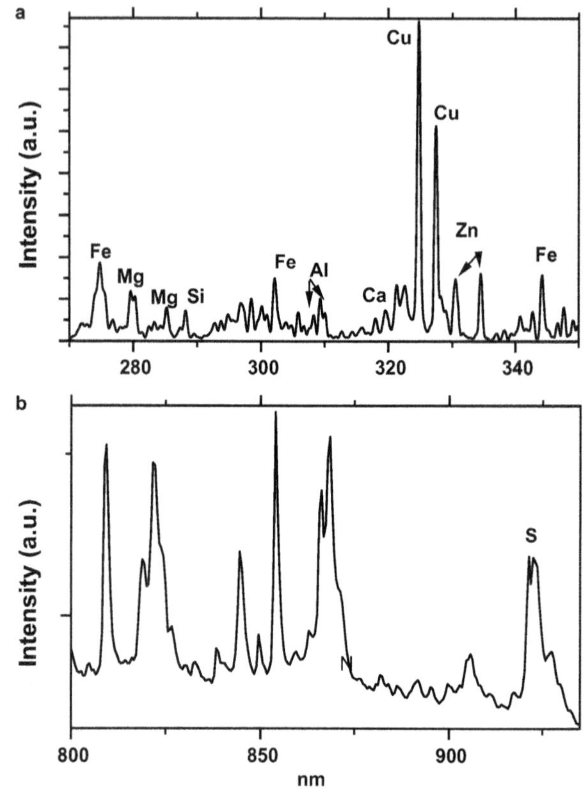

Fig. 8.32 (**a**–**b**) Typical breakdown spectra of the copper bearing ore in UV (**a**) and IR (**b**) spectral ranges

possibility of process control in industrial online LIBS application (Gaft et al. 2014a). In another task, Fe/Ni concentration ratio and MgO content were the aims of online analyses. All elements may be analyzed by online LIBS analyzer (Fig. 8.34b) and the correlation between laboratory data and laboratory-scale LIBS conveyer tests gave R^2 of 0.97 and 0.94, correspondingly (Gaft et al. 2014a).

8.3.4.3 Molybdenum Lump Sorting

The task was to evaluate the possibility for individual sorting of Mo bearing particles for high, average and low Mo products. Figure 8.35a–c demonstrates breakdown spectra received in online mode demonstrating clear analytical lines of Mo at different Mo contents. Correlation between LIBS and laboratory data demonstrates that individual sorting by LIBS is quite possible (Fig. 8.35d).

8.3.4.4 Tin Bearing Minerals: Cassiterite

In its oldest use tin was alloyed with copper and zinc to make bronze. Tin is still valuable today in alloys and coatings for steel. Cassiterite, a tin oxide, is the only

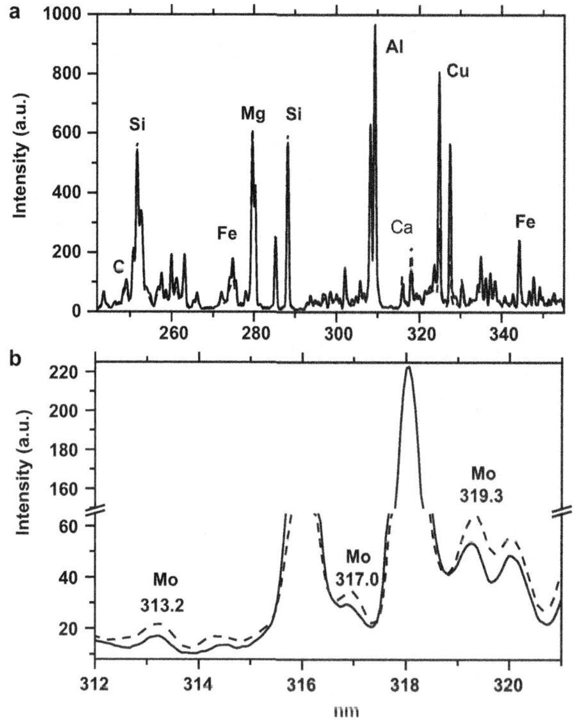

Fig. 8.33 (**a–b**) Typical breakdown spectra of Mo bearing copper ore (**a**) with weak but distinctive emission lines of Mo (**b**)

important ore mineral. Under powerful laser excitation cassiterite very often is characterized by intensive luminescence. The emission band consists of the main band at 560 nm with decay time of 200–250 ns, the weaker band at 650 nm with decay time of approximately 300 ns and the weak band peaking at 450 nm with very short decay time of 50 ns. Using shortwave part of the excitation band the blue emission becomes relatively stronger, while under longwave part the red emission dominates. The main background luminescence in cassiterite case is connected with the sorption of organic materials, but its extremely short decay of less than 20 ns enables effective sorting using delay of approximately 75 ns and gate of 1–5 μs.

Tin has many intensive lines in breakdown spectrum, the strongest ones at 284, 286, and 303 nm (Fig. 8.36) which enables LIBS sorting.

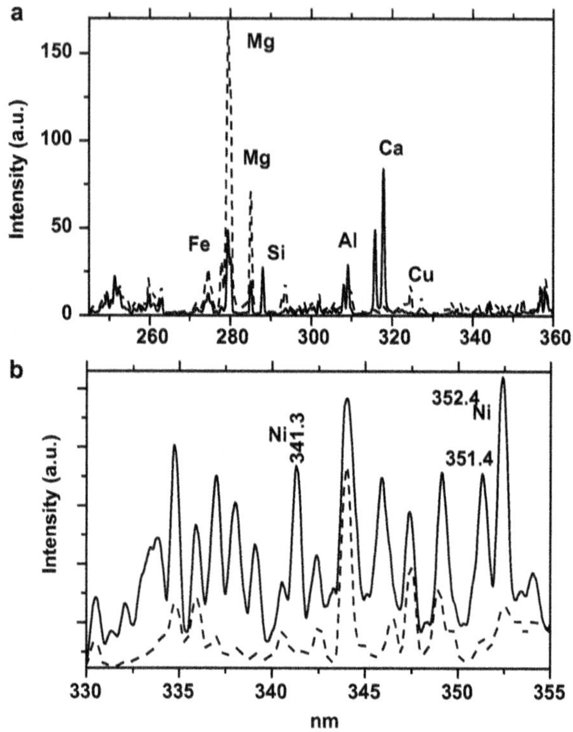

Fig. 8.34 (**a–b**) Typical breakdown spectra of Cu (**a**) and Ni (**b**) in Ni bearing ore

8.3.5 Non-ferrous Metals

8.3.5.1 Scheelite (W)

Tungsten is mainly used in lamp filaments and steel-alloy cutting tools. Scheelite and wolframite are the major ores of tungsten. At present time even deposits with very low tungsten concentrations (down to 0.1 %) are considered as industrial.

Scheelite was the second (after diamond) mineral which was sorted using X-ray excited luminescence enabling to select the pebbles with elevated W contents. Using this method it was possible to sort the tails with 0.03 % of WO_3 and to select the product with 0.12 % of WO_3. Nevertheless, in certain cases such sorting may be difficult. For example, fluorite, apatite and calcite under X-ray excitation also have luminescence in the selective windows used for scheelite separation (Gaft et al. 1989). Laser-induced time-resolved luminescence may be much more effective, using different excitation spectra and decay times of corresponding minerals. Luminescence and excitation spectra demonstrate that excitation at 240 nm is optimal for scheelite, while Mn^{2+}, Ce^{3+} and Eu^{2+} in barren rock are substantially diminished. The strong difference in decay times enables to lower the background further, because delay of approximately 1 μs enables to quench the short lived Ce^{3+}, Eu^{2+} and organic matter, while the gate of 20 μs liquidates long lived Mn^{2+}

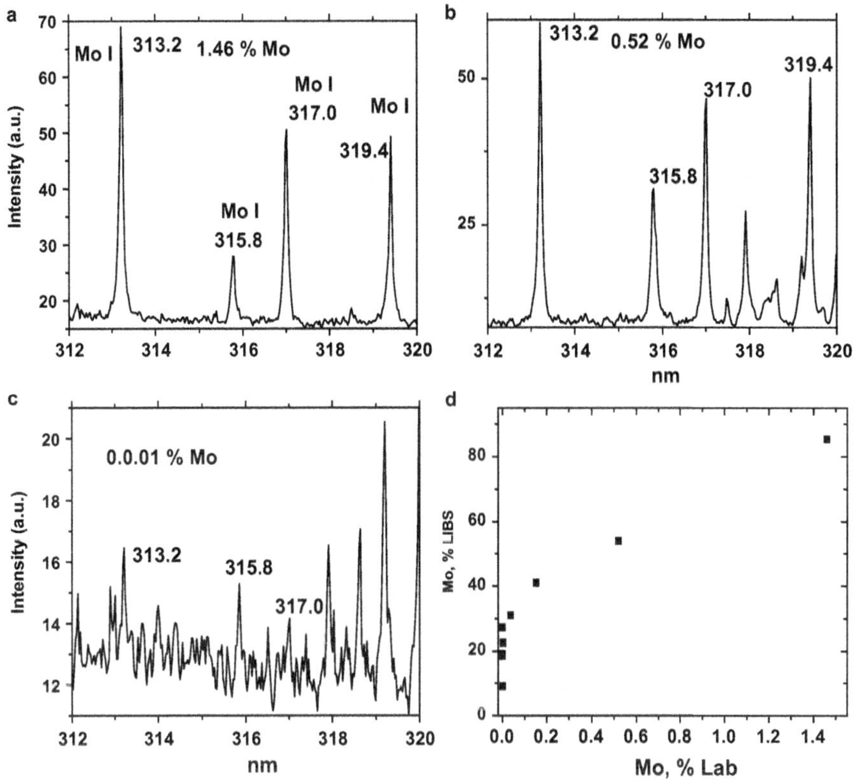

Fig. 8.35 (**a–d**) Breakdown spectra of Mo bearing particles (**a–c**) and correlation between Mo concentration based on LIBS and laboratory data

background. From the other hand, at such time window the scheelite emission with decay time of 5–6 μs will be strong.

Tungsten has many intensive lines in breakdown spectrum, the best one at 401 nm (Fig. 8.37) which enables potential LIBS sorting.

8.3.5.2 Zircon and Baddeleyite (Zr)

The most important zirconium minerals are zircon and baddeleyite, often associated with rutile and ilmenite deposits. Zircon and baddeleyite are characterized by strong luminescence under powerful laser excitation. The main background emission is connected with organic matter sorption and sometimes with fluorite. Emission spectra of zircon and baddeleyite together with their decay times demonstrate that delay time of 200 ns and gate of 10–50 μs combined with excitation near 300 nm enable to remove background signal drastically.

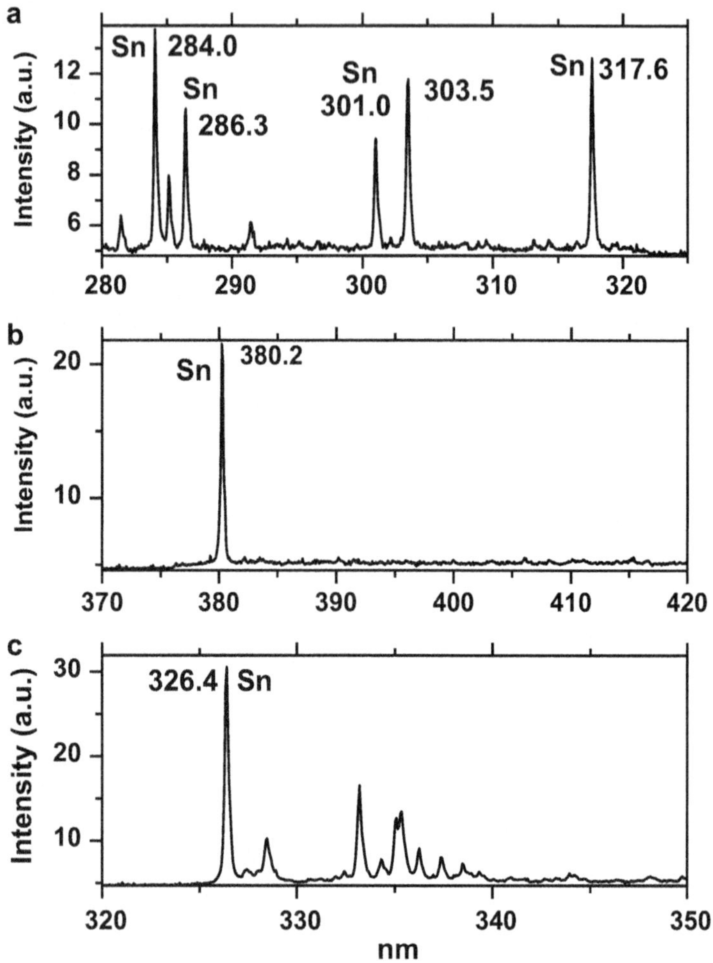

Fig. 8.36 (**a–c**) Time-resolved LIBS of cassiterite with Sn emission lines in different spectral ranges

The task was to analyze brown colored sand type Zr bearing products by LIBS in order to evaluate Zr, Ti, Al and Fe contents in online mode. Figure 8.38 demonstrates that all relevant elements may be analyzed by online LIBS analyzer. Correlation of LIBS determined Ti concentrations with laboratory data is also good.

One of the features of Zr breakdown emission is strong Zr III and Zr IV emissions (Gaft et al. 2013) (Fig. 8.39a) and characteristic molecular emission of ZrO (Fig. 8.39b), which may be effectively used for analytical purposes, especially when high contents of Fe takes place in the analyzed material. For example, unique Fe-P deposit in Cola Peninsula (Russia) accompanied by high levels of baddeleyite (Fig. 8.39c).

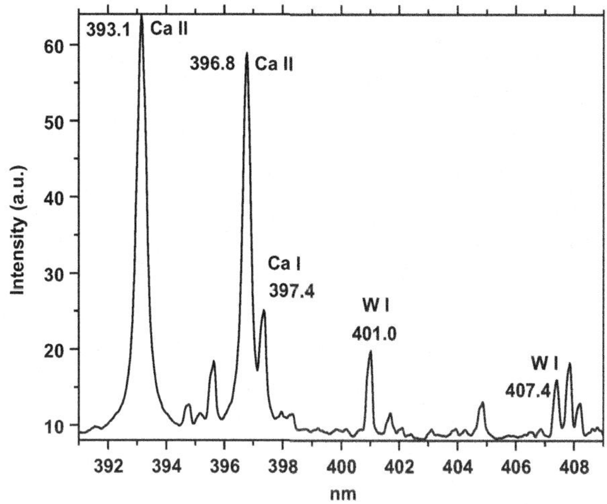

Fig. 8.37 Time-resolved LIBS of scheelite with W emission lines

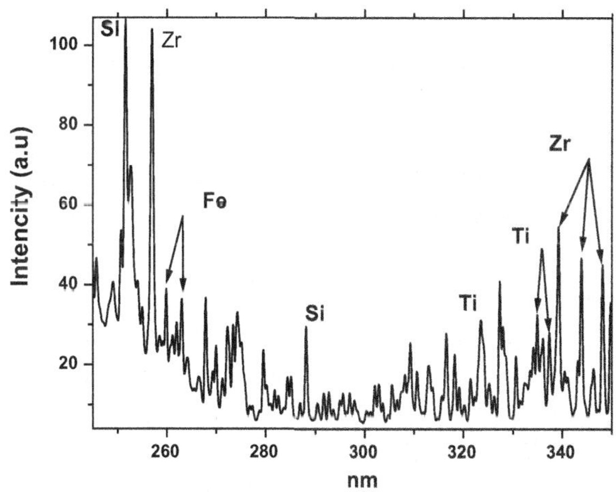

Fig. 8.38 Typical breakdown spectrum of ZrO2 heavy sand

8.3.5.3 Eucryptite and Spodumen (Li)

These minerals have been successfully separated by X-ray excited luminescence, but the serious problem was orange emission of carbonate host rocks (Gorobets et al. 1997). Eucryptite is characterized by intensive red emission at 780 nm of Fe^{3+} (Prokofiev et al. 1979) with very long decay time of 4–5 ms. Because Mn^{2+} emission in carbonates has also long decay, the main effect may be in using the differences in excitation and emission spectra. Spodumen under laser excitation is often characterized by blue emission with decay time of 50–60 μs and such relatively long decay enables effective sorting on the background of short-lived fluorite luminescence.

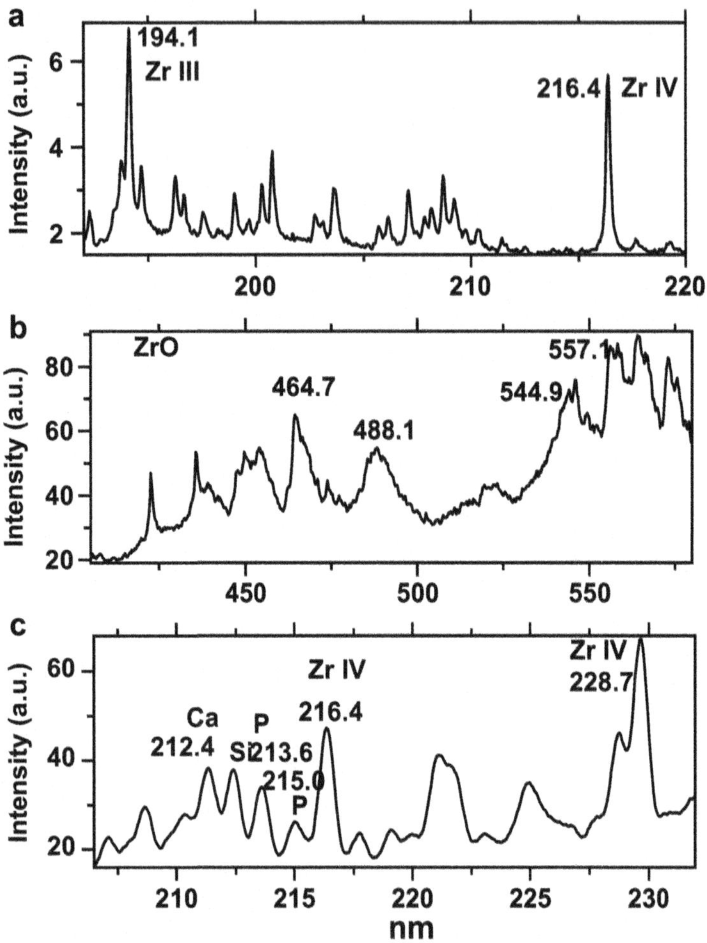

Fig. 8.39 (**a–c**) Breakdown spectra of Zr III, Zr IV (**a**) and ZrO molecules (**b**). Breakdown spectra of P-Fe-Zr deposit, Cola Peninsula, Russia

It may be proposed that LIBS method may be used for on-line real-time Li bearing minerals identification because of strong and characteristic lines of Li emission at 671 and 610 nm (Fig. 8.40).

8.3.6 *Precious Metal*

8.3.6.1 Gold Bearing Minerals

Gold is used extensively in the manufacture of electronics, in medicine, and as the world's monetary standard. This valuable metal is often found in nature in the

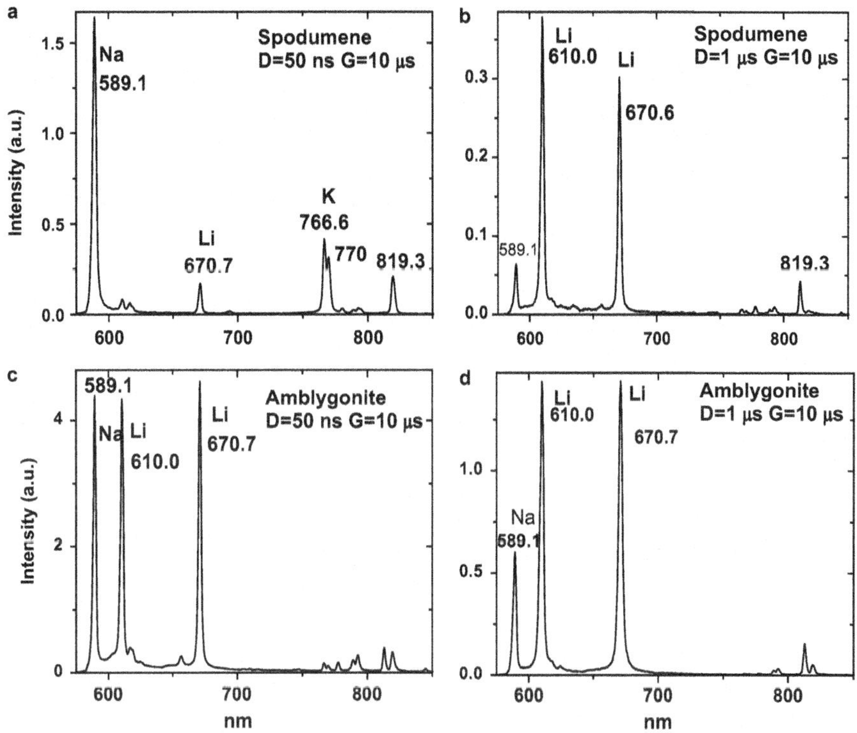

Fig. 8.40 (**a–d**) Breakdown spectra of several Li bearing minerals

native form, usually associated with quartz or pyrite. When deposits containing gold are eroded, the heavy gold may become concentrated in placer deposits in the gravel of stream beds. Only rarely is visible gold found in good ore. It does occur as a compound with tellurium in minerals such as sylvanite and calaverite. On average, only five parts per million of every ton of ore mined are actually gold. It is, for example, necessary to separate the precious metal from the more than 100 million tons of ore milled each year in South Africa. This is carried out in the mine's gold plant. Initially, ore hoisted from the mine is broken into smaller pieces by a primary crusher; secondary crushers break it down further, followed by a milling stage to produce a fine rock dust. At this point most of the tiny particles of gold contained in the ore have been exposed. A cyanide solution chemically dissolves away the exposed gold particles and the resultant liquid follows one of two routes to recover the gold now held in solution. Gold bearing ores are now sorted by photometric and radioactive or X-ray fluorescence methods. First one is used for gold containing quartz, while accompanying rocks have dark color. The second method detects the certain elements, which are correlated with Au (Mokrousov and Lileev 1979). All these methods use selective features, which are not directly connected with gold contents. Luminescent sorting of gold containing ores was never used because native gold and its minerals are not luminescent. In order to overcome this problem, the methods of artificial luminescence creation on the surface of gold bearing minerals have been developed (Gaft et al. 1991a). Different kinds of the chemical and physical treatments induce strong yellow luminescence (Fig. 8.41) on the surface of gold minerals, while home rock remains non-luminescent. This gives an opportunity to discriminate and to separate gold bearing lumps. Laser excitation plays a decisive role, because in the case of extremely thin luminescent layers which are formed on the surface, X-rays and UV lamps are not effective excitation sources. The best types in the gold case are conglomeratic, vein and stockwork deposits: mainly quartz, quartz-pyrite-chalcopyrite, quartz-silicate, barite and quartz-arsenopyrite compositions; low oxidized and with low and middle sulfides concentrations; with mainly gravitational and coarse Au; emulsion Au in sulfides with content in mineral ~0.01 wt% (Moroshkin et al. 1997). Regular LIBS online analyzer is not suitable for trace elements with extremely low concentrations, as in the gold case. Au mostly will be not on the particles surface and surely will be missed. Consequently the direct native Au analysis is not possible. From the other side, it possibly may be used for the gold ore by indirect way. For example, gold containing quartz may be quantitatively analyzed or pyrite and other sulfides which is often associated with gold. Besides, very often good correlation takes place between gold and other major or minor elements, such as Cu, Pb, As, U and so on. Thus gold concentration may be evaluated in real time using online analysis of those elements with substantially higher levels.

8.3.6.2 Silver Bearing Minerals

Man has used this metal for centuries. It sometimes is found in native silver, but is more commonly found combined with sulfur as argentenite. Most of the world supply of silver is obtained as by-product when other metals such as lead, zinc, and copper are processed. Silver is used in photography, in medicines, and in coins.

Natural silver halogenides, which are secondary ores for silver production, are characterized by intensive luminescence under powerful laser excitation. The main source of background luminescence is the sorption of organic matter. The relatively long decay time of their luminescence may enable to lower the short-lived background emission by several orders of magnitude.

Native silver and argentenite are not luminescent at natural form, but as in the case of gold, it is possible to induce the artificial emission (Fig. 8.41, Gaft et al. 1989).

Silver has many intensive lines in breakdown spectrum, the strongest ones at 328 and 338 nm in the UV part of the spectrum and at 520 and 546 nm in the visible (Fig. 8.42a, b), which it may be potentially used in radiometric sorting. Silver is relatively uniformly distributed in massive sulphide ores (average values for the deposits 10–50 ppm), but local enrichments occur (up to 150–300 ppm of Ag). Silver usually has higher concentration comparing to gold and in certain cases may be analyzed by online LIBS. For example, we have positive laboratory scale experience with 10 samples of copper bearing rocks from Peru. It is very rare

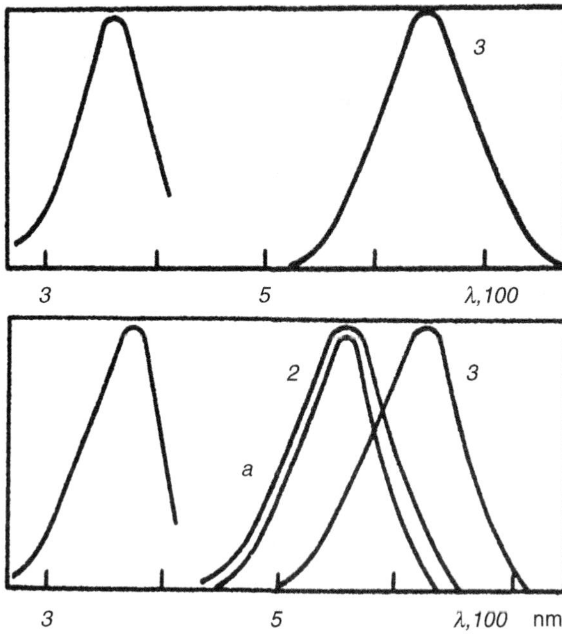

Fig. 8.41 Artificially induced luminescence on Au (*upper*) and Ag bearing minerals (1 – silver and argentenite; 2 – galenite) (Gaft et al. 1989c)

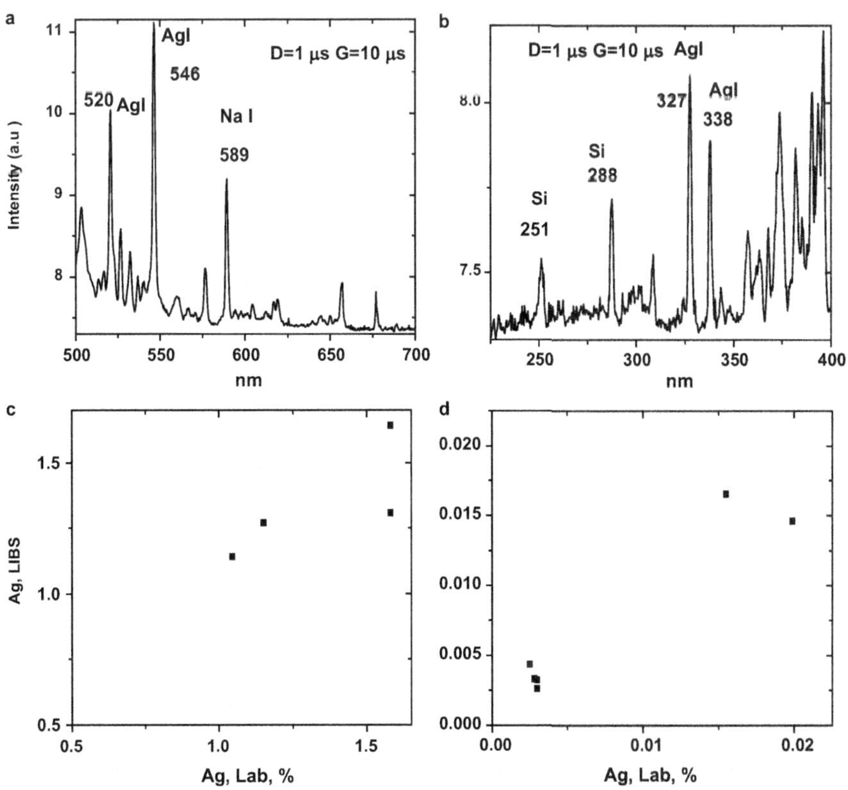

Fig. 8.42 (**a–d**) LIBS spectra of Ag containing minerals (**a**, **b**) and correlation between laboratory and online LIBS determined Ag concentrations for middle and low concentrations ranges (**c**, **d**)

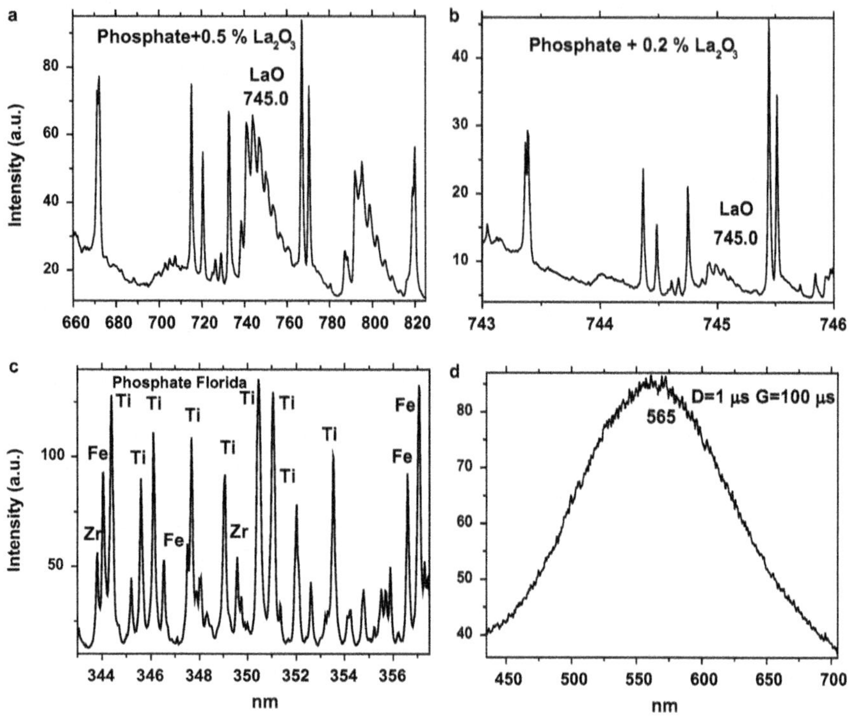

Fig. 8.43 (**a–d**) Breakdown spectra of LaO molecules in phosphates with 0.5 (**a**) and 0.2 % (**b**) levels. Emission lines of Zr and Ti in natural phosphate (**c**) and luminescence of zircon impurity inside phosphate

case where we are capable to detect starting from approximately 25 ppm level (Fig. 8.42c, d).

8.3.7 *Rare Earth Elements (REE)*

LIBS is mainly characterized by narrow emission lines of individual atoms and ions. But in case of REE it is not effective because of their usual very low concentrations and the presence of many accompanying elements with strong plasma emission, which covers the relatively weak REE lines. From the other side, emission from simple molecules, formed as the atoms recombine in the cooling plasma, may be effectively used (Fig. 6.11).

In order to evaluate the potential ability of LIBS for REE detection in natural phosphates, their artificial mixtures with different concentrations of La_2O_3 have been studied. It was found that LaO molecules are confidently detected in natural phosphates at 0.5 % and 0.2 % levels (Fig. 8.43a, b). Additional efforts are needed

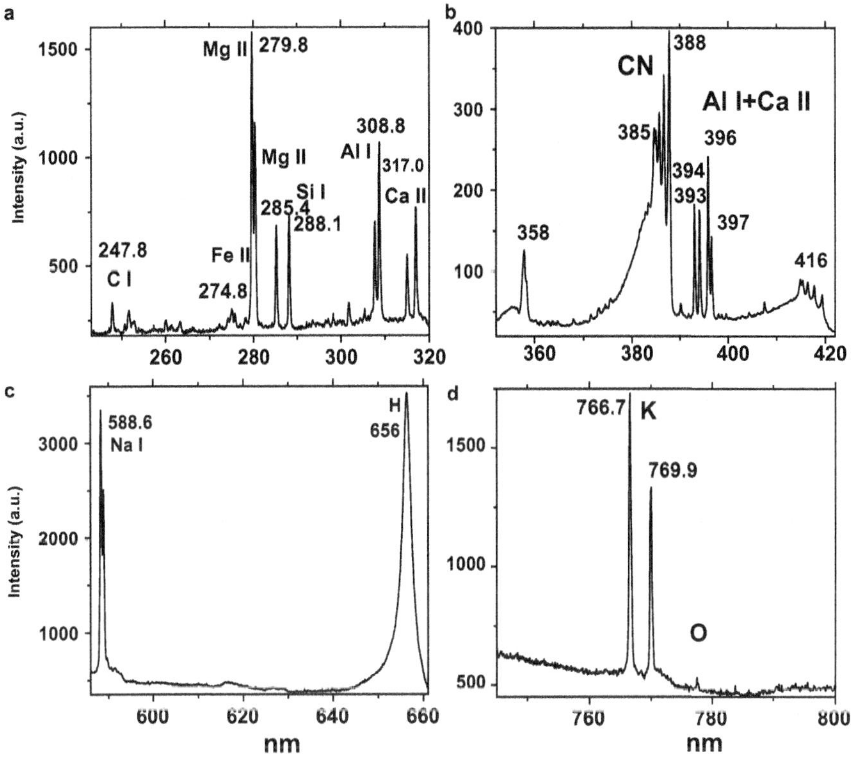

Fig. 8.44 (a–d) Typical online breakdown spectrum of coal

in order to earn another order of magnitude sensitivity, which are relevant to natural abundance of REE in natural phosphates. Regular LIBS study revealed that specific phosphate with elevated REE concentration contains Ti and Zr emission lines (Fig. 8.43c). Luminescence study confirmed that zircon impurity takes place in this phosphate, which evidently contains REE (Fig. 8.43d).

8.3.8 Coal

8.3.8.1 Ash Content Evaluation

Online impurities content evaluation enables raw and washed coal quality monitoring; control of sorting, blending and stockpiling actions; beneficiation plant optimization; product compliance monitoring; boiler feed control; coke oven and furnace feed control; consignment and batch monitoring; and integrated coal quality management systems. Presently mainly PGNAA machines have been used for this task.

In coal analysis detectable elements include the key inorganic components of coal – such as Al, Si, Mg, Ca, Fe, Na, and K – in addition to C and H. Ash content may be determined from these element analyses, similar to the approach used in PGNAA. Figure 8.44 demonstrates typical online breakdown spectra of coal determined in spectral range from 250 to 340 nm. The lines of C, Mg, Al, Si, Ca, Fe, Na, K and organic matter were detected. A weighted sum of all detected elements was sought for best correlation with ash content; hypothetically, each element has its specific coefficient of the strength of its plasma emission relative to its participation in ash formation. Approximately 500 spectra were accumulated for one analytical point. The actual analytical task was as follows. The material on the conveyer belt consists of a mixture of 2 Seam and 4 Seam coals, with ratio changing continuously. The coal product ash content demands stringent control. The ash analyses standard error should be less than 0.55 % mean absolute error however a higher accuracy is desirable.

As it was presented above, it was found that the correlation between LIBS and PGNAA online analyzers is good, the average standard error being only 0.32 % mean absolute error, while the correlations between LIBS and Coalscan with laboratory data are worse, 0.5 % and 0.38 % mean absolute error, correspondingly

8.3.8.2 Volatiles Content Evaluation

Besides ash, another important parameter for coal quality evaluation is its calorific value (CV), which mainly depends on ash content, moisture and volatiles. Online moisture analyzers are well known and may be easily used together with LIBS machine. Thus, to enable CV real time evaluation by LIBS it is very important to develop LIBS technique for volatiles analysis. Volatiles in coal refer to the components of coal, except moisture, which are liberated at high temperature in the absence of the air. This is usually attributed to a mixture of organics as hydrocarbons chains, aromatic hydrocarbons, some sulfur groups etc. Volatile matter of coal usually is determined in laboratories using heating the coal samples to 900–950 °C. Such method is time consuming and not suitable for real time control.

The aim was to evaluate if LIBS approach is principally applicable for volatile matter evaluation. It is known that LIBS technique is capable to analyze the organic impurities in coal, such as H, O, N and C. Recently laboratory LIBS ability for non-metallic impurities evaluation was studied on a series of coal samples with large-scale variety of volatiles matter (Zhang et al. 2008). We tried to determine what organic elements may be selected as quantitative indicators of volatile matter presence and if breakdown spectra of those elements are good enough for their quantitative analyses. This stage was accomplished in laboratory conditions with the final aim to evaluate LIBS perspective for volatiles online control in coal on moving belt conveyer in real industrial conditions. Analyzing known literature data on the correlation between different organic elements and volatile contents, it was found that only hydrogen exhibits clear linear correlation with volatile matter. We studied the hydrogen-volatile matter correlation using representative sample of

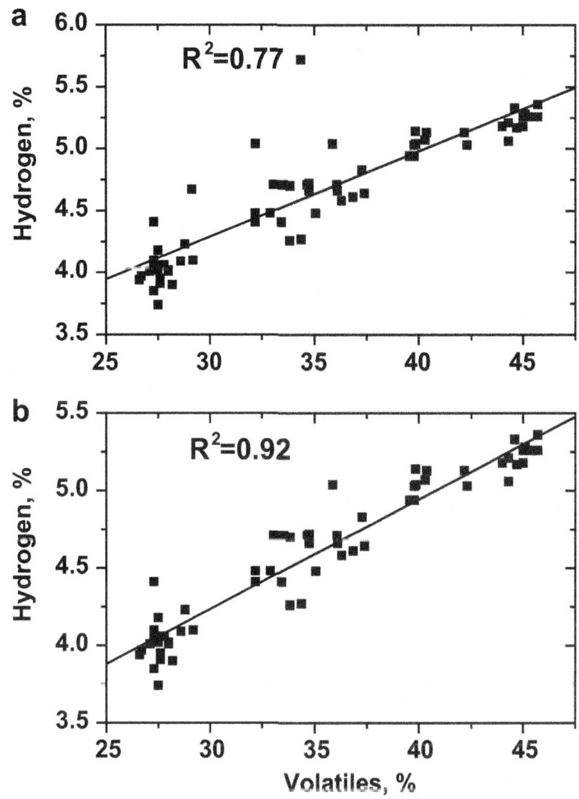

Fig. 8.45 (a–b) Correlations between hydrogen impurities and volatile matter in coal used by IEC during 2008–2009 years

coals used by Israeli Electrical Company during 2008–2009 years. Among 80 samples only 3 revealed substantial deviation from the main trend. After those samples exclusion, the hydrogen-volatiles correlation looks quite linear with very high linear correlation coefficient of $R^2 = 0.92$ (Fig. 8.45). Thus it may be concluded that hydrogen-volatiles correlation exists and H may be selected as quantitative volatile matters indicator.

The next step was LIBS evaluation as a technique for hydrogen quantitative analyses. Figure 8.46 demonstrates coal breakdown baseline corrected spectra in visible spectral range. The delay $\Delta t = 100$ ns and gate width $W = 500$ ns were optimized to get the best conditions for H, C, O, and N lines detection. The lines at 495.8 nm and 559.2 nm (Fig. 8.46a) are the second orders of corresponding ultraviolet lines of C and Mg, enabling to detect C at 495.8 nm avoiding the necessity of two spectral measurements in UV and VIS spectral ranges. Those lines disappear when 320 nm long pass filter blocks emission at wavelength shorter than 300 nm (Fig. 8.46b). The intensities at 656.3, 746.8, and 777.3 nm lines were measured for H(I), N(I) and O(I) evaluation, respectively.

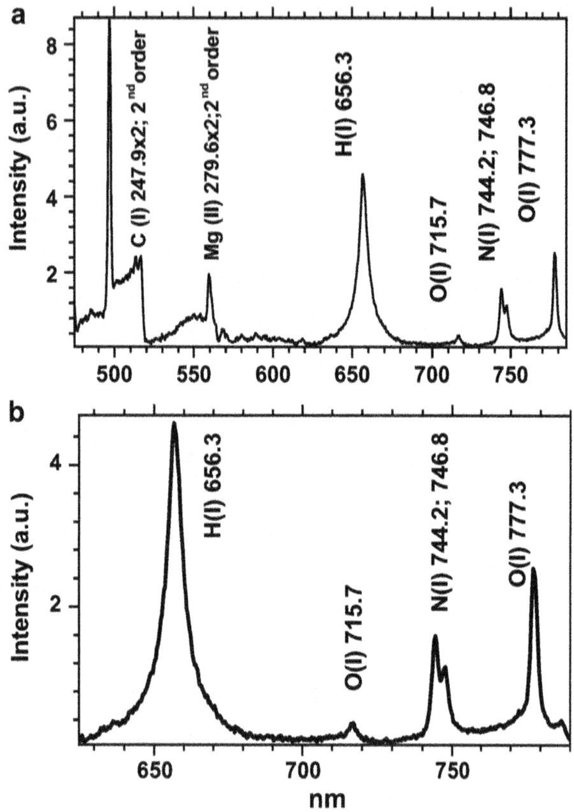

Fig. 8.46 (a–b) Breakdown spectra of coal sample with background subtracting. $\Delta t = 100$ ns and $W = 500$ ns (**a**); the same with long pass 320 nm filter placed in front of the entrance end of the optical fiber (**b**)

In laboratory with constant measurements conditions (laser intensity, laser spot diameter and so on) the line intensity (I_i) of the certain element is proportional only to its concentration (N_i) in sample. Thus having number of the reference samples with known concentrations of desired elements one can construct the calibration dependence ($I_i \propto N_i$). For online analyses of materials on moving belt conveyer the measurement conditions are constantly changing and it is obligatory to construct the calibration dependence for each element as relation ($I_i/I_o \propto N_i$) where (I_0) is the line intensity of the element, which concentration is constant during the measurements, thus (I_i/I_o) will depends only on N_i and not on experimental conditions.

Carbon may be a potential candidate for such reference, but it presents in coal in a fixed form and as a part of volatiles, making such normalization doubtful. Such conclusion is confirmed by the fact that C emission intensity is strongly influenced by volatiles quantity. Oxygen in coal also exists as component of organics and as different oxides of inorganic elements (Si, Al, Mg, Ca) and also may occur as results of air breakdown on the surface of the samples. From the other side, the contents of nitrogen N are usually low in coal and still the N(I) line doublet (744.2

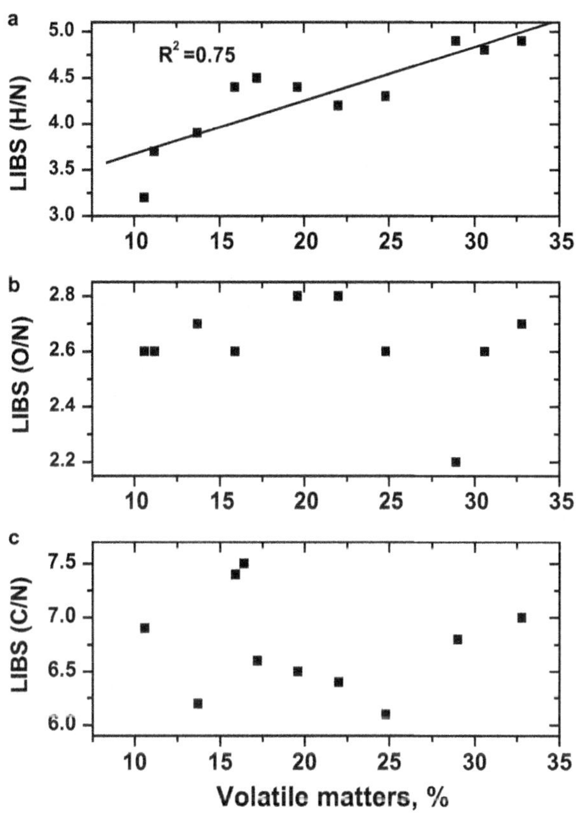

Fig. 8.47 (a–c) Correlations between hydrogen (**a**), oxygen (**b**) and carbon (**c**) emissions normalized to ambient nitrogen emission and volatile matter in coal

and 746.8 nm) observed in our LIBS is relatively strong. Thus it may be proposed that those lines are mostly due to the air breakdown on the surface of the samples and may be used as reference signal in our measurements. Similarly, the N(I) line at 746.8 nm in LIBS spectra of anthracite coal was used as a reference signal for organic oxygen determination (Zhang et al. 2008).

Figure 8.47 demonstrates lines intensities ratio dependences for O(I)/N(I) (triangles), C(I)/N(I) (circles) and H(I)/N(I) (squares) on volatile matters. As it was expected, O(I)/N(I) and C(I)/N(I) line intensities ratios are independent on volatile contents within accuracy of our experiments. By the way, almost constant O(I)/N (I) lines ratio confirms that N(I) lines as well as O(I) lines in LIBS of our samples are mostly due to air breakdown nearby to the surface of the samples. From the other side, H (I)/N(I) lines ratio reveals real dependence on volatile matter. This relation exhibits linear dependence: H(I)/N(I) $= A + BX$, where $A = 3.1$, $B = 0.06$ and X is volatile matters in % with correlation linearity coefficient $R^2 = 0.72$.

It has to be noted that all those experiments have been conducted on dry or slightly wet coal samples. Our preliminary tests of coals with high moisture contents revealed that hydrogen emission depends on moisture content and its correlation with volatiles becomes less evident. The solution is to separate the hydrogen emission connected with moisture from emission connected with volatiles by using independent moisture analyzer simultaneously with LIBS measurements.

8.3.8.3 Sulfur Content Evaluation

The most intense sulfur plasma lines are in VUV peaking at 180.7, 182.03 and 182.6 nm. In this region atmospheric oxygen has strong absorption and detection of LIBS signal becomes difficult. From the first glance, such approach is not suitable for on-line measurements in ambient conditions, but our failure to have the good results for LIBS detection of low sulfur levels in other spectral regions stimulated deeper research in this direction (Gaft et al. 2009c). First experiments were accomplished when the sample was very close to spectrometer entrance. Plasma emission was dispersed in an imaging spectrograph, which was purged with nitrogen in order to prevent atmospheric absorption inside the spectrograph. Nitrogen was also purged between the sample and the slit of the spectrometer. It was found that single pulse IR excitation is suitable for the task (1064 nm, 30 mJ/pulse, D = 100 ns, G = 3 μs). It was found that in such conditions definite sulfur emission lines may be detected in natural coal samples with 1.2–2.7 % of S. Good signal/noise ratio evidently enables to go down to 0.1–0.2 % of S which is actually our desirable detection limit for online analyses.

During the next stage the experiments have been accomplished using a spectrograph purged with nitrogen, while the samples were 1 cm from the slit in ambient conditions without nitrogen purging. Figure 8.48 demonstrates that at such conditions not only pure S and artificial mixtures of coal with sulfur may be analyzed, but also natural coal samples with very low sulfur concentrations.

It is possible to evaluate what is the maximum distance between the sample and the spectrometer in ambient conditions, at which it is possible to detect sulfur emission lines in this spectral range using a N2-filled spectrometer. This spectral range is opaque in air due to O_2 absorption. Using the absorption cross section for O_2 molecule and the ideal gas equation: PV = NkT, where P = 105 Pa is the normal atmosphere pressure, $V = 10^{-6}$ m^3 is the volume = 1 cm^{-3}, and $k = 1.38 \times 10^{-23}$ J K^{-1} is the Boltzmann constant, 24×10^{18} molecules exist in 1 cm^3 of air at 300 K. The air consists 20 % of O_2, so in 1 cm^3 of air there exist 4.8×10^{18} O_2 molecules. The absorption coefficient κ of the air at normal pressure and at 300 K in around the 180 nm spectral region is 4.810^{-20} cm^2, yielding a working distance of 0.23 $cm^{-1} = 4.8 \times 10^{-20} \times 4.8 \times 10^{18}$ cm^{-3}. This is equivalent to a penetration depth (signal diminishing in e times) of the light at 180 nm of about 5 cm. Using an optical system from LiF material, which is transparent down to 110 nm, it is possible to move the sulfur sample as far as 20 cm from the detector with the emission intensity stronger than those received from 1 cm distance without collecting optics. It confirms the possibility of using a N_2 – filled spectrometer for an on-line analysis of sulfur in ambient conditions with sample-detector distances

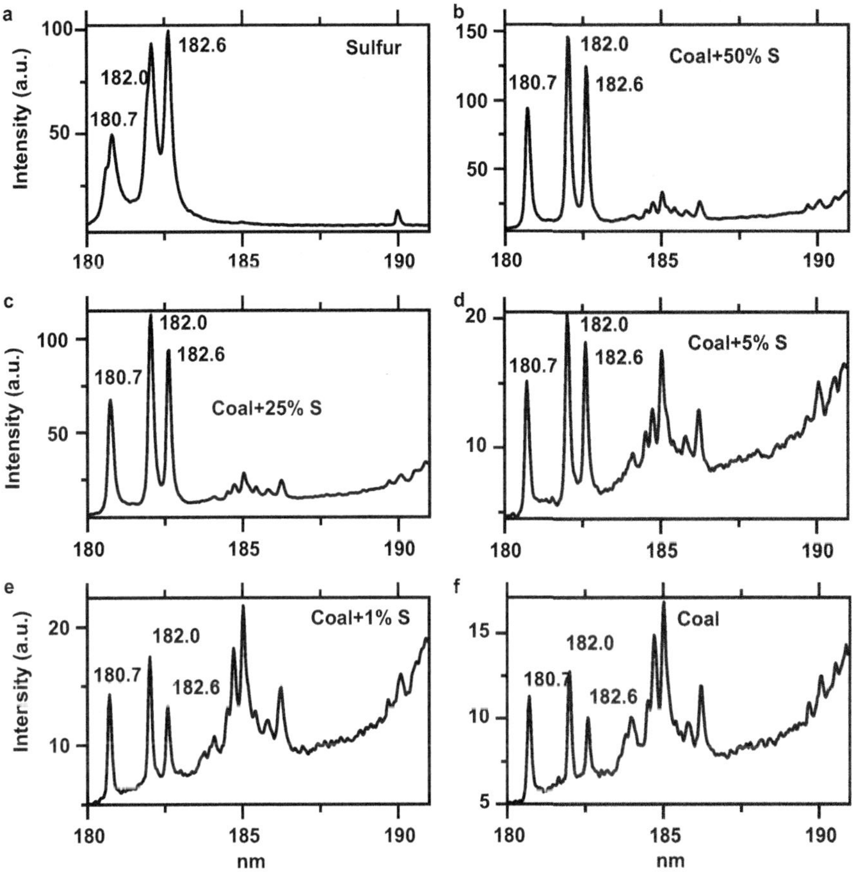

Fig. 8.48 (a–f) Single-pulsed LIBS of mixtures of coal with sulfur in VUV with nitrogen flashed spectrometer situated 1 cm from spectrograph slit in ambient conditions

up to about 20 cm. The sulfur emission lines detected from 20 cm distance with collimating optics from natural coal samples containing 0.5 % S. Thus, even low sulfur concentrations may be quantified.

We checked such possibility by analyzing two sets of coal samples from South Africa and Australia with known sulfur concentrations in static laboratory conditions (1064 nm, 30 mJ/pulse, D = 100 ns, G = 3 μs). Ten shots have been accumulated and averaged from each sample and such procedure was repeated five times. Two different correlations were detected between sulfur concentrations and corresponding LIBS intensity for coal from South Africa and Australia (Fig. 8.49a). They are evidently connected with different matrix effects, a well-known LIBS phenomenon, which was often detected by us during ash content analysis in coal by LIBS. Correspondingly two different algorithms were used in order to recalculate LIBS intensity to weight % of sulfur (Fig. 8.49b). Linear

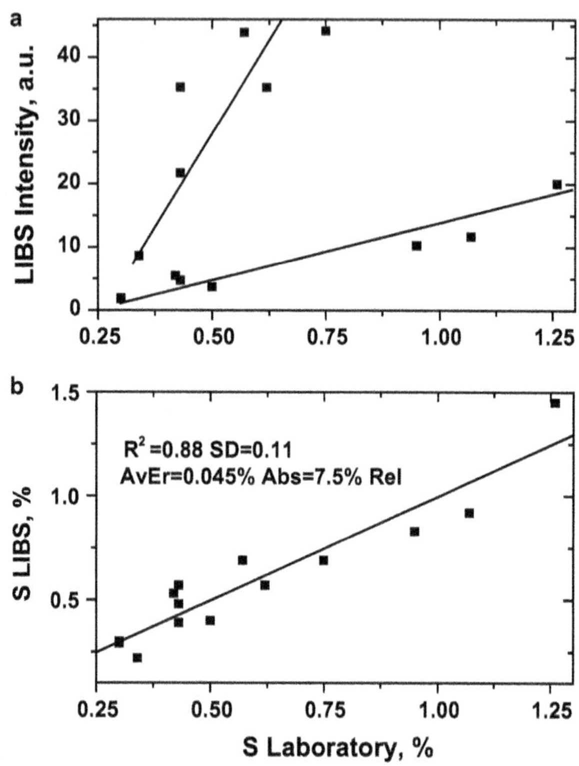

Fig. 8.49 (a–b) Correlation between sulfur laboratory data and LIBS intensity in natural coals in VUV in air with nitrogen flashed spectrometer using single-pulse IR excitation (**a**). Two different trends represent two coal types and result from matrix effect (**b**)

regression $R^2 = 0.88$ standard deviation (SD = 0.11) and relative average error (AvEr = 7.5 % Rel) look quite satisfactory, yet statistic is very limited and those data may be considered as potentially promising but only preliminary. Next step will be laboratory prototype development and tests in dynamic regime with much more representative sampling.

8.3.9 Cement

8.3.9.1 LIBS Elemental Control

Figure 8.50 demonstrates typical breakdown spectra of cement with clear analytical lines of Ca, Al, Si, Mg, Fe and Ti. The correlation between LIBS and control laboratory data for those elements in representative raw mix samples are quite satisfactory and promise successful industrial implementation (Gaft et al. 2014c).

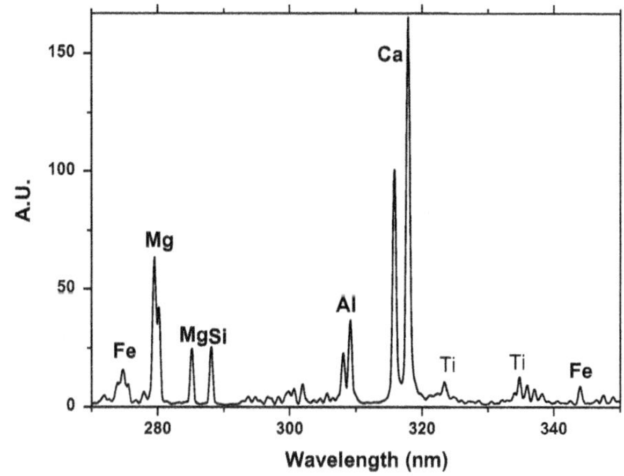

Fig. 8.50 Typical breakdown spectrum of cement raw mix with clear analytical lines of Ca, Al, Si, Fe, Mg and Ti

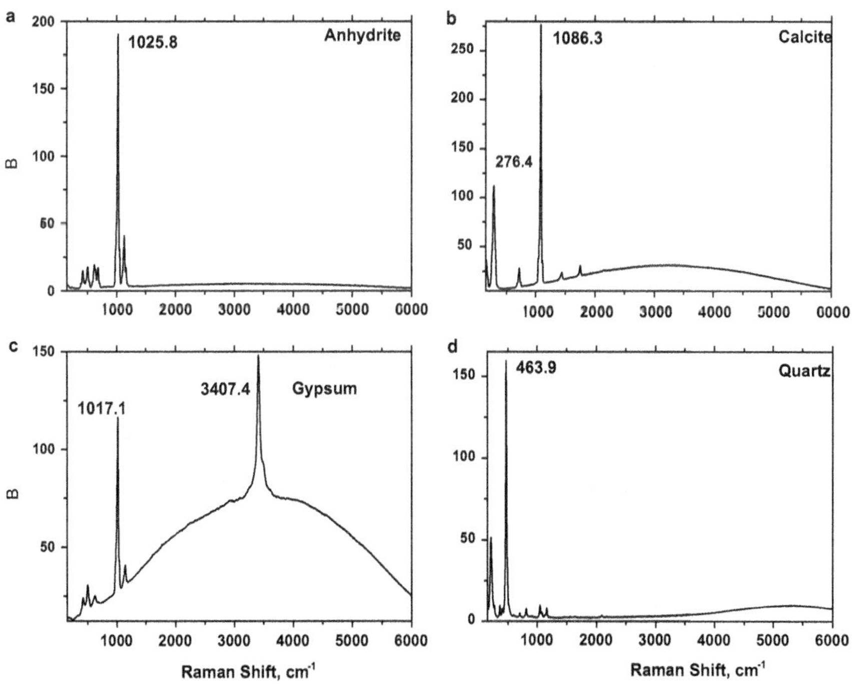

Fig. 8.51 (**a–d**) Gated Raman spectra with excitation by 532 nm of calcite, anhydrite, gypsum and quartz relevant for cement industry

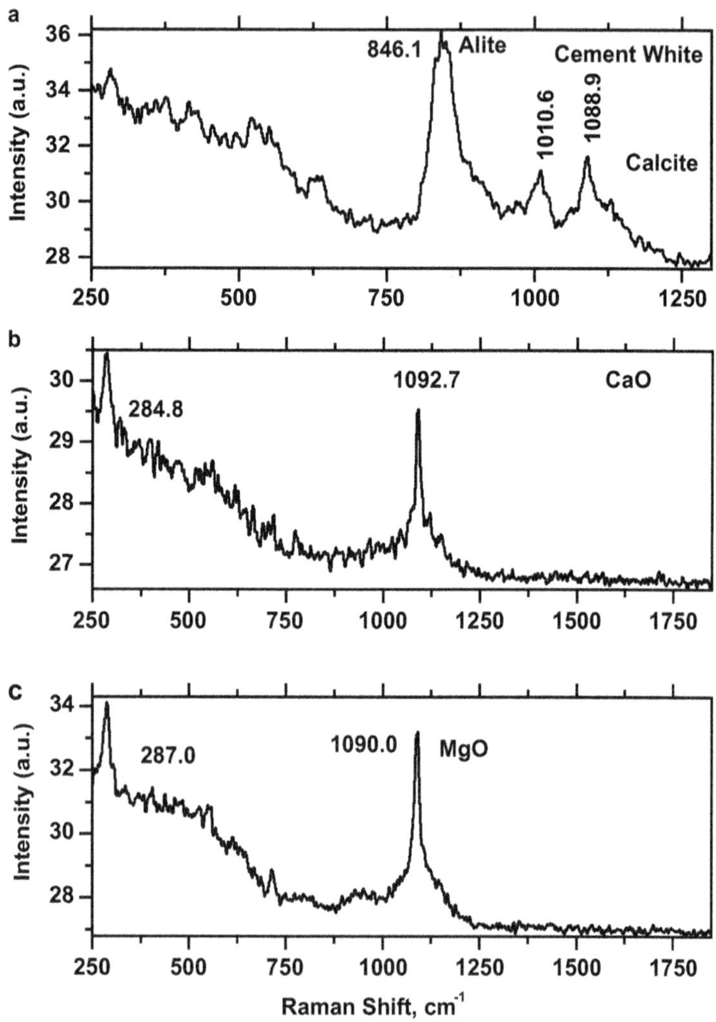

Fig. 8.52 (**a–c**) Gated Raman spectra with excitation by 532 nm of alite, CaO and MgO

8.3.9.2 LIBS-Raman for Online Mineral Analysis

Figure 8.51 presents the Raman spectra of several phases important for the cement industry received with excitation by 532 nm, which is the second harmonic of industrial Nd-YAG laser and may be effectively used for simultaneous detection of LIBS and Raman signatures. The major and mostly analyzed mineral phases of cement clinker are alite and belite. Alite is characterized by the strong Raman band peaking at 845 cm^{-1} range, while belite at 805 cm^{-1}. Tricalcium aluminate and ferrite have also been studied and exhibit the strongest Raman bands peaking at

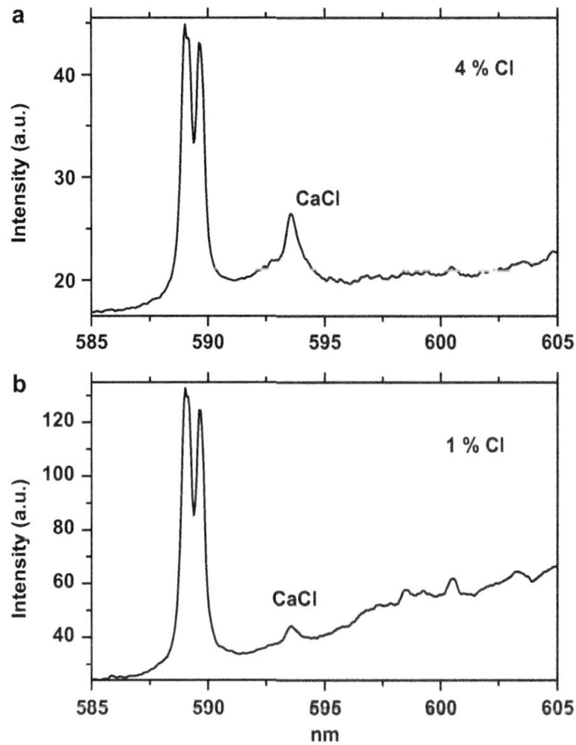

Fig. 8.53 (a–b) Breakdown spectra of CaCl in concrete with different Cl concentrations

754 and 737 cm^{-1}, correspondingly. Hydrous cementitious materials, sulfates and carbonates also have characteristic Raman signals. Figure 8.52 presents Raman spectra of alite, CaO and MgO received in online mode. It is a proof that online Raman may be a potential tool for cement production online control.

8.3.9.3 Chlorine Detection in Concrete

Chlorine detection in concrete is very important for its quality evaluation. Concrete contains a lot of calcium, thus CaCl molecular emission is a potential tool for Cl quantitative analysis in concrete. Figure 8.53 presents the data on CaCl molecular emission in concrete with 4 and 1 % of Cl. It is confidently detected at such levels and it seems that the 0.2 % of Cl concentrations may be detected by this method.

References

Cutmore N, Eberhardt J (2002) The future of ore sorting in sustainable processing. Cairns: s.n., Processing Conference, pp 287–289

Cremers D, Radziemski R (2006) Handbook of laser-induced breakdown spectroscopy. Wiley, New York

Frick-Begemann C, Noll R, Wotruba H, Schmidts C (2010) Fast and flexible – laser assisted sorting of minerals. AT Miner Proc 51:2–7

Gaft M (1989) Luminescence of minerals under laser excitation. Ministry of Geology, Moscow (in Russian)

Gaft M, Scorobogatova N, Rassulov V, Moroshkin V (1989) The use of natural silver halogens luminescence for mineral prospecting. Miner J 11:58–64 (in Russian)

Gaft M, Nagli L, Panczer G, Reisfeld R (2002) Laser-induced luminescence and breakdown spectroscopies evaluations of phosphates with high dolomite content. In: Zhang P, El-Shall H, Somasundran P, Stana R (eds) Beneficiation of phosphates, fundamentals and technology. SME, Littleton, pp 145–152

Gaft M, Reisfeld R, Panczer G (2005) Modern luminescence spectroscopy of minerals and materials. Springer, Berlin/New York

Gaft M, Sapir-Sofer I, Modiano H, Stana R (2007) Laser induced breakdown spectroscopy for bulk minerals online analyses. Spectrochim Acta B 62:1496–1503

Gaft M, Dvir E, Modiano H, Schone U (2008) Laser induced breakdown spectroscopy machine for online ash analyses in coal. Spectrochim Acta B 63:1177–1182

Gaft M, Nagli L, Fasaki I et al (2009b) Laser-induced breakdown spectroscopy for sulfur on-line analyses of minerals in ambient conditions. Spectrochim Acta B 64:1098–1104

Gaft M, Nagli L, Groisman Y (2011) Luminescence excited by laser induced plasma. Opt Mater 34:368–375

Gaft M, Nagli L, Groisman Y, Barishnikov A (2014a) Industrial online raw materials analyzer based on laser-induced breakdown spectroscopy. Appl Spectrosc 68:7–18

Gaft M, Nagli L, Groisman Y et al (2014c) LIBS-Raman analyzer for online process control. World Cement September: 83–87

Gao Z, Zheng S, Gu Z (2002) Review of beneficiation technology for Florida high dolomite pebbel. In: Zhang P, El-Shall H, Somasundran P, Stana R (eds) Beneficiation of phosphates, fundamentals and technology. SME, Littleton, pp 247–259

Gesing A (2007) Elemental analysis and chemical composition based material separation and blending. TMS (The Minerals, Metals & Materials Society), Annual meeting and exhibition, elemental analysis and chemical composition based material separation and blending, pp 101–110

Gorobets B, Rogojine A (2001) Luminescent spectra of minerals. Handbook. RPC VIMS, Moscow

Gorobets B, Litvintsev E, Rogojine A (1997) Luminescence sorting of nonmetallic raw materials. In: ICAM'96, 5th international on congress applied mineralogy, Warsaw, Programme and Abstract, 2–5 June 1996, pp 229–233

Groisman Y, Gaft M (2010) Online LIBS analyses in potassium fertilizers industry. Spectrochim Acta B 65:744–749

Gudaev O, Kanaev I, Shlyufman E (1999) Laqser based separation of diamonds from the rock. Detect Syst 3:19–23

Lamprecht G, Himan H, Snyman L (2007) Detection of diamond in ore using pulsed laser Raman spectroscopy. J Miner Proc 84:262–273

Meisner L (1994) Nonlinear characteristics of minerals. In: Marfunin A (ed) Composition, structure and properties of mineral matter. Springer, Berlin, pp 496–497

Meisner L, Kuz'min V (1986) Theoretical aspects of nonlinear optical methods of separation. Obogashenie rud 5:22–25 (in Russian)

Mokrousov V, Lileev V (1979) Radiometric beneficiation of not radioactive ores. Nedra, Moscow (in Russian)

Moroshkin V, Evdokimenko E, Gaft M et al (1997) Method of artificial luminescent films – a new method of luminescence analysis of minerals and rocks. Rocks Metals 3:63–72

Nagli L, Gaft M, Gornushkin I (2012) Fraunhofer type absorption lines in double pulse laser induced plasma. Appl Opt 51:201–212

Noll R, Fricke-Begemann C, Brunk M et al (2014) Laser-induced breakdown spectroscopy expands into industrial applications. Spectrochim Acta B 93:41–51

Ostapenko M (1990) Technological evaluation of the mineral raw materials. Nedra, Moscow (in Russian)

Palanco S, Laserna J (2004) Remote sensing instrument for solid samples based on open-path atomic emission spectrometry. Rev Sci Instrum 75:2068–2075

Prokofiev I, Gorobets B, Shuriga T et al (1979) Origin of the fluorescence of lithium minerals. Izv Akad Nauk SSSR Ser Geol 3:88–94 (in Russian)

Rosenwasser S, Asimellis G, Bromley B et al (2001) Development of a method for automated quantitative analysis of ores using LIBS. Spectrochim Acta B 56:707–723

Sabsabi M, Cielo P (1995) Quantitative analyses of aluminium alloys by laser induced breakdown spectroscopy and plasma characterization. Appl Spectrosc 49:499–507

Sallé B, Mauchien P, Maurice S (2007) Laser-induced breakdown spectroscopy in open-path configuration for the analysis of distant objects. Spectrochim Acta B 62:739–768

Salter J, Wyatt N (1991) Sorting in the minerals industry: past, present and future, vol 4, Miner Engin. Pergamon Press, Great Britain, pp 779–796

White W (1984) Separation or concentration of magnesium-bearing minerals by induced fluorescence. US Patent 4,423814

Winefordner J (2000) Laser induced breakdown spectroscopy for elemental process monitoring of slurry streams. Florida Institute of Phosphate Research (FIPR), publication N 04-057-169

Wotruba H, Harbeck H (2010) Sensor-based sorting. In: Ullamann's Encyclopedia of industrial chemistry, John Wiley and Sons Eds (32nd edn)

Zhang L, Dong L, Dou H et al (2008) Laser-induced breakdown spectroscopy for determination of the organic oxygen content in anthracite coal under atmospheric conditions. Appl Spectrosc 62:458–463

Chapter 9
Different Applications

9.1 Gemology

Gemology is the study of gemstones. At the heart of gemmology is gem identification and treatments (color and purity enhancement by thermal treatments, irradiation…). For example, some rubies, spinels and garnets are impossible to tell apart from each other by observation, but their values are considerably different. A precise and accurate means to tell them apart is absolutely necessary. When dealing with whole crystals, the ruby and garnet are easy to distinguish, but most of the material that gets cut into gems is not found in whole crystals, but in broken pieces. Using the techniques of mineralogy, they are easily distinguished from each other. Scratch tests, where various substances scratch the unknown, will determine its hardness. Other useful tests are the reaction to acids and the flame of a blowtorch. These are categorized as destructive tests and are obviously inappropriate for cut gems.

A method needed to be devised where cut gems could be identified without damage. To this end scientists began to first, identify the measurable physical and optical properties of the gems. Next they devised instruments to measure these properties. That is not to say that it doesn't require substantial education to identify gems. It is a large and complex subject that is continuing to increase in complexity as new gems are discovered and new ones are created in the laboratory.

Simulants or imitations are anything that is posing as something else. For example, a white topaz posing as a diamond is an imitation. With the continued development of crystal growth technology, various synthetic and imitation gem materials are increasingly encountered in the jewelry industry. For example, synthetic cubic zirconia, and synthetic moissanite (SiC-6H) have a very similar appearance to diamonds, and are used as imitation diamonds. Because of their very different chemical and physical properties, it is relatively simple to determine the imitation. Raman spectroscopy and LIBS may be useful for express gemstones identification, combining in-situ chemical and structural analysis (Table 9.1).

M. Gaft et al., *Modern Luminescence Spectroscopy of Minerals and Materials*, Springer Mineralogy, DOI 10.1007/978-3-319-24765-6_9

Table 9.1 Luminescence, Raman and LIBS parameters of several gemstones

Mineral	Laser-based spectroscopy		
	Luminescence	Raman, cm^{-1}	LIBS, nm
Diamond	N3, S2, S3, S4, A	1332	C (248, 406, 427)
Spinel	Cr^{3+}	407, 665, 766	Mg (280, 383, 518) Al (309, 394, 396)
Topaz	Cr^{3+}	407, 459, 553, 843, 928	Al (309, 394, 396) Si (251, 287)
Danburite	Eu^{2+}, Yb^{2+} (77 K)	611	Ca (393, 396, 422) B (251)
Forsterite	Mn^{2+}	824, 855	Mg (280, 383) Si (251, 287)
Fluorite	Eu^{2+}, Yb^{2+} (77K)	320	Ca (393, 396, 422) Ca F (603) K (766, 770)
Corundum	Cr^{3+}	420, 750, 812	Al (308, 394, 396)
Scapolite	$S^{2}\_$	461, 538, 769	Ca (393, 396, 422) Na (589) K (766, 770) Al (309, 394, 396) Si (251, 287)
Apophyllite	Mn^{2+}	432, 583, 663, 1061	Ca (393, 396, 422) Si (251, 287) CaF (603) K (766, 770)
Rhodonite	Mn^{2+}	671, 975, 999	Mn (258, 280) Si (251, 287)
Rhodochrosite	Mn^{2+}	1086	Mn (258, 280)
Datolite	Ce^{3+}, Eu^{2+}, Yb^{2+} (77 K)	597, 697, 918, 987, 1082, 1175	Ca (393, 396, 422) B (251)
Scheelite	WO_4, MoO_4	798, 911, 1010, 1076, 1305	Ca (393, 396, 422) W (295, 401, 407)
Emerald	Cr^{3+}	398, 686, 1012, 1069	Al (309, 394, 396) Si (252, 252) Mg (280, 383)
Alexandrite	Cr^{3+}	932	Be (318) Al (309, 394, 396) Si (252, 252)
Garnet	Cr^{3+}	371, 549, 825, 879	Al (309, 394, 396) Si (252, 252) Mg (280, 383)
Apatite	Ce^{3+}, Eu^{2+}, Sm^{3+}, Mn^{2+}	432, 584, 965	Ca (393, 396, 422) CaF (603)
Kyanite	Cr^{3+}	441, 490, 896, 957	Al (308, 394, 396) Si (251, 287)
Benitoite	Ti^{4+}, Ti^{3+}	535, 573, 932	Ba (413, 456, 493) Ti (307, 317, 323, 334) Si (252, 252)
Charoite	Ce^{3+}, Eu^{2+}	640, 677, 1141	Ca (393, 396, 422) Si (251, 287) Na (589) Ca F (603) K (766, 770)
Zircon	$(MeO_m)^{n-}$, Dy^{3+}, Fe^{3+}	974, 1008	Zr (268, 339, 348)
Titanite	Sm^{3+}, Cr^{3+}, Ti^{3+}	470, 611, 861	Ca (393, 396, 422) Ti (307, 317, 323, 334)
Tourmaline	Mn^{2+}, Cr^{3+}, Fe^{3+}	375, 413, 643, 716	Al (308, 394, 396) Si (251, 287, 507) Mg (280, 383)

"Synthetic" gem refers to materials that duplicate their natural counterparts. Emerald, sapphires and spinel are common synthetics (Armstrong et al. 2000). One of the most critical areas of innovation and challenge involves alexandrite. This material is a beryllium aluminum oxide that typically exhibits a red to green color change as lighting environments are altered. This rare variety of chrysoberyl occurs in numerous places around the world, including Russia, Sri Lanka, Africa, Brazil, and India. Some laboratory-grown alexandrites have become so similar to their natural counterparts that there is a confidence crisis in the marketplace. Synthetic or laboratory-grown materials duplicate natural gems in physical, optical, and chemical terms. Indeed, synthetic materials grown under controlled laboratory conditions can exhibit higher overall quality than their natural counterparts. The significant advantage of gemstone synthesis is consistently higher yields of fine quality material, which is a stark contrast to the relatively finite percentages of higher-quality gems that emerge from the production of mines around the world. As a result, gem synthesis consistently produces larger quantities of fine and uniform gem material at a fraction of the cost of comparable natural alternatives.

While natural and man-made materials can share the same physical and optical properties, there are still considerably differences, the main one being rarity. A natural gem takes considerable time to form and is usually millions of years old. Plus, many feel they have aesthetic qualities not found in mass-produced materials. While natural and man-made materials appear nearly identical, their values vary considerably. For this reason it is important to be able to distinguish between the two. As crystallization techniques continue to improve, the technical challenges facing gem-testing laboratories will increase dramatically. However, regardless of the similarities between natural and laboratory grown products, the differences between these growth environments can be entrapped in the chemical and physical natures of each material. Whether laboratory-grown or natural, each growth system leaves its indelible fingerprint on the final product. As the technical differences diminish between the natural and laboratory-grown gemstones, the degree of laboratory sophistication necessary to nondestructively distinguish between these materials must increase significantly. Technical superiority is essential to maintain the long-term stability of the gemstone market and the credibility of laboratory testing facilities (Armstrong et al. 2000).

Because chemical and structural properties of natural and artificial gems are very similar in this case, the possibilities of Raman and LIBS methods are rather limited. It was found that other laser-based techniques could be very effective for rapid spectroscopic discrimination between natural and synthetic emeralds, rubies, and alexandrite (Armstrong et al. 2000). The first one is DRIFTS (Diffuse Reflectance Fourier Transformed Infra-Red Spectroscopy) and the second one is NIR (Near Infra-Red) Spectroscopy. In some cases it was even possible to discriminate between gems made by different synthetic processes. Once again, there is a significant benefit to having two independent methods available.

Luminescence properties may also be very effective. The main two differences are the following (Gorobets et al. 2003).

1. Luminescence intensity with very rare exceptions is much higher in artificial gemstones compare to natural counterparts. It may be explained by the fact that the activator contents are usually higher in laboratory made gems, while the quenching centers concentrations are lower.
2. Fine and super-fine structure of luminescence lines of Cr^{3+} and TR^{3+} are narrower in artificial gems. The reason is that in natural precious stones many optically active centers usually present, while in artificial ones only one or two occur.

Luminescence appears to be effective for differentiation between the artificial and natural diamonds. For example, fluorescent characteristics are used in diamonds quality gemological certificates, for distinction between natural and synthetic, irradiated and enhanced gem diamonds. Most type IIa (and rare type IaB) diamonds transmit short-wave ultraviolet (SWUV) light while most of the diamonds (type Ia) do not transmit SWUV. As well, recording of the sample's photoluminescence spectrum at 77 K provides an excellent way to identified high pressure-high temperature treated (HTHP) diamond (GE POL, General Electric). Emission peaks at 637 and 575 nm under continuous 514 nm Ar laser excitation indicates the presence of a small amount of N-V centers (single nitrogen linked to a carbon vacancy) in the gemstone. While most but not all natural, i.e. non-treated colorless IIa diamonds also show an N-V (637 nm) and the N-V0 (575 nm) feature, the peak height ratio 637/575 is considered indicative for treatment identification. Diamonds with high ratios >2.8 are HTHP treated, whereas diamonds with low ratios <1.6 are natural (Chalain et al. 2000). Lindblom et al. (2003, 2005) studied differentiation of natural and synthetic gem-quality diamonds by luminescence properties. It was found that time-resolved luminescence spectra of natural and synthetic diamonds display clear mutual differences. The spectra of natural diamonds revealed emission bands caused by complex nitrogen-vacancy aggregates whereas the bands of synthetic diamonds reflect simple nitrogen-vacancy aggregates and nickel-containing defects. The deep-UV excited time-resolved luminescence spectra of natural diamonds showed fast-decaying donor-acceptor pair recombination bands at around 410 nm. With longer delay and gates the broad band shifted to 435 nm. All synthetic diamonds display luminescence from nickel-related defects centered at 480 and 530 nm at 300 and 77 K, respectively, and all synthetic diamonds had a sharp luminescence bands at 694 nm caused by the Cr^{3+} ion impurity in corundum inclusions.

As crystallization techniques continue to improve, the technical challenges facing gem-testing laboratories will increase dramatically. However, regardless of the similarities between natural and laboratory grown products, the differences between these growth environments can be entrapped in the chemical and physical natures of each material. Whether laboratory-grown or natural, each growth system leaves its indelible fingerprint on the final product. As the technical differences diminish between the natural and laboratory-grown gemstones, the degree of laboratory sophistication necessary to nondestructively distinguish between these materials must increase significantly. Technical superiority is essential to maintain the long-term stability of the gemstone market and the credibility of laboratory testing facilities (Armstrong et al. 2000).

9.2 Micro-LIBS Multi-element Mapping of Mo-rich Glass-Ceramic

Based on the principle of the Raman microscope, a regular transmission and reflex ion optical microscope has been modified to unable tunable pulsed laser injection as well as fluorescence collecting via an optic fiber up to a intensified CCD camera allowing time resolved records. For this purpose, all optics was replaced to support high-energy laser pulses. Such equipment allows the following (Panczer et al. 2003):

- micro identification of the zone of interest in heterogeneous samples under white polarized light or UV illumination;
- a spatial resolution inferior to 5 μm of the analyzed zone, an monochromatic pulsed excitation (from 266 to 650 nm) and a time-resolved spectral analysis of the emission signal. It allows also in situ microanalyses at low and high temperature (Linkham type furnace −196 to 1500 °C) as well as high pressure (sapphire or diamond anvil cells).

To conduct efficient mapping, it is essential to detect as many elements as possible in a given spectral range, especially for the trace rare earth elements. Figure 9.1 shows LIBS spectra of the glassy phase and crystalline phase of the Mo-rich glass-ceramic sample in the range of 368–398 nm recorded with a gate delay of 150 ns and a gate width of 1 μs (atomic emission lines are labeled with "I" and ionic emission lines are labeled with "II"). Mo, Ca, Sr, REEs, Al, Fe and Zr emission lines were found. All the detected elements were validated by LA-ICP-MS technique. Emission lines of the detected elements are listed in Table 9.1, which are assigned according to the NIST database. Strong lines such as Mo I at 379.8, 386.4 and 390.3 nm, Ca II at 393.3 and 396.9 nm, and Al I at 394.4 and 396.2 nm, the spectral intensities change dramatically from glassy phase to crystalline one. Line intensities of Mo I and Ca II are much higher for crystal spectrum than for glass. By comparison, Al I 394.4 and 396.2 nm lines are very intense in the glass spectrum but almost disappear in the crystal one. This observation indicates that Mo and Ca are mainly located in the crystal while Al remains in the glass phase. This result is consistent with the statement that partial crystallization occurs in the Mo-rich borosilicate during thermal treatment and precipitates the $CaMoO_4$ powellite crystallites. Eu II at 390.7 nm, Nd II at 384.7 nm, La II at 394.8 nm, Pr II at 391.8 nm as well as Sr I at 394.1 nm lines are present in the spectrum from the crystal whereas Fe I at 372.0, 373.5, 373.7 and 374.5 nm, and Zr I at 388.5 and 389.1 nm lines are well observed in the glass. It can be preliminary inferred that REEs (Eu, Nd, La and Pr) and Sr are preferentially incorporated into the $CaMoO_4$ crystal while Fe and Zr prefer to stay in the glass. Spectral line intensities of REEs (Eu, Nd, La and Pr), Sr, Fe and Zr were relatively weak. However, the intensity variation of these species between crystal and glass was still large enough to obtain elemental maps. Lines selected for multi-elemental mapping are indicated in Table 9.2.

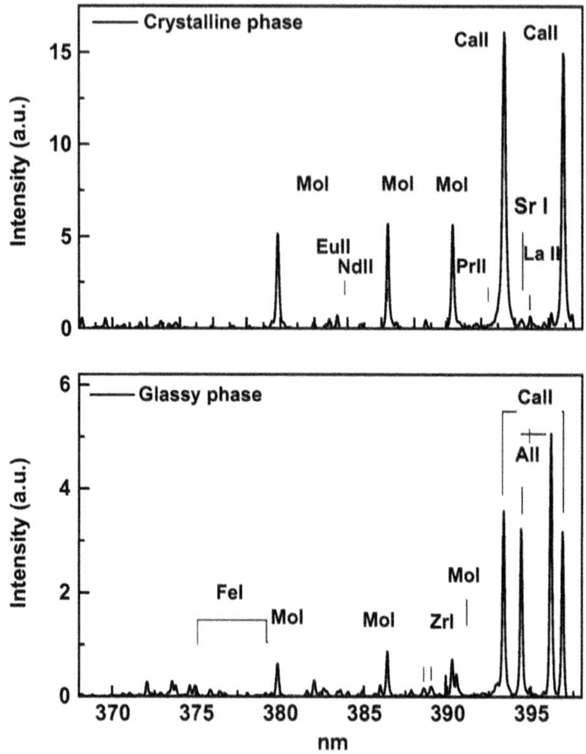

Fig. 9.1 Highly resolved LIBS spectra of glassy and $CaMoO_4$ powellite crystalline phases of a Mo-rich borosilicate glass-ceramic in the near UV range (368–398 nm); atomic lines are labeled with "I" and ionic lines are labeled with "II"

Table 9.2 Spectral lines of detected elements in the range of 364–398 nm (atomic lines are labeled with "I" and ionic lines are labeled with "II")

Species	Line position (nm)				
Mo I	379.8[a]	386.4	390.3		
Ca II	393.3[a]	396.8			
Sr I	394.1[a]				
Eu II	390.7[a]				
La II	394.8[a]				
Nd II	384.7[a]				
Pr II	391.8[a]				
Al I	394.4[a]	396.2			
Fe I	372.1	373.6	373.8	374.5[a]	374.9
Zr I	388.6	389.1[a]			

[a]Lines selected for μLIBS mapping

A 1200 × 1600 μm^2 (i.e. 1.2 × 1.6 mm^2) rectangular area on the polished surface containing both residual glass and crystallites was selected to conduct μLIBS mapping. Sequences of single-laser-shots covering the expected scanning area were defined by the software based on Labview program. Successive single-shot ablations were performed point by point with a step of 10 μm between neighboring craters. Consequently, the total mapping area consists of 120 × 160 points, which requires a durationof about 1 h. For each point in the scanned area, the corresponding line intensity of every species (marked with "*") in Table 9.1 was retrieved from the spectra with background subtracted. Intensity maps of each element were therefore constructed, as shown in Fig. 9.2. Maps of Mo and Ca are full-sized images which correspond exactly to the scanned area on the sample surface. Bright zones in these two maps present very intense emission from Mo I 379.8 nm and Ca II 393.3 nm lines, which indicate the regions rich in Mo and Ca on the sample surface. Dark areas correspond to small amount of Mo and Ca. Distributions of Mo and Ca coincide well with the localizations of the crystallites shown in the optical microscope image. It is important to know that only crystallites exposed at the sample surface can be detected by μLIBS. Since crystallites are randomly located in the glass matrix with various tilt angles with respect to the surface, only certain part of some crystallites was exposed after polishing. Nevertheless, most of the crystallites are deeply embedded inside the glass matrix. Taking into account the result of AFM measurement on crater depth, this part of crystallites locates at least 1.2 μm below the sample surface and was not detected by μLIBS mapping. Consequently, all the bright areas correspond to crystallites lying on the surface and the dark areas to glassy surface. As shown in the Mo and Ca maps, surface crystallites distribute on the right side of the scanned area. The white rectangle in the optical microscope image (Fig. 9.2) shows an area rich in crystals on the scanned sample surface. Zoomed maps in this area for Sr, REEs (Eu, La, Nd and Pr), Al, Fe and Zr are displayed in Fig. 9.2. All the spectral lines which were used to extract the elemental map of each element are listed in Table 9.2. According to these multi-elemental maps, distributions of elements are very clear. Sr and REEs are concentrated in powellite crystallites since their maps are similar to those of Mo and Ca. However, Al, Fe and Zr are barely observed in the crystallite areas. They have a complementary distribution with respect to REEs, which indicates that Al, Fe and Zr prefer to locate in glass rather than in crystallites. The green and yellow spots on the Al, Fe and Zr maps are artifact due to the increase of image contrast. These results confirm that Sr and REEs (La, Nd, Eu and Pr) preferentially incorporate $CaMoO_4$ crystal during its crystallization while Al, Fe and Zr remain in the glass matrix. Such distribution of the elements between glassy and crystalline phases can also be illustrated easily by plotting the intensity correlations between Mo I 379.8 nm and the other species, as shown in Fig. 9.3. It indicates that Mo is correlated to Ca, Sr and REEs but anti-correlated to Al, Fe and Zr.

Signal from the border of crystallites and glass can be influenced by both crystalline and glassy phases. Thus, the size of crystallites shown in the multi-elemental maps is more or less different from the real one. The determination of the real size of a crystallite is affected by space resolution of the mapping system. However, the distributions of all detected elements are clear enough to show us the information about their fractionation behaviors after the nuclear glass thermal

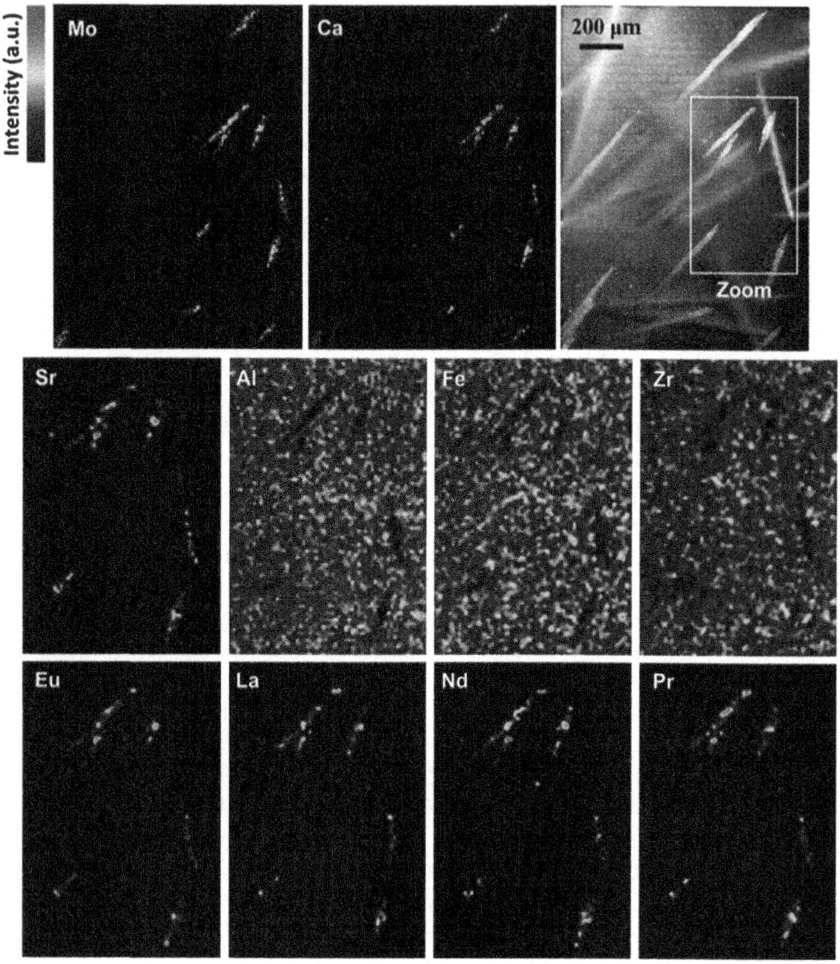

Fig. 9.2 LIBS mapping of multi-element of the Mo-rich borosilicate glass-ceramic sample and a optical microscope image of the mapping area; Mo and Ca mappings are full-sized images of the scanned area while all other mappings are images of the area enclosed by the white rectangle in the optical microscope image

treatment. Such information provides clue for investigations of the other nuclear waste glass systems or minerals which contain micro-structures.

It has to be noticed here that quantitative elemental analysis by μLIBS is still on progress for the moment. The images in Fig. 9.2 are only normalized intensity maps of multi elements. The intensity variation of element in the crystalline phase and in the residual glass is due to the increase of image contrast, which is less likely linked to inhomogeneous element distribution of in each phase. To further improve the ability of quantitative analysis by LIBS, more experiments are required, including calibration of element concentrations and determination of analyzed sample

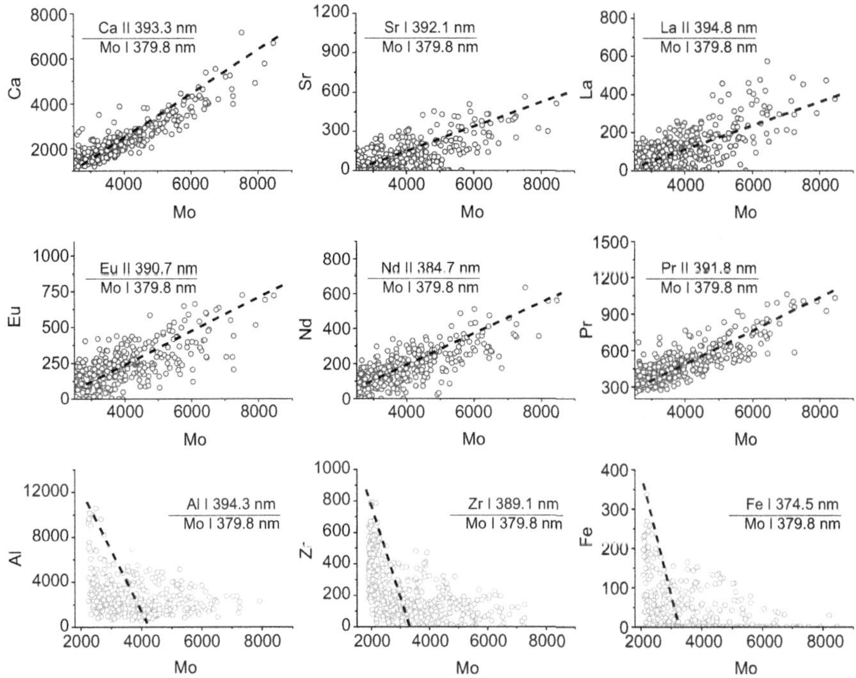

Fig. 9.3 Line intensity correlation between Mo I 379.8 nm line and the lines of other species in the mapping area of Mo-rich borosilicate glass-ceramic (around 300 points for each figure)

volume, etc. However, it is still worth taking advantage of LIBS for the detection and analysis of nuclear materials, since it is sensitive to most of the elements (particularly to low-content REEs) and can response to multi-elements fast and instantaneously.

9.3 Waste Storage Geomaterials

In the framework of nuclear waste, and especially long-lived high-level radioactive wastes disposal, multiple barriers matrices such as waste package, man-made barriers and geological barrier are studied (Ewing 2001). Therefore specific doped vitreous and crystalline phases are elaborated and their capacity to trap fission products (FP) and minor actinides, to resist to dissolution and to auto-irradiation is evaluated. At present, the uranium and plutonium are extracted from the general waste flow during spent-fuel reprocessing, the ultimate waste (C type), consisting of minor actinides (Np, Am and Cm) and fission products FP (Tc, I, Cs and Zr), is currently vitrified by incorporation in molten glass. Vitrification remains the most suitable process for all the high-level waste. As a matter of fact, the amorphous structure of glass allows accommodating a wide range of elements.

Because of the similarity in their chemical features and electronic configuration, REE have been chosen for modelling the trivalent minor actinides. Their luminescent properties allow recognize whether the elements are incorporated in the bulk of the matrix, or adsorbed on the surface where they can be subjected to leaching out.

9.3.1 Borosilicate Glass

Figure 9.4 presents the emission lines of the $^5D_0 \rightarrow {}^7F_1$ transition revealing two kind of environment in Eu^{3+}-doped complex nuclear borosilicate glass. There are attributed to (1) a borate environment and (2) a silicate one, by comparison of the ligand field parameters B_2 with referenced glassy matrices (Ollier et al. 2002). Furthermore, specific mineral matrices for use in conditioning separated radio nuclides are also elaborated and studied on the basis of mineral structures, known in nature for their selective-insertion capabilities with respect to certain radio nuclides and their stability over time. Among those phases, hollandite, zirconolite, apatite and monazite are particularly studied.

9.3.2 Hollandite, Britholite and Monazite

Hollandite structure derived from the quadratic rutile TiO_2 one with general formula $A_2B_8O_{16}$. In nature, there is a whole range of minerals that crystallize in this structure. The difference between them lies in the chemical elements inserted as a substitute for titanium Ti in the B-site: hollandite (Mn totally substituted for Ti), priderite (Fe^{3+}), henrymeyerite (Fe^{2+}), redledgeite (Cr), ankangite (Cr, V). FP can be fixed in $BaCs_{0.1}Al_{1.5}Ti_{6.5}O_{16}$, with A-site occupied by Ba^{2+} and Cs^+ while Al^{3+}, Ti^{3+} and Ti^{4+} are located in the B-site. Zirconolite structure $Ca^{2+}Zr_x{}^{4+}Ti_{(3\text{-}x)}{}^{4+}O_7$ with $0.8 < x < 1.35$, is able to accommodate lanthanides, hafnium Hf and tri- and tetravalent actinides by insertion into the calcium Ca and zirconium Zr sites as it was shown in the nature. Charge compensations are ensured by substituting trivalent cations (Al^{3+}, Ti^{3+}) in the titanium Ti^{4+} site.

Phospho-silicate britholites, $Ca_9Nd(PO_4)_5(SiO_4)F_2$, with apatite structure present high chemical durability, thermal stability and withstand radiation damage, are therefore consider as good candidates for actinide retention. Figures 9.5 and 9.6 show how the $^5D_0 \rightarrow {}^7F_0$ transition in trivalent europium-doped britholites along the phosphate-silicate solid solution allows to distinguish different sites identified as Ca(II), Ca(I) + La(I) and La(II) sites (El Ouenzerfi et al. 2001). Phospho- and phospho-vanadate apatites are also consider as well to trap fission products ($Ca_8NdCs(PO_4)_6F_2$ and $Pb_{10}(VO_4)_{4.8}(PO_4)_{1.2}I_2$).

Monazite, a rare earth orthophosphate $LaCePO_4$, has in the nature the ability to incorporate large amounts of U and Th which lead sometime to the $CaTh(PO_4)_2$ brabantite phase. It remains crystalline despite high α-decay doses even in the case

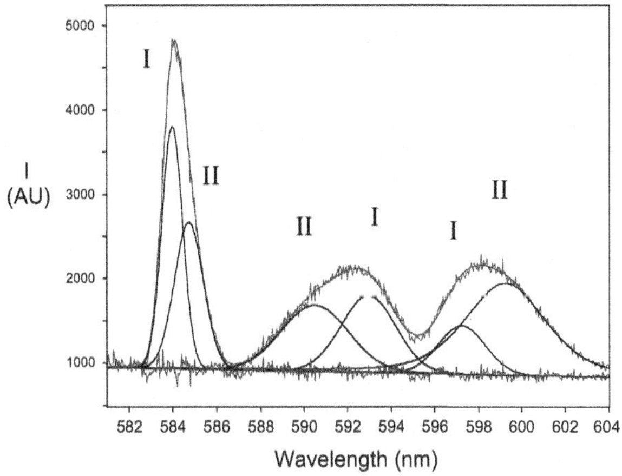

Fig. 9.4 Deconvolution of the $^5D_0 \rightarrow {}^7F_1$ transition of Eu^{3+} revealing two kinds of sites in Eu-doped complex nuclear borosilicate glass (577 nm excitation, 14 K)

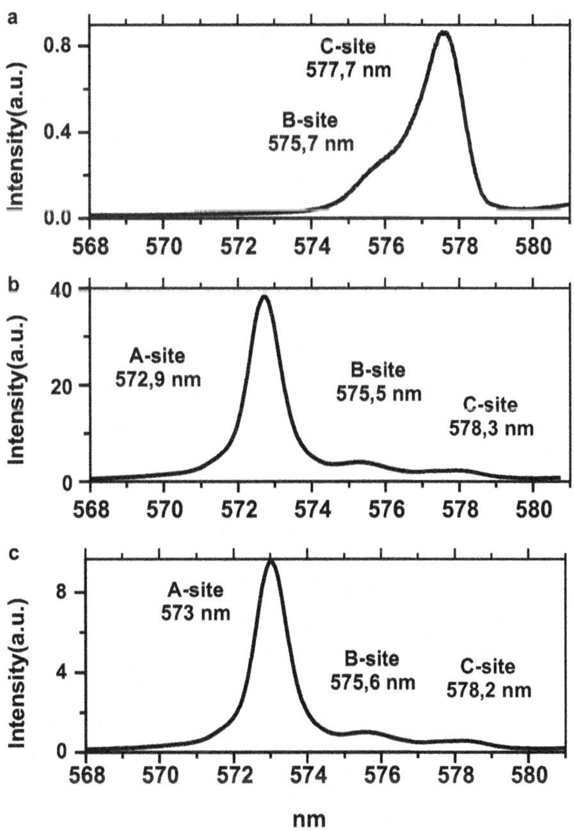

Fig. 9.5 (**a–c**) Emission of the $^5D_0 \rightarrow {}^7F_0$ of Eu^{3+}-doped britholithes $Ca_2La_8(SiO_4)_6O_2$ (1), $Ca_{4,79}La_{4,89}(SiO_4)_{3,12}(PO_4)_{2,88}O_x$ (2), $Ca_{8,06}La_{2,11}(PO_4)_6O_x$ (3). The A, B and C-sites correspond to the Ca(II), Ca(I) + La(I) and La(II) sites

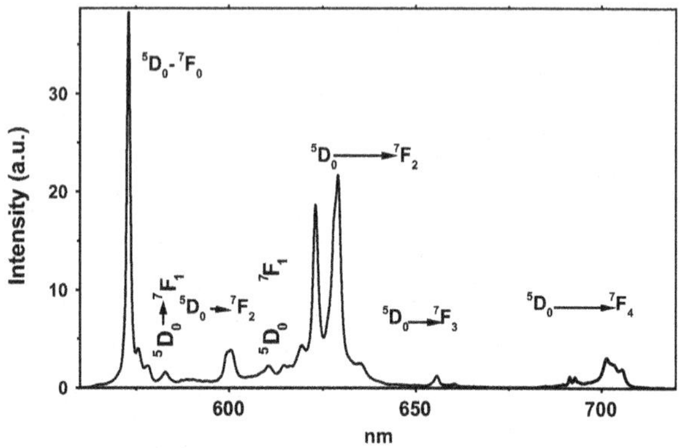

Fig. 9.6 Emission of the Eu^{3+}-doped (2 wt.%) britholithe

of very old crystals (over 2 billiard years). Cathodoluminescence allowed estimate the degree of auto irradiation received by natural monazite (Nasdala et al. 2002). Photoluminescence was used as well on natural monazites from different ages and different Th and U content, to follow thermal healing of radiation-induced defects. The full-width at half-maximum (FWHM) of Nd emission lines which indicates the degree of disorder, appears to decrease linearly with the treatment temperature (Fig. 9.7). The data collected from different natural monazites are compared in Table 9.3 While the Madagascar monazite present the largest FWHM (6.1 nm) and can be consider as having undergone the strongest metamictization, the modern alpine monazite are the most perfectly ordered sample (FWHM 2.1 nm). The evolution trend with temperature of the Madagascar sample, which present before annealing the largest FWHM, is used to estimate the equivalent annealing fictive temperature of the other samples. The degree of disorder appears to be proportional to the thorium content and not to the age of the samples (Paquette et al. 1994; Schärer and Deutsch 1990; Seydoux-Guillaume et al. 2002; Schärer et al. 1994).

9.3.3 *Powellite*

Powellite $CaMoO_4$, is the main crystalline phase which nucleates in low waste loading nuclear glass (R7T7). $CaMoO_4$ has I41/a space-group symmetry and C4h point-group symmetry. Due to the preferential incorporation of radioactive fission products (FP), the structure of powellite could be modified after a certain dose of self-irradiation (Wang et al. 2013). Hence, it is important and interesting to study the behaviors (swelling, structural damage, sites modification, etc.) of powellite under irradiations. Eu and Nd were doped into powellite as FP surrogates. External Ar-ion beam bombardment was used to simulate the internal self-irradiation (Mendoza et al. 2009, 2011). Different damage levels, in unit of displacement per atom (DPA) were selected in order to follow the evolution of structural behavior (Wang et al. 2014). Micro-Raman spectroscopy was used to investigate the medium range

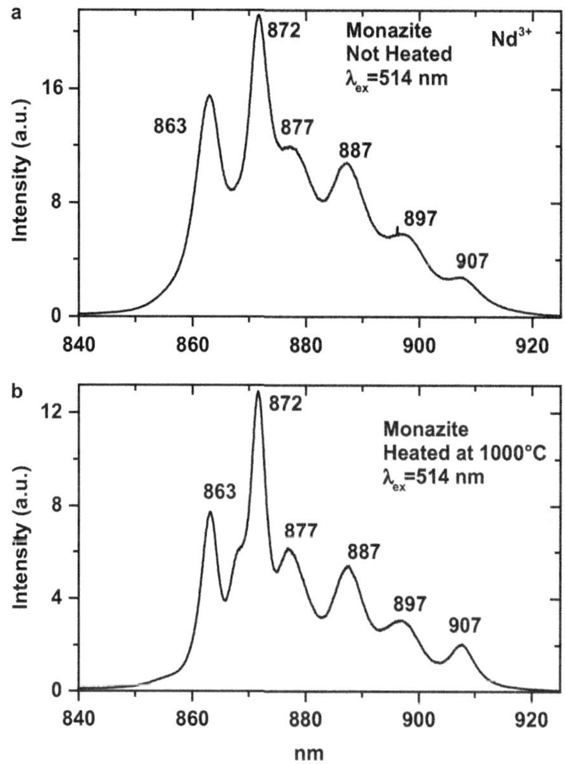

Fig. 9.7 (**a–b**) Emission spectra of Nd^{3+} ($^4F_{3/2} \rightarrow {}^4I_{9/2}$ transition) of natural monazite from Madagascar before (**a**) and after (**b**) treatment at 1000 °C

order modification of tetrahedral powellite structure after irradiation while photoluminescence (532 nm cw excitation) of europium probes the local environment distortions. The fine analyses of the $^5D_0$-$^7F_1$ transition allowed identifying the contribution of Eu^{3+} in S_4 and C_{2V} local symmetries (Fig. 9.8) along the irradiated lamella. A complete mapping of the lamellae (C_{2V}/S_4 contribution ratio in the $^5D_0$-$^7F_1$ transition) reveals the local modifications of the powellite structure for 0.012 and 5.0 displacements per atom (dpa) with a micrometer scale resolution. At low dose and fluence (0.012 dpa) the highest irradiated zone is mainly of asymmetric C_{2V} symmetry while at high dose and fluence (5.0 dpa), it evolves to a S_4 more symmetric one.

9.4 Luminescent Bio-minerals

Hydroxyapatite $Ca_5(PO_4)_3OH$ is the main mineral constituent of bones and teeth. The chemical composition of biological apatite is somewhat different from the above given formula. They are usually defined as non-stoichiometric apatites. This non- stoichiometry is caused by the presence of different ions in very small or in

Table 9.3 Comparison of full-width at half-maximum (nm) of two Nd emission lines with Th concentration of different natural monazites and determination of equivalent annealing fictive temperature (Madagascar sample considered as reference material). (nd: not determined) (age: Paquette et al. 1994; Schärer and Deutsch 1990; Seydoux-Guillaume et al. 2002; Schärer et al. 1994)

Monazites	Madagascar (reference)	Norway	Canada	Brazil	China	Alps Trimouns	Alps Lauzière
Age (Ma)	545	nd	1928	474	24	nd	nd
Th (wt.%)	13.25	nd	9.03–10.33	6.92	5.74–15.60	nd	nd
FWHM 3 (nm)	4.6(1)	4.4(5)	4.4(3)	3.9(4)	3.4(7)–3.6(9)	2.3(5)	2.1(5)
Fictive T (°C)	25	160	178	670	921–1140	$\gg$1200	$\gg$1200
FWHM 1 (nm)	6.1(1)	5.7(9)	5.5(1)	5.4(2)	4.5(6)–5.1(6)	3.2(3)	2.9(3)
Fictive T (°C)	25	211	400	460	680–1025	$\gg$1200	$\gg$1200

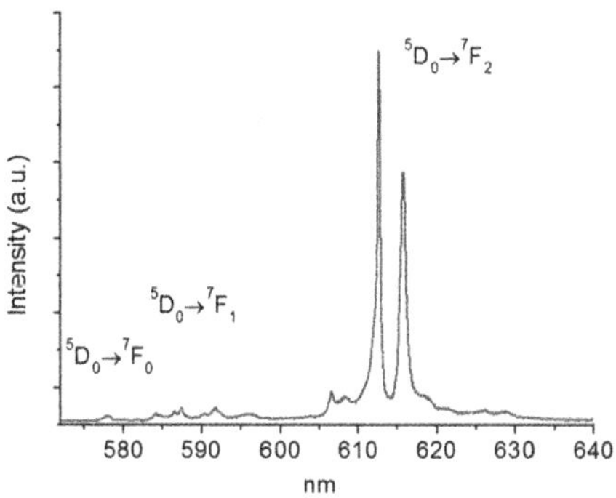

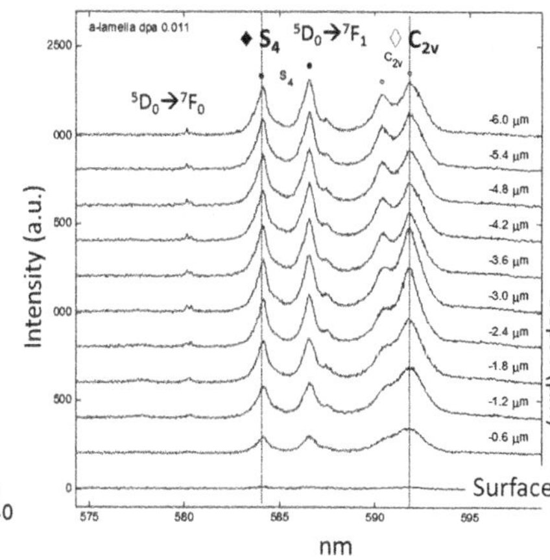

Fig. 9.8 (**a–b**) Luminescence of Eu^{3+} doped $CaMoO_4$ powellite (**a**) and microluminescence investigation of the $^5D_0$-$^7F_1$ transition (S_4 and C2v local symmetries) of irradiated powellite lamellae (0.012 dpa) in function of depth (μm) from irradiated surface (excitation by 532 nm)

trace amounts. Foreign ions might have an effect on the chemical or physical properties of hydroxyapatite. Trace ions might affect the mineralisation process or the dissolution of biological minerals, teeth and bones. The effect of trace ions is also important in the investigation of hydroxyapatite as a bone implant or as coating material for Ti or other implants. In this cases trace ions might influence the stability of the implant, the chemical interaction at the bone and the implant interface or mechanical properties. Whether foreign ions can substitute Ca^{2+} in one of its crystallographical sites in the hydroxyapatite lattice ions is a principal question in the preparation of mineral with foreign ions. The size of the ions, their chemical properties and the preparation conditions are among the most important factors in this respect.

La^{3+} is of interest in relationship to biological hydroxyapatite because of their inhibitory effect on the demineralisation of dental enamel. La^{3+} presence form was not clear, namely surface adsorption or lattice incorporation. In attempt to clarify it, laser-induced luminescence has been used (Mayer et al. 1999). La^{3+} is not luminescent Gd-containing samples were prepared and studied. Figure 9.9 demonstrates Gd luminescence spectra with 266 nm laser excitation before and after heating at 800 °C. It is clearly seen that luminescence intensity is drastically stronger after thermal treatment. Thus precipitated samples must be heated to 800 °C to enable Gd^{3+} replace Ca^{2+} and becomes luminescent and its incorporation form is surface adsorption.

Another example is LIBS application for real-time identification of carious teeth (Samek et al. 2001). In the dental practice, usually more healthy tissue is removed than ultimately necessary. Carious and healthy tooth material can be identified through the decrease of matrix elements Ca and P in hydroxyapatite and/or the increase of non-matrix elements, typically Li, Sr, Ba, Na, Mg, Zn and C, using pattern recognition algorithms. A fibre-based LIBS assembly was successfully used

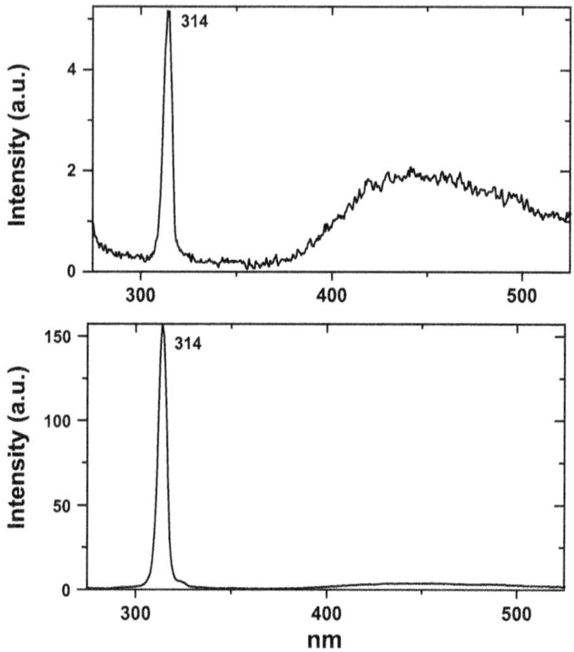

Fig. 9.9 Laser-induced luminescence of Gd^{3+} in precipitated hydroxyapatite before and after heating at 800 °C

for this task. As for the case of phosphate ores evaluation, the efforts aimed to normalising the spectrum collection conditions and procedures, so that the spectra are sufficiently reproducible for precise quantitative analysis, down to detection of a few ppm, are not required. The combination of LIBS and discrimination analysis is a useful tool for real time in vivo/in vitro caries identification during the drilling process when luminous short laser pulses create plasma.

9.5 LED Applications

Mellilite rich glass-ceramics investigated for white LED were obtained by thermally treating calcium aluminosilicate glass at 900 °C at various durations (Bouchouicha et al. 2015). Melilite which was identify by Raman and XRD, crystalizes with tetragonal symmetry of the P-421 m space group and belongs to the solid solution akermanite ($Ca_2MgSi_2O_7$)-gehlenite ($Ca_2Al_2SiO_7$) with Ak/Geh = 25/75 (mol %). The emission spectra of the Eu^{3+} doped and treated samples under continuous 532 nm excitation are shown in Fig. 9.10. The strongest emission peak near 613 nm, assigned to the transition $^5D_0$-$^7F_2$ of Eu^{3+}, is dominant in all of the spectra. This peak appears when Eu^{3+} occupies non-centrosymmetric lattice center. After heat treatment, emission spectra of Eu^{3+} particularly the $^5D_0$-$^7F_2$ transition changed remarkably compared to the spectra of glass sample. This change is the result of the incorporation of Eu^{3+} in an ordered crystalline

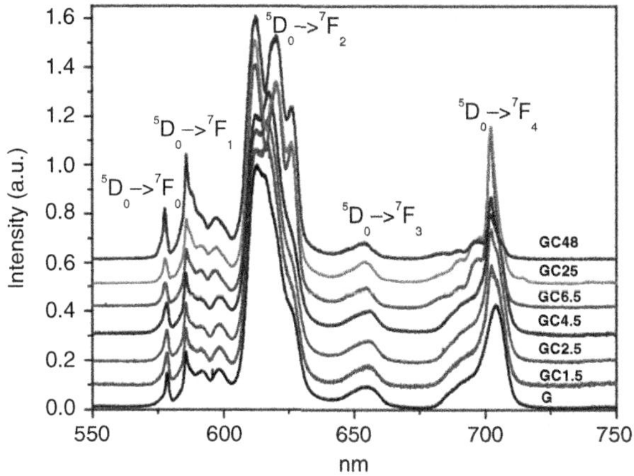

Fig. 9.10 Emission spectra ($\lambda_{ex} = 532$ nm) of Eu^{3+} transitions in the initial glass G and melilite-rich glass-ceramics in function of thermal duration (in hours) at 900 °C. Spectra are normalized to the maximum $^5D_0$-$^7F_2$

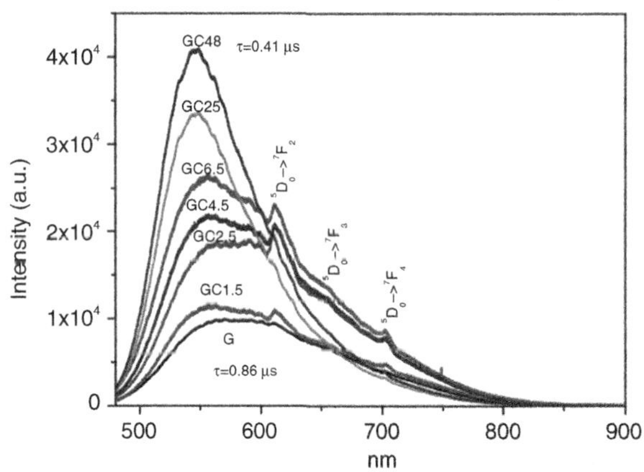

Fig. 9.11 Emission spectra ($\lambda_{ex} = 473$ nm) of Eu^{2+} and Eu^{3+} in the initial glass G and melilite-rich glass-ceramics in function of thermal duration (hours) at 900 °C (a)

environment. Prominent Stark splitting is observed particularly in the $^5D_0$-$^7F_2$ transition of Eu^{3+} suggesting that Eu^{3+} is efficiently incorporated in the melilite crystalline phase. Under 473 nm cw excitation new feature appears revealing the broad band in the 480–900 nm region attributed to 4f65d – 4f7 transition of Eu^{2+} (Fig. 9.11). For the G glass, the broad emission band of Eu^{2+} with a FWHM of 4568 cm^{-1} is centered at 580 nm. Furthermore, it is possible to notice the presence of narrower peaks between 570 and 750 nm due to the $^5D_0$-^7F_J (J = 1–4) of Eu^{3+}. However the Eu^{2+} emission is much more effective compared to Eu^{3+} one. This effect can be explained by the preparation method of the initial glass sample which was conducted under vacuum atmosphere and consequently, europium is mainly under its divalent form. Moreover, the influence of the silicon content on the Eu^{2+}/Eu^{3+} ratio was more deeply investigated by XANES. After annealing, the Eu^{3+} peaks drop down while, intensity of Eu^{2+} emission enhanced remarkably and shifted to the lower wavelength (blue-shifted) with increase of heat-treatment duration.

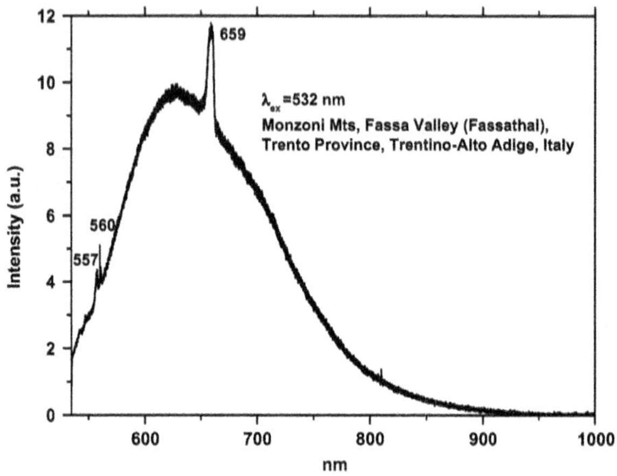

Fig. 9.12 Laser induced cw 532 nm excited luminescence of natural gehlenite with evidently Mn^{2+} emission

In natural gehlenite the luminescence of Eu was not detected, while the broad band peaking at approximately 630 nm may be preliminary attributed to Mn^{2+} luminescence (Fig. 9.12).

References

Armstrong D, Wang X, Beesley C, Rabinovitz R (2000) Rapid spectroscopic discrimination between natural and synthetic emeralds, rubies, and alexandrite. Am Lab 30:26–34

Bouchouicha H, Panczer G, de Ligny D et al (2015) Synthesis and luminescent properties of Eu^{3+}/Eu^{2+} co-doped calcium aluminosilicate glass-ceramics. J Lumin (Pre published online)

Chalain J-P, Fritsch E, Hanni H (2000) Identification of GE POL diamonds. J Gemm 27:73–78

El Quenzerfi R, Panczer G, Goutaudier C et al (2001) Relationship between structural and luminescence properties in Eu^{3+}-doped oxyphosphate-silicate apatite $Ca_{2-x}La_{8-x}$ $(SiO_4)_{6-x}(PO_4)_xO_2$. Opt Mater 16:301–310

Ewing R (2001) The design and evaluation of nuclear-waste forms: clues from mineralogy. Can Mineral 39:697–715

Gororbets B, Kononov O, Rogojine A, Kvitko T (2003) Application of luminescence in gemology. Vest Gem 3(10):38–56 (in Russian)

Lindblom J, Hösla J, Papunen H et al (2003) Differentiation of natural and synthetic gem-quality diamonds by luminescence properties. Opt Mater 24:243–253

Lindblom J, Hölsä J, Papunen H (2005) Luminescence study of defects in synthetic as-grown and HPHT diamonds compared to natural diamonds. Am Miner 90:428–440

Mayer I, Layani J, Givan A et al (1999) La ions in precipitated hydroapatites. J Inorg Biochem 73:221–226

Mendoza C, Panczer G, de Ligny D et al (2009) $CaMoO_4$ in a molybdenum rich borosilicate glass-ceramic: a spectroscopic study. proceeding of the 8th PACRIM conference on ceramic and glass technology. Vancouver Can Ceram Transact 217:43–55

Mendoza C, de Ligny D, Panczer G et al (2011) Behaviour of the Eu^{3+} $^5D_0 \rightarrow {}^7F_0$ transition in $CaMoO_4$ powellite type ceramics under Ar and Pb ions implantation. Opt Mater 34:386–390

Nasdala L, Lengauer C, Hanchar J et al (2002) Annealing radiation damage and the recovery of cathodoluminescence. Chem Geol 191:121–140

Ollier N, Panczer G, Champagnon B et al (2002) Europium as a luminescent probe of an aluminoborosilicate nuclear glass and its weathering Gels. J Lumin 94–95:197–201

Panczer G, Ollier N, Champagnon B, Gaft M et al (2003) New development in time-resolved micro-photoluminescence spectroscopy of heterogeneous natural and synthetic inorganic materials. Opt Mater 24:253–257

Paquette J, Nedelec A, Moine B, Rakotondrazafi M (1994) U-Pb, single zircon Pb-evaporation, and Sm-Nd isotopic study of a granulite domain in SE Madagascar. J Geol 102:523–538

Samek O, Telle H, Beddows D (2001) Laser-induced breakdown spectroscopy: a tool for real-time, in vitro and in vivo identification of carious teeth. BMC Oral Health 1:1–12

Schärer U, Deutsch A (1990) Isotop systematics and shoch-wave metamorphism II: U-Pb and Rb-Sr in naturalkly shocked rocks, the Haughton impact structure. Geochim Cosmochim Acta 54:3435–3447

Schärer U, Zhang L, Tapponnier P (1994) Duration of strike-slip movements in large shear zones – the red river belt, China. Earth Planet Sci Lett 126:379–397

Seydoux-Guillaume A, Wirth R, Nasdala L et al (2002) An XRD, TEM and Raman study of experimentally annealed natural monazite. Phys Chem Miner 29:240–253

Wang X, Motto-Ros V, Panczer G et al (2013) Mapping of rare earth elements in nuclear waste glass-ceramic using microlaser-induced breakdown spectroscopy. Spectrochim Acta B 87:139–146

Wang X, Panczer G, de Ligny D et al (2014) Irradiated rare-earth-doped powellite single crystal probed by confocal Raman mapping and transmission electron microscopy. J Raman Spectrosc 45:383–391

Conclusions

In the proposed book there is an emphasis on luminescence lifetime, which is a measure of the transition probability and non-radiative relaxation from the emitting level. Luminescence in minerals is observed over a time interval of nanoseconds to milliseconds. It is therefore a characteristic and a unique property and no two luminescence emissions will have exactly the same decay time. The best way for a combination of the spectral and temporal nature of the emission can be determined by time-resolved spectra. Such techniques can often separate overlapping features, which have different origins and therefore different luminescence lifetimes. The method involves recording the intensity in a specific time "window" at a given delay after the excitation pulse where both delay and gate width have to be carefully chosen. The added value of the method is the energetic selectivity of a laser beam, which enables to combine time-resolved spectroscopy with powerful individual excitation.

The book deals mainly with theoretical approach, experimental results and their interpretation of laser-induced time-resolved spectroscopy of minerals in the wide spectral range from 250 to 2000 nm, which enables to reveal new luminescence previously hidden by more intensive centers. Artificial activation by potential luminescence centers has been accomplished in many cases, which makes the sure identification possible. The mostly striking example is mineral apatite, which has been extremely well studied by many scientists using practically all known varieties of steady-state luminescence spectroscopy: photoluminescence with lamp and laser excitations, X-ray excited luminescence, cathodoluminescence, ionoluminescence and thermoluminescence. Nevertheless, time-resolved spectroscopy revealed that approximately 50 % of luminescence information remained hidden. The mostly important new information is connected with luminescence of trivalent rare-earth elements in minerals, such as Eu^{3+}, Pr^{3+}, Ho^{3+}, Tm^{3+}, Er^{3+}, and Nd^{3+} (Table 1).

Luminescence information on transition elements is substantially improved. Besides well-known Mn^{2+} centers, emission of Mn^{5+} was found and Mn^{4+} proposed as possible participant in minerals luminescence. Luminescence

M. Gaft et al., *Modern Luminescence Spectroscopy of Minerals and Materials*,
Springer Mineralogy, DOI 10.1007/978-3-319-24765-6

Table 1 REE luminescence centers detected in minerals

Center	λ_{lum}	λ_{ex}	τ, μs	Transition	Minerals
Ce^{3+}	360–400	308	0.02	5d^1-4f^1 ($^2F_{5/2}$)	Apatite, barite, anhydrite, calcite, feldspars, danburite, datolite, esperite, apophyllite, charoite, fluorite, hardystonite
Pr^{3+}	228 239 268 480 480 600 650 858	460	10 100 20	$^1S_0$-$^3H_5$ $^1S_0$-$^3H_6$ $^1S_0$-$^1G_4$ $^1D_2$-$^3H_6$ $^3P_0$-$^3H_4$ $^1D_2$-$^3H_6$ $^3P_0$-$^3F_2$ $^3P_0$-$^1G_4$	Apatite, scheelite, zircon, anhydrite, zircon, titanite
Nd^{3+}	890 1070 1340 395 416 446 588 672 743 846	360 355	300 10–30	$^4F_{3/2}$-$^4I_{9/2}$ $^4F_{3/2}$-$^4I_{11/2}$ $^4F_{3/2}$-$^4I_{13/2}$ $^2F_{5/2}$-$^4F_{5/2}$ $^2F_{5/2}$-$^4F_{9/2}$ $^2P_{1/2}$-$^4I_{9/2}$ $^4G_{7/2}$-$^4I_{11/2}$ $^4G_{7/2}$-$^4I_{12/2}$ $^4G_{7/2}$-$^4I_{13/2}$ $^4G_{7/2}$-$^4I_{15/2}$	Apatite, scheelite, fluorite, barite, titanite, zircon, anhydrite, calcite, rhodonite, feldspars
Sm^{3+} Sm^{2+}	565 599 645 630 688 700 734	308	2250	$^4G_{5/2}$-$^6H_{5/2}$ $^4G_{5/2}$-$^6H_{7/2}$ $^4G_{5/2}$-$^6H_{9/2}$ 5d-4f $^7D_0$-$^7F_0$ $^7D_0$-$^7F_1$ $^7D_0$-$^7F_2$	Apatite, zircon, scheelite, monazite, fluorite, anhydrite, titanite, zircon, hardystonite, pyromorphite Anhydrite, fluorite, apatite
Eu^{3+} (I) Eu^{3+} (II) Eu^{2+}	579 590 618 653 700 574 601 630 711 390–450	384 266 337	1800 0.6– 0.8	$^5D_0$-$^7F_0$ $^5D_0$-$^7F_1$ $^5D_0$-$^7F_2$ $^5D_0$-$^7F_3$ $^5D_0$-$^7F_4$ $^5D_0$-$^7F_0$ $^5D_0$-$^7F_1$ $^5D_0$-$^7F_2$ $^5D_0$-$^7F_4$ 4f5d-4f	Apatite, zircon, scheelite, danburite, datolite, fluorite, calcite, anhydrite, xenotime, orangite, titanite Fluorite, apatite, barite, danburite, feldspars, anhydrite, zoisite, charoite
Gd^{3+}	312	266	2000	$^6P_{7/2}$-$^8S_{7/2}$	Fluorite, zircon, anhydrite, hardystonite
Tb^{3+}	380 415 437 545	308	600 2400	$^5D_3$-$^7F_6$ $^5D_3$-$^7F_5$ $^5D_3$-$^7F_4$ $^5D_4$-$^7F_5$	Apatite, scheelite, zircon, calcite, fluorite, zoisite, monazite, feldspars, calcite
Dy^{3+}	480 575 663 750	354	720	$^4F_{9/2}$-$^6H_{15/2}$ $^4F_{9/2}$-$^6H_{13/2}$ $^4F_{9/2}$-$^6H_{11/2}$ $^4F_{9/2}$-$^6H_{9/2}$	Apatite, scheelite, zircon, calcite, fluorite, zoisite, monazite, titanite
Er^{3+}	545 854 978 1540	308 532 532	23 2100	$^4S_{3/2}$-$^4I_{15/2}$ $^4S_{3/2}$-$^4I_{9/2}$ $^4I_{11/2}$-4$I_{15/2}$ $^4I_{13/2}$-4$I_{15/2}$	Apatite, fluorite, scheelite, zircon, feldspars, titanite

(continued)

Table 1 (continued)

Center	λ_{lum}	λ_{ex}	τ, μs	Transition	Minerals
Tm^{3+}	364 452 796	360	17	$^1D_2$-$^3H_6$ $^1D_2$-$^3H_4$ $^1G_4$-$^3F_5$	Apatite, scheelite, zircon, calcite, fluorite, anhydrite, aragonite, hardystonite
Yb^{3+}	993	308	3100	$^2F_{5/2}$-$^2F_{7/2}$	Apatite, scheelite, zircon
Ho^{3+}	540	360	5	$^5S_2$-$^5I_8$	Zircon, scheelite, fluorite

characteristics of Mn^{6+}, Cr^{5+}, Ti^{3+}, V^{2+}, V^{3+}, Ni^{+}, Sb^{3+} are presented and their possible role in minerals luminescence is considered. Luminescence of Pb^{2+}, Pb^{+}, Bi^{3+}, Bi^{2+}, Tl^{+}, Ag^{+}, Cu^{+}, Ni^{+} has been found. Reabsorption lines of molecular H_2O and O_2 in apatite and diamond have been identified.

The possibilities of time-resolved laser based spectroscopies have been demonstrated, combining such techniques as luminescence, Raman, breakdown and second-harmonic generation. New type of luminescence excitation mechanism, Plasma Induced Luminescence, was found. UV Gated Raman spectroscopy proved to be an effective tool for minerals Raman detection on the strong luminescence background.

Online real time process control in mining industry based on different laser spectroscopies was implemented. Spectroscopic properties of minerals, rocks and foliage were considered as an homeland security aspects.

Bibliography

Agekyan V, Shiryapov I, Oreshnikova L (1974) Exciton-phonon transitions in tin oxide. FTT 16:2473–2475 (in Russian)

Cutmore N, Eberhard M. www.min-eng.com/protected/me02ex

Czaja M, Bodył Gajowska S, Mazurak Z (2013) Steady-state luminescence measurement for qualitative identification of rare earth ions in minerals. J Miner Petrol Sci 108:47–54

de Boisbaudran L (1885) Sur la fluorescence des terres rares. C R Acad Sci Paris 101:552–588

Gaft M, Panczer G (2013) Laser induced time resolved luminescence spectroscopy of minerals – a powerful tool for studying the nature of emission centers. Miner Petrol 107:363–372

Gaft M, Reisfeld R, Panczer G et al (1998a) Laser-induced time-resolved luminescence of minerals. Spectrochim Acta A 54:2163–2175

Gaft M, Panczer G, Uspensky E, Reisfeld R (1999b) Laser-induced time-resolved luminescence of rare-earth elements in scheelite. Miner Mag 63:199–210

Gaft M, Nagli L, Fasaki I et al (2009b) Laser-induced breakdown spectroscopy for sulfur on-line analyses of minerals in ambient conditions. Spectrochim Acta B 64:1098–1104

Mao X, Bol'shakov A, Choi I et al (2011) Laser ablation molecular isotopic spectrometry: strontium and its isotopes. Spectrochim Acta B 66:767–775

Matsyuk S, Zintchouk N (2001) Optical spectroscopy of the upper mantle minerals. Nedra, Moscow (in Russian)

Measures R, Houston W, Stephenson D (1974) Laser induced fluorescent decay spectra – a new form of environmental signature. Opt Eng 13:494–501

Moroshkin V, Evdokimenko E, Gaft M et al (1997) Method of artificial luminescent films – a new method of luminescence analysis of minerals and rocks. Rocks Metals 3:63–72

NIST Atomic Spectra Database: http://physics.nist.gov/PhysRefData/ASD/lines_form.html

Ollier N, Fuchs Y, Cavani O et al (2015) Influence of impurities on Cr^{3+} luminescence properties in Brazilian emeralds and alexandrite minerals. Eur Mineral J (Pre published online)

Panczer G (2001) La photoluminescence resolue en temps, nouvel outil pour la mineralogy. Habilitation thesis, Universite Claude Bernard, Lyon (in French)

Peixoto J, Oort A (1992) The physics of climate. American Institute of Physics, New York

Prokic M (1977) Analysis of the thermoluminescence glow curves of natural barite. J Phys Chem Solid 38:617–622

Radziemski L, Loree T, Cremes D, Hoffman N (1983) Time resolved laser induced breakdown spectrometry of aerosols. Anal Chem 55:1246–1253

Reiche I, Vignand C, Champagnon B et al (2001) From mastodon ivory to gemstone: The origin of turquoise color in odontolite. Am Miner 86:1519–1524

M. Gaft et al., *Modern Luminescence Spectroscopy of Minerals and Materials*,
Springer Mineralogy, DOI 10.1007/978-3-319-24765-6

Reisfeld R (1991) Luminescence and nonradiative processes in porous glasses. In: Di Bartolo B, Chen X (eds) Proceedings of advanced study Institute on advances in nonradiative processes in solids. Plenum, NATO ASI Series (B) Physics 249:397–425

Sallé B, Mauchien P, Maurice S (2007) Laser-induced breakdown spectroscopy in open-path configuration for the analysis of distant objects. Spectrochim Acta B 62:739–768

Sands H, Hayward I, Kirkbride T et al (1998) UV excited resonance Raman spectroscopy of narcotics and explosives. J Forens Sci 43:509–513

Shoval S, Gaft M, Panczer G (2003) Luminescence of Cr^{3+} in natural and calcined diaspore. J Therm Anal Calorim 71:699–706

Sidike A, Kusachi I, Yamashita N (2003) Natural fluorite emitting yellow fluorescence under UV light. Phys Chem Miner 30:478–485

Tarashchan A, Waychunas G (1995) Luminescence of minerals. Interpretation of the luminescence spectra in terms of band theory, crystal field theory, sensitization and quenching. In: Marfunin A (ed) Advanced mineralogy 2, methods and instrumentation. Springer, Berlin/Heidelberg/New York, pp 124–135

Index

M. Gaft et al., *Modern Luminescence Spectroscopy of Minerals and Materials*, Springer Mineralogy, DOI 10.1007/978-3-319-24765-6

The manufacturer's authorised representative in the EU is Springer Nature Customer Service Centre GmbH, Europaplatz 3, 69115 Heidelberg, Germany. If you have any concerns regarding our products, please contact ProductSafety@springernature.com

Printed and bound by CPI Group (UK) Ltd, Croydon, CR0 4YY
01/10/2026
02225860-0018